Bayesian Analysis for the Social Sciences

WILEY SERIES IN PROBABILITY AND STATISTICS

Established by WALTER A. SHEWHART and SAMUEL S. WILKS

A complete list of the titles in this series appears at the end of this volume.

Bayesian Analysis for the Social Sciences

Simon Jackman

Department of Political Science, Stanford University, USA.

WILEY

A John Wiley and Sons, Ltd., Publication

This edition first published 2009

Registered office
John Wiley & Sons Ltd, The Atrium, Southern Gate, Chichester, West Sussex, PO19 8SQ, United Kingdom

For details of our global editorial offices, for customer services and for information about how to apply for permission to reuse the copyright material in this book please see our website at www.wiley.com.

Library of Congress Cataloguing-in-Publication Data

Record on file

A catalogue record for this book is available from the British Library.

ISBN: 978-0-470-01154-6

Typeset in 10/12pt Times Roman by Laserwords Pvt Ltd, Chennai, India

To Janet

Contents

List of figures

List of Tables

Preface

Markov chain Monte Carlo methods prompted an explosion of interest in Bayesian statistical inference through the decade of the 1990s. Today, Bayesian approaches are almost *de rigueur* in the social sciences. 'MCMC-enabled' Bayesian inference is well and truly part of the social science methodological toolkit, and one can find applications of Bayesian statistical inference in virtually every field of the social sciences. It is reasonable to expect that a 'well trained', quantitatively inclined social scientist is at least familiar with Bayesian statistical inference. And even the terms 'MCMC' and 'hierarchical model' have acquired a certain cachet.

The incentive structure of graduate education encourages students to explore the latest methodological tools. The reasons are clear enough. Methodological precocity makes collaboration with professors more likely, can boost professional visibility and may enhance the prospect of peer-refereed publications, all of which can help launch a successful academic career. But my experience – based on teaching and lecturing to many social science audiences – is that many students are attracted to Bayesian modeling and the power of MCMC techniques without a sufficient understanding of how Bayesian inference actually works. It is not uncommon to find students beginning to use MCMC methods while being less than clear about some key concepts: e.g. the distinctions between a posterior density, the likelihood function and the sampling distribution of a statistic; the role of priors in a Bayesian analysis; how hierarchical models work; predictive inference via prior and posterior predictive densities; how to summarize a prior or posterior density, and so on. Moreover, MCMC techniques for exploring posterior densities are often regarded in a 'black box' way, with an overly casual concern for what these methods are actually doing, or how well they do it.

Accordingly, in this book I have tried to provide a reasonably self-contained introduction to Bayesian statistical inference, with an emphasis on fundamentals in addition to focusing on social science applications. We consider 'the basics' of Bayesian inference in considerable detail in Part I, deferring the consideration of simulation-based approaches via MCMC to Part II. Part III of the book is driven by examples, complemented by `WinBUGS/JAGS` programs and `R` code. Appendices supply some mathematical and statistical background, covering vectors and matrices (Appendix A), the foundations of probability (Section B.1), some of the widely used probability mass functions (Section B.2.1) and densities (Section B.2.2) and proofs of some key results concerning the conjugate analysis of normal data (Section C.2).

In Part I of the book I provide a detailed introduction to Bayesian inference, starting with fundamental concepts in the Bayesian approach: e.g., subjective probability (Section 1.1), Bayes Theorem itself (Sections 1.3, 1.4 and 1.10), exchangeability (Section 1.9), the contrast with frequentist inference (Section 1.5), how to summarize and communicate features of the posterior density, (Section 1.6), and Bayesian hypothesis testing (Section 1.8). Chapter 2 presents the 'pre-simulation', conjugate, Bayesian analysis of some key – if simple – inferential problems encountered in statistical work in the social sciences: e.g. inference for rates and proportions (Section 2.1), two-sample binomial problems (Section 2.2), count data (Section 2.3), inference for the mean and variance in normal data problems (Section 2.4) and regression (Section 2.5).

Part II introduces simulation methods, considering the two 'MC's of Markov chain Monte Carlo in reverse order, as it were. Chapter 3 introduces the Monte Carlo method and algorithms for sampling from arbitrary densities. Chapter 4 provides some background and key results on Markov chains, leading up to a statement of the Ergodic Theorem (Section 4.4). In Chapter 5 we bring the two 'MCs' together to consider the workhorses of MCMC: the Metropolis algorithm (Section 5.1) and the Gibbs sampler (Section 5.2). In Chapter 6 we consider implementing Markov chain Monte Carlo, looking at strategies for effective Bayesian computation via the freely available, general purpose computer programs for Bayesian statistical inference `WinBUGS` and its close cousin `JAGS`. A series of applications demonstrating MCMC at work appears in Section 6.5. In both Parts I and II I have tried to provide a reasonably rigorous treatment of the underlying issues, exposing my target readers in the social sciences to the relevant literature in statistics. I will not provide proofs of some key propositions (at least not in the body of the text), and in many cases the reader is referred to the wonderful books by Bernardo and Smith (1994) and Robert and Casella (2004).

The remainder of the book – Part III – is devoted to applications in the social sciences. These include hierarchical models of various sorts (Chapter 7) and regression models for binary outcomes (Section 8.1), ordinal (Section 8.2) and multinomial outcomes (Section 8.4). In Chapter 9 we consider Bayesian analysis of latent variable models, including a presentation of the factor analysis model as a (Bayesian) latent variable model (Section 9.2), item-response theory (IRT) models for binary indicators (Section 9.3) and state-space models with dynamic latent variables (Section 9.4). Hierarchical models appear repeatedly throughout these applications: e.g. as extensions to the ordinal response model (Example 8.6), the factor analysis model (Example 9.1) and IRT models (Example 9.2).

In emphasizing the foundations of Bayesian inference, I have opted not to cover some important, advanced topics. At the risk of writing the preface to a second edition, a list of topics that 'could have, should have' appeared here includes: Bayesian model selection when using MCMC (e.g. computing marginal likelihoods and Bayes factors with the output of MCMC algorithms; the Deviance Information Criterion); a Bayesian treatment of 'econometric' problems such as endogenous predictors and sample selection; 'reversible jump' Metropolis algorithms (e.g. so to induce a posterior mass function over the number of components in a finite mixture problem); the relatively new but very interesting world of non-parametric Bayesian inference via Dirichlet process priors, and Bayesian inference for social network problems. These omissions are regrettable, but I have been surprised at how long it has taken to cover the introductory and intermediate level material presented here, and how many pages a thorough treatment of these topics has consumed.

The book has many examples, with the examples in Part III complete with `WinBUGS/JAGS` and `R` code. Almost all the data used in the examples is available in my `R` package, `pscl` (Jackman 2008b). My source for the mother-child sexual orientation data (Example 2.8) and the sex partner count data (Example 2.10) is Agresti and Finlay (1997). I am grateful to Orley Ashenfelter for sharing the data used in Example 2.13. Bruce Western originally built the data set used in Example 2.16. Jonathan Katz and Gary King were the original collectors of the U.K. House of Commons electoral results analyzed in Example 6.9. Don Green and Lynn Vavreck graciously shared the data that appears in Example 7.9. Jonathan Nagler shared the data analyzed in Example 8.1. Matt Levendusky organized the data from the American National Election Studies analyzed in Example 8.6. R. Michael Alvarez and Jonathan Nagler organized the data from the American National Election Studies that is analyzed in Examples 8.7 and 8.8. Some of the data used in Example 1.2 and Example 6.8 was collected by Matt Levendusky; Josh Clinton collected some of the data analyzed in Example 9.2; Andrea Abel helped me collect some of the data used in Example 9.3.

SIMON JACKMAN

Sydney, Australia
2009

Acknowledgments

This book was a long time in the making. I thank everyone at Wiley for their patience and indulgence, but especially Susan Barclay, Beth Dufour and Heather Kay.

I owe Larry Bartels many professional and intellectual debts. Here I will thank him for a lecture he gave in 1990 at the University of Rochester, my first real introduction to Bayesian inference. Larry's lecture helped change the way I thought about statistical inference, and started me on a path that leads to this moment, many years later, finally finishing this book on Bayesian statistical inference in the social sciences.

An ongoing, twenty-five year conversation with Bruce Western on statistical practice in the social sciences has been especially valuable. Bruce and I were exposed to Bayesian ideas at similar stages in our respective graduate careers, with our "zeal-of-a-convert" enthusiasm for Bayesian inference appearing in a co-authored publication (Western and Jackman 1994). The argument I present in the Introduction as to the utility of Bayesian analysis in social science settings has its origins in that article. The section in the Introduction titled "Investigating Causal Heterogeneity..." draws from the title of Bruce's 1998 article.

Hank Heitowit invited me to deliver a week-long series of lectures on Bayesian inference at the ICPSR Summer School at the University of Michigan for several years beginning in the late 1990s. Preparing for these lectures was my first, serious attempt at organizing an introduction to Bayesian inference in a way that would be accessible to a broad spectrum of social science graduate students. Week-long workshops at Yale University and Princeton University were similarly valuable; I thank Don Green and Larry Bartels, respectively. I gave two weeks of lectures on Bayesian inference in the social sciences at the ECPR Summer School at the University of Essex in 2008, providing me a wonderful opportunity to give much of the book a test run; I am very grateful to the students who took my class, and to my hosts at Essex (Anthony McGann, Thomas Plümper, Vera Troeger and the Summer School staff). Similarly, I thank my fellow members of the Society for Political Methodology for the opportunity to present numerous applications at various SPM summer meetings: Chris Achen, Neal Beck, Henry Brady, John Freeman, Jeff Gill, John Jackson, and Gary King have offered many helpful suggestions on presenting Bayesian inference to an audience of social science researchers. I thank Herbert Kitschelt for the opportunity to present material from Chapter 1 at Duke University, Harold Clarke for hosting a two-day workshop at UT Dallas, and workshop participants at UC Davis.

I thank my Stanford colleagues David Laitin, Doug Rivers and Jonathan Wand for feedback on drafts of Chapters 1 and 2, along with Jonathan Katz (CalTech) and Lynn Vavreck (UCLA) and my former students Sarah Anderson, John Bullock, Stephen Jessee, and Christina Maimone. Even before I went to Stanford in 1997 – but especially since – Doug Rivers has been a terrific resource and source of encouragement; we team-taught a Bayesian statistics "topics" class in Stanford's Political Science Department in the late 1990s that was enormously valuable, helping to shape my "Bayesian pedagogy", in addition to breaking ground on several projects that have since seen their way into print or into examples presented in this book. James Holland Jones and I team-taught a class on Bayesian inference for the social sciences in the Spring of 2007, which was especially helpful, making me realize that some important topics were missing from the book; I thank him and our students in that class. David Rogosa has been especially helpful with references and suggestions on multi-level modeling.

Persi Diaconis introduced me to some of the very elegant mathematics that underlies Markov chain Monte Carlo, almost all of which is beyond the level of exposition in this book. Nonetheless, my exposure to this material inspired me to provide a more rigorous treatment of Markov chains and the properties of MCMC algorithms (Chapters 4 and 5) than I would have otherwise. Persi's writing on exchangeability and de Finetti's Theorem (reproduced here as Proposition 1.9) – much of it with the late David Freedman – has also had a profound impact on my thinking about statistical modeling; I have tried to convey the important role of exchangeability and prediction in Bayesian inference in Section 1.9.

More than a few Stanford students have suffered through the evolution of my thinking about Bayesian inference and the disorganized state of early drafts of this book: the list is long – and my memory short – but I single out Andrea Abel, Marc André Bodet, Josh Clinton, Wendy Gross, Xiaojun Li, Matthew Levendusky, Jeremy Pope, Alex Tahk, Chris Tausanovitch, Shawn Treier, and Chris Warshaw.

I spent seven weeks in 2007 as a guest of the Econometrics group at the University of Sydney, as a recipient of a New South Wales Government Expatriate Return Fellowship. This was an extremely productive time and I am very grateful for the opportunity, and to the hospitality and collegiality of Richard Gerlach. Between January and August of 2009 I was a Visiting Professor at the United States Studies Centre at the University of Sydney. I thank Geoff Garrett and the staff of the Centre for allowing me to dedicate the first few months of my visit to completing the book, and in a wonderfully productive environment.

Finally, I thank my family: my wife Janet, my daughter Josephine and my son Thomas. Janet gets the dedication page, but it isn't nearly enough. For my part, countless evenings and weekend afternoons in the office are bound up in this book: as they often say, "you don't get those hours back". But what they don't often say is that it is usually the other parent – in our case, Janet – who has to pick up the slack with respect to home and family. For those thousands of hours and your hard work, Janet, "thank you".

S.D.J

Introduction

Bayesian statistical analysis relies on Bayes's Theorem, which tells us how to update *prior beliefs* about parameters and hypotheses in light of *data*, to yield *posterior beliefs*. Bayes Theorem[1] itself is utterly uncontroversial and follows directly from the conventional definition of conditional probability. A derivation and discussion of Bayes Theorem is provided in Chapter 1. For now, it is sufficient to consider the following stylized, graphical rendition of Bayesian inference:

$$\begin{array}{c}\text{prior}\\\text{beliefs}\end{array} \longrightarrow \text{data} \longrightarrow \begin{array}{c}\text{posterior}\\\text{beliefs}\end{array}$$

or equivalently,

$$p(\theta) \longrightarrow y \longrightarrow p(\theta|y)$$

That is, if θ is some object of interest, but subject to uncertainty – a parameter, a hypothesis, a model, a data point – then Bayes Theorem tells us how to rationally revise prior beliefs about θ, $p(\theta)$, in light of the data y, to yield posterior beliefs $p(\theta|y)$.

In this way Bayes Theorem provides a solution to the general problem of *induction* (e.g. Hacking 2001). In the specific case of statistical inference, Bayes Theorem provides a solution to the problem of how to *learn from data*. Thus, in a general sense, Bayesian statistical analysis is remarkably simple and even elegant, relying on this same simple recipe in each and every application.

Why be Bayesian?

This book is about the application of the Bayesian recipe in a specific context: statistical inference in social science settings. As I see it, there are three major reasons why we ought to adopt a Bayesian approach to statistical inference, and in the social sciences in particular:

Bayesian inference is simple and direct

The result of a Bayesian analysis is a *posterior probability statement*, 'posterior' in the literal sense, in that such a statement characterizes beliefs *after* looking at data. Examples include

[1] Throughout the book I will drop the 'apostrophe-s' in 'Bayes's Theorem'.

- the posterior probability that a regression coefficient is positive, negative or lies in a particular interval;
- the posterior probability that a subject belongs to a particular latent class;
- the posterior probability that a hypothesis is true; or,
- the posterior probability that a particular statistical model is true model among a family of statistical models.

Note that the posterior probability statements produced by a Bayesian analysis are probability statements over the quantities or objects of direct substantive interest to the researcher (e.g. parameters, hypotheses, models, predictions from models).

We will contrast the Bayesian approach to inference with *frequentist inference*, the method of statistical inference most commonly taught, and the method almost all social scientists encounter early in their methodological training. The contrast between the two approaches to induction is stark, simple to grasp, and important. Bayesian procedures condition on the data at hand to produce posterior probability statements about parameters and hypotheses. Frequentist procedures do just the reverse: one conditions on a null hypothesis to assess the plausibility of the data one observes (and more 'extreme' data sets that one did not observe but we might have had we done additional sampling), with another step of reasoning required to either reject or fail to reject the null hypothesis. Thus, compared to frequentist procedures, Bayesian procedures are simple and straightforward, at least conceptually. When compared to Bayesian procedures, frequentist procedures seem to be geared towards computing the wrong conditional probability.

Say what you mean, mean what you say

Many students and colleagues in the social sciences think about and talk about the end product of their (frequentist) analyses in *subjective* terms that can only be justified in a Bayesian approach. For instance, a small p-value is often thought to justify a statement of the sort 'having looked at these data, I believe that hypothesis H_0 is quite implausible'. Of course, and as I detail below, this is *not* what a p-value is: the p-value is a statement about the plausibility of the data (and other, more extreme, but hypothetical/unobserved data sets) conditional on a null hypothesis. Adopting an explicitly Bayesian approach would resolve a recurring source of confusion for these researchers, letting them say what they mean and mean what they say. In addition, a Bayesian approach would also help researchers who know the correct interpretation of a frequentist hypothesis test, but are unsure how to apply that reasoning in the case of where the data can not be considered a sample from some larger population, a fairly common situation in the social sciences (e.g. Berk, Western and Weiss 1995). I have more to say about this problematic feature of frequentist inference, below.

Investigating causal heterogeneity: hierarchical modeling

The prior density $p(\theta)$ plays a key role in Bayesian inference. Critics of Bayesian inference often point to this as weakness of Bayesian inference, in that two researchers who fit the same likelihood to the same data, but possess different prior beliefs over the parameters, will have different sets of posterior beliefs (e.g. Efron 1986). This is true, and we will return to this specific criticism later.

But the prior density also provides a way for model expansion when we work with data sets that pool data over multiple units and/or time periods. Data sets of this sort abound in the social sciences. Individuals live in different locations, with environmental factors that are constant for anyone within that location, but vary across locations; students attend different schools; voters live in different constituencies; firms operate in different types of markets; politicians compete under different sets of electoral rules, and so on. Understanding the relative weight of micro-level variables and macro-level or 'contextual' variables – and critically, the interaction between them – is central to many research programs throughout the social sciences. A key question in research of this type is how the causal structure that operates at one level of analysis (e.g. individuals) varies across a 'higher' level of analysis (e.g. localities or time periods).

The Bayesian approach to statistical inference is extremely well-suited to answering this question. Recall that in the Bayesian approach parameters are always random variables, typically (and most basically) in the sense that the researcher is unsure as to their value, but can characterize that uncertainty in the form of a prior density $p(\theta)$. But it is no great stretch to replace the prior $p(\theta)$ with a stochastic model formalizing the researcher's assumptions about the way that parameters θ might vary across groups $j = 1, \ldots, J$, perhaps as a function of observable characteristics of the groups; e.g., $\theta_j \sim f(z_j, \gamma)$, where now γ is a set of unknown *hyperparameters*. That is, the model is now comprised of a nested *hierarchy* of stochastic relations: the data from unit j, $\mathbf{y}_j$, are modeled as a function of covariates and parameters θ_j, while cross-unit heterogeneity in the θ_j is modeled as function of unit-specific covariates z_j and hyperparameters γ.

Models of this sort are known to Bayesians as *hierarchical* models, but go by many different names in different parts of the social sciences depending on the specific form of the model and the estimation strategy being used (e.g. 'random' or 'varying' coefficients models, 'multilevel' or 'mixed' models). A detailed consideration of Bayesian hierarchical modeling is deferred until Chapter 7, and hierarchical modeling will appear repeatedly in various applications throughout Part III of this book. But we will introduce some key supporting concepts before then, especially the key concept of *exchangeability* (e.g. Section 1.9.6).

It is no exaggeration to say that hierarchical modeling is one of the great successes of the Bayesian revival of the last twenty years. Simple models of causal heteroegenity – in which parameters are assumed to vary in a non-deterministic way across units – are substantively attractive and yet difficult to estimate in a classical setting. The Bayesian approach easily accommodates the notion of 'random parameters'. But in the wake of Markov chain Monte Carlo revolution (see below), Bayesian computation for these models has also become rather simple.

Bayesian simulation methods make 'hard' statistical problems easier

An important attraction of the Bayesian framework rests on more pragmatic considerations. In recent decades, the advent of cheap, fast computing power has seen simulation-based approaches to statistical inference become feasible for the typical social scientist. In particular, throughout the 1990s, there was something of an explosion of interest in Bayesian approaches, almost entirely driven by the fact that cheap computing power makes it feasible to do simulation-based, Bayesian statistical data analysis. A suite of algorithms known as Markov chain Monte Carlo (MCMC) makes the Bayesian apporach not just a theoretical curiosity, but a practical reality for applied researchers; I introduce these methods in Part II of the book.

MCMC algorithms have proven themselves amazingly powerful and flexible, and have brought wide classes of models and data sets out of the 'too hard' basket. We have already mentioned hierarchical models. But other examples include data sets with lots of missing data, or models with lots of parameters, model with latent variables, mixture models, and flexible semi- and non-parametric models, many of which we will consider in Part III of this book. Quite simply, the combination of fast, cheap computers and MCMC algorithms let researchers estimate models that they simply could not estimate previously, or at least not easily. Thus, whatever one's view of the intellectual or philosophical virtues of the Bayesian approach, by adopting a Bayesian approach to inference, researchers can avail themselves of MCMC algorithms, an extraordinarily powerful set of techniques for estimation and inference. By the end of the 1990s, statistics journals were teeming with Bayesian applications, as applied researchers exploited the power and flexibility of simulation-based analysis via MCMC methods. In Part III of the book I show how Bayesian analysis via MCMC algorithms can be extremely valuable for a wide variety of social science problems.

Bayesian inference in social science settings

The advantages of Bayesian inference just described are quite general. But, in the specific context of the social sciences, the Bayesian approach has some particular strengths that warrant elaboration.

What do you mean when you say something is 'statistically significant'?

Let me make a claim that I hope you find provocative: for many inferential problems in the social sciences, *the conventional, textbook approach to statistical inference makes no sense*. As mentioned earlier, the 'conventional, textbook approach' to statistical inference is *frequentist* inference. With only a few exceptions, every social scientist I know (and almost every statistician I know) was introduced to statistics from this frequentist perspective, although seldom is the adjective 'frequentist' used in introductory statistics classes.

Frequentist and Bayesian inference differ in several fundamental ways. Frequentist statistics does not deny the truth of Bayes Theorem, but does not rely on it for statistical inference. In fact, frequentist inference proceeds by answering a different question to that answered by a Bayesian, posterior probability. Frequentist inference asks 'assuming hypothesis H_0 is true, how often would we obtain a result at least as extreme as the result actually obtained?', where 'extreme' is relative to the hypothesis being tested. If results such as the one obtained are sufficiently rare under hypothesis H_0 (e.g. generate a sufficiently small p value), then we conclude that H_0 is incorrect, rejecting it in favor of some alternative hypothesis. Indeed, we teach our students to say that when the preceding conditions hold, we have a 'statistically significant' result.

The simplicity of this two word phrase relative to its actual, substantive content is both powerful and deceptive. We learn and repeat the simple two word phrase – 'statistically significant' – rather than the considerably longer description of the relative frequencies of extreme results in repeated sampling under a maintained null hypothesis. My experience is that in substituting the phrase 'statistical significance' for the much longer textbook definition, students quickly forget the frequentist underpinnings of what it is they are *really* asserting, and, hence seldom question whether the appeal to the long-run, repeated sampling properties of a statistical procedure is logical or realistic.

Again, contrast the Bayesian position. In the Bayesian approach we condition on the data at hand to assess the plausibility of a hypothesis (via Bayes Rule), while the frequentist approach conditions on a hypothesis to assess the plausibility of the data (or more extreme data sets), with another step of reasoning required to either reject or fail to reject hypotheses.

This difference may not seem important. But some key conceptual issues are at stake and are especially relevant for the social sciences. For a frequentist, the key, inferential question is the 'how often' question, as in 'how often would we obtain a result at least as extreme as the result actually obtained?' Hence the 'frequentist' moniker: the frequentist p-value is the relative *frequency* of obtaining a result at least as extreme as the result actually obtained, assuming hypothesis H_0 to be true, where the *sampling distribution* of the result tells us how to assess relative frequencies of possible different results, under H_0. But what about cases where *repeated sampling makes no sense, even as a thought experiment?*

Data that are not samples

Consider researchers analyzing cross-national data in economics, political science, or sociology, say, using national accounts data from the OECD. Alternatively, consider researchers in American politics analyzing all trade votes in the United States Senate in a given period, or an analysis of the determinants of civil war, using an authoritative, comprehensive listing of all civil wars in the world since World War Two.

In what sense do data such as these comprise a sample from a population? Repeating the data collection exercise would not yield a new sample from the population, but rather, save for coding error or other errors in transmission or transcription of the data, 'repeated sampling' would yield exactly the same data set. Thus, the situation encountered in many social science settings is very different from the thought experiments of introductory statistics: e.g. flip a coin n times, record $\hat{\pi}_{(n)}$, the proportion of the n flips that came up heads, repeating the process *ad infinitum*, thereby generating the sampling distribution of $\hat{\pi}_{(n)}$ through brute force (albeit imaginary) random sampling. No, for the analyst of the OECD national accounts data, repeated 'sampling' will not yield different values for the variables in the analysis, nor for the parameter estimates produced by model fitting. In cases such as these *there is no uncertainty due to variation in repeated sampling from a population*: the data available for analysis exhaust the population of substantive interest.

Nonetheless, one can feed the OECD data to a computer program, and have standard errors and significance tests reported for various statistics (e.g. means, correlations, regressions) as usual. But what do those standard errors *mean* in this context? What is a cross-national researcher really saying when they report that such-and-such a regression coefficient in their cross-national statistical analysis is 'statistically significant'?

At this point several counter-arguments come to mind. To be sure, coding error might see the numbers changing over repetitions of the data collection process, say, if the data were being manually entered into a spreadsheet. But this form of measurement error may have quite different operational characteristics than classical sampling error: e.g. what is the population of possible data sets that might be generated via coding error, and how are we (randomly?) sampling from that population? Are those coding errors distributed randomly, and if so, via what distribution? In short, does a central limit theorem apply to the resulting estimator and standard errors? That is, what assumptions need to hold such that can one could interpret the standard errors produced by our statistical software as a measure of the variability of estimates across repetitions of the data collection process? Almost surely not: as $n \rightarrow \infty$, frequentist sampling error vanishes (and, for independent data, at rate $\sqrt{n}$), but this hardly seems likely for coding errors.

Nor is it satisfying or particularly helpful to say that the data are just one of many possible data sets that could have been generated if 'history were to be replayed many times over'. What is the sampling mechanism that selected the history we did happen to observe? No one knows, or can know. And frankly, it is astonishing to listen to well trained, quantitatively literate, professional social scientists attempt to rationalize a repeated sampling, frequentist approach to inference when working with non-repeatable data (or, at least when they are confronted the conundrum of their data being non-repeatable). Such is the power of the simple little phrase, 'statistically significant'. Frankly, adhering to a frequentist conception of probability in the face of non-repeatable data risks intellectual embarrassment.

The Bayesian approach provides a coherent basis for inference when working with non-repeatable data. As I detail in Sections 1.1 and 1.2, almost all Bayesian statisticians adopt a *subjective* approach to probability. That is, uncertainty attaches to the researcher's state of knowledge about quantities of interest (data, parameters, etc.). As the name suggests, *frequentist* probability is conceived of as a relative frequency. But, as the previous discussion indicates, this frequentist conception of probability runs into difficulties when the thought experiment of repeated sampling is inappropriate. In turn, this undermines the substantive meaning ascribed to standard errors and hypothesis tests in a frequentist analysis.

Moreover, it is not clear how to attach frequentist probabilities to past or non-repeatable events about which I may be uncertain (e.g. the probability that an asteroid impact caused the K-T mass extinction), or future events (e.g. the probability that Democrats will hold a majority of seats in the US. House of Representatives after the next congressional elections). That is, there is a large class of events of great interest to social scientists for which the frequentist notion of probability is inappropriate.

Understanding the confidence interval

Suppose we administer a survey and find that the mean income reported by the respondents is \$45 000, with a standard error of \$2500. With a large sample, a 95 % confidence interval for the mean level of income in the population, μ, is \$40 000 to \$50 000. But what, exactly, is the interpretation of this confidence interval? Again, my experience is that many colleagues and students are confused on this score. One often heard interpretation of the 95 % confidence interval in this instance is 'there is a .95 probability that μ lies between \$40 000 and \$50 000'.

But this is not the correct frequentist interpretation. Recall that in the frequentist approach, parameters are fixed characteristics of populations, so μ either lies in the interval or it doesn't. The correct interpretation of a frequentist confidence interval concerns the repeated sampling characteristics of a sample statistic. In the case of a 95 % confidence interval, the correct frequentist interpretation is that 95 % of the 95 % confidence intervals one would draw in repeated samples will include μ. Now, is the 95 % confidence interval that one constructs from the data set at hand one of the lucky 95 % that actually contains μ, or not? No ones knows.

This example highlights how the repeated sampling, frequentist notion of probability is simply not how ordinary researchers use and understand the term 'probability'. From the frequentist perspective, the statement 'there is a .95 probability that μ lies between \$40 000 and \$50 000' is valid, since for frequentists 'probability' is at least tacitly understood to mean 'relative frequency in repeated sampling'. But this is not how most practioners use confidence intervals. Rather, subjective statements of the sort 'I am 95 % sure that $\mu \in$ [\$40 000, \$50 000]' are quite typical of the way confidence intervals are interpreted by colleagues and students in the social sciences. Alas, the correct frequentist interpretation is the less helpful statement about the performance of the 95 % confidence interval in repeated sampling. This leads to considerable confusion, and frankly, makes teaching and learning statistics harder than it should be.

For instance, many readers may recall their first encounter with (frequentist) statistics as follows:

> When I first learned a little statistics, I felt confused ... Not because the mathematics was difficult ... but because I found it difficult to follow the logic by which inferences were arrived from data .... the statement that a 95 % confidence interval for an unknown parameter ran from −2 to +2 sounded as if the parameter lay in that interval with 95 % probability and yet I was warned that all I could say was that if I carried out similar procedures time after time then the unknown parameters would lie in the confidence intervals I constructed 95 % of the time. It appeared that the books I looked at were not answering the questions that would naturally occur to a beginner ... (Lee 2004, xiii).

Indeed, statements about the performance of a statistical procedure in repeated sampling are only of so much use to the researcher or policy maker who typically wants to know what they should believe about a parameter given relevant data. And bear in mind that data collection in the social sciences is often a single-shot affair: repeated sampling is unrealistic, even in the contexts where it is hypothetically possible (such as survey research of large populations). Again, it is in this sense that Bayesian procedures offer a more direct path to inference; as I put it earlier, the Bayesian approach lets researchers mean what they say and say what they mean. For instance, the statement, 'having looked at the data, I am 95 % sure that $\mu \in$ [\$40 000, \$50 000]' is a natural product of a Bayesian analysis, a characterization of the researcher's *beliefs* about a parameter in formal, probabilistic terms, rather than a statement about the repeated sampling properties of a statistical *procedure*.

But is it scientific?

Some readers might find the last few pages diconcerting. My experience is that more than a few social scientists find the very notion that there are different ways of 'doing statistics' disquieting. After all, isn't it statistics that helps put the 'science' in 'social science'? Isn't it through statistical analysis of data that our hunches and conjectures about social processes are either falsified or not? Well, yes, but as anyone who has ever used statistical methods will recognize, data analysis requires many 'judgement calls' by the researcher. Familiar examples include the perennial question 'what variables go into my analysis?', say, on the right-hand side of a regression equation (e.g. Leamer 1978). Is, say, ordinary least squares appropriate for the model being estimated, given the data on hand? To what extent are the variables measured with error? Are the predictors in my regression equation exogenous? The data themselves can help supply answers to some of these questions, but it is often the case that social scientists are stuck with small data sets that were generated under less than ideal conditions, and subjective judgements about model specification and model adequacy are unavoidable.

So then, let us acknowledge that subjectivity is inherent to the scientific exercise. The Bayesian approach rests on a subjective notion of probability, but demands that subjective beliefs conform to the laws of probability. Put differently, in the Bayesian approach, the subjectivity of scientists is acknowledged, but simultaneously insists that subjectivity be rational, in the sense that when confronted with evidence, subjective beliefs are updated rationally, in accord with the axioms of probability.

Doing Bayesian statistics

As we shall see, for all its virtues, Bayesian analysis is often more easily talked about or described than actually *done*, or at least this was the case up until recently. The mathematics and computation underlying Bayesian analysis has been dramatically simplified via a suite of algorithms known collectively as Markov chain Monte Carlo (MCMC), to be discussed in Chapter 4. The combination of the popularization of MCMC and vast increases in the computing power available to social scientists means that Bayesian analysis is now well and truly part of the mainstream of quantitative social science, as we will see in some detail in later chapters. Indeed, whatever one's view of the philosophical or intellectual virtues of Bayesianism, there is no doubting that Bayesian modeling, say, post-1990, is tremendously powerful, useful, and popular. Models and data sets that were once consigned to the 'too hard' basket can now be feasibly tackled by workaday social scientists, and this is one of my chief motivations for writing this book.

Despite these important pragmatic reasons for adopting the Bayesian approach, it is important to remember that MCMC algorithms are *Bayesian* algorithms: they are tools that simplify the computation of posterior densities. So, before we can fully and sensibly exploit the power of MCMC algorithms, it is important that we understand the foundations of Bayesian inference, and this is the subject of this first part of the book.

Part I

Introducing Bayesian Analysis

1

The foundations of Bayesian inference

In this chapter I elaborate on the overview of Bayesian statistical inference provided in the introduction. I begin by reviewing the fundamental role of probability in statistical inference. In the Bayesian approach, probability is usually interpreted in subjective terms, as a formal, mathematically rigorous characterization of beliefs. I distinguish the subjective notion of probability from the classical, objective or frequentist approach, before stating Bayes Theorem in the various forms it is used in statistical settings. I then review how Bayesian data analysis is actually done. At a high level of abstraction, Bayesian data analysis is extremely simple, following the same, basic recipe: via Bayes Rule, we use the data to update prior beliefs about unknowns. Of course, there is much to be said on the implementation of this procedure in any specific application, and these details are the subjects of later chapters. The discussion in this chapter deals with some general issues. For instance, how does Bayesian inference differ from classical inference? Where do priors come from? What is the result of a Bayesian analysis, and how does one report those results? How does hypothesis testing work in the Bayesian approach? What kinds of considerations motivate model specification in the Bayesian approach?

1.1 What is probability?

As a formal, mathematical matter, the question 'what is probability?' is utterly uncontroversial. The following axioms, known as the Kolmogorov (1933) axioms, constitute the conventional, modern, mathematical defintion of probability, which I reproduce here (with measure-theoretic details omitted; see the Appendix for a more rigorous set of definitions). If Ω is a set of events, and $P(A)$ is a function that assigns real numbers to events $A \subset \Omega$, then $P(A)$ is a probability measure if

1. $P(A) \geq 0, \forall A \subset \Omega$ (probabilities are non-negative)

Bayesian Analysis for the Social Sciences S. Jackman

2. $P(\Omega) = 1$ (probabilities sum to one)
3. If A and B are disjoint events, then $P(A \cup B) = P(A) + P(B)$ (the joint probability of disjoint events is the sum of the probabilities of the events).

On these axioms rests virtually all of contemporary statistics, including Bayesian statistics. This said, one of the ways in which Bayesian statistics differs from classical statistics is in the *interpretation* of probability. The very idea that probability is a concept open to interpretation might strike you as odd. Indeed, Kolmogorov himself ruled out any questions regarding the interpretation of probabilities:

> The theory of probability, as a mathematical discipline, can and should be developed from axioms in exactly the same way as Geometry and Algebra. This means that after we have defined the elements to be studied and their basic relations, and have stated the axioms by which these relations are to be governed, all further exposition must be based exclusively on these axioms, independent of the usual concrete meaning of these elements and their relations (Kolmogorov 1956, 1).

Nonetheless, for anyone actually deploying probability in a real-world application, Kolmogorov's insistence on a content-free definition of probability is quite unhelpful. As Leamer (1978, 24) points out:

> These axioms apply in many circumstances in which no one would use the word probability. For example, your arm may contain 10 percent of the weight of your body, but it is unlikely that you would report that the probability of your arm is .1.

Thus, for better or worse, probability is open to interpretation, and has been for a long time. Differences in interpretation continue to be controversial (although less so now than, say, 30 years ago), are critical to the distinction between Bayesian and non-Bayesian statistics, and so no book-length treatment of Bayesian statistics can ignore it. Most thorough, historical treatments of probability identify at least *four* interpretations of probability (e.g., Galavotti 2005). For our purposes, the most important distinction is between probability as it was probably (!) taught to you in your first statistics class, and probability as interpreted by most Bayesian statisticians.

1.1.1 Probability in classical statistics

In classical statistics probability is often understood as a property of the phenomenon being studied: for instance, the probability that a tossed coin will come up heads is a characteristic of the coin. Thus, by tossing the coin many times under more or less identical conditions, and noting the result of each toss, we can estimate the probability of a head, with the precision of the estimate monotonically increasing with the number of tosses. In this view, probability is the limit of a long-run, relative frequency; i.e. if A is an event of interest (e.g. the coin lands heads up) then

$$\Pr(A) = \lim_{n \to \infty} \frac{m}{n}$$

is the probabilty of A, where m is the number of times we observe the event A and n is the number of repetitions. Given this definition of probability, we can understand why classicial statistics is sometimes referred to as

1. *frequentist*, in the sense that it rests on a definition of probability as the long-run relative *frequency* of an event;
2. *objectivist*, in the sense that probabilities are characteristics of objects or things (e.g. the staples of introductory statistics, such as cards, dice, coins, roulette wheels); this position will be contrasted with a *subjectivist* interpretation of probability.

One of the strongest statements of the frequentist position comes from Richard von Mises:

> we may say at once that, up to the present time [1928], no one has succeeded in developing a complete theory of probability without, sooner or later, introducing probability by means of the relative frequencies in long sequences.

Further,

> The rational concept of probability, which is the only basis of probability calculus, applies only to problems in which either the same event repeats itself again and again, or a great number of uniform elements are involved at the same time... [In] order to apply the theory of probability we must have a practically unlimited sequence of observations (quoted in Barnett 1999, 76).

As we shall see, alternative views long pre-date von Mises' 1928 statement and it is indeed possible to apply the theory of probability without a 'practically unlimited' sequence of observations. This is just as well, since many statistical analyses in the social sciences are conducted without von Mises' 'practically unlimited' sequence of observations.

1.1.2 Subjective probability

Most introductions to statistics are replete with examples from games of chance, and the naïve view of the history of statistics is that interest in games of chance spurred the development of probability (e.g. Todhunter 1865), and, in particular, the frequentist interpretation of probability. That is, for simple games of chance it is feasible to enumerate the set of possible outcomes, and hence generate statements of the likelihood of particular outcomes in relative frequency terms (e.g. the 'probability' of throwing a seven with two dice, an important quantity in craps). But historians of science stress that at least two notions of probability were under development from the late 1600s onwards: the objectivist view described above, and a subjectivist view. According to Ian Hacking, the former is 'statistical, concerning itself with stochastic laws of chance processes', while the other notion is 'epistemological, dedicated to assessing reasonable degrees of belief in propositions' (Hacking 1975, 12).

As an example of the latter, consider Locke's *Essay Concerning Human Understanding* (1698). Book IV, Chapter XV of the *Essay* is titled 'On Probability', in which Locke notes that 'most of the propositions we think, reason, discourse – nay, act upon, are such that we cannot have undoubted knowledge of their truth.' Moreover, there are 'degrees' of belief, 'from the very neighborhourhood of certainty and demonstration, quite down to improbability and unlikeliness, even to the confines of impossibility'. For Locke, 'Probability is likeliness to be true', a definition in which (repeated) games of chance play no part.

The idea that one might hold different degrees of belief over different propositions has a long lineage, and was apparent in the theory of proof in Roman and canon law, in which judges were directed to employ an 'arithmetic of proof', assigning different weights to various pieces of evidence, and to draw distinctions between 'complete proofs' or 'half proofs' (Daston 1988, 42–43). Scholars became interested in making these notions more rigorous, with Leibniz perhaps the first to make the connection between the qualitative use of probabilistic reasoning in jurisprudence with the mathematical treatments being generated by Pascal, Huygens, and others.

Perhaps the most important and clearest statement linking this form of jurisprudential 'reasoning under uncertainty' to 'probability' is Jakob Bernoulli's posthumous *Ars conjectandi* (1713). In addition to developing the theorem now known as the weak law of large numbers, in Part IV of the *Ars conjectandi* Bernoulli declares that 'Probability is degree of certainty and differs from absolute certainty as the part differs from the whole', it being unequivocal that the 'certainty' referred to is a state of mind, but, critically, (1) varied from person to person (depending on one's knowledge and experience) and (2) was quantifiable. For example, for Bernoulli, a probability of 1.0 was an absolute certainty, a 'moral certainty' was nearly equal to the whole certainty (e.g., 999/1000, and so a morally impossible event has only $1 - 999/1000 = 1/1000$ certainty), and so on, with events having 'very little part of certainty' still nonetheless being possible.

In the early-to-mid twentieth century, the competition between the frequentist and subjectivist interpretations intensified, in no small measure reflecting the competition between Bayesian statistics and the then newer, frequentist statistics being championed by R. A. Fisher. Venn (1866) and later von Mises (1957) made a strong case for a frequentist approach, apparently in reaction to 'a growing preoccupation with subjective views of probability' (Barnett 1999, 76). During this period, both the objective/frequentist and subjective interpretations of probability were formalized in modern, mathematical terms – von Mises formalizing the frequentist approach, and Ramsey (1931) and de Finetti (1974, 1975) providing the formal links between subjective probability and decisions and actions.

Ramsey and de Finetti, working independently, showed that subjective probability is not just *any* set of subjective beliefs, but beliefs that conform to the axioms of probability. The Ramsey-de Finetti Theorem states that if $p_1, p_2, \ldots$ are a set of betting quotients on hypotheses $h_1, h_2, \ldots$, then if the p_j do not satisfy the probability axioms, there exists a betting strategy and a set of stakes such that whoever follows this betting strategy will lose a finite sum whatever the truth values of the hypotheses turn out to be (e.g. Howson and Urbach 1993, 79). This theorem is also known as the Dutch Book Theorem, a Dutch book being a bet (or a series of bets) in which the bettor is guaranteed to lose.

In de Finetti's terminology, subjective probabilities that fail to conform to the axioms of probability are *incoherent* or *inconsistent*. Thus, subjective probabilities are whatever a particular person believes, provided they satisfy the axioms of probability. In particular, the Dutch book results extend to the case of *conditional probabilities*, meaning that if I do not update my subjective beliefs in light of new information (data) in a manner consistent with the probability axioms, and you can convince me to gamble with you, you have the opportunity to take advantage of my irrationality, and are guaranteed to profit at my expense. That is, while probability may be subjective, Bayes Rule governs how rational people should update subjective beliefs.

1.2 Subjective probability in Bayesian statistics

Of course, it should come as no suprise that the subjectivist view is almost exclusively adopted by Bayesians. To see this, recall the proverbial coin tossing experiment of introductory statistics. And further, recall the goal of Bayesian statistics: to update probabilities in light of evidence, via Bayes' Theorem. But which probabilities? The objective sense (probability as a characteristic of the coin) or the subjective sense (probability as degree of belief)? Well, almost surely we do not mean that the coin is changing; it is conceivable that the act of flipping and observing the coin is changing the tendency of the coin to come up heads when tossed, but unless we are particularly violent coin-tossers this kind of physical transformation of the coin is of an infintisimal magnitude. Indeed, if this occured then both frequentist and Bayesian inference gets complicated (multiple coin flips no longer constitute an independent and identically distributed sequence of random events). No, the probability being updated here can only be a subjective probability, the observer's degree of belief about the coin coming up heads, which may change while observing a sequence of coin flips, via Bayes' Theorem.

Bayesian probability statements are thus about states of mind over states of the world, and not about states of the world *per se*. Indeed, whatever one believes about determinism or chance in social processes, the meaningful uncertainty is that which resides in our brains, upon which we will base decisions and actions. Again, consider tossing a coin. As Emile Borel apparently remarked to de Finetti, one can guess the outcome of the toss while the coin is still in the air and its movement is perfectly determined, or even after the coin has landed but before one reviews the result; that is, subjective uncertainty obtains irrespective of 'objective uncertainty (however conceived)' (de Finetti 1980b, 201). Indeed, in one of the more memorable and strongest statements of the subjectivist position, de Finetti writes

> **PROBABILITY DOES NOT EXIST**
>
> The abandonment of superstitious beliefs about...Fairies and Witches was an essential step along the road to scientific thinking. Probability, too, if regarded as something endowed with some kind of objective existence, is not less a misleading misconception, an illusory attempt to exteriorize or materialize our true probabilistic beliefs. In investigating the reasonableness of our own modes of thought and behaviour under uncertainty, all we require, and all that we are reasonably entitled to, is consistency among these beliefs, and their reasonable relation to any kind of relevant objective data ('relevant' in as much as subjectively deemed to be so). This is Probability Theory (de Finetti 1974, 1975, x).

The use of subjective probability also means that Bayesians can report probabilities without a 'practically unlimited' sequence of observations. For instance, a subjectivist can attach probabilities to the proposition 'Andrew Jackson was the eighth president of the United States' (e.g. Leamer 1978, 25), reflecting his or her degree of belief in the proposition. Contrast the frequentist position, in which probability is defined as the limit of a relative frequency. What is the frequentist probability of the truth of the proposition 'Jackson was the eighth president'? Since there is only one relevant experiment for this problem, the frequentist probability is either zero (if Jackson was not the eighth president) or one (if Jackson was the eighth president). Non-trivial frequentist probabilities, it seems, are reserved for phenomena that are standardized and repeatable (e.g. the exemplars of introductory statistics such as coin tossing and cards, or, perhaps, random sampling in survey research). Even greater difficulties for the frequentist position arise when considering events that have not yet occured, e.g.

- What is the probability that the Democrats win a majority of seats in the House of Representatives at the next Congressional elections?
- What is the probability of a terrorist attack in the United States in the next five years?
- What is the probability that over the course of my life, someone I know will be incarcerated?

All of these are perfectly legitimate and interesting social-scientific questions, but for which the objectivist/frequentist position apparently offers no helpful answer.

With this distinction between objective and subjective probability firmly in mind, we now consider how Bayes Theorem tells us how we should rationally update subjective, probabilistic beliefs in light of evidence.

1.3 Bayes theorem, discrete case

Bayes Theorem itself is uncontroversial: it is merely an accounting identity that follows from the axioms of probability discussed above, plus the following additional definition:

Definition 1.1 (Conditional probability). *Let A and B be events with $P(B) > 0$. Then the conditional probability of A given B is*

$$P(A|B) = \frac{P(A \cap B)}{P(B)} = \frac{P(A, B)}{P(B)}.$$

Although conditional probability is presented here (and in most sources) merely as a definition, it need not be. de Finetti (1980a) shows how coherence requires that conditional probabilities behave as given in Definition 1.1, in work first published in 1937. The thought experiment is as follows: consider selling a bet at price $P(A) \cdot S$, that pays S if event A occurs, but is annulled if event B does not occur, with $A \subseteq B$. Then unless your conditional probability $P(A|B)$ conforms to the definition above, someone could collect arbitrarily large winnings from you via their choice of the stakes S; Leamer (1978, 39–40) provides a simple retelling of de Finetti's argument.

Conditional probability is derived from more elementary axioms (rather than presented as a definition) in the work of Bernardo and Smith (1994, ch. 2). Some authors work with a set of probability axioms that are explicitly conditional, consistent with the notion that there are no such things as unconditional beliefs over parameters; e.g. Press (2003, ch. 2) adopts the conditional axiomization of probability due to Rényi (1970) and see also the treatment in Lee (2004, ch. 1).

The following two useful results are also implied by the probability axioms, plus the definition of conditional probability:

Proposition 1.1 (Multiplication rule)

$$P(A \cap B) = P(A, B) = P(A|B)P(B) = P(B|A)P(A)$$

Proposition 1.2 (Law of total probability)

$$\begin{aligned} P(B) &= P(A \cap B) + P(\sim A \cap B) \\ &= P(B|A)P(A) + P(B|\sim A)P(\sim A) \end{aligned}$$

Bayes Theorem can now be stated, following immediately from the definition of conditional probability:

Proposition 1.3 (Bayes Theorem). *If A and B are events with $P(B) > 0$, then*

$$P(A|B) = \frac{P(B|A)P(A)}{P(B)}$$

Proof. By proposition 1.1 $P(A, B) = P(B|A)P(A)$. Substitute into the definition of the conditional probability of $P(A|B)$ given in Definition 1.1. ◁

Bayes Theorem is much more than an interesting result from probability theory, as the following re-statement makes clear. Let H denote a hypothesis and E evidence (data), then we have

$$\Pr(H|E) = \frac{\Pr(E \cap H)}{\Pr(E)} = \frac{\Pr(E|H)\Pr(H)}{\Pr(E)}$$

provided $\Pr(E) > 0$. In this version of Bayes Theorem, $\Pr(H|E)$ is the probability of H *after* obtaining E, and $\Pr(H)$ is the *prior* probability of H before considering E. The conditional probability on the left-hand side of the theorem, $\Pr(H|E)$, is usually referred to as the *posterior* probability of H. Bayes Theorem thus supplies a solution to the general problem of inference or induction (e.g. Hacking 2001), providing a mechanism for learning about the plausibility of a hypothesis H from data E.

In this vein, Bayes Theorem is sometimes referred to as the *rule of inverse probability*, since it shows how a conditional probability B given A can be 'inverted' to yield the conditional probability A given B. This usage dates back to Laplace (e.g. see Stigler 1986b), and remained current up until the popularization of frequentist methods in the early twentieth century – and, importantly, criticism of the Bayesian approach by R. A. Fisher (Zabell 1989a).

I now state another version of Bayes Theorem, that is actually more typical of the way the result is applied in social-science settings.

Proposition 1.4 (Bayes Theorem, multiple discrete events). *Let $H_1, H_2, \dots H_k$ be mutually exclusive and exhaustive hypotheses, with $P(H_j) > 0 \; \forall j = 1, \dots, k$, and let E be evidence with $P(E) > 0$. Then, for $i = 1, \dots, k$,*

$$P(H_i|E) = \frac{P(H_i)P(E|H_i)}{\sum_{j=1}^{k} P(H_j)P(E|H_j)}.$$

Proof. Using the definition of conditional probability, $P(H_i|E) = P(H_i, E)/P(E)$. But, again using the definition of conditional probability, $P(H_i, E) = P(H_i)P(E|H_i)$. Similarly, $P(E) = \sum_{j-1}^{k} P(H_j)P(E|H_j)$, by the law of total probability (proposition 1.2). ◁

■ Example 1.1

Drug testing. Elite athletes are routinely tested for the presence of banned performance-enhacing drugs. Suppose one such test has a false negative rate of .05 and a false positive rate of .10. Prior work suggests that about 3 % of the subject pool uses a particular prohibited drug. Let H_U denote the hypothesis 'the subject uses the prohibited substance'; let $H_{\sim U}$ denote the contrary hypothesis. Suppose a subject is drawn randomly from the subject pool for testing, and returns a positive test, and denote this event as E. What is the posterior probability that the subject uses the substance? Via Bayes Theorem in Proposition 1.4,

$$\begin{aligned} P(H_U|E) &= \frac{P(H_U)P(E|H_U)}{\sum_{i\in\{U,\sim U\}} P(H_i)P(E|H_i)} \\ &= \frac{.03 \times .95}{(.03 \times .95) + (.97 \times .10)} \\ &= \frac{.0285}{.0285 + .097} \\ &\approx .23 \end{aligned}$$

That is, in light of (1) the positive test result (the evidence, E), (2) what is known about the *sensitivity* of the test, $P(E|H_U)$, and (3) the *specificity* of the test, $1 - P(E|H_{\sim U})$, we revise our beliefs about the probability that the subject is using the prohibited substance from the baseline or prior belief of $P(H_U) = .03$ to $P(H_U|E) = .23$. Note that this posterior probability is still substantially below .5, the point at which we would say it is more likely than not that the subject is using the prohibited substance.

■ Example 1.2

Classifying Congressional districts. The United States House of Representatives consists of 435 Congressional districts. Even a casual, visual inspection of district level election results suggests that there are $J = 3$ 'clumps' or classes of districts: Republican seats ($T_i = 1$), Democratic seats ($T_i = 2$), and a small cluster of extremely Democratic seats ($T_i = 3$); see Figure 6.6 in Example 6.8. Let y_i be the proportion of the two-party vote won by the Democratic candidate for Congress in district i, and λ_j be the proportion of districts in class j (i.e. $\sum_{j=1}^{J} \lambda_j = 1$). We will assume that the distribution of the y_i within each of the $J = 3$ classes is well approximated by a normal distribution, i.e. $y_i|(T_i = j) \sim N(\mu_j, \sigma_j^2)$.

Analysis of data from the 2000 U.S. Congressional elections ($n = 371$ contested districts) suggests the following values for μ_j, σ_j and λ_j (to two decimal places, see Example 6.8 for details):

Class	μ_j	σ_j	λ_j
1. Republican	.35	.08	.49
2. Democratic	.66	.10	.46
3. Extremely Democratic	.90	.03	.05

By Bayes Theorem (as stated in Proposition 1.4), the probability that district i belongs to class j is

$$P(T_i = j|y_i) = \frac{P(T_i = j) \cdot P(y_i|T_i = j)}{\sum_{k=1}^{J} \left[P(T_i = k) \cdot P(y_i|T_i = k)\right]}$$

$$= \frac{\lambda_j \cdot \phi(y_i; \mu_j, \sigma_j^2)}{\sum_{k=1}^{J} \left[\lambda_k \cdot \phi([y_i - \mu_k]/\sigma_k)\right]} \qquad (1.1)$$

where $\phi(y; \mu, \sigma)$ is the normal probability density function (see Definition B.30).

In 2000, California's 15th congressional district was largely comprised of Silicon Valley suburbs, at the southern end of the San Francisco Bay Area, and some of the wealthy, neighboring suburban communities running up into the Santa Cruz mountains. The incumbent, Republican Tom Campbell, had been re-elected in 1998 with over 61 % of the two-party vote, but vacated the seat in order to run for the US Senate: according to the *Almanac of American Politics* (Barone, Cohen and Ujifusa 2002, 198),

> the authorities at Stanford Law School had told him [Campbell] he would lose tenure if he stayed in Congress, so instead of winning another term in the House as he could easily have done, he decided to gamble and win either the Senate or Stanford. Predictably, Stanford won.

In the parlance of American politics, CA-15 was an 'open seat' in 2000. An interesting question is the extent to which Campbell's incumbency advantage had been depressing

Democratic vote share. With no incumbent contesting the seat in 2000, it is arguable that the 2000 election would provide a better gauge of the district's type. The Democratic candidate, Mike Honda, won with 56 % of the two-party vote. So, given that $y_i = .56$, to which class of congressional district should we assign CA-15? An answer is given by substituting the estimates given in the above table into the version of Bayes Theorem given in Equation 1.1: to two decimal places we have

$$P(T_i = 1|y_i = .56) = \frac{.49 \times \phi(.56; \mu = .35, \sigma^2 = .07^2)}{.49 \times \phi(.56; \mu = .35, \sigma^2 = .07^2) + .46 \times \phi(.56; \mu = .66, \sigma^2 = .10^2) + .05 \times \phi(.56; \mu = 90, \sigma^2 = .03^2)}$$

which is approximately .07 (to two decimal places). Similarly, we obtain $P(T_i = 2|y_i = .56) = .93$ and $P(T_i = 3|y_i = .56) \approx 0$.

That is, the posterior probability that CA-15 belongs to the 'Democratic' class is .96. Note that the result in CA-15, $y_i = .56$ lies a long way from the 'extremely Democratic' class ($\mu_3 = .90, \sigma_3 = .03$) and so the probability of assigning CA-15 to that class is virtually zero.

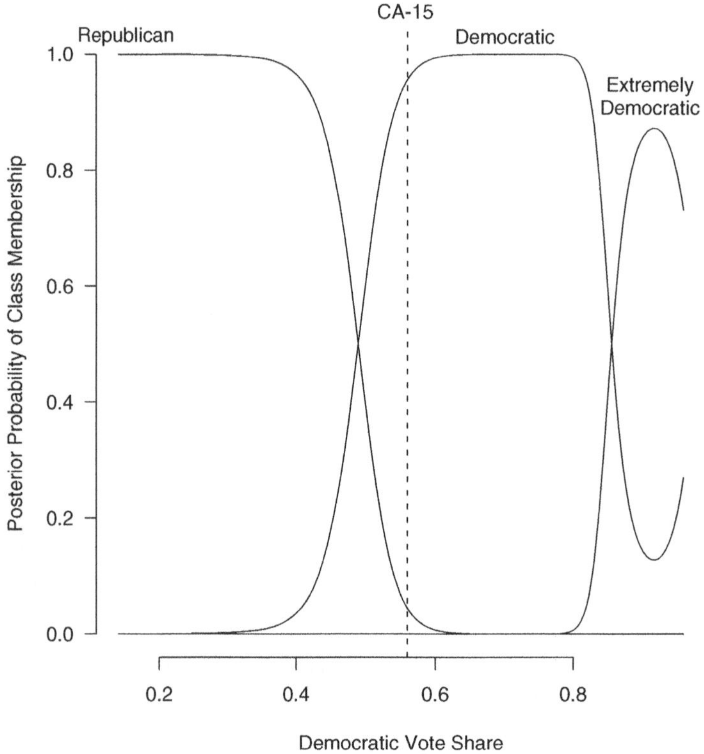

Figure 1.1 Posterior probability of class membership, Congressional districts. The probability that CA-15 ($y_i = .56$) belongs to the 'Democratic' class is .96.

This calculation can be repeated for any plausible value of y_i, and hence over any range of plausible values for y_i, showing how posterior classification probabilities change as a function of y_i. Figure 1.1 presents a graph of the posterior probability of membership in each of three classes of congressional district, as Democratic congressional vote share ranges over the values observed in the 2000 election. We will return to this example in Chapter 5.

1.4 Bayes theorem, continuous parameter

In most analyses in the social sciences, we want to learn about a continuous parameter, rather than the discrete parameters considered in the discussion thus far. Examples include the mean of a continuous variable, a proportion (a continuous parameter on the unit interval), a correlation, or a regression coefficient. In general, let the unknown parameter be θ and denote the data available for analysis as $\mathbf{y} = (y_1, \ldots, y_n)'$. In the case of continuous parameters, beliefs about the parameter are represented as *probability density functions* or pdfs (see Definition B.12); we denote the prior pdf as $p(\theta)$ and the posterior pdf as $p(\theta|\mathbf{y})$.

Then, Bayes Theorem for a continuous parameter is as follows:

Proposition 1.5 (Bayes Theorem, continuous parameter).

$$p(\theta|\mathbf{y}) = \frac{p(\mathbf{y}|\theta)p(\theta)}{\int p(\mathbf{y}|\theta)p(\theta)d\theta}$$

Proof. By the multiplication rule of probability (Proposition 1.1),

$$p(\theta, y) = p(\theta|y)p(y) = p(y|\theta)p(\theta), \qquad (1.2)$$

where all these densities are assumed to exist and have the properties $p(z) > 0$ and $\int p(z)dz = 1$ (i.e. are *proper* probability densities, see Definitions B.12 and B.13). The result follows by re-arranging the quantities in Equation 1.2 and noting that $p(y) = \int p(y, \theta)d\theta = \int p(y|\theta)p(\theta)d\theta$. ◁

Bayes Theorem for continuous parameters is more commonly expressed as follows, perhaps the most important formula in this book:

$$p(\theta|\mathbf{y}) \propto p(\mathbf{y}|\theta)p(\theta), \qquad (1.3)$$

where the constant of proportionality is

$$\left[\int p(\mathbf{y}|\theta)p(\theta)d\theta\right]^{-1}$$

i.e. ensuring that the posterior density integrates to one, as a proper probability density must (again, see Definitions B.12 and B.13).

The first term on the right hand side of Equation 1.3 is the *likelihood function* (see Definition B.16), the probability density of the data $\mathbf{y}$, considered as a function of θ. Thus, we can state this version of Bayes Theorem in words, providing the 'Bayesian mantra',

> *the posterior is proportional to the prior times the likelihood.*

This formulation of Bayes Rule highlights a particularly elegant feature of the Bayesian approach, showing how the likelihood function $p(\mathbf{y}|\theta)$ can be 'inverted' to generate a probability statement about θ, given data $\mathbf{y}$.

Figure 1.2 shows the Bayesian mantra at work for a simple, single-parameter problem: the success probability, $\theta \in [0, 1]$, underlying a binomial process, an example which we will return to in detail in Chapter 2. Each panel shows a combination of a prior, a likelihood, and a posterior distribution (with the likelihood re-normalized to be comparable to the prior and posterior densities).

The first two panels in the top row of Figure 1.2 have a uniform prior, $\theta \sim \text{Unif}(0, 1)$, and so the prior is absorbed into the constant of proportionality, resulting in a posterior density over θ that is proportional to the likelihood; given the normalization of the likelihood I use in Figure 1.2, the posterior and the likelihood graphically coincide. In these cases, the mode of the posterior density is also that value of θ that maximizes the likelihood function. For the special case considered in Figure 1.2, the prior distribution $\theta \sim \text{Unif}(0, 1)$ corresponds to an *uninformative prior* over θ, the kind of prior we might specify when we have no prior information about the value of θ, and hence no way to *a priori* prefer one set of values for θ over any other. Of course, there is another way to interpret this result: from a Bayesian perspective, likelihood based analyses of data assume prior ignorance, although seldom is this assumption made explicit, even if it were plausible. In the examples we encounter in later chapters, we shall see circumstances in which prior ignorance is plausible, and cases in which it is not. We will also consider the priors that generate 'the usual answer' for well-known problems (e.g. estimating a mean, a correlation, regression coefficients, etc.).

Posterior densities as precision-weighted combination of prior information and likelihood

The other panels in Figure 1.2 display how Bayesian inference works with more or less *informative priors* for θ. In the top right of Figure 1.2 we see what happens when the prior and the likelihood more or less coincide. In this case, the likelihood is a little less diffuse than the prior, but the prior and the likelihood have the same mode. Application of Bayes Theorem in this instance yields a posterior distribution that has the same mode as the prior and the likelihood, but is more precise (less diffuse) than both the prior and the likelihood. In the other panels of Figure 1.2, this pattern is more or less repeated, except that the mode of the prior and the likelihood are not equal. In these cases, the mode of the posterior distribution lies between the mode of the prior distribution and the mode of the likelihood. Specifically, the mean of the posterior distribution is a *precision-weighted average* of the prior and the likelihood, a feature that we will see repeatedly in this book, a consequence of working with so-called *conjugate* priors in the exponential family, which we define in the next section. Many standard statistical models are in the exponential

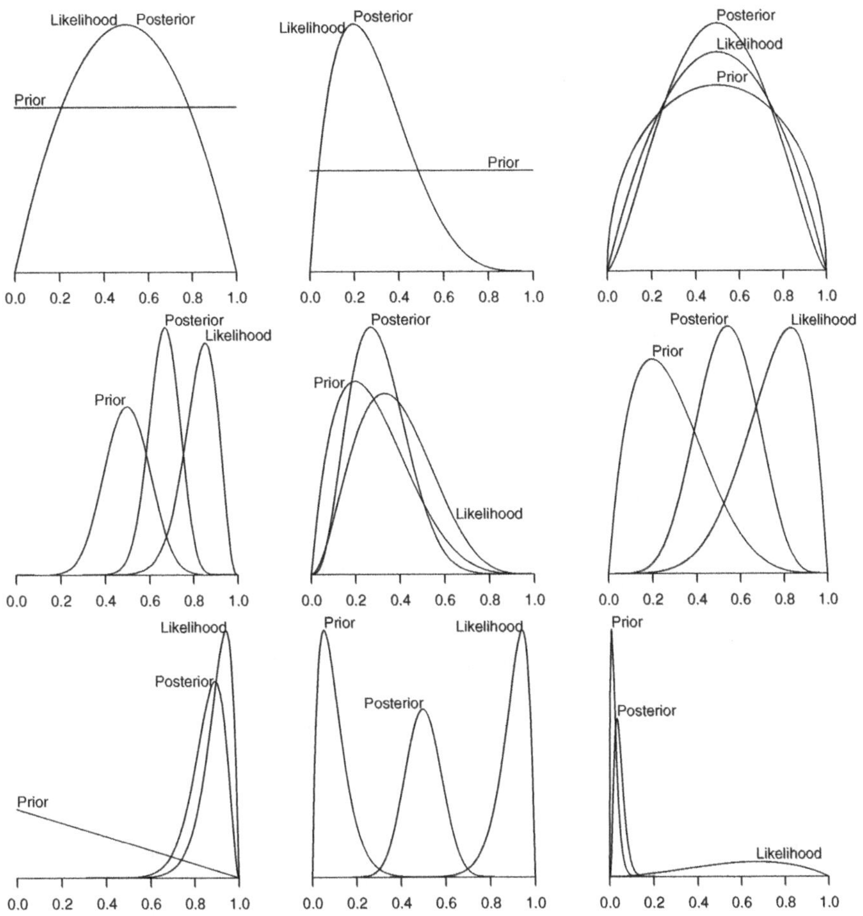

Figure 1.2 Priors, likelihoods and posterior densities. Each panel shows a prior density, a likelihood, and a posterior density over a parameter $\theta \in [0, 1]$. In the top two panels on the left the posterior and the likelihood coincide, since the prior is uniform over the parameter space.

family (but not all), for which conjugate priors are convenient ways of mathematically representing prior beliefs over parameters, and make Bayesian analysis mathematically and computationally quite simple.

1.4.1 Conjugate priors

Since conjugacy is such an important concept in Bayesian statistics, it is worth pausing to sketch a definition:

Definition 1.2 *Suppose a prior density $p(\theta)$ belongs to a class of parametric densities, $\mathcal{F}$. Then the prior density is said to be conjugate with respect to a likelihood $p(\mathbf{y}|\theta)$ if the posterior density $p(\theta|\mathbf{y})$ is also in $\mathcal{F}$.*

Of course, this definition rests on the unstated definition of a 'class of parametric densities', and so is not as complete as one would prefer, but a thorough explanation involves more technical detail than is warranted for now. Examples are perhaps the best way to illustrate the simplicity that conjugacy brings to a Bayesian analysis. And to this end, all the examples in Chapter 2 use priors that are conjugate with respect to their respective likelihoods.

In particular, the examples in Figure 1.2 show the results of a Bayesian analysis of binomial data (n independent realizations of a binary process, also known as Bernoulli trials, such as coin flipping), for which the unknown parameter is $\theta \in [0, 1]$, the probability of a 'success' on any given trial. For the likelihood function formed with binomial data, any Beta density (see Definition B.28) over θ is a conjugate prior: that is, if prior beliefs about θ can be represented as a Beta density, then after those beliefs have been updated (via Bayes Rule) in light of the binomial data, posterior beliefs about θ are also characterized by a Beta density. In Section 2.1 we consider the Bayesian analysis of binomial data in considerable detail.

For now, one of the important features of conjugacy is the one that appears graphically in Figure 1.2: for a wide class of problems (i.e. when conjugacy holds), Bayesian statistical inference is equivalent to *combining information*, marrying the information in the prior with the information in the data, with the relative contributions of prior and data to the posterior being proportional to their respective precisions. That is, Bayesian analysis with conjugate priors over a parameter θ is equivalent to taking a *precision-weighted average* of prior information about θ and the information in the data about θ.

Thus, when prior beliefs about θ are 'vague', 'diffuse', or, in the limit, uninformative, the posterior density will be dominated by the likelihood (i.e. the data contains much more information than the prior about the parameters); e.g. the lower left panel of Figure 1.2. In the limiting case of an uninformative prior, the *only* information about the parameter is that in the data, and the posterior has the same shape as the likelihood function. When prior information is available, the posterior incorporates it, and rationally, in the sense of being consistent with the laws of probability via Bayes Theorem. In fact, when prior beliefs are quite precise relative to the data, it is possible that the likelihood is largely ignored, and the posterior distribution will look almost exactly like the prior, as it should in such a case; e.g. see the lower right panel of Figure 1.2. In the limiting case of a degenerate, infinitely-precise, 'spike prior' (all prior probability concentrated on a point), the data are completely ignored, and the posterior is also a degenerate 'spike' distribution. Should you hold such a dogmatic prior, no amount of data will ever result in you changing your mind about the issue.

1.4.2 Bayesian updating with irregular priors

Figure 1.3 displays a series of prior and posterior densities for less standard cases, where the prior densities are not simple unimodal densities. In each instance, Bayes Rule applies as usual, with the posterior density being proportional to the prior density times the likelihood, and appropriately normalized such that the posterior density encloses an area equal to one. In the left-hand series of panels, the prior has two modes, with the left mode more dominant than the right mode. The likelihood is substantially less dispersed than the prior, and attains a maximum at a point with low prior probability. The resulting posterior density clearly represents the merger of prior and likelihood: with a mode just

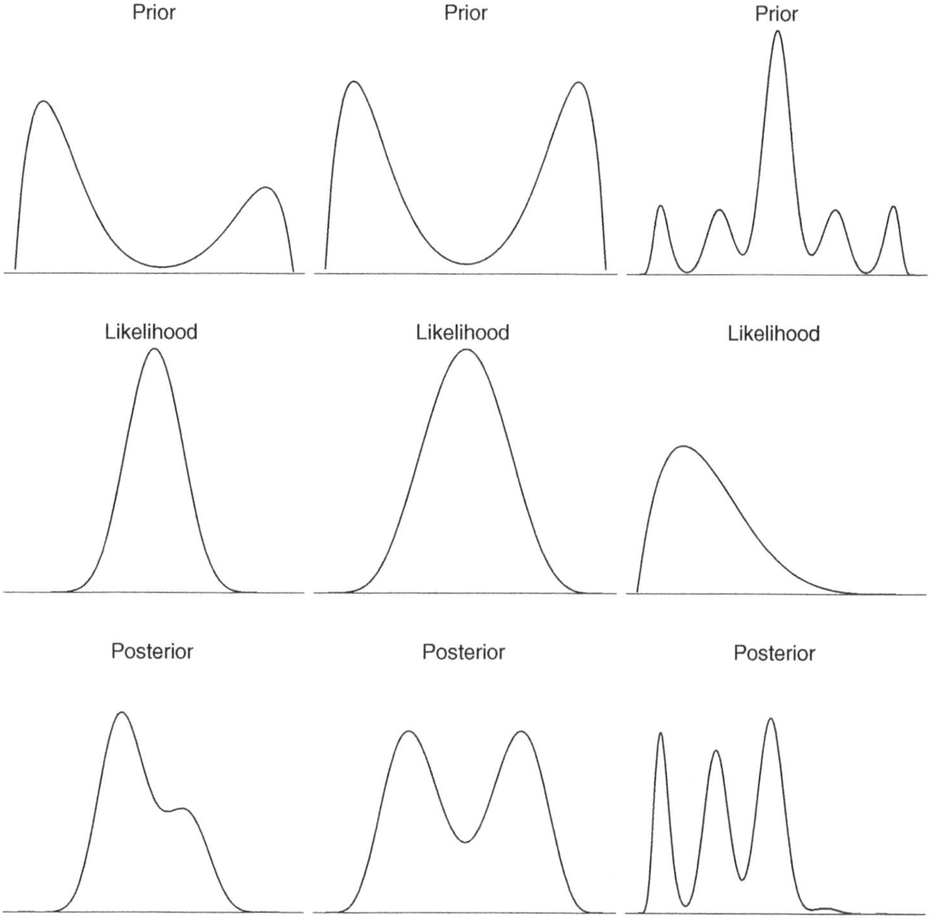

Figure 1.3 Priors, likelihoods and posterior densities for non-standard cases. Each column of panels shows the way Bayes Rule combines prior information (top) with information in the data (characterized by the likelihood, center) to yield a posterior density (lower panels).

to the left of the mode of the likelihood function, and a smaller mode just to the right of the mode of the likelihood function. The middle column of panels in Figure 1.3 shows a symmetric case: the prior is bimodal but symmetric around a trough corresponding to the mode of the likelihood function, resulting in a bimodal posterior distribution, but with modes shrunk towards the mode of the likelihood. In this case, the information in the data about θ combines with the prior information to reduce the depth of the trough in the prior density, and to give substantially less weight to the outlying values of θ that receive high prior probability. In the right-hand column of Figure 1.3 an extremely flamboyant prior distribution (but one that is nonetheless symmetric about its mean) combines with the skewed likelihood to produce the trimodal posterior density, with the posterior modes located in regions with relatively high likelihood. Although this prior (and posterior) are somewhat fanciful (in the sense that it is hard to imagine those densities corresponding to beliefs over a parameter), the central idea remains the same: Bayes Rule governs the mapping from prior to posterior through the data. Implementing Bayes Rule may be difficult when the prior is not conjugate to the likelihood, but, as we shall see, this is where modern computational tools are particularly helpful (see Chapter 3).

1.4.3 Cromwell's Rule

Note also that via Bayes Rule, if a particular region of the parameter space has zero prior probability, then it also has zero posterior probability. This feature of Bayesian updating has been dubbed 'Cromwell's Rule' by Lindley (1985). After the English deposed, tried and executed Charles I in 1649, the Scots invited Charles' son, Charles II, to become king. The English regarded this as a hostile act, and Oliver Cromwell led an army north. Prior to the outbreak of hostilities, Cromwell wrote to the synod of the Church of Scotland, 'I beseech you, in the bowels of Christ, consider it possible that you are mistaken'. The relevance of Cromwell's plea to the Scots for our purposes comes from noting that a prior that assigns zero probability to a hypothesis can never be revised; likewise, a hypothesis with prior weight of 1.0 can never be refuted.

The operation of Cromwell's Rule is particularly clear in the left-hand column of panels in Figure 1.4: the prior for θ is a uniform distribution over the left half of the support of the likelihood, and zero everywhere else. The resulting posterior assigns zero probability to values of θ assigned zero prior probability, and since the prior is uniform elsewhere, the posterior is a re-scaled version of the likelihood in this region of non-zero prior probability, where the re-scaling follows from the constraint that the area under the posterior distribution is one. The middle column of panels in Figure 1.4 shows a prior that has positive probability over all values of θ that has non-zero likelihood, and a discontinuity in the middle of the parameter space, with the left-half of the parameter space supporting having half as much probability mass as the right-half. The resulting posterior has a discontinuity at the point where the prior does, but since the prior is otherwise uniform, the posterior inherits the shape of the likelihood on either side of the discontinuity, subject to the constraint (implied by the prior) that the posterior has twice as much probability mass to the right of the discontinuity than to the left, and integrates to one. The right-hand column of Figure 1.4 shows a more elaborate prior, a step function over the parameter space, decreasing to the right. The resulting posterior has discontinuities at the discontinuities in the prior, and some that are quite abrupt, depending on the conflict between the prior and likelihood in any particular segment of the prior.

The point here is that posterior distributions can sometimes look quite unusual, depending on the form of the prior and the likelihood for a particular problem. The fact that a posterior distribution may have a peculiar shape is of no great concern in a Bayesian analysis: provided one is updating prior beliefs via Bayes Rule, all is well. Unusual looking posterior distributions might suggest that one's prior distribution was poorly specified, but, as a general rule, one should be extremely wary of engaging this kind of procedure. Bayes Rule is a procedure for generating posterior distributions over parameters in light of data. Although one can always re-run a Bayesian analysis with different priors (and indeed, this is usually a good idea), Bayesian procedures should not be used to hunt for priors that generate the most pleasing looking posterior distribution,

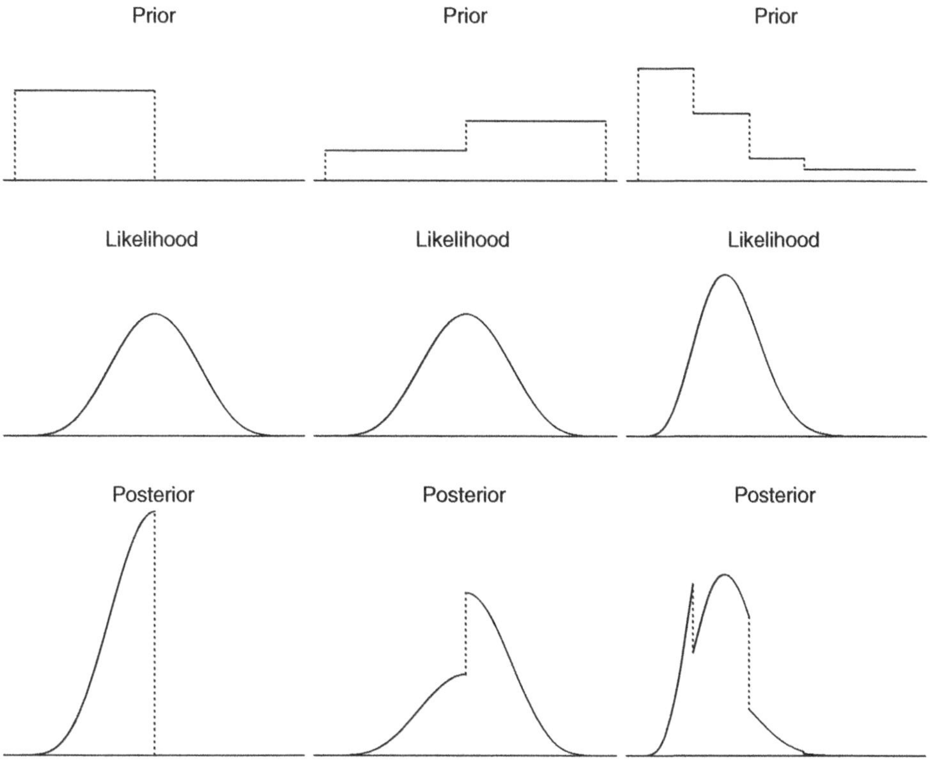

Figure 1.4 Discontinuous prior and posterior densities. Each column of panels shows the way Bayes Rule combines prior information (top) with information in the data (characterized by the likelihood, center) to yield a posterior density (lower panels). The dotted lines indicate discontinuities.

given a particular data set and likelihood. Indeed, such a practice would amount to an inversion of the Bayesian approach: i.e. if the researcher has strong ideas as to what values of θ are more likely than others, aside from the information in the data, then that auxiliary information should be considered a prior, with Bayes Rule providing a procedure for rationally combining that auxiliary information with the information in the data.

1.4.4 Bayesian updating as information accumulation

Bayesian procedures are often equivalent to combining the information in one set of data with another set of data. In fact, if prior beliefs represent the result of a previous data analysis (or perhaps many previous data analyses), then Bayesian analysis is equivalent to pooling information. This is a particularly compelling feature of Bayesian analysis, and one that takes on special significance when working with conjugate priors. In these cases, Bayesian procedures *accumulate information* in the sense that the posterior distribution is more precise than either the prior distribution or the likelihood alone. Further, as the amount of data increases, say through repeated applications of the data generation process, the posterior precision will continue to increase, eventually overwhelming any non-degenerate prior; the upshot is that analysts with different (non-degenerate) prior beliefs over a parameter will eventually find their beliefs coinciding, provided they (1) see enough data and (2) update their beliefs using Bayes Theorem (Blackwell and Dubins 1962). In this way Bayesian analysis has been proclaimed as a model for scientific practice (e.g. Howson and Urbach 1993; Press 2003) acknowledging that while reasonable people may differ (at least prior to seeing data), our views will tend to converge as scientific knowledge accumulates, provided we update our views rationally, consistent with the laws of probability (i.e. via Bayes Theorem).

■ Example 1.3

Drug testing, Example 1.1, continued. Suppose that the randomly selected subject is someone you know personally, and you strongly suspect that she does not use the prohibited substance. Your prior over the hypothesis that she uses the prohibited substance is $P(H_U) = 1/1000$. I have no special knowledge regarding the athlete, and use the baseline prior $P(H_U) = .03$. After the positive test result, my posterior belief is $P(H_U|E) = .23$, while yours is

$$\begin{aligned} P(H_U|E) &= \frac{P(H_U)P(E|H_U)}{\sum_{i\in\{U,\sim U\}} P(H_i)P(E|H_i)} \\ &= \frac{.001 \times .95}{(.001 \times .95) + (.999 \times .10)} \\ &= \frac{.00095}{.000095 + .0999} \\ &\approx .009 \end{aligned}$$

A second test is performed. Now, our posteriors from the first test become the priors with respect to the second test. Again, the subject tests positive, which we denote as the event E'. My beliefs are revised as follows:

$$\begin{aligned} P(H_U|E') &= \frac{.23 \times .95}{(.23 \times .95) + (.77 \times .10)} \\ &= \frac{.2185}{.2185 + .077} \\ &= .74, \end{aligned}$$

while your beliefs are updated to

$$\begin{aligned} P(H_U|E') &= \frac{.009 \times .95}{(.009 \times .95) + (.991 \times .10)} \\ &= \frac{.00855}{.00855 + .0991} \\ &\approx .079. \end{aligned}$$

At this point, I am reasonably confident that the subject is using the prohibited substance, while you still attach reasonably low probability to that hypothesis. After a 3rd positive test your beliefs update to .45, and mine to .96. After a 4th positive test your beliefs update to .88 and mine to .996, and after a 5th test, your beliefs update to .99 and mine to .9996. That is, given this stream of evidence, common knowledge as to the properties of the test, and the fact that we are both rationally updating our beliefs via Bayes Theorem, our beliefs are converging.

In this case, given the stream of postitive test results, our posterior probabilities regarding the truth of H_U are asymptotically approaching 1.0, albeit mine more quickly than yours, given the low *a priori* probability you attached to H_U. Note that with my prior, I required just two consecutive positive test results to revise my beliefs to the point where I considered it more likely than not that the subject is using the prohibited substance, whereas you, with a much more skeptical prior, required four consecutive postive tests.

It should also be noted that the specific pattern of results obtained in this case depend on the properties of the test. Tests with higher sensitivity and specificity would see our beliefs be revised more dramatically given the sequence of positive test results. Indeed, this is the objective of the design of diagnostic tests of various sorts: given a prior $P(H_U)$, what levels of sensitivity and specificity are required such that after just one or two positive tests, $P(H_U|E)$ exceeds a critical threshold where an action is justified. See Exercise 1.2.

1.5 Parameters as random variables, beliefs as distributions

One of the critical ways in which Bayesian statistical inference differs from frequentist inference is immediately apparent from Equation 1.3 and the examples shown in Figure 1.2: the result of a Bayesian analysis, the posterior density $p(\theta|\mathbf{y})$ is just that, a probability density. Given a subjectivist interpretation of probabilty that most Bayesians adopt, the 'randomness' summarized by the posterior density is a reflection of the researcher's uncertainty over θ, conditional on having observed data $\mathbf{y}$.

Contrast the frequentist approach, in which θ is not random, but a fixed (but unknown) property of a population from which we randomly sample data $\mathbf{y}$. Repeated applications of the sampling process, if undertaken, would yield different $\mathbf{y}$, and different sample based estimates of θ, denoted $\hat{\theta} = \hat{\theta}(\mathbf{y})$, this notation reminding us that estimates of parameters are functions of data. In the frequentist scheme, the $\hat{\theta}(\mathbf{y})$ vary randomly across data sets (or would, if repeated sampling was undertaken), while the parameter θ is a constant feature of the population from which data sets are drawn. The distribution of values of $\hat{\theta}$ that would result from repeated application of the sampling process is called the *sampling distribution*, and is the basis of inference in the frequentist approach; the standard deviation of the sampling distribution of $\hat{\theta}$ is the *standard error* of $\hat{\theta}$, which plays a key role in frequentist inference.

The Bayesian approach does not rely on how $\hat{\theta}$ might vary over repeated applications of random sampling. Instead, Bayesian procedures center on a simple question: "what should I believe about θ in light of the data available for analysis, $\mathbf{y}$?" The quantity $\hat{\theta}(\mathbf{y})$ has no special, intrinsic status in the Bayesian approach: as we shall see with specific examples in Chapter 2, a least squares or maximum likelihood estimate of θ is a feature of the data that is usually helpful in *computing* the posterior distribution for θ. And, under some special circumstances, a least squares or maximum likelihood estimate of θ, $\hat{\theta}(\mathbf{y})$, will correspond to a Bayes estimate of θ (see Section 1.6.1). But the critical point to grasp is that in the Bayesian approach, the roles of θ and $\hat{\theta}$ are reversed relative to their roles in classical, frequentist inference: θ is random, in the sense that the researcher is uncertain about its value, while $\hat{\theta}$ is fixed, a feature of the data at hand.

1.6 Communicating the results of a Bayesian analysis

In a Bayesian analysis, all relevant information about θ after having analyzed the data is represented by the posterior density, $p(\theta|\mathbf{y})$. An important and interesting decision for the Bayesian researcher is how to communicate posterior beliefs about θ.

In a world where journal space was less scarce than it is, researchers could simply provide pictures of posterior distributions: e.g. density plots or histograms, as in Figure 1.2. Graphs are an extremely efficient way of presenting information, and, in the specific case of probability distributions, let the researcher and readers see the location, dispersion and shape of the distribution, immediately gauging what regions of the parameter space are more plausible than others, if any. This visualization strategy works well when θ is a scalar, but quickly becomes more problematic when working with multiple parameters, and so the posterior density is a *multivariate* distribution: i.e. we have

$$p(\boldsymbol{\theta}|\mathbf{y}) = p(\theta_1, \ldots, \theta_k|\mathbf{y}) \propto p(\boldsymbol{\theta})p(\mathbf{y}|\boldsymbol{\theta}) \tag{1.4}$$

Direct visualization is no longer feasible once $k > 2$: density plots or histograms have two-dimensional counterparts (e.g. contour or image plots, used throughout this book, and perspective plots), but we simply run out of dimensions at this point. As the dimension of the parameter vector increases, we can graphically present one or two dimensional slices of the posterior density. For problems with lots of parameters, this means that we may have lots of pictures to present, consuming more journal space than even the most sympathetic editor may be able to provide.

Thus, for models with lots of parameters, graphical presentation of the posterior density may not be feasible, at least not for all parameters. In these cases, numerical summaries of the posterior density (or the marginal posterior densities specific to particular parameters) are more feasible. Moreover, for most standard models, and if the researcher's prior beliefs have been expressed with conjugate priors, the analytic form of the posterior is known (indeed, as we shall see, this is precisely the attraction of conjugate priors!). This means that for these standard cases, almost any interesting feature of the posterior can be computed directly: e.g., the mean, the mode, the standard deviation, or particular quantiles. For non-standard models, and/or for models where the priors are not congujate, modern computational power lets us deploy Monte Carlo methods to compute these features of posterior densities; see Chapter 3. Finally, it should be noted that with large sample sizes, provided the prior is not degenerate, the posterior densities are usually well approximated by normal densities, for which it is straightforward to compute numerical summaries (see Section 1.7). In this section I review proposals for summarizing posterior densities.

1.6.1 Bayesian point estimation

If a Bayesian point estimate is required – reducing the information in the posterior distribution to a single number – this can be done, although some regard the attempt to reduce a posterior distribution to a single number as misguided and *ad hoc*. For instance,

> While it [is] easy to demonstrate examples for which there can be no satisfactory point estimate, yet the idea is very strong among people in general and some statisticians in particular that there is a need for such a quantity. To the idea that people like to have a single number we answer that usually they shouldn't get it. Most people know they live in a statistical world and common parlance is full of words implying uncertainty. As in the case of weather forecasts, statements about uncertain quantities ought to be made in terms which reflect that uncertainty as nearly as possible (Box and Tiao 1973, 309–10).

This said, it is convenient to report a point estimate when communicating the results of a Bayesian analysis, and, so long as information summarizing the dispersion of the posterior distribution is also provided (see Section 1.6.2, below), a Bayesian point estimate is quite a useful quantity to report.

The choice of which point summary of the posterior distribution to report can be rationalized by drawing on (Bayesian) decision theory. Although we are interested in the specific problem of choosing a single-number summary of a posterior distribution, the question of how to make rational choices under conditions of uncertainty is quite general, and we begin with a definition of loss:

Definition 1.3 (Loss Function). *Let Θ be a set of possible states of nature θ, and let $a \in \mathcal{A}$ be actions available to the researcher. Then define $l(\theta, a)$ as the loss to the researcher from taking action a when the state of nature is θ.*

Recall that in the Bayesian approach, the researcher's beliefs about plausible values for θ are represented with a probability density function (or a probability mass function, if θ take discrete values), and, in particular, after looking at data $\mathbf{y}$, beliefs about θ are represented by the posterior density $p(\theta|\mathbf{y})$. Generically, let $p(\theta)$ be a probability density over θ, which in turn induces a density over losses. Averaging the losses over beliefs about θ yields the Bayesian expected loss (Berger 1985, 8):

Definition 1.4 (Bayesian expected loss). *If $p(\theta)$ is the probability density for $\theta \in \Theta$ at the time of decision making, the Bayesian expected loss of an action a is*

$$\varrho(p(\theta), a) = E[l(\theta, a)] = \int_\Theta l(\theta, a) p(\theta) d\theta.$$

A special case is where the density p in Definition 1.4 is a posterior density:

Definition 1.5 (Posterior expected loss). *Given a posterior density for θ, $p(\theta|\mathbf{y})$, the posterior expected loss of an action a is $\varrho(p(\theta|\mathbf{y}), a) = \int_\Theta l(\theta, a) p(\theta|\mathbf{y}) d\theta$.*

A Bayesian rule for choosing among actions $\mathcal{A}$ is to select $a \in \mathcal{A}$ so to minimize posterior expected loss. In the specific context of point estimation, the decision problem is to choose a Bayes estimate, $\tilde{\theta}$, and so actions $a \in \mathcal{A}$ now index feasible values for $\tilde{\theta} \in \Theta$. The problem now is that since there are plausibly many different loss functions one might adopt, there are plausibly many Bayesian point estimates one might choose to report. If the chosen loss function is convex, then the corresponding Bayes estimate is unique (DeGroot and Rao 1963), so the choice of what Bayes estimate to report usually amounts to what (convex) loss function to adopt. We briefly consider some well-studied cases.

Definition 1.6 (Quadratic loss). *If $\theta \in \Theta$ is a parameter of interest, and $\tilde{\theta}$ is an estimate of θ, then $l(\theta, \tilde{\theta}) = (\theta - \tilde{\theta})^2$ is the quadratic loss arising from the use of the estimate $\tilde{\theta}$ instead of θ.*

With quadratic loss, we obtain the following useful result:

Proposition 1.6 (Posterior mean as a Bayes estimate under quadratic loss). *Under quadratic loss the Bayes estimate of θ is the mean of the posterior density, i.e. $\tilde{\theta} = E(\theta|\mathbf{y}) = \int_\Theta \theta p(\theta|\mathbf{y}) d\theta$.*

Proof. Quadratic loss (Definition 1.6) implies that the posterior expected loss is

$$\varrho(\theta, \tilde{\theta}) = \int_\Theta (\theta - \tilde{\theta})^2 p(\theta|\mathbf{y}) d\theta.$$

and we seek to minimize this expression with respect to $\tilde{\theta}$. Expanding the quadratic yields

$$\begin{aligned} \varrho(\theta, \tilde{\theta}) &= \int_\Theta \theta^2 p(\theta|\mathbf{y}) d\theta + \tilde{\theta}^2 \int_\Theta p(\theta|\mathbf{y}) d\theta - 2\tilde{\theta} \int_\Theta \theta p(\theta|\mathbf{y}) d\theta \\ &= \int_\Theta \theta^2 p(\theta|\mathbf{y}) d\theta + \tilde{\theta}^2 - 2\tilde{\theta} E(\theta|\mathbf{y}), \end{aligned}$$

Differentiate with respect to $\tilde{\theta}$, noting that the first term does not involve $\tilde{\theta}$. Then set the derivative to zero and solve for $\tilde{\theta}$ to establish the result. ◁

This result also holds for the case of performing inference with respect to a parameter vector $\boldsymbol{\theta} = (\theta_1, \ldots, \theta_K)'$. In this more general case, we define a multidimensional quadratic loss function as follows:

Definition 1.7 (Multidimensional quadratic loss). *If $\boldsymbol{\theta} \in \mathbb{R}^K$ is a parameter, and $\tilde{\boldsymbol{\theta}}$ is an estimate of $\boldsymbol{\theta}$, then the (multidimensional) quadratic loss is $l(\boldsymbol{\theta}, \tilde{\boldsymbol{\theta}}) = (\boldsymbol{\theta} - \tilde{\boldsymbol{\theta}})'\mathbf{Q}(\boldsymbol{\theta} - \tilde{\boldsymbol{\theta}})$ where $\mathbf{Q}$ is a positive definite matrix.*

Proposition 1.7 (Multidimensional posterior mean as Bayes estimate). *Under quadratic loss (Definition 1.7), the posterior mean $E(\boldsymbol{\theta}|\mathbf{y}) = \int_{\boldsymbol{\Theta}} \boldsymbol{\theta} p(\boldsymbol{\theta}|\mathbf{y}) d\boldsymbol{\theta}$ is the Bayes estimate of $\boldsymbol{\theta}$.*

Proof. The posterior expected loss is $\varrho(\boldsymbol{\theta}, \tilde{\boldsymbol{\theta}}) = \int_{\Theta}(\boldsymbol{\theta} - \tilde{\boldsymbol{\theta}})'\mathbf{Q}(\boldsymbol{\theta} - \tilde{\boldsymbol{\theta}})p(\boldsymbol{\theta}|\mathbf{y})d\boldsymbol{\theta}$. Differentiating with respect to $\tilde{\boldsymbol{\theta}}$ yields $2\mathbf{Q}\int_{\Theta}(\boldsymbol{\theta} - \tilde{\boldsymbol{\theta}})p(\boldsymbol{\theta}|\mathbf{y})d\boldsymbol{\theta}$. Setting the derivative to zero and re-arranging yields $\int_{\Theta}(\boldsymbol{\theta} - \tilde{\boldsymbol{\theta}})p(\boldsymbol{\theta}|\mathbf{y})d\boldsymbol{\theta} = 0$ or $\int_{\Theta}\boldsymbol{\theta}p(\boldsymbol{\theta}|\mathbf{y})d\boldsymbol{\theta} = \int_{\Theta}\tilde{\boldsymbol{\theta}}p(\boldsymbol{\theta}|\mathbf{y})d\boldsymbol{\theta}$. The left-hand side of this expression is just the mean of the posterior density, $E(\boldsymbol{\theta}|\mathbf{y})$, and so $E(\boldsymbol{\theta}|\mathbf{y}) = \int_{\Theta}\tilde{\boldsymbol{\theta}}p(\boldsymbol{\theta}|\mathbf{y})d\boldsymbol{\theta} = \tilde{\boldsymbol{\theta}}\int_{\Theta}p(\boldsymbol{\theta}|\mathbf{y})d\boldsymbol{\theta} = \tilde{\boldsymbol{\theta}}$. ◁

Remark. This result holds irrespective of the specific weighting matrix **Q**, provided **Q** is positive definite.

The mean of the posterior distribution is a popular choice among researchers seeking to quickly communicate features of the posterior distribution that results from a Bayesian data analysis; we now understand the conditions under which this is a rational point summary of one's beliefs over θ. Specifically, Proposition 1.6 rationalizes the choice of the mean of the posterior density as a Bayes estimate.

Of course, other loss functions rationalize other point summaries. Consider linear loss, possibly asymmetric around θ:

Definition 1.8 (Linear loss). *If $\theta \in \Theta$ is a parameter, and $\tilde{\theta}$ is a point estimate of θ, then the linear loss function is*

$$l(\theta, \tilde{\theta}) = \begin{cases} k_0(\theta - \tilde{\theta}) & \text{if } \tilde{\theta} < \theta \\ k_1(\tilde{\theta} - \theta) & \text{if } \theta \leq \tilde{\theta} \end{cases}$$

Loss in absolute value results when $k_0 = k_1 = 1$, a special case of a class of symmetric, linear loss functions (i.e. $k_0 = k_1$). Asymmetric linear loss results when $k_0 \neq k_1$.

Proposition 1.8 (Bayes estimates under linear loss). *Under linear loss (definition 1.8), the Bayes estimate of θ is the $k_1/(k_0 + k_1)$ quantile of $p(\theta|\mathbf{y})$, the $\tilde{\theta}$ such that $P(\theta \leq \tilde{\theta}) = k_0/(k_0 + k_1)$.*

Proof. Following Bernardo and Smith (1994, 256), we seek the $\tilde{\theta}$ that minimizes

$$\varrho(\theta, \tilde{\theta}) = \int_{\Theta} l(\theta, \tilde{\theta})p(\theta|\mathbf{y})d\theta = k_0 \int_{\{\tilde{\theta}<\theta\}} (\theta - \tilde{\theta})p(\theta|\mathbf{y})d\theta + k_1 \int_{\{\theta\leq\tilde{\theta}\}} (\tilde{\theta} - \theta)p(\theta|\mathbf{y})d\theta.$$

Differentiating this expression with respect to $\tilde{\theta}$ and setting the result to zero yields

$$k_0 \int_{\{\tilde{\theta}<\theta\}} p(\theta|\mathbf{y})d\theta = k_1 \int_{\{\theta\leq\tilde{\theta}\}} p(\theta|\mathbf{y})d\theta$$

Adding $k_0 \int_{\{\theta\leq\tilde{\theta}\}} p(\theta|\mathbf{y})d\theta$ to both sides yields $k_0 = (k_0 + k_1) \int_{\{\theta\leq\tilde{\theta}\}} p(\theta|\mathbf{y})d\theta$ and so re-arranging yields $\int_{\{\theta<\tilde{\theta}\}} p(\theta|\mathbf{y})d\theta = k_0/(k_0 + k_1)$. ◁

Note that with symmetric linear loss, we obtain the median of the posterior density as the Bayes estimate. Asymmetric loss functions imply using quantiles other than the median.

■ **Example 1.4**

Graduate Admissions. A professor reviews applications to a Ph.D. program. The professor assumes that each applicant $i \in \{1, \ldots, n\}$ possesses ability θ_i. After reviewing the applicants' files (i.e. encountering data, or $\mathbf{y}$), the professors's beliefs regarding each θ_i can be represented as a distribution $p(\theta_i|\mathbf{y})$. The professor's loss function is asymmetric, since the professor has determined that it is 2.5 times as costly to overestimate an applicant's ability than it is to underestimate ability: i.e.

$$\varrho(\theta, \tilde{\theta}) = \begin{cases} \theta - \tilde{\theta} & \text{if } \theta > \tilde{\theta} \\ 2.5(\tilde{\theta} - \theta) & \text{if } \theta \leq \tilde{\theta} \end{cases}$$

Ability is measured on an arbitrary scale, normalized to have mean zero and standard deviation one across the applicant pool. Suppose that for applicant i, $p(\theta_i|\mathbf{y}) \approx N(1.8, 0.4^2)$, while for applicant j, $p(\theta_j|\mathbf{y}) \approx N(2.0, 1.0^2)$; i.e. there is considerably greater posterior uncertainty as to the ability of applicant j. Given the professors's loss function, the Bayes estimate of θ_i is the $1/(1 + 2.5) = .286$ quantile of the $N(1.8, 0.4^2)$ posterior density, or 1.57; for applicant j, the Bayes estimate is the .286 quantile of a $N(2.0, 1.0^2)$ density, or 1.43. Thus, although $E(\theta_j|\mathbf{y}) > E(\theta_i|\mathbf{y})$, the greater uncertainty associated with applicant j, when coupled with the asymmetric loss function, results in the professor assigning a higher Bayes estimate to applicant i than to applicant j (i.e. $\tilde{\theta}_j < \tilde{\theta}_j$).

1.6.2 Credible regions

Bayes estimates are an attempt to summarize beliefs over θ with a single number, providing a rational, best guess as to the value of θ. But Bayes estimates do not convey information as to the researcher's uncertainty over θ, and indeed, this is why many Bayesian statisticians find Bayes estimates fundamentally unsatisfactory. To communicate a summary of prior or posterior uncertainty over θ, it is necessary to somehow summarize information about the location and shape of the prior or posterior distribution, $p(\theta)$. In particular, what is the set or region of more plausible values for θ? More formally, what is the region $C \subseteq \Omega$ that supports proportion α of the probability under $p(\theta)$? Such a region is called a *credible region*:

Definition 1.9 (Credible region). *A region $C \subseteq \Omega$ such that $\int_C p(\theta)d\theta = 1 - \alpha$, $0 \leq \alpha \leq 1$ is a $100(1 - \alpha)$% credible region for θ.*

For single-parameter problems (i.e. $\Omega \subseteq \mathbb{R}$), if C is not a set of disjoint intervals, then C is a credible interval.

If $p(\theta)$ is a (prior/posterior) density, then C is a (prior/posterior) credible region.

There is trivially only one 100 % credible region, the entire support of $p(\theta)$. But non-trivial credible regions may not be unique. For example, suppose $\theta \sim N(0, 1)$: it is obvious that there is no unique $100(1 - \alpha)$% credible region for any $\alpha \in (0, 1)$: any interval spanning $100(1 - \alpha)$ percentiles will be such an interval. A solution to this problem comes from restricting attention to credible regions that have certain desirable properties, including minimum volume (or, for a one dimensional parameter problem, minimum length) in the set of credible regions induced by a given choice of α, for a specific $p(\theta)$. This kind of optimal credible region is called a *highest probability density region*, sometimes referred to as a HPD region or a 'HDR'. The following definition of a HPD region is standard and appears in many places in the literature, e.g. Box and Tiao (1973, 123) or Bernardo and Smith (1994, 260):

Definition 1.10 (Highest probability density region). *A region $C \subseteq \Omega$ is a* $100(1 - \alpha)\%$ *highest probability density region for θ under $p(\theta)$ if*

1. $P(\theta \in C) = 1 - \alpha$
2. $P(\theta_1) \geq P(\theta_2), \forall\, \theta_1 \in C, \theta_2 \notin C$

A $100(1 - \alpha)\%$ HPD region for a symmetric, unimodal density is obviously unique and symmetric around the mode. In fact, if $p(\theta)$ is a univariate normal density, a HPD is the same as a interval around the mean:

■ Example 1.5

Suppose $p(\theta) \equiv N(a, b^2)$. Then a $100(1 - \alpha)\%$ HPD region is the interval

$$(a - |z_\alpha|b, a + |z_\alpha|b)$$

where z_α is the α quantile of the standard normal density. With $\alpha = .05$, $|z_\alpha| \approx 1.96$, and a 95 % HPD corresponds to a 95 % interval; see Figure 1.5.

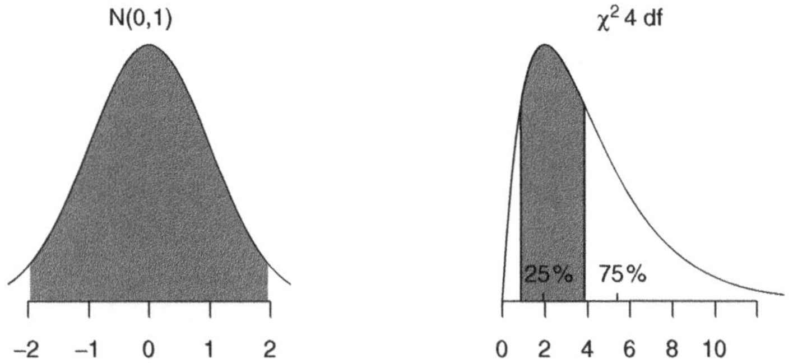

Figure 1.5 95 % HPD intervals for standard normal (left panel) and a 50 % HPD for a χ^2_4 density.

Note that the correspondence between intervals and HPD intervals does not hold for non-symmetric densities, as we demonstrate with a simple example.

■ Example 1.6

The right panel of Figure 1.5 shows a χ^2 density with 4 degrees of freedom, and its 50 % HPD interval. Notice that the 50 % HPD interval is more concentrated around the mode of the density, and has shorter length than the interval based on the 25th to 75th percentiles of the density.

As the next two examples demonstrate, (1) the HPD need not be a connected set, but a collection of disjoint intervals (say, if $p(\theta)$ is not unimodal), and (2) the HPD need not be unique.

■ Example 1.7

Extreme missingness in bivariate normal data. Consider the data in Table 1.1, where two variables (y_1 and y_2) are observed subject to a pattern of severe missingness, but are otherwise assumed to be distributed bivariate normal each with mean zero, and an unknown covariance matrix. These manufactured data have been repeatedly analyzed to investigate the properties of algorithms for handling missing data (e.g. Murray 1977; Tanner and Wong 1987).

Table 1.1 Twelve observations from a bivariate normal distribution.

y_1:	1	1	−1	−1	2	2	−2	−2	NA	NA	NA	NA
y_2:	1	−1	1	−1	NA	NA	NA	NA	2	2	−2	−2

Given the missing data pattern, what should we conclude about the correlation ρ between y_1 and y_2? For this particular example, with an uninformative prior for the covariance matrix of $\mathbf{Y} = (\mathbf{y}_1, \mathbf{y}_2)$, the posterior density for ρ is bimodal, as shown in Figure 1.6. The shaded areas represent half of the posterior density for ρ; the intervals supporting the shaded areas together constitute a 50 % HPD region for ρ, and are the disjoint intervals (−.914, −.602) and (.602, .914). We return to this example in more detail in Examples 5.11 and 6.6.

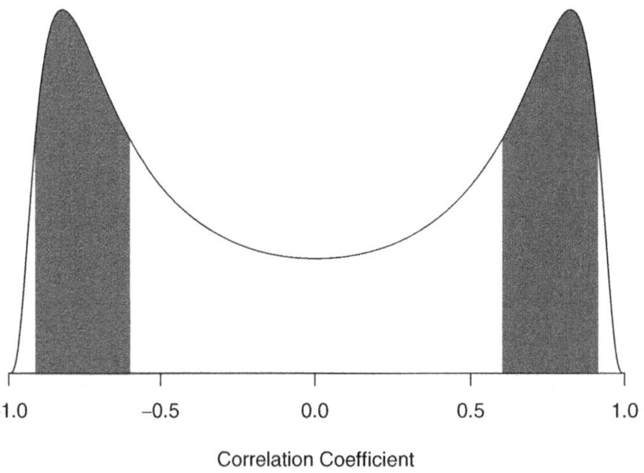

Figure 1.6 Bimodal posterior density for a correlation coefficient, and 50 % HPD.

■ **Example 1.8**

Non-unique HDRs. Suppose $\theta \sim \text{Uniform}(0, 1)$. Then any HPD region of content α is not unique, $\forall\ 0 < \alpha < 1$. See Figure 1.7. The shaded regions are both supported by 25 % HPDs, as are any other intervals of width .25 we might care to draw.

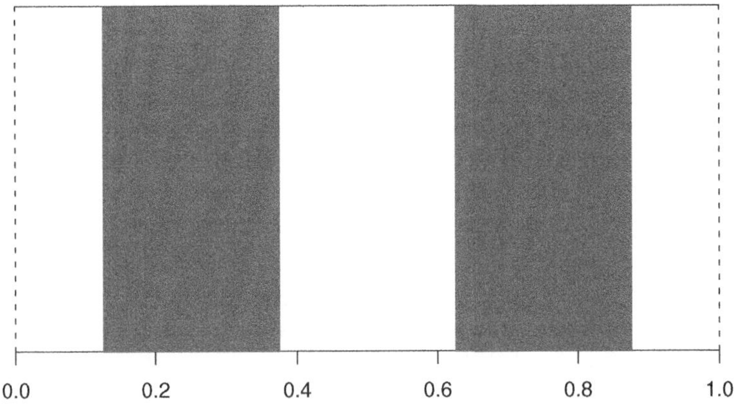

Figure 1.7 Uniform density and (non-unique) 25 % HPDs.

For higher dimensional problems, the HPD is a *region* in a parameter space and numerical approximations and/or simulation may be required to compute it. For some simple cases, such as multiple regression analysis with conjugate priors, although the posterior distribution is multivariate, it has a well known form for which it is straightforward to compute HPDs; see Proposition 2.13.

1.7 Asymptotic properties of posterior distributions

As we have seen, Bayes Rule tells us how we ought to revise our prior beliefs in light of data. In Section 1.4 we saw that as the precision of one's prior beliefs tends to zero, posterior beliefs are increasingly dominated by the data (through the likelihood). This also occurs as the data set 'gets larger': subject to an exception to be noted below, for a given prior, as the size of the data set being analyzed grows without bound, the usual result is that the resulting sequence of posterior densities collapses to a spike on the true values of the parameters in the model under consideration.

Of course, some Bayesians find such thinking odd: in a Bayesian analysis, we condition on the data at hand, updating beliefs via Bayes Rule. Unlike frequentist inference, Bayesian inference does not rest on the repeated sampling and/or asymptotic properties of the statistical procedures being used. Many Bayesians consider asking what would happen as one's data set gets infinitely large as an interesting mathematical exercise, but not particularly relevant to the inferential task at hand. This view holds that provided we update our beliefs via Bayes Rule in light of *this* data set, and with a model/likelihood appropriate to the data at hand (not a trivial matter), we are behaving rationally, and the repeated sampling or asymptotic properties of our inferences are second order concerns. Some Bayesians even go further, arguing that models and parameters have no objective, exterior reality, but are mathematical fictions we conjure so as to help us make probability assignments over data (we explore this 'subjectivist' position further in §1.9), and so questions such as consistency are moot.

My own position – echoing that of Diaconis and Freedman (1986a, 11) – is that even subjectivist Bayesians ought to consider asymptotic properties of Bayes estimates, since if Bayesian inference is to be a model of scientific practice, we should be able to establish the convergence of (initially disparate) opinions as relevant evidence accumulates.

So what can we say about Bayesian inferences, asymptotically? The key idea here is that subject to some regularity conditions, as the data set grows without bound, the posterior density is increasingly dominated by the contribution from the data through the likelihood function, and the standard asymptotic properties of maximum likelihood estimators apply to the posterior density. These properties include

- consistency, at least in the sense that the posterior density is increasingly concentrated around the true parameter value as $n \to \infty$; or, in the additional sense of Bayes point estimators of θ (Section 1.6.1) being consistent;
- asymptotic normality, i.e. $p(\theta|\mathbf{y})$ tends to a normal distribution as $n \to \infty$.

There is a large literature establishing the conditions under which frequentist and Bayesian procedures coincide, at least asymptotically. These results are too technical to be reviewed in any detail in this text; see, for instance, Bernardo and Smith (1994, ch. 5) for statements of necessary regularity conditions and proofs of the main results and references to the literature. Diaconis and Freedman (1986a,b) provide some counter-examples to the consistency results; the 'incidental parameters' problem (Neyman and Scott 1948) is one such counter-example which we briefly return to in Section 9.1.2. I provide a brief illustration of 'Bayesian consistency' with two examples, below, and sketch a proof of a 'Bayesian central limit theorem' in the Appendix.

Bayesian consistency works as follows. Suppose the true value of θ is θ^*. Then provided the prior distribution $p(\boldsymbol{\theta})$ does not place zero probability mass on θ^* (say, for a discrete parameter), or on a neigborhood of θ^* (say, for a continuous parameter), then as $n \to \infty$, the posterior will be increasingly dominated by the contribution from the likelihood, which, under suitable regularity conditions, tends to a spike on θ^*.

Figures 1.8 and 1.9 graphically demonstrate the Bayesian version of consistency as described above. In each case, the prior is held constant as the sample size increases, leading to a progressively tighter correspondence between the posterior and the likelihood. Even with modest amounts of data, the multimodality of the priors are being overwhelmed by the information in the data, and the likelihood and posterior are collapsing to a spike on θ^*.

Although the scale used in Figures 1.8 and 1.9 doesn't make it clear, the likelihoods and posterior in the Figures are also tending to normal distributions: re-scaling by the usual $\sqrt{n}$ would make this clear. The fact that posterior densities start to take on a normal shape as $n \to \infty$ is particularly helpful. The normal is an extremely well-studied distribution, and completely characterized by its first two moments. This can drastically

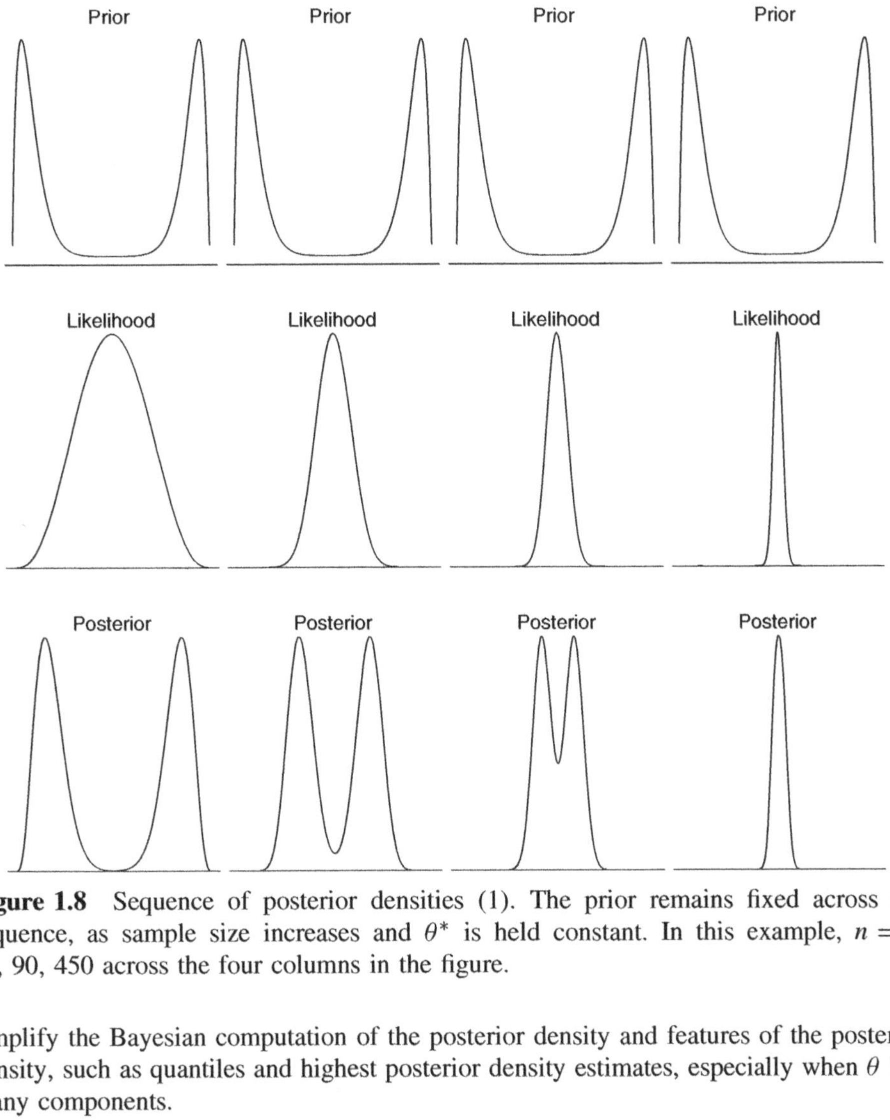

Figure 1.8 Sequence of posterior densities (1). The prior remains fixed across the sequence, as sample size increases and θ^* is held constant. In this example, $n = 6$, 30, 90, 450 across the four columns in the figure.

simplify the Bayesian computation of the posterior density and features of the posterior density, such as quantiles and highest posterior density estimates, especially when θ has many components.

1.8 Bayesian hypothesis testing

The posterior density of θ also provides the information necessary to test hypotheses about θ. At the outset, it is worth stressing that Bayesian hypothesis testing and frequentist hypothesis testing differ starkly. The most common hypothesis test of classical statistics, $H_0 : \theta = 0$, is untestable in the Bayesian approach if θ is a continuous parameter; to see this, note that if a continuous parameter $\theta \in \Omega \subseteq \mathbb{R}$ has the posterior distribution $p(\theta|\mathbf{y})$, then a 'point null' hypothesis such as $H_0 : \theta = c$ has zero probability, since c is a one-point set with measure zero (see Definition B.3). This difficulty also afflicts

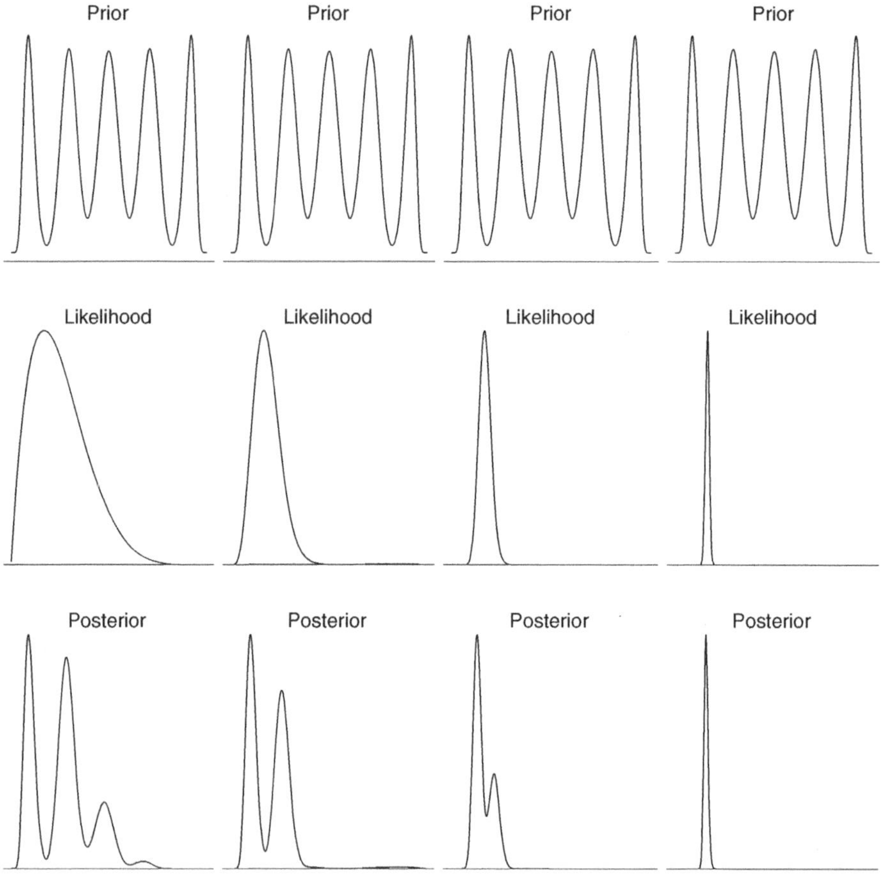

Figure 1.9 Sequence of posterior distributions (2). The prior remains fixed across the sequence, as sample size increases and θ^* is held constant. In this example, $n = 6$, 30, 150, 1500 across the four columns in the figure.

hypothesis testing in the frequentist world: with respect to a continuous parameter, *all point null hypotheses are false*, as the researcher would eventually discover if they were to successively test a point null hypothesis at a pre-specified, non-zero significance level, with increasing amounts of data (a fact that is typically ignored in introductory statistics classes). By concentrating attention on the posterior density, $p(\theta|\mathbf{y})$, the Bayesian approach helps to make clear the logical deficiencies of point null hypothesis testing. Thus, at least for continuous parameters, we don't test point null hypotheses in the Bayesian approach, and for that matter nor should a frequentist.

Instead, suppose we have a continuous parameter $\theta \in \mathbb{R}$, then two, exculsive, exhaustive and non-trivial (non-point) hypotheses are $H_0 : \theta < c$ and the alternative hypothesis $H_1 : \theta \geq c$. Posterior probabilities for these hypotheses are defined as follows:

$$\Pr(H_0|\mathbf{y}) = \Pr(\theta < c|\mathbf{y}) = \int_{-\infty}^{c} p(\theta|\mathbf{y})d\theta$$

and

$$\Pr(H_1|\mathbf{y}) = \Pr(\theta \geq c|\mathbf{y}) = \int_c^\infty p(\theta|\mathbf{y})d\theta.$$

For standard models, where conjugate priors have been deployed, these posterior probabilities are straightforward to compute; in other cases, modern computing power means Monte Carlo methods can be deployed to assess these probabilities, as we will see in Chapter 3.

The posterior probability of a hypothesis is something that only makes sense in a Bayesian framework. There is no such corresponding quantity in a frequentist framework, although this is how a frequentist p-value is often misinterpreted. For a frequentist, θ is a fixed but unknown number, and so hypotheses about θ are either true or false, and $\Pr(H_0|\mathbf{y}) = 1$ if H_0 is true, and zero if it is not. As such, for a frequentist, the falsity or truth of a hypothesis does not depend on the data, and so a quantity such as $\Pr(H_0|\mathbf{y})$ is meaningless. In contrast, for the Bayesian, θ is not fixed, but subject to (subjective prior/posterior) uncertainty, and so too is H_0, and so the posterior probability $\Pr(H_0|\mathbf{y})$ is quite useful. Indeed, one might argue that those types of posterior probability statement are exactly what one wants from a data analysis, letting us make statements of the sort 'how plausible is hypothesis H_0 in light of these data?' A frequentist p-value answers a different question: 'how frequently would I observe a result at least as extreme as the one obtained if H_0 were true?', which is a statement about the plausibility of the data *given* the hypothesis. Turning this assessment into an assessment about the hypothesis requires another step in the frequentist chain of reasoning (e.g. conclude H_0 is false if the p-value falls below some preset level). Contrast the Bayesian procedure, which lets us assess the plausibility of H_0 directly. A long line of papers contrasts p-values with Bayesian posterior probabilities, arguing (as I have here) that many analysts interpret the former as the latter, but that these two quantities can often be very different from one another; especially helpful papers on this score include Dickey (1977), Berger and Sellke (1987) and Berger (2003).

The following example provides a demonstration of Bayesian hypothesis testing using data from a survey. To help understand how Bayesian and frequentist approaches to hypothesis testing differ, a frequentist analysis is also provided.

■ Example 1.9

Attitudes towards abortion. Agresti and Finlay (1997, 133) report that in the 1994 General Social Survey, 1934 respondents were asked

> Please tell me whether or not you think it should be possible for a pregnant woman to obtain a legal abortion if the woman wants it for any reason.

Of the 1934 respondents, 895 reported 'yes' and 1039 said 'no'. Let θ be the unknown population proportion of respondents who agree with the proposition in the survey item, that a pregnant woman should be able to obtain an abortion if the woman wants it for any reason. The question of interest is whether a majority of the population supports the proposition in the survey item.

Frequentist approach. The survey estimate of θ is $\hat{\theta} = 895/1934 \approx .46$, the approximation coming via rounding to two significant digits. Although the underlying data are binomial (independent Bernoulli trials), with this large sample, the normal distribution provides an excellent approximation to the frequentist sampling distribution of $\hat{\theta}$; binomial data are considered in detail in Chapter 2. Suppose interest focuses on whether the unknown population proportion $\theta = .5$. A typical frequentist approach to this question is to test the null hypothesis $H_0 : \theta = .5$ against all other alternatives $H_A : \theta \neq .5$, or a one-sided alternative $H_B : \theta > .5$. We would then ask how unlikely it is that one would see the value of $\hat{\theta}$ actually obtained, or an even more extreme value if H_0 were true, by centering the sampling distribution of $\hat{\theta}$ at the hypothesized value. The standard deviation of the normal sampling distribution (the standard error of $\hat{\theta}$) under H_0 is

$$\text{se}(\hat{\theta}_{H_0}) = \sqrt{\frac{\theta_{H_0}(1-\theta_{H_0})}{n}} = \sqrt{\frac{.50 \times (1-.50)}{1934}} \approx .011.$$

The realized value of $\hat{\theta}$ is $(.5 - .46)/.011 \approx 3.64$ standard errors away from the hypothesized value. Under a normal distribution, this an extremely rare event. Over repeated applications of random sampling, only a small proportion of estimates of θ will lie 3.64 or more standard errors away from the hypothesized mean of the sampling distribution. This proportion is

$$2 \times \int_{3.64}^{\infty} \phi(z)dz = 2 \times [1 - \Phi(3.64)] \approx .00028,$$

where $\phi(\cdot)$ and $\Phi(\cdot)$ are the normal pdf and cdfs, respectively. Given this result, most (frequentist) analysts would reject the null hypothesis in favor of either alternative hypothesis, reporting the p-values for H_0 against H_A as .00028 and for H_0 against H_B as .00014.

Bayesian approach. The unknown parameter is $\theta \in [0, 1]$ and suppose we bring little or no prior information to the analysis. In such a case, we know that the posterior density has the same shape as the likelihood, which with the large sample used here is well approximated by a normal density (the details of Bayesian estimation and inference for a sample proportion are presented in Chapter 2), specifically, a normal distribution centered on the maximum likelihood estimate of .46 with standard deviation .011; i.e. $p(\theta|y) \approx N(.46, .011^2)$ and inferences about θ are based on this distribution. We note immediately that most of the posterior probability mass lies below .5, suggesting that the hypothesis $\theta > .5$ is not well-supported by the data. In fact, the posterior probability of this hypothesis is

$$\Pr(\theta > .5|y) = \int_{.5}^{\infty} p(\theta|y)d\theta = \int_{.5}^{\infty} \phi\left(\frac{\theta - .46}{.011}\right) d\theta = .00014.$$

That is, there is an apparent symmetry between the frequentist and Bayesian answers: in both instances, the 'answer' involved computing the same tail area probability of a normal distribution, with the probability of H_0 under the Bayesian posterior distribution corresponding with the p-value in the frequentist test of H_0 against the one-sided alternative H_B; see Figure 1.10. But this similarity really is only superficial. The Bayesian probability is a statement about the researcher's beliefs about θ, obtained via application

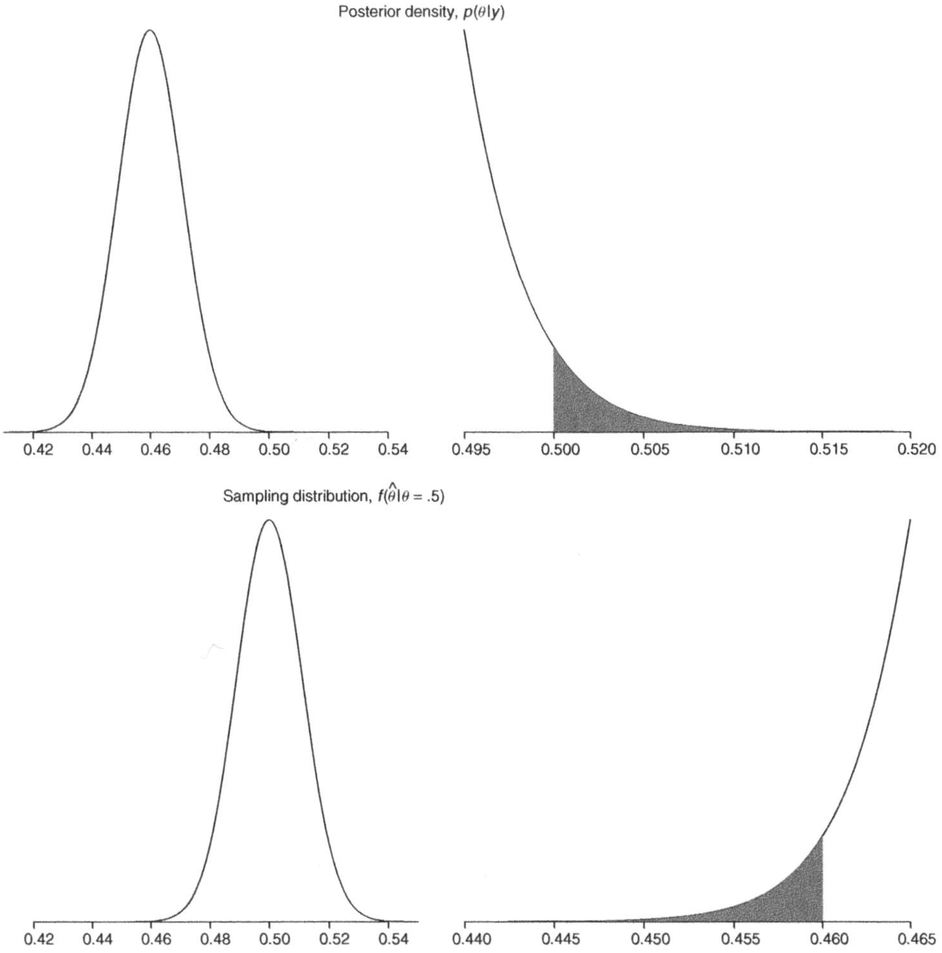

Figure 1.10 Posterior density contrasted with sampling distribution under $H_0 : \theta = .5$, for Example 1.9. The top right panel shows the posterior density in the neighborhood of $\theta = .5$, with the shaded region corresponding to the posterior probability $p(\theta > .5|\mathbf{y}) = \int_{.5}^{\infty} p(\theta|\mathbf{y})d\theta = .00014$. The lower right panel shows the sampling distribution in the neighborhood of $\hat{\theta} = .46$, with the shaded region corresponding to the proportion of times one would observe $\hat{\theta} \leq .46$ if $H_0 : \theta = .5$ were true, corresponding to .00014 of the area under the sampling distribution.

of Bayes Rule, and is $\Pr(H_0|y)$, obtained by computing the appropriate integral of the posterior distribution $p(\theta|y)$. The frequentist p-value is obtained via a slightly more complex route, and has a quite different interpretation than the Bayesian posterior probability, since it conditions on the null hypothesis; i.e. the sampling distribution is $f(\hat{\theta}|H_0)$ and the p-value for H_0 against the one-sided alternative, the proportion of $\hat{\theta} < .46$ we would see under repeated sampling, with the sampling distribution given by the null hypothesis.

1.8.1 Model choice

Applying Bayes Rule produces a posterior density, $f(\theta|\text{y})$, not a point estimate or a binary decision about a hypothesis. Nonetheless, in many settings the goal of statistical analysis is to inform a discrete decision problem, such as choosing the 'best model' from a class of models for a given data set. We now consider Bayesian procedures for making such a choice.

Let M_i index models under consideration for data $\mathbf{y}$. What may distinguish the models are parameter restrictions of various kinds. A typical example in the social sciences is when sets of predictors are entered or dropped from different regression-type models for $\mathbf{y}$; if j indexes candidate predictors, then dropping x_j from a regression corresponds to imposing the parameter restriction $\beta_j = 0$. Alternatively, the models under consideration may not nest or overlap. For example, consider situations where different theories suggest disjoint sets of predictors for some outcome y. In this case two candidate models M_1 and M_2 may have no predictors in common.

Consider a closed set of models, $\mathcal{M} = \{M_1, \ldots, M_J\}$; i.e. the researcher is interested in choosing among a distinct number of models, rather than the (harder) problem of choosing a model from an infinite set of possible models. In the Bayesian approach, the researcher has prior beliefs as to which model is correct, which are formulated as prior probabilities, denoted $P(M_i)$ with i indexing the set of models $\mathcal{M}$. The goal of a Bayesian analysis is to produce posterior probabilities for each model, $P(M_i|\mathbf{y})$, and to inform the choice of a particular model. This posterior probability comes via application of Bayes Rule for multiple discrete events, which we encountered earlier as Proposition 1.4. In the specific context of model choice, we have

$$P(M_i|\mathbf{y}) = \frac{P(M_i)p(\mathbf{y}|M_i)}{\sum_{j=1}^{J} P(M_j)p(\mathbf{y}|M_j)}. \tag{1.5}$$

The expression $p(\mathbf{y}|M_i)$ is the *marginal likelihood*, given by the identity

$$p(\mathbf{y}|M_i) = \int_{\Theta_i} p(\mathbf{y}|\theta_i, M_i)p(\theta_i)d\theta_i \tag{1.6}$$

i.e. averaging the likelihood for $\mathbf{y}$ under M_i over the prior for the parameters θ_i of M_i.

As we have seen in the discussion of Bayes estimates, the mapping from a researcher's posterior distribution to a particular decision depends on the researcher's loss function. To simplify the model choice problem, suppose that one of the models in $\mathcal{M}$ is the 'best model', M^*, and the researcher possesses the following simple loss function

$$l(M_i, M^*) = \begin{cases} 0 & \text{if} \quad M_i = M^* \\ 1 & \text{if} \quad M_i \neq M^* \end{cases}$$

For each model, $P([M_i \neq M^*]|\mathbf{y}) = 1 - P(M_i|\mathbf{y})$, and so the expected posterior loss of choosing model i is $1 - P(M_i|\mathbf{y})$. Thus, the loss minimizing choice is to choose the model with highest posterior probability.

■ **Example 1.10**

Attitudes towards abortion, Example 1.9, continued. The likelihood for these data is approximated by a normal distribution with mean .46 and standard deviation .011. We consider the following hypotheses: $H_0 : .5 \leq \theta \leq 1$ and $H_1 : 0 \leq \theta < .5$, which generate priors

$$p_0(\theta_0) \equiv \text{Uniform}(.5, 1) = \begin{cases} 2 & \text{if } .5 \leq \theta_0 \leq 1 \\ 0 & \text{otherwise} \end{cases}$$

and

$$p_1(\theta_1) \equiv \text{Uniform}(0, .5) = \begin{cases} 2 & \text{if } 0 \leq \theta_1 < .5 \\ 0 & \text{otherwise} \end{cases}$$

respectively. We are *a priori* neutral between the two hypotheses, setting $P(H_0) = P(H_1)$ to 1/2. Now, under H_0, the marginal likelihood is

$$\begin{aligned} p(y|H_0) &= \int_{.5}^{1} p(y|H_0, \theta_0) p_0(\theta_0) d\theta_0 = 2 \int_{.5}^{1} p(y|H_0, \theta_0) d\theta_0 \\ &= 2\left(\Phi\left(\frac{1 - .46}{.011}\right) - \Phi\left(\frac{.5 - .46}{.011}\right)\right) = .00028 \end{aligned}$$

Under H_1, the marginal likelihood is

$$\begin{aligned} p(y|H_1) &= \int_{0}^{.5} p(y|H_1, \theta_1) p_1(\theta_1) d\theta_1 = 2 \int_{0}^{.5} p(y|H_1, \theta_1) d\theta_1 \\ &= 2\left(\Phi\left(\frac{.5 - .46}{.011}\right) - \Phi\left(\frac{-.46}{.011}\right)\right) = 2. \end{aligned}$$

Thus, via Equation 1.5:

$$\begin{aligned} P(H_0|y) &= \frac{\frac{1}{2} \times .00028}{(\frac{1}{2} \times .00028) + (\frac{1}{2} \times 2)} = \frac{.00014}{.00014 + 1} = .00014 \\ P(H_1|y) &= \frac{1}{.00014 + 1} = .99986 \end{aligned}$$

indicating that H_1 is much more plausible than H_0.

1.8.2 Bayes factors

For any pairwise comparison of models or hypotheses, we can also rely on a quantity known as the Bayes factor. Before seeing the data, the *prior odds* of M_1 over M_0 are $p(M_1)/p(M_0)$, and after seeing the data we have the *posterior odds* $p(M_1|\mathbf{y})/p(M_0|\mathbf{y})$. The ratio of these two sets of odds is the Bayes factor:

Definition 1.11 (Bayes Factor). *Given data y and two models M_0 and M_1, the Bayes factor*

$$B_{10} = \frac{p(y|M_1)}{p(y|M_0)} = \left\{\frac{p(M_1|y)}{p(M_0|y)}\right\} \Big/ \left\{\frac{p(M_1)}{p(M_0)}\right\} \tag{1.7}$$

is a summary of the evidence for M_1 against M_0 provided by the data.

The Bayes factor provides a measure of whether the data have altered the odds on M_1 relative to M_0. For instance, $B_{10} > 1$ indicates that M_1 is now more plausible relative to M_0 than it was *a priori*.

The Bayes factor plays something of an analagous role to a likelihood ratio. In fact, twice the logarithm of B_{10} is on the same scale as the deviance and likelihood ratio test statistics for model comparisons. For cases where the models are labelled by point restrictions on θ, the Bayes factor is a likelihood ratio. However, unlike the likelihood ratio test statistic, in the Bayesian context there is no reference to a sampling distribution with which to assess the particular statistic obtained in the present sample. In the Bayesian approach, all inferences are made conditional on the data at hand (not with reference to what might happen over repeated applications of random sampling). Thus, the Bayes factor has to be interpreted as a summary measure of the information in the data about the relative plausibility of models or hypotheses, rather than offering a formulaic way to choose between those model or hypotheses. Jeffreys (1961) suggests the following scale for interpreting the Bayes factor:

B_{10}	$2 \log B_{10}$	Evidence for M_1
<1	<0	negative (support M_0)
1 to 3	0 to 2	barely worth mentioning
3 to 12	2 to 5	positive
12 to 150	5 to 10	strong
>150	>10	very strong

Good (1988) summarizes the history of the Bayes factor, which long predates likelihood ratio as a model comparison tool.

■ **Example 1.11**

Attitudes towards abortion, Example 1.9, continued. We computed the marginal likelihoods under the two hypotheses in Example 1.10, which we now use to compute the Bayes factor,

$$B_{10} = \frac{p(y|H_1)}{p(y|H_0)} = \frac{2}{.00028} = 7142,$$

again indicating that the data strongly favor H_1 over H_0.

1.9 From subjective beliefs to parameters and models

Earlier in this chapter I introduced Bayes Theorem with some simple examples. But in so doing I have brushed over some important details. In particular, the examples are all *parametric* (as are almost all statistical models in the social sciences), in the sense that the probability distribution of the data is written as a function of an unknown parameter θ (a scalar or vector). This approach to statisical inference – expressing the joint density of the data as a function of a relatively small number of unknown parameters – will be familiar to many readers, and may not warrant justification or elaboration. But given the subjectivist approach adopted here the question of how and why parameters and models enter the picture is not idle.

Recall that in the subjectivist approach championed by de Finetti (and adopted here), the idea that probability is a property of a coin, a die, or any other object under study, is regarded as metaphysical nonsense. All that is real is the data at hand. We may also possess knowledge (or at least beliefs) about how the data were generated. For instance, are the data real at all, or are they output of a computer simulation? Were the data produced via an experiment with random assignment to treatment and control groups, by random sampling from a specific population, or are the data a complete enumeration of a population? But everything else is a more or less convenient fiction created in the mind of the researcher *including parameters and models*.

To help grasp the issue a little more clearly, consider the following example. A coin is flipped n times. The possible set of outcomes is $\mathcal{S} = \{\{H, T\}_1 \times \ldots \times \{H, T\}_n\}$, with cardinality 2^n. Assigning probabilities over the elements of $\mathcal{S}$ is a difficult task, if only because for any moderate to large value of n, 2^n is a large number. Almost instinctively, we start falling back on familar ways to simplify the problem. For example, reaching back to our introductory statistics classes, we would probably inquire 'are the coin flips independent?' If satisfied that the coin flips are independent, we would then fit a binomial model to the data, modeling the r flips coming up heads as a function of a 'heads' probability θ, given the n flips. In a Bayesian analysis we would also have a prior density $p(\theta)$ as part of the model, and we would report the posterior density over $p(\theta|r, n)$ as the result of the analysis.

I now show that this procedure – using parameteric models to simplify data analysis – can be justified by recourse to a deeper principle called *exchangeability*. In particular, if data are 'infinitely exchangeable', then a Bayesian approach to modeling the data is not only possible or desirable, but is actually *implied* by exchangeability. That is, prior distributions over parameters are not merely a 'Bayesian addition' to an otherwise classical analysis, but *necessarily* arise when one believes that the data are exchangeable. This is the key insight of one of the most important theorems in Bayesian statistics – de Finetti's Representation Theorem – which we will also encounter below.

1.9.1 Exchangeability

We begin with a definition:

Definition 1.12 (Finite exchangeability). *The random quantities $y_1, \ldots, y_n$ are finitely exchangeable if their joint probability density (or mass function, for discrete y),*

$$p(y_1, \ldots, y_n) = p(y_{z(1)}, \ldots, y_{z(n)})$$

for all permutations z of the indices of the y_i, $\{1, \ldots, n\}$.

Remark. An infinite sequence of random quantities $y_1, y_2, \ldots$ is infinitely exchangeable if every finite subsequence is finitely exchangeable.

Exchangeability is thus equivalent to the condition that the joint density of the data $\mathbf{y}$ remains the same under any re-ordering or re-labeling of the indices of the data.

Similarly, exchangeability is often interpreted as the Bayesian version of the 'iid assumption' that underlies much statistical modeling, where 'iid' stands for 'independently and identically distributed'. In fact, if data are exchangeable they are conditionally iid, where the conditioning is usually on a parameter, θ (but contrast Problem 1.8). Indeed, this is one of the critical implications of de Finetti's Representation Theorem.

As we shall now see, de Finetti's Theorem shows that beliefs about data being infinitely exchangeable imply a belief about the data having 'something in common', a 'similiarity' or 'equivalence' (de Finetti's original term) such that I can swap y_i for y_j in the sequence without changing my beliefs that either y_i or y_j will be one or zero (i.e. there is nothing special about y_i having the label i, or appearing in the i-th position in the sequence). That is, under exchangeability, two sequences, each with the same length n, and the same proportion of ones, would be assigned the same probability. As Diaconis and Freedman (1980a) point out: 'only the number of ones in the ... trials matters, not the location of the ones'.

de Finetti's Thoerem takes this implication a step further, showing that if I believe the data are infinitely exchangeable, then it is as if there is a parameter θ that drives a stochastic model generating the data, *and* a density over θ that doesn't depend on the data. This density is interpretable as a prior density, since it characterizes beliefs about θ that are not conditioned on the data. That is, the existence of a prior density over a parameter is a *result* of de Finetti's Representation Theorem, rather than an assumption.

We now state this remarkable theorem, referring interested readers elsewhere for a proof.

Proposition 1.9 (de Finetti Representation Theorem, binary case). *If $y_1, y_2, \ldots$ is an infinitely exchangeable sequence, with $y_i \in \{0, 1\}, \forall\, i = 1, 2, \ldots$, then there exists a probability density function P such that the joint probability mass function for n realizations of y_i, $P(y_1, \ldots, y_n)$ can be represented as follows,*

$$P(y_1, \ldots, y_n) = \int_0^1 \prod_{i=1}^{n} \theta^{y_i}(1-\theta)^{1-y_i}\, dF(\theta)$$

where $F(\theta)$ is the limiting distribution of θ, i.e.

$$F(\theta) = \lim_{n\to\infty} \Pr(n^{-1}\sum_{i=1}^{n} y_i \leq \theta).$$

Proof. See de Finetti (1931; 1937), Heath and Sudderth (1976). ◁

Remark. Hewitt and Savage (1955) proved the uniqueness of the representation.

Since this theorem is so important to the subjectivist, Bayesian approach adopted here, we pause to examine it in some detail. First, consider the object on the left-hand side of the equality in the proposition. Given that $y_i \in \{0, 1\}$, $P(y_1, \ldots, y_n)$ is an assignment of probabilities to all 2^n possible realizations of $\mathbf{y} = (y_1, \ldots, y_n)$. It is daunting to consider allocating probabilities to all 2^n realizations, but an implication of de Finetti's Representation Theorem is that *we don't have to*. The proposition shows that probability assignments to $y_1, \ldots, y_n$ (a finite subset of an infinitely exchangeable sequence) can be made in terms of a single parameter θ, interpretable as the limiting value of the proportion of ones in the infinite, exchangeable sequence $y_1, y_2, \ldots$. This is extraordinarily convenient, since under exchangeability, the parameter θ can become the object of statistical modeling, rather than much more cumbersome object $P(y_1, \ldots, y_n)$. Thus, in the subjectivist approach, parameters feature in statistical modeling not necessarily because they are 'real' features of the world, but because they are part of a convenient, mathematical representation of probability assignments over data.

Perhaps more surprisingly, di Finetti's Representation Theorem also implies the existence of a prior probability density over θ, $F(\theta)$, in the sense that it is a density over θ that does not depend on the data. If $F(\theta)$ in Proposition 1.9 is absolutely continuous, then we obtain the probability density function for θ, $p(\theta) = dF(\theta)/d\theta$. In this case, the identity in the proposition can be re-written as

$$P(y_1, \ldots, y_n) = \int_0^1 \prod_{i=1}^n \theta^{y_i}(1-\theta)^{1-y_i} p(\theta) d\theta. \tag{1.8}$$

We recognize the first term on the right-hand-side of equation 1.8 as the likelihood for a series of Bernoulli trials, distributed independently conditional on a parameter θ, i.e. under independence conditional on θ,

$$\mathcal{L}(\theta; \mathbf{y}) \equiv f(\mathbf{y}|\theta) = \prod_{i=1}^n f(y_i|\theta)$$

where

$$f(y_i|\theta) = \begin{cases} \theta = \theta^{y_i} & \text{if } y_i = 1 \\ (1-\theta) = (1-\theta)^{1-y_i} & \text{if } y_i = 0 \end{cases}$$

The second term in Equation 1.8, $p(\theta)$, is a prior density for θ, The integration in equation 1.8 is how we obtain the marginal density for $\mathbf{y}$, as a weighted average of the likelihoods implied by different values of $\theta \in [0, 1]$, where the prior density $p(\theta)$ supplies the weights.

That is, a simple assumption such as (infinite) exchangeability implies the existence of a parameter θ and a prior over θ, and hence a justification for adopting a Bayesian approach to inference:

> This [de Finetti's Representation Theorem] is one of the most beautiful and important results in modern statistics. Beautiful, because it is so general and yet so simple. Important, because exchangeable sequences arise so often in practice. If there are, and we are sure there will be, readers who find $p(\theta)$ distasteful, remember it is only as distasteful as exchangeability; and is that unreasonable? (Lindley and Phillips 1976, 115)

1.9.2 Implications and extensions of de Finetti's Representation Theorem

The parameter θ considered in Proposition 1.9 is recognizable as a success probability for independent Bernoulli trials. But other parameters and models can be considered. A simple example comes from switching our focus from the individual zeros and ones to $S = \sum_{i=1}^{n} y_i$, the number of ones in the sequence $\mathbf{y} = (y_1, \ldots, y_n)$, with possible values $s \in \{0, 1, \ldots, n\}$. Since there are $\binom{n}{s}$ ways of obtaining $S = s$ successes in n trials, de Finetti's Representation Theorem implies that probability assignments for S represented as

$$\Pr(S = s) = \binom{n}{s} \int_0^1 \theta^s (1 - \theta)^{n-s} dF(\theta).$$

where $F(\theta) = \lim_{n\to\infty} \Pr(n^{-1}S \leq \theta)$ is the limiting probability distribution function for θ. Put differently, conditional on θ and n (the number of trials), the number of successes S is distributed following the binomial probability point mass function.

A general form of de Finneti's Representation Theorem exists, and here I re-state a relatively simple version of the general form, due to Smith (1984, 252):

Proposition 1.10 (Representation Theorem for Real-Valued Random Quantities). *If* $\mathbf{y}_n = (y_1, \ldots . y_n)$ *are realizations from an infinitely exchangeable sequence, with* $-\infty < y_i < \infty$ *and with probability measure* P*, then there exists a probability measure* μ *over* $\mathcal{F}$*, the space of all distribution functions on* $\mathbb{R}$ *such that the joint distribution function of* $\mathbf{y}_n$ *has the representation*

$$P(y_1, \ldots, y_n) = \int_{\mathcal{F}} \prod_{i=1}^{n} F(y_i) d\mu(F)$$

where

$$\mu(F) = \lim_{n\to\infty} P(F_n)$$

and where F_n *is the empirical distribution function for* $\mathbf{y}$ *i.e.*

$$F_n(y) = n^{-1}[I(y_1 \leq y) + I(y_2 \leq y) + \ldots + I(y_n \leq y)]$$

where $I(\cdot)$ *is an indicator function evaluating to one if its argument is true and zero otherwise.*

Proof. See de Finetti (1937; 1938). The result is a special case of the more abstract situation considered by Hewitt and Savage (1955) and Diaconis and Freedman (1980b). ◁

Note for this general case that the Representation Theorem implies a *nonparametric*, or, equivalently, a infinitely-dimensional parametric model. That is, the F_n in Proposition 1.10 is the unknown distribution function for y, a series of asymptotically-diminishing step functions over the range of y. Conditional on this distribution, it is as if we have independent data. The distribution μ is equivalent to a prior over what F_n would look like in a large sample.

Thus, the general version of de Finetti's Representation Theorem is only so helpful, at least as a practical matter. What typically happens is that the infinite-dimensional F_n is approximated with a distribution indexed by a finite parameter vector $\boldsymbol{\theta}$; e.g., consider $\boldsymbol{\theta} = (\mu, \sigma^2)$, the mean and variance of a normal density, respectively. Note the *parametric, modeling* assumptions being made here. This said, the use of particular parameteric models is not completely *ad hoc*. There is much work outlining the conditions under which exchangeability plus particular *invariance* assumptions imply particular parameteric models (e.g., under what conditions does a belief of exchangeability over real-valued quantities justify a normal model, under what conditions does a belief of exchangeability over positive, integer-valued random quantities justify a geometric model, and so on). Bernardo and Smith (1994, §4.4) provide a summary of these results.

1.9.3 Finite exchangeability

Note that both Propositions 1.9 and 1.10 rest on an assumption of infinite exchangeability: i.e. that the (finite) data at hand are part of a infinite, exchangeable sequence. de Finetti type theorems do not hold for finitely exchangeable data; see Diaconis (1977) for some simple but powerful examples. This seems problematic, especially in social science settings, where it is often not at all clear that data can be considered to be a subset of an infinite, exchangeable sequence. Happily, finitely exchangeable sequences can be shown to be approximations of infinitely exchangeable sequences, and so de Finetti type results hold approximately for finitely exchangeable sequences. Diaconis and Freedman (1980b) bound the error induced by this approximation for the general case, in the sense that the de Finneti type representation for $P(\mathbf{y}_n)$ under finite exchangeability differs from the representation obtained under an assumption of infinite exchangeability by a factor that is smaller than a constant times $1/n$. Thus, for large n, the 'distortion' induced by assuming infinite exchangeability is vanishingly small. A precise definition of this 'distortion' and sharp bounds for specific cases are reported in Diaconis and Freedman (1981), Diaconis, Eaton and Lauritzen (1992), and Wood (1992).

Still, n can be quite small in social science settings, and exchangeability judgements themselves may not be enough to justify parameteric modeling. In these cases models arise not so much as a consequence of having exchangeable data via de Finetti's Representation Theorem, but exist in the mind of the researcher prior to the analysis. This is perfectly fine, and indeed, corresponds to the way many social scientists go about their business: we look for data sets to tests theories and models, rather than (as may happen more often in statistics) we look for models to fit to the data we've been given to analyze. That is, one can (and ought!) to adopt a Bayesian approach even in the absence of exchangeable data; the point here is that models and prior densities are necessarily implied by accepting that one's data is exchangeable.

1.9.4 Exchangeability and prediction

Exchangeability also makes clear the close connections between prediction and Bayes Rule, and between parameters and observables. Consider tossing a coin n times with the outcomes, $\mathbf{y}$, infinitely exchangeable. Arbitrarily, let $y_i = 1$ for a head. We observe r heads out of n tosses. Then we consider the next toss of the coin, with the outcome denoted $\tilde{y}$, conditional on the observed sequence of r heads in n flips, $\mathbf{y}$. The probability density (or mass function) we form over this future outcome $\tilde{y}$ is known as the posterior predictive density (or mass function). In this case,

$$P(\tilde{y} = 1|\mathbf{y}) = \frac{P(\tilde{y} = 1, \mathbf{y})}{P(\mathbf{y})}$$

and, by exchangeability,

$$= \frac{\int_0^1 \theta^{r+\tilde{y}}(1-\theta)^{n+1-r-\tilde{y}} p(\theta)d\theta}{\int_0^1 \theta^r (1-\theta)^{n-r} p(\theta)d\theta}$$

$$= \frac{\int_0^1 \theta^{\tilde{y}}(1-\theta)^{1-\tilde{y}} \theta^r (1-\theta)^{n-r} p(\theta)d\theta}{\int_0^1 \theta^r (1-\theta)^{n-r} p(\theta)d\theta}$$

$$= \frac{\int_0^1 \theta^{\tilde{y}}(1-\theta)^{1-\tilde{y}} \mathcal{L}(\theta; \mathbf{y}) p(\theta)d\theta}{\int_0^1 \mathcal{L}(\theta; \mathbf{y}) p(\theta)d\theta},$$

since up to a constant multiplicative factor (that will cancel across numerator and denominator) $\mathcal{L}(\theta; \mathbf{y}) = \theta^r(1-\theta)^{n-r}$. But, by Bayes Rule (Proposition 1.5),

$$p(\theta|\mathbf{y}) = \frac{\mathcal{L}(\theta; \mathbf{y})p(\theta)}{\int_0^1 \mathcal{L}(\theta; \mathbf{y})p(\theta)d\theta}$$

and so $P(\tilde{y} = 1|\mathbf{y}) = \int_0^1 \theta^{\tilde{y}}(1-\theta)^{1-\tilde{y}} p(\theta|\mathbf{y})d\theta = \int_0^1 \theta p(\theta|\mathbf{y})d\theta = E(\theta|\mathbf{y})$. That is, under exchangeability (and via Bayes Rule), beliefs about the outcome of the next realization of the binary sequence corresponds to beliefs about the parameter θ. It is provocative to note that θ need not correspond to anything in the physical world; indeed, the parameter θ may well be nothing more than a convenient fiction we conjure up to make a prediction problem tractable. The general point here is that we will rely on this property of modeling under exchangeability quite frequently, with parameters providing an especially useful way to summarize beliefs not only about the data at hand, but future realizations of the (exchangeable) data.

1.9.5 Conditional exchangeability and multiparameter models

Once again, consider the simple case in Proposition 1.9, where $\mathbf{y} = (y_1, \ldots, y_n)$ is a sequence of zeros and ones. In this case, without any other information about the data, exchangeability seems quite plausible. That is, probability assignments over the data conform to the form given in Proposition 1.9, in which the data are considered independent Bernoulli trials, conditional on the parameter θ, and $p(\theta)$ is a prior density over θ.

But consider a different situation. What if instead of (canonical) coin flips, we had asked survey respondents if they had ever engaged in political protest, for instance, a street demonstration. The data are coded $y_i = 1$ if survey respondent i responds 'Yes' and 0 otherwise. But we also know that the data come from J different countries: let $j = 1, \ldots, J$ index the countries covered by the survey, and let $C_i = j$ if respondent i is in country j. Suppose for a moment that the country information is given to us only in the most minimal form: a set of integers, i.e. $C_i \in \{1, \ldots, J\}$. That is, we know that the data come from different countries, but that is all.

Even with this little amount of extra information I suspect most social scientists would not consider the entire sequence of data $\mathbf{y} = (y_1, \ldots, y_n)'$ as exchangeable, since there are good reasons to suspect levels of political protest vary considerably by country. We would want to condition any assignment of a zero or a one to y_i on the country label of case i, C_i. *Within* any given country, and absent any other information, the data might be considered exchangeable. Data with this feature are referred to as *partially exchangeable* or *conditionally exchangeable* (e.g. Lindley and Novick 1981). In this example, exchangeability within each country implies that each country's data can be modeled via country-specific, Bernoulli models: i.e. for $j = 1, \ldots, J$,

$$\begin{aligned} y_i|C_i = j &\sim \text{Bernoulli}(\theta_j) \quad \text{(likelihoods)} \\ \theta_j &\sim p_j(\theta_j) \quad \text{(priors)} \end{aligned}$$

or, equivalently, since the data are exchangeable within a country, we can model the number of respondents reporting engaging in political protest in a particular country r_j via a binomial model, conditional on θ_j and the number of respondents in that country n_j:

$$\begin{aligned} r_j|\theta_j, n_j &\sim \text{Binomial}(\theta_j; n_j) \quad \text{(likelihood)} \\ \theta_j &\sim p_j(\theta_j) \quad \text{(priors)} \end{aligned}$$

1.9.6 Exchangeability of parameters: hierarchical modeling

The hypothetical multi-country example just considered takes a step in the direction of 'hierarchical models'. That is, idea of exchangeability applies not just to data, but to parameters as well: i.e. note the deliberate use of the general term 'random quantities' rather than 'data' in Propositions 1.9 and 1.10.

Consider the example again. We know that data span J different countries. But that is all we know. Under these conditions, the θ_j can be considered exchangeable: i.e. absent any information to distinguish the countries from one another, the probability assignment $p(\theta_1, \ldots, \theta_J)$ is invariant to any change of the labels of the countries (see Definition 1.12). Put simply, the country labels j do not meaningfully distinguish the countries with respect to their corresponding θ_j. In this case, de Finetti's Representation Theorem implies that the joint density of the θ_j has the representation

$$p(\boldsymbol{\theta}) = p(\theta_1, \ldots, \theta_J) = \int \prod_{j=1}^{m} p(\theta_j|\nu)p(\nu)d\nu \tag{1.9}$$

where ν is a *hyperparameter*. That is, under exchangeability at the level of countries, it is *as if* we have the following two-stage or *hierarchical* prior structure over the θ_j:

$$\begin{aligned} \theta_j|\nu &\sim p(\theta_j|\nu) \quad \text{(hierarchical model for } \theta_j) \\ \nu &\sim p(\nu) \quad \text{(prior for hyperparameter } \nu) \end{aligned}$$

For the example under consideration – modeling country-specific proportions – we might employ the following choices for the various densities:

$$r_j|\theta_j, n_j \sim \text{Binomial}(\theta_j; n_j)$$

$$\theta_j|\nu \sim \text{Beta}(\alpha, \beta)$$

$$\alpha \sim \text{Exponential}(2)$$
$$\beta \sim \text{Exponential}(2)$$

with $\nu = (\alpha, \beta)$ the hyperparameters for this problem. Details on the specific densities come later: e.g. in Chapter 2 we discuss models for proportions in some detail, and a hierarchical model for binomial data is considered in Example 7.9. At this stage, the key point is that exchangeability is a concept that applies not only to data, but to parameters as well.

We conclude this brief introduction to hierarchical modeling with an additional extension. If we possess more information about the countries other than case labels, then exchangeability might well be no longer plausible. Information that survey respondents were located in different countries prompted us to revise a belief of exchangeability for them; similarly, information allowing us to distinguish countries from one another might lead us to revisit the exchangeability judgement over the θ_j parameters. In particular, suppose we have variables at the country level, $\mathbf{x}_j$, measuring factors such as the extent to which the country's constitution guarantees rights to assembly and freedom of expression, and the repressiveness of the current regime. In this case, exchangeability might hold *conditional* on a unique combination of those country-level predictors. A statistical model that exploits the information in $\mathbf{x}_j$ might be the following *multi-level* hierarchical model:

$$\begin{aligned} r_j|\theta_j, n_j &\sim \text{Binomial}(\theta_j; n_j) \\ z_j &= \log\left(\frac{\theta_j}{1-\theta_j}\right) \\ z_j|\mathbf{x}_j &\sim N(\mathbf{x}_j\boldsymbol{\beta}, \sigma^2) \\ \boldsymbol{\beta}|\sigma^2 &\sim N(\mathbf{b}, \sigma^2\mathbf{B}) \\ \sigma^2 &\sim \text{Inverse-Gamma}\left(\frac{\nu_0}{2}, \frac{\nu_0\sigma_0^2}{2}\right). \end{aligned}$$

Again, details on the specific models and densities deployed here come in later chapters; Example 7.10 provides a detailed consideration of a multi-level model. The key idea is that the information in $\mathbf{x}_j$ enters as the independent variables in a regression model for z_j, the log-odds of each country's θ_j. In this way *contextual* information about country j is incorporated into a model for the survey responses. These types of exchangeability judgements will play an important role in the discussion of hierarchical models in Chapter 7.

1.10 Historical note

Bayes Theorem is named for the Reverend Thomas Bayes, who died in 1761. The result that we now refer to as Bayes Theorem appeared in an essay attributed to Bayes and communicated to the Royal Society after Bayes death by his friend, Richard Price (Bayes 1763). This famous essay has been republished many times since (e.g. Bayes 1958).

Several authors have noted that there is some doubt that Bayes actually discovered the theorem named for him; see, for instance, Stigler (1999, Ch 14) and the references in Fienberg (2006). Nor is it clear that Bayes himself was a 'Bayesian' in the sense that we use the term today (e.g. Stigler 1982).

The subject of Bayes *Essay towards solving a problem in the doctrine of chances* was what we would today recognize as a binomial problem: given x successes in n independent binary trials, what should we infer about π, the underlying probability of success? Bayes himself studied the binomial problem with a uniform prior. In 1774 Laplace (apparently unaware of Bayes work) stated Bayes theorem in its more general form, and also considered non-uniform priors (Laplace 1774). Laplace's article popularized what would later become known as 'Bayesian' statistics. Perhaps because of Laplace's work on the subject, Bayes' essay itself 'was ignored until after 1780 and played no important role in scientific debate until the twentieth century' (Stigler 1986b, 361). Additional historical detail can be found in Bernardo and Smith (1994, ch. 1), and Stigler (1986a, ch. 3). We return to the relatively simple statistical problem considered by Bayes (drawing inferences given binomial data) in Chapter 2.

The adjective 'Bayesian' did not enter the statistical vernacular until the 20th century. Fienberg (2006) reviews the 'neo-Bayesian revival' of the 20th century, and, via a review by Edwards (2004), traces the first use of 'Bayesian' as an adjective to Fisher (1950), in an introduction to a paper originally written in 1921. Unsurprisingly, Fisher's use of the term was not flattering, since he was at pains to contrast his approach to statistical inference from the subjectivism he disliked in the Bayesian approach. In contrast with Fisher's pejorative use of the term, Fienberg (2006) provides a detailed exposition of how Bayesians themselves came to adopt the 'Bayesian' moniker in the 20th century.

Problems

1.1 Consider a cross-national study of economic development, where the data comprise all OECD countries in 2000. A researcher argues that while these data are the population of OECD countries in 2000, they are nonetheless a random sample from the histories of these countries. Discuss.

1.2 Consider the drug testing problem given in Example 1.1. Consider the false negative rate and the false positive rate of the drug test as two variables.

1. Construct a grid of hypothetical values for these two variables. At each point on the grid, compute the posterior probability of H_U, the hypothesis 'the subject uses the prohibited substance' given the prior on this hypothesis of $P(H_u) = .03$ and a postitive test result. Use a graphical technique such as a contour plot or an image plot to summarize the results.

2. What values for the two error rates of the test give rise to a posterior probability on H_U that exceeds 0.5?

3. Repeat this exercise, but now considering a run of 3 positive tests: what values of the test error rates give rise to a posterior probability for H_U in excess of 0.95?

1.3 Suppose $p(\theta) \equiv \chi^2_2$. Compute a 50 % highest density region for θ. Compare this region with the inter-quartile range of $p(\theta)$.

1.4 Consider a density $p(\theta)$. Under what conditions can a HDR for θ of content α be determined by simply noting the $(1-\alpha)/2$ and $1-(1-\alpha)/2$ quantiles of $p(\theta)$? That is, what must be true about $p(\theta)$ so as to let us compute a HDR this way?

1.5 Consider Example 1.4. Repeat the analysis in the example assuming that it is (a) two times and (b) five times as costly to overestimate applicant ability than it is to underestimate ability.

1.6 A poll of 500 adults in the United States taken in the Spring of 2008 finds that just 29 % of respondents approve of the way that George W. Bush is handling his job as president.

1. Report the posterior probabilities of $H_0 : \theta > .33$ and $H_1 : \theta < .33$. The threshold $\theta = .33$ has some politically interest, say, if we assume that (up to a rough approximation) the electorate is evenly partitioned into Democrat, Independent, and Republican identifiers.
2. Report a Bayes factor for H_0 vs H_1. Comment briefly on your finding.
3. Contrast how a frequentist approach would distinguish between these two hypotheses.

1.7 Consider the poll data in the previous question. Suppose you had the following uniform prior for θ, $\theta \sim \text{Unif}(.4, .6)$. What is your posterior density, given the polling data?

1.8 Is exchangeability merely a Bayesian way of saying 'iid'? That is, establish whether statistical independence is a necessary and sufficient condition for exchangeability. In particular, can you come up with an example where exchangeability holds, but independence does not?

2

Getting started: Bayesian analysis for simple models

2.1 Learning about probabilities, rates and proportions

Perhaps the first problem ever considered in Bayesian statistics – and probably the first problem you encountered in your first statistics class – is the problem of learning from binomial data. Bayes opened his famous 1763 *Essay towards solving a problem in the doctrine of chances* with the following statement:

PROBLEM:

> *Given* the number of times in which an unknown event has happened and failed: *Required* the chance that the probability of its happening in a single trial lies somewhere between any two degrees of probability that can be named.

In modern terms, we recognize this as the problem of trying to learn about the parameter θ governing a binomial process, having observed r successes in n independent binary trials (also known as Bernoulli trials). Estimation and inference for the binomial success probability is a fundamental problem in statistics (and a relatively easy one) and so provides a convenient context in which to gain exposure to the mechanics of Bayesian inference.

Moreover, binary data abound in the social sciences. Examples include micro-level phenomena such as

- voter turnout ($y_i = 1$ if person i turned out to vote, and 0 otherwise), purchasing a specific product, using the Internet, owning a computer;

Bayesian Analysis for the Social Sciences S. Jackman

- life events such as graduating from high school or college, marriage, divorce, having children, home ownership, employment, or going to jail, and even death itself;
- self-reports of attitudes (e.g. approval or disapproval of a policy or a politician; reporting that one is happy or satisifed with one's life, intending to vote for a specific candidate);
- correctly responding to a test item (e.g. $y_{ij} = 1$ if test-taker i correctly answers test item j);
- legislative behavior (e.g. $y_{ij} = 1$ if legislator i votes "Aye" on proposal j).

Aggregate-level binary data are also common in social-science settings, coding for

- the presence or absence of interstate or civil war,
- whether a country is a democracy or not,
- if a congressional district is home to a military base,
- features of a country's policy of social welfare (universalist or not) or industrial relations regime (centralized bargaining or not).

In general, the observed data consist of a string of binary data: $y_i \in \{0, 1\}$, with $i = 1, \ldots, n$. Let r be the number of times $y_i = 1$. In section 1.9 we saw that if binary data can be considered exchangeable, then it is *as if* the data were generated as n independent Bernoulli trials with unknown parameter θ, i.e.

$$y_i \stackrel{\text{iid}}{\sim} \text{Bernoulli}(\theta), \quad i = 1, \ldots, n, \tag{2.1}$$

and with a prior density $p(\theta)$ over θ. Similarly, under exchangeability it is *as if*

$$r \sim \text{Binomial}(\theta; n) \tag{2.2}$$

and again $p(\theta)$ is a prior density over θ. Note that in either model, θ has the interpretation as the proportion of ones we would observe in an arbitrarily long, exchangeable sequence of zeros and ones. In most applications involving binary data in the social sciences, a model such as Equation 2.1 or 2.2 is adopted without recourse to exchangeability, the researcher presuming that θ exists in some meaningful sense and that the data at hand were generated as independent Bernoulli trials, conditional on θ. A typical example is the analysis of survey data, where, via random sampling, respondents are considered to be providing binary responses independently of one another, and θ is interpreted as a characteristic of the population, the *proportion* of the population with $y_i = 1$.

■ Example 2.1

Attitudes towards abortion, Example 1.9, continued. 895 out of 1934 survey respondents reported that they thought it should be possible for a pregnant woman to obtain a legal abortion if the woman wants it for any reason. Under the assumptions of independence and random sampling (operationally equivalent to exchangeability), we can represent the data with the binomial model in Equation 2.2, i.e.

$$895 \sim \text{Binomial}(\theta; n = 1934)$$

The corresponding binomial likelihood function is

$$\mathcal{L}(\theta; r = 895, n = 1934) = \binom{1934}{895} \theta^{895}(1-\theta)^{1934-895}$$

and the maximum likelihood estimate is $\hat{\theta}_{\text{MLE}} = 895/1934 \approx .463$ with standard error .011.

2.1.1 Conjugate priors for probabilities, rates and proportions

Any Bayesian analysis requires a specification of the prior. In this case, the parameter θ is a probability (or, under exchangeability, a population proportion), and is restricted to the [0, 1] unit interval. Any proper prior density over θ, $p(\theta)$ is thus a function with the following properties:

1. $p(\theta) \geq 0$, $\theta \in [0, 1]$.
2. $\int_0^1 p(\theta)d\theta = 1$.

Obviously, many functions have these properties: in fact, all the prior densities in Figures 1.2, 1.3 and 1.4 satistfy these conditions.

I initially consider *conjugate priors*: priors with the property that when we apply Bayes Rule – multiplying the prior by the likelihood – the resulting posterior density is of the same parametric type as the prior; see Definition 1.2. Thus, whether a prior is conjugate or not is a property of the prior density with respect to the likelihood function. In this way a conjugate prior is sometimes said to be "closed under sampling", in the sense that after sampling data and modifying the prior via Bayes Rule, our posterior beliefs over θ can still be characterized with a probability density function in the same parameteric class as we employed to characterize our prior beliefs (i.e. the class of probability densities representing beliefs is not expanded by considering the information about θ in the likelihood).

It is worth stressing that conjugacy is not as restrictive as it might sound. Conjugate priors are capable of characterizing quite a wide array of beliefs. This is especially true if one considers *mixtures* of conjugate priors (Diaconis and Ylvisaker 1979). In fact, mixtures of conjugate priors were used to generate the odd looking prior distributions in Figures 1.3 and 1.9.

As we now see, Beta densities are conjugate prior densities with respect to a binomial likelihood. The Beta density is quite a flexible probability density, taking two shape parameters as its arguments, conventionally denoted with α and β, and with its support confined to the unit interval. A uniform density on [0, 1] is a special case of the Beta density, arising when $\alpha = \beta = 1$. Symmetric densities with the mode/mean/median at .5 are generated when $\alpha = \beta$ for $\alpha, \beta > 1$. Unimodal densities with positive skew are generated by $\alpha > \beta > 1$; negative skew arises with $\beta > \alpha > 1$. Other features of the Beta density are simple functions of α and β:

- the mean: $\dfrac{\alpha}{\alpha + \beta}$

- the mode: $\dfrac{\alpha - 1}{\alpha + \beta - 2}$
- the variance: $\dfrac{\alpha\beta}{(\alpha + \beta)^2(\alpha + \beta + 1)}$

Further details on the Beta density appear at Definition B.28 in Appendix B. The Beta density is

$$p(\theta; \alpha, \beta) = \frac{\Gamma(\alpha + \beta)}{\Gamma(\alpha)\Gamma(\beta)}\theta^{\alpha-1}(1 - \theta)^{\beta-1}, \tag{2.3}$$

where $\theta \in [0, 1]$, $\alpha, \beta > 0$ and $\Gamma(\cdot)$ is the Gamma function (again, see Appendix B). Note that the leading terms involving the Gamma functions do not involve θ, and so Equation 2.3 can be rewritten as

$$p(\theta; \alpha, \beta) \propto \theta^{\alpha-1}(1 - \theta)^{\beta-1}. \tag{2.4}$$

This prior is conjugate for θ with respect to the binomial likelihood in Equation 2.2:

Proposition 2.1 (Conjugacy of Beta prior, binomial data). *Given a binomial likelihood over r successes in n Bernoulli trials, each independent conditional on an unknown success parameter* $\theta \in [0, 1]$, *i.e.*

$$\mathcal{L}(\theta; r, n) = \binom{n}{r}\theta^r(1 - \theta)^{n-r}$$

then the prior density $p(\theta) = Beta(\alpha, \beta)$ *is conjugate with respect to the binomial likelihood, generating the posterior density* $p(\theta|r, n) = Beta(\alpha + r, \beta + n - r)$.

Proof. By Bayes Rule,

$$p(\theta|r, n) = \frac{\mathcal{L}(\theta; r, n)p(\theta)}{\int_0^1 \mathcal{L}(\theta; r, n)p(\theta)d\theta} \propto \mathcal{L}(\theta; r, n)p(\theta)$$

The binomial coefficient $\binom{n}{r}$ in the likelihood does not depend on θ and can be absorbed into the constant of proportionality. Similiarly, we use the form of the Beta prior given in Equation 2.4. Thus

$$\begin{aligned} p(\theta|r, n) &\propto \underbrace{\theta^r(1 - \theta)^{n-r}}_{\text{likelihood}} \; \underbrace{\theta^{\alpha-1}(1 - \theta)^{\beta-1}}_{\text{prior}} \\ &= \theta^{r+\alpha-1}(1 - \theta)^{n-r+\beta-1} \end{aligned}$$

which we recognize as the kernel of a Beta distribution (a definition of the kernel of a probability density function is provided in Definition B.15). That is, $p(\theta|r, n) = c\theta^{r+\alpha-1}(1 - \theta)^{n-r+\beta-1}$ where c is the normalizing constant

$$\frac{\Gamma(n + \alpha + \beta)}{\Gamma(r + \alpha)\Gamma(n - r + \beta)},$$

i.e. $\int_0^1 \theta^{r+\alpha-1}(1-\theta)^{n-r+\beta-1} = c^{-1}$, see Definition B.19. In other words,

$$\theta | r, n \sim \text{Beta}(\alpha + r, \beta + n - r) \tag{2.5}$$

$\triangleleft$

As is the case for most conjugate priors, the form of the posterior density has a simple interpretation. It is *as if* our prior distribution represents the information in a sample of $\alpha + \beta - 2$ independent Bernoulli trials, in which we observed $\alpha - 1$ "successes" in which $y_i = 1$. This interpretation of the parameters in the conjugate Beta prior in "data equivalent terms" makes it reasonably straightforward to specify conjugate priors for probabilities, rates and proportions.

■ Example 2.2

Attitudes towards abortion, Example 1.9, continued. A researcher has no prior information regarding θ, the proportion of the population agreeing with the proposition that it should be possible for a pregnant woman to obtain a legal abortion if the woman wants it for any reason. Any value of $\theta \in [0, 1]$ is *a priori* as likely as any other, implying the researcher's prior is a uniform distribution. The uniform distribution on the unit interval is a special case of the Beta distribution with $\alpha = \beta = 1$ (i.e. corresponding to possessing a prior set of observations of zero successes in zero trials). Recall that in these data $r = 895$ and $n = 1934$. Then, applying the result in Proposition 2.1, the posterior density for θ is

$$p(\theta | r = 895, n = 1934) = \text{Beta}(\alpha^*, \beta^*),$$

where

$$\begin{aligned} \alpha^* &= \alpha + r = 1 + 895 = 896 \\ \beta^* &= \beta + n - r = 1 + 1934 - 895 = 1040 \end{aligned}$$

Given the "flat" or "uninformative" prior, the location of the the mode of the posterior density corresponds *exactly* with the maximum likelihood estimate of θ: i.e.

$$\begin{aligned} \text{posterior mode of } p(\theta | r, n) &= \frac{\alpha^* - 1}{\alpha^* + \beta^* - 2} = \frac{896 - 1}{896 + 1040 - 2} \\ &= \frac{895}{1934} = \frac{r}{n} = .463 = \hat{\theta}_{\text{MLE}} \end{aligned}$$

Moreover, when α^* and β^* are large (as in this example), the corresponding Beta density is almost exactly symmetric and so the mode is almost exactly the same as the mean and the median. That is, because of the near-perfect symmetry of the posterior density, Bayes estimates of θ such as the posterior mean, median or mode (considered in Section 1.6.1) almost exactly coincide. In addition, with the "flat"/uniform prior employed in this example, these Bayes estimate of θ all almost exactly coincide with the maximum likelihood estimate. For instance, the mean of the posterior density is at $896/(896 + 1040) \approx .463$, which corresponds to the maximum likelihood estimate to three decimal places.

The posterior variance of θ is

$$\begin{aligned} V(\theta|r, n) &= \frac{\alpha^* \beta^*}{(\alpha^* + \beta^*)^2(\alpha^* + \beta^* + 1)} \\ &= \frac{896 \times 1040}{(896 + 1040)^2 \times (896 + 1040 + 1)} \\ &= \frac{931\,840}{3\,748\,096 \times 1937} \approx 0.000128, \end{aligned}$$

and so the posterior standard deviation of θ is $\sqrt{.000128} = .0113$, corresponding to the standard error of the maximum likelihood estimate of θ.

Finally, note that with α^* and β^* both large (as they are in this example), the Beta density is almost exactly symmetric, and so a very accurate approximation to a 95 % highest posterior density region for θ can be made by simply computing the 2.5 and 97.5 percentiles of the posterior density, the Beta(α^*, β^*) density. With $\alpha^* = 896$ and $\beta^* = 1040$, these percentiles bound the interval [.441, .485]. Computing the 95 % HPD directly shows this approximation to be very good in this case: the exact 95 % HPD corresponds to the approximation based on the percentiles, at least out to 4 significant digits.

Of course, the results of a Bayesian analysis will not coincide with a frequentist analysis if the prior density is informative about θ, relative to the information in the data about θ. This can happen even when priors over θ are "flat" or nearly so, if the data set is small and/or relatively uninformative about θ. Alternatively, even with a large data set, if prior information about θ is plentiful, then the posterior density will reflect this, and will not be proportional to the likelihood alone.

■ Example 2.3

Combining information for improved election forecasting. In early March 2000, Mason-Dixon Polling and Research conducted a poll of voting intentions in Florida for the November presidential election. The poll considered George W. Bush and Al Gore as the presumptive nominees of their respective political parties. The poll had a sample size of 621, and the following breakdown of reported vote intentions: Bush 45 % ($n = 279$), Gore 37 % (230), Buchanan 3 % (19) and undecided 15 % (93). For simplicity, we ignore the undecided and Buchanan vote share, leaving Bush with 55 % of the two-party vote intentions, and Gore with 45 %, and $n = 509$ respondents expressing a preference for the two major party candidates. We assume that the survey responses are independent, and (perhaps unrealistically) that the sample is a random sample of Floridian voters. Then, if θ is the proportion of Floridian voters expressing a preference for Bush, the likelihood for θ given the data is (ignoring constants that do not depend on θ),

$$\mathcal{L}(\theta; r = 279, n = 509) \propto \theta^{279}(1 - \theta)^{509-279}.$$

The maximum likelihood estimate of θ is $\hat{\theta}_{\text{MLE}} = r/n = 279/509 = .548$ with a standard error of $\sqrt{(.548 \times (1 - .548))/509} = .022$. Put differently, this poll provides very strong evidence to suggest that Bush was leading Gore in Florida at that relatively early stage of the 2000 presidential race.

But how realistic is this early poll result? Is there other information available that bears on the election result? Previous presidential elections are an obvious source of information. Even a casual glance at state-level presidential election returns suggests one can do a fairly good job at predicting state-level election outcomes by looking at previous elections in that state, say, by taking a weighted average of previous election outcomes via regression analysis. This procedure is then used to generate a forecast for Republican presidential vote share in Florida in 2000. This prediction is 49.1 %, with a standard error of 2.2 percentage points.

We can combine the information yielded by this analysis of previous elections with the survey via Bayes' Theorem. We can consider the prediction from the analysis as supplying a prior, and the survey as "data", although mathematically, it doesn't matter what label we give to each piece of information.

To apply Bayes' Theorem in this case, and to retain conjugacy, I first characterize the information from the regression forecast as a Beta distribution. That is, we seek values for α and β such that

$$E(\theta; \alpha, \beta) = \alpha/(\alpha + \beta) = .491$$

$$V(\theta; \alpha, \beta) = \frac{\alpha\beta}{(\alpha + \beta)^2(\alpha + \beta + 1)} = .022^2$$

which yields a system of equations in two unknowns. Leonard and Hsu (1999, 100) suggest the following strategy for solving for α and β. First define $\theta_0 = E(\theta; \alpha, \beta) = \alpha/(\alpha + \beta)$ and then note that $V(\theta; \alpha, \beta) = \theta_0(1 - \theta_0)/(\gamma + 1)$, with $\gamma = \alpha + \beta$ interpretable as the size of a hypothetical, prior sample. Then,

$$\begin{aligned}
\gamma &= \frac{\theta_0(1 - \theta_0)}{V(\theta; \alpha, \beta)} - 1 \\
&= \frac{.491 \times (1 - .491)}{.022^2} - 1 = 515.36, \\
\alpha &= \gamma\theta_0 = 515.36 \times .491 = 253.04, \\
\beta &= \gamma(1 - \theta_0) = 515.36 \times (1 - .491) = 262.32.
\end{aligned}$$

That is, the information in the previous elections is equivalent to having ran another poll with $n \approx 515$ in which $r \approx 253$ respondents said they would vote for the Republican presidential candidate.

Now, with a conjugate prior for θ it is straightforward to apply Bayes Rule. The posterior density for θ is a Beta density with parameters $\alpha^* = 279 + 253.04 = 532.04$ and $\beta^* = 509 - 279 + 262.32 = 492.32$. With α^* and β^* both large, this Beta density is almost exactly symmetric, with the mean, mode and median all equal to .519 (to three significant digits). A 95 % HPD is extremely well approximated by the 2.5 and 97.5 percentiles of the Beta density, which bound the interval [.489, .550]. Figure 2.1 displays the prior, the posterior and the likelihood (re-scaled so as to be comparable to the prior and posterior densities). The posterior density clearly represents a compromise between the two sources of information brought to bear in this instance, with most of the posterior probability mass lying between the mode of the prior density and the mode of the

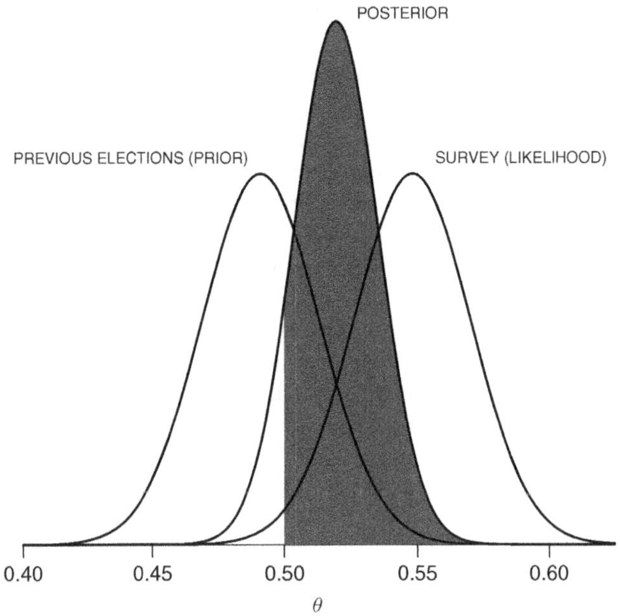

Figure 2.1 Florida election polling. The shaded area under the posterior density represents the posterior probability that Bush leads Gore, and is .893.

likelihood function. It is also clear that at least in this case, the variance of the posterior density is smaller than either the variance of the prior or the dispersion of the likelihood function. This reflects the fact that in this example the prior and the likelihood are not in wild disagreement with one another, and the posterior density "borrows strength" from each source of information.

Other features of the posterior density are worth reporting in this case. For instance, given the combination of the survey data with the information supplied by the historical election data, the probability that Bush leads Gore in Florida is simply the posterior probability that $\theta > 1/2$, i.e.

$$\begin{aligned}
P(\theta > 1 - \theta | r, n, \alpha, \beta) &= P(\theta > .5 | r, n, \alpha, \beta) \\
&= \int_{.5}^{1} p(\theta | r, n, \alpha, \beta) d\theta \\
&= \int_{.5}^{1} p(\theta; \alpha^* = 532.04, \beta^* = 492.32) d\theta \\
&= 1 - \int_{0}^{.5} p(\theta; \alpha^* = 532.04, \beta^* = 492.32) d\theta.
\end{aligned}$$

This probability is .893, suggesting that the probability of a Bush victory was still quite high, but not as high as near-certainty implied by the survey estimate.

■ Example 2.4

Will the sun rise tomorrow? In his *Essai philopsophique sur les probabilités*, Laplace considered the following problem: "If we place the dawn of history at 5000 years before the present date, then we have 1 826 213 days on which the sun has constantly risen in each 24 hour period" (Laplace 1825, 11). Given this evidence, what is the probability that the sun will rise tomorrow? Bayes also considered this problem in his famous posthumous *Essay* (1763). Bayes and Laplace computed their solution to this problem supposing (in Bayes' words), "a previous total ignorance of nature". Bayes (or at least Richard Price, who edited and communicated Bayes' essay to the *Royal Society*) imagined "a person just brought forth into this world", who, after witnessing sunrise after sunrise is gradually revising their assessment that the sun will rise tomorrow:

> this odds [of the sun rising] would increase, ... with the number of returns to which he was witness. But no finite number of returns would be sufficient to produce absolute or physical certainty (Bayes 1958, 313).

Laplace's solution to this problem is known as Laplace's "rule of succession", and is one of the first, correct Bayesian results to be widely published.

In a more modern statistical language, both Bayes and Laplace considered the problem of what to believe about a binomial success probability θ, given n successes in n trials, and a uniform prior on θ: i.e.

$$\mathcal{L}(\theta; n) = \binom{n}{n} \theta^n (1-\theta)^{n-n} = \theta^n \quad \text{(likelihood)}$$

$$p(\theta) = \begin{cases} 1 & \text{if } 0 \le \theta \le 1 \\ 0 & \text{otherwise} \end{cases} \quad \text{(prior)}$$

By Bayes Rule, the posterior distribution for θ is

$$p(\theta|n) \propto \mathcal{L}(\theta; n) p(\theta) \propto \theta^n$$

which we recognize as the kernel of a Beta(α^*, β^*) distribution with $\alpha^* = n + 1$ and $\beta = 1$; i.e. $\theta|n \sim \text{Beta}(n + 1, 1)$. This posterior distribution has a mode at $\theta = 1$, but the posterior mean is

$$E(\theta|n \text{ success in } n \text{ trials}) = \frac{n+1}{n+2} = 1 - \frac{1}{n+2}$$

As we saw in section 1.9.4, exchangeability and Bayes Rule imply that

$$P(y_{n+1} = 1 | \sum_{i=1}^{n} y_i = r) = E(\theta | \sum_{i=1}^{n} y_i = r),$$

for any $r = 0, 1, \ldots, n$. Here $r = n$ and Laplace considered $n = 1\,826\,213$. Thus, the probability of the sun rising tomorrow, conditional on a uniform prior over this probability and 5000 years of exchangeable, daily sunrises, is $1 - 1.826215 \times 10^{-6}$ or 0.999998173785. That is, it is not an absolute certainty that the sun will rise tomorrow, nor should it be.

Remark. This example highlights some interesting differences between the frequentist and Bayesian approaches to inference. I paraphrase Diaconis (2005) on this result:

> if this seems wild to you, compare it with the frequentist estimate of θ given n successes in n trials. It is $\hat{\theta} = 1$ (even if $n = 2$).

That is, suppose we flip a coin twice. It comes up heads both times. The maximum likelihood estimate of θ is 1.0, which is also the frequentist estimate of the probability of a head on the next toss. With a uniform prior over θ, the Bayesian estimate of the probability of a head on the next toss is $1 - 1/4 = .75$. It is an interesting exercise to see if in these circumstances a frequentist would be prepared to make a wager based on the estimate $\hat{\theta} = 1$.

Remark. Laplace fully understood his example was contrived, ignoring relevant knowledge such as orbital mechanics in the solar system, the life-cycles of stars, and so on. Immediately after concluding that under the stipulated conditions of a uniform prior over θ and 5000 years of sun rises "we may lay odds of 1 826 214 to 1 that it [the sun] will rise again tomorrow", Laplace writes

> But this number would be incomparably greater for one who, perceiving in the coherence of phenomena the principle regulating days and seasons, sees that nothing at the present moment can check the sun's course (Laplace 1825, 11)

Further discussion on the historical and scientific background surrounding Laplace's use of this example, and Laplace's Bayesian solution is provided in two essays by Zabell (1988; 1989b). The latter essay makes clear the links between Laplace's rule of succession and the concept of exchangeability (see Section 1.9).

2.1.2 Bayes estimates as weighted averages of priors and data

Proposition 2.1 shows that with a Beta(α, β) prior for a success probability θ and a binomial likelihood over the r successes out of n trials, the posterior density for θ is a Beta density with parameters $\alpha^* = \alpha + r$ and $\beta^* = \beta + n - r$. We have also seen that under these conditions, the posterior mean of θ is

$$E(\theta|r,n) = \frac{\alpha^*}{\alpha^* + \beta^*} = \frac{\alpha + r}{\alpha + \beta + n} = \frac{n_0\theta_0 + n\hat{\theta}}{n_0 + n} \tag{2.6}$$

where $\hat{\theta} = r/n$ is the maximum likelihood estimate of θ, $n_0 = \alpha + \beta$ and $\theta_0 = \alpha/(\alpha + \beta) = E(\theta)$ is the mean of the prior density for θ. Equation 2.6 can be rewritten as

$$E(\theta|r,n) = \gamma\theta_0 + (1 - \gamma)\hat{\theta}$$

where $\gamma = n_0/(n_0 + n)$, and since $n_0, n > 0$, $\gamma \in [0, 1]$. Alternatively,

$$E(\theta|r,n) = \hat{\theta} + \gamma(\theta_0 - \hat{\theta})$$

These simple, linear form of these three expressions reflect the fact that with a conjugate Beta prior, the posterior mean – a Bayes estimate of θ – is a *weighted average* of the prior mean θ_0 and the maximum likelihood estimate $\hat{\theta}$. This means that with conjugate priors, the posterior mean for θ will be a convex combination of the prior mean and the maximum likelihood estimate, and so lies somewhere between the prior mean and the value supported by the data. In this way, the Bayes estimate is sometimes said to *shrink* the posterior towards the prior, with $0 < \gamma < 1$ being the shrinkage factor. With a increasingly precise set of prior beliefs – a prior based on an increasing number of hypothetical prior observations – the posterior mean will be shifted increasingly closer to the prior mean. On the other hand, as the precision of the prior information degrades, n is large relative to n_0, and $\hat{\theta}$ dominates the weighted average in Equation 2.6.

■ Example 2.5

Combining information for improved election forecasting, Example 2.3 continued. Previous election results supply prior information for θ, George W. Bush's share of the two-party vote for president in Florida in 2000, represented by a Beta(253.04, 262.32) density. But it is reasonable to consider other priors. In particular, one might think that while the previous election results supply a reasonable "best guess" as to θ, there is actually quite a lot of uncertainty as to the relevance of previous elections for the problem of making a forecast for the current election. Put differently, it might be reasonable to doubt that presidential election outcomes in Florida are exchangeable. In fact, it could well be that precisely because Florida was expected to be a pivotal state in the 2000 election (if not *the* pivotal state), the campaigns devote a disproportionate amount of resources there, changing the nature of the election there, such that the 2000 election is "special" and no longer exchangeable with past elections. In this case we might consider a range of priors over θ, reflecting more or less confidence in the prior over θ dervied from previous elections.

The top panel of Figure 2.2 shows how the location of the posterior mean, $E(\theta|r, n)$ changes over more or less stringent versions of the prior derived from previous elections. The parameter λ governs the precision of the prior, playing the role of a user-adjustable *hyperparameter*. That is, the prior is $\theta \sim \text{Beta}(253.04/\lambda, 262.32/\lambda)$ with $\lambda > 0$. For all values of λ, the prior mean remains constant, i.e. $E(\theta; \lambda) = (\alpha/\lambda)/(\alpha/\lambda + \beta/\lambda) = \alpha/(\alpha + \beta)$. But as $\lambda \to 0$, the precision of the prior is increasing, since $n_0 = (\alpha + \beta)/\lambda$ is growing large; conversely, as $\lambda \to \infty$, $n_0 \to 0$ and the prior precision decreases. In turn, since Equation 2.6 shows that the mean of the posterior density for θ has the weighted average form

$$E(\theta|r, n) = \frac{n\hat{\theta} + n_0\theta_0}{n + n_0},$$

it follows that

$$\lim_{\lambda \to \infty} E(\theta|r, n) = \hat{\theta}$$

$$\lim_{\lambda \to 0} E(\theta|r, n) = \theta_0$$

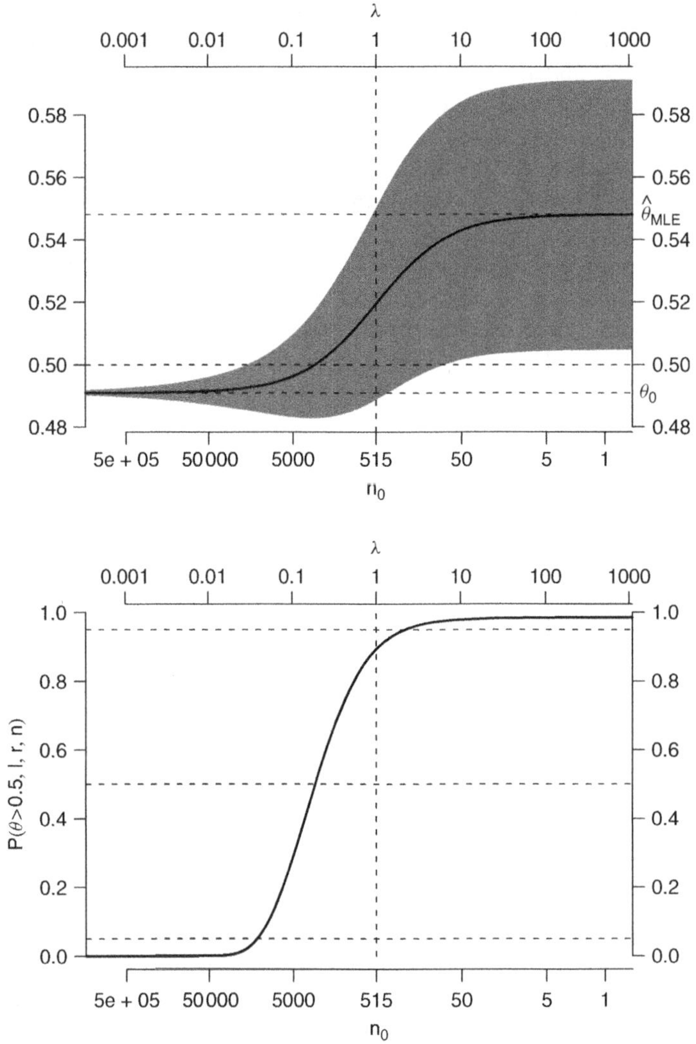

Figure 2.2 Posterior mean as a convex combination of prior and data for binomial success parameter, θ. In the top panel, the solid line traces out a series of posterior means, $E(\theta|r, n)$, generated using the prior, $p(\theta; \alpha, \beta, \lambda) = \text{Beta}(\alpha/\lambda, \beta/\lambda)$. The shaded area shows the width of a posterior 95 % HPD for θ, also varying as a function of λ. The lower panel shows the predicted probability of a Bush victory in Florida, the posterior probability that $\theta > .5$, as a function of λ.

In the top panel of Figure 2.2 λ increases from left to right, on the log scale. On the left, as $\lambda \to 0$, the posterior is completely dominated by the prior; on the right of the graph, as $\lambda \to \infty$, the prior receives almost zero weight, and the posterior mean is almost exactly to the maximum likelihood estimate $\hat{\theta}$.

The top panel of Figure 2.2 also displays a 95 % HDR around the locus of posterior means. As $\lambda \to 0$, the 95 % HDR collapses around the prior mean, since the prior is becoming infinitely precise, and the posterior tends towards a degenerate density with point mass on θ_0. On the other hand, as $\lambda \to \infty$, the prior has virtually no impact on the posterior, and the posterior 95 % HDR widens, to correspond to a classical 95 % confidence interval around $\hat{\theta}$.

The lower panel of Figure 2.2 shows the posterior probability that Bush wins Florida in 2000 (the posterior probability that $\theta > .5$). Clearly, as the posterior density is shrunk towards θ_0 as $\lambda \to 0$, this posterior probability can be expected to vary. The survey data along suggest that this posterior probability is quite high: the 95 % HDR for θ with $\lambda \to \infty$ does not overlap .5 (top panel), and the posterior probability that $\theta > .5$ is almost 1.0 (lower panel). With the prior suggested by the historical election data ($\lambda = 1$), the posterior probability of Bush winning Florida has fallen below the conventional 95 % level of confidence, but Bush winning Florida is over 4 times as likely as the alternative outcome. With $\lambda \approx .19$, $E(\theta|r, n) \approx .5$ and the posterior probability that $\theta > .5$ falls to around .5. For values of $\lambda < .19$, the prior is sufficiently stringent that the posterior probability that Bush carries Florida is below .5. With $\lambda \approx .19$, $n_0 \approx 2712$, or equivalently, it is as if the information in the historical election data is as precise as that obtained from a survey of around 2 700 respondents. That is, we would have to place considerable weight on the historical election data in order to contradict the implication of the survey data that Bush would defeat Gore in Florida in 2000.

As it turned out, Bush won 50 % of the two-party vote in Florida, in famously controversial circumstances, and Florida was pivotal in deciding the 2000 presidential election. Bush's 50 % performance in Florida in 2000 is only one percentage point above the level implied by the prior, well within a prior 95 % HDR and the posterior 95 % HDR generated using the historical data as a prior (i.e. with $\lambda = 1$), and outside the 95 % HDR implied by the February survey data. With the benefit of hindsight, it would seem that combining the February survey data with the prior information supplied by historical election data was prudent. Bayes Rule tell us how to perform that combination of information.

This relatively simple, weighted average form for the mean of the posterior density is not unique to the use of conjugate Beta prior and a binomial likelihood. As we shall see below, we obtain similar weighted average forms for the mean of the posterior density in other contexts, when employing prior densities that are conjugate with respect to a likelihood. In fact, the weighted average form of the posterior mean is a general feature of conjugacy, and indeed, as been proposed as a *definition* of conjugacy by Diaconis and Ylvisaker (1979).

2.1.3 Parameterizations and priors

In the examples considered above, prior ignorance about θ has been expresed via a uniform prior over the unit interval, which is actually a conjugate Beta prior with respect to the binomial likelihood. That is, absent any knowledge to the contrary, any value of θ should be considered as likely as any other. Bayes and Laplace employed this prior. Bayes motivated his 1763 *Essay* with reference to the location of a ball thrown onto a square table or a bounded plane, for which "there shall be the same probability that it rests upon any one equal part of the plane as another". In this case, and in many others, Bayes thought this postulate of prior ignorance appropriate:

> in the case of an event concerning the probability of which we absolutely know nothing antecedently to any trials made concerning it, seems to appear from the following consideration; viz. that concerning such an event I have no reason to think that, in a certain number of trials, it should rather happen any one possible number of times than another.

The more general principle – in the absence of *a priori* information, give equal weight to all possible outcomes – is known as the *principle of insufficient reason*, and is usually attributed to Jakob Bernoulli in *Ars conjectandi*, but may well be due to Leibniz (Hacking 1975, ch. 14). Laplace embraced this postulate, apparently accepting it as "an intuitively obvious axiom" (Stigler 1986b, 359), whereas Bayes offered a quite elaborate justification of what we today would call a uniform prior; of course, Laplace also considered non-uniform priors.

Even the simple binomial problem considered above can become complicated if we consider a *reparameterization*. That is, suppose we are interested not in the success probability θ, but a parameter that is a function of θ, say $g(\theta)$. R. A. Fisher (1922) famously pointed to a shortcoming of the simple use of uniform priors over θ in this case. With n trials and r successes, the maximum likelihood estimate is $\hat{\theta} = r/n$. But, Fisher notes, "we might equally have measured probability upon an entirely different scale" (Fisher 1922, 325), and considers the re-parameterization $\sin q = 2\theta - 1$. By the invariance property of maximum likelihood (Proposition B.1), the MLE of q corresponds to the MLE of θ, once we map $\hat{q}_{\text{MLE}}$ back to θ, as we now demonstrate. Assuming conditional independence of the observations, the likelihood function is (omitting constants that do not depend on q)

$$\mathcal{L}(q; r, n) = (1 + \sin q)^r (1 - \sin q)^{n-r}$$

with log-likelihood

$$\log \mathcal{L}(q; r, n) = r \log(1 + \sin q) + (n - r) \log(1 - \sin q),$$

and, after differentiating with respect to q, the MLE is given by solving

$$\frac{r \cos q}{1 + \sin q} = \frac{(n - r) \cos q}{1 - \sin q},$$

yielding $\sin q = 2(r/n) - 1$ as the MLE. This implies that even if we maximize the likelihood with respect to q, we still find that $\hat{\theta}_{\text{MLE}} = r/n$, the same result that we get from solving for the MLE of θ directly.

Fisher noted that in a Bayesian analysis of this problem, a uniform prior over q did not imply a uniform prior over a parameter such as θ, and that two sets of Bayes estimates for θ could be obtained, depending on which "uniform prior" one adopted. That is, the results of the Bayesian analysis are sensitive to choice of parameterization and prior in a way that a maximum likelihood analysis is not. If we have the model

$$q \sim \text{Uniform}\left(-\frac{\pi}{2}, \frac{\pi}{2}\right)$$
$$\theta = (1 + \sin q)/2$$
$$r \sim \text{Binomial}(\theta; n)$$

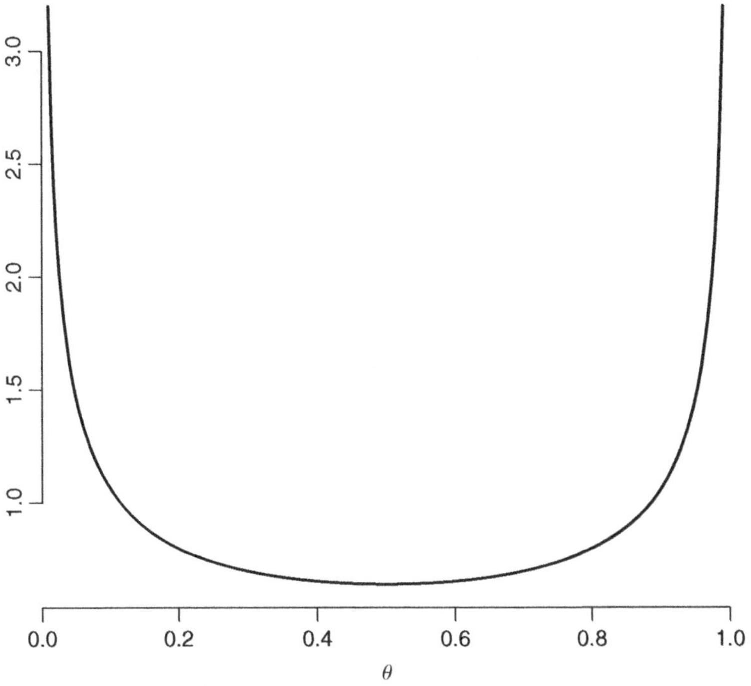

Figure 2.3 The Beta($\frac{1}{2}$, $\frac{1}{2}$) density.

then the posterior density for θ is Beta$(r + \frac{1}{2}, n - r + \frac{1}{2})$ density, with a mode at $(r - \frac{1}{2})/(n - 1)$. This differs from the maximum likelihood estimate of r/n by an amount that vanishes as $n \to \infty$. On the other hand, a uniform prior on θ yields a posterior density that has the same shape as the likelihood function.

Using the results in Proposition B.7, the uniform prior on $q \in [\frac{-\pi}{2}, \frac{\pi}{2}]$ induces a Beta($\frac{1}{2}$, $\frac{1}{2}$) prior density on θ, which integrates to one over the unit interval even though the density is infinite at zero and one. As Figure 2.3 displays, the Beta($\frac{1}{2}$, $\frac{1}{2}$) density is somewhat odd, in that it assigns high probability to relatively extreme values of θ: e.g. a 50 % highest density region for $\theta \sim$ Beta($\frac{1}{2}$, $\frac{1}{2}$) is the set of disjoint intervals $\{[0, .15], [.85, 1]\}$. Nonetheless, this is the prior density that arises when specifying a uniform prior over a (non-linear) transformation of θ.

The points here are (a) the prior is important in a Bayesian analysis; (b), expressing prior ignorance may not be as straightforward as you might think. A prior density that is uninformative with respect to one parameter (or parameterization) may not be uninformative with respect to some other parameter of interest. What to do? For one thing, if prior densities are expressing prior beliefs, then analysts need to decide precisely what is the object under study. If the binomial success probability, θ, is an object of interest and for which it is easy to mathematically express one's prior beliefs, then one should go ahead and use that prior. But we should not be surprised, let alone dismayed, that, say, a uniform prior on θ implies an unusual or surprising distribution over some parameter $g(\theta)$, or vice-versa. Likewise, if interest focuses on $g(\theta)$, then the analyst should formulate a prior with respect to that parameter.

2.1.4 The variance of the posterior density

With a Beta(α, β) prior over θ, and a binomial likelihood for r successes in n Bernoulli trials (independent conditional on θ), the variance of the posterior density for θ is

$$V(\theta|r, n) = \frac{(\alpha + r)(\beta + n - r)}{(\alpha + \beta + n)^2(\alpha + \beta + n + 1)}$$

It is not always the case that the posterior variance is smaller than the prior variance

$$V(\theta) = \frac{\alpha\beta}{(\alpha + \beta)^2(\alpha + \beta + 1)}.$$

It is not obvious from inspecting the previous two equations as to when posterior uncertainty about θ can exceed prior uncertainty. But if the prior information about θ and the data starkly contradict one another, then it is possible for the posterior uncertainty to exceed the prior uncertainty. Conversely, when the prior and the likelihood are in relative agreement, then the posterior density will be less dispersed than the prior, reflecting the gain in information about θ from the data.

■ **Example 2.6**

Will I be robbed at Stanford? Laura has just arrived at Stanford University for graduate study. She guesses the probability that she will be the victim of property theft while on campus during the course of her studies to be 5 %; i.e. $E(\theta) = .05$. Laura has heard lots of good things about Stanford: e.g. Stanford is a private university surrounded by relatively affluent neighborhoods, with a high proportion of graduate students to undergraduates. These factors not only lead Laura to make a low *a priori* estimate of θ, but to make Laura reasonably confident in her guess. While Laura's *a priori* probability estimate is subject to uncertainty, she figures that the probability of her falling victim to theft is no greater than 10 %, but almost surely no lower than 1 %. These prior beliefs are well approximated by a Beta distribution over θ, with its mean at .05, and with $\Pr(\theta > .10) \approx .01$ and $\Pr(\theta < .01) \approx .01$. That is, if $\theta \sim \text{Beta}(\alpha, \beta)$, then Laura's belief that $E(\theta) = .05$ implies the constraint $\alpha/(\alpha + \beta) = .05$ or $\beta = 19\alpha$. Further, the additional constraints that $\Pr(\theta > .10) \approx .01$ and $\Pr(\theta < .01) \approx .01$ imply that $\alpha \approx 7.4$ and $\beta \approx 140.6$.

Over the course of conversations with twenty other students (more or less randomly selected), Laura raises the topic of theft on campus. It transpires that twelve of the twenty students she speaks to report having been the victim of theft while at Stanford. Laura is reasonably satisfied that the experiences of the twenty students are independent of one another (i.e. the twenty students constitute part of an exchangeable sequence of Stanford students). Accordingly, Laura summarizes the new information with a binomial likelihood for the $r = 12$ "successes" in $n = 20$ trials, considered independent conditional on θ. Via Bayes Rule, Laura updates her beliefs about θ, yielding the posterior density

$$\theta|r, n \sim \text{Beta}(\alpha + r, \beta + n - r) = \text{Beta}(19.4, 148.6)$$

Laura's prior "best guess" (prior mean for θ) of .05 is revised upwards in light of the fact that 60 % of the 20 students she spoke to reported being a victim of theft. Laura's *a posteriori* estimate of θ (the mean of the posterior density) is $19.4/(19.4 + 148.6) \approx .12$.

But what it is also interesting is what has happened to the *dispersion* or *precision* of Laura's beliefs. The variance of Laura's prior beliefs was

$$\begin{aligned} V(\theta) &= \frac{\alpha\beta}{(\alpha+\beta)^2(\alpha+\beta+1)} \\ &= \frac{7.4 \times 140.6}{(7.4+140.6)^2(7.4+140.6+1)} \\ &\approx .00032 \end{aligned}$$

but the variance of the posterior density over θ is

$$\begin{aligned} V(\theta|r,n) &= \frac{(\alpha+r)(\beta+n-r)}{(\alpha+\beta+n)^2(\alpha+\beta+n+1)} \\ &= \frac{(7.4+12)\times(140.6+8)}{(7.4+140.6+20)^2(7.4+140.6+20+1)} \\ &\approx .00060, \end{aligned}$$

or roughly twice the size of the prior variance. That is, while Laura has revised her estimate of θ upwards, she is in some senses more uncertain about θ than she was previously. The 95 % highest density region for Laura's prior density is [.018, .086], spanning .068, while the corresponding region for the posterior density is [.069, .164], spanning .095.

The previous example demonstrates that in some circumstances posterior uncertainty can be larger than prior uncertainty. But this result does not hold if we work with an alternative parameterization. In particular, for conjugate analysis of binomial problems, if the analysis is performed with respect to the log of the odds ratio

$$\Lambda = \text{logit}(\theta) = \ln\left(\frac{\theta}{1-\theta}\right)$$

then posterior density of Λ has the property that it always has variance no greater than the variance of the prior density for Λ. If $\theta \sim \text{Beta}(\alpha,\beta)$, the using the change of variables result in Proposition B.7,

$$f(\Lambda) = f_\theta\left(\frac{\exp(\Lambda)}{1+\exp(\Lambda)};\alpha,\beta\right)\frac{\exp(\Lambda)}{(1+\exp(\Lambda))^2} \tag{2.7}$$

where f_θ is a Beta(α,β) density. For even moderately-sized α and β (say, $\alpha,\beta \geq 5$), this density is extremely well-approximated by a normal distribution with mean $\log(\alpha/\beta)$ and variance $\alpha^{-1}+\beta^{-1}$; see Lindley (1964, 1965) and Bloch and Watson. Since the variance of $f(\Lambda)$ is approximately $\alpha^{-1}+\beta^{-1}$, it is obvious that the posterior variance will be smaller than the prior variance; i.e. if $\theta \sim \text{Beta}(\alpha,\beta)$ and $\theta|r,n \sim \text{Beta}(\alpha+r,\beta+n-r)$, then $V(\Lambda|r,n) \approx (\alpha+r)^{-1}+(\beta+n-r)^{-1} < V(\Lambda) \approx \alpha^{-1}+\beta^{-1}$, since for any data set $r \geq 0$ and $n > 0$.

■ Example 2.7

Will I be robbed at Stanford? (Example 2.6, continued). Figure 2.4 shows Laura's prior and posterior densities, on both the unit probability interval, and on the log-odds scale. Visual inspection confirms that on the log-odds scale, the posterior variance is smaller than the prior variance. Using the Lindley (1965) normal approximation to $f(\Lambda)$, we obtain $V(\Lambda) \approx .14$, and $V(\Lambda | r = 12, n = 20) \approx .058$. The normal approximations to the prior and posterior densities for Λ are overlaid on Figure 2.4 as dotted lines.

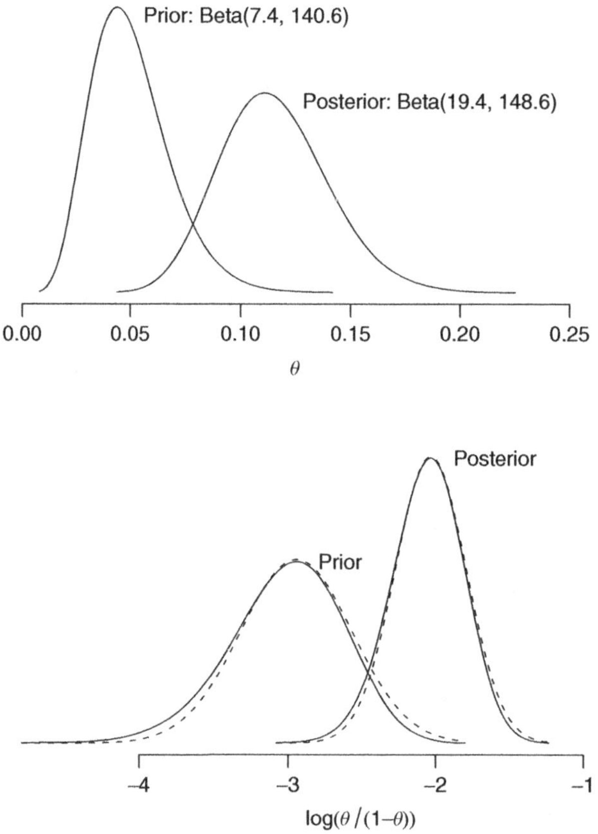

Figure 2.4 Prior and posterior densities for Example 2.6, probability and log-odds scale. The dotted lines on the log-odds scale indicate the Lindley (1965) normal approximation.

Again, this example highlights that re-parameterizations are not necessarily innocuous in a Bayesian setting. We saw earlier that a prior that may be "flat" or "uninformative" with respect to θ may not be uninformative with respect to $g(\theta)$; Example 2.6 highlights that application of Bayes Rule can properly lead to a *decrease* in precision with respect to θ, but an *increase* in precision with respect to $\Lambda = \text{logit}(\theta)$. There is nothing "wrong" with these results, or with the underlying Bayesian analyses. But analysts should be aware of the possibility that sometimes, rationally updating one's beliefs via Bayes Rule can lead to a decrease in precision, depending on the parameterization employed and the degree of conflict between the data and the prior.

2.2 Associations between binary variables

Analysis of the two-by-two crosstabulation has long been a staple of data analysis in the social sciences. Just as many outcomes of interest in social-scientific analysis are binary variables, so too are plausible predictors of those outcomes. Examples include analyzing a binary outcome such as voter turnout by race (white/non-white), or examining whether democractic regimes are more likely to go to war with one another than with autocratic regimes.

Bayesian approaches to testing for association between binary variables builds on the ideas presented in the previous section. There we assumed that the data for analysis consisted of a sequence of n binary outcomes, $y_i \in \{0, 1\}$ that could be considered independent conditional on a parameter $\theta \in (0, 1) = E(y_i) = \Pr(y_i = 1)$. Given independence, the number of "successes", $r = \sum_{i=1}^{n} y_i$, can be modeled as binomial random variable, conditional on the unknown success probability θ and the number of outcomes n. We now generalize this model to consider two sequences of binary outcomes, y_{i0} when a predictor $x_i = 0$, and y_{i1} when a predictor $x_i = 1$. Each sequence of binary data is presumed to be exchangeable, with success probabilities θ_0 and θ_1, respectively. We assume that the binary realizations are independent within and across the two groups defined by $x_i \in \{0, 1\}$, and so the likelihood for the two streams of binary data is just

$$\begin{aligned}\mathcal{L}(\theta_0, \theta_1; \mathbf{y}) &= \mathcal{L}(\theta_0; \mathbf{y})\,\mathcal{L}(\theta_1; \mathbf{y}) \\ &= \prod_{\{i:x_i=0\}} \theta_0^{y_i}(1-\theta_0)^{1-y_i} \prod_{\{i:x_i=1\}} \theta_1^{y_i}(1-\theta_1)^{1-y_i}\end{aligned}$$

Given independence, we can also model these two data as two binomial processes, and ignoring constants that do not depend on θ_0 or θ_1 we have

$$\mathcal{L}(\theta_0, \theta_1; \mathbf{y}) \propto \theta_0^{r_0}(1-\theta_0)^{n_0-r_0}\theta_1^{r_1}(1-\theta_1)^{n_1-r_1} \tag{2.8}$$

where r_0 and r_1 are the numbers of "successes" in the two groups defined by $x_i = 0$ and $x_i = 1$ respectively; see Table 2.1. Since the likelihood for θ_0 and θ_1 in equation 2.8 is a function solely of the four quantities r_0, r_1, n_0 and n_1, the likelihood is in effect a model for the two-by-two table in Table 2.1.

Table 2.1 Two-by-two cross table.

	x_i		
y_i	0	1	
0	$n_0 - r_0$	$n_1 - r_1$	
1	r_0	r_1	
	n_0	n_1	n

As in any Bayesian analysis, we formalize our prior beliefs over unknown parameters with probability densities. In this case, the parameter space is the unit square, since both $0 \leq \theta_0, \theta_1 \leq 1$; more formally, $\boldsymbol{\theta} = (\theta_0, \theta_1)' \in \Theta = [0, 1] \times [0, 1]$. Specifying a prior for $\boldsymbol{\theta}$ is simplified when prior information about the parameters θ_0 and θ_1 is independent, since in that case $p(\boldsymbol{\theta}) = p(\theta_0)p(\theta_1)$, i.e. for independent random variables, the joint probability density is equal to the product of the marginal densities. But this need not be the case. For example, if *a priori*, a belief that θ_0 is more likely to be large than small is also informative about θ_1, then θ_0 and θ_1 are not independent. In this case, we can not simply factor the joint prior density over $\boldsymbol{\theta}$ as the product of the two marginal prior densities for θ_0 and θ_1. For now, I focus on the simple case, where prior information about θ_0 and θ_1 is independent, and represented with Beta densities, which are conjugate with respect to the binomial likelihood in Equation 2.8: i.e. if $\theta_0 \sim \text{Beta}(\alpha_0, \beta_0)$ and $\theta_1 \sim \text{Beta}(\alpha_1, \beta_1)$ then under the assumption that θ_0 and θ_1 are independent *a priori*

$$\begin{aligned} p(\boldsymbol{\theta}) &= p(\theta_0)\, p(\theta_1) \\ &\propto \theta_0^{\alpha_0 - 1}(1 - \theta_0)^{\beta_0 - 1} \theta_1^{\alpha_1 - 1}(1 - \theta_1)^{\beta_1 - 1}. \end{aligned}$$

Via Bayes Rule, the posterior is proportional to the prior times a likelihood, and so

$$\begin{aligned} p(\theta_0, \theta_1 | r_0, r_1, n_0, n_1) &\propto p(\theta_0)p(\theta_1)\mathcal{L}(\theta_0, \theta_1; \mathbf{y}) \\ &\propto \theta_0^{\alpha_0 - 1}(1 - \theta_0)^{\beta_0 - 1}\, \theta_1^{\alpha_1 - 1}(1 - \theta_1)^{\beta_1 - 1} \\ &\qquad \times \theta_0^{r_0}(1 - \theta_0)^{n_0 - r_0}\, \theta_1^{r_1}(1 - \theta_1)^{n_1 - r_1} \\ &\propto \theta_0^{\alpha_0 + r_0 - 1}(1 - \theta_0)^{\beta_0 + n_0 - r_0 - 1} \\ &\qquad \times \theta_1^{\alpha_1 + r_1 - 1}(1 - \theta_1)^{\beta_1 + n_1 - r_1 - 1}. \end{aligned}$$

The terms involving θ_0 we recognize as the kernel of a Beta density with parameters $\alpha_0^* = \alpha_0 + r_0$ and $\beta_0^* = \beta_0 + n_0 - r_0$, and similarly for θ_1: i.e. $p(\boldsymbol{\theta}|r_0, r_1, n_0, n_1) = p(\theta_0|r_0, r_1)p(\theta_1|r_1, n_1)$. Thus, for this problem not only are θ_0 and θ_1 independent *a priori*, they are also independent *a posteriori*, and in particular,

$$\begin{aligned} \theta_0|(r_0, r_1, n_0, n_1) &\sim \text{Beta}(\alpha_0^*, \beta_0^*) \\ \theta_1|(r_0, r_1, n_0, n_1) &\sim \text{Beta}(\alpha_1^*, \beta_1^*) \end{aligned}$$

The question of social-scientific interest with data such as these is whether the outcome $y_i = 1$ is more or less likely depending on whether $x_i = 0$ versus $x_i = 1$. Let the quantity of interest be $q = \theta_1 - \theta_0$. Then, in a Bayesian approach, we need to compute the posterior density for q, a function of the unknown parameters. In particular, interest centers on how much of the posterior density for q lies above zero, since this is the posterior probability that $\theta_1 > \theta_0$. Since the posterior densities for θ_0 and θ_1 are independent Beta densities, the posterior density of q is the density of the *difference* of two independent Beta densities. The analytical form of such a density has been derived by Pham-Gia and Turkkan (1993), but is extremely cumbersome and I do not reproduce it here.

Other functions of θ_0 and θ_1 are also of interest in the analysis of two-by-two tables. For instance, if we define $w = \theta_1/\theta_0$, then the posterior probability that $\theta_1 > \theta_0$ is the posterior probability that $w > 1$. The distribution of w, a ratio of independent Beta densities, has been derived by Weisberg (1972), and again, this has a particularly cumbersome form that I will not reproduce here.

A popular measure of association in two-by-two tables is the odds ratio: if

$$\mathrm{OR}_0 = \frac{\theta_0}{1-\theta_0} \text{ and } \mathrm{OR}_1 = \frac{\theta_1}{1-\theta_1}$$

then the ratio

$$\mathrm{OR} = \frac{\mathrm{OR}_1}{\mathrm{OR}_0} = \frac{\theta_1\,(1-\theta_0)}{(1-\theta_1)\,\theta_0} \tag{2.9}$$

is a measure of association between y_i and x_i. In particular, if $\theta_1 > \theta_0$, then $\mathrm{OR} > 1$. Under the assumptions that $y_i|x_i = 0$ and $y_i|x_i = 1$ are independent Bernoulli processes of length n_0 and n_1, and assuming conjugate priors for θ_0 and θ_1, Altham (1969) showed that the posterior probability that $\mathrm{OR} < 1$ can be computed as a finite sum of hypergeometric probabilities. In fact, Altham (1969) demonstrated a close connection between the Bayesian posterior probability and the p-value produced by Fisher's (1935) exact test, the standard frequentist tool for testing associations in cross-tabulations with small samples. As we shall see below, relative to the posterior probabilities generated from a Bayesian analysis with uniform priors over θ_0 and θ_1, Fisher's exact test is too conservative, in the sense that it generates p-values that are larger (in favor of the null hypothesis of no association) than the corresponding Bayesian posterior probabilities.

In addition, many analysts use the log of the odds-ratio as a measure of association in two-by-two tables. If

$$\Lambda_0 = \log\left(\frac{\theta_0}{1-\theta_0}\right) \text{ and } \Lambda_1 = \log\left(\frac{\theta_1}{1-\theta_1}\right)$$

then the log of the odds ratio

$$\begin{aligned}\Lambda &= \Lambda_1 - \Lambda_0\\ &= \log\left(\frac{\theta_1}{1-\theta_1}\right) - \log\left(\frac{\theta_0}{1-\theta_0}\right)\\ &= \log\left(\frac{\theta_1(1-\theta_0)}{(1-\theta_1)\theta_0}\right).\end{aligned} \tag{2.10}$$

is positive when $\theta_1 > \theta_0$ (since the log-odds transformation is a strictly increasing function). The posterior density for Λ is well approximated by a normal density, with the approximation being quite good when all the entries in the two-by-two crosstabulation are at least 5. This result follows from the fact that the distributions of Λ_0 and Λ_1 are approximately normal, as noted in Section 2.1.3, and the fact that the difference (or sum) of two normal distributions is a normal distribution. In particular,

$$\Lambda|(r_0, r_1, n_0, n_1) \sim N\left(\log\left(\frac{(\alpha_1^* - \frac{1}{2})(\beta_0^* - \frac{1}{2})}{(\beta_1^* - \frac{1}{2})(\alpha_0^* - \frac{1}{2})}\right), \frac{1}{\alpha_1^*} + \frac{1}{\beta_1^*} + \frac{1}{\alpha_0^*} + \frac{1}{\beta_0^*}\right) \tag{2.11}$$

Closed-form expressions for the densities of some of these estimands are either extremely cumbersome or even unknown. In addition to the articles cited earlier, see Nurminen and Mutanen (1987) or Pham-Gia and Turkkan (2002) for a taste of the difficulties involved in deriving posterior densities for various summaries of the relationships in two-by-two tables. See Problem 2.8 for some simple cases. The simulation methods to be discussed in Chapter 3 provide a simple way to characterize posterior densities with non-standard or cumbersome analytical forms. A mixture of simulation and analytical methods were used to compute and graph the prior and posterior densities reported in the following example; again, the details await Chapter 3.

■ Example 2.8

Do parents influence the sexual orientation of their children? Golombok and Tasker (1996) studied whether there is any link between the sexual orientation of parents and children. Twenty-five children of lesbian mothers ($x_i = 1$) and a control group of 21 children of heterosexual single mothers ($x_i = 0$) were first seen at age 9.5 years (on average), and again at 23.5 years (on average). In the second interview, children were asked about their sexual identity, responding as either as bisexual/lesbian/gay ($y_i = 1$) or heterosexual ($y_i = 0$); one of the participants did not supply information on this variable.

Table 2.2 Young adults sexual identity, by sexual orientation of mother.

	Mother (x_i)	
Child's identity (y_i)	Heterosexual ($x_i = 0$)	Lesbian ($x_i = 1$)
Heterosexual ($y_i = 0$)	20	23
Bisexual/lesbian/gay ($y_i = 1$)	0	2
Total	20	25

Table 2.2 reproduces the cross-tabulation of mother's sexual orientation by child's sexual identity, reported in Table 2 of Golombok and Tasker (1996). All of the children of the heterosexual mothers report a heterosexual identity ($r_0 = 0; n_0 = 20$), while almost all of the children of the lesbian mothers report a heterosexual identity ($r_1 = 2; n_1 = 25$). Thus, the maximum likelihood estimates of θ_0 and θ_1 are $\hat{\theta}_0 = r_0/n_0 = 0$ and $\hat{\theta}_1 = r_1/n_1 = 0.08$. Golombok and Tasker (1996) report a frequentist test of the null hypothesis that $\theta_0 = \theta_1$, presumably against the two-sided alternative hypothesis $\theta_0 \neq \theta_1$ with Fisher's (1935) exact test, and report the p-value of the test as "ns" for "not significant". Given the data in Table 2.2, the actual p-values for Fisher's exact text can be easily computed. Let $H_0 : \theta_1 = \theta_0 \iff \text{OR} = 1$, where OR is the odds-ratio defined in Equation 2.9 (note that the sample estimate of the odds ratio is infinite). Then, the p-values yielded by Fisher's exact test against the following alternatives are

H_A	p
$\text{OR} \neq 1 \iff \theta_1 \neq \theta_0$	.49
$\text{OR} > 1 \iff \theta_1 > \theta_0$	.30
$\text{OR} < 1 \iff \theta_1 < \theta_0$	1.0

A Bayesian analysis proceeds as follows. The data are presumed to be two realizations from two independent binomial processes, of lengths n_0 and n_1 respectively, with success probabilities θ_0 and θ_1, respectively. I posit uniform priors for both θ_0 and θ_1, i.e. $\theta_0 \sim \text{Beta}(1, 1)$ and $\theta_1 \sim \text{Beta}(1, 1)$. Then the posterior densities are $\theta_0 \sim \text{Beta}(1, 21)$ and $\theta_1 \sim \text{Beta}(3, 24)$.

Figure 2.5 shows prior densities (dotted lines) and posterior densities for various parameters: θ_0, θ_1, the difference $\theta_1 - \theta_0$, the ratio θ_1/θ_0, the odds ratio (Equation 2.9) and the log of the odds ratio (Equation 2.10). Recall that the question of scientific interest is whether $\theta_1 > \theta_0$; or more generally, is the probability of reporting a L/B/G

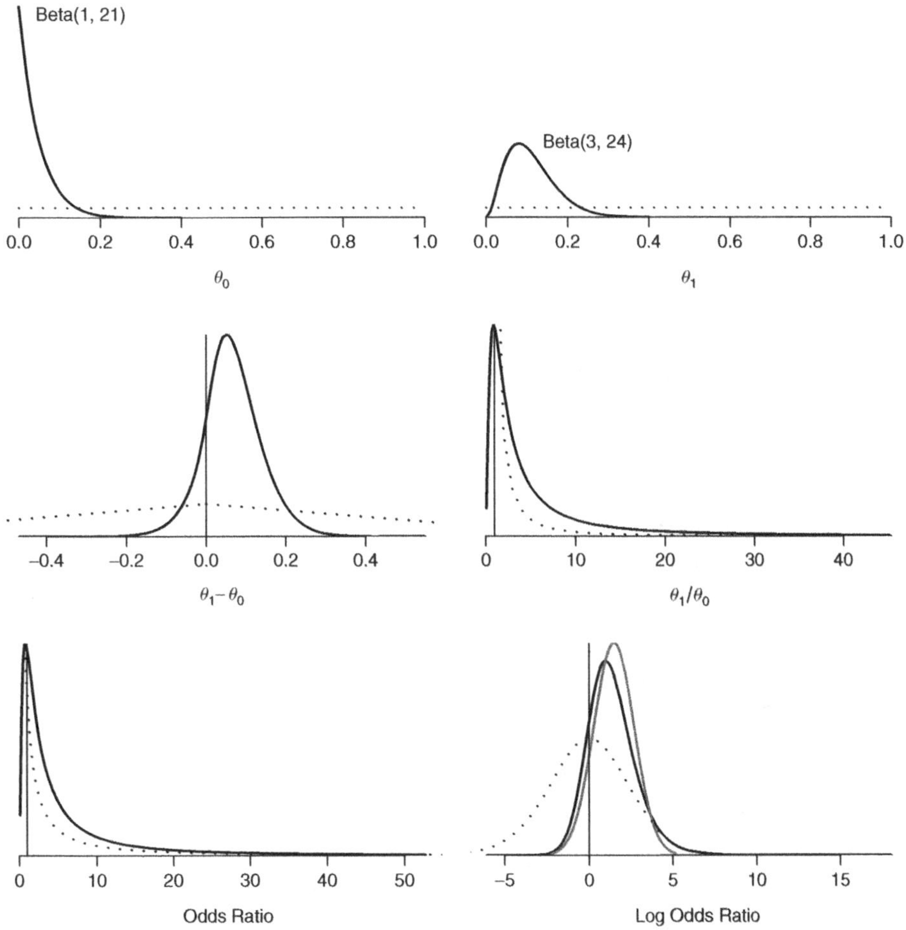

Figure 2.5 Posterior densities for Example 2.8. The top two panels show the posterior densities of θ_0 and θ_1, which are Beta densities; the remaining panels show the posterior densities for various functions of θ_0 and θ_1: the difference $\theta_1 - \theta_0$, the ratio θ_1/θ_0, the odds ratio and the log odds ratio. The vertical lines indicate the point where $\theta_1 = \theta_0$; the area under the various curves to the right of this line is the posterior probability that $\theta_1 > \theta_0$. The dotted line in each panel is the prior density for the particular parameter, implied by the uniform priors on θ_0 and θ_1.

sexual identity associated with whether one's mother is lesbian or heterosexual? The four posterior densities in the lower four panels of Figure 2.5 indicate that the sample information does suggest $\theta_1 > \theta_0$, but the evidence is not strong. For the difference of θ_1 and θ_0, their ratio, their odds ratio, and the log of their odds ratio, the posterior probability that $\theta_1 > \theta_0$ is the area under the respective posterior density to the right of the vertical line. In each instance, this probability is .84, not overwhelming evidence, and falling short of conventional standards of statistical significance. However, this posterior probability is substantially higher than the probability implied by Fisher's exact test, which yields a p-value of .30, for the test of H_0 : OR = 1 against the one-sided alternative OR > 1, suggesting that the alternative hypothesis is considerably much less likely *a posteriori* relative to the conclusions of the Bayesian procedure (with uniform priors on θ_0 and θ_1).

Finally, note the normal approximation to the posterior density of the log-odds ratio, the gray line in the lower right panel of Figure 2.5, corresponding to the density in Equation 2.11. With the small cell counts observed for this example, it is well known that the normal approximation is not accurate, and is shifted to the right relative to the actual posterior for the log of the odds ratio, Λ. Under the normal approximation, the estimated posterior probability that $\Lambda > 0 \iff \theta_1 > \theta_0$ is .89, reasonably close to the actual result of .84, but a substantial overestimate nonetheless.

Fisher's exact test is well known to produce p-values that are too conservative relative to the analogous quantities obtained from a Bayesian analysis of two binomial proportions that places uniform priors over each unknown parameters. Specifically, Altham (1969) showed that the p-values against the null hypothesis produced by Fisher's test can be obtained from a Bayesian procedure – again, approaching the problem as the difference of two binomial proportions – that employs improper Beta(0,1) and Beta(1,0) priors over the two unknown binomial success parameters. In addition to not being proper probability densities (see Definitions B.12 and B.13), this prior is somewhat odd, in that it corresponds to a strong prior belief in negative association between the two variables.

■ **Example 2.9**

War and revolution in Latin America. Sekhon (2005) presents a reanalysis of a cross-tabulation originally appearing in Geddes (1990), examining the relationship between foreign threat and social revolution in Latin America, in turn, testing a claim advanced by Skocpol (1979). Sekhon updates the data presented by Geddes (1990), generating the following 2-by-2 table:

	Revolution	No revolution
Defeated and invaded or lost territory	1	7
Not defeated for 20 years	2	74

Each observation is a 20-year period for each Latin American country, with any given Latin American country contributing multiple observations (this makes exchangeability seem questionable, but like the original authors we ignore this potential complication). The sole observation in the top left of the cross-tabulation is Bolivia: it suffered a military defeat in 1935, and a social revolution in 1952. The two observations in the lower left of the table are Mexico (revolution in 1910) and Nicaragua (revolution in 1979). We are interested in the conditional, binomial probabilities:

- $\theta_0 = \Pr(y_i = 1|x_i = 0)$, the probability of revolution conditional on no military defeat within 20 years;
- $\theta_1 = \Pr(y_i = 1|x_i = 1)$, the probability of revolution conditional on having suffered a military loss.

The MLEs are $\hat{\theta}_0 = 2/(2 + 74) = .026$ and $\hat{\theta}_1 = 1/(7 + 1) = .125$, suggesting that revolutions are much more likely conditional on military defeat than conditional on not having experienced a military defeat. However, Fisher's test of the null hypothesis of no association against an unspecified alternative yields a p-value of .262, which is the same p-value obtained with the one-sided alternative $H_A : \theta_1 > \theta_0$.

With the uniform priors $\theta_j \sim \text{Beta}(1, 1)$, $j = 0, 1$, the posterior densities are $\theta_0 \sim \text{Beta}(3, 75)$ and $\theta_1 \sim \text{Beta}(2, 8)$. Sekhon (2005) reports the results of a simulation-based analysis of the posterior density of the quantity $q = \theta_1 - \theta_0$ (see Example 3.3), finding the posterior probability that $\theta_1 > \theta_0$ to be .9473, which is quite strong evidence in favor of the hypothesis. Thus, the "conservatism" of Fisher's exact test relative to this Bayesian approach is readily apparent.

But, consider the improper priors $\theta_0 \sim \text{Beta}(1, 0)$ and $\theta_1 \sim \text{Beta}(0, 1)$; see Definitions B.12 and B.13. These improper priors have the properties $\lim_{\theta_0 \to 0} p(\theta_0) = \lim_{\theta_1 \to 1} p(\theta_1) = +\infty$ and have the effect of being favoring the null hypothesis that $\theta_1 = \theta_0$. With these priors the posterior densities are $\theta_0 \sim \text{Beta}(3, 74)$ and $\theta_1 \sim \text{Beta}(1, 8)$, and the posterior probability that $\theta_1 > \theta_0$ is .738, corresponding to result given by Fisher's exact test. That is, Fisher's exact test has a p-value of $1 - .738 = .262$, which frequentists interpret as the relative frequency (and hence, "probability") of observing a result at least as extreme as the result actually observed, in repeated sampling of 2-by-2 tables with the same row and column marginals as the table actually observed, if the null hypothesis were true.

Two conclusions follow: (1) the priors that "rationalize" Fisher's exact test for a Bayesian are improper (i.e. any elicitation of coherent prior beliefs could not lead to these priors) and (2) lead to the conclusions that are "too conservative" (in favor of the null hypothesis of no association), at least in 2-by-2 tables with small cell counts.

2.3 Learning from counts

Count data are another frequently encountered type of data in the social sciences. Examples include counts of the number of militarized interstate disputes the United States engages in per year (Gowa 1998); the number of executive agencies created by a US president (Howell and Lewis 2002); homicides in St Louis' census tracts (Kubrin and Weitzer 2003); overvoting and undervoting in elections in Los Angeles County (Sinclair and Alvarez 2004); the production of scholarly articles (Keith *et al.* 2002); patent applications by companies (Cameron and Trivedi 1998); civil rights protests (Isaac and Christiansen 2002); claims of human rights abuses by countries (Cole 2006); cites to the Federalist papers in Supreme Court opinions (Corley, Howard and Nixon 2005).

The most frequently deployed model for count data uses the Poisson probability mass function (pmf). The use of the Poisson pmf can be rationalized as follows. Suppose we observe n counts, say generated by observing n individuals over some fixed time period t, or by observing n spatial regions. Let $i = 1, \ldots, n$ index the observations, and denote the counts as $y_i \in \{0, 1, 2, \ldots\}$. If the observations are from an infinitely exchangeable sequence (in the sense of Section 1.9.1) – say because we have no information with which to distinguish either the n individuals or n regions – then it is "as if" the counts were generated by a Poisson process with unknown intensity parameter $\lambda > 0$. That is, the probability mass function for a count $y_i \in \{0, 1, 2, \ldots\}$ is a Poisson pmf, indexed by an intensity parameter λ, i.e.

$$p(y_i|\lambda) = \frac{e^{-\lambda}\lambda^{y_i}}{y_i!}. \tag{2.12}$$

The intensity parameter λ is the mean of a Poisson pmf, and it is a curious property of the Poisson mass function that λ is also the variance (see Definition B.23). This feature of the Poisson is often too restrictive for many count processes generated by social or political processes, and alternative, more flexible models often provide a better fit to the data, as we shall see momentarily. For now we assume that exchangeability holds, and that the Poisson is an appropriate model for the counts.

Under exchangeability, the likelihood function for count data is the simply joint probability of $\mathbf{y} = (y_1, \ldots, y_n)'$, given by the n-fold product of the observation-specific probabilities:

$$p(\mathbf{y}|\lambda) = e^{-n\lambda}\frac{\lambda^S}{\prod_{i=1}^n y_i!}$$

where $S = \sum_{i=1}^n y_i$. Up to a constant of proportionality, the log-likelihood is

$$\ln \mathcal{L}(\lambda; \mathbf{y}) \propto -n\lambda + S \ln \lambda$$

with the first-order condition

$$\frac{\partial \ln \mathcal{L}}{\partial \lambda} = -n + S/\lambda = 0$$

implying that the maximum likelihood estimate of λ, $\hat{\lambda}_{\text{MLE}}$, is simply $\bar{y} = S/n$.

A Bayesian analysis of count data uses the likelihood to update prior beliefs about the intensity parameter, λ, via Bayes Rule. For a Poisson likelihood for λ, the conjugate prior is a Gamma density (see Definition B.34). We state this result formally in the following proposition:

Proposition 2.2 (Conjugate analysis of Poisson data). *If $y_i \sim Poisson(\lambda)$, $i = 1, \ldots, n$, with prior beliefs over λ, $p(\lambda)$, represented with a $Gamma(a, b)$ density, then the posterior density over λ, $p(\lambda|\mathbf{y})$, is a $Gamma(a^*, b^*)$ density, where $a^* = a + S$ and $b^* = b + n$, and where $S = \sum_{i=1}^n y_i$.*

Proof. By Bayes Rule, and from the definition of a Gamma density (Definition B.34),

$$\begin{aligned}
p(\lambda|\mathbf{y}) &\propto p(\mathbf{y}|\lambda)p(\lambda) \\
&\propto \underbrace{\frac{\lambda^S}{\prod_{i=1}^n y_i!}\exp(-n\lambda)}_{\text{Poisson likelihood}} \underbrace{\frac{b^a}{\Gamma(a)}\lambda^{a-1}\exp(-b\lambda)}_{\text{Gamma prior}} \\
&\propto \lambda^{S+a-1}\exp(-\lambda[n+b])
\end{aligned}$$

which we recognize as the kernel of a Gamma density over λ with parameters $a^* = a + S$ and $b^* = b + n$. ◁

As is usually the case with conjugate priors, the hyper-parameters of the conjugate prior density are interpretable in data equivalent terms. In this case, the parameter a in the Gamma prior for λ is equivalent to a prior sum of counts, and b is equivalent to a prior sample size. Moreover, since a/b is the mean of a Gamma(a, b) density, the information in the prior Gamma density is the same as that which would arise from a sample of b exchangeable counts in which the average count was a/b. This convenient, data equivalent form of the prior hyperparameters makes specifying a Gamma prior density for a Poisson intensity parameters easier than it might otherwise be.

Further, the simple form of the posterior density in Proposition 2.2 makes clear the circumstances in which the prior will have more or less influence on the posterior density. As the prior hyperparameter $a, b \to 0$, then clearly the posterior hyperparameters converge on their analogs in the data: i.e.

$$\lim_{a\to 0} a^* = S = \sum_{i=1}^n y_i$$

$$\lim_{b\to 0} b^* = n, \qquad \text{and}$$

$$\lim_{a,b\to 0} E(\lambda|\mathbf{y}) = a^*/b^* = (a+S)/(b+n) = S/n = \bar{y} = \hat{\lambda}_{\text{MLE}}.$$

That is, as the prior density over λ becomes less precise (in the data equivalent sense that it is based on a diminishing number of prior observations) then the resulting posterior density for λ is increasingly dominated by the likelihood. Moreover, a Bayes estimate for λ, such as the posterior mean $E(\lambda|\mathbf{y})$ tends towards the maximum likelihood estimate of λ, $\bar{y}$, as $a, b \to 0$. See also Problem 2.15. We demonstrate these features of conjugate Bayesian analysis of count data in the following example.

■ Example 2.10

Reports of number of sexual partners. In 1991, the General Social Survey asked respondents the following question: "How many sex partners have you had in the last 12 months'?" The responses are summarized in Table 2.3.

Table 2.3 Self-reported number of sex partners for Example 2.10.

Count	Men	Women	Count	Men	Women
0	44	102	7	1	0
1	195	233	8	1	0
2	20	18	9	0	0
3	3	9	10	0	0
4	3	2	⋮	⋮	⋮
5	5	1	14	1	0
6	3	0			

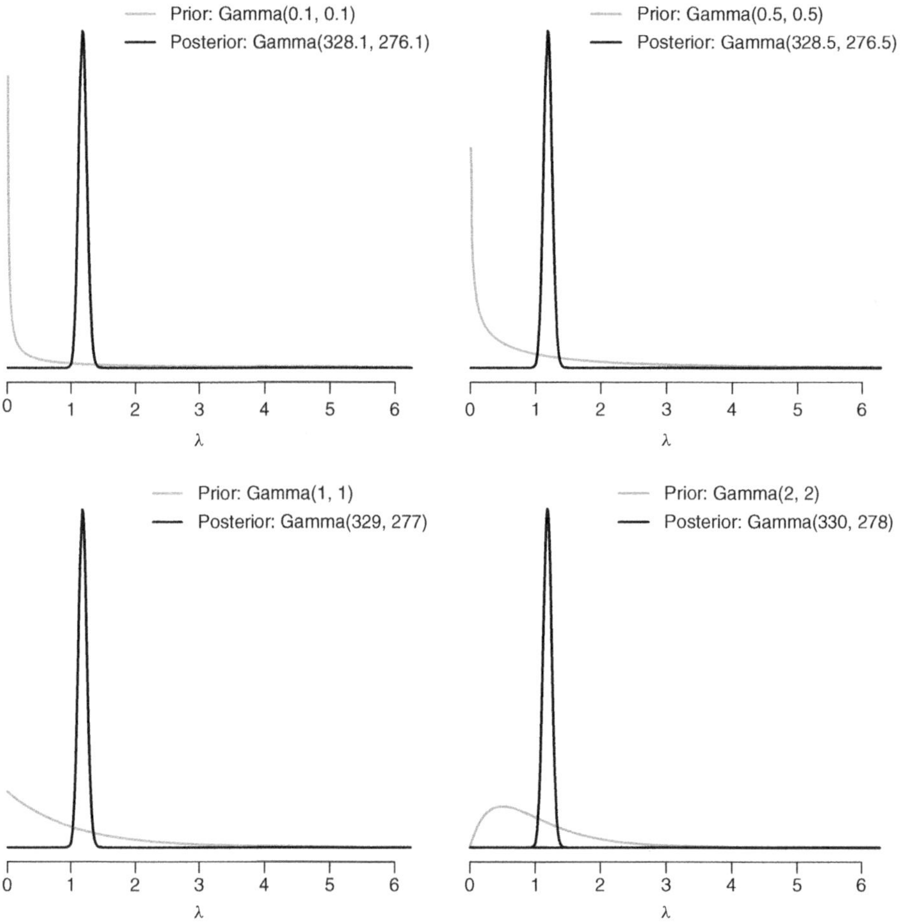

Figure 2.6 Prior and posterior densities for Poisson intensity parameter, sexual partners data, male respondents, General Social Survey.

There are two questions of substantive interest: (1) given the information in the data, what should we infer about the numbers of sexual partners for men and women in the population; (2) are the reported numbers of sexual partners the same for women and men? For now, we proceed with the information in the table, with no information with which to distinguish the respondents beyond their sex. Conditional on sex, we treat the counts as exchangeable and fit two Poisson models, one for men and one for women. That is, for observations $i = 1, \ldots, n$ and counts $y_i \in \{0, 1, \ldots, \}$, then

$$p(y_i|\lambda_j) = \frac{\lambda_j^{y_i}}{y_i!} e^{-\lambda_j}, \; j \in \{\underline{\text{M}}\text{en}, \underline{\text{W}}\text{omen}\}$$

Further, we assume that the survey responses are generated independently, a plausible assumption for the General Social Survey.

Priors. We have little prior information to bring to bear on this problem. On the one hand, the mean number of sex partners in a closed, heterosexual, two-sex population must be the same. This would suggest that both prior distributions should have at least the same expected value, or values that are close to one another. On the other hand, it is well known that men consistently report more sex partners in sexual history surveys than do women (e.g. Morris 1993). This would suggest using a prior in which the mean for men exceeded that for women. For expository purposes, I use four different vague, conjugate Gamma priors for both λ_M and λ_W in this analysis: i.e. $\lambda_j \sim \text{Gamma}(b, b)$, with $b \in \{.1, .5, 1, 2\}$.

Posterior densities. Each prior generates a unique posterior density over λ, as shown in Figure 2.6, and displayed numerically in Table 2.4. The graphs and the table demonstrate that the information about λ in the priors are weak relative to the information in the data,

Table 2.4 Comparison of prior and posterior densities, sexual partners data, male respondents. Each row of the table corresponds to a different Gamma(b, b) prior over the Poisson intensity parameter λ. Cell entries are prior/posterior means; ranges in brackets are the 2.5 and 97.5 percentiles of the respective prior/posterior Gamma densities (quantities reported as zero are less then 10^{-3}). The maximum likelihood estimate of λ is 1.19 (to three significant digits).

b	Prior	Posterior
.1	1 [0, 9.78]	1.19 [1.06, 1.32]
.5	1 [0, 5.02]	1.19 [1.06, 1.32]
1	1 [0.02, 3.69]	1.19 [1.06, 1.32]
2	1 [0.12, 2.79]	1.19 [1.06, 1.32]

such that more or less than same posterior density results in each instance. Since the prior information is weak relative to the data, a Bayes estimate such as the posterior mean, $E(\lambda|\mathbf{y})$, is close to the maximum likelihood estimate, at least to a reasonable number of significant digits. Uncertainty over the Poisson intensity parameter λ is quite tightly concentrated in a small region for each prior: the four 95 % HPDs for λ reported in Table 2.4 range from about 1.06 to 1.32.

2.3.1 Predictive inference with count data

Given that in a conjugate Bayesian analysis the Poisson intensity parameter λ follows a Gamma density (both *a priori* and *a posteriori*) it is interesting to consider the predictions that such a model makes for the counts themselves. That is, if we were to use the conjugate Poisson-Gamma model for predictive purposes, then what is the resulting marginal probability mass function over the counts themselves? Put concretely, if $y|\lambda \sim \text{Poisson}(\lambda)$, then we know that $E(y) = V(y) = \lambda$, and it is straightforward to compute interesting facts about y (e.g. the probability that someone will report 3 or more sexual partners). But since λ itself is subject to prior/posterior uncertainty (characterized with a Gamma density), what does this imply for our beliefs over the counts? The answer is that after averaging over a Gamma density for λ, the probability mass function of counts follows a negative binomial density:

Proposition 2.3 (Prior/posterior predictive mass function, Poisson count data, conjugate Gamma prior). *If $y_i \sim \text{Poisson}(\lambda)$, $i = 1, \ldots, n$ with prior/posterior beliefs over λ represented with a Gamma density with parameters a and b, then the prior/posterior predictive mass function for a future observation $\tilde{y}$, $p(\tilde{y}|\mathbf{y})$ is a negative binomial density.*

Proof.

$$
\begin{aligned}
p(\tilde{y}|\mathbf{y}) &= \int p(\tilde{y}|\lambda, \mathbf{y})p(\lambda|\mathbf{y})d\lambda \\
&= \int_0^\infty \underbrace{\frac{\lambda^{\tilde{y}}}{\tilde{y}!}\exp(-\lambda)}_{\text{Poisson}(\lambda)} \underbrace{\frac{b^a}{\Gamma(a)}\lambda^{a-1}\exp(-b\lambda)\,d\lambda}_{\text{Gamma}(a,b)} \\
&= \frac{b^a}{\tilde{y}!\Gamma(a)}\int_0^\infty \lambda^{\tilde{y}+a-1}\exp[-\lambda(b+1)]d\lambda.
\end{aligned}
$$

Letting $z = \lambda(b+1)$, we have

$$
\begin{aligned}
p(\tilde{y}|\mathbf{y}) &= \frac{b^a}{\tilde{y}!\Gamma(a)}\int_0^\infty \frac{z^{\tilde{y}+a-1}}{(b+1)^{\tilde{y}+a-1}}\exp(-z)\frac{d\lambda}{dz}dz \\
&= \frac{b^a}{\tilde{y}!\Gamma(a)}\frac{1}{(b+1)^{\tilde{y}+a-1}}\frac{1}{(b+1)}\int_0^\infty z^{\tilde{y}+a-1}\exp(-z)dz \\
&= \frac{b^a}{\tilde{y}!\Gamma(a)}\frac{1}{(b+1)^{\tilde{y}+a-1}}\frac{1}{(b+1)}\Gamma(\tilde{y}+a) \\
&\propto \frac{\Gamma(\tilde{y}+a)}{\Gamma(a)\tilde{y}!}\frac{1}{(b+1)^{\tilde{y}}}
\end{aligned}
$$

Now let $\theta = 1 - 1/(b+1)$ such that $1 - \theta = 1/(b+1)$. Then

$$p(\tilde{y}|\mathbf{y}) \propto \frac{\Gamma(\tilde{y}+a)}{\Gamma(a)\tilde{y}!}(1-\theta)^{\tilde{y}}$$

which we recognize as proportional to a negative binomial probability mass function over $\tilde{y}$ with success probability θ and success count a (see Definition B.24). ◁

■ Example 2.11

Reports of number of sexual partners, Example 2.10, continued. Using the prior $\lambda \sim \text{Gamma}(1, 1)$, we obtain the posterior density $\lambda|\mathbf{y} \sim \text{Gamma}(329, 277)$. By

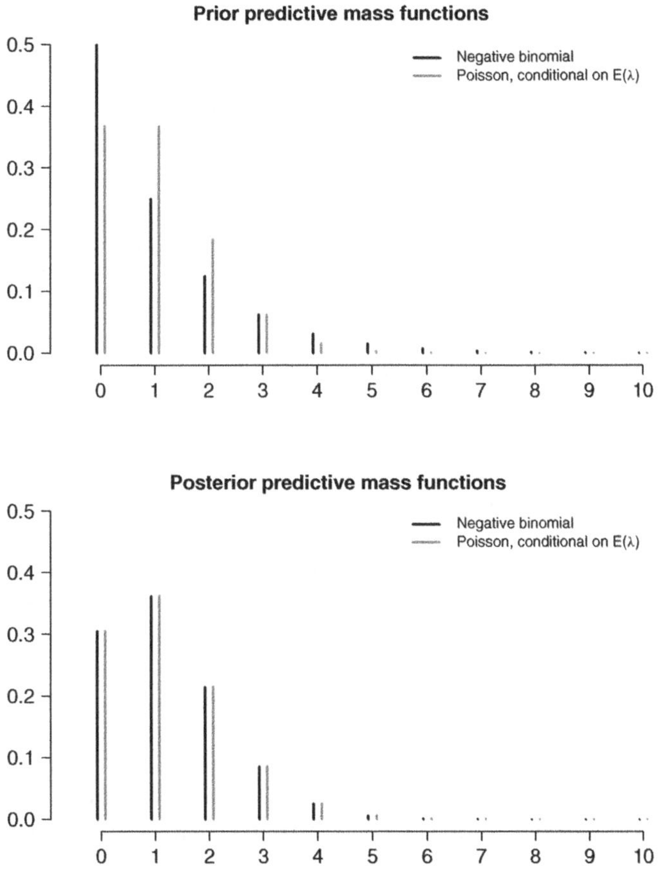

Figure 2.7 Prior and posterior predictive mass functions for counts of reported sex partners. The gray bars show a Poisson pmf with intensity parameters set to the prior/posterior expected value of λ, given a prior Gamma (1,1) density, or the posterior Gamma(329, 277) density; the correct predictive mass function, taking into account (prior/posterior) uncertainty in λ is the negative binomial. Note that when there is little uncertainty over λ as in the posterior analysis in this example, the Poisson and negative binomial pmfs are practically indistinguishable.

Proposition 2.3 this implies that for a future observation $\tilde{y}$, $p(\tilde{y}|\mathbf{y})$ is a negative binomial pmf with success probability $\theta = 1 - 1/(b+1)$ where $b = 277$ and success count $a = 329$.

Figure 2.7 contrasts the prior and predictive mass functions for this example. The Gamma(1,1) prior over λ results in a prior predictive mass function (a negative binomial pmf) that assigns probability .5 to a reported count of zero sex partners; *a posteriori* the probability of a reported zero count is .31. Ignoring the uncertainty over λ results in a Poisson pmf as the predictive pmf over the counts (gray bars in Figure 2.7); *a posteriori* there is very little uncertainty in λ and the negative binomial and Poisson predictive mass functions are practically indistinguishable.

2.4 Learning about a normal mean and variance

Much quantitative work in the social sciences involves making inferences about the mean of a population given sample data. When the variable being analyzed is continuous (or approximately so), the normal distribution is often used as an underlying probability model for the sample observations: e.g. $y_i \sim N(\mu, \sigma^2)$, with interest centering on the unknown mean parameter μ. Examples include analysis of surveys measuring incomes and working hours. Sometimes the normal distribution is used when the data being analyzed are discrete, but take on a relatively large number of categories and so are approximately continuous (e.g. 7 point scales in surveys measuring political attitudes). Regression analysis, one of the most widely used tools in the social sciences, is a special case of this approach to modeling, where the mean of continuous response y_i is modeled as a function of predictors $\mathbf{x}_i$; for instance, consider the normal, linear regression model $y_i|\mathbf{x}_i \sim N(\mathbf{x}_i\boldsymbol{\beta}, \sigma^2)$, with $\boldsymbol{\beta}$ a vector of unknown parameters. We return to the regression model in Section 2.5. For data for which normal distributions are implausible, other distributions are required: examples include binary data, considered in the previous sections, or counts, for which the Poisson or negative binomial distributions may be more appropriate. This said, the normal distribution is enormously popular in statistical modeling, and many variables that do not appear to have a normal distribution can be transformed to better approximate normality; e.g. taking the logarithm of a variable that is highly skewed, such as income. In addition, as sample size increases, discrete distributions such as the binomial or the Poisson are well approximated by the normal distribution. Thus, the normal distribution is often used as a substitute for these discrete distributions in large samples; a common example is the analysis of a binary variable in a large sample (e.g. estimating vote intentions from surveys with large numbers of respondents).

2.4.1 Variance known

Consider the case where we have n observations on a variable y, yielding a vector of observations $\mathbf{y} = (y_1, \ldots, y_n)'$. Conditional on the unknown mean parameter μ, we consider the y_i exchangeable, and we also assume that the variance σ^2 is known. If each (exchangeable) observation is modeled as $y_i \sim N(\mu, \sigma^2)$, with σ^2 known, then the likelihood is

$$\mathcal{L}(\mu; \mathbf{y}, \sigma^2) \equiv f(\mathbf{y}|\mu) \propto \prod_{i=1}^{n} \exp\left[\frac{-(y_i - \mu)^2}{2\sigma^2}\right], \tag{2.13}$$

i.e. the maximum likelihood estimate of μ is simply the sample mean, i.e. $\hat{\mu} = \bar{y} = n^{-1}\sum y_i$.

Bayesian inference for μ in this case proceeds as usual: we multiply the likelihood for μ by a prior density for μ, $p(\mu)$, to obtain the posterior density, $p(\mu|\mathbf{y})$. For this likelihood, generated from the normal densities for each y_i, a conjugate prior density is the normal density. That is, if prior beliefs about μ are represented with a normal density, then given a normal model for the data, the posterior density for μ is also a normal density. Specifically,

Proposition 2.4 *Let* $y_i \stackrel{iid}{\sim} N(\mu, \sigma^2)$, $i = 1, \ldots, n$, *with* σ^2 *known, and* $\mathbf{y} = (y_1, \ldots, y_n)'$. *If* $\mu \sim N(\mu_0, \sigma_0^2)$ *is the prior density for* μ, *then* μ *has posterior density*

$$\mu|\mathbf{y} \sim N\left(\frac{\mu_0\sigma_0^{-2} + \bar{y}\frac{n}{\sigma^2}}{\sigma_0^{-2} + \frac{n}{\sigma^2}},\ \left(\sigma_0^{-2} + \frac{n}{\sigma^2}\right)^{-1}\right).$$

Proof. See Proposition C.3 in the Appendix. ◁

This result draws on familiar result in conjugate, Bayesian analysis: the mean of the posterior density is a *precision-weighted average* of the mean of the prior density, and the maximum likelihood estimate. In this case, the mean of the prior density is μ_0 and has precision σ_0^{-2}, equal to the inverse of the variance of the prior density. The maximum likelihood estimate of μ is $\bar{y}$, which has precision equal to n/σ^2, the inverse of the variance of $\bar{y}$. Inspecting the result in Proposition 2.4 makes it clear that as prior information about μ becomes less precise ($\sigma_0^{-2} \to 0, \sigma_0^2 \to \infty$), the posterior density is dominated by the information about μ in the likelihood. In this case popular Bayes estimates such as the posterior mean (or mode) tend to the maximum likelihood estimate, $\bar{y}$, and the variance of the posterior density tends to σ^2/n, the variance of the sampling distribution of $\hat{\mu}$ that would arise via a frequentist approach to inference for this problem. In Problem 2.23 you are invited to extend this to the proposition to the multivariate case.

Note that just as we saw in the case of inference for a binomial success probability, a Bayes estimate can be expressed as the sample estimate, plus an offset that reflects the influence of the prior. In this case, the mean of the posterior density can be expressed as

$$E(\mu|\mathbf{y}) = \bar{y} + \lambda(\mu_0 - \bar{y})$$

where

$$\lambda = \frac{\sigma_0^{-2}}{\sigma_0^{-2} + \frac{n}{\sigma^2}}$$

is the precision of the prior relative to the total precision (the prior precision plus the precision of the data). As $\lambda \to 1$, $E(\mu|\mathbf{y}) \to \mu_0$, while as $\lambda \to 0$, say, when prior information about μ is sparse relative to the information in the data, $E(\mu|\mathbf{y}) \rightarrow \bar{y}$. And in general, the Bayes estimate $E(\mu|\mathbf{y})$ is a convex combination of the prior mean μ_0 and the MLE $\bar{y}$.

■ Example 2.12

Combining information from polls, via the normal approximation to the Beta density. In Example 2.3, information from a poll and from regression analysis of previous election returns was combined via Bayes Rule to generate a posterior density over George Bush's likely share of the two-party vote in Florida in the 2000 presidential election. We now revisit this example, but representing both sources of information with normal densities, since the Beta can be well approximated by a normal as the parameters of the Beta get large (as is the case, say, when the Beta density is being used to represent uncertainty about a population parameter generated by a survey based on a large sample). That is, if $\theta \sim \text{Beta}(\alpha, \beta)$, then as $\alpha, \beta \rightarrow \infty$, $\theta \sim N(\tilde{\theta}, \sigma^2)$, where $\tilde{\theta} = E(\theta; \alpha, \beta) = \alpha/(\alpha + \beta)$ and $\sigma^2 = V(\theta; \alpha, \beta) = \alpha\beta/[(\alpha + \beta)^2(\alpha + \beta + 1)]$.

So, in Example 2.3, 279 respondents said they would vote for Bush in the November 2000 election, of 509 respondents expressing a preference for either Bush or Gore; i.e. $\alpha = 279$ and $\alpha + \beta = 509$, so $\tilde{\theta} = 279/509 = .548$ and $\sigma^2 = [279 \times (509 - 279)]/[509^2 \times 510] = .000486$. The regression analysis of historical election returns produces a predictive distribution for θ, specifically, $\theta \sim N(.491, .000484)$. For the purposes of modeling, we treat the predictive distribution from the historical analysis as a prior over θ, and the information in the poll as the likelihood; that is, the poll result $\tilde{\theta}$ is observed data, generated by a sampling process governed by θ and the poll's sample size (n.b., $n = \alpha + \beta$). Specifically, using the normal approximation, $\tilde{\theta} \sim N(\theta, \sigma^2)$ or $.548 \sim N(\theta, .000486)$.

We are now in a position to use the result in Proposition 2.4, as follows:

$$\begin{array}{rcll} \theta & \sim & N(.491, .000484) & \text{(prior)} \\ .548 & \sim & N(\theta, .000486) & \text{(data/likelihood)} \\ \theta|\text{data} & \sim & N(\theta^*, \sigma^{2*}) & \text{(posterior)} \end{array}$$

where

$$\theta^* = \left(\frac{.491}{.000484} + \frac{.548}{.000486}\right)\left(\frac{1}{.000484} + \frac{1}{.000486}\right)^{-1} = .519$$

$$\sigma^{2*} = \left(\frac{1}{.000484} + \frac{1}{.000486}\right)^{-1} = .000243$$

Notice that the resulting posterior density has smaller variance than either the prior variance or the variance of the distribution representing the data; the situation is well-represented by Figure 2.1, which accompanies Example 2.3. This elegant feature is property of conjugate analysis of normal-based likelihoods when the variance term is known, and is consistent with Bayesian analysis being equivalent to combining information from multiple sources. However, contrast the way that application of Bayes Rule led to a loss of precision in Example 2.6.

2.4.2 Mean and variance unknown

The more typical case encountered in practice is that both the mean *and* the variance are unknown parameters. That is, the typical situation is that we have data $\mathbf{y} = (y_1, \ldots, y_n)'$ with $y_i \overset{\text{iid}}{\sim} N(\mu, \sigma^2)$, and interest focuses on the *vector* of parameters $\boldsymbol{\theta} = (\mu, \sigma^2)'$. The prior density is now a bivariate, joint density, $p(\boldsymbol{\theta}) = p(\mu, \sigma^2)$, as is the posterior density $p(\mu, \sigma^2|\mathbf{y})$. The likelihood is defined with respect to the two unknown parameters, i.e.

$$\mathcal{L}(\mu, \sigma^2; \mathbf{y}) = \prod_{i=1}^{n} \frac{1}{\sqrt{2\pi\sigma^2}} \exp\left[\frac{-(y_i - \mu)^2}{2\sigma^2}\right]. \tag{2.14}$$

Note that the likelihood for this problem is a function defined over a two dimensional parameter space. More precisely, the parameter space for this problem is the real half plane, i.e. $-\infty < \mu < \infty$ and $\sigma^2 > 0$. And of course, just as the likelihood has support over the half-plane, so too do the prior and posterior densities.

To obtain a conjugate prior for $\boldsymbol{\theta} = (\mu, \sigma^2)'$ with respect to the normal likelihood in Equation 2.14 we factor the joint prior density $p(\mu, \sigma^2)$ as the product of a normal, conditional prior density for μ, $p(\mu|\sigma^2)$ and a marginal, inverse-Gamma prior density over σ^2, $p(\sigma^2)$. That is, $p(\mu, \sigma^2) = p(\mu|\sigma^2)p(\sigma^2)$ where

$$\mu|\sigma^2 \sim N(\mu_0, \sigma^2/n_0) \tag{2.15}$$

$$\sigma^2 \sim \text{Inverse-Gamma}(\nu_0/2, \nu_0\,\sigma_0^2/2) \tag{2.16}$$

where the functional form of the inverse-Gamma density is given in the Appendix (see definition B.35) and where

- $\mu_0 = E(\mu|\sigma^2) = E(\mu)$ is the mean of the prior density for μ;
- σ^2/n_0 is the variance of the prior density for μ, conditional on σ^2, with n_0 interpretable as a "prior sample size"; i.e. as $n_0 \to 0$ the variance of the prior gets larger, reflecting the fact that a prior belief based on less information (or "prior observations") will be less precise than a prior based on more information;
- $\nu_0 > 0$ is a prior "degrees of freedom" parameter;
- $\nu_0\,\sigma_0^2$ is equivalent to the sum of squares one obtains from a (previously observed) data set of size ν_0, i.e. $\sigma_0^2 > 0$.

The marginal inverse-Gamma density has a χ^2 shape, and indeed, a χ^2 density is a special case of a Gamma density, and an inverse-χ^2 density is a special case of an inverse-Gamma density (see the defintions of these densities in Appendix B). In particular, if $z \sim \chi^2_{\nu_0}$ then the quantity $\nu_0\sigma_0^2/z$ has the inverse-Gamma $(\nu_0/2, \nu_0\sigma_0^2/2)$ density, and so some authors refer to the inverse-Gamma density as a "scaled-inverse-χ^2 density."

Examples of inverse-Gamma$(\nu_0/2, \nu_0\,\sigma_0^2/2)$ densities appear in Figure 2.8. The mean, $E(\sigma^2)$, is $\nu_0\sigma_0^2/(\nu_0 - 2)$, provided $\nu_0 > 2$, otherwise the mean is undefined, and the mode occurs at $\nu_0\sigma_0^2/(\nu_0 + 2)$. Thus, the mean and the mode of the inverse-Gamma density will tend to coincide as $\nu_0 \to \infty$; i.e. the inverse-Gamma density tends to a (symmetric) normal density as $\nu_0 \to \infty$, but otherwise is skewed right.

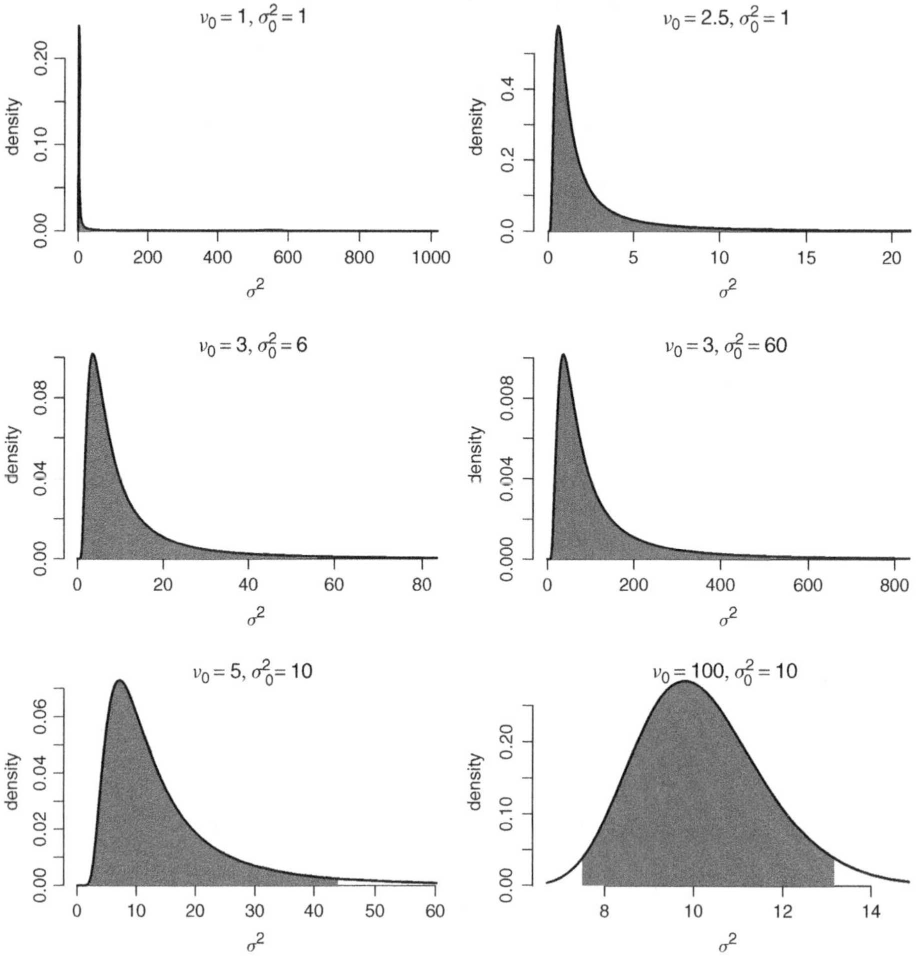

Figure 2.8 Inverse-Gamma densities. The shaded areas are 95 % highest density regions. Note the different scales on the horizontal axes.

A specification of the normal/inverse-Gamma prior requires specifying values for the four parameters, μ_0, n_0, ν_0 and σ_0^2 appearing in Equations 2.15 and 2.16. Graphical depictions of normal/inverse-Gamma densities via contour plots appear in Figure 2.9, using shading to visualize how the density varies over its support on the half-plane. The graphs show that the joint density over (μ, σ^2) has the skewed, χ^2 shape in the vertical σ^2 direction, and is symmetric in the horizontal μ direction around $E(\mu) = \mu_0 = 0$. The effect of a greater number of prior observations (n_0) is immediately apparent in Figure 2.9: the top panel shows a normal/inverse-Gamma density with $n_0 = 5$ that is much more dispersed in the μ direction than is the density in the lower left panel where $n_0 = 50$. Of course, the dispersion of the density in the μ direction is not solely a function of n_0; recall that the normal/inverse-Gamma prior is the product of a conditional normal density for μ (conditional on σ^2) and a marginal inverse-Gamma density for σ^2, and so uncertainty over σ^2 will propagate into uncertainty over μ.

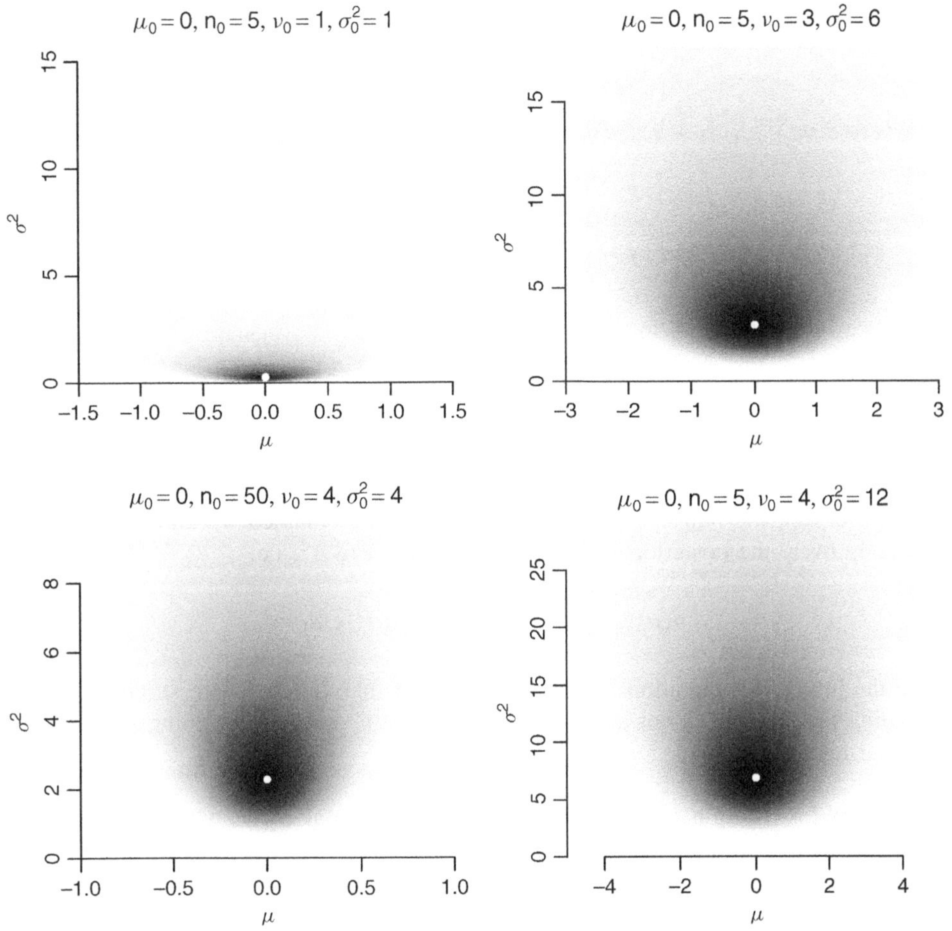

Figure 2.9 Normal/inverse-Gamma densities. Darker colors indicate regions with higher density; the mode is shown with a white circle. Note that the axes are scaled differently across the four panels.

The fact that the normal/inverse-Gamma prior over (μ, σ^2) is conjugate with respect to the likelihood for iid normal data is stated in the following proposition, with the proof relegated to the Appendix:

Proposition 2.5 (Conjugate priors for mean and variance, normal data). *Let* $y_i \overset{iid}{\sim} N(\mu, \sigma^2)$, *$i = 1, \ldots, n$, and let* $\mathbf{y} = (y_1, \ldots, y_n)'$. *If* $\boldsymbol{\theta} = (\mu, \sigma^2)'$ *has a normal/inverse-Gamma prior density with parameters* μ_0, n_0, ν_0 *and* σ_0^2, *then the posterior density of* $\boldsymbol{\theta}$ *is also a normal/inverse-Gamma density with parameters* μ_1, n_1, ν_1 *and* σ_1^2, *where*

$$\mu_1 = \frac{n_0\mu_0 + n\bar{y}}{n_0 + n}$$

$$n_1 = n_0 + n$$

$$\nu_1 = \nu_0 + n$$
$$\nu_1 \sigma_1^2 = \nu_0 \sigma_0^2 + S + \frac{n_0 n}{n_0 + n}(\mu_0 - \bar{y})^2$$

and where $S = \sum_{i=1}^{n}(y_i - \bar{y})^2$. *That is,*

$$\mu | \sigma^2, \mathbf{y} \sim N(\mu_1, \sigma^2/n_1)$$
$$\sigma^2 | \mathbf{y} \sim \textit{inverse-Gamma}\left(\frac{\nu_1}{2}, \frac{\nu_1 \sigma_1^2}{2}\right)$$

Proof. See Proposition C.4 in the Appendix. ◁

The conditional posterior density for μ is a normal density in which σ^2 appears in the expression for the variance of the conditional posterior density. But usually, interest focuses on μ itself, and we need to confront the fact that σ^2 is itself subject to posterior uncertainty, and that uncertainty should rightly propagate into inferences about μ itself. That is, we seek the *marginal* posterior density for μ, obtained by integrating out or "averaging over" the posterior uncertainty with respect to σ^2; i.e.

$$p(\mu|\mathbf{y}) = \int_0^\infty p(\mu|\sigma^2, \mathbf{y}) p(\sigma^2|\mathbf{y}) d\sigma^2$$

where the limits of integration follow from the fact that variances are strictly positive. Evaluating this integral is not as hard as it might seem, and indeed, doing so generates a familiar result.

Proposition 2.6 (Marginal posterior density of the normal mean with conjugate priors). *Under the conditions of Proposition C.4, the marginal posterior density of μ is a student-t density (Definition B.37), with location parameter μ_1, scale parameter $\sqrt{\sigma_1^2/n_1}$ and ν_1 degrees of freedom, where $n_1 = n_0 + n$,*

$$\mu_1 = \frac{n_0 \mu_0 + n\bar{y}}{n_1},$$

$\sigma_1^2 = S_1/\nu_1$, $S_1 = \nu_0 \sigma_0^2 + (n-1)s^2 + \frac{n_0 n}{n_1}(\bar{y} - \mu_0)^2$, $\nu_1 = \nu_0 + n$, $s^2 = (n-1)^{-1} \sum_{i=1}^{n}(y_i - \bar{y})^2$ *and* $\bar{y} = n^{-1}\sum_{i=1}^{n} y_i$.

Proof. See Proposition C.7 in the Appendix. ◁

That is, when fitting a normal model to data, after we account for uncertainty in the variance parameter, uncertainty about the mean follows a t-density; analogously, in frequentist statistics, the t-density is used for inference about the normal mean when both the mean and variance are unknown. We also state another familiar and related result, again deferring the proof to the Appendix:

Proposition 2.7 (Posterior predictive density, normal data, conjugate priors, mean and variance unknown). *If $y_i \overset{iid}{\sim} N(\mu, \sigma^2)$, $i = 1, \ldots, n$, with prior beliefs over $\boldsymbol{\theta} = (\mu, \sigma^2)$ represented with a normal/inverse-Gamma density with parameters $\mu_0, n_0, \nu_0, \nu_0\sigma_0^2$, then* *the posterior predictive density for a future observation y^*, $p(y^*|\mathbf{y})$, is a student-t density, with location parameter*

$$E(y^*|\mathbf{y}) = E(\mu|\mathbf{y}) = \mu_1 = \frac{n_0\mu_0 + n\bar{y}}{n_0 + n},$$

scale parameter $\sigma_1\sqrt{(n_1+1)/n_1}$ and $\nu_1 = n + \nu_0$ degrees of freedom, where

$$n_1 = n_0 + n$$
$$\sigma_1^2 = S_1/\nu_1$$
$$S_1 = \nu_0\,\sigma_0^2 + (n-1)s^2 + \frac{n_0\,n}{n_0+n}(\mu_0 - \bar{y})^2$$

Proof. See Proposition C.8 in the Appendix. ◁

We now put these results to work in an example.

■ Example 2.13

Suspected voter fraud in Pennsylvania. In November 1993 Pennsylvania conducted elections for its state legislature. The result in the Senate election in the 2nd district (based in Philadelphia) was challenged in court, and ultimately overturned. The Democratic candidate won 19 127 of the votes cast by voting machine, while the Republican won 19 691 votes cast by voting machine, giving the Republican a lead of 564 votes. However, the Democrat won 1396 absentee ballots, while the Republican won just 371 absentee ballots, more than offsetting the Republican lead based on the votes recorded by machines on election day. The Republican candidate sued, claiming that many of the absentee ballots were fraudulent. The judge in the case solicited expert analysis from Orley Ashenfelter, an economist at Princeton University. Ashenfelter examined the relationship between absentee vote margins and machine vote margins in 21 previous Pennsylvania Senate elections in seven districts in the Philadelphia area over the preceding decade (Ashenfelter 1994).

A simple version of Ashenfelter's analysis is as follows (e.g., Ashenfelter. Levine and Zimmerman 2003, 113): for each election one can compute the difference between the Democratic margin recorded via voting machines, and note how unusual the result in the 2nd district is relative to the distribution of the historical data. That is, let $i = 1, \ldots, 21$ index the decade's worth of previous elections, and define $y_i = a_i - m_i$ where a_i is Democratic percentage of the two-party vote cast via absentee ballots, and m_i is the Democratic percentage of the two-party vote recorded on election day by voting machines. For the 2nd district in 1993 (subject to court challenge), $a_{22} = 1396/(1396 + 371) = 79.0\,\%$ and $m_{22} = 19127/(19127 + 19691) = 49.3\,\%$, and so $y_{22} = 79 - 49.3 = 29.7$. We model the historical data as exchangable realizations from a normal density with an unknown mean and variance, i.e. $y_i \sim N(\mu, \sigma^2)$. Since the observations are actually repeated observations from seven districts, we note that the exchangeablility assumption underlying the simple normal model is perhaps suspect; a more plausible model might be a hierarchical model in which the data are exchangeable within each district, with the district-specific means modeled as exchangeable. In the interests of simplicity, we ignore this possibility for now.

Under the assumed normal model, the likelihood is given by Equation 2.14. Maximizing the likelihood with respect to the unknown parameters yields the MLEs

$$\hat{\mu} = \bar{y} = n^{-1}\sum_{i=1}^{n} y_i = -5.8 \qquad \text{and} \qquad \hat{\sigma}^2 = n^{-1}\sum_{i=1}^{n}(y_i - \bar{y})^2 = 55.3.$$

Eliciting a prior density. In a conjugate Bayesian analysis, it is necessary to specify a normal/inverse-Gamma prior over $\boldsymbol{\theta} = (\mu, \sigma^2)'$. Here, y_i is the difference of two percentages and so has a range of $(-100, 100)$, as must μ. This suggests that a uniform prior for μ over $(-100, 100)$ would be one way to express prior ignorance for this problem: this uniform prior can be well approximated with a normal density with a suitably large variance, with the minor inconvenience that the normal density has unbounded support, and will attach probability mass to values of μ outside the $(-100, 100)$ interval. In addition, we might bring some substantive knowledge to bear on the problem, noting that y_i is not merely the difference between two percentages, but the differences between two vote shares, recorded in the same district, in the same election, but via two voting methods (voting machines and absentee balloting). Absent something very wrong (such as massive fraud), it would be unusual to think that these two methods of voting could generate large differences in vote shares. One could speculate that Republicans might be more likely to use absentee balloting than Democrats, in the sense that factors that predispose someone towards Republican partisanship (e.g. income and socio-economic status) also determine who is likely to engage in the relatively costly act of applying for and completing an absentee ballot, and indeed, there is considerable support for this proposition in the literature on voter turnout (e.g. Karp and Banducci 2001). On the other hand, the parties themselves do a lot of the "leg work" in helping their partisans get absentee ballots (Oliver 1996), meaning that a candidate's ability to win absentee ballots will be somewhat determined by the strength of their party in that district (Patterson and Caldeira 1985), meaning that the difference between election day votes shares and absentee vote ought to be small. On balance, it seems that *a priori* we should not expect massive differences between vote shares recorded via voting machine and by absentee balloting. Zero difference between the two methods seems a reasonable *a priori* expectation, although this prior belief is rather flimsy, as say, might arise from an analysis with very few observations. In terms of the parameters of the normal/inverse-Gamma prior density, I specify $\mu_0 = E(\mu) = 0$ and $n_0 = 5$.

To elicit a prior for the variance σ^2, I reason as follows: *a priori* I expect 50 % of the y_i to lie within $\pm\tau$ of the prior mean of zero (i.e. τ is a prior guess as to the 75-th percentile of the y_i). My prior 95 % confidence on τ ranges from a low of 3, and has an upper limit of 10. That is, I attach 2.5 % probability to the events (a) 50 % of the y_i lie inside the $(-3, 3)$; (b) 50 % of the y_i lie outside the $(-10, 10)$ interval). These characterizations of my prior beliefs are sufficient to induce a prior density over the variance of y_i, σ^2, since under the assumption that y_i follows a normal distribution, there is a straightforward mapping from a prior density over a quantile such as τ to a prior density over the variance σ^2. In particular, with τ known and with $E(y_i) = 0$, we have

$$\int_{-\infty}^{\tau} \phi\left(\frac{y_i}{\sigma}\right) dy_i = \Phi\left(\frac{\tau}{\sigma}\right) = .75$$

where ϕ and Φ are the standard normal probability density and distribution functions, respectively. Thus, $\sigma = \tau/\Phi^{-1}(.75)$ (i.e. if τ is the 75-th percentile of a normal density with mean zero, then what is the standard deviation of that normal density?). Thus $\tau = 3 \iff \sigma = 4.45$ and $\tau = 10 \iff \sigma = 14.83$. From this we infer that the 2.5th and 97.5th percentiles of the prior density for σ^2 are $(3/\Phi^{-1}(.75))^2 = 19.78$ and $(10/\Phi^{-1}(.75))^2 = 219.81$, respectively. Since the prior density for σ^2 is an inverse-Gamma density, with shape and rate parameters $\nu_0/2$ and $\nu_0\sigma_0^2/2$, respectively, we complete the specification of the prior by finding the inverse-Gamma density with 2.5 and 97.5 percentiles of 19.78 and 219.81, respectively. I use a simple computer program to do this, yielding $\nu_0 = 6.20$ and $\sigma_0^2 = 47.07$. This prior density has a mean at $\nu_0\sigma_0^2/(\nu_0 - 2) = 69.4$.

So, to summarize, my prior density over the parameter vector for this problem, $\boldsymbol{\theta} = (\mu, \sigma^2)'$ is a normal/inverse-Gamma density with parameters $(\mu_0, n_0, \nu_0, \sigma_0^2) = (0, 5, 6.20, 47.07)$.

Posterior density. Using the result in Proposition 2.5, the posterior density for $\boldsymbol{\theta} = (\mu, \sigma^2)'$ is a normal/inverse-Gamma density with parameters

$$\begin{aligned}
\mu_1 &= E(\mu|\sigma^2, \mathbf{y}) \\
&= \frac{n_0\mu_0 + n\bar{y}}{n_0 + n} \\
&= \frac{(5 \times 0) + (21 \times -5.8)}{5 + 21} = -4.69 \\
n_1 &= n_0 + n = 5 + 21 = 26 \\
\nu_1 &= \nu_0 + n = 6.20 + 21 = 27.20
\end{aligned}$$

and, since $s^2 = n^{-1}\sum_{i=1}^{n}(y_i - \bar{y})^2 = 58.08$,

$$\begin{aligned}
\nu_1\sigma_1^2 &= \nu_0\sigma_0^2 + (n-1)s^2 + \frac{n_0 n}{n_0 + n}(\mu_0 - \bar{y})^2 \\
&= (6.20 \times 47.07) + (20 \times 58.08) + \left(\frac{5 \cdot 21}{5 + 21} \times (0 + 5.81)^2\right) \\
&= 291.83 + 1161.51 + \frac{105}{26} \cdot 33.76 = 1589.47
\end{aligned}$$

and so $\sigma_1^2 = 1589.47/27.20 = 58.44$.

A graphical summary of this posterior density appears in Figure 2.10, with the prior density also shown for comparison. The posterior density is quite concentrated around its mode, at least relative to the dispersion of the prior density, indicating that the data are supplying a considerable amount of information about $\boldsymbol{\theta} = (\mu, \sigma^2)'$. The maximum likelihood estimates are indicated with a white square in the top right panel of Figure 2.10, and are located close to the posterior mode; the fact that the posterior and the likelihood are not that dissimilar again stems from the fact that the prior information about (μ, σ^2) is relatively weak relative to the information in the data. Figure 2.10 also displays the prior and posterior marginal, inverse-Gamma densities for σ^2, with 95 % highest density regions shaded gray. A visual comparison of these prior and posterior densities also

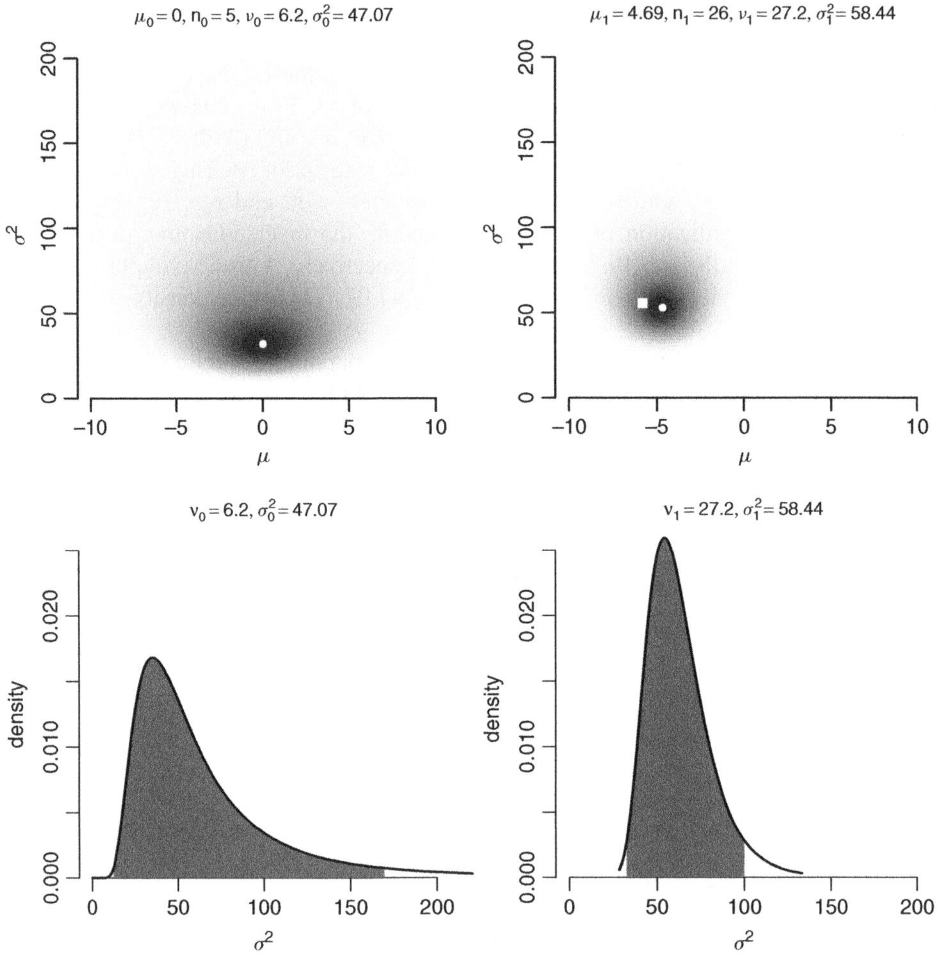

Figure 2.10 Prior and posterior normal/inverse-Gamma densities. Prior densities on the left; posterior densities on the right. Normal/inverse-Gamma densities for (μ, σ^2) are in the upper panels, with darker colors indicating regions of higher density, the circle indicating the mode, and for the posterior density, the square indicating the location of the maximum likelihood estimates. Marginal inverse-Gamma densities for σ^2 appear in the lower panels, with the shaded area corresponding to a 95 % highest density region.

reveals that beliefs about σ^2 have been dramatically modified by the data. *A priori*, a 95 % HPD for σ^2 ranged from 12.7 to 169.2, with the mean of the marginal prior density at 69.4; after revising beliefs about σ^2 in light of the data, the 95 % HPD has shrunk to (32.8, 99.9), the mean of the marginal posterior density is 63.08, and the mode of the marginal posterior of σ^2 is 54.4, close to the maximum likelihood estimate of 55.3.

Marginal posterior density for μ. The marginal prior and posterior densities for μ appear in Figure 2.11. Proposition 2.6 establishes that these marginal densities are (unstandardized) t densities, with ν_0 and ν_1 degrees of freedom for the prior and posterior densities,

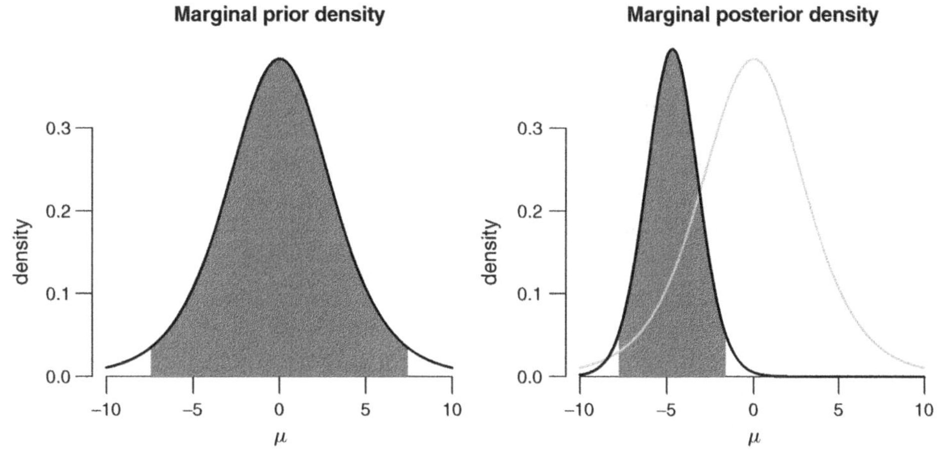

Figure 2.11 Marginal prior and posterior densities for μ. Shaded regions are 95 % HDRs. The gray curve superimposed on the right panel is the prior density.

respectively. The scale parameter of the prior marginal density for μ is 3.07, while the corresponding parameter in the posterior density is 1.50. In addition, since $\nu_0 = 6.2$ and $\nu_1 = 27.2$, the prior marginal density for μ is reasonably heavy-tailed relative to the posterior marginal density. The prior marginal density for μ is centered over $E(\mu) = \mu_0 = 0$, with 95 % HDR spanning the interval (−7.4, 7.4). Constrast the posterior marginal density for μ, centered over the posterior mean for μ, −4.69, with 95 % HDR spanning the interval (−7.8, −1.6), which does not overlap zero.

Hypothesis testing. We can use the marginal posterior density of μ to attach probabilities to conjectures of interest. For instance, the posterior probability that $\mu > 0$,

$$\Pr(\mu > 0|\mathbf{y}) = \int_0^\infty p(\mu|\mathbf{y})d\mu$$

is the posterior probability that on average, Democratic candidates outperform Republican candidates on absentee ballots relative to their performance on election day. Since the marginal posterior density is a t density, computing this posterior probability is rather straightforward: any statistical computer program can produce tail areas for t densities. In this case, $\Pr(\mu > 0|\mathbf{y}) = .023$; note that *a priori*, this probability was .5, so the data are quite informative on this question, strongly suggesting that Republicans enjoy an advantage in absentee balloting over Democrats. More formally, if we consider the two hypotheses $H_1 : \mu < 0$ (Republican advantage) and $H_0 : \mu \geq 0$ (Democratic advantage) it is straightforward to compute the Bayes factor summarizing the strength of the evidence for H_1 against H_0 (see Definition 1.11): i.e.

$$\begin{aligned} B_{10} &= \left\{\frac{p(H_1|\mathbf{y})}{p(H_0|\mathbf{y})}\right\} \Big/ \left\{\frac{p(H_1)}{p(H_0)}\right\} \\ &= \left\{\frac{1 - .023}{.023}\right\} \Big/ \left\{\frac{.5}{.5}\right\} \\ &= 42.5 \end{aligned}$$

which constitutes strong evidence in favor of the hypothesis of a Republican advantage in absentee ballots over the alternative hypothesis of a Democratic advantage.

Predictive density for a future observation. Proposition 2.7 establishes that the predictive density for a future observation y^*, $p(y^*|\mathbf{y})$, is an unstandardized t density, centered at the mean of the posterior density for μ, $\mu_1 = -4.69$, with scale parameter $\sigma_1\sqrt{(n_1+1)/n_1} = 7.64 \times \sqrt{27/26} = 7.79$ and $\nu_1 = 27.2$ degrees of freedom. The 95 % HDR for this predictive density is $(-20.67, 11.29)$.

Assessing the unusual result in the 2nd district in 1993. The posterior predictive density characterizes beliefs over any yet-to-be-realized observation. We treat the observed result in the 2nd district in 1993 as a observation "in the future" by excluding it from $\mathbf{y}$, the data used in the analysis above. In this way, the posterior predictive density is actually a prior of sorts, a characterization of our beliefs over y^* prior to seeing it. In any event, the predictive density provides a way of quantitatively assessing the result in the 2nd district in 1993, relative to the data and our prior beliefs. Recall that the Democratic candidate won 29.7 percentage points more of the two-party vote absentee ballots than on machine ballots. The probability of observing a Democratic advantage this large or larger given the data, model and prior used above is just a tail area for the predictive density; specifically,

$$\Pr(y^* \geq 29.7|\mathbf{y}) = \int_{29.7}^{\infty} p(y^*|\mathbf{y})dy^*$$

which is simply the tail area of an unstandardized t density, and easily computed. Given that a 95 % HDR for $p(y^*|\mathbf{y})$ ranges from $(-20.67, 11.29)$, it comes as no surprise that $\Pr(y^* \geq 29.7|\mathbf{y})$ is quite small, just 7.2×10^{-5}. That is, the Democratic margin on absentee ballots observed in the Senate race in the 2nd district in 1993 is extremely unusual, so much so that one would probably question the assumption of conditional exchangeability if it were included in the data set. In other words, it seems quite likely that the process that generated the result in the 2nd district in 1993 is different to the process generating the rest of the data, which is what the judge in the case found.

2.4.3 Conditionally conjugate prior

Propositions 2.5, 2.6 and 2.7 deal with the case where the prior over $\boldsymbol{\theta} = (\mu, \sigma^2)'$ has a rather special form. The normal/inverse-Gamma prior for $\boldsymbol{\theta}$ used above exploits the fact that the joint prior (and posterior) density for $\boldsymbol{\theta}$ can be expressed as $p(\mu, \sigma^2) = p(\mu|\sigma^2)p(\sigma^2)$, with $p(\mu|\sigma^2)$ a normal density and the marginal prior (and posterior density) for σ^2 an inverse-Gamma density.

We will consider a subtle modification of this prior, assuming prior independence between μ and σ^2 while retaining a normal prior for μ and an inverse-Gamma prior for σ^2. These priors are said to be *conditionally conjugate* with respect to the normal likelihood for $\mathbf{y}$. The modifier "conditionally" here stems from the fact that although we can insist on prior independence for μ and σ^2, we will nonetheless obtain a normal posterior density for μ *conditional* on σ^2; some authors describe these priors as *semi-conjugate*. We state this result in the following proposition.

Proposition 2.8 (Conditionally Conjugate Prior for the Normal Mean). Let $y_i \overset{iid}{\sim} N(\mu, \sigma^2)$, $i = 1, \ldots, n$. If $p(\mu, \sigma^2) = p(\mu)p(\sigma^2)$, where $p(\mu) \equiv N(\mu_0, \omega_0^2)$ and $p(\sigma^2) \equiv$ inverse-Gamma$(\nu_0/2, \nu_0\sigma_0^2/2)$ then

$$\mu|(\sigma^2, \mathbf{y}) \sim N(\mu_1, \omega_1^2) \quad \text{and} \quad \sigma^2|(\mu, \mathbf{y}) \sim \text{inverse-Gamma}\left(\frac{\nu_1}{2}, \frac{\nu_1\sigma_1^2}{2}\right)$$

where

$$\mu_1 = \frac{\frac{n}{\sigma^2}\bar{y} + \frac{1}{\omega_0^2}\mu_0}{\frac{n}{\sigma^2} + \frac{1}{\omega_0^2}}, \qquad \omega_1^2 = \left(\frac{n}{\sigma^2} + \frac{1}{\omega_0^2}\right)^{-1},$$

$\bar{y} = n^{-1}\sum_{i=1}^n y_i$, $\nu_1 = \nu_0 + n$, $\nu_1\sigma_1^2 + S$ and $S = \sum_{i=1}^n (y_i - \mu)^2$.

Proof. See Proposition C.10 in the Appendix. ◁

Note the (by now) familiar form of the mean of the conditional posterior density of μ; i.e. $\mu_1 \equiv E(\mu|\mathbf{y})$ is a precision weighted average of the mean of the prior, μ_0, and the maximum likelihood estimate of μ, $\bar{y}$.

This way of expressing the prior for the parameters in normal data problems is especially popular when working with the simulation-based MCMC methods that are our focus later in this book. With this conditionally conjugate approach, we only obtain the posterior for μ *conditional* on σ^2; when working with simulation methods we will obtain the marginal posterior density for μ by sampling from the posterior density for σ^2, and then sampling from the normal density for μ conditional on the sampled value of σ^2 (i.e. a simple instance of the *method of composition* and central to the simulation methods we will examine later). We resort to this two-step, simulation-based procedure in this case, since the marginal posterior density of μ has no closed form expression.

In addition, many practitioners are attracted to the conditionally conjugate prior specification since it is slightly easier to formulate beliefs about location parameters (here, μ, or regression coefficients $\boldsymbol{\beta}$) with beliefs about dispersion parameters (here, σ^2) put to one side. The use of simulation-based methods (and fast desktop computing) means that the fact that the resulting marginal posterior density for the location parameters has no closed form expression is of no great practical consequence.

2.4.4 An improper, reference prior

A Bayesian analysis can be made to produce a posterior density that corresponds with the results of a frequentist analysis. In the case of normal data with unknown mean and variance, consider the improper prior $p(\mu, \sigma^2) \propto 1/\sigma^2$. This prior density is uniform with respect to $\mu \in \mathbb{R}$, generating a so-called marginal *reference prior* $p(\mu) \propto 1$ and

$p(\sigma^2) \propto 1/\sigma^2, \sigma^2 \in \mathbb{R}^+$. Clearly this density can not integrate to one over its support; i.e. $(\mu, \sigma^2) \in \boldsymbol{\Theta} = \mathbb{R} \times \mathbb{R}^+$. Such a density is said to be improper; see Definitions B.12 and B.13.

This improper prior is completely uninformative with respect to μ, but attaches greater prior density to smaller values of σ^2 than larger values; however, the prior is uniform with respect to $\log \sigma^2$. Note also that unlike the conjugate prior considered above, μ and σ^2 are independent *a priori* – the improper, joint prior density is simply the product of the two marginal prior densities – consistent with the notion that if we are quite ignorant about μ and σ^2, then gaining some information over one of these parameters should not affect beliefs about the other parameter (e.g. Lee 2004, 62–63). Multiplying this prior by the normal likelihood in Equation 2.14 yields:

Proposition 2.9 *If* $y_i \overset{iid}{\sim} N(\mu, \sigma^2)$ *and* $p(\mu, \sigma^2) \propto 1/\sigma^2$ *then*

$$p(\mu, \sigma^2|\mathbf{y}) \propto (\sigma^2)^{\frac{-n}{2}-1} \exp\left[\frac{-1}{2\sigma^2}\left(S + n(\bar{y} - \mu)^2\right)\right], \tag{2.17}$$

where $S = \sum_{i=1}^{n}(y_i - \bar{y})^2$ *and* $\bar{y} = n^{-1}\sum_{i=1}^{n} y_i$. *Equivalently,*

$$\mu|\sigma^2, \mathbf{y} \sim N(\bar{y}, \sigma^2/n) \tag{2.18}$$

$$\sigma^2|\mathbf{y} \sim \textit{inverse-Gamma}\left(\frac{n-1}{2}, \frac{S}{2}\right) \tag{2.19}$$

$$\frac{\mu - \bar{y}}{\sqrt{S/((n-1)n)}} \sim t_{n-1} \tag{2.20}$$

Proof. See Exercise 2.19. ◁

Equation 2.18 states that with the improper reference prior, $p(\mu, \sigma^2) \propto 1/\sigma^2$, the conditional posterior density for μ is a normal density centered over the conventional estimate of μ, $\bar{y}$. Moreover, the conditional posterior density for μ has variance σ^2/n, equal to the variance of frequentist sampling distribution of $\bar{y}$ the conventional unbiased estimator of μ. Equation 2.19 provides the marginal posterior density of σ^2, an inverse-Gamma density with mean $S/(n-3)$ and mode $S/(n+1)$; compare these quantities with the unbiased frequentist estimate for σ^2 of $S/(n-1)$ and the maximum likelihood estimate of S/n. Equation 2.20 provides the marginal posterior density of μ in standardized form, a (standardized) student-t distribution with $n-1$ degrees of freedom, identical to the density arising in a frequentist analysis when testing hypotheses about μ. These results – generated with an improper, reference prior – can be considered as a limiting case of the results obtained with a proper normal/inverse-Gamma prior with $n_0, \nu_0 \to 0$ (i.e. the amount or precision of the prior information approaches zero).

A new observation $\tilde{y}$, has a posterior predictive density $p(\tilde{y}|\mathbf{y})$, which we first considered in Proposition 2.7, finding $p(\tilde{y}|\mathbf{y})$ to be a t-density. With the improper, reference prior considered here, the posterior predictive density $p(\tilde{y}|\mathbf{y})$ remains a t-density, identical to the density a frequentist would use in testing hypotheses about a predicted value of y:

Proposition 2.10 *Under the conditions of Proposition 2.9, the posterior predictive density for a future observation $\tilde{y}$, $p(\tilde{y}|\mathbf{y})$ is a student-t density, with location parameter $\bar{y}$, scale parameter $s\sqrt{\frac{n+1}{n}}$ and degrees of freedom parameter $n - 1$, where*

$$s^2 = (n-1)^{-1}\sum_{i=1}^{n}(y - \bar{y})^2 \qquad \textit{and}$$

$$\bar{y} = n^{-1}\sum_{i=1}^{n} y_i.$$

Proof. See Proposition C.9 in the Appendix. ◁

We now return to Example 2.13 using the improper, reference prior $p(\mu, \sigma^2) \propto 1/\sigma^2$, using the results of Propositions 2.9 and 2.10. Given these results, we will find that a Bayesian analysis with the improper, reference prior produces posterior densities over μ and future values $\tilde{y}$ that correspond to the frequentist sampling distributions of the estimators $\hat{\mu}$ and $\hat{y}$.

■ Example 2.14

Suspected Pennsylvania voter fraud; Example 2.13, continued. Suppose we adopt the improper, reference prior $p(\mu, \sigma^2) \propto 1/\sigma^2$. Note that $\bar{y} = \hat{\mu} = -5.8$, $n = 21$ and $s^2 = (n-1)^{-1}\sum_{i=1}^{n}(y_i - \bar{y})^2 = 58.1$. Thus, using Equations 2.18 to 2.20 of Proposition 2.9,

$$\mu|\sigma^2, \mathbf{y} \sim N(-5.8, \sigma^2/21)$$

$$\sigma^2|\mathbf{y} \sim \text{inverse-Gamma}(10, 581)$$

$$\frac{\mu + 5.8}{1.44} \sim t_{20}$$

and via Proposition 2.10, a future observation $\tilde{y}|\mathbf{y} \sim t_{20}(-5.8, 7.8)$. Note that with this posterior predictive density, the probability that $\tilde{y} \geq 29.7|\mathbf{y}$ is 3.2×10^{-4}. This is an extremely unlikely event, and the substantive conclusion does not differ from that obtained with the proper prior used in Example 2.13; that is, it is extremely likely that a different process generated the result in the 2nd district in 1993 than the process presumed to have generated the results in the earlier data.

Figure 2.12 shows a sensitivity analysis for the analysis of these data, demonstrating how features of the posterior predictive density $p(\tilde{y}|\mathbf{y})$ change as the normal/inverse-Gamma prior used in Example 2.13 tends towards impropriety (as $n_0 \to 0$) and, at the other extreme, as the prior becomes increasingly dogmatic (as $n_0 \to \infty$). That is, the improper, reference prior $p(\mu, \sigma^2) \propto 1/\sigma^2$ can be considered a limiting case of the proper, normal/inverse-Gamma prior, obtained as $n_0, \nu_0 \to 0$. Note that in the sensitivity analysis shown in Figure 2.12, only n_0 varies; $\mu_0 = 0$, $\nu_0 = 6.2$ and $\sigma_0^2 = 47.07$, and so the marginal prior density for σ^2 remains unchanged. The top panel of Figure 2.12 shows the posterior predictive density becoming more diffuse and centered on the maximum likelihood estimate $\hat{\mu}_{\text{MLE}} = \bar{y} = -5.8$ as the

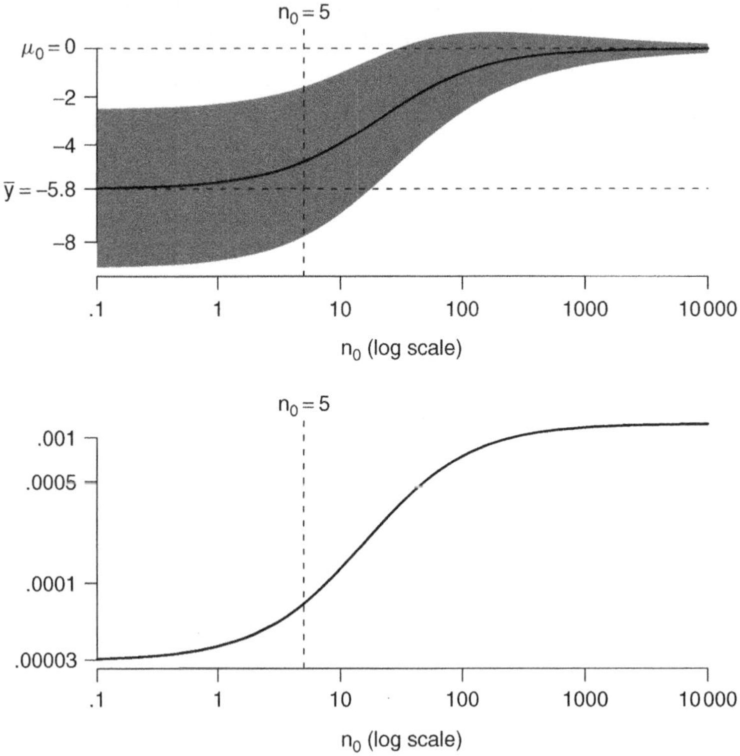

Figure 2.12 Sensitivity analysis, suspected Pennslyvania voter fraud. The top panel shows the posterior predictive density $p(\tilde{y}|\mathbf{y})$ changing as a function of n_0, the number of "prior observations"; the solid line is the mean of the posterior predictive desnity, $E(\tilde{y}|\mathbf{y})$, and the shaded region shows the width of a 95 % HDR. The lower panel shows the posterior probability that $\tilde{y} \geq 29.7$, the value of y observed in the 2nd district in 1993.

prior tends towards impropriety (i.e. as $n_0 \to 0$), and collapsing to a spike on $\mu_0 = 0$ as the prior becomes more dogmatic, ignoring the data (as $n_0 \to \infty$). The proper, normal/inverse-Gamma prior employed in Example 2.13 is obtained with $n_0 = 5$, which does not differ dramatically from the results obtained with the MLEs under the improper, reference prior (see also Figure 2.10). The lower panel Figure 2.12 shows the posterior probability of observing a result at least as pro-Democratic as that obtained in the 2nd district in 1993, i.e. $\Pr(\tilde{y} \geq 29.7|\mathbf{y})$. This posterior probability also changes as n_0 is allowed to vary, but again we see that the qualitative conclusion is extremely robust to the variation in the prior demonstrated here: observing a Democratic advantage of some 29.7 percentage points on absentee ballots is *extremely* unlikely, irrespective of how many "prior observations" one may possess about μ.

Of course, there are priors that can overwhelm this conclusion and Figure 2.13 shows what such priors might be, providing a mapping from the "prior space" given by combinations of μ_0 and n_0 to $\mu_1 = E(\tilde{y}|\mathbf{y})$ (the mean of the posterior predictive density, top panel) and to $\Pr(\tilde{y} \geq 29.7|\mathbf{y})$ (the posterior probabilty of observing an event at least

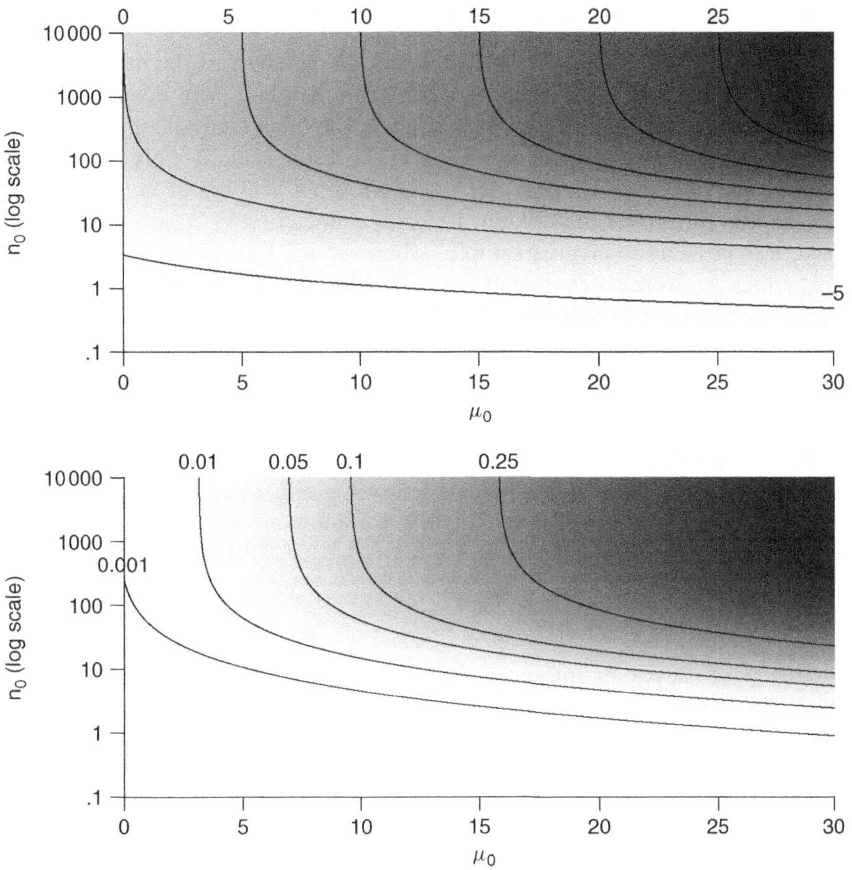

Figure 2.13 Sensitivity analysis, suspected Pennsylvania voter fraud. The top panel shows how the mean of the posterior predictive density, $E(\tilde{y}|\mathbf{y}) = \mu_1$ varies as a function of both the mean of the prior for μ, μ_0, and the precision of the prior, as reflected in the number of prior observations, n_0; darker regions indicate higher values of μ_1. The lower panel shows how the posterior probability $\Pr(\tilde{y} \geq 29.7|\mathbf{y})$ varies as a function of both μ_0 and n_0; darker regions indicate the prior space generating higher levels of the posterior probability, such that reasonably stringent prior beliefs are required so as to make the event $\tilde{y} \geq 29.7$ reasonably likely, given the data.

as extreme as the outcome in the 2nd district in 1993, lower panel). For instance, if one's prior beliefs corresponded to $\mu_0 > 15$ and $n_0 > 10$, then the event $\tilde{y} \geq 29.7$ is still an unlikely event, but with probability greater than 0.05. Any prior more agnotsic than this is more or less swamped by the data (i.e. $\bar{y} = -5.8, n = 21$), rendering the event $\tilde{y} \geq 29.7$ quite unlikely. In sum, this analysis suggests that strong to very strong prior beliefs about a Democratic advantage in absentee ballots would be required so as to be "unsurprised" by observing $y \geq 29.7$. Absent such prior beliefs, the substantive conclusion remains that a Democratic advantage of 29.7 percentage points is highly unlikely and starkly at odds with the rest of the data.

2.4.5 Conflict between likelihood and prior

With the variance σ^2 unknown, it is possible that the variance of the marginal posterior density for μ can be larger than the variance of the marginal prior density. This would seem to contradict a widespread view of conjugate Bayesian analysis, in which posterior precision is no less than the prior precision. This result actually only holds for some special cases, such as when the variance is known in an analysis of normal data with a conjugate normal prior over the unknown mean parameter.

Indeed, it is possible to create examples where we are less certain about a parameter *a posteriori* than we were *a priori*, and even in the relatively innoucuous context considered here: normal data and conjugate normal/inverse-Gamma priors over $\boldsymbol{\theta} = (\mu, \sigma^2)'$. A necessary condition for this is that the prior for μ be in sufficient conflict with the data. In Proposition 2.6 the marginal posterior density of μ is a t density, with scale parameter $\sqrt{\sigma_1^2/n_1}$ and ν_1 degrees of freedom. The term $n_0 n/n_1(\bar{y} - \mu_0)$ makes a contribution to σ_1^2, and taps the conflict between the prior mean of μ, μ_0 and the MLE of μ, $\bar{y}$.

Problem 2.5 invites the reader to create an example in which we can look at data $\mathbf{y}$ and come away "knowing less" about μ than we did *a priori*. See also Figure 3.2 in O'Hagan (2004).

2.4.6 Non-conjugate priors

The discussion of Bayesian inference for a normal mean (and variance) above has explored either conjugate priors or improper, reference priors. For the conjugate case, the conditional posterior density of μ is a normal density, and similarly, in the case of an improper reference prior, the posterior density of μ is also normal. This set of results is especially convenient, since is it quite easy to both elict and report beliefs about a random quantity if those (prior or posterior) beliefs follow a normal density. The normal density is symmetric, unimodal, completely characterized by its mean and variance, and it is trivial to compute quantiles, tail area probabilities and highest density regions for a normal density.

But with only a slight change to the form of the prior, some surprising results can arise. In particular, if we use a student-t density as the prior for μ, it is possible for the posterior density to be bimodal. Proposition C.11 in the Appendix establishes that with a t prior for μ and a normal likelihood for the data, the marginal prior density of μ is proportional to the product of two t densities. That is, if we have the prior densities

$$\mu \sim t_{\nu_0}(\mu_0, \omega_0^2)$$

$$\sigma^2 \sim \text{inverse-Gamma}\left(\frac{a}{2}, \frac{a\sigma_0^2}{2}\right)$$

for the model $y_i \overset{\text{iid}}{\sim} N(\mu, \sigma^2)$, $i = 1, \ldots, n$, then the marginal posterior density for μ is

$$p(\mu|\mathbf{y}) \propto \left[1 + \frac{1}{\nu_0}\frac{(\mu - \mu_0)^2}{\omega_0}\right]^{-(\nu_0+1)/2} \times \left[1 + \frac{1}{\nu_1}\frac{(\bar{y} - \mu)^2}{\tau^2}\right]^{-(\nu_1+1)/2} \tag{2.21}$$

where $\nu_1 = n + a - 1$ and τ^2 is defined in the proof to Proposition C.11. This density arises in a number of statistical contexts and has been studied by a long line of scholars:

in particular, Dickey (1975) showed this density to be a mixture of one-dimensional t densities. Box and Tiao (1973, §9.2.1) studied the conditions under which the density is bimodal. The basic result is that if the data ($\bar{y}$) are in sufficient conflict with the prior μ_0, and the degrees of freedom parameters are sufficiently small (so the t densities have reasonably heavy tails), then the density in Equation 2.21 will be bimodal. As the degrees of freedom parameters get large, the t densities in Equation 2.21 tend to normal densities, and the standard conjugate result from Proposition 2.6 applies (i.e. the marginal posterior density of μ is a t density, and hence symmetric and unimodal).

2.5 Regression models

It is difficult to overstate the importance of the regression model to quantitative social science. In any given year, literally thousands of regression models are reported in the pages of social science journals, and thousands more regressions are run but are unreported. The regression model appears almost whenever a continuous variable is the outcome of substantive interest, with the regression model used to estimate the effects of independent variables on the continuous response. The regression model occupies center stage in research methods classes, particularly at the graduate level: if there is one statistical model that we expect Ph.D. students to understand, it is the regression model. Advanced statistical training in the social sciences also relies on the regression model as its point of departure. The regression model is typically the context in which students are exposed to treatments of properties of statistical estimators (e.g. bias, consistency). More practically, the regression model is often the "launch pad" for a treatment of models for binary, ordinal or nominal responses, panel data, counts, durations, variance components, hierarchical models, or time series models.

As will be familiar to many readers, the linear regression model relates a continuous response variable, $\mathbf{y} = (y_1, \ldots, y_n)'$ and a n-by-k matrix of predictors $\mathbf{X}$ via the following linear model for the conditional mean of $\mathbf{y}$,

$$E(\mathbf{y}|\mathbf{X}, \boldsymbol{\beta}) = \mathbf{X}\boldsymbol{\beta}, \tag{2.22}$$

where $\boldsymbol{\beta}$ is a k-by-1 vector of unknown regression coefficients. The most common goal of regression analysis is inference for $\boldsymbol{\beta}$, since, as is clear from Equation 2.22, these parameters tap how change in the predictors $\mathbf{X}$ maps into change in the average value of $\mathbf{y}$. In addition, the regression model is used to predict values of $\mathbf{y}$ that might arise from particular configuations of values on the predictors; this particular use of regression modeling is especially important in applied policy analysis, where the researcher faces a decision (or is supplying advice to decision-makers) on how or whether to manipulate particular elements of $\mathbf{X}$ so as to bring about change in $\mathbf{y}$.

Here I outline the assumptions that underlie the standard Bayesian treatment of the linear regression model:

1. **weak exogeneity:** Let $p(y_i, \mathbf{x}_i)$ be the joint probability density of y_i and $\mathbf{x}_i$, where $i = 1, \ldots, n$ indexes the observations. That is, both $\mathbf{y}$ and $\mathbf{X}$ are considered random variables. Since a joint density can always be factored as the product of a conditional density and a marginal density, we have

$$p(y_i, \mathbf{x}_i|\boldsymbol{\theta}) = p(y_i|\mathbf{x}_i, \boldsymbol{\theta})p(\mathbf{x}_i|\boldsymbol{\theta}).$$

The assumption of weak exogeneity consists of the following two restrictions:

$$p(y_i|\mathbf{x}_i, \boldsymbol{\theta}) = p(y_i|\mathbf{x}_i, \boldsymbol{\theta}_{y|x})$$

and

$$p(\mathbf{x}_i|\boldsymbol{\theta}) = p(\mathbf{x}_i|\boldsymbol{\theta}_x),$$

where $\boldsymbol{\theta} = (\boldsymbol{\theta}_{y|x}, \boldsymbol{\theta}_x)'$. That is, without loss of generality, the parameter vector $\boldsymbol{\theta}$ can be decomposed into two components, one indexing the conditional density $p(y_i|\mathbf{x}_i)$ and the other indexing the marginal density $p(\mathbf{x}_i)$. An implication of these assumptions (and this is the substantive content of weak exogeneity) is that knowledge of the parameters $\boldsymbol{\theta}_x$ indexing $p(\mathbf{x}_i)$ provides no additional information about y_i beyond that in $\mathbf{x}_i$ and $\boldsymbol{\theta}_{y|x}$. In this sense, the fact that $\mathbf{x}_i$ is a random variable is of no consequence when we turn to the problem of trying to learn about $\boldsymbol{\theta}_{y|x}$. When these conditions hold, $\mathbf{x}_i$ is said to be *weakly exogenous* for $\boldsymbol{\theta}_{y|x}$, as Engle, Hendry and Richard (1983) define the term.

2. **conditional independence:** y_j and $\mathbf{x}_j$ convey no additional information about y_i beyond the information in $\mathbf{x}_i$ and the parameters $\boldsymbol{\theta}_{y|x}$, $\forall\ i \neq j$. That is, for all $i, j \in \{1, \ldots, n\}$, the definition of conditional probability and our assumption of weak exogeneity lets us write

$$p(y_i, y_j, \mathbf{x}_i, \mathbf{x}_j|\boldsymbol{\theta}) = p(y_i, y_j|\mathbf{x}_i, \mathbf{x}_j, \boldsymbol{\theta}_{y|x})p(\mathbf{x}_i, \mathbf{x}_j, \boldsymbol{\theta}_x).$$

The conditional independence assumption lets us simplify the first term on the right-hand side of the equality as

$$p(y_i, y_j|\mathbf{x}_i, \mathbf{x}_j, \boldsymbol{\theta}_{y|x}) = p(y_i|\mathbf{x}_i, \boldsymbol{\theta}_{y|x})p(y_j|\mathbf{x}_j, \boldsymbol{\theta}_{y|x}),$$

factoring the joint density for (y_i, y_j) given $(\mathbf{x}_i, \mathbf{x}_j, \boldsymbol{\theta}_{y|x})$ as the product of the two conditional densities, $p(y_i|\mathbf{x}_i, \boldsymbol{\theta}_{y|x})$ and $p(y_j|\mathbf{x}_j, \boldsymbol{\theta}_{y|x})$. Note that $\mathbf{x}_i$ appears only as a conditioning argument for y_i, and $\mathbf{x}_j$ appears only as a conditioning argument for y_j. That is, by the definition of conditional probability we have

$$\begin{aligned} p(y_i, y_j|\mathbf{x}_i, \mathbf{x}_j, \boldsymbol{\theta}_{y|x}) &= p(y_i|y_j, \mathbf{x}_i, \mathbf{x}_j\boldsymbol{\theta}_{y|x})p(y_j|\mathbf{x}_i, \mathbf{x}_j, \boldsymbol{\theta}_{y|x}) \\ &= p(y_j|y_i, \mathbf{x}_i, \mathbf{x}_j\boldsymbol{\theta}_{y|x})p(y_i|\mathbf{x}_i, \mathbf{x}_j, \boldsymbol{\theta}_{y|x}), \end{aligned}$$

and the effect of the conditional independence assumption is that

$$p(y_i|y_j, \mathbf{x}_i, \mathbf{x}_j, \boldsymbol{\theta}_{y|x}) = p(y_i|\mathbf{x}_i, \boldsymbol{\theta}_{y|x})$$

and, vice-versa,

$$p(y_j|y_i, \mathbf{x}_i, \mathbf{x}_j, \boldsymbol{\theta}_{y|x}) = p(y_j|\mathbf{x}_j, \boldsymbol{\theta}_{y|x})$$

again, $\forall\ i \neq j$. To re-state the effect of the assumption, after we know $\mathbf{x}_i$ and the parameters $\boldsymbol{\theta}_{y|x}$, there is no additional information about y_i in any of the $\mathbf{x}_j$ or y_j.

Note the importance of the conditioning on $\boldsymbol{\theta}_{y|x}$. It is not that the case that y_j and $\mathbf{x}_j$ do not contain information about $p(y_i|\mathbf{x}_i)$. Quite the contrary: the vector $\mathbf{y}_{-i}$ (all observations on y except observation i) and the matrix $\mathbf{X}_{-i}$ supply information about the parameters $\boldsymbol{\theta}_{y|x}$, and through those parameters supply information about $p(y_i|\mathbf{x}_i, \boldsymbol{\theta}_{y|x})$.

One way that data can be generated in a manner consistent with this conditional independence assumption is via *random sampling*. In fact, the conditional independence relations described above constitute a formal definition of "random sampling"; recall the discussion of exchangeability in Section 1.9. One way in which the assumption of conditional independence will not hold is when a lagged value of y_i appears in $\mathbf{x}_i$, as in many regression models for data that form a time series.

Note also that the assumptions of conditional independence assumption and weak exogeneity together imply *strict exogeneity*. That is, weak exogeneity permits a factorization of the joint density $p(y_i, \mathbf{x}_i|\boldsymbol{\theta})$ as the product $p(y_i|\mathbf{x}_i, \boldsymbol{\theta}_{y|x})p(\mathbf{x}_i|\boldsymbol{\theta}_x)$. But via conditional independence, that factorization can be applied for all $\mathbf{x}_i$ and across all $\mathbf{y}_i$: i.e. conditional independence lets us assert

$$p(y_i|\mathbf{x}_i, \boldsymbol{\theta}_{y|x}) = p(y_i|\mathbf{X}, \boldsymbol{\theta}_{y|x}),$$

and so

$$p(y_i, \mathbf{X}|\boldsymbol{\theta}) = p(y_i|\mathbf{X}, \boldsymbol{\theta}_{y|x})p(\mathbf{X}|\boldsymbol{\theta}_x)$$

and in turn,

$$p(\mathbf{y}, \mathbf{X}|\boldsymbol{\theta}) = p(\mathbf{y}|\mathbf{X}, \boldsymbol{\theta}_{y|x})p(\mathbf{X}|\boldsymbol{\theta}_x).$$

As we will soon see, this factorization is especially helpful when it comes time to specify the likelihood for the normal, linear regression model.

3. **normal densities:** $p(y_i|\mathbf{x}_i, \boldsymbol{\theta}_{y|x})$ is a normal density, $\forall\ i = 1, \ldots, n$. Since a normal density is completely characterized by its mean and variance parameters, the specific role of the parameters $\boldsymbol{\theta}_{y|x}$ is now clearer: i.e. $\mathbf{X}$ and $\boldsymbol{\theta}_{y|x}$ most together define (a) the mean and (b) the variance of each normal density $p(y_i|\mathbf{x}_i, \boldsymbol{\theta}_{y|x}), i = 1, \ldots, n$. The next two assumptions specify what these two quantities are. Moreover, the assumption of conditional independence means that $p(\mathbf{y}|\mathbf{X}, \boldsymbol{\theta}_{y|x})$ is a multivariate normal density (see Definition B.31) with a very simple form to its n-by-n variance-covariance matrix, $\boldsymbol{\Sigma}$. An implication of conditional independence is that $\text{cov}(y_i, y_j|\mathbf{X}, \boldsymbol{\theta}_{y|x}) = 0, \forall\ i \neq j$, and so the variance-covariance matrix $\boldsymbol{\Sigma}$ is a diagonal matrix (see Definition A.7).

4. **linearity of the conditional expectation (regression) function:** $E(y_i|\mathbf{x}_i, \boldsymbol{\theta}_{y|x}) = \mathbf{x}_i\boldsymbol{\beta}$. The assumption that the predictors enter the model in a linear, additive fashion is not as restrictive as it might first seem. Readers familiar with the linear regression model will recognize that the "linearity" here is only a restriction on the parameters, $\boldsymbol{\beta}$, not on the way that predictor variables enter the model. For

instance, the researcher can always non-linearly transform the predictors of substantive interest to "trick" the linear regression model into estimating non-linear functions of interest: i.e. the predictors that enter the linear model, $\mathbf{X}$, may actually be non-linear transformations of predictors of substantive interest. Likewise, $\mathbf{y}$ may actually be a non-linear transformation of a response variables of interest. No class on the linear regression model is complete without showing students how logarithmic transformations, polynomials, taking reciprocals or square roots, or interactions can be used to estimate non-linear relationships between $\mathbf{y}$ and predictors of interest.

5. **conditional homoskedasticity:** $\text{var}(y_i|\mathbf{X}) = \sigma^2, i = 1, \ldots, n$. That is, given the predictors, the y_i have constant variance, σ^2, with σ^2 an unknown parameter.

Given that the y_i have normal densities (Assumption 3, above) and are conditionally independent (Assumption 2), it follows that the y_i have zero conditional covariances (see Proposition B.2). Combined with the assumption of conditional homoskedasticity (Assumption 5), we can now write the regression with the following compact expression,

$$\mathbf{y}|\mathbf{X}, \boldsymbol{\beta}, \sigma^2 \sim N(\mathbf{X}\boldsymbol{\beta}, \sigma^2\mathbf{I}_n), \tag{2.23}$$

where $\mathbf{I}_n$ is a n-by-n identity matrix (see Defintion A.9). Also, note that $\boldsymbol{\theta}_{y|x} = (\boldsymbol{\beta}, \sigma^2)$; i.e. the dependence of $\mathbf{y}$ on $\mathbf{X}$ is captured by the linear, normal model parameterized by the regression parameters $\boldsymbol{\beta}$ and conditional variance parameter σ^2.

2.5.1 Bayesian regression analysis

The regression model in Equation 2.23 is a *conditional model* for $\mathbf{y}$. That is, we model $\mathbf{y}$ conditional on the predictors $\mathbf{X}$ and in particular, we adopt the model for the conditional mean of y_i given in Assumption 4, above (and see also Equation 2.22). From the perspective of Bayesian modeling, this does require a little additional justification, as I now elaborate.

Thus far, Bayesian analysis has been presented as following the same, simple recipe: i.e. via Bayes Rule, the posterior density is proportional to the density of "the data" times the prior density. It is important to remember that in the context of regression, "the data" comprise both $\mathbf{y}$ and the predictors $\mathbf{X}$. This would seem to imply that a Bayesian analysis would use the joint density $p(\mathbf{y}, \mathbf{X}|\boldsymbol{\theta})$ to learn about $\boldsymbol{\theta}$; i.e. via Bayes Rule

$$p(\boldsymbol{\theta}|\mathbf{y}, \mathbf{X}) \propto p(\mathbf{y}, \mathbf{X}|\boldsymbol{\theta})p(\boldsymbol{\theta}).$$

Indeed, applying Bayes Rule in this way lets us learn about $\boldsymbol{\theta}$, but the goal of regression analysis is to learn about the (linear) dependence of $\mathbf{y}$ on $\mathbf{X}$, which, given the assumptions above, is captured in the parameters $\boldsymbol{\theta}_{y|x} = (\boldsymbol{\beta}, \sigma^2)$. Now, recall that in the previous section we used the assumptions of conditional independence and weak exogeneity to factor the joint density for $(\mathbf{y}, \mathbf{X})$ as $p(\mathbf{y}|\mathbf{X}, \boldsymbol{\theta}_{y|x})p(\mathbf{X}|\boldsymbol{\theta}_x)$. We now make the additional assumption that prior beliefs about $\boldsymbol{\theta}_{y|x}$ and $\boldsymbol{\theta}_x$ are independent: i.e.

$$p(\boldsymbol{\theta}) = p(\boldsymbol{\theta}_{y|x}, \boldsymbol{\theta}_x) = p(\boldsymbol{\theta}_{y|x})p(\boldsymbol{\theta}_x) = p(\boldsymbol{\beta}, \sigma^2)p(\boldsymbol{\theta}_x). \tag{2.24}$$

Now consider the posterior density for all the unknown parameters introduced thus far, $p(\boldsymbol{\beta}, \sigma^2, \boldsymbol{\theta}_x)$: via Bayes Rule and our assumptions,

$$
\begin{aligned}
p(\boldsymbol{\beta}, \sigma^2, \boldsymbol{\theta}_x|\mathbf{y}, \mathbf{X}) &= \frac{p(\mathbf{y}, \mathbf{X}|\boldsymbol{\beta}, \sigma^2, \boldsymbol{\theta}_x)p(\boldsymbol{\beta}, \sigma^2, \boldsymbol{\theta}_x)}{p(\mathbf{y}, \mathbf{X})} \\
&= \frac{p(\mathbf{y}|\mathbf{X}, \boldsymbol{\beta}, \sigma^2, \boldsymbol{\theta}_x)p(\mathbf{X}|\boldsymbol{\beta}, \sigma^2, \boldsymbol{\theta}_x)p(\boldsymbol{\beta}, \sigma^2)p(\boldsymbol{\theta}_x)}{p(\mathbf{y}|\mathbf{X})p(\mathbf{X})} \\
&= \frac{p(\mathbf{y}|\mathbf{X}, \boldsymbol{\beta}, \sigma^2)p(\boldsymbol{\beta}, \sigma^2)}{p(\mathbf{y}|\mathbf{X})} \times \frac{p(\mathbf{X}|\boldsymbol{\theta}_x)p(\boldsymbol{\theta}_x)}{p(\mathbf{X})} \\
&= p(\boldsymbol{\beta}, \sigma^2|\mathbf{y}, \mathbf{X})p(\boldsymbol{\theta}_x|\mathbf{X}). \qquad (2.25)
\end{aligned}
$$

This product form of this factorization highlights that our assumptions let us break the Bayesian analysis of data $(\mathbf{y}, \mathbf{X})$ into two independent pieces: one part that lets us make posterior inferences about the parameters $\boldsymbol{\theta}_{y|x} = (\boldsymbol{\beta}, \sigma^2)$ that characterize the (linear) relationship between $\mathbf{y}$ and $\mathbf{X}$, and another piece that lets us make inferences about $\boldsymbol{\theta}_x$ given data $\mathbf{X}$. Our interest lies with $\boldsymbol{\theta}_{y|x}$.

It is worth noting that assumptions about $\mathbf{X}$ being fixed or stochastic have no bearing on a Bayesian analysis of the regression model, since the same posterior density results in either case. In the case of "fixed" (or non-stochastic) regressors, $\mathbf{X}$ is not drawn from a density $p(\mathbf{X})$, and so $p(\mathbf{X})$ and $\boldsymbol{\theta}_x$ drop out of the analysis. In the case of stochastic regressors, our assumptions mean that a Bayesian regression analysis can focus on inference for the parameters of substantive inference, $\boldsymbol{\theta}_{y|x} = (\boldsymbol{\beta}, \sigma^2)'$, with the likelihood $p(\mathbf{y}|\mathbf{X}, \boldsymbol{\beta}, \sigma^2)$ containing the information in $\mathbf{y}$ and $\mathbf{X}$ about $\boldsymbol{\beta}$ and σ^2. To restate the main point, our assumptions maintain that given $\mathbf{X}$, there is no additional information about $\boldsymbol{\theta}_{y|x} = (\boldsymbol{\beta}, \sigma^2)$ to be had from knowing $\boldsymbol{\theta}_x$.

2.5.2 Likelihood function

Given the assumptions above, the likelihood for the normal linear model has a simple form, i.e.

$$
p(\mathbf{y}; \mathbf{X}, \boldsymbol{\beta}, \sigma^2) = (2\pi\sigma^2)^{-n/2} \exp\left[\frac{-(\mathbf{y} - \mathbf{X}\boldsymbol{\beta})'(\mathbf{y} - \mathbf{X}\boldsymbol{\beta})}{2\sigma^2}\right] \qquad (2.26)
$$

which follows directly from the form of the regression model given in Equation 2.23 and the definition of a multivariate normal density (Definition B.31). If we define

$$
\epsilon_i = y_i - E(y_i|\mathbf{x}_i, \boldsymbol{\beta}) = y_i - \mathbf{x}_i\boldsymbol{\beta}
$$

then $(\mathbf{y} - \mathbf{X}\boldsymbol{\beta})'(\mathbf{y} - \mathbf{X}\boldsymbol{\beta}) = \sum \epsilon_i^2$, and so an estimate of $\boldsymbol{\beta}$ that maximizes this likelihood in 2.26 also minimizes the sum of the squared residuals. In other words, under our maintained assumptions, the maximum likelihood estimator of $\boldsymbol{\beta}$ is also the least squares estimator, or $\hat{\boldsymbol{\beta}}_{\text{MLE}} = \hat{\boldsymbol{\beta}}_{\text{LS}} = (\mathbf{X}'\mathbf{X})^{-1}\mathbf{X}'\mathbf{y}$.

2.5.3 Conjugate prior

A conjugate prior density for the parameter vector $\boldsymbol{\theta} = (\boldsymbol{\beta}, \sigma^2)'$ is a generalization of the conjugate normal/inverse-Gamma prior density considered for the case of inference for a normal mean and variance (Section 2.4.2). There, the prior for μ, conditional on σ^2, is a univariate normal density. In the current context of the normal linear regression model, the prior for $\boldsymbol{\beta}$, conditional on σ^2, is a *multivariate* normal density (see Definition B.31). The following proposition summarizes the result.

Proposition 2.11 (Conjugate priors for regression parameters and variance, normal regression model). *Let* $y_i|\mathbf{x}_i \stackrel{iid}{\sim} N(\mathbf{x}_i\boldsymbol{\beta}, \sigma^2)$, $i = 1, \ldots, n$ *and* $\mathbf{y} = (y_1, \ldots, y_n)'$, *where* $\mathbf{x}_i$ *is a 1-by-k vector of predictors,* $\boldsymbol{\beta}$ *is a k-by-1 vector of unknown regression parameters and* σ^2 *is an unknown variance parameter. If* $\boldsymbol{\beta}|\sigma^2 \sim N(\mathbf{b}_0, \sigma^2\mathbf{B}_0)$ *is the conditional prior density for* $\boldsymbol{\beta}$ *given* σ^2, *and* $\sigma^2 \sim$ *inverse-Gamma*$(\nu_0/2, \nu_0\sigma_0^2/2)$ *is the prior density for* σ^2, *then*

$$\begin{aligned}\boldsymbol{\beta}|\sigma^2, \mathbf{y}, \mathbf{X} &\sim N(\mathbf{b}_1, \sigma^2\mathbf{B}_1),\\ \sigma^2|\mathbf{y}, \mathbf{X} &\sim \textit{inverse-Gamma}(\nu_1/2,\ \nu_1\sigma_1^2/2)\end{aligned}$$

where

$$\begin{aligned}\mathbf{b}_1 &= (\mathbf{B}_0^{-1} + \mathbf{X}'\mathbf{X})^{-1}(\mathbf{B}_0^{-1}\mathbf{b}_0 + \mathbf{X}'\mathbf{X}\hat{\boldsymbol{\beta}})\\ \mathbf{B}_1 &= (\mathbf{B}_0^{-1} + \mathbf{X}'\mathbf{X})^{-1}\\ \nu_1 &= \nu_0 + n \qquad \textit{and}\\ \nu_1\sigma_1^2 &= \nu_0\sigma_0^2 + S + r,\end{aligned}$$

and where

$$\begin{aligned}\hat{\boldsymbol{\beta}} &= (\mathbf{X}'\mathbf{X})^{-1}\mathbf{X}'\mathbf{y}\\ S &= (\mathbf{y} - \mathbf{X}\hat{\boldsymbol{\beta}})'(\mathbf{y} - \mathbf{X}\hat{\boldsymbol{\beta}}) \qquad \textit{and}\\ r &= (\mathbf{b}_0 - \hat{\boldsymbol{\beta}})'(\mathbf{B}_0 + (\mathbf{X}'\mathbf{X})^{-1})^{-1}(\mathbf{b}_0 - \hat{\boldsymbol{\beta}}).\end{aligned}$$

Proof. See Proposition C.12 in the Appendix. ◁

An important element of the result in Proposition 2.11 generalizes a point made in Section 2.1.2: the mean of the conditional posterior density for $\boldsymbol{\beta}$ is a precision-matrix weighted average of the prior mean for $\boldsymbol{\beta}$, $\mathbf{b}_0$, and the maximum likelihood estimate, $\hat{\boldsymbol{\beta}} = (\mathbf{X}'\mathbf{X})^{-1}\mathbf{X}'\mathbf{y}$. Since both $\mathbf{b}_0$ and $\hat{\boldsymbol{\beta}}$ are k-by-1 vectors, the precision of each is k-by-k precision matrices, $\sigma^{-2}\mathbf{B}_0^{-1}$ and $\sigma^{-2}\mathbf{X}'\mathbf{X}$, respectively. Thus, as prior knowledge about $\boldsymbol{\beta}$ becomes less and less precise, $\mathbf{B}_0^{-1} \rightarrow \mathbf{0}$ and the relative contribution of the prior to the posterior tends to zero. On the other hand, should the prior be sufficiently informative about $\boldsymbol{\beta}$ relative to the information in the data, then the posterior can be dominated by the prior, with the data contributing very little, if anything, to posterior inferences about the regression parameters.

In particular, some or all of the variables in $\mathbf{X}$ may be highly collinear, or even perfectly collinear. In this case, the data are relatively uninformative about the corresponding elements of $\boldsymbol{\beta}$ (recall that in a classical setting, the effect of multicollinearity is to inflate the sampling variances of particular elements of $\hat{\boldsymbol{\beta}}$). In the case of perfect collinearity, the data are completely uninformative about particular elements of $\boldsymbol{\beta}$; $\mathbf{X}'\mathbf{X}$ is singular, $(\mathbf{X}'\mathbf{X})^{-1}$ does not exist, and $\hat{\boldsymbol{\beta}}$ can not be computed (and this is the sense in which the data are uninformative about $\boldsymbol{\beta}$). More formally, perfect multicollinearity is a case in which the model parameters are *unidentified*: i.e. when there are linear dependencies among the columns of $\mathbf{X}$, the likelihood function in Equation 2.26 is unidentified with respect to the corresponding elements of $\boldsymbol{\beta}$ (see Definition B.17). In general, the fact that the likelihood function is not identified poses no formal problem in a Bayesian analysis: via Bayes Rule, the posterior density is proportional to the prior density times the likelihood, irrespective of whether the likelihood is identified or not. In the specific case of linear regression, note that the expression for $\mathbf{b}_1$ in Proposition 2.11 can be rewritten as

$$\mathbf{b}_1 = (\mathbf{B}_0^{-1} + \mathbf{X}'\mathbf{X})^{-1}(\mathbf{B}_0^{-1}\mathbf{b}_0 + \mathbf{X}'\mathbf{y})$$

which exists if and only if $\mathbf{B}_0^{-1}$ is non-singular; i.e. even if $\mathbf{X}'\mathbf{X}$ is not invertible, the matrix $\mathbf{B}_0^{-1} + \mathbf{X}'\mathbf{X}$ will be invertible provided $\mathbf{B}_0$ is positive definite (see Definition A.17). That is, the posterior density for $\boldsymbol{\beta}$ exists so long as the prior density for $\boldsymbol{\beta}$ is proper (i.e. so long as $\mathbf{B}_0$ is invertible), irrespective of whether $\mathbf{X}'\mathbf{X}$ is singular or non-singular. Any Bayesian analysis combines prior and sample information about parameters; in the case of perfect multicollinearity in a regression setting, the prior density supplies information about particular elements of $\boldsymbol{\beta}$ for which the data is uninformative. Of course, similar but less dramatic results hold for the far more common case of less than perfect multicollinearity. See Leamer (1978) or Box and Tiao (1973) for more detail on this particular aspect of Bayesian analysis of the linear regression model.

Marginal Posterior Density for $\boldsymbol{\beta}$

While the conditional posterior density $p(\boldsymbol{\beta}|\sigma^2, \mathbf{y}, \mathbf{X})$ is a multivariate normal density, the marginal posterior density

$$p(\boldsymbol{\beta}|\mathbf{y}, \mathbf{X}) = \int p(\boldsymbol{\beta}|\sigma^2, \mathbf{y}, \mathbf{X})p(\sigma^2|\mathbf{y}, \mathbf{X})d\sigma^2$$

is a multivariate t density (see Definition B.38). This result is a generalization of Proposition 2.6, which we encountered in Section 2.4.2 in considering the conjugate, Bayesian analysis of normal data with unknown mean and variance. Again, a proof of the proposition is relegated to the Appendix.

Proposition 2.12 (Marginal Posterior Density for Regression Parameters, Conjugate Priors, Normal Regression). *Assume the conditions of Proposition 2.11. Then the marginal posterior density for* $\boldsymbol{\beta}$, $p(\boldsymbol{\beta}|\mathbf{y}, \mathbf{X})$, *is a multivariate t density, with location parameter* $\mathbf{b}_1$, *squared scale parameter* $\sigma_1^2\mathbf{B}_1$ *and* ν_1 *degrees of freedom, where* $\mathbf{b}_1$ *and* $\mathbf{B}_1$ *are as defined in Proposition 2.11.*

Proof. See Proposition C.14 in the Appendix. ◁

Highest density region for $\boldsymbol{\beta}$

The preceeding proposition establishes that in a conjugate analysis, the marginal prior/posterior density of $\boldsymbol{\beta}$, denoted $p(\boldsymbol{\beta})$, is a multivariate t density. Recall that if $\boldsymbol{\beta} \in \mathbb{R}^k$, a highest density region of content α ($0 < \alpha < 1$) is a set $\mathcal{C}^\alpha \subset \mathbb{R}^k$, such that (1) $\int_{\mathcal{C}^\alpha} p(\boldsymbol{\beta}) d\boldsymbol{\beta} = \alpha$, and (2) $p(\boldsymbol{\beta}) \geq p(\tilde{\boldsymbol{\beta}}), \forall \boldsymbol{\beta} \in \mathcal{C}^\alpha, \tilde{\boldsymbol{\beta}} \in \mathbb{R}^k \setminus \mathcal{C}^\alpha$. The following proposition provides an expression for the HDR $\mathcal{C}^\alpha$, showing a connection between the size of $\mathcal{C}^\alpha$ and the α quantile of Snedecor's F density:

Proposition 2.13 (Highest density region for $\boldsymbol{\beta}$ under conjugacy). *Suppose* $\boldsymbol{\beta} \sim t_\nu(\mathbf{b}, \mathbf{B})$ *(a prior or posterior multivariate t density),* $\boldsymbol{\beta} \in \mathbb{R}^k$. *Suppose further that*

$$\sigma^2 \sim \textit{inverse-Gamma}(\nu/2, \nu s^2/2)$$

is a prior or posterior density for σ^2. *Then the highest density region (HDR) of* $\boldsymbol{\beta}$ *of content* α ($0 < \alpha < 1$) *is a set* $\mathcal{C}^\alpha \subset \mathbb{R}^k$, *specifically,*

$$\mathcal{C} = \{\boldsymbol{\beta} : (\boldsymbol{\beta} - \mathbf{b})'\mathbf{B}^{-1}(\boldsymbol{\beta} - \mathbf{b}) \leq c\}$$

where $c = ks^2F$ *and* F *is the* α *quantile of Snedecor's* F *density with* k *and* ν *degrees of freedom.*

Proof. Since $\boldsymbol{\beta} \sim t_\nu(\mathbf{b}, \mathbf{B})$, and $\sigma^2 \sim$ inverse-Gamma$(\nu/2, \nu s^2/2)$ it follows that $\boldsymbol{\beta}|\sigma^2 \sim N(\mathbf{b}, \sigma^2\mathbf{B})$ (Proposition 2.12) and $q = \sigma^{-2}(\boldsymbol{\beta} - \mathbf{b})\mathbf{B}^{-1}(\boldsymbol{\beta} - \mathbf{b}) \sim \chi^2_k$ (see the remarks appearing immediately after Definition B.36). By definition, $q|\sigma^2$ is independent of σ^2. Since $\sigma^2 \sim$ inverse-Gamma$(\nu/2, \nu s^2/2)$, $\nu s^2 \sigma^{-2} \sim \chi^2_\nu$ (again, see the remarks appearing immediately after Definition B.36). Then by Definition B.39,

$$\frac{q/k}{\nu s^2 \sigma^{-2}/\nu} = \frac{(\boldsymbol{\beta} - \mathbf{b})\mathbf{B}^{-1}(\boldsymbol{\beta} - \mathbf{b})}{ks^2} \sim F_{k,\nu}$$

Now consider a region $\mathcal{C}^* = \{\boldsymbol{\beta} : (\boldsymbol{\beta} - \mathbf{b})\mathbf{B}^{-1}(\boldsymbol{\beta} - \mathbf{b}) \leq c^*\}$. For any such region we have $\Pr(\boldsymbol{\beta} \in \mathcal{C}^*) = \Pr[z \leq c^*/(ks^2)]$, where $z \sim F_{k,\nu}$. To obtain a HDR for $\boldsymbol{\beta}$ of content α, we set c^* so that $\Pr[z \leq c^*/(ks^2)] = \alpha$, i.e. $c^* = ks^2F$, where F is the α quantile of Snedecor's F density with k and ν degrees of freedom. ◁

The HDR for $\boldsymbol{\beta}$ is therefore enclosed by an ellipsoid in $\mathbb{R}^k$, centered on $\mathbf{b}$, the mean of the prior/posterior density of $\boldsymbol{\beta}$, with a shape or orientation determined by the elements of $\mathbf{B}$. The extent of the HDR is easily determined since the solution for its boundary is a quadratic form in $\boldsymbol{\beta}$, with the α-quantile of an F density appearing in the solution; for $k = 2$ or for two-dimensional, marginal "slices" of $p(\boldsymbol{\beta})$, it is possible to display the HDR graphically (e.g. see Figures 2.15 and 2.16).

Note also that the HDR with content α for a single element of $\boldsymbol{\beta}$, β_j, is easily derived. Under conjugacy, the marginal prior/posterior density of β_j is a $t_\nu(b_j, B_j)$ density. Then the HDR is simply an interval on the real line, with its range given by the $.5 \pm \alpha/2$ quantiles of a $t_\nu(b_j, B_j)$ density, or, equivalently, a F density with 1 and ν degrees of freedom, exploiting the fact that if $\beta_j \sim t_\nu(b_j, B_j)$, then $(\beta_j - b_j)^2/B_j \sim F_{1,\nu}$.

Predictive density for a new observation

We also state a generalization of Proposition 2.7, establishing that the posterior predictive density for a new observation $\tilde{y}$ is a t distribution:

Proposition 2.14 (Posterior predictive density, normal regression, conjugate priors). *Assume the conditions of Proposition C.12. Then the posterior predictive density for q new observations $\tilde{\mathbf{y}}$ given predictors taking on values $\mathbf{X}$ (a q-by-k matrix), is a multivariate t density with location $\mathbf{X}\mathbf{b}_1$, squared scale parameter $\sigma_1^2(\mathbf{X}\mathbf{B}_1\mathbf{X}' + \mathbf{I}_q)$ and ν_1 degrees of freedom, where $\mathbf{b}_1$ and $\mathbf{B}_1$ are as defined in Proposition 2.11.*

Proof. See Proposition C.15 in the Appendix. ◁

2.5.4 Improper, reference prior

In Section 2.4.4 we saw that for the simple model $y_i \sim N(\mu, \sigma^2)$, the reference prior $p(\mu, \sigma^2) \propto 1/\sigma^2$ yields a posterior distribution for μ and σ^2 equal to the sampling distribution of the least squares estimates $\hat{\mu}$ and $\hat{\sigma}^2$, stated as Proposition 2.9. We have an analogous result in the case of regression modeling:

Proposition 2.15 *Assume the conditions of Proposition 2.11, assigning the parameters $(\boldsymbol{\beta}, \sigma^2)'$ the improper prior density $p(\boldsymbol{\beta}, \sigma^2) \propto 1/\sigma^2$. Further assume that $\mathbf{X}'\mathbf{X}$ is non-singular. Then*

$$\boldsymbol{\beta}|\sigma^2, \mathbf{y}, \mathbf{X} \sim N\left(\hat{\boldsymbol{\beta}}, \sigma^2(\mathbf{X}'\mathbf{X})^{-1}\right)$$

$$\sigma^2|\mathbf{y}, \mathbf{X} \sim \textit{inverse-Gamma}\left(\frac{n-k}{2}, \frac{S}{2}\right)$$

$$\boldsymbol{\beta}|\mathbf{y}, \mathbf{X} \sim t_{n-k}\left(\hat{\boldsymbol{\beta}}, S/(n-k)(\mathbf{X}'\mathbf{X})^{-1}\right)$$

where $\hat{\boldsymbol{\beta}} = (\mathbf{X}'\mathbf{X})^{-1}\mathbf{X}'\mathbf{y}$ *and* $S = (\mathbf{y} - \mathbf{X}\hat{\boldsymbol{\beta}})'(\mathbf{y} - \mathbf{X}\hat{\boldsymbol{\beta}})$.

Proof. See Proposition C.16 in the Appendix. ◁

In other words, with the (improper) prior beliefs $p(\boldsymbol{\beta}, \sigma^2) \propto 1/\sigma^2$, the Bayesian posterior density of $\boldsymbol{\beta}$ (conditional on σ^2) is a multivariate normal density centered over the maximum likelihood estimate $\hat{\boldsymbol{\beta}} = (\mathbf{X}'\mathbf{X})^{-1}\mathbf{X}'\mathbf{y}$ with variance-covariance matrix equal to the variance-covariance matrix of the frequentist sampling distribution of $\hat{\boldsymbol{\beta}}$ (again, conditional on σ^2), $\sigma^2(\mathbf{X}'\mathbf{X})^{-1}$. It is worth stressing that this "correspondence" between a Bayesian regression analysis and a frequentist analysis can be obtained without resorting to an improper prior: if one adopts a vague-but-proper, conjugate, normal/inverse-Gamma prior for $(\boldsymbol{\beta}, \sigma^2)$ then the resulting normal/inverse-Gamma posterior density will closely approximate the posterior density of Proposition 2.15. Of course, whether such a prior actually is an honest representation of one's prior beliefs is up to the researcher.

We now see how these general propositions about Bayesian regression analysis operate via two worked examples.

■ Example 2.15

Suspected voter fraud in Pennsylvania, Example 2.13, continued. Regression analysis provides another way to assess the plausibility of the outcome in the disputed Pennsylvania state senate election. For the purposes of this analysis we treat the Democratic lead (expressed as share of the two party vote, in percentage points) among absentee ballots as the dependent variable (y_i), and the Democratic lead in the two-party vote among votes cast on voting machines as the predictor (x_i). Again, we omit the disputed election outcome from the analysis. The regression analysis provides a baseline against which to assess any "future" outcome, such as the disputed state senate election in the 2nd district in 1993.

Model. With just the single predictor, x_i, the regression model is $y_i|x_i \stackrel{\text{iid}}{\sim} N(\beta_0 + \beta_1 x_i, \sigma^2)$, and so the unknown parameters are $\boldsymbol{\theta} = (\boldsymbol{\beta}, \sigma^2)' = (\beta_0, \beta_1, \sigma^2)'$. As in Example 2.15, the model assumes that the data are exchangeable, ignoring the fact that the 21 data points in the "historical" data are sets of 3 observations from 7 districts; specifically, the possibility that the regression relationship between x_i and y_i might vary across the 7 districts is ignored. That is, given an x_i, our beliefs about y_i are the same irrespective of the district that produced the particular x_i.

Prior densities. A Bayesian analysis of a regression problem requires specification of priors for the unknown parameters $\boldsymbol{\theta} = (\boldsymbol{\beta}, \sigma^2)$. To keep the analysis simple, I will employ a conjugate prior for $\boldsymbol{\theta}$, which in this case (see Proposition 2.11) I represent as $p(\boldsymbol{\theta}) = p(\boldsymbol{\beta}, \sigma^2) = p(\boldsymbol{\beta}|\sigma^2)p(\sigma^2)$, where

$$\begin{aligned} \boldsymbol{\beta}|\sigma^2 &\sim N(\mathbf{b}_0, \sigma^2\mathbf{B}_0) \\ \sigma^2 &\sim \text{inverse-Gamma}(\nu_0/2, \nu_0\sigma_0^2/2). \end{aligned}$$

Given this conjugate normal/inverse-Gamma prior, a formal specification of the prior amounts to choosing values of

- $\mathbf{b}_0$, a vector of length 2, containing the mean of the multivariate normal prior density for $\boldsymbol{\beta} = (\beta_0, \beta_1)'$.
- $\mathbf{B}_0$, a 2 by 2 matrix, such that conditional on σ^2, $\sigma^2\mathbf{B}_0$ is the variance-covariance matrix for the multivariate normal prior density for $\boldsymbol{\beta}$.
- $\nu_0 > 0$, a prior "degrees of freedom" parameter
- $\sigma_0^2 > 0$, where $\nu_0\sigma_0^2$ is a "prior sum of squares" (i.e. as would be obtained from a data set of size ν_0).

As I argued in Example 2.13, absent any knowledge to the contrary, a reasonable presumption is that, on average, the Democratic margin among absentee ballots (y_i) is the same as the Democratic margin among machine ballots (x_i), implying the regression relationship $E(y_i|x_i) = x_i$, and so I set $\mathbf{b}_0 = (0, 1)'$. Of course, we don't expect this relationship to hold exactly, and, in any event, I do not place a great deal of confidence in my "prior guess" that $\beta_0 = 0$ and $\beta_1 = 1$. This uncertainty in $\boldsymbol{\beta}$ will be reflected in the values of $\mathbf{B}_0$ and the parameters characterizing prior beliefs over σ^2, ν_0 and σ_0^2. I use the same values for ν_0 and σ_0^2 as in Example 2.13, since the quantity being modeled there corresponds to $y_i - x_i$ in the current example, or $y_i - E(y_i|x_i)$; thus, σ^2 from Example 2.13 is $V(y_i - x_i) = V(y_i|x_i)$ in the current example. That is, I set $\nu_0 = 6.2$ and $\sigma_0^2 = 47.07$.

To set $\mathbf{B}_0$, I first assume that my prior information about the intercept parameter β_0 is independent of my prior information about the slope parameter β_1; i.e. $\mathbf{B}_0$ is a 2-by-2 diagonal matrix, with the zero on the off-diagonal meaning that there is no *a priori* covariance between β_0 and β_1. I then make use of the result in Proposition 2.12: i.e. if $\boldsymbol{\beta}|\sigma^2$ follows a multivariate normal density, then the marginal density for $\boldsymbol{\beta}$ is a multivariate t density. Under the assumption of prior independence of β_0 and β_1, it follows that each of these parameters have marginal densities that are univariate t densities. Specifically, if $\boldsymbol{\beta}|\sigma^2 \sim N(\mathbf{b}_0, \sigma^2\mathbf{B}_1)$ and $\sigma^2 \sim$ inverse-Gamma$(\nu_0/2, \nu_0\sigma_0^2/2)$ with $\mathbf{b}_0 = (b_0, b_1)'$ and $\mathbf{B}_1 = \text{diag}(\lambda_0, \lambda_1)$ then $\beta_j \sim t_{\nu_0}(b_j, \sigma_0^2\lambda_j)$, $j = 0, 1$. Moreover, we can also exploit the fact that in standardized form, for $j = 0, 1$,

$$(\beta_j - b_j)/\sqrt{\sigma_0^2\lambda_j} \sim t_{\nu_0}.$$

This is convenient, since given $\mathbf{b} = (b_0, b_1)$, ν_0 and σ_0^2, we can recover λ_0 and λ_1 as follows. Suppose q_{ν_0} is the q-quantile of the (unstandardized) t_{ν_0} density for β_0. Let $\tilde{q}_{\nu_0}$ be the q-quantile of the standardized t_{ν_0} density for β_j, $j = 0, 1$. Since it is straightforward to compute critical quantiles of standardized t densities, we know that $q_{\nu_0} = \tilde{q}_{\nu_0} \times \sqrt{\sigma_0^2\lambda_j}$ or, re-arranging,

$$\lambda_j = \left(\frac{q_{\nu_0}}{\tilde{q}_{\nu_0}}\right)^2 / \sigma_0^2 \qquad (2.27)$$

Here I posit that a 95 % prior interval for the intercept parameter β_0 ranges from -25 to 25, i.e. for β_0, q_{ν_0} is 25 with $\nu_0 = 6.2$ and so $\tilde{q}_{\nu_0}$ is the .975 quantile of a t distribution with 6.2 degrees of freedom, or 2.43. Given that $\sigma_0^2 = 47.07$, Equation 2.27 tells us that $\lambda_0 = 2.25$. For the slope parameter, it is almost certain that the relationship between y_i and x_i will be positive (as the Democratic vote margin recorded via voting machines increases, so too should the vote margin recorded among absentee ballots), and so I assign prior probability .01 to the event $\beta_1 \leq 0$. That is, the .01 percentile of the prior marginal density for β_1 is 0, meaning that $q_{\nu_0} = 1$, $\tilde{q}_{\nu_0} = 3.11$ is the 99th percentile of a t distribution with $\nu_0 = 6.2$ degrees of freedom, and so via Equation 2.27, $\lambda_1 = .0022$. Thus,

$$\mathbf{B}_0 = \text{diag}(\lambda_0, \lambda_1) = \begin{bmatrix} 2.25 & 0 \\ 0 & .0022 \end{bmatrix}$$

and thus my prior for $\boldsymbol{\theta} = (\boldsymbol{\beta}, \sigma^2)$ is

$$\boldsymbol{\beta}|\sigma^2 \sim N\left(\begin{bmatrix} 0 \\ 1 \end{bmatrix}, \sigma^2\begin{bmatrix} 2.25 & 0 \\ 0 & .0022 \end{bmatrix}\right)$$

$$\sigma^2 \sim \text{inverse-Gamma}\left(\frac{6.2}{2}, \frac{6.2 \times 47.07}{2}\right)$$

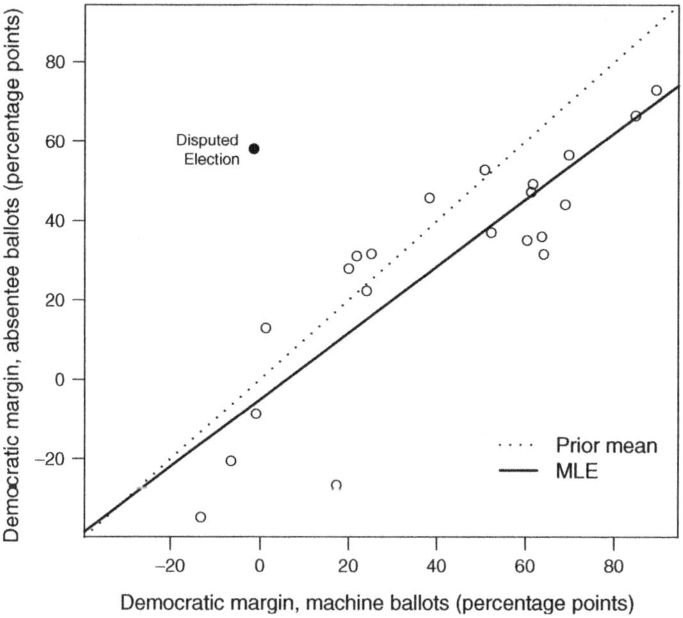

Figure 2.14 Pennsylvania State Senate Election data. The regression line corresponding to the mean of the posterior density for $\boldsymbol{\beta}$ is visually indistinguishable from the MLEs.

Data and likelihood. The data are shown in Figure 2.14, with the disputed election highlighted. The disputed election is excluded from the analysis of the 21 historical data points. I make the standard regression assumptions (see Section 2.1.2), such that conditional on x_i, $\boldsymbol{\beta}$ and σ^2, the y_i are iid normal with mean $\beta_0 + \beta_1 x_i$ and variance σ^2. The corresponding likelihood function is provided in Equation 2.26, for which the maximum likelihood estimate of $\boldsymbol{\beta}$ is $\hat{\boldsymbol{\beta}}_{\text{MLE}} = (\mathbf{X}'\mathbf{X})^{-1}\mathbf{X}'\mathbf{y} = (-5.08, 0.84)'$ with

$$(\mathbf{X}'\mathbf{X})^{-1} = \begin{bmatrix} .136 & -.00217 \\ -.00217 & .0000532 \end{bmatrix}.$$

The sum of the squared residuals is $S = (\mathbf{y} - \mathbf{X}\hat{\boldsymbol{\beta}})'(\mathbf{y} - \mathbf{X}\hat{\boldsymbol{\beta}}) = 4162.4$, and so the MLE of σ^2 is $S/n = 4162.4/21 = 198.21$. The frequentist standard errors of $\hat{\boldsymbol{\beta}}_{\text{MLE}}$ come from taking the square root of the diagonal of the estimated variance-covariance matrix $\hat{\sigma}^2(\mathbf{X}'\mathbf{X})^{-1}$, yielding $\text{se}(\hat{\beta}_0) = 5.19$ and $\text{se}(\hat{\beta}_1) = .10$. That is, a maximum likelihood analysis would conclude that the intercept parameter is not distinguishable from zero at conventional levels of statistical significance, while the slope parameter of .84 is about 1.6 standard errors less than 1.0, suggesting that as Democractic vote share increases, increasingly less of the Democratic advantage as recorded via voting machines is apparent in the absentee ballots. All in all, the MLEs do not conflict with the relatively vague prior information about $\boldsymbol{\beta}$, and, as we shall now see, this will mean that the shape of the posterior density is largely determined by the likelihood function.

Posterior density. With the conjugate, normal/inverse-Gamma prior given above, and the normal regression model assumed for the data, then via Proposition 2.11, the posterior density for $\boldsymbol{\beta}$ is, conditional on σ^2, a multivariate normal density with mean vector

$$\begin{aligned}
\mathbf{b}_1 &= (\mathbf{B}_0^{-1} + \mathbf{X}'\mathbf{X})^{-1}(\mathbf{B}_0^{-1}\mathbf{b}_0 + \mathbf{X}'\mathbf{X}\hat{\boldsymbol{\beta}}) \\
&= \left(\begin{bmatrix} 2.25 & 0 \\ 0 & .0022 \end{bmatrix}^{-1} + \begin{bmatrix} 21 & 854 \\ 854 & 53\,500 \end{bmatrix} \right)^{-1} \\
&\quad \times \left(\begin{bmatrix} 2.25 & 0 \\ 0 & .0022 \end{bmatrix}^{-1} \begin{bmatrix} 0 \\ 1 \end{bmatrix} + \begin{bmatrix} 21 & 854 \\ 854 & 53\,500 \end{bmatrix} \begin{bmatrix} -5.08 \\ 0.84 \end{bmatrix} \right) \\
&= \begin{bmatrix} -4.95 \\ 0.84 \end{bmatrix},
\end{aligned}$$

and variance-covariance matrix

$$\begin{aligned}
\sigma^2\mathbf{B}_1 &= \sigma^2(\mathbf{B}_0^{-1} + \mathbf{X}'\mathbf{X})^{-1} \\
&= \sigma^2 \left(\begin{bmatrix} 2.25 & 0 \\ 0 & .0022 \end{bmatrix}^{-1} + \begin{bmatrix} 21 & 854 \\ 854 & 53\,500 \end{bmatrix} \right)^{-1} \\
&= \sigma^2 \begin{bmatrix} .126 & -.002 \\ -.002 & 5.01 \times 10^{-5} \end{bmatrix}.
\end{aligned}$$

Proposition 2.11 also lets us determine that the marginal posterior density for σ^2 is an inverse-Gamma density with parameters $\nu_1/2$ and $\nu_1\sigma_1^2/2$, where

$$\begin{aligned}
\nu_1 &= \nu_0 + n = 6.2 + 21 = 27.2 \\
\nu_1\sigma_1^2 &= \nu_0\sigma_0^2 + S + r = 6.2 \cdot 47.02 + 4162.4 + 22.95 = 4477.2,
\end{aligned}$$

recalling from Proposition 2.11 that $r = (\mathbf{b}_0 - \hat{\boldsymbol{\beta}})'(\mathbf{B}_0 + (\mathbf{X}'\mathbf{X})^{-1})^{-1}(\mathbf{b}_0 - \hat{\boldsymbol{\beta}})$. From Definition B.35, we can compute the mean of this marginal posterior density for σ^2,

$$E(\sigma^2|\mathbf{y}, \mathbf{X}) = \frac{\nu_1\sigma_1^2/2}{\nu_1/2 - 1} = \frac{\nu_1\sigma_1^2}{\nu_1 - 2} = \frac{4477.2}{27.2 - 2} = 177.7,$$

and its mode,

$$\frac{\nu_1\sigma_1^2}{\nu_1 + 2} = \frac{4477.5}{27.2 + 2} = 153.3.$$

The mode of the marginal posterior density for σ^2 is slightly smaller than the maximum likelihood estimate $\hat{\sigma}^2 = 198.21$, reflecting the effect of the prior density for σ^2, which puts more relatively weight on values of σ^2 closer to zero than the likelihood (see the lower left panel of Figure 2.15).

Proposition 2.12 states that the marginal posterior density for $\boldsymbol{\beta} = (\beta_0, \beta_1)'$, $p(\boldsymbol{\beta}|\mathbf{y}, \mathbf{X})$ is a bivariate t density, as is the marginal prior density $p(\boldsymbol{\beta})$. These two densities are represented graphically in the top two panels of Figure 2.15, with darker colors indicating regions of higher density. The marginal posterior density (top right panel) is much more concentrated than the prior marginal density, since the information about the model

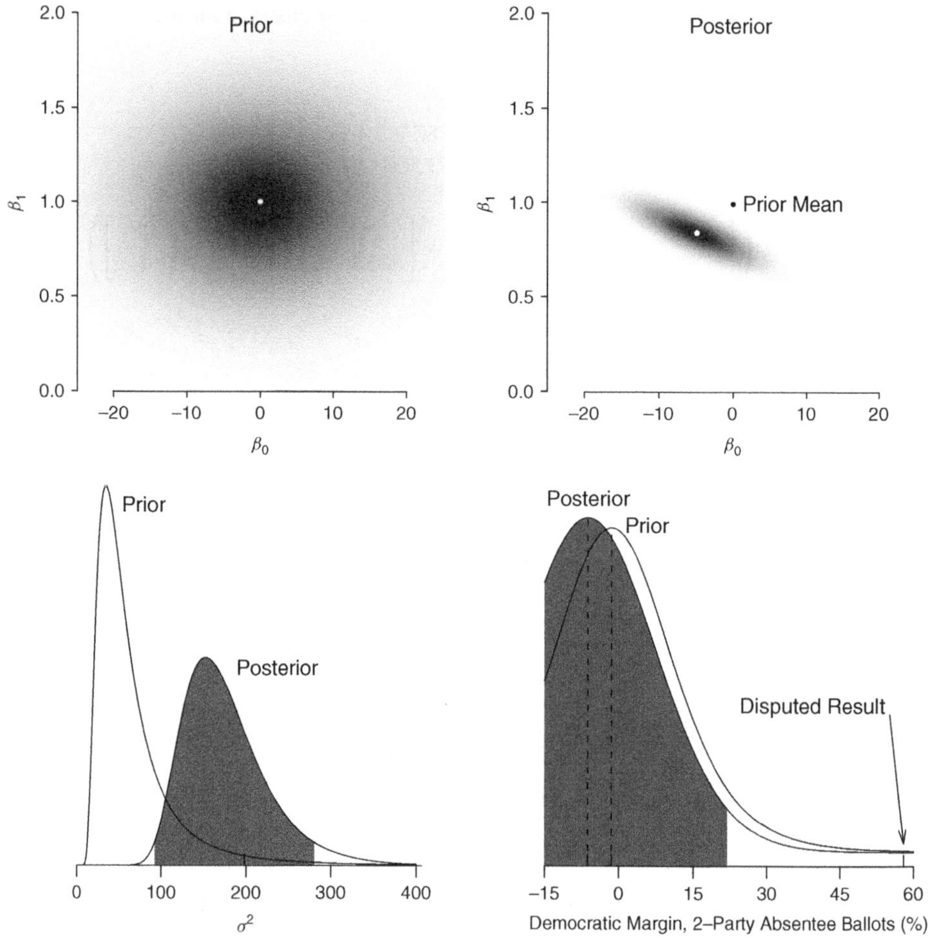

Figure 2.15 Prior and posterior densities, regression analysis of Pennsylvania State Senate Data. The top left panel shows the marginal prior density for the intercept (β_0) and slope (β_1) parameters (a bivariate t density), with darker colors indicating regions of higher density and the mode indicated with a white dot. The top right panel shows the marginal posterior density for β_0 and β_1, again, a bivariate t density. The lower left panel shows the marginal prior and posterior densities for σ^2, with a posterior 95 % highest density region shaded gray. The lower right panel shows the prior and posterior predictive densities given with $x = -1.45$, the value recorded in the disputed election; the Democratic absentee vote margin recorded in the disputed election lies far outside the 95 % highest density region of the posterior predictive density.

parameters contained in the data is much more precise than the prior information. In fact, the MLEs and the posterior mode of $\boldsymbol{\beta}$ are extremely close to one another, again reflecting the facts that the posterior density is almost completely determined by the likelihood, and that the prior information is relatively imprecise. Note also that although prior information about β_0 and β_1 is independent (the shading in the top left panel of Figure 2.15 has a circular appearance), the information in the data about these parameters is not independent (i.e. $\mathbf{X}'\mathbf{X}$ is not a diagonal matrix). Consequently, the marginal posterior density for $\boldsymbol{\beta}$ i(top right panel) has a pronounced elliptical shape, with the "top-left to bottom-right" orientation of the principal axis of the ellipse reflecting the negative covariance between β_0 and β_1 in the likelihood (i.e. the off-diagonal element of $\mathbf{X}'\mathbf{X}$ is positive, equal to the $\sum_{i=1}^{n} x_i$, and the off-diagonal element of $(\mathbf{X}'\mathbf{X})^{-1}$ is negative).

A tabular, numerical summary of the prior and posterior densities for this problem appears in Table 2.5. The posterior densities obtained via the proper, normal/inverse-Gamma prior described above closely correspond to those produced by the improper, reference prior $p(\boldsymbol{\beta}, \sigma^2) \propto 1/\sigma^2$; the means of the posterior densities for $\boldsymbol{\beta}$ are almost identical. The proper prior for σ^2 is reasonably informative, putting more weight on lower values of σ^2 than does the likelihood function, and so the marginal posterior densities for σ^2 differ slightly.

Table 2.5 Numerical summaries of prior and posterior densities, regression analysis of Pennsylvania State Senate Elections. Cell entries are the means of the corresponding densities, with the range of the respective 95 % marginal HPDs in brackets. The improper prior is $p(\boldsymbol{\beta}, \sigma^2) \propto 1/\sigma^2$ (see Section 2.5.4), and produces posterior means for $\boldsymbol{\beta}$ equal to the maximum likelihood estimates $\hat{\boldsymbol{\beta}} = (\mathbf{X}'\mathbf{X})^{-1}\mathbf{X}'\mathbf{y}$.

	Proper prior	Posterior (proper prior)	Posterior (improper prior)
Intercept (β_0)	0 [−25.0, 25.0]	−4.95 [−14.3, 4.40]	−5.08 [−16.5, 6.33]
Slope (β_1)	1 [.22, 1.78]	.84 [.65, 1.02]	.84 [.61, 1.07]
σ^2	56.1 [12.7, 169.0]	177.7 [92.5, 281.0]	231.2 [108.6, 420.5]
Prediction for disputed election ($\tilde{x} = -1.45$)	−1.45 [−31.5, 28.6]	−6.16 [−34.2, 21.8]	−6.30 [−39.4, 26.8]

Predictions for the disputed election. Of particular interest is what the historical data tells us about the outcome in the disputed election. We do this by using the fitted regression model to generate predictions for a new data point, where the value of the predictor is set to the value of x observed in the disputed election. Recall that in the disputed election, the Democratic candidate narrowly trailed based on the ballots cast by voting machine: $\tilde{x} = -1.45$. We use this value of x to generate a posterior density for the predicted Democratic margin among absentee ballots, $\tilde{y}|\tilde{x} = -1.45$, posterior in the sense of after the regression analysis of the historical data, $\mathbf{y}$ and $\mathbf{X}$. Proposition 2.14 tells us that the posterior density for $\tilde{y}$ given a new observation $\tilde{\mathbf{x}}$ is a t density with mean $\tilde{\mathbf{x}}\mathbf{b}_1$, squared scale parameter $\sigma_1^2(1 + \tilde{\mathbf{x}}\mathbf{B}_1\tilde{\mathbf{x}}')$ and ν_1 degrees of freedom, where $\tilde{\mathbf{x}} = (1, -1.45)$ is the vector formed by concatenating a 1 for the intercept term and the value of -1.45 for x recorded in the disputed election, and $\mathbf{b}_1$, $\mathbf{B}_1$, σ_1^2 and ν_1 are the parameters indexing the posterior density for $\boldsymbol{\theta} = (\boldsymbol{\beta}, \sigma^2)$, as discussed above. This posterior predictive density is shown in the lower right panel of Figure 2.15. With $\mathbf{b}_1 = (-4.95, 0.84)'$, the posterior predictive density for $\tilde{y}|\tilde{x} = -1.45$ is centered at -6.16 %, while the actual y value recorded in the disputed election is 58.0 %. A 95 % highest density region for the posterior predictive density extends up to 21.8 %, indicating that the result actually recorded is highly improbable given the regression modeling of the historical data. The posterior probability of observing a 58 % (or higher) Democratic margin among absentee ballots given a -1.45 % margin

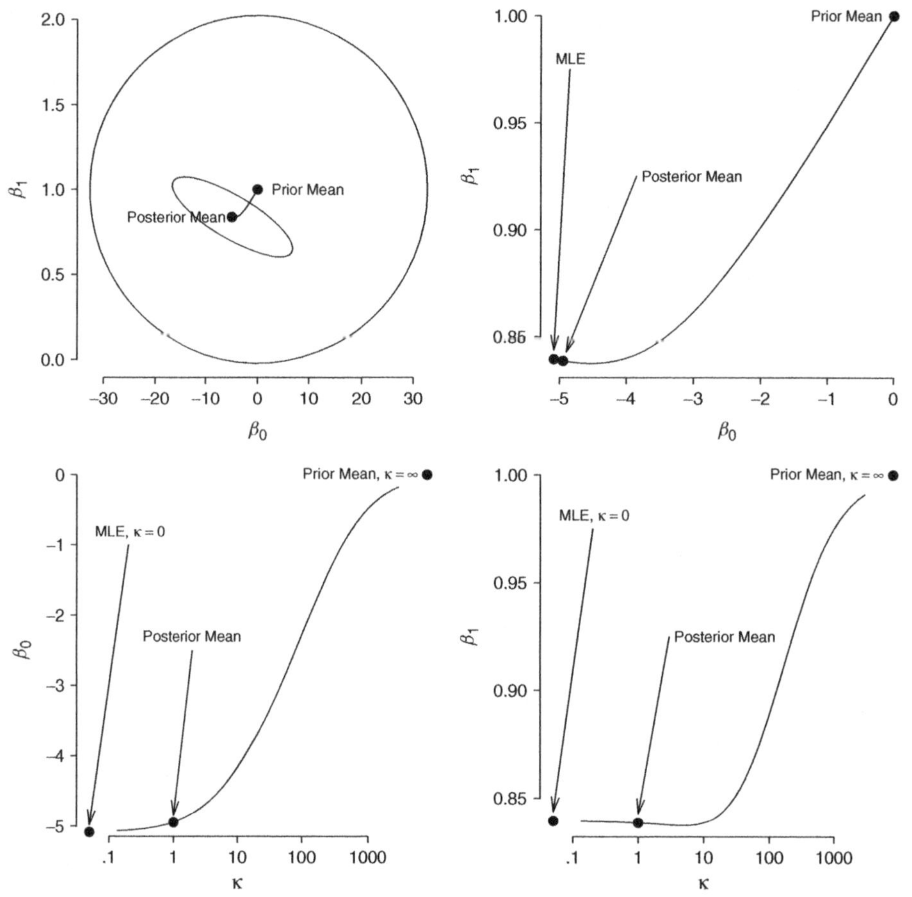

Figure 2.16 Sensitivity analysis, regression analysis of Pennsylvania State Senate Election Data. The top left panel shows 95 % highest density regions for the prior and posterior marginal densities of $\boldsymbol{\beta} = (\beta_0, \beta_1)'$, with a line connecting the mean of the prior marginal density and the mean of the posterior marginal density. The prior marginal HDR is the larger, circular region; the posterior marginal HDR is the smaller ellipse (see also Figure 2.15). The top right panel "zooms in" on the line connecting the prior mean and the maximum likelihood estimates of $\boldsymbol{\beta}$, showing the line to be a quadratic function, with the posterior mean located on this line, but close to the MLEs. The lower panels show the results of a sensitivity analysis, letting the means of the marginal posterior densities for β_0 and β_1 vary as a function of the precision of the prior density, controlled via a hyper-parameter κ. As $\kappa \to 0$, the prior becomes less precise, and the MLEs dominate the posterior density; as $\kappa \to \infty$, the prior swamps the data. The prior used in Example 2.15 results with $\kappa = 1$.

among machine ballots is just 3.35×10^{-5}, strongly suggesting that this disputed absentee ballot margin was generated by a process other than the one that generated the historical data.

Sensitivity analysis. We conclude this example with some brief examination of how the results reported above change given a range of prior densities. Recall that conditional on σ^2, the prior density of $\boldsymbol{\beta}$ is a $N(\mathbf{b}_0, \sigma^2\mathbf{B}_0)$ density, where $\mathbf{B}_0$ is a diagonal matrix. For the purposes of the sensitivity analysis, I consider more or less precise versions of the prior density used above, replacing the (diagonal) matrix $\mathbf{B}_0$ with the matrix $1/\kappa\mathbf{B}_0$, where $0 < \kappa < \infty$. As $\kappa \to 0$, the prior density for $\boldsymbol{\beta}$ becomes less precise, and the posterior density is dominated by the contribution from the likelihood function; as $\kappa \to \infty$, the prior becomes increasingly more precise, and overwhelms the information in the data. Thus, for any given value of κ, the mean of the posterior density for $\boldsymbol{\beta}$ is a precision-matrix weighted average of the maximum likelihood estimates and the expectation of $\boldsymbol{\beta}$ under the prior, as demonstrated in the various panels of Figure 2.16; as κ varies, so too does the resulting posterior mean of $\boldsymbol{\beta}$, tracing out the quadratic function shown in the top right panel of Figure 2.16. The prior actually deployed in the analysis above corresponds to $\kappa = 1$, and produces a posterior mean for $\boldsymbol{\beta}$ very close to the MLEs: i.e. in this case, the prior information is not especially stringent and does not conflict with the information in the likelihood (see especially the top left panel of Figure 2.16).

There always exists a prior that could overwhelm the information in the data; in this case, one could find a prior that would make the result in the disputed election seem plausible, but the prior adopted here does not do that. As the lower right panel of Figure 2.15 makes clear, even under the (reasonably vague) prior density, the result in the disputed election is extremely improbable. It would seem that the only way to make the disputed election result seem *a posterior* plausible would be via a very precise set of prior beliefs that there is no relationship between absentee ballot margins and voting machine margins.

In the next example, we see how Bayesian analysis provides a way for non-quantitative information to be formally introduced into a regression analysis, via a conjugate prior density for $\boldsymbol{\beta}$ and σ^2. Unlike the previous example, the data are relatively uninformative about the parameters of interest, and accordingly, the posterior is quite sensitive to the prior.

■ Example 2.16

Cross-national analysis of trade union density. Western and Jackman (1994) present a Bayesian re-analysis of an exchange between Stephens and Wallerstein (1991) concerning the determinants of cross-national variation in trade union density (the percentage of a work force that are members of trade unions). This variable is of particular interest in the field of comparative political economy, since it is linked to outcomes such as income equality, the strength of political parties of the Left, and welfare state expenditure, to name just a few (e.g. Stephens and Wallerstein 1991, 941). Both authors in the debate have a key predictor variable, motivated by theoretical and historical arguments. For Wallerstein, the size of the civilian labor force is key to understanding cross-national variation in union density. Wallerstein argues that the benefits to the union movement from organizing (union bargaining power in industrial disputes) depend on the proportion of the labor force being organized, but that the cost of organizing is an increasing function of the number of new members recruited, implying that the optimal union density is inversely related to the absolute size of the labor force. Thus, bigger countries with bigger labor forces can be expected to have relatively smaller union densities, while smaller countries can be expected to have relatively larger union densities. Stephens proposes a different mechanism. Population size largely determines the size of the domestic market for a country's economic output, which, in turn, is the main determinant of industrial concentration. It is this variable – industrial concentration – that determines union density, since it is less costly to organize workers when the labor force is relatively concentrated in a smaller number of sectors and/or firms. Additional detail appears in Stephens (1979) and Wallerstein (1989).

The data set available for analysis is small, comprising just 20 observations, an almost exhaustive subset of the 21 advanced industrial countries that have experienced a continuous history of democracy since World War Two (Luxembourg is excluded for want of data on union density). Immediately we recognize that these data are not a sample in the conventional sense of being randomly selected from a population. Frequentist inference – which rests on the repeated sampling properties of estimators – is simply not appropriate in this case. Since the data set is so small, it is replicated in Table 2.6 (sorted by levels of union density, in descending order), along with a listing of data sources and definitions. Regression analysis is used to assess the effects of the various, hypothesized sources of variation in union density. Both Stephens and Wallerstein agree that left governments are an important determinant of union density and include it in their regression analyses. The dispute between the authors concerns the two other variables: size of the labor force (Wallerstein's preferred predictor) and economic concentration (Stephen's preferred predictor). Size of the labor force enters the analysis in logged form, and so the regression model is actually non-linear with respect to absolute labor force size: if y is union density, and $\beta < 0$ is the regression coefficient on $\log x$, the log of labor force size, then $\partial y/\partial x = \beta/x$, which, since $\beta < 0$ is negative, but approaching zero as $x \to \infty$. Put differently, as labor force size increases, and other determinants are held constant, the expected union density decreases, but at a diminishing marginal rate. Economic concentration enters the regression model "as is", but noting that the variable is measured as a ratio relative to economic concentration in the United States (see Table 2.6). That is, the regression model to be considered is

$$E(y_i|L_i, S_i, C_i) = \beta_0 + \beta_1 L_i + \beta_2 S_i + \beta_3 C_i \qquad (2.28)$$

where i indexes the $n = 20$ countries in the data set, y_i is union density, L_i is the measure of left government, S_i is logged labor-force size, and C_i is the economic concentration ratio.

A noteworthy feature of the data is presented in Figure 2.17. The two predictors at the source of the Stephens-Wallerstein exchange correlate at −.92. According to Stephens, "because of multicollinearity, economic concentration and the size of the labor force could not be entered in the same equation" (Stephens and Wallerstein 1991, 945–6). There is no statistical or mathematical justification for Stephens' statement: only in the case of *perfect* multicollinearity does regression analysis become computationally infeasible, with the $\mathbf{X}'\mathbf{X}$ matrix singular such that $(\mathbf{X}'\mathbf{X})^{-1}$ does not exist. Stephens is simply observing

Table 2.6 Cross-national data on union membership and its determinants, as presented in Stephens and Wallerstein (1991). Union density is defined as a percentage of the total number of wage and salary earners plus the unemployed, measured between 1975 and 1980, with most of the data drawn from 1979. Left Government is an index tapping the extent to which parties of the left have controlled governments since 1919, due to Wilensky (1981). Size of the labor force is the number of wage and salary earners plus the unemployed, in thousands. Economic concentration is the percentage of employment, shipments, or production accounted for by the four largest enterprises in a particular industry, averaged over industries (with weights proportional to the size of the industry) and the resulting measure is normalized such that the United States scores a 1.0, and is due to Pryor (1973). Economic concentration data marked with an asterisk indicates an imputed value, generated using procedures described in Stephens and Wallerstein (1991, 945). The data set is supplied as `unionDensity` in the `R` package `pscl`.

Country	Union density	Left government	Size of labor force	Economic concentration
Sweden	82.4	111.84	3931	1.55
Israel	80.0	73.17	997	1.71*
Iceland	74.3	17.25	81	2.06*
Finland	73.3	59.33	2034	1.56*
Belgium	71.9	43.25	3348	1.52
Denmark	69.8	90.24	2225	1.52*
Ireland	68.1	0.0	886	1.75*
Austria	65.6	48.67	2469	1.53*
NZ	59.4	60.0	1050	1.64*
Norway	58.9	83.08	1657	1.58*
Australia	51.4	33.74	5436	1.37*
Italy	50.6	0.0	15 819	0.86
UK	48.0	43.67	25 757	1.13
Germany	39.6	35.33	23 003	0.92
Netherlands	37.7	31.50	4509	1.25
Switzerland	35.4	11.87	2460	1.68
Canada	31.2	0.0	10 516	1.35
Japan	31.0	1.92	39 930	1.11
France	28.2	8.67	18 846	0.95
USA	24.5	0.0	92 899	1.00

what many of us have recognized when working with correlated predictors: the data provide relatively little information about the effect of each predictor, and the estimated coefficients may conflict with prior expectations as to their signs and magnitudes.

Prior information. Western and Jackman (1994) note that both Stephens and Wallerstein make informal use of prior information in their regression analyses of these data (here I focus on the prior information relied upon by Stephens). For instance, a common "solution" to the "problem" of multicollinearity is to exclude one of the correlated predictors, tantamount to a Bayesian analysis in which the researcher has strong prior information that the effect of the included variable is not zero, and that the effect of the excluded

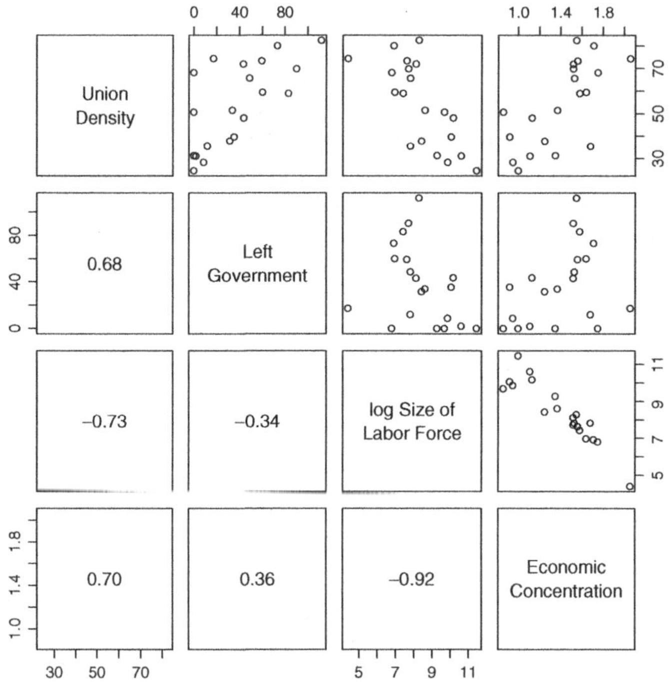

Figure 2.17 Scatterplot and correlation matrix, union density and its determinants. Pearson correlations are reported below the diagonal; pairwise scatterplots appear above the diagonal. Note the strong linear relationship between economic concentration and log of labor force size ($r = -.92$).

variables *is* zero. Wallerstein develops expectations about the effect of labor-force size from more primitive propositions about relations between firms, unions and workers in labor markets. Stephens develops his expectation as to the effect of economic concentration from historical accounts of the trajectory of union densities over time in several countries, particularly those provided by Kjellberg (1983). Indeed, Stephens notes that

> [t]he impasse in the statistical analysis [that is, the high level of collinearity between economic concentration and log of labor-force size] suggests it is necessary to examine comparative historical evidence on the causes of cross-national differences in unionization" (Stephens and Wallerstein 1991, 948).

That is, Stephens sees the historical narratives and case studies as providing information about the impact of economic concentration on union densities beyond that in the quantitative, cross-national data set used in the regression analysis. Western and Jackman (1994) attempt to formalize this supplemental, historical information as a prior density over the regression parameters. Stephens (and Wallerstein) believe the historical record strongly suggests that governments comprised of parties of the Left promote union growth, and Western and Jackman quantify that prior belief by suggesting that it implies that one additional year of left-wing government would increase union density by about 1 percentage point, and this regression effect has only a small probability of being non-positive (just

2.5 %). That is, returning to the notation introduced in the regression model in Equation (2.28), Western and Jackman adopt a prior density on β_1 that is centered on .3 (one extra year of left wing government increases the Left Government index by a factor of 3.33) but the prior density assigns only probability .025 to $\beta_1 \leq 0$. The historical data Stephens relies on has relatively little to say about the role of labor force size: according to Stephens, Kjellberg's (1983) comparative historical analysis indicates that "any direct influence of the industrial structure (or [labor force] size) came before 1930". In Western and Jackman's reanalysis, they quantify this absence of prior information with a vague prior density for β_2, centered on 0, but with a massive variance, 10^{12}, essentially supplying *no* prior information about that parameter.

Quantifying the historical information about the effect of economic concentration is more difficult, but Western and Jackman look at the way union density declined in the 1980s in response to delincing levels of economic concentration.

> Measuring economic concentration as the average size of British manufacturing establishments as a ratio of the size of American firms at a fixed point in time, economic concentration in Britain declined by about .3 from the late 1970s to the late 1980s (See the United Nations Yearbook for Industrial Statistics for 1982 and 1991). Union density in this period [in the United Kingdom] declined by about 15 percentage points. If economic concentration generated about a fifth, or 3 percentage points, of the decline, a plausible prior mean would $3/.3 = 10\ldots$ (Western and Jackman 1994, 418)

That is, Western and Jackman suggest that the prior mean of β_3 is 10, and assign only .025 prior probability to $\beta_3 < 0$. Westerm and Jackman also use a vague prior density for the intercept parameter, β_0, assigning that prior density a large variance.

This information is not sufficient to specify a conjugate normal/inverse-Gamma prior for all the parameters in the regression analysis. A regression analysis with three predictors and an intercept term has *five* parameters, $\boldsymbol{\theta} = (\boldsymbol{\beta}, \sigma^2) = (\beta_0, \ldots, \beta_3, \sigma^2)$, and a conjugate, normal/inverse-Gamma prior density is $\boldsymbol{\beta}|\sigma^2 \sim N(\mathbf{b}_0, \sigma^2\mathbf{B}_0)$ and $\sigma^2 \sim$ inverse-Gamma$(\nu_0/2, \nu_0\sigma_0^2/2)$ where in this case

- $\mathbf{b}_0$ is a 4-by-1 vector of parameters, the mean of the prior density for $\boldsymbol{\beta}$, which according to Western and Jackman is $\mathbf{b}_0 = (0, .3, 0, 10)'$;
- $\mathbf{B}_0$ is a 4-by-4 symmetric matrix and so has as many as 10 unique parameters;
- ν_0 is a prior degrees of freedom parameter;
- $\nu_0\sigma_0^2$ is equivalent to a prior "sum of squared errors".

A helpful simplification is to assume that prior information about the regression parameters is uncorrelated, such that $\mathbf{B}_0$ is a diagonal matrix (i.e. has zeroes on the off-diagonal elements, see Definition A.7). This means that once we have specified values for ν_0 and σ_0^2, we can use the quantitative information provided by Western and Jackman about the dispersion of the marginal prior densities of $\beta_0, \ldots, \beta_3$ to complete the specification of the normal/inverse-Gamma prior. I proceed as in Example 2.15, first specifying values for ν_0 and σ_0^2 and then using the fact that the marginal prior density of β_j, $j = 0, \ldots, 3$, is a t density with ν_0 degrees of freedom, centered over b_{0j} and with scale $\sigma_0\sqrt{\lambda_j}$, where $\mathbf{B}_0 = \text{diag}(\lambda_0, \ldots, \lambda_3)$.

I assume that the historical information relied on by Stephens is somewhat imprecise, coming from the equivalent of a regression run on a relatively small number of cases: Kjellberg (1983) examined the histories of 12 countries, and with 4 regression parameters, I equate this $\nu_0 = 8$ degrees of freedom. I specify a value for σ_0^2 by assuming that the "prior regression analysis" corresponding to Kjellberg (1983) comparative historical analysis provided only a moderate fit to the union density data, with, say, 95 % of the data within plus or minus 30 percentage points of the fitted regression surface (plus or minus 30 percentage points would seem, *a priori*, to span a large range of the values for union densities, given that the densities are expressed as percentages). The .975 quantile of a t density with $\nu_0 = 8$ degrees of freedom is $q = 2.31$, and so this judgment about the fit of the "prior regression" amounts to the assessment that $q\sigma_0 = 30$, or that $\sigma_0 = 30/q = 13.01$, or that $\sigma_0^2 = 169.2$. Thus, given that β_1 has a marginal prior 95 % interval ranging from 0 to .6, we infer that the scale of the marginal prior t density for β_1 is $.3/q = \sigma_0\sqrt{\lambda_1}$; since $\sigma_0 = 30/q$ it follows that $\lambda_1 = (.3/30)^2$ or 10^{-4}. The corresponding calculation for β_3, the coefficient on industrial concentration, yields $\lambda_3 = 1/9$. For the elements of $\boldsymbol{\beta}$ with vague, uninformative priors, β_0 and β_2, I set $\lambda_0 = \lambda_2 = 10^8$. To reiterate, my normal/inverse-Gamma prior for this problem is characterized by the parameters

- $\mathbf{b}_0 = (0, .3, 0, 10)'$
- $\mathbf{B}_0 = \text{diag}(10^8, 10^{-4}, 10^8, 1/9)$
- $\nu_0 = 8$ and $\sigma_0^2 = 169.2$.

Posterior density. Assuming a normal likelihood for these data, Proposition 2.11 tells us that the posterior density for $\boldsymbol{\theta} = (\boldsymbol{\beta}, \sigma^2)'$ is

$$\boldsymbol{\beta}|\sigma^2, \mathbf{y}, \mathbf{X} \sim N(\mathbf{b}_1, \sigma^2\mathbf{B}_1)$$

$$\sigma^2 \sim \text{inverse-Gamma}\left(\frac{\nu_1}{2}, \frac{\nu_1\sigma_1^2}{2}\right)$$

where $\mathbf{b}_1 = (69.47, 0.28, -4.70, 9.69)'$,

$$\mathbf{B}_1 = \begin{bmatrix} 2.683 & -0.003 & -0.247 & -0.313 \\ -0.003 & 0.000 & 0.000 & -0.000 \\ -0.247 & 0.000 & 0.025 & 0.019 \\ -0.313 & -0.000 & 0.019 & 0.108 \end{bmatrix},$$

$\nu_1 = 28$ and $\sigma_1^2 = 111.02$.

To grasp what this posterior density is saying about the inferential problem at hand, it is more instructive to examine the prior and posterior marginal highest density regions for each regression coefficient. Table 2.7 presents numerical summaries of the prior density described above, and two posterior densities: one resulting from an improper, reference prior density (such that the mean of the posterior density for $\boldsymbol{\beta}$ is equal to the maximum likelihood estimates, $\hat{\boldsymbol{\beta}} = (\mathbf{X}'\mathbf{X})^{-1}\mathbf{X}'\mathbf{y}$), and the posterior density resulting from the proper, normal/inverse-Gamma prior density described above. The difficulty posed by the strong correlation (−.92) between the log of Labor Force Size and Economic

Table 2.7 Numerical summaries of prior and posterior densities, regression analysis of cross-national rates of union density. Cell entries are the means of the corresponding densities, with the range of the respective 95 % marginal HPDs in brackets. The improper prior is $p(\boldsymbol{\beta}, \sigma^2) \propto 1/\sigma^2$ (see Section 2.5.4), and produces posterior means for $\boldsymbol{\beta}$ equal to the maximum likelihood estimates $\hat{\boldsymbol{\beta}} = (\mathbf{X}'\mathbf{X})^{-1}\mathbf{X}'\mathbf{y}$.

	Prior	Posterior (improper prior)	Posterior (proper prior)
Intercept (β_0)	0.0 $[-3 \times 10^5, 3 \times 10^5]$	97.6 [−24.3, 219]	69.5 [34.1, 105]
Left Government (β_1)	0.3 [0, .6]	.27 [.11, .43]	.28 [.15, .40]
log Labor Force Size (β_2)	0 $[-3 \times 10^5, 3 \times 10^5]$	−6.46 [−14.5, 1.58]	−4.70 [−8.12, −1.28]
Economic Concentration (β_3)	10 [0, 20]	.351 [−40.5, 41.2]	9.69 [2.61, 16.8]
σ^2	225 [54, 501]	122 [50, 220]	120 [63, 188]

Concentration is evident in the 2nd column of Table 2.7: the HPDs for β_2 and β_3 both overlap zero, and, in particular, β_3 has a massive 95 % HPD. That is, absent any prior information, regression analysis of the data yields no meaningful information as to the relative contributions of log labor force size (β_2) and economic concentration (β_3).

The effect of incorporating the prior information (summarized in column one of Table 2.7) is apparent from inspecting column three. All three predictors have posterior densities that imply sizeable effects. The left government coefficient (β_1) has a posterior mean of .28 and a marginal posterior 95 % HDR ranging from .15 to .40 (i.e. we attach virtually zero posterior probability to the possibility that $\beta_1 \leq 0$); these results closely correspond to the results obtained without any prior information (column 2 of Table 2.7), reflecting the fact that the proper prior largely agrees with the likelihood. But the story is starkly different for the parameters β_2 and β_3. For these two parameters, the introduction of prior information is especially consequential. Even though the proper prior supplies virtually zero information about the marginal density of β_2, the posterior density for β_2 is centered at −4.7, and has a 95 % HPD that does not overlap zero. Most interestingly, *the posterior density for β_3 almost coincides with the prior*, again, a consequence of the fact that there is next to no information about the effect of β_3 in the data: the mean of the prior density for β_3 is 10 while the mean of the posterior density is 9.69, and the marginal prior 95 % HPD ranges from zero to 20, while the marginal posterior 95 % HPD ranges from 2.6 to 16.8. Here, then, is an example where the posterior density is *heavily* determined by the prior, as will always be the case when the data are as uninformative as they in this case (i.e. a small number of observations, with two or more highly collinear predictors).

To recap, the sample data can not resolve the relative contributions of two "competing" predictors. A typical "solution" is to drop one or other of the predictors, equivalent to employing a "spike" prior at zero for the coefficient on the excluded variable, but these seems scientifically unrealistic and heavy-handed. Frankly, a better solution is to simple acknowledge the multicollinearity, to concede that the available data can not resolve the competition between the two predictors, and to think about acquiring more, or different data. But, as is often the case in a setting like comparative politics, additional data collection is problematic: the data available for analysis exhausts the relevant population (e.g. the set of advanced industrial democracies with union movements), and extending the analysis back in time is infeasible for want of useable measures. To the extent that additional, earlier data does exist, it is in the form of historical narratives and series of case studies, of the sort provided by Kjellberg (1983) and relied on by Stephens (1991). But, as this example demonstrates, the Bayesian approach offers the possibility of systematically exploiting that additional information, by converting into probability densities (priors) over the parameters of the regression model.

Sensitivity analysis. The role of the prior information is critical in this example. Without incorporating the supplemental, historical data via the prior density, the regression analysis yields extremely imprecise estimates of the effects of log labor force size (β_2) and economic concentration (β_3). With the historical information incorporated via a prior density and Bayes Rule, we conclude that both predictors have large effects: as labor forces get larger, rates of union density decline (other things being held constant), and as economic concentration increases, rate of union density increase (again, other thing being held constant). But, as in any Bayesian analysis, it is worthwhile exploring the sensitivity of the conclusions to the specification of the prior.

This type of sensitivity analysis seems all the more pressing in the current example, where Western and Jackman's quantification of Kjellsberg's (1983) historical analysis can hardly be considered authoritative. That is, different people could well differ on the *relevance* of the Kjellberg (1983) analysis, and in particular, as to the particular normal/inverse-Gamma prior over $\boldsymbol{\theta} = (\boldsymbol{\beta}, \sigma^2)'$ implied by the case studies examined by Kjellberg. If seemingly small changes to the parameters of the prior density produce large changes in the posterior density, then this should be clearly noted. As in Example 2.15, I produce a sensitivity analysis by making the prior density over $\boldsymbol{\beta}$ more or less precise via a tuning parameter, $\kappa > 0$, where the marginal prior density for $\boldsymbol{\beta}$ becomes a multivariate t density with ν_0 centered over the prior mean, $\mathbf{b}_0$, but with squared scale matrix $\sigma_0^2/\kappa\mathbf{B}_0$. As $\kappa \to \infty$, the prior becomes increasingly precise, collapsing to a "spike" on $\mathbf{b}_0$, as will the posterior density. On the other hand, as $\kappa \to 0$, the prior tends towards impropriety, with the marginal prior variances for the elements of $\boldsymbol{\beta}$ tending towards ∞, and so posterior inferences about $\boldsymbol{\beta}$ will be completely driven by the information in the data.

Figure 2.18 displays how the marginal posterior densities for β_2 and β_3 change as a function of the "prior precision hyperparameter", κ. These marginal posterior densities are t densities with ν_1 degrees of freedom, but with location and scale parameters that depend on κ. The areas shaded gray are marginal posterior 95 % highest density regions, that become smaller as $\kappa \to \infty$, reflecting the gains in posterior precision that result from using an increasingly precise prior density (recall that in a conjugate Bayesian analysis, posterior precision is usually the sum of prior precision and data precision). The results reported in column three of Table 2.7 correspond to $\kappa = 1$. Figure 2.18 makes clear that as the prior precision is gradually relaxed (as κ takes on values smaller

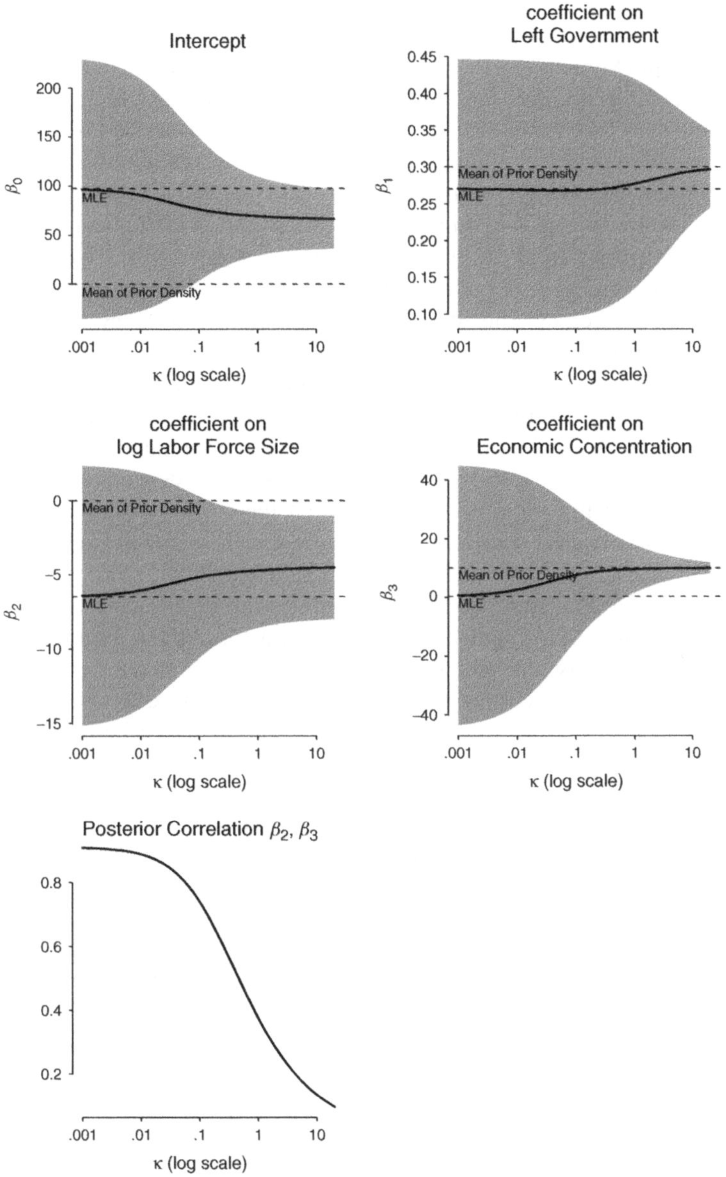

Figure 2.18 Sensitivity analysis, regression analysis of cross-national rates of union density. The solid lines in the top four panels show how the means of the marginal posterior densities for the four regression coefficients vary as a function of the precision of the prior density, controlled via κ. As $\kappa \to 0$, the prior becomes less precise, and the MLEs dominate the posterior density; as $\kappa \to \infty$, the prior swamps the data. The prior used in Example 2.16 results with $\kappa = 1$, producing the results in column three of Table 2.7; the results in column two of Table 2.7 are obtained with $\kappa = 0$. The gray areas vertically span 95 % highest density regions. The lower panel shows how the posterior correlation between β_2 and β_3 varies as a function of κ.

than 1.0), posterior precision quickly diminishes. For β_3, the coefficient on economic concentration, at $\kappa \approx .7$, the marginal posterior 95 % HPD on β_3 begins to overlap zero and grows quite quickly as $\kappa \rightarrow 0$. For β_2, the coefficient on log labor force size, recall that there is essentially no prior information about this coefficient. But because the prior information about β_3 is independent of the vague prior information about β_2, the result is that there is less "posterior multicollinearity" than the multicollinearity that results from relying on the data alone (i.e. the posterior scale matrix $\mathbf{B}_1$ is "more diagonal" than $(\mathbf{X}'\mathbf{X})^{-1}$); see the lower panel of Figure 2.18, where the posterior correlation between β_2 and β_3 quickly grows from .37 with $\kappa = 1$, to .74 with $\kappa \approx .1$. That is, as the prior is made less informative ($\kappa \rightarrow 0$), the posterior covariance between β_2 and β_3 increases, posterior precision decreases, and the marginal posterior 95 % HDR for β_3 expands to overlap zero at $\kappa \approx .1$.

In short, it seems that with just a mild re-specification of the prior (e.g., $\kappa \approx .7$ or smaller), the strong multicollinearity of the data reasserts itself, making it difficult to gauge the independent contributions of the log labor force size and economic concentration as determinants of union density. The results do seem to depend sharply on how we choose to quantify the historical information about the role of economic concentration in the Kjellberg (1983) case studies, and in particular, how much precision we attach to the information in those case studies. This conclusion might seem a little unsatisfying. But the Bayesian approach used here has the virtue of allowing the supplemental, non-quantitative information to enter the analysis in a formal way, via the specification of the prior, and sensitivity analysis has the virtue of transparancy, making it very clear that when the data are as weak as they are in this case, the conclusions drawn from a data analysis will depend strongly on prior assumptions. Bayesian analysis offers no "magic bullet" to the problem of weak, collinear data, but it does lay bare the *process* by which prior assumptions will inevitably structure the results in such a case. Contrast the typical regression analysis in this case, where an appeal to "theory" and/or historical data is cited as justification for dropping one of the collinear variables from the analysis entirely, amounting to a use of prior information that manages to be both heavy-handed and informal: recall that dropping a variable from a regression analysis is equivalent to a dogmatic prior belief that its coefficient is exactly zero. If only because the Bayesian approach requires researchers to specify prior densities over the model parameters, it surely represents an improvement over less formal "solutions" to the problems posed by weak, collinear data.

2.6 Further reading

The results on Bayesian analysis of the linear regression model are not at all new. For instance, the result in Proposition 2.15 – Bayesian analysis under an improper reference prior – was stated by Jeffreys (1961) and Savage (1962). The result in Proposition 2.12 – under conjugacy, the marginal posterior density of a vector of regression coefficients in a normal regression model is a multivariate t density – can be found in Savage (1961). Box and Tiao (1973) remains a valuable reference on Bayesian analysis of regression models. Zellner (1971) and Leamer (1978) were influential texts for my teachers – and hence for me – with many results and insights on the application of the Bayesian approach to econometrics, especially the linear regression model.

Problems

2.1 I present to you a coin that I claim is fair.

1. Assign a prior over $\theta \in [0, 1]$, the probability that any given toss of the coin will result in a head. Given this prior, what is the prior odds ratio for the two hypotheses $H_0 : \theta < .5$ and $H_1 : \theta \geq .5$.

2. The coin is flipped and comes up heads. What is your posterior density over θ? What is your posterior odds ratio for the two hypotheses?

3. Use a graphical technique to display how your posterior beliefs are modified over a sequence of flips all coming up heads. That is, plot a Bayes estimate of θ against the number of coin flips (starting with zero flips). Generate another graph showing how the odds ratio is similarly evolving in response to the sequence of coin flips coming up heads.

2.2 What is the probability that your vote will be pivotal when you are voting over two alternatives with $2n$ other voters? Consider two scenarios: (a) each voter is assumed to vote for either alternative with probability $\pi = 1/2$; (b) π is subject to uncertainty, expressed with a uniform prior $\pi \sim \text{Unif}(0, 1)$. If you get stuck, see Chamberlain and Rothschild (1981). Comment on how the "prior makes a difference" for this problem, for various values of n.

2.3 Three polls are fielded on the weekend prior to a election, each using simple random sampling, producing the following estimates of vote intention for the incumbent candidate.

Poll	Support (%)	Sample size
A	52	350
B	49	500
C	51	650

1. Let the proportion of the electorate intending to vote for the incumbent be θ. Write out a plausible statistical model for each poll's results in which θ appears as a parameter.

2. Suppose you have no prior information over θ. Taking each poll separately, what is the posterior density for θ given the information in each poll? Generate a graphical summary of each posterior density.

3. What is the posterior density for θ given the information in *all* the polls?

4. With the posterior density calculated in answering the previous question, report the posterior probability that $\theta > .5$.

5. Is the mean of the posterior density the same as the average of the poll results? Why or why not?

2.4 (Suggested by John Bullock.) A voter V is unsure as to John McCain's age. The voter represents this uncertainty with a normal distribution with a mean of 70 and variance σ_0^2. V has a friend, A, who reports that his beliefs as to McCain's age can be represented with a normal distribution with a mean of 80 and variance σ_1^2. Another friend, B, reports that his beliefs about McCain's age can be represented with a normal distribution with a mean of 60 and variance σ_1^2.

1. If V treats his friends' self-reports of their beliefs as to McCain's age as "data", then what are V's posterior beliefs as to McCain's age?

2. Suppose V were to hear from B before he heard from A. Would this change V's posterior beliefs about McCain's age? Why or why not?

3. Suppose V thinks A's information about McCain's age is not as reliable as B's. V decides to nonetheless incorporate A's information but by "inflating" the variance term of A's information, σ_1^2, to $\lambda\sigma_1^2$, where $\lambda > 1$. Use a graphical or tabular summary to demonstrate the sensitivity Vs posterior beliefs about McCain's age to beliefs about the reliability of A's information.

2.5 Consider the discussion in Section 2.4.5. Construct a conjugate, normal data example in which the marginal variance of the mean, μ, actually increases after looking at data. This can be a "fake" example, and you can formulate the problem in terms of the parameters of the normal/inverse-Gamma prior μ_0, n_0, σ_0^2 and ν_0 and the sufficient statistics $\bar{y}$, n, and $s^2 = \sum_{i=1}^n (y_i - \bar{y})^2$. See Proposition 2.6 for the form of the marginal posterior density of μ in this case. Graph the marginal prior for μ, the likelihood and the marginal posterior for μ.

2.6 Let $\text{OR} = \dfrac{\theta_1(1-\theta_0)}{(1-\theta_1)\theta_0}$ be the odds-ratio of θ_1 to θ_0. Show that $\text{OR} > 1 \iff \theta_1 > \theta_0$.

2.7 Suppose $\theta \sim \text{Unif}(0, 1)$.

1. Derive the density of the quantity $y = -\log\theta$. Does this density have a recognizable form? Consult the densities listed in the Appendix.

2. Derive the density of the quantity $\text{OR} = \theta/(1-\theta)$.

3. Derive the density of the quantity $\log\left(\frac{\theta}{1-\theta}\right)$.

2.8 Suppose $\theta_0 \sim \text{Unif}(0, 1)$ and $\theta_1 \sim \text{Unif}(0, 1)$.

1. Derive the density of the quantity $w = \theta_1/\theta_0$.

2. Derive the density of the quantity $\delta = \theta_1 - \theta_0$.

2.9 We are interested in the quantity $\delta = \theta_1 - \theta_0$, where $\theta_j \in [0, 1]$, $j = 0, 1$, and so $\delta \in [-1, 1]$. Suppose we have a uniform prior on δ, i.e. $\delta \sim \text{Unif}(-1, 1)$.

1. Is this prior on δ sufficient to induce prior densities over both θ_j?

2. Suppose $\theta_0 \sim \text{Unif}(0, 1)$ and $\delta \sim \text{Unif}(-1, 1)$. What is $p(\theta_1)$, the density of $\theta_1 = \theta_0 + \delta$? Indeed, with this setup, is it the case that $p(\theta_1) = 0, \forall\, \theta_1 \notin [0, 1]$?

2.10 Prove that if $\theta \sim \text{Beta}(\alpha, \beta)$, then $E(\theta) = \alpha/(\alpha + \beta)$.

2.11 Show that the density corresponding to a Beta(0, 1) density is improper.

2.12 Prove that if $\theta \sim \text{Gamma}(a, b)$, then $E(\theta) = a/b$.

2.13 Show that the Poisson probability mass function sums to one over its support. That is, show that $\sum_{i=0}^{\infty} p_i = 1$, where $p_i = \Pr(y = i)$, where $y \sim \text{Poisson}(\lambda)$, $\lambda > 0$.

2.14 Prove that if $\theta \sim \text{Poisson}(\lambda)$ then $E(\theta) = \lambda$.

2.15 Recall Proposition 2.2 on the conjugate analysis of Poisson count data. Show that the posterior mean of the Poisson intensity parameter $E(\lambda|\mathbf{y}) = \alpha E(\lambda) + (1 - \alpha)\hat{\lambda}_{\text{MLE}}$, where $E(\lambda)$ is the mean of the conjugate Gamma prior over λ, $\hat{\lambda}_{\text{MLE}}$ is the maximum likelihood estimate of the Poisson parameter, and $0 < \alpha < 1$. That is, show that the mean of the posterior for λ is a convex combination of the mean of the prior, and the maximum likelihood estimate.

2.16 Consider the sex partners data reported in Example 2.10. Generate the equivalents of Table 2.4 and Figure 2.6 using the data from female respondents.

2.17 Suppose the vector $\boldsymbol{\theta} = (\mu, \sigma^2)'$ has a normal/inverse-Gamma density with parameters $(\mu_0, n_0, \nu_0, \sigma_0^2)$. Where is the joint posterior mode of this density?

2.18 The data used to examine irregularities in Pennsylvania state senate elections (Example 2.15) appears as part of the author's `R` package, `pscl`. In the `pscl` package, the data frame is called `absentee`. The disputed election is the last election in the data set. Replicate the analysis in Example 2.15, over a range of priors. That is, what kind of prior beliefs would one have to hold in order to find that the disputed election is not particularly unusual? That is, perform a sensitivity analysis of the sort presented graphically in Figures 2.16, but where the output of interest is the posterior predictive density for the disputed election result.

2.19 Prove Proposition 2.9. That is, suppose $y_i \overset{\text{iid}}{\sim} N(\mu, \sigma^2)$ with μ and σ^2 unknown, and $i = 1, \ldots, n$. If the (improper) prior density for (μ, σ^2) is $p(\mu, \sigma^2) \propto 1/\sigma^2$, prove that

1. $\mu|\sigma^2, \mathbf{y} \sim N(\bar{y}, \sigma^2/n)$
2. $\sigma^2|\mathbf{y} \sim \text{inverse-Gamma}\left(\frac{n-1}{2}, \frac{S}{2}\right)$

where $S = \sum_{i=1}^{n}(y_i - \bar{y})^2$. Consult the definitions of the normal and inverse-Gamma densities provided in Appendix B.

2.20 Use your results from the last two questions to find the mode of the joint posterior density for (μ, σ^2) given the improper reference prior $p(\mu, \sigma^2) \propto 1/\sigma^2$. Does the mode of the joint posterior density correspond to the maximum likelihood estimates of (μ, σ^2)?

2.21 Extend Proposition 2.8 to the multiple regression setting. That is, given

$$\mathbf{y}_i|(\mathbf{X}, \boldsymbol{\beta}, \sigma^2) \overset{\text{iid}}{\sim} N(\mathbf{x}_i\boldsymbol{\beta}, \sigma^2) \quad \forall\, i = 1, \ldots, n$$

with $\boldsymbol{\beta} \sim N(\mathbf{b}_0, \mathbf{B}_0)$ and $\sigma^2 \sim$ inverse-Gamma$(\nu_0/2, \nu_0\sigma_0^2/2)$, derive the posterior densities $p(\sigma^2|\mathbf{y}, \mathbf{X})$ and $p(\boldsymbol{\beta}|\sigma^2, \mathbf{y}, \mathbf{X})$. Verify that that the *marginal* posterior density of $\boldsymbol{\beta}$,

$$p(\boldsymbol{\beta}|\mathbf{y}, \mathbf{X}) = \int_0^\infty p(\boldsymbol{\beta}, \sigma^2|\mathbf{y}, \mathbf{X})d\sigma^2$$

is not recognizable as a standard density (but we can sample from this density using techniques to be introduced in Part II).

2.22 Suppose we have two sets of data, labelled 1 and 2, and a normal linear regression model holds in each. Consider the model that results from "pooling" the two sets of data, i.e.

$$\begin{bmatrix} \mathbf{y}_1 \\ \mathbf{y}_2 \end{bmatrix} = \begin{bmatrix} \mathbf{X}_1 \\ \mathbf{X}_2 \end{bmatrix} \boldsymbol{\beta} + \begin{bmatrix} \boldsymbol{\epsilon}_1 \\ \boldsymbol{\epsilon}_2 \end{bmatrix}$$

where the $\boldsymbol{\epsilon}_j$ are stochastic disturbances with $E(\boldsymbol{\epsilon}_j|\mathbf{X}_j) = \mathbf{0}$ and $V(\boldsymbol{\epsilon}_j|\mathbf{X}_j) = \sigma_j^2\mathbf{I}_{n_j}$, $j = 1, 2$. Show that the least squares estimate of $\boldsymbol{\beta}$ formed by *pooling* data sets 1 and 2 is equivalent to a Bayesian analysis, in which the information about $\boldsymbol{\beta}$ in one data set is treated as prior information.

2.23 Generalize the result in Proposition 2.4 to the case where $\mathbf{y}_i \sim N(\boldsymbol{\mu}, \boldsymbol{\Sigma})$, where $\mathbf{y}_i$ is a k-by-1 vector, $\boldsymbol{\mu}$ is a unknown vector of mean parameters, and $\boldsymbol{\Sigma}$ is a known k-by-k covariance matrix. Consult Definition B.31 to obtain the likelihood function given n iid observations $\mathbf{Y} = \{\mathbf{y}_i\}$. You should be able to factor the likelihood into a term that is a function of $\mathbf{y}_i$ and $\bar{\mathbf{y}}$ and another term that is a function of $\bar{\mathbf{y}}$ and $\boldsymbol{\mu}$ (hint: write $\mathbf{y}_i - \boldsymbol{\mu}$ as $\mathbf{y}_i - \bar{\mathbf{y}} + \bar{\mathbf{y}} - \boldsymbol{\mu}$). What is the conjugate prior and posterior density for $\boldsymbol{\mu}$? Express the mean of the posterior density for $\boldsymbol{\mu}$ as a function of the MLE of $\boldsymbol{\mu}$ and the mean of the prior conjugate density.

Part II

Simulation Based Bayesian Analysis

The models and examples presented in Part I are reasonably simple: e.g. inference for an unknown success probability, a proportion, the mean of a normal, linear regression where the y_i is conditionally iid given the predictors. For these problems, the likelihood functions are standard, and, if one uses conjugate priors, deriving the posterior density poses no great computational burden (indeed, this is why conjugate prior densities are widely employed in Bayesian analysis). Recall that in Section 2.1, we saw that for n conditionally independent Bernoulli trials, given a success probability θ with a conjugate prior $\theta \sim \text{Beta}(\alpha, \beta)$, the posterior density is also a Beta density with parameters $\alpha + r$ and $\beta + n - r$, where r is the number of 'successes' in the n trials. In this case, Bayesian computation amounts to nothing more than simple addition!

But Bayesian computation quickly becomes more challenging when working with more complicated models, or when one uses non-conjugate priors. In fact, we do not have to venture far beyond the examples considered in Part I to encounter problems where conjugacy can't be exploited, and characterizing the posterior density becomes a non-trivial exercise. Many statistical models commonly deployed in the social sciences have this feature, as the following example demonstrates.

■ Example II.1

Regression models for binary data. Suppose we observe n independent binary variables $\mathbf{y} = (y_1, \ldots, y_n)'$, where $y_i \sim \text{Bernoulli}(\pi_i)$, with π_i related to predictors $\mathbf{x}_i$ as follows, $\pi_i = F(\mathbf{x}_i\boldsymbol{\beta})$, where $\boldsymbol{\beta} \in \mathcal{B} \subseteq \mathbb{R}^k$ is a vector of unknown parameters, and $F : \mathbb{R} \rightarrow [0, 1]$. A 'probit' model results if F is the cumulative distribution function (CDF) of the standard normal, and a 'logit' model results if F is the logistic CDF. Interest centers on inference for $\boldsymbol{\beta}$, parameters tapping the way that change in the predictors maps into change in π. A Bayesian analysis proceeds by specifying a prior for $\boldsymbol{\beta}$, $p(\boldsymbol{\beta})$, and applying Bayes' Rule to obtain the posterior density: i.e. $p(\boldsymbol{\beta}|\mathbf{y}, \mathbf{X}) \propto p(\mathbf{y}|\mathbf{X}, \boldsymbol{\beta})p(\boldsymbol{\beta})$, where $p(\mathbf{y}|\mathbf{X}, \boldsymbol{\beta})$ is the likelihood. Under conditional independence, the likelihood is the product of the observation-specific Bernoulli densities, i.e.

$$p(\mathbf{y}|\mathbf{X}, \boldsymbol{\beta}) = \prod_{i=1}^{n} p(y_i|\mathbf{x}_i, \boldsymbol{\beta}) = \prod_{i=1}^{n} \pi_i^{y_i}(1-\pi_i)^{1-y_i}$$
$$= \prod_{i=1}^{n} F(\mathbf{x}_i\boldsymbol{\beta})^{y_i}[1 - F(\mathbf{x}_i\boldsymbol{\beta})]^{1-y_i}.$$

The maximum likelihood estimate of $\boldsymbol{\beta}$ is the value of $\boldsymbol{\beta}$ that maximizes this likelihood. It is well known that (at least for both logit and probit models) there is no closed form, analytical solution to this maximization problem, and MLEs of $\boldsymbol{\beta}$ are found by iterative techniques (e.g. McCullagh and Nelder 1989).

Bayesian analysis of this model fares no better in terms of computational simplicitly. By Bayes' Rule, the posterior density is

$$p(\boldsymbol{\beta}|\mathbf{y}, \mathbf{X}) = \frac{p(\boldsymbol{\beta}) \prod_{i=1}^{n} F(\mathbf{x}_i\boldsymbol{\beta})^{y_i}[1 - F(\mathbf{x}_i\boldsymbol{\beta})]^{1-y_i}}{\int_B p(\boldsymbol{\beta}) \prod_{i=1}^{n} F(\mathbf{x}_i\boldsymbol{\beta})^{y_i}[1 - F(\mathbf{x}_i\boldsymbol{\beta})]^{1-y_i} d\boldsymbol{\beta}} \tag{II.1}$$

Depending on the form of the prior, $p(\boldsymbol{\beta})$, and whether we employ a probit or a logit likelihood (or something else), it is not at all obvious that this posterior density corresponds to any standard, parametric density. Writing in 1984, Zellner and Rossi remarked that analytically evaluating this posterior density 'seems very difficult' (p. 367). Albert and Chib (1993) describe the posterior density in equation II.1 as 'largely intractable' (p. 669). That said, Zellner and Rossi considered some approximations. For instance, for the case where the prior for $\boldsymbol{\beta}$ is an improper density, $p(\boldsymbol{\beta}) \propto c$, a constant, the posterior density is proportional to the likelihood, and under standard regularity conditions, the posterior density is approximately normal in a large sample. In addition, Taylor series expansions can be used to provide approximations for the case where $p(\boldsymbol{\beta})$ is a multivariate normal density. Zellner and Rossi note that the k-dimensional integration in the denominator of the posterior density poses a formidable computational challenge, at least given statistical computing as it looked in 1984. Thus, for all practical purposes, a Bayesian approach to logit or probit regression modeling was in the 'too hard' basket for the typical social scientist, at least until about 1990.

As we shall see in the following chapters, much has changed since Zellner and Rossi's 1984 article. Since that time, there has been an explosion in the computing power available to social scientists. Expressions of the sort in Equation II.1, with multi-dimensional integrals, can now be evaluated with relative ease, via computationally intensive simulation methods.

It is hard to understate the impact of this massive increase in computer power on statistical practice. In particular, the Bayesian approach has become feasible for not just the simple problems we encounted in Part I, but for virtually *any* statistical problem. Indeed, quite aside from the logical or philosophical attractiveness of the Bayesian approach, many researchers are attracted to Bayesian modeling because modern computational tools make simulation-based Bayesian approaches relatively straightforward, flexible, and capable of generating inferences where other methods fail or are intractable.

It is no stretch to state that there has been a Bayesian revolution in the statistics profession: the number of papers in statistics journals and presented at conferences that utilize Bayesian methods soared through the 1990s, to the point where the Bayesian approach can no longer be considered exotic or out of the mainstream, standing in the wings while center stage belongs to a frequentist orthodoxy. The contrast with the position of Bayesian ideas in the statistics profession *circa* the mid-20th-century could not be more stark. In 1992, the Royal Statistical Society convened a meeting on 'The Gibbs sampler and other Markov chain Monte Carlo methods', at which three papers were read, showing how computationally intensive algorithms and increases in computing power were rapidly expanding the scope of the Bayesian approach (Besag and Green 1993; Gilks *et al.* 1993; Smith and Roberts 1993a). As is customary for the Royal Statistical Society, a formal discussion followed and was also reported in the Society's *Journal*. In opening the discussion, Peter Clifford quipped that

> Bayesianism has obviously come a long way. It used to be that you could tell a Bayesian by his tendency to hold meetings in isolated parts of Spain and his obsession with coherence, self-interrogation and other manifestations of paranoia. Things have changed... (Clifford 1993)

In this part of the book I detail how this transformation took place, highlighting the role of simulation in modern Bayesian computation. These chapters serve as a vital precursor to Part III, where we will deploy simulation-based Bayesian methods to tackle inferential problems at the research frontiers of the social sciences. The suite of techniques that make modern, Bayesian computation possible are collectively known as Markov chain Monte Carlo algorithms, often referred to by the acronym 'MCMC'. I discuss each 'MC' in turn, starting with the 'Monte Carlo' principle in Chapter Three, before turning to see how Markov chains can be used to let us characterize (possibly high-dimensional) posterior densities in Chapter Four. In Chapters Five and Six we put these ideas to work, examining how to deploy MCMC algorithms in Bayesian statistical analysis.

3

Monte Carlo methods

Modern Bayesian computation makes extensive use of a simple idea, *the Monte Carlo principle*:

> *anything we want to know about a random variable θ can be learned by sampling many times from $f(\theta)$, the density of θ.*

Moreover, the *precision* with which we learn about θ is limited only by the number of samples from $f(\theta)$ we are willing to generate, store, and summarize. Hence, as computer processing speeds have gotten faster and as computer processors have become cheaper – along with computer memory and storage devices – learning about random variables via Monte Carlo methods has become increasingly feasible. Note also that in the specific context of Bayesian statistical inference, the density $f(\theta)$ referred to above is usually a posterior density for θ, $p(\theta|\mathbf{y})$, For the next few pages at least, the discussion will be quite general, and I will refer to a generic density of interest, $f(\theta)$.

Monte Carlo methods are not new. And it is worth noting that the methods can be applied to deterministic problems as well as stochastic, statistical problems. Some of the earliest examples of Monte Carlo long pre-date the advent of electronic computers: e.g. using a long run of Buffon's needle experiment to estimate the value of π (e.g. Hall 1873); see also Exercise 3.2. According to a review in a Los Alamos National Laboratory technical report (X-5 Monte Carlo Team 2003, Ch. 2), Lord Kelvin appears to have used random sampling to aid in evaluating some integrals that appear in the kinetic theory of gasses and 'acknowledged his secretary for performing calculations for more than 5000 collisions'. In the twentieth century, but still prior to the advent of electronic computing, Enrico Fermi used 'statistical sampling' to study the 'moderation of neutrons', using a small, mechanical adding machine in Italy in the 1930s (Segré 1980). Fermi, of course, was part of the team of scientists that constituted the Manhattan

Bayesian Analysis for the Social Sciences S. Jackman

Project, the US effort to develop nuclear weapons during World War Two. Nicholas Metropolis and Stanislaw Ulam were also Manhattan Project scientists, and were key proponents of using then-nascent computing power to solve the high-dimensional integral equations underlying weapons design. Yet Ulam's interest in Monte Carlo methods had more mundane beginnings:

> The first thoughts and attempts I made to practice [the Monte Carlo method] were suggested by a question which occurred to me in 1946 as I was convalescing from an illness and playing solitaires. The question was what are the chances that a Canfield solitaire laid out with 52 cards will come out successfully? After spending a lot of time trying to estimate them by pure combinatorial calculations, I wondered whether a more pratical method than 'abstract thinking' might not be lay it out say one hundred times and simply observe and count the number of successful plays. This was already possible to envisage with the beginning of the new era of fast computers, and I immediately thought of problems of neutron diffusion and other questions of mathematical physics... Later... [in 1946, I] described the idea to John von Neumann and we began to plan actual calculations (quoted in Eckhardt 1987, p. 131)

In particular, post-war work on developing the 2nd generation of nuclear weapons – thermonuclear devices – posed a formidable analytical challenge, giving rise to a problem involving many unknown, interdependent, random quantities (e.g. the location and velocities of neutrons). According to Metropolis and Ulam (1949, 338),

> [t]he classical methods for dealing with these equations [characterizing neutron diffusion and multiplication] are extremely laborious and incomplete in the sense that solutions in 'closed form' are unobtainable. The idea of using a statistical approach ... is sometimes referred to as the Monte Carlo method.

We now consider some simple examples, and present some important results about Monte Carlo methods.

3.1 Simulation consistency

A key property of Monte Carlo estimates is *simulation consistency*, which I explain informally and motivate with an example before stating the result as a proposition. Suppose we are interested in a random quantity θ, which has density $p(\theta)$, and we possess a computing technology that can provide arbitrarily many draws from $p(\theta)$, store those draws, and, manipulate them so as to form summary statistics, or summaries of transformations of the draws. Then, as the number of draws grows arbitrarily large, via the strong law of large numbers, the summary statistics of the sampled values are *simulation-consistent*, in the following sense: summaries of the sampled values are Monte Carlo *estimates* of a particular feature of $p(\theta)$ and have the property that they are estimates as the number of sampled values increases. Here we are making use of one of the most fundamental results in all of statistics: e.g. that the quality of a sample statistic increases with sample size. In short, the more Monte Carlo samples we can draw, store and summarize from $p(\theta)$, the better.

Attentive readers will note a subtle irony here: having forcefully argued against the utility of a frequentist/repeated-sampling approach to statistical inference in Part I, we will now use sampling (i.e. Monte Carlo methods) as a way to characterize subjective, posterior beliefs.

In this chapter we will restrict attention to the case of having a series of independent draws from $p(\theta)$. When the sequence of draws $\{\theta_t\}$, $0 \leq t \leq T$, are not independent – as is almost always the case in Bayesian analysis – we will use tools from Markov chain theory to characterize the statistical properties of the $\{\theta_t\}$ and summaries of the $\{\theta_t\}$, the subject of Chapter 4.

■ Example 3.1

Learning about a uniform random variable. For the sake of argument, pretend you've forgotten almost everything you've ever learned about probability and statistics! In particular, suppose that $\theta \sim \text{Unif}(0, 1)$ but we don't know how to use that information to compute functions of θ, nor how to derive facts such as $E(\theta) = .5$, $\text{sd}(\theta) = \sqrt{1/12}$, or even that $\Pr(\theta \leq q) = q$, for some $q \in (0, 1)$. Monte Carlo methods provide a way forward. Almost all statistical computer programs contain a *pseudo-random* number generator, providing approximately independent draws from a Unif(0, 1) density. Suppose we take T draws from uniform distribution, $\theta_1, \theta_2, \ldots, \theta_T$, where T is an arbitrarily large number. Then any function of the sampled values will be arbitrarily close to the corresponding feature of $f(\theta)$: e.g. in this case, given that $\theta \sim \text{Unif}(0, 1)$,

1. the average of the sampled values, $\bar{\theta}^{(T)} = T^{-1} \sum_{t=1}^{T} \theta_t$, will be very close to $E(\theta) = .5$. Importantly, $\bar{\theta}^{(T)}$ gets arbitrarily close to .5 as $T \rightarrow \infty$.
2. likewise, the standard deviation of the sampled values
$$\text{sd}(\theta)^{(T)} = \left[(T-1)^{-1} \sum_{t=1}^{T} (\theta_t - \bar{\theta}^{(T)})^2 \right]^{1/2}$$
will get arbitrarily close to $\sqrt{1/12}$ as $T \rightarrow \infty$.
3. the p-th quantile of the sampled values, $q_p^{(T)}$, $p \in (0, 1)$, will be very close to the p-th quantile of $f(\theta)$, which, in the case of a uniform density on the unit interval, is p. Again, as $T \rightarrow \infty$, $q_p^{(T)}$ will get arbitrarily close to p.

Figure 3.1 shows three summary statistics of draws from a uniform distribution as T increases to ten million: the three summary statistics are those discussed above, the mean, the standard deviation, and a quantile (the 95th percentile). Clearly, as the number of samples increases, the simulation-based estimates of all three quantities get closer to their known values, and are essentially indistinguishable from their known values for large values of T, at least on the scale shown in Figure 3.1. To better see simulation consistency, Figure 3.2 plots the absolute error of these three Monte Carlo estimates as a function of Monte Carlo sample size, showing the error generally decreasing as sample size increases.

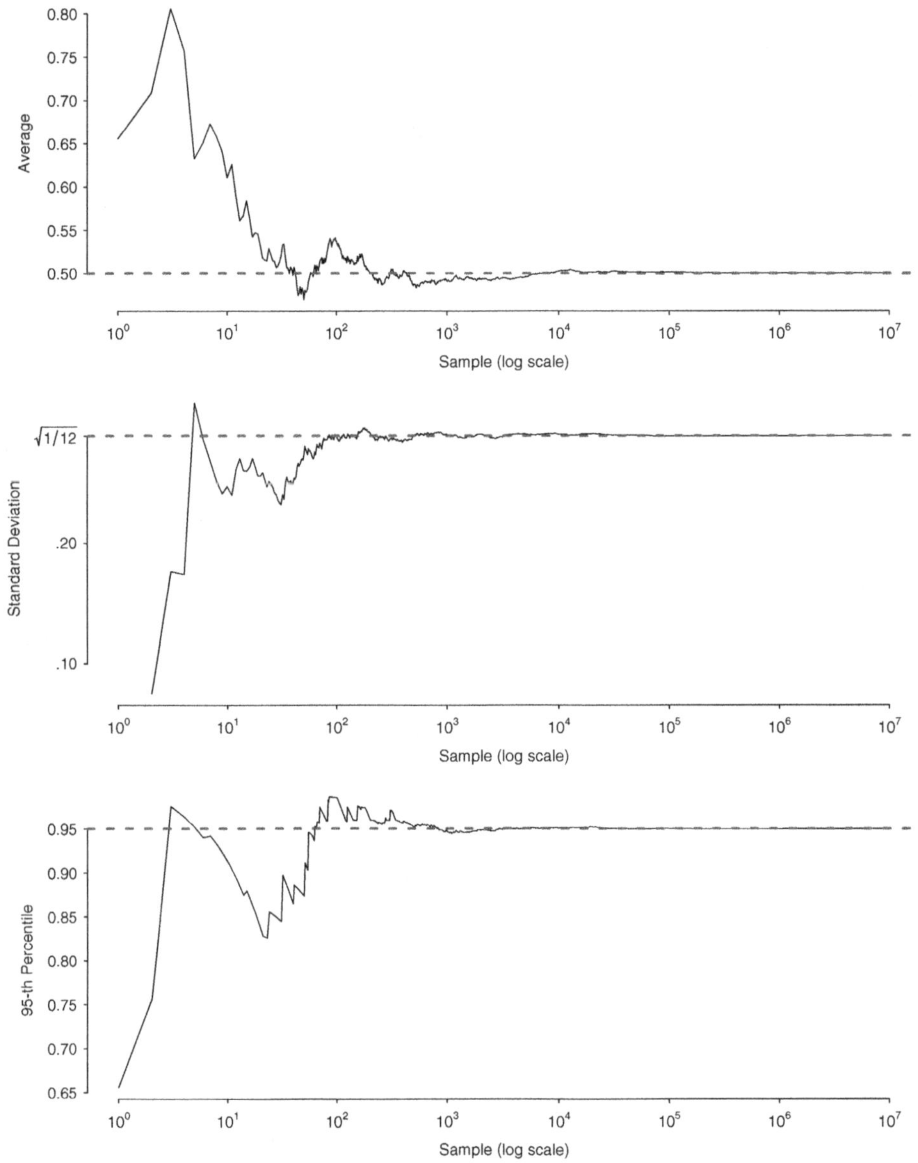

Figure 3.1 History of Monte Carlo simulation, various functions of a uniform random variable. The top panel shows the mean of the sampled values, the middle panel shows the standard deviation, and the lower panel shows the 95th percentile. Dotted horizontal lines indicate the actual value of the mean, the standard deviation and 95th percentile of $\theta \sim \text{Unif}(0, 1)$. The sample count is plotted on a logarithmic scale on the horizontal axis.

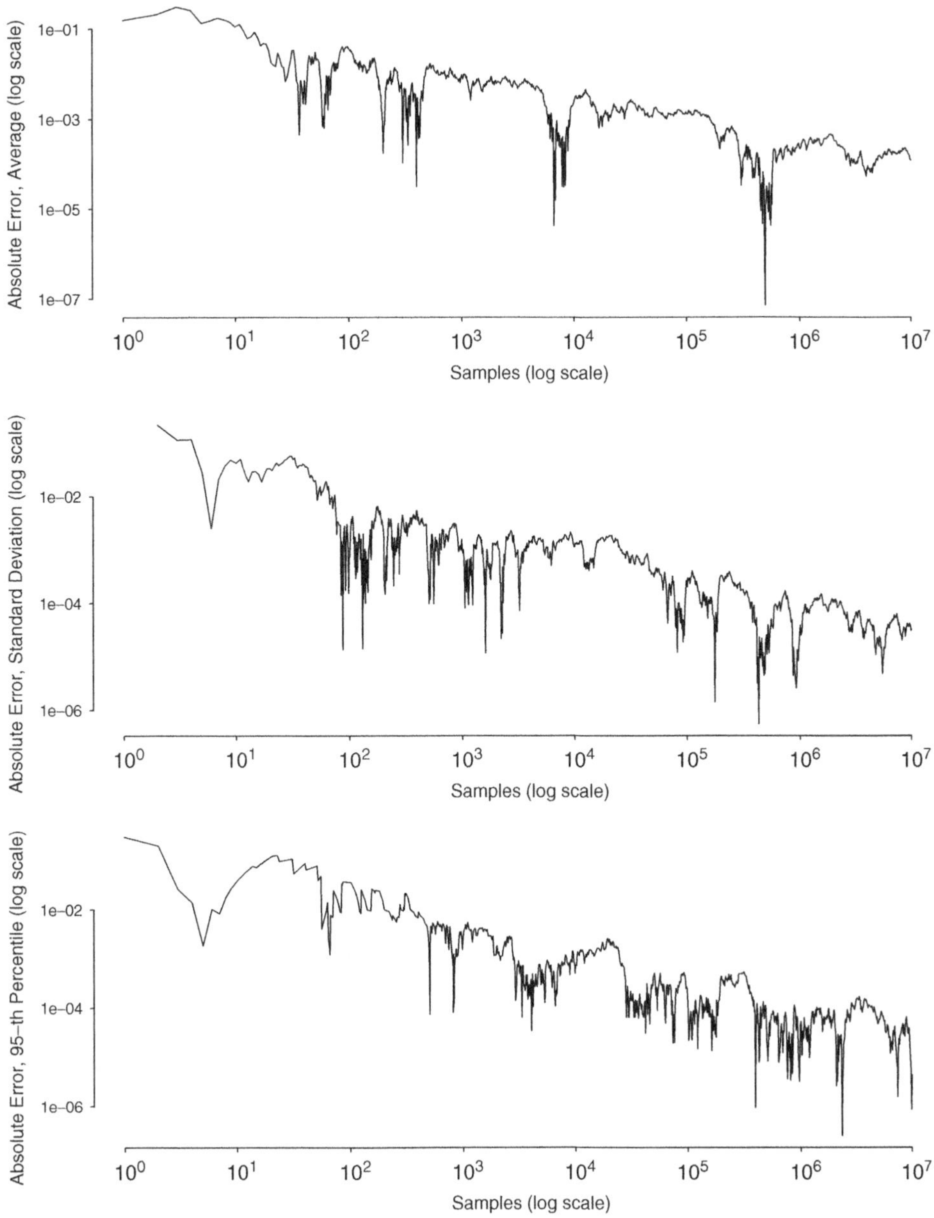

Figure 3.2 Error of Monte Carlo estimates, various functions of a uniform random variable. The absolute value of the Monte Carlo estimate of the mean (top panel), standard deviation (middle panel) and 95th percentile (lower panel) of $\theta \sim \text{Unif}(0, 1)$ are plotted against the sample count; both axes are scaled logarithmically.

The phenomenon observed in Figure 3.1 – 'simulation consistency' – is stated formally in the following proposition. The first part of the proposition considers the simulation consistency of an estimator of a scalar, $h(\boldsymbol{\theta})$, where $\boldsymbol{\theta}$ has (a possibly multivariate) density $f(\boldsymbol{\theta})$ from which we can sample. The second part of the proposition deals with an estimate of a quantile of $h(\boldsymbol{\theta})$.

Proposition 3.1 (Simulation consistency, independent draws). *Suppose* $\{\boldsymbol{\theta}^{(t)}\}$ *is a sequence of independent draws from the density* $f(\boldsymbol{\theta})$, *with* $\boldsymbol{\theta} \in \mathbb{R}^k$ *and* $h : \mathbb{R}^k \to \mathbb{R}$.

1.

$$\bar{h}^{(T)} = T^{-1} \sum_{t=1}^{T} h(\boldsymbol{\theta}^{(t)}) \stackrel{a.s.}{\to} E[h(\boldsymbol{\theta})]$$

2. *Further, suppose that for* $p \in (0, 1)$, *there is a unique* q_p *such that* $Pr[h(\boldsymbol{\theta}) \leq q_p] \geq p$ *and* $Pr[h(\boldsymbol{\theta}) \geq q_p] \geq 1 - p$ *are both true. Consider* $q_p^{(T)} \in \mathbb{R}$ *such that*

$$T^{-1} \sum_{t=1}^{T} \mathcal{I}(-\infty < h(\boldsymbol{\theta}^{(t)})) < q_p^{(T)} \geq p$$

where $p \in (0, 1)$ *and* $\mathcal{I}(\cdot)$ *is a binary indicator function, equal to 1 if its argument is true, and zero otherwise. Then* $q_p^{(T)} \stackrel{a.s.}{\to} q_p$.

Proof. (1). The claim is a restatement of the strong law of large numbers (Proposition B.10). (2) Geweke (2005, Theorem 4.1.1); Rao (1973, 423); van der Vaart (1998, 305). ◁

That is, via the law of large numbers, as we draw more samples from $f(\boldsymbol{\theta})$, our simulation-based estimate of a particular feature of $\boldsymbol{\theta}$, say, $h(\boldsymbol{\theta})$, gets better and better. Note that the second part of the proposition establishes simulation consistency for estimators of a quantile of the density of $h(\boldsymbol{\theta})$.

We can also state a result as to the *rate* at which the error of a Monte Carlo based estimate tends to zero, applying the central limit theorem to the case of independent samples realized via Monte Carlo simulation:

Proposition 3.2 (Simulation Central Limit Theorem, independent draws). *Assume the conditions of Proposition 3.1 and let* $E[h(\boldsymbol{\theta})] = \bar{h}$ *and* $var[h(\boldsymbol{\theta})] = \sigma^2$. *Further, let* $\bar{h}^{(T)} = T^{-1} \sum_{t=1}^{T} h(\boldsymbol{\theta}^{(t)})$ *and* $\sigma^{2\,(T)} = T^{-1} \sum_{t=1}^{T} \left[h(\boldsymbol{\theta}^{(t)}) - \bar{h}^{(T)}\right]^2$. *Then*

1. $T^{1/2}(\bar{h}^{(T)} - \bar{h}) \stackrel{d}{\to} N(0, \sigma^2)$
2. $\sigma^{2\,(T)} \stackrel{a.s.}{\to} \sigma^2$

Proof. The result follows via the Lindberg-Lévy central limit theorem. See Geweke (2005, Theorem 4.1.1 (c)). ◁

That is, we obtain a familiar $\sqrt{T}$ rate of convergence when estimating features of a density $p(\theta)$ with a sequence of T independent samples from $p(\theta)$. Restating this result, let the *error* of a Monte Carlo estimate of $\bar{h}$ based on T independent draws from $p(\boldsymbol{\theta})$ be $e^{(T)} = \bar{h}^{(T)} - \bar{h}$. As $T \to \infty$, the density of $e^{(T)}$ converges to a normal density, with mean zero and variance σ^2/T, where $\sigma^2 = \text{var}[(h(\boldsymbol{\theta})]$. Any particular Monte Carlo estimate of $\bar{h}$ will never *exactly* equal $\bar{h}$, but will get arbitrarily close to $\bar{h}$ as $T \to \infty$. More precisely, for large T, the fact that $e^{(T)}$ has a normal density means that we know a good deal about the performance of the Monte Carlo estimator. For example, for large T, the probability that $\bar{h}^{(T)}$ lies more than some distance c away from $\bar{h}$ is just the probability of observing a draw from a standard normal density larger than $c\sqrt{T}/\sigma$ in absolute value, for the case of the independent Monte Carlo samplers contemplated here. Again observe that for a given c and σ, this tail area probability diminishes as $T \to \infty$.

■ Example 3.2

Learning about a uniform random variable, Example 3.1, continued. We know that if $\theta \sim \text{Unif}(0, 1)$, then $E(\theta) = .5$ and $\text{var}(\theta) = 1/12$. Suppose $h(\theta)$ is an identity, such that $h(\theta) = \theta$ and therefore $\bar{h} = E[h(\theta)] = E(\theta) = .5$ and $\sigma^2 = 1/12$. Then, by Proposition 3.2,

$$T^{1/2}(\bar{h}^{(T)} - .5) \xrightarrow{d} N(0, 1/12).$$

In this example h is an identity, and so $\bar{h}^{(T)} = T^{-1} \sum_{t=1}^{T} \theta^{(t)} = \bar{\theta}^{(T)}$, where $\theta^{(1)}, \theta^{(2)}, \ldots$ is a sequence of independent samples from $p(\theta) \equiv \text{Unif}(0, 1)$. Thus we have $T^{1/2}\bar{\theta}^{(T)} \xrightarrow{d} N(.5, 1/12)$, and equivalently, $\bar{\theta}^{(T)} \xrightarrow{d} N(.5, T^{-1}/12)$. We can use this result to infer that after sampling 1 000 000 independent samples from $p(\theta) \equiv \text{Unif}(0, 1)$, we are 95 % confident that our Monte Carlo estimate $\bar{\theta}^{(T)}$ lies in the interval $.5 \pm 1.96\sqrt{T^{-1}/12} = .5 \pm .000566 = (.499434, .500566)$, i.e. very close to the true value of $E(\theta) = .5$. By way of contrast, note that with a sample of $T = 100$ independent samples, we can be 95 % confident that the Monte Carlo estimate will lie in the interval (.4434, .5566), an interval some $\sqrt{10^6/10^2} = 100$ times wider than the interval obtained with $T = 1\,000\,000$ samples.

Figure 3.3 shows the behavior of the Monte Carlo based estimate of the mean of a uniform density, used in Example 3.1, and replicates the top panel of Figure 3.2. The dotted line corresponds to the nominal 50 % error bound from applying the result in Proposition 3.2. That is, a Monte Carlo estimator of the mean of a random variable with a Unif(0, 1) density based on T independent samples has absolute error no greater than $.67\sqrt{T^{-1}/12}$ with probability .5. The performance of the one Monte Carlo estimator shown in Figure 3.3 is close to this nominal rate.

Remark. Note that these results follow from classical sampling theory, and, in particular, the central limit theorem. However, note that in the context of Monte Carlo methods we are *actually* sampling many times from a density of interest, rather than *imagining* what would happen if one were to sample an estimator many times from its sampling distribution. And, perhaps most importantly, unlike most applications of classical sampling theory, the researcher controls the number of samples, guaranteeing that these desirable

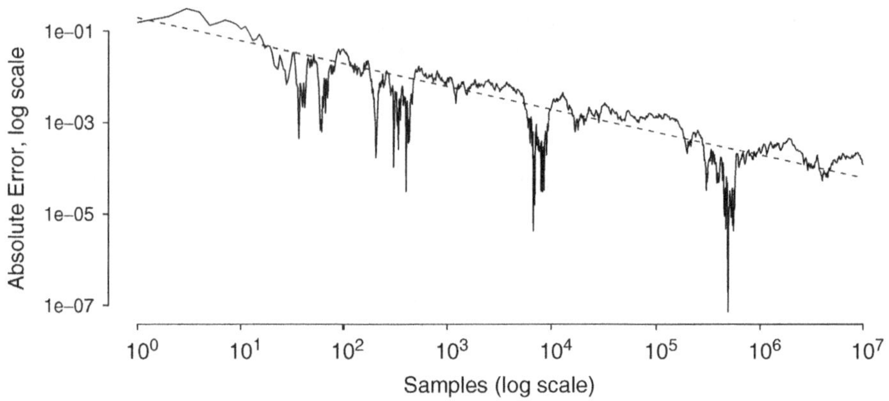

Figure 3.3 Absolute value of the error of a Monte Carlo based estimate of the mean of a Unif(0, 1) random variable. The dotted line indicates a 50 % upper-bound on the error, using the result in Proposition 3.2; i.e. over repeated implementations of the Monte Carlo procedure, for any given number of samples, half of the implementations would lie above the line, given by $.67\sqrt{T^{-1}/12}$, and half of the implementations would lie below. The error decreases at rate $\sqrt{T}$; the logarithmic scaling used on the axes makes the rate of convergence appear linear in the graph.

asymptotic properties of simulation-based estimates actually will hold in any particular application.

3.2 Inference for functions of parameters

Monte Carlo methods can be very useful when interest centers on functions of parameters which do not have standard probability densities. That is, suppose a statistical model for data $\mathbf{y}$ is parameterized in terms of a parameter $\boldsymbol{\theta}$, but the estimand of interest is $h(\boldsymbol{\theta})$. Even though the posterior density for $\boldsymbol{\theta}$, $p(\boldsymbol{\theta}|\mathbf{y})$, may be a standard, well-studied density, the implied posterior density for $h(\boldsymbol{\theta})$, $p(h(\boldsymbol{\theta})|\mathbf{y})$, may not be. In such a case Propositions 3.1 and 3.2 tell us that a Monte Carlo methods are an attractive way to proceed: samples from the posterior for $\boldsymbol{\theta}$ can be transformed to yield $h(\boldsymbol{\theta})$; storing and summarizing an arbitrarily large number of $h(\boldsymbol{\theta})$ lets us learn about any feature of the posterior density of $h(\boldsymbol{\theta})$, via the Monte Carlo principle.

This general strategy – using Monte Carlo methods to learn about non-standard functions of parameters – is actually widely used in the social sciences. The following example provides a simple demonstration of the technique.

■ Example 3.3

War and revolution in Latin America, Example 2.9, continued. Substantive interest centers on the difference of two independent binomial success probabilities, $q = \theta_1 - \theta_0$. For the case considered by Sekhon (2005) – a study of the probability of revolution in Latin American countries conditional on whether the country has experienced

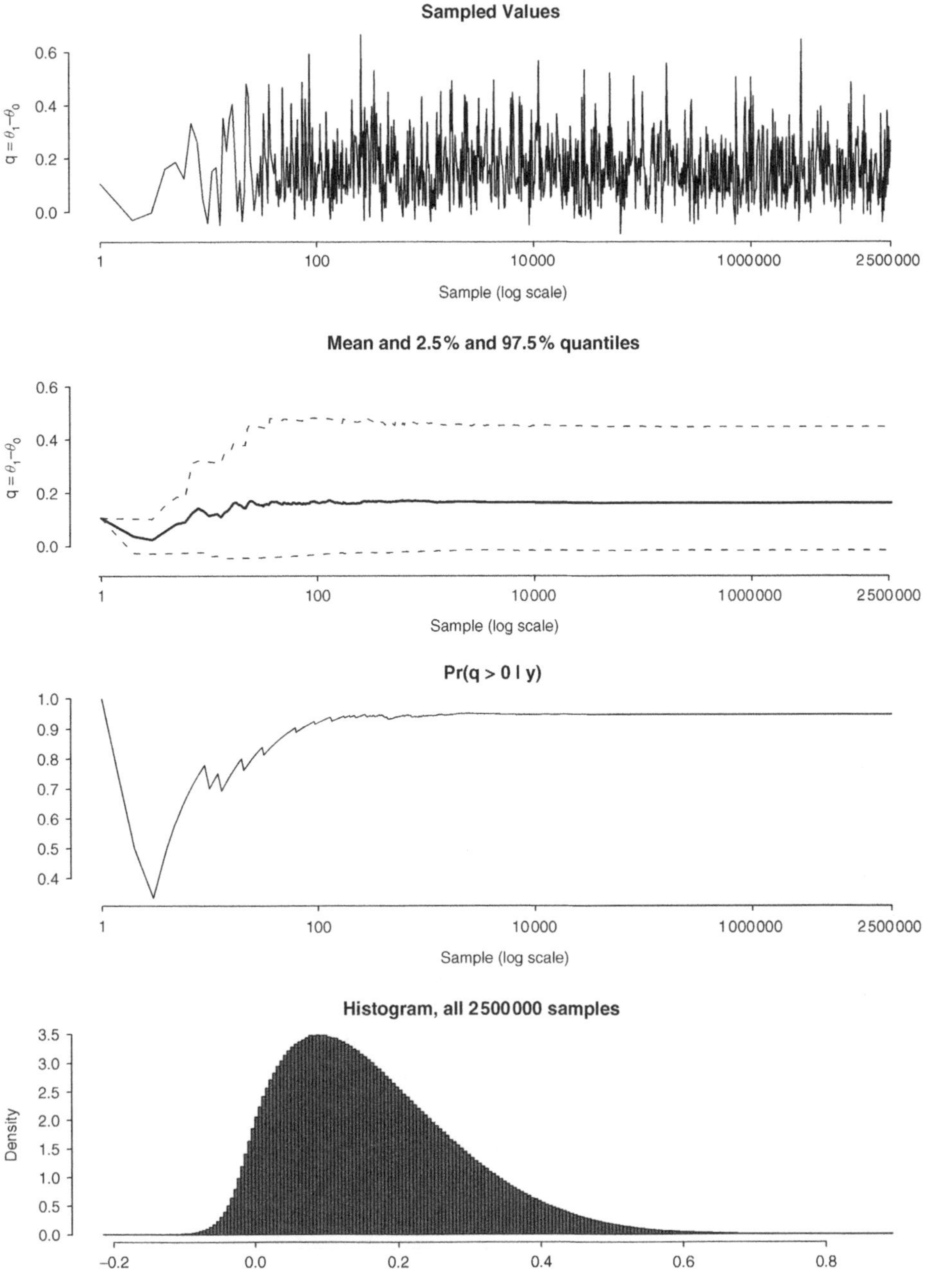

Figure 3.4 Monte Carlo simulation for Example 3.3. The top panel shows the 2 500 000 sampled values for $q = \theta_1 - \theta_0$ plotted as a time series; the 2nd panel from the top shows the mean (solid line) and 2.5 % and 97.5 % quantiles (dotted lines); the 3rd panel shows the probability that $q > 0$, again as a time series; the lower panel displays a histogram for q.

a military defeat in the last 20 years – the posterior densities are $\theta_0 \sim \text{Beta}(3, 75)$ and $\theta_1 \sim \text{Beta}(2, 8)$; θ_0 is the probability of a revolution conditional on no military defeat with the last 20 years, while θ_1 is the probability of a revolution conditional on a military defeat. Given that θ_0 and θ_1 have independent Beta densities, the density of $q = \theta_1 - \theta_0$ is surprisingly complicated, has only recently been derived (Pham-Gia and Turkkan 1993), and remains difficult to compute. Sekhon (2005) uses simulation to compute the posterior density of q, via the following algorithm:

1: **for** $t = 1$ to T **do**
2: sample $\theta_1^{(t)}$ from $p(\theta_1|\mathbf{y}) \equiv \text{Beta}(2, 8)$
3: sample $\theta_0^{(t)}$ from $p(\theta_0|\mathbf{y}) \equiv \text{Beta}(3, 75)$
4: $q^{(t)} \leftarrow \theta_1^{(t)} - \theta_0^{(t)}$
5: **end for**

Equivalently,

1: $\tilde{\boldsymbol{\theta}}_1 \leftarrow T$ independent draws from $p(\theta_1|\mathbf{y}) \equiv \text{Beta}(2, 8)$
2: $\tilde{\boldsymbol{\theta}}_0 \leftarrow T$ independent draws from $p(\theta_0|\mathbf{y}) \equiv \text{Beta}(3, 75)$
3: $\tilde{\mathbf{q}} \leftarrow \tilde{\boldsymbol{\theta}}_1 - \tilde{\boldsymbol{\theta}}_0$

where $\tilde{\boldsymbol{\theta}}_1, \tilde{\boldsymbol{\theta}}_0$ and $\tilde{\mathbf{q}}$ are vectors of length T. Implementing this algorithm with T set to 2 500 000 produced the graphical summaries depicted in Figure 3.4. Since the sampled values of q are all independent of one another, we observe no serial dependence when we visually inspect the sequence of sampled q (top panel of Figure 3.4). The trajectory of the simulation-based estimate of the mean of the posterior density for q appears to stabilize after a couple of thousand sampled values at .16, and the estimates of the 2.5 % and 97.5 % quantiles display no appreciable movement after approximately 10 000 iterations, at least on the scale of the graph shown in the 2nd panel of Figure 3.4. The 3rd panel of Figure 3.4 shows the simulation history of a substantively important quantity, the posterior probability that $q > 0$; the simulation-based estimate of this probability is initially quite unstable, but converges on about .95 after a couple of thousand samples. The histogram in the lower panel of Figure 3.4, based on all 2 500 000 iterations, shows $p(q|\mathbf{y})$ to be right-skewed: most of the posterior density lies above zero, with just a small tail lying below zero. In fact, the posterior probability that $\theta_1 > \theta_0$ is .95, suggesting that revolution is more likely given a military defeat than given no military defeat.

The key methodological points here are: (1) the posterior density $p(q|\mathbf{y})$ is difficult to work with analytically; (2) we can't sample directly from $p(q|\mathbf{y})$; (3) nonetheless, we can sample from the posterior densities $p(\theta_1|\mathbf{y})$ and $p(\theta_0|\mathbf{y})$, from which we construct samples from the target posterior density, $p(q|\mathbf{y})$.

3.3 Marginalization via Monte Carlo integration

A common use of Monte Carlo methods is *Monte Carlo integration*. Perhaps the simplest examples of Monte Carlo integration involve computing the mean of a random variable or tail areas of the density of a random quantity, as in Example 3.3. For instance, if we are interested in computing the mean of θ given that it has density $p(\theta)$, i.e. $E_p(\theta) = \int_\Theta p(\theta)\theta d\theta$, we can use the following Monte Carlo algorithm:

Algorithm 3.1 Monte Carlo estimate of the mean.

1: **for** $t = 1$ to T **do**
2: sample $\theta^{(t)}$ from $p(\theta)$
3: **end for**
4: estimate $E(\theta)$ with $T^{-1}\sum_{t=1}^{T}\theta^{(t)}$

Alternatively, consider computing the tail area of a density, such as the probability that a random quantity θ, $-\infty < \theta < \infty$, is greater than some constant c. If θ has density $p(\theta)$, then the relevant quantity is $\Pr(\theta > c) = \int_c^\infty p(\theta)d\theta$, which we can estimate via the following Monte Carlo algorithm:

Algorithm 3.2 Monte Carlo estimate of tail area probability.

1: **for** $t = 1$ to T **do**
2: sample $\theta^{(t)}$ from $p(\theta)$
3: $z^{(t)} \leftarrow \mathcal{I}(\theta^{(t)} > c)$, where $\mathcal{I}(\cdot)$ is a binary indicator function
4: **end for**
5: estimate $\Pr(\theta > c)$ with $T^{-1}\sum_{t=1}^{T} z^{(t)}$

In each of these two cases, the computational problem of evaluating the integral of the density of a random variable has been solved by applying the Monte Carlo principle.

Another extremely useful application of the Monte Carlo principle involves another application of integration: computing marginal densities given joint densities. That is, suppose a model involves multiple parameters. For simplicity, assume that we have just two parameters, $\boldsymbol{\theta} = (\theta_1, \theta_2)$, with $\theta_j \in \Theta_j \subseteq \mathbb{R}$, $j = 1, 2$. The posterior density of $\boldsymbol{\theta}$ given data $\mathbf{y}$ is $p(\boldsymbol{\theta}|\mathbf{y})$. Suppose we are interested in computing the marginal posterior density of θ_1, $p(\theta_1|\mathbf{y})$. Obtaining the marginal posterior density involves integration: i.e. $p(\theta_1|\mathbf{y}) = \int_{\Theta_2} p(\theta_1, \theta_2|\mathbf{y})d\theta_2 = \int_{\Theta_2} p(\theta_1|\theta_2, \mathbf{y})p(\theta_2|\mathbf{y})d\theta_2$. Monte Carlo methods are an attractive way to perform this integration, via the following algorithm:

Algorithm 3.3 Monte Carlo marginalization (method of composition).

1: **for** $t = 1$ to T **do**
2: sample $\theta_2^{(t)}$ from $p(\theta_2|\mathbf{y})$
3: sample $\theta_1^{(t)}$ from $p(\theta_1|\theta_2^{(t)}, \mathbf{y})$.
4: **end for**

The sampled values, $\theta_1^{(t)}, t = 1, \ldots, T$ constitute a sample from the marginal posterior density $p(\theta_1|\mathbf{y})$, while the sampled pairs $\boldsymbol{\theta}^{(t)} = (\theta_1^{(t)}, \theta_2^{(t)}), t = 1, \ldots, T$ are samples from the joint posterior density $p(\boldsymbol{\theta}|\mathbf{y}) = p(\theta_1|\theta_2, \mathbf{y})p(\theta_2|\mathbf{y})$. This method of obtaining a sample from a marginal density for θ_1 – sampling from the marginal for θ_2 and then the conditional density for θ_1 given the sampled value of θ_2 – is known as the *method of composition* (e.g. Tanner 1996, 52).

■ Example 3.4

Sampling from a t density. In Sections 2.4 and 2.5 we saw that the t density arises in numerous contexts in the conjugate, Bayesian analysis of normal data: e.g. Proposition 2.12 establishes that in a normal, linear regression model with a conjugate normal/inverse-Gamma prior for $\boldsymbol{\theta} = (\boldsymbol{\beta}, \sigma^2)$, $\boldsymbol{\beta} \in \mathbb{R}^k, 0 < \sigma^2 < \infty$, the marginal prior and posterior densities of the regression coefficients $\boldsymbol{\beta}$ are multivariate t densities. That is, if $p(\theta)$ is the (prior/posterior) density for $\boldsymbol{\theta} = (\boldsymbol{\beta}, \sigma^2)$, then the marginal (prior/posterior) density for $\boldsymbol{\beta}$ is

$$p(\boldsymbol{\beta}) = \int_0^\infty p(\boldsymbol{\beta}, \sigma^2) d\sigma^2 = \int_0^\infty p(\boldsymbol{\beta}|\sigma^2) p(\sigma^2) d\sigma^2$$

where $p(\boldsymbol{\beta}|\sigma^2)$ is the conditional (prior/posterior) density of $\boldsymbol{\beta}$ given σ^2, and $p(\sigma^2)$ is the marginal (prior/posterior) density of σ^2. More specifically, Proposition C.6 establishes that if $p(\sigma^2)$ is an inverse-Gamma density (see Definition B.35) and $p(\boldsymbol{\beta}|\sigma^2)$ is a normal density (see Definition B.31), then the marginal density $p(\boldsymbol{\beta})$ is a t density.

Monte Carlo methods offer a way to sample from this marginal t density via the method of composition. That is, suppose $\sigma^2 \sim \text{inverse-Gamma}(\nu/2, \nu s^2/2)$ and $\boldsymbol{\beta}|\sigma^2 \sim N(\mathbf{b}, \sigma^2\mathbf{B})$, such that $\boldsymbol{\beta} \sim t_\nu(\mathbf{b}, s^2\mathbf{B})$. Then T draws from $p(\boldsymbol{\beta})$ can be obtained via Algorithm 3.4.

The T sampled values $\sigma^{2(1)}, \ldots, \sigma^{2(T)}$ produced by Algorithm 3.4 are independent draws from the marginal (prior/posterior) density for σ^2, while the T sampled values $\boldsymbol{\beta}^{(1)}, \ldots, \boldsymbol{\beta}^{(T)}$ comprise T independent draws from the marginal t (prior/posterior) density for $\boldsymbol{\beta}$. The key idea is that the draw from the marginal density for σ^2 is used in forming the distribution from which we sample the $\boldsymbol{\beta}$.

Algorithm 3.4 Sampling from a t density via method of composition.

1: **for** $t = 1$ to T **do**
2: sample $\sigma^{2(t)}$ from $p(\sigma^2) \equiv \text{inverse-Gamma}(\nu/2, \nu s^2/2)$
3: sample $\boldsymbol{\beta}^{(t)}$ from $p(\boldsymbol{\beta}|\sigma^{2(t)}) \equiv N(\mathbf{b}, \sigma^{2(t)}\mathbf{B})$
4: **end for**

Incidentally, this Monte Carlo based approach to sampling from a t distribution provides a vivid demonstration of the fact that the t density is a 'scale-mixture' of normal densities: uncertainty over σ^2 propagates into uncertainty over $\boldsymbol{\beta}$, meaning that the resulting marginal density for $\boldsymbol{\beta}$ is the heavier-tailed t density, relative to the normal *conditional* density for $\boldsymbol{\beta}$ that results when σ^2 is treated as given.

Further, this approach – sampling via the method of composition – is an easily implemented solution to the problem of characterizing the marginal posterior density of $\boldsymbol{\beta}$ in regression problems when we use priors that are only conditionally conjugate (recall Proposition 2.8 and Problem 2.11).

The next example exploits Monte Carlo methods in two respects: (1) to induce posterior densities over various, non-standard estimands of interest; (2) to sample from the marginal posterior density of regression coefficients using the algorithm introduced in Example 3.4.

■ Example 3.5

Estimating partisan bias in electoral systems. A *seats-votes curve* is a mapping from a party's vote share in election i, $v_i \in [0, 1]$ into its parliamentary seat share $s_i \in [0, 1]$. One popular and simple method for estimating seats-votes curves is the 'multi-year' method (e.g. Niemi and Fett 1986; Niemi and Jackman 1991): i.e. aggregate election results over multiple elections are pooled, yielding n elections, and then a log-odds on log-odds regression is used to estimate the seats-vote curves as follows (e.g. Tufte 1973):

$$\log\left(\frac{s_i}{1-s_i}\right) = \beta_0 + \beta_1 \log\left(\frac{v_i}{1-v_i}\right) + \epsilon_i \tag{3.1}$$

where $\boldsymbol{\beta} = (\beta_0, \beta_1)$ are unknown parameters to be estimated and ϵ_i is a zero-mean, independently and identically distributed error term with variance σ^2, usually assumed to follow a normal density. Note also the technical restriction implied by the log-odds transformation, $v_i, s_i \in (0, 1)$ respectively, ruling out cases where a party wins all seats and/or all votes.

This approach to estimating seats-votes curves has some obvious drawbacks: in order to get a reasonable sample of n elections, the data collection pools over large and potentially quite heterogeneous collections of elections through history and/or across jurisdictions, and so the resulting seats-votes curve provides an abstract or general characterization of the different electoral systems and redistricting plans that were used over the set of elections used in the analysis. See Gelman and King (1990) and King and Gelman (1991) for approaches that generate estimates of seats-votes curves specific to a single election based on district-level results.

The 'log-odds on log-odds' functional form in Equation 3.1 has the property that it generates seats-votes curves with the sensible properties that $\lim_{v_i \to 0} s_i = 0$ and $\lim_{v_i \to 1} s_i = 1$. Moreover,

$$E(s_i|v_i = .5) = \frac{\exp(\beta_0)}{1+\exp(\beta_0)} \tag{3.2}$$

is the proportion of seats the political party can expect to win conditional on winning half of the votes. If $E(s_i|v_i = .5) = .5$ then the electoral system is said to be *unbiased*, and the quantity

$$q = \frac{\exp(\beta_0)}{1+\exp(\beta_0)} - .5 \tag{3.3}$$

is a measure of the extent to which the electoral system is biased in favor of the political party. Note that $\beta_0 = 0 \iff q = 0$. The parameter β_1 taps the *responsiveness* of the electoral system. Pure proportional representation corresponds to $\boldsymbol{\beta} = (\beta_0, \beta_1)' = (0, 1)'$. 'Winner-take-all' electoral systems have $\beta_1 \to \infty$, such that the party winning 50 % of the votes plus one additional vote wins all the seats (e.g. the way Electoral College votes are allocated in most of the American states). The 'cube law' (Kendall and Stuart 1950) holds that $\beta_1 = 3$, and is silent on the possibility of electoral bias, implicitly constraining $\beta_0 = 0$.

Given data $\mathbf{v} = (v_1, \ldots, v_n)'$ and $\mathbf{s} = (s_1, \ldots, s_n)'$, inference for $\boldsymbol{\beta} = (\beta_0, \beta_1)'$ is straightforward, amounting to nothing more than the ordinary least squares regression in Equation 3.1. But, because of the non-linearity of Equation 3.1, inference for quantities of interest in this model is not so straightforward. For instance, the quantity q (the partisan bias of the electoral system) is a non-linear function of the parameters. Interest also centers on the following quantity

$$v_{50} = \frac{\exp(-\beta_0/\beta_1)}{1 + \exp(-\beta_0/\beta_1)}, \tag{3.4}$$

the vote share required such that $E(s|v = v_{50}) = .5$, i.e. the vote share required such that the party can expect to win half of the parliamentary seats. This is a politically important quantity in two-party, parliamentary democracies, where typically, the party that wins a majority of the seats forms a government. Note that the quantity v_{50} has an analog in bioassay: the LD50 (where 'LD' stands for 'lethal dose'), that value of the dose (e.g. a poison) where 50 % of the exposed subjects can be expected to die.

Data. Table 3.1 presents the results of 23 elections for Australia's House of Representatives, commencing with the 1949 election. The 1949 election is widely considered to be the first election of the contemporary Australian party system, in which the Australian Labor Party (ALP) competes with a coalition of two conservative parties, the Liberal Party and the rural-based National Party. Australian House of Representatives elections utilize single transferable voting (also known as the 'alternative' vote, or 'instant-runoff' voting), and so the vote shares are reported as share of the two-party preferred vote (or 2PP, essentially counting the proportion of the electorate that ranked ALP candidates ahead of coalition candidates). The bulk of the data is broadly consistent with the notion that the electoral systems that have governed Australian House of Representatives elections is 'fair': in 12 of 23 elections, the ALP won less than 50 % of the 2PP and won less than 50 % of the seats; in another 6 elections, the ALP won more than 50 % of the 2PP and more than 50 % of the seats; in 4 elections (1954, 1961, 1969 and 1998) the ALP did not win sufficient seats to form a government despite winning more than 50 % of the 2PP, and in 1990 the ALP was able to form a government with only 49.9 % of the 2PP. These 5 'off-diagonal' elections constitute 22 % of the data, and in 4 cases, the conservative coalition parties are the beneficiaries of the mismatch between vote share and seat share.

The regression model in equation 3.1 is fit to these data, using an improper prior density $p(\boldsymbol{\beta}, \sigma^2) \propto 1/\sigma^2$, and assuming a normal, linear regression model for the log-odds transformations of seats and vote shares. With this prior and likelihood for the data, we use Proposition 2.15 to see that the posterior density of $(\boldsymbol{\beta}, \sigma^2)$ is

$$\sigma^2|\mathbf{v}, \mathbf{s} \sim \text{inverse-Gamma}\left(\frac{n-k}{2}, \frac{S}{2}\right)$$

$$\boldsymbol{\beta}|\sigma^2, \mathbf{v}, \mathbf{s} \sim N\left(\hat{\boldsymbol{\beta}}, \frac{S}{n-k}(\mathbf{X}'\mathbf{X})^{-1}\right)$$

Table 3.1 Election results, Australian House of Representatives, 1949–2004. Vote share is two-party preferred. Source: Australian Electoral Commission. The data are available as `AustralianElections` in the `R` package `pscl`.

Date	ALP vote share	ALP seat share
10-12-49	49.0	38.8
28-4-51	49.3	43.0
29-5-54	50.7	46.7
10-12-55	45.8	38.5
22-11-58	45.9	36.9
9-12-61	50.5	49.2
30-11-63	47.4	41.0
26-11-66	43.1	33.1
25-10-69	50.2	47.2
2-12-72	52.7	53.6
18-5-74	51.7	52.0
13-12-75	44.3	28.3
10-12-77	45.4	30.6
18-10-80	49.6	40.8
5-3-83	53.2	60.0
1-12-84	51.8	55.4
11-7-87	50.8	58.1
24-3-90	49.9	52.7
13-3-93	51.4	54.4
2-3-96	46.4	33.1
3-10-98	51.0	45.3
10-11-01	49.0	43.3
9-10-04	47.3	40.0

where $n = 23$, $k = 2$, $S = .5285$ (the sum of the squared residuals from the log-odds on log-odds regression), $\hat{\boldsymbol{\beta}} = (\mathbf{X}'\mathbf{X})^{-1}\mathbf{X}'\mathbf{s} = (-.11, 3.04)'$, and

$$(\mathbf{X}'\mathbf{X})^{-1} = \begin{bmatrix} .0497 & .151 \\ .151 & 3.67 \end{bmatrix},$$

with $\mathbf{X}$ the 'regressor matrix' formed by concatenating a unit column vector to the vector of vote shares $\mathbf{v}$.

The seats-votes curve implied by this regression analysis is overlaid on the scatterplots of the data in Figure 3.5. The fitted seats-votes curve runs just below the $(v, s) = (.5, .5)$ point, indicating that over the 55 years of Australian political history covered by these data, Australian electoral systems appear to be have been biased slightly in favor of the coalition parties. This said, the data also appear to be reasonably dispersed around the seats-vote curve implied the regression analysis, consistent with the notion that in any given election, this historical, average, seats-votes curve is of only so much use in predicting how a given ALP vote share maps into an actual seat shares.

Numerical summaries of the regression analysis are provided in Table 3.2, providing the posterior mean and 95 % highest density regions of the marginal posterior densities of β_0, β_1 and σ^2. In computing these summaries we rely on the fact that the marginal

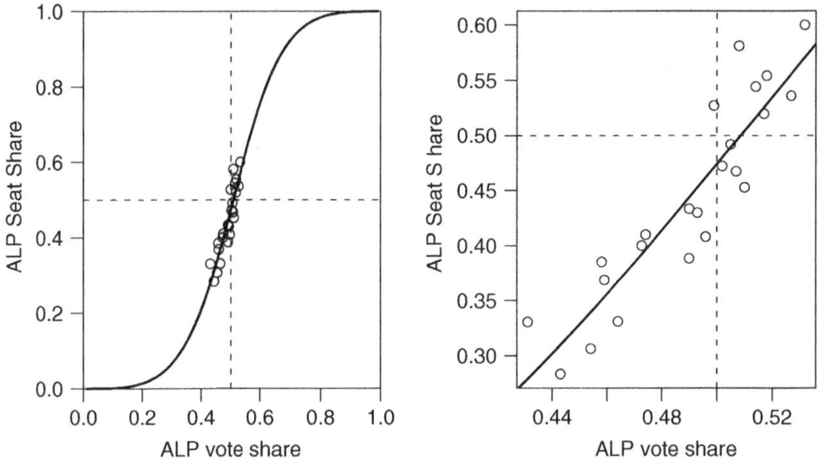

Figure 3.5 Seats-votes curve, Australian House of Representatives, 1949–2004. The left panel shows the seats-votes curve plotted over the entire unit interval (with the 23 data points superimposed), showing the constraints inherent in the functional form of log-odds on log-odds regression: the seats-votes curve passes through the points $(0, 0)$ and $(1, 1)$. The sigmoidal seats-vote curve is the curve corresponding to the ordinary least squares regression estimates of Equation 3.1. The right panel 'zooms in' on the range of the actual data, showing the fitted seats-vote curve to run just below the (.5, .5) point.

Table 3.2 Regression analysis, seats-votes curve for Australian House of Representatives elections, numerical summaries of marginal posterior densities. The Monte Carlo approximation uses $T = 2\,500\,000$ independent draws from the posterior density of the model parameters. The two sets of results are indistinguishable (at least to the level of precision reported in the table), indicative of the quality of the Monte Carlo approximation with this many draws.

	Analytic result		Monte Carlo approximation	
Parameter	Mean	95% HDR	Mean	95% HDR
Intercept (β_0)	−.11	(−.18, −.03)	−.11	(−.18, −.03)
Slope (β_1)	3.04	(2.41, 3.67)	3.04	(2.41, 3.67)
σ^2	.028	(.013, .047)	.028	(.013, .047)
r^2	.83			

posterior densities of both β_0 and β_1 are t densities centered on the maximum likelihood estimates $\hat{\boldsymbol{\beta}}$ and easily summarized; similarly, the marginal posterior density of σ^2 is an inverse-Gamma density and is also easily summarized. In what follows below, we will use Monte Carlo methods to generate summaries of these posterior densities and of the posterior densities of interesting functions of the model parameters. Most of the marginal posterior probability mass for β_0 lies below zero (and the 95 % posterior HDR for β_0 excludes zero), consistent with the notion that on average, Australian electoral systems exhibit anti-ALP bias. The posterior mean for β_1 is centered close to 3.0, the value predicted by the 'cube law', with the hypothesized value of 3.0 lying comfortably inside the 95 % posterior HDR.

Marginal Posterior Density of $\boldsymbol{\beta}$. Proposition 2.9 establishes that the marginal posterior density of $\boldsymbol{\beta}$ is a multivariate t density with $n - k = 21$ degrees of freedom, location vector $\hat{\boldsymbol{\beta}}$ and squared scale parameter $S/(n-k)(\mathbf{X}'\mathbf{X})^{-1}$. We use Monte Carlo methods

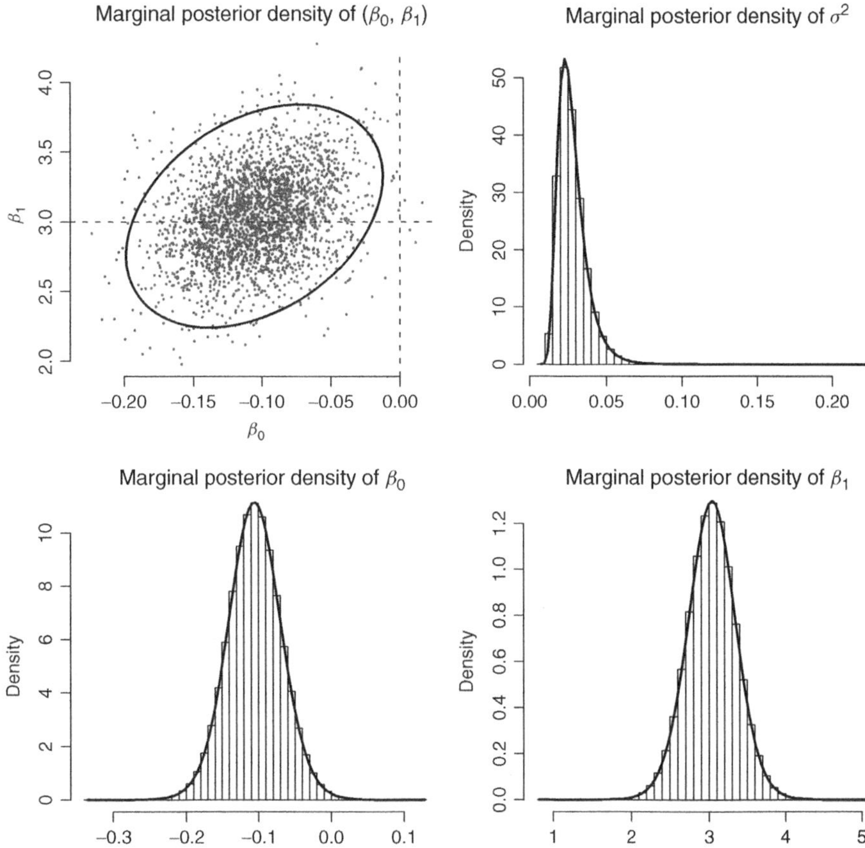

Figure 3.6 Graphical summaries, marginal posterior densities, regression analysis of Australian House of Representatives Elections. The top left panel shows 2500 samples from the marginal posterior density for $\boldsymbol{\beta} = (\beta_0, \beta_1)'$, generated using the method of composition; the ellipse marks the edge of a 95 % highest density region. The remaining panels present histograms summarizing $T = 2\,500\,000$ draws from the marginal posterior densities of σ^2, β_0 and β_1 (with the method of composition used to generate samples for β_0 and β_1). The solid lines superimposed on the histograms are the known, marginal posterior densities of each quantity (an inverse-Gamma density for σ^2, and t densities for β_0 and β_1); the fact that the histogram based summaries and the densities are visually indistinguishable indicates that the Monte Carlo approximation (based on $T = 2\,500\,000$ independent draws) is extremely good.

to generate an arbitrarily large sample of size T from this density via the following scheme, using the method of composition (see Algorithms 3.3 and 3.4):

1: **for** $t = 1$ to T **do**
2: sample $\sigma^{2(t)}$ from $p(\sigma^2|\mathbf{v}, \mathbf{s}) \equiv$ inverse-Gamma $((n-k)/2, S/2)$
3: sample $\boldsymbol{\beta}^{(t)}$ from $p(\boldsymbol{\beta}|\sigma^{2(t)}, \mathbf{v}, \mathbf{s}) \equiv N\left(\hat{\boldsymbol{\beta}}, \sigma^{2(t)}(\mathbf{X}'\mathbf{X})^{-1}\right)$
4: **end for**

where n, k, S, $\hat{\boldsymbol{\beta}}$ and $(\mathbf{X}'\mathbf{X})^{-1}$ are as given above. Figure 3.6 graphically summarizes the results of running this algorithm for $T = 2\,500\,000$ iterations. Histograms are used to summarize the draws from the marginal posterior densities of β_0 and β_1, with the actual, known densities (t densities) overlaid for comparison. With $T = 2\,500\,000$ independent samples, the Monte Carlo based approximation is extremely good: the analytic results (the known t densities for each parameter) are visually indistinct from the histogram-based summary of the Monte Carlo approximation. In fact, a comparison of the analytic and Monte Carlo results reported in Table 3.2 shows that the Monte Carlo approximation can not be distinguished from the analytic results, at least up to the 2 or 3 significant digits of precision supported by these data.

We can use the Monte Carlo approximation to report any function of the parameters that may be of interest. For instance, the posterior probability that the Australian electoral system has an anti-Labor bias is $\Pr(\beta_0 < 0)$ which we can estimate by simply noting the proportion of the Monte Carlo samples of β_0 that lie below zero: 2 491 122 of the $T = 2\,500\,000$ draws lie below zero, meaning that the Monte Carlo estimate of $\Pr(\beta_0 < 0)$ is $2\,491\,122/2\,500\,000 = .996$ (to three digits). The analytic estimate of this probability, generated using statistical software to evaluate the tail-area of a t density, is just one minus the p-value from the frequentist hypothesis test that $\beta_0 = 0$ against the one-sided alternative that $\beta_0 < 0$, and is also .996 (to three digits).

Inference for auxiliary quantities of interest. Monte Carlo methods are also especially useful in making inferences about the more complex functions of the parameters discussed earlier: e.g. the expected seat share conditional on $v = .5$ (Equation 3.2) and the vote share, v_{50}, needed to set the expected seat share to .5 (Equation 3.4). Random samples from the posterior densities of these quantities can be obtained by manipulating the samples from the posterior density of β_0, β_1 and σ^2. In addition, we can also generate samples from the posterior predictive density for seat shares $\tilde{s}$ conditional on $v_i = .5$, $p(s_i|v_i = .5, \mathbf{v}, \mathbf{s})$. The following algorithm is used to generate samples from these densities:

1: **for** $t = 1$ to T **do**
2: sample $\sigma^{2(t)}$ from $p(\sigma^2|\mathbf{v}, \mathbf{s}) \equiv$ inverse-Gamma $((n-k)/2, S/2)$
3: sample $\boldsymbol{\beta}^{(t)}$ from $p(\boldsymbol{\beta}|\sigma^{2(t)}, \mathbf{v}, \mathbf{s}) \equiv N\left(\hat{\boldsymbol{\beta}}, \sigma^{2(t)}(\mathbf{X}'\mathbf{X})^{-1}\right)$
4: $\bar{s}^{(t)}|v = .5 \leftarrow \dfrac{\exp(\beta_0^{(t)})}{1+\exp(\beta_0^{(t)})}$
5: sample $\tilde{g}^{(t)} \sim N(\beta_0^{(t)}, \sigma^{2(t)})$.
6: $\tilde{s}^{(t)} \leftarrow \dfrac{\exp(\tilde{g}^{(t)})}{1+\exp(\tilde{g}^{(t)})}$,

7: $v_{50}^{(t)} \leftarrow \dfrac{\exp(-\beta_0^{(t)}/\beta_1^{(t)})}{1+\exp(-\beta_0^{(t)}/\beta_1^{(t)})}$

8: **end for**

where n, k, S, $\hat{\boldsymbol{\beta}}$ and $(\mathbf{X}'\mathbf{X})^{-1}$ are as given above.

Figure 3.7 presents summaries of the posterior densities of the three quantities generated by this Monte Carlo algorithm: (1) $\bar{s}|v = .5$ (top panel, the expected ALP seat share conditional on winning 50 % of the 2PP vote); (2) $\tilde{s}$ (middle panel, ALP seat shares conditional on winning 50 % of the 2PP vote); (3) v_{50} (lower panel). The quantity $\bar{s}$ (top panel) is the mean of the predictive density for $\tilde{s}$ (middle panel): $\tilde{s}$ has a more dispersed posterior density than does $\bar{s}$, reflecting the additional uncertainty due to the fact that the seat share in any particular election result will lie some distance above or below the predicted seat share $\bar{s}$. This extra uncertainty is reflected in the additional sampling at step 5 of the algorithm listed above, and converted to a predicted seat share at step 6.

This extra uncertainty is consequential. If the ALP wins 50 % of the 2PP, then, averaged over 55 years of Australian political history, it can expect to win 47.4 % of the seats in the House of Representatives, which is not sufficient to form government in its own right (conversely, the Liberal/National coalition can expect to form a government if it wins 50 % of the 2PP, or even less). Of course, this assessment is subject to uncertainty. The posterior probability that $\bar{s} < .5$ is also the posterior probability that $\beta_0 < .5$, which we computed as .996; that is, we can be extremely confident that the *average* ALP seat share conditional on $v = .5$ is less than 50 %. But because the regression analysis does not provide a perfect fit to the data (i.e. actual election results vary around the average, long-run result), the prediction about the outcome in a *particular* election in which the ALP won 50 % of the 2PP is reasonably vague. Our 'best guess' (the mean of the posterior density for $\tilde{s}$) remains that the ALP would win 47.4 % of the seats, but as the middle panel of Figure 3.7 reveals, there is considerable uncertainty around this estimate. A 95 % HDR for $\tilde{s}$ ranges from 39.1 % to 55.8 %, and the probability that $\tilde{s} < .5$ is .74, far less than the corresponding probability for $\bar{s}$. That is, there is a 26 % chance that the ALP could win a majority of seats with 50 % of the 2PP; this constitutes strong, but not overwhelming evidence of a long-run, anti-ALP bias in Australian electoral systems.

Finally, the lower panel of Figure 3.7 summarizes $T = 2\,500\,000$ draws from the posterior density of v_{50}. This density has a mean of 50.9 % (2PP) and is skewed right, with a 95 % HDR ranging from 50.2 % to 51.7 % of the 2PP. The posterior probability that $v_{50} > .5$ is again equal to the posterior probability $\beta_0 < 0$, which we computed earlier to be .996. This finding is again consistent with the conclusion that over the long-run, Australian electoral systems have displayed a small to moderate amount of anti-Labor bias.

The advantage of the Monte Carlo approach adopted here is that we can induce posterior densities over quantities of political interest that are readily interpretable, such as v_{50}. That is, we see that the lesson of 55 years of Australian political history is that the ALP can not count on 'falling over the finish line' (that is, winning a slim majority of the 2PP vote). Rather, to be reasonably confident of winning government, the ALP needs to win 51 % of the 2PP or more. This finding is consistent with either or both of the following possibilities: (1) gerrymandering or malapportionment generating bias

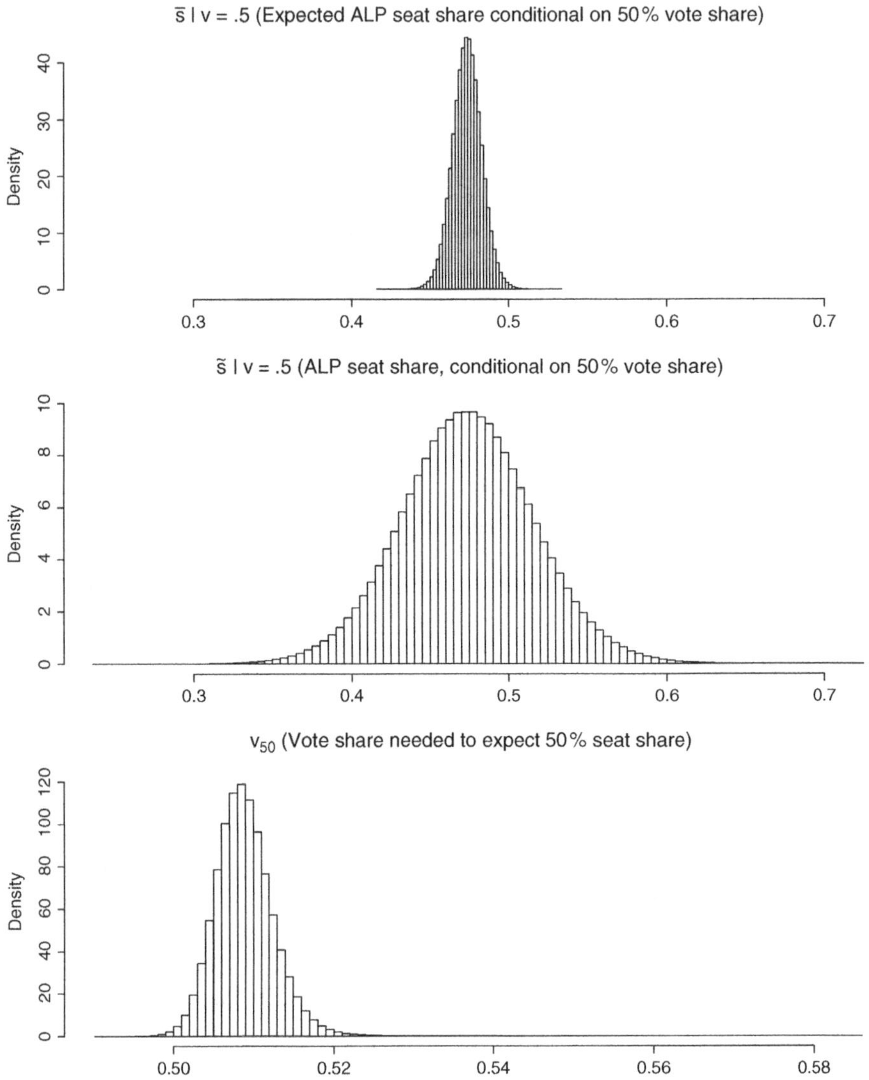

Figure 3.7 Graphical summaries, posterior densities for auxiliary quantities, regression analysis of Australian House of Representatives elections. The top panel presents a histogram summarizing T samples from the posterior density for $\bar{s}|(v = .5) = E(s|v = .5) = \exp(\beta_0)/(1 + \exp(\beta_0))$. The middle panel summarizes T samples from the predictive density for $s|v = .5$. The lower panel summarizes T samples from the posterior density of the vote share v_{50} such that $E(s|v = v_{50}) = .5$. $T = 2\,500\,000$ in each case.

through the way electoral divisions are drawn; (2) the Liberal/National coalition parties being more efficient in the way their vote share is spread across seats (e.g. winning more marginal seats than the ALP). Further discussion of Australian electoral systems (at least up through 1993) appears in Jackman (1994).

3.4 Sampling algorithms

Generating samples from 'standard' densities – those densities routinely encountered in conjugate, Bayesian analysis, such as the normal, the Beta, the inverse-Gamma, and so on – is easy to do. Modern computer packages provide support for random number generation from these densities, and many others. Indeed, if interest centers on estimands that follow these densities, then Monte Carlo methods are redundant: again, today's statistical software packages all contain algorithms that generate extremely precise estimates of quantiles or tail area probabilities of commonly encountered densities. To be sure, and as we saw in the previous section, even in this case Monte Carlo methods are useful when we need to summarize posterior beliefs about *functions* of parameters, say, some estimand $h(\theta)$ that does not follow a 'standard' density, even though the posterior density of θ is itself 'standard'.

In this section we consider the problem of sampling from some arbitrary density, $p(\theta)$. This type of problem arises frequently in Bayesian analysis, as soon as we leave reasonably simple, conjugate problems behind and we begin to tackle more interesting models, such as the probit or logit model described in Example 2.1. To reiterate, we face the following problem. Bayes' Rule tells us that the posterior density for θ, $p(\theta|\mathbf{y})$ is proportional to the prior density $p(\theta)$ times the likelihood $p(\mathbf{y}|\theta)$. The particular form of the prior and the likelihood means that their product – the function proportional to the posterior density – may not correspond to any density known to statistics. How, then, to characterize this posterior density (e.g. its mean, its dispersion, critical quantiles, a highest density region)? Moreover, how do we characterize posterior beliefs for estimands that are functions of θ? We know that via the Monte Carlo principle that if we can generate random samples from $p(\theta)$, we can compute simulation-consistent characterizations of any interesting feature of $p(\theta)$. In this way the hard/impossible mathematics problem of integrating $p(\theta)$ then becomes the considerably less difficult problem of how to draw a large number of samples from $p(\theta)$.

Here I briefly review three common algorithms for sampling from arbitrary densities: (1) the inverse-CDF method (Section 3.4.1); (2) importance sampling (Section 3.4.2); (3) accept-reject sampling (Section 3.4.3). We defer a consideration of a very useful sampling algorithm known as the 'slice sampler' until Section 5.2.7.

3.4.1 Inverse-CDF method

Suppose we wish to sample from $p(\theta)$, where $\theta \in \mathbb{R}$. Denote the cumulative distribution function (CDF) of θ as $F(q)$; i.e. $F(q) = \Pr(\theta \leq q) = \int_{-\infty}^{q} p(\theta)d\theta$ is a function that maps from $\mathbb{R}$ into the unit probability interval. Suppose further that the inverse-CDF exists, i.e. $F^{-1} : (0, 1) \rightarrow \mathbb{R}$, and, more importantly, is computable. Then the problem of sampling T values from $p(\theta)$ is fairly trivial, as described in the following algorithm:

Algorithm 3.5 Sampling via the Inverse-CDF.

1: **for** $t = 1$ to T **do**
2: sample $q^{(t)} \sim \text{Unif}(0, 1)$
3: $\theta^{(t)} \leftarrow F^{-1}(q^{(t)})$
4: **end for**

That is, provided we possess a computing technology that can (1) generate random draws from a uniform distribution; (2) compute F^{-1}, we have all we need to sample from $p(\theta)$.

It is actually reasonably rare that we can compute that inverse-CDF of a random quantity, but not know its mean or how to compute (or approximate) critical quantiles of its density, etc. Nonetheless, in many situations the inverse-CDF method is especially useful, as the following example demonstrates.

■ Example 3.6

Sampling from a truncated normal density. In the Bayesian analysis of discrete choice models, a helpful computational step involves sampling from a truncated density. For instance, consider a probit model for binary response data, $y_i \sim \text{Bernoulli}(\pi_i)$, $\pi_i = \Phi(\mathbf{x}_i\boldsymbol{\beta})$, where $i = 1, \ldots, n$, $\mathbf{x}_i$ is a vector of predictors, $\boldsymbol{\beta}$ are coefficients to be estimated, and $\Phi(\cdot)$ is the normal CDF. A helpful, alternative representation of this model is in terms of an unobserved or *latent* variable, y_i^*, where

$$y_i^*|y_i = 0 \sim N(\mathbf{x}_i\boldsymbol{\beta}, \sigma^2)\mathcal{I}(y_i^* \leq c)$$
$$y_i^*|y_i = 1 \sim N(\mathbf{x}_i\boldsymbol{\beta}, \sigma^2)\mathcal{I}(y_i^* > c)$$

where the indicator function $\mathcal{I}$ evaluates to 1 if its argument is true and 0 otherwise, and so in this case, induces a *truncation* of the normal densities for y_i^*. As it appears above, this representation of the probit model is not identified: the conventional identifying restrictions are to set $\sigma^2 = 1$ and $c = 0$.

Given $\mathbf{x}_i$ and $\boldsymbol{\beta}$, how then to sample the latent variable y_i^*? One strategy is simply to sample from the regular, untruncated normal density $N(\mathbf{x}_i\boldsymbol{\beta}, 1)$ until a draw of y_i^* is made that satifies the truncation constraint implied by the value of y_i: i.e. suppose $y_i = 0$,

---

Algorithm 3.6 Truncated normal sampling, naïve rejection method.

1: **repeat**
  $z \sim N(\mathbf{x}_i\boldsymbol{\beta}, 1)$
2: **until** $z \leq 0$ {truncation constraint}

This strategy works quite well, save for cases where the mean of the untruncated density lies some distance from the truncation point $c = 0$, in which case many draws may be generated before obtaining one that satisfies the truncation constraint.

A more direct approach exploits the inverse-CDF approach. Again, consider the case where $y_i = 0$, so the latent variable is $-\infty < y_i^* \leq 0$. Let $p(y_i^*|y_i = 0)$ be the density of y_i^* we wish to sample from, a $N(\mathbf{x}_i\boldsymbol{\beta}, 1)$ density with support truncated to the non-positive half-line. In order for $p(y_i^*|y_i = 0)$ to be a density, it must be the case that $\int_{-\infty}^{0} p(y_i^*|y_i = 0)dy_i^* = 1$. That is,

$$p(y_i^*|y_i = 0) = \begin{cases} \dfrac{\phi(y_i^* - \mathbf{x}_i\boldsymbol{\beta})}{\Phi(-\mathbf{x}_i\boldsymbol{\beta})} & \text{if} \quad y_i^* \leq 0 \\ 0 & \text{otherwise,} \end{cases}$$

where $\phi(\cdot)$ is the standard normal probability density function and $\Phi(-\mathbf{x}_i\boldsymbol{\beta})$ is the (unconditional) probability $\Pr(y_i^* \leq 0)$. The CDF of the truncated y_i^* is

$$F(z) = \int_{-\infty}^{z \leq 0} p(y_i^*|y_i = 0)dy_i^* = \frac{1}{\Phi(-\mathbf{x}_i\boldsymbol{\beta})} \int_{-\infty}^{z \leq 0} \phi(y_i^* - \mathbf{x}_i\boldsymbol{\beta})dy_i^* = \frac{\Phi(z - \mathbf{x}_i\boldsymbol{\beta})}{\Phi(-\mathbf{x}_i\boldsymbol{\beta})}.$$

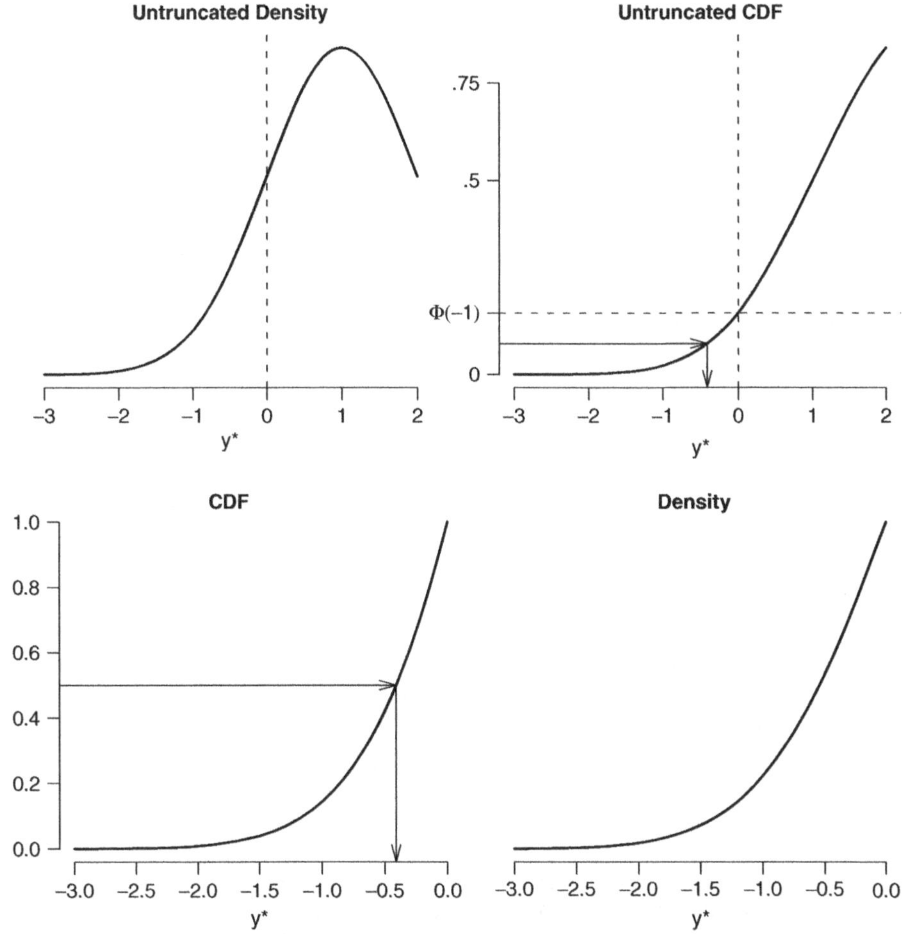

Figure 3.8 Truncated normal sampling. The goal is to sample y^* from a $N(1, 1)$ density, truncated to lie below zero. The upper panels show the untruncated $N(1, 1)$ density and distribution function, with the truncation point shown with the dotted vertical line. The lower panels show the truncated density and CDF. Samples from the truncated density can be generated via the inverse-CDF method, sampling $d \sim \text{Unif}(0, 1)$, then rescaling the d to lie between 0 and $b = \Phi(-1) \approx .16$. The CDF in the lower left panel results from rescaling the portion of the untruncated CDF that lies to the left of the truncation point in the top right panel. Draws from the density in the lower right panel are obtained by simply projecting draws from the uniform distribution through the inverse-CDF function, as indicated by the arrows.

Let $\Phi^{-1}(\cdot)$ denote the inverse of the normal CDF, i.e. $\Phi^{-1} : (0, 1) \mapsto \mathbb{R}$. Then an algorithm for generating a random draw $\tilde{y}^* \sim p(y_i^*|y_i = 0)$ is:

Algorithm 3.7 Inverse-CDF algorithm for truncated normal sampling.

1: $b \leftarrow \Phi(-\mathbf{x}_i\boldsymbol{\beta})\{b \in (0, 1)\}$
2: $d \sim \text{Unif}(0, 1)$
3: $q \leftarrow bd$
4: $z \leftarrow \Phi^{-1}(q)$
5: $\tilde{y}^* \leftarrow z + \mathbf{x}_i\boldsymbol{\beta}$

Figure 3.8 depicts this algorithm graphically. The untruncated density is $y^* \sim N(1, 1)$, and the truncation constraint is $y^* \leq 0$. The inverse-CDF method produces draws by first sampling a uniform deviate, d, but then transformed to meet the truncation constraint (contrast the top left and lower right panels of Figure 3.8). The arrows in Figure 3.8 show the inverse-CDF method at work, with the transformed uniform deviate being mapped 'back through' the CDF into the set of values that satisfies the truncation constraint. The arrows are drawn to correspond to a sampled y^* exactly at the median of the truncated density: i.e. the arrows correspond to $d = .5$, but since $b = \Phi(-1) \approx .16$, $q = bd \approx .08$, and so $z = \Phi^{-1}(q) \approx -1.41$ and $y^* = -1.41 + \mu = -.41$.

Remark. The algorithm used in this example makes use of the inverse normal CDF. Of course, neither the normal CDF nor the inverse-CDF are available in closed form: statistical software relies on high-order polynomials that approximate these functions, but are extremely precise. For instance, the inverse normal CDF function in `R` (R Development Core Team 2009), `qnorm`, uses combinations of polynomials varying from 2nd to 7th order, using an algorithm of Wichura (1988). This algorithm generates accuracy to 'about 16 figures' over the range $10^{-316} < \min(p, 1 - p)$, recalling that the inverse normal CDF maps $p \in (0, 1) \mapsto \mathbb{R}$. This approximation is so accurate that the inverse-CDF method has been selected as the default method of generating (untruncated) random normal deviates in `R`.

3.4.2 Importance sampling

In the discussion that follows, let $p(\theta)$ be the *target* density, the density we seek to characterize via repeated sampling. Assume the following:

1. we can evaluate the target density at any given point in its support; i.e. we can compute $p(\theta) \; \forall \; \theta \in \Theta$.
2. we do not possess an algorithm for direct sampling from the target density;
3. we *can* sample from a density $s(\theta)$, where $s(\theta)$ has the property that $p(\theta) > 0 \Rightarrow s(\theta) > 0, \forall \; \theta \in \Theta$.

If these properties hold, then we can exploit the following identity: if we are interested in an estimand that is some function of θ, $h(\theta)$, then

$$\text{E}[h(\theta)] = \int_{\Theta} h(\theta)p(\theta)d\theta = \int_{\Theta} h(\theta)p(\theta)/s(\theta)s(\theta)d\theta.$$

Algorithm 3.8 Importance sampling.

1: **for** $t = 1$ to T **do**
2: sample $\theta^{(t)} \sim s(\theta)$.
3: $w^{(t)} \leftarrow p(\theta^{(t)})/s(\theta^{(t)})$
4: **end for**
5: $\bar{h}^{(T)} \leftarrow T^{-1}\sum_{t=1}^{T} h(\theta^{(t)})w^{(t)}$

This expectation can be estimated via the *importance sampling* algorithm (Algorithm 3.8). The ratios $w^{(t)}$ (line 3 of the algorithm) are *importance weights*, literally telling us how the samples from the approximating density $s(\theta)$ should be weighted so as to make them representative of the target density. Note also that as a special case, the transformation $h(\cdot)$ could be an identity, in which case this algorithm provides a way of estimating $E(\theta)$.

■ Example 3.7

Sampling from an inverse-Gamma density. We have encountered the inverse-Gamma density numerous times (see Definition B.35): it is, among other things, the conjugate prior density for the variance parameter in the Bayesian analysis of normal data. Suppose we wish to know $E(\theta)$, where $\theta \sim$ inverse-Gamma$(\nu/2, \nu s^2/2)$, $0 < \theta < \infty$. Of course, via Definition B.35, we know that $E(\theta) = \nu s^2/(\nu - 2)$, but let us momentarily put that result to one side.

Suppose we possess a technology for sampling from a normal density, generating arbitrarily many independent draws $z^{(t)} \sim N(\mu, \sigma^2)$ where $-\infty < z^{(t)} < \infty$, $t = 1, \ldots, T$. Further, consider the random variable $v = \log\theta$, or conversely $\theta = \exp(v)$; i.e. $-\infty < v < \infty$. Via Proposition B.7 – the 'change of variables' proposition – we deduce that if $\theta \sim$ inverse-Gamma$(\nu/2, \nu s^2/2)$, then $p(v) = p_\theta(\exp(v))\exp(v)$ where p_θ is an inverse-Gamma density. We note immediately that $p(v)$ is positive over its support (the real line), as is the normal density. We can thus use importance sampling to estimate $E(\theta)$ as follows:

Algorithm 3.9 Importance sampling, expectation of an inverse-Gamma variate.

for $t = 1$ to T **do**
sample $z^{(t)} \sim N(\mu, \sigma^2)$
$q^{(t)} \leftarrow p_\theta(\exp(z^{(t)}))\exp(z^{(t)})$
$\phi^{(t)} \leftarrow \frac{1}{\sqrt{2\pi\sigma^2}} \exp\left[\frac{-(z^{(t)}-\mu)^2}{2\sigma^2}\right]$
$w^{(t)} \leftarrow q^{(t)}/\phi^{(t)}$
end for
$\bar{\theta}^{(T)} \leftarrow T^{-1}\sum_{t=1}^{T} \exp(z^{(t)})w^{(t)}$

Consider the case where $\theta \sim$ inverse-Gamma$(2, 5)$. In this case, $E(\theta) = 5/(2-1) = 5$. I implement the algorithm above with the parameters of the normal sampling density set to $\mu = 0$ and $\sigma^2 = 4$, and I draw $T = 10^7$ samples. Figure 3.9 compares the target density, $p(\log(\theta))$ and the normal sampling density, showing the densities themselves (top panel) and log densities (middle panel). The importance weighting function – the

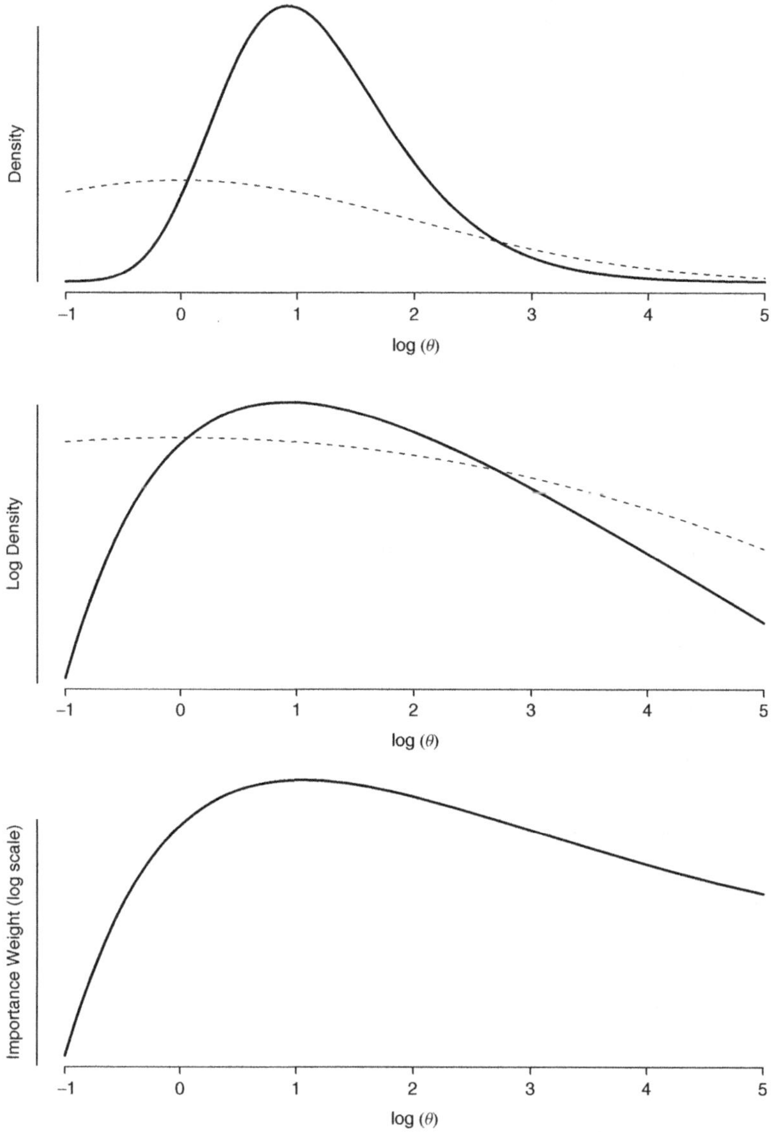

Figure 3.9 Target density and sampling density for importance sampling. The target density is the density of $\log(\theta)$ where $\theta \sim$ inverse-Gamma(2, 5) and the sampling density (dotted line) is $N(0, 4)$. The middle panel compares the densities on the log-density scale, highlighting that much of the support of the normal sampling density (the region below -1) is assigned virtually zero weight (see the lower panel).

ratio of the two densities – is shown in the lower panel of Figure 3.9. The key point here is that much of the support of the normal sampling density gets very low weight: in turn, this means that a high proportion of the samples from the normal density are given almost zero importance weights $w^{(t)}$, and hence contribute virtually nothing to the Monte Carlo estimate of $E(\theta)$. That is, importance sampling with this particular sampling density will be reasonably inefficient, requiring a relatively large number of draws to attain a given level of accuracy than if we had used a sampling density more closely approximating the target density.

This said, with modern computing power, it is trivial to generate, store and summarize a massive number of draws from a normal density, letting us overcome the inefficiency of the sampling density through sheer volume of samples. In this case, with ten million draws from the normal sampling density, we obtain an excellent estimate of $E(\theta)$, 4.998. Figure 3.10 shows how the importance sampling estimate converges on the known, true value of $E(\theta) = 5.0$ as $T \to \infty$.

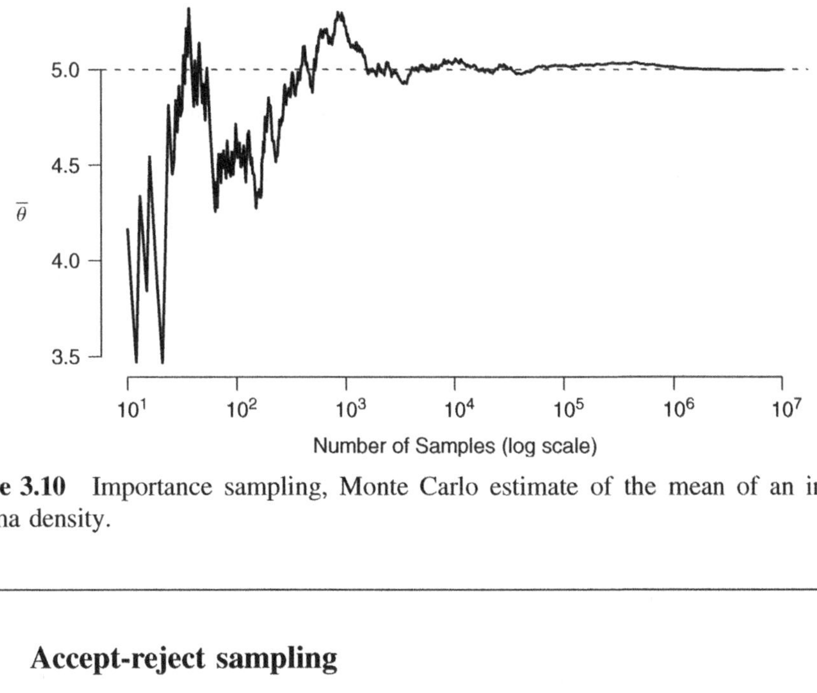

Figure 3.10 Importance sampling, Monte Carlo estimate of the mean of an inverse-Gamma density.

3.4.3 Accept-reject sampling

It is often the case that none of the aforementioned algorithms are applicable, either because there exists no 'off-the-shelf' technique for direct sampling, say, because the CDF of the density isn't known. and so the inverse-CDF method can't be deployed. In these cases a technique very similar to importance sampling can be used, to generate not just estimates of functions of a random quantity θ, but samples from its density $p(\theta)$. The ability to generate samples from any arbitrary density will prove especially useful when we consider Markov chain Monte Carlo methods in the next chapter.

Accept-reject sampling is due to von Neumann (1951) and works as follows. We wish to sample from the 'target' density $p(\theta)$, $\theta \in \mathbb{R}$. We can evaluate the density $p(\theta)$, but we can't sample 'directly' from that density. However, we do possess a technology for sampling from a *majorizing function* $g(\theta)$, that is, where $g(\theta) > p(\theta), \forall\ \theta$. What makes accept-reject sampling so attractive is that the majorizing function can be obtained by simply rescaling an *instrumental* or *proposal* density, i.e. $g(\theta) = cm(\theta)$, where $m(\theta)$ is the instrumental or proposal density. A judicious choice of $m()$ and c can prove especially helpful in trying to sample from $g(\theta)$.

Algorithm 3.10 Accept-reject sampling.

```
1: for t = 1 to T do
2:    sample z ~ m(θ)
3:    sample u ~ Unif(0, 1)
4:    r ← p(z)/(cm(z))
5:    if u ≤ r then
6:       θ^(t) ← z {"accept"}
7:    else
8:       go to 2 {"reject"}
9:    end if
10: end for
```

The accept-reject algorithm is sketched in Algorithm 3.10. We now prove that this algorithm indeed does produce samples from the target density p.

Proposition 3.3 (Accept-Reject Sampling). *The accept-reject algorithm described above produces random deviates $\tilde{\theta} \in \mathbb{R}$ whose density is the target density p.*

Proof. The proof is straightforward and appears in many places in the literature; e.g. Devroye (1986, 40–42), Ripley (1987, 60–62), or Robert and Casella (2004, 50).

Consider the cumulative distribution function of the random deviates produced by the accept-reject algorithm:

$$\Pr(\tilde{\theta} \leq q) = \Pr\left(z \leq q | u \leq \frac{p(z)}{cm(z)}\right) = \frac{\Pr\left(z \leq q, u \leq \frac{p(z)}{cm(z)}\right)}{\Pr\left(u \leq \frac{p(z)}{cm(z)}\right)}.$$

Re-write these probabilities as integrals, noting that $u \sim \text{Unif}(0, 1)$, to yield

$$\Pr(\tilde{\theta} \leq q) = \frac{\displaystyle\int_{-\infty}^{q} \int_{0}^{p(z)/(cm(z))} du\, m(z)\, dz}{\displaystyle\int_{-\infty}^{\infty} \int_{0}^{p(z)/(cm(z))} du\, m(z)\, dz}.$$

Since

$$\int_{0}^{p(z)/(cm(z))} du = \frac{1}{c} \frac{p(z)}{m(z)}$$

we have

$$\Pr(\tilde{\theta} \leq q) = \frac{\displaystyle\frac{1}{c} \int_{-\infty}^{q} p(z) dz}{\displaystyle\frac{1}{c} \int_{-\infty}^{\infty} p(z) dz} = \int_{-\infty}^{q} p(z) dz$$

which, by definition, implies that $\tilde{\theta}$ has density p. ◁

Some important features of the accept-reject algorithm warrant elaboration:

1. The target density p need only be known up to a factor of proportionality. All that matters is that we can sample from a function that majorizes the target density, and we can control that through the scaling constant, c. This is especially useful in the context of Bayesian analysis, where it is often the case that posterior densities are only known up to an (unknown) proportionality constant: i.e. via Bayes Rule, the posterior density is *proportional to* the prior density times the likelihood. In a sense, the unknown proportionality constant can be absorbed into the user-controlled scaling constant in the accept-reject algorithm.

2. An accept-reject algorithm produces potentially many draws that are rejected, and hence the algorithm can be computationally inefficient. This inefficiency will be mitigated to the extent that the majorizing function g closely approximates the target density, p. In such a case the ratio $r = p/g$ will be approximately one, and almost all draws will be accepted at lines 5 and 6 of the accept-reject algorithm, above. This highlights that careful choices of the proposal density and the scaling constant c can increase the efficiency of an accept-reject algorithm.

3. In the typical case where both the target density p and the proposal density are in fact densities, the scaling constant must be greater than 1.0. Note however, that since the majorizing function simply re-scales the proposal density, if the proposal density has thinner tails than the target density, then no matter the choice of $c > 1$, the re-scaled proposal density will never fully envelope the target. Thus, it is impossible to use an accept-reject algorithm to sample from a heavy-tailed density like the Cauchy using a normal density as the proposal density; however, the reverse can indeed work (Robert and Casella 2004, Problem 2.34).

■ Example 3.8

Truncated normal sampling, Example 3.6, continued. Accept-reject sampling is also an attractive way to generate samples from a truncated normal distribution. Suppose the target density is a normal density, truncated to the region $\theta > k$. Without loss of generality, suppose the normal density has mean zero and unit variance, so the target density is

$$p(\theta) = \phi(\theta)/(1 - \Phi(k)) \propto \exp(-\theta^2/2),$$

if $\theta > k$, and 0 otherwise. Robert (1995) considered using a *translated exponential density* as a majorizing function for this problem, i.e.

$$m(\theta) = \alpha \exp\left[-\alpha(\theta - k)\right], \tag{3.5}$$

(contrast the standard exponential density given in Definition B.33). Note that the parameter α in the translated exponential density acts as the scale parameter c in the accept-region algorithm. The ratio of the target density to the majorizing function is

$$\begin{aligned} r = p(\theta)/m(\theta) &\propto \alpha^{-1} \exp(-\theta^2/2)/\exp[-\alpha(\theta - k)] \\ &= \alpha^{-1} \exp\left[-\theta^2/2 + \alpha(\theta - k)\right]. \end{aligned}$$

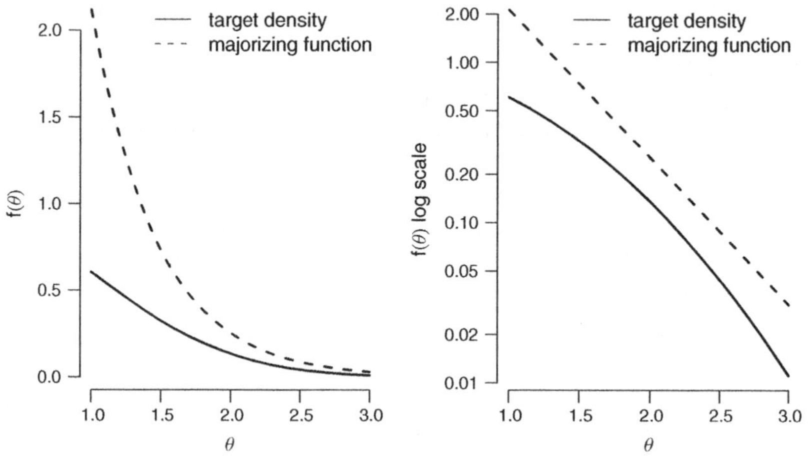

Figure 3.11 Target and proposal densities, Example 3.8. The target density $p(\theta)$ is a standard normal density, truncated to the region $\theta > 1$, i.e. $p(\theta) \propto \exp(-\theta^2/2)\mathcal{I}(\theta > 1)$. The majorizing function is a translated exponential density, $m(\theta) = \alpha \exp[-\alpha(\theta - k)]$, with $\alpha \approx 1.62$ and $k = 1$ (the truncation point). The panel on the right shows the two densities on a logarithmic scale, highlighting the heavier tails of the majorizing translated exponential density relative to the Gaussian target density.

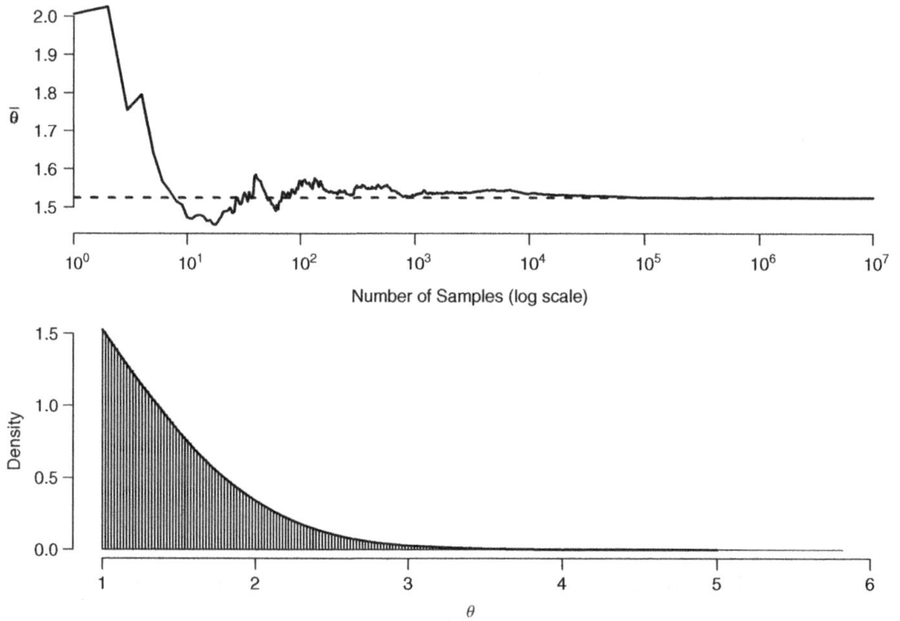

Figure 3.12 Accept-reject sampling, Example 3.8. The top panel shows a running Monte Carlo estimate of the mean of the target density (n.b., the sample count is plotted on a logarithmic scale); the known mean, $\phi(1)/(1 - \Phi(1)) \approx 1.52$ is shown with a dotted horizontal line. The lower panel is a histogram of the 10^7 sampled values, with the actual target density, $\phi(\theta)/(1 - \Phi(1))$ overlaid as a solid line.

Optimizing this expression with respect to α generates the optimal proposal density in the class of translated exponential densities. Differentiating and solving for α yields $\alpha = (k \pm \sqrt{k^2 + 4})/2$ as the optimal choices of α, again recalling that k is the truncation point. We take the positive root as the solution, ensuring $\alpha > 0$; i.e. we require the majorizing function (Equation 3.5) to be everywhere positive, such that it envelopes the target. See also Liu, (2001, 25).

As a specific example, consider the case of sampling from a standard normal density, truncated to lie above $k = 1$. The discussion above implies that the optimal choice of α is $\tilde{\alpha} = (1 + \sqrt{1^2 + 4})/2 \approx 1.62$.

Figure 3.11 compares the target and proposal densities for this specific example, with the right panel comparing the densities on the logarithmic scale, highlighting the heavier tails of the translated exponential density relative to the Gaussian target density. The top panel of Figure 3.12 shows the Monte Carlo estimate of the mean of the truncated normal density over 10^7 iterations of the accept-reject algorithm. As expected, the Monte Carlo estimate converges on the mean of the truncated normal density, $\phi(1)/(1 - \Phi(1)) \approx 1.52$ reasonably quickly. The lower panel of Figure 3.12 shows a histogram of the sampled values, with the actual target density, $\phi(\theta)/(1 - \Phi(1))$ overlaid as a solid line. Clearly, the accept-reject algorithm produces an extremely good approximation to the target density.

3.4.4 Adaptive rejection sampling

Adaptive rejection sampling is an especially useful sampling technique, generalizing the accept-reject method presented in Section 3.4.3. As the name suggests, the technique adapts the proposal density over the course of the iterations, building a proposal that gives a successively better approximation to the target density.

The algorithm is especially valuable when the density is of some arbitrary or unknown form such that choosing a majorizing proposal density is difficult. In fact, the algorithm is initialized without any proposal density at all, with the adaptive nature of the algorithm seeing it build up a piecewise series of exponential densities as the proposal density. The only technical restriction is that the target density be log-concave and continuously differentiable, which is the case for an extremely wide set of distributions encountered in applied Bayesian analysis (e.g. Dellaportas and Smith 1993). In this way the algorithm has tremendously wide applicability, and formed an essential ingredient in early versions of 'general purpose' Bayesian simulation software such as BUGS (Thomas, Spiegelhalter and Gilks 1992).

The algorithm is due to Gilks (1992) and Gilks and Wild (1992) with the ARS algorithm (as is it known) appearing as Algorithm AS 287 in the *Applied Statistics* statistical algorithm series (Wild and Gilks 1993). Variants of the algorithm are available where the target density fails the log-concavity restriction (e.g. see Section 5.1.2), or where the target is not continuously differentiable (the so-called 'DFARS' algorithm, for 'derivative-free' adaptive rejection sampling).

For completeness, we provide a formal definition of log-concavity before discussing the algorithm. The discussion here focuses on the derivative-free version of the algorithm, since in practice the derivatives of the target density may be costly to derive and program, even if they exist.

Definition 3.1 *A density $p(\boldsymbol{\theta})$, $\boldsymbol{\theta} \in \mathbb{R}^k$, is log-concave if the determinant of*

$$\mathbf{H} = \begin{pmatrix} \frac{\partial^2 \log p}{\partial \theta_1 \partial \theta_1} & \frac{\partial^2 \log p}{\partial \theta_1 \partial \theta_2} & \cdots & \frac{\partial^2 \log p}{\partial \theta_1 \partial \theta_k} \\ \frac{\partial^2 \log p}{\partial \theta_2 \partial \theta_1} & \frac{\partial^2 \log p}{\partial \theta_2 \partial \theta_2} & \cdots & \frac{\partial^2 \log p}{\partial \theta_2 \partial \theta_k} \\ \vdots & \vdots & \ddots & \vdots \\ \frac{\partial^2 \log p}{\partial \theta_k \partial \theta_1} & \frac{\partial^2 \log p}{\partial \theta_k \partial \theta_2} & \cdots & \frac{\partial^2 \log p}{\partial \theta_k \partial \theta_k} \end{pmatrix}$$

is non-positive.

Remark. For the case where $k = 1$, the density $p(\theta)$ is log-concave if $\partial^2 \log p/\partial\theta^2 \leq 0$.

The key thing about the focus on log-concave densities is that (a) the operation of taking logs turns many densities into concave functions (e.g. see part 2 of Problem 3.10); (b) any strictly concave function has a unique maximum. Knowing that the log target density is concave makes it easy to not only find the mode of the target density, but to build an envelope over the target density, as we now describe.

Suppose a function $f(\theta)$ is log-concave, $\theta \in \Theta \subseteq \mathbb{R}$. Consider a set of $J \geq 3$ ordered points $\mathcal{D}$ such that $\theta_1 < \theta_2 < \ldots < \theta_J$. Let $h(\theta) \propto \log f(\theta)$, and let $\mathcal{P} = \{P_j = (\theta_j, h(\theta_j))' \in \mathbb{R}^2 : j = 1, \ldots, J\}$. Assume that at least two points in $\mathcal{D}$ lie on either side of the unique mode of f i.e. the set $\mathcal{D}$ brackets the mode of f. We can verify that $\mathcal{D}$ does in fact bracket the mode of f if the chord joining the two left most points in $\mathcal{P}$ has a positive gradient and the chord joining the two right most points has a negative gradient. Observe that computing these chords requires only evaluations of $h \propto \log f$.

By the concavity of h, the chords form a *lower envelope* of h, which we denote as $\underline{h}_J$; the notation and much of the exposition here follows Robert and Casella (2004, §2.4.2). If we project the chords beyond the points where they intersect h, the resulting lines lie above h. Let $L_{j,j+1}$ be the line that is the extension of the chord connecting P_j and P_{j+1}. We can form an *upper envelope* $\bar{h}$ of h by connecting segments of the $L_{j,j+1}$ that most closely envelope h. That is, at the abscissa θ^*, the ordinate of the upper envelope is given by finding

$$\bar{h}(\theta^*) \equiv \min\{L_{j,j+1}(\theta^*) : L_{j,j+1}(\theta^*) \geq h(\theta^*), j = 1, \ldots, J-1\},$$

where $L_{j,j+1}(\theta^*)$ denotes the ordinate on $L_{j,j+1}$ corresponding to abscissa θ^*. Generally, for $\theta^* \in [\theta_j, \theta_{j+1}]$, $\bar{h}(\theta^*)$ will be given by either $L_{j-1,j}(\theta^*)$ or $L_{j+1,j+2}(\theta^*)$, save for some special handling of $\theta^* \notin [\theta_1, \theta_J]$, as we will see with the examples in Figure 3.13. The lower envelope is simply $\underline{h}(\theta) = L_{j,j+1}(\theta)$, $\theta \in [\theta_j, \theta_{j+1}]$. In terms of the original target density f, we can construct the lower envelope $\underline{f} = \exp \underline{h}$ and an upper envelope $\bar{f} = \exp(\bar{h})$.

Figure 3.13 shows the resulting upper and lower envelopes generated by this construction for three target densities: a $N(0, 1)$ density (top row), a Beta(3, 1.3) density (middle row) and a Gamma(1.5,1.15) density (bottom row). Panels on the left show the log target density h (thick line), with the extended chords $L_{j,j+1}$ (thin lines) and the

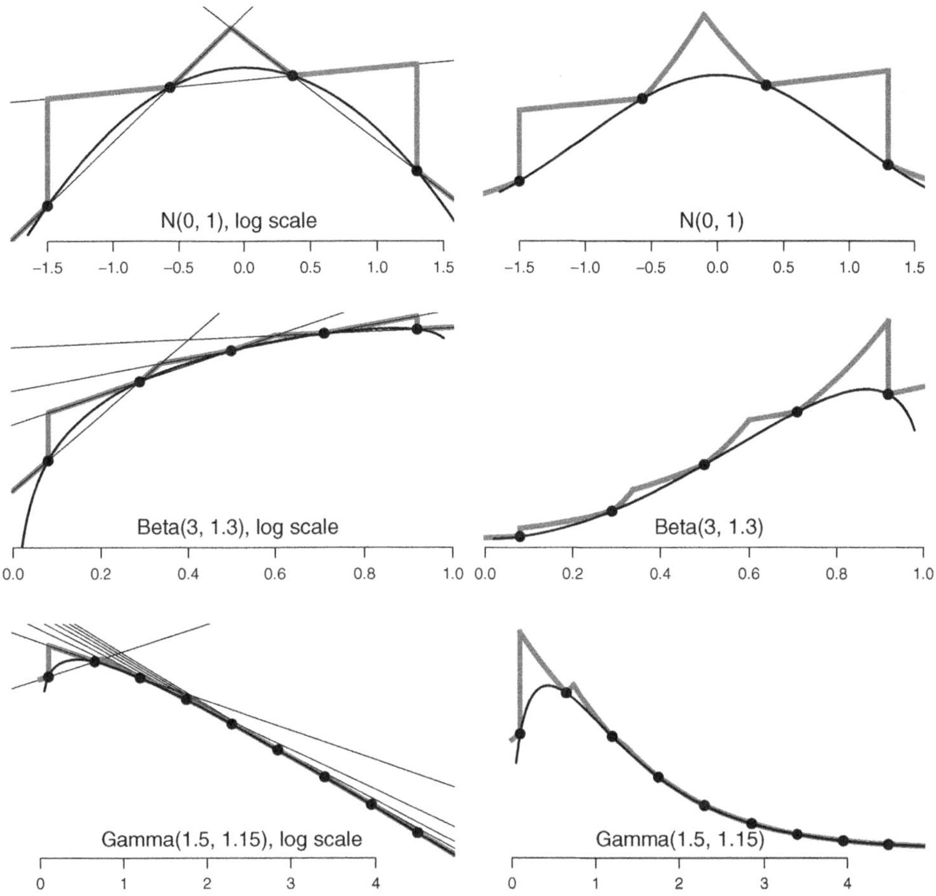

Figure 3.13 Upper and lower envelopes for derivative-free adaptive rejection sampling, three densities. Panels on the left show the log target densities $h(\theta)$; panels on the right show the target densities. Thin lines (left panels) show the chords joining evaluation points P_j. $L_{j,j+1}$ is the line formed by extending the $P_j P_{j+1}$ chord. Segments of the $L_{j,j+1}$ combine to generate the upper envelopes $\bar{h}$ (gray lines).

resulting upper envelope $\bar{h}$ (thick gray line). Panels on the right show the target density f and the upper envelope $\bar{f} = \exp(\bar{h})$. For the normal density we have chosen just $J = 4$ evaluation points $\mathcal{D}$, shown as points on the target density; for the Beta and Gamma densities we have $J = 5$ and $J = 9$ evaluation points, respectively. These different values of J were chosen to illustrate that as J increases the envelopes tend to the target density. Note that in Figure 3.13 the upper envelopes of the Beta and Gamma densities are almost visually indistinguishable from the respective target densities, in part due to the way that these functions are approximately linear over reasonably large ranges on the log scale.

We handle the extension to the limits of the support of the target density of f as follows. For f with bounded support (on either or both sides), we can always set x_1 to any known lower bound and x_J to any known upper bound. In some cases the target density f may be zero at the boundary and $h = \log f = -\infty$ (e.g. the Beta density example in Figure 3.13). Alternatively, f may have unbounded support. In these cases the envelope $\bar{h}$ can be defined as an increasing linear function of θ for $\theta \leq \theta_1$ and/or as a decreasing linear function of θ for $\theta \geq \theta_J$.

The upper envelope $\bar{h}$ is a series of r_J piecewise linear functions, each one an element of $\mathcal{L} = \{L_{j,j+1} : j = 1, \ldots, J-1\}$ operating on $[\theta_i, \theta_{i+1}]$, $i = 0, \ldots, r_J$. Note that $r_J > J$ in the examples in Figure 3.13; e.g. $J = 4$ but $r_J = 6$ in the normal example in the top panel. Each linear component of $\bar{h}$ can be written as $\bar{h}_i(\theta) = \alpha_i + \beta_i\theta$.

We can now define a proposal density as the piecewise exponential function

$$g_J(\theta) = G_J^{-1}\left[\sum_{i=0}^{r_J} \exp(\beta_i\theta + \alpha_i) \times \mathcal{I}(\theta \in [\theta_i, \theta_{i+1}])\right] \tag{3.6}$$

where θ_0 and θ_{r_J+1} are the (possibly infinite) limits of the support of f, $\mathcal{I}$ is an indicator function returning one if its argument is true and zero otherwise and G_J is the normalizing constant

$$\sum_{i=0}^{r_J} \int_{\theta_i}^{\theta_{i+1}} \exp(\beta_i\theta + \alpha_i)d\theta = \sum_{i=0}^{r_J} \exp(\alpha_i)\frac{\exp(\beta_i\theta_{i+1}) - \exp(\beta_i\theta_i)}{\beta_i}. \tag{3.7}$$

This normalizing constant takes a slightly different form in the case of unbounded support ($\theta_0 = -\infty$ and/or $\theta_{r_J+1} = +\infty$); see Robert and Casella (2004, 70). Note that the function $G_J g_J(\theta)$ is the piecewise exponential majorizing function with respect to f (the gray lines in the right-hand column of panels in Figure 3.13).

A key step in the ARS algorithm involves sampling from the proposal density g_J, which can be done using the following piecewise version of an inverse-CDF approach (Section 3.4.1):

Algorithm 3.11 Sampling from piecewise exponential density (Robert and Casella 2004, 70–71).

Require: $\theta_i, \theta_{i+1}, \alpha_i, \beta_i, i = 0, \ldots, r_J$ as in equation 3.6; G_J as in equation 3.7.

$p_i \leftarrow \exp(\alpha_i)\dfrac{\exp(\beta_i\theta_{i+1}) - \exp(\beta_i\theta_i)}{G_J\beta_i}, i = 0, \ldots, r_J.$

sample i^* from Multinomial($\mathbf{p}$, 1), $\mathbf{p} = (p_0, \ldots, p_{r_J})'$.

sample $U \sim$ Unif(0, 1)

$\theta \leftarrow \beta_{i^*}^{-1} \log[\exp(\beta_{i^*}\theta_{i^*}) + U(\exp(\beta_i\theta_{i^*+1}) - \exp(\beta_i\theta_{i^*}))]$.

We can now sketch the ARS algorithm itself; see Algorithm 3.12. Note that ARS exploits the fact that we have computed a lower envelope, to augment traditional accept-reject sampling. Line 6 implements what is known as a 'squeezing' step, utilizing the information in the lower envelope; see Marsaglia (1977), Gilks (1996) and Robert and Casella (2004, § 2.4.1).

Algorithm 3.12 Adaptive rejection sampling.

1: $t \leftarrow 0$
2: Initialize $J \geq 3$.
3: Initialize evaluation points $\mathcal{D}$ and chords $\mathcal{L}$

Require: $\mathcal{D}$ brackets the mode of h
4: sample $\theta \sim g_J(\theta)$, see Algorithm 3.11, above.
5: sample U $\sim$ Unif(0, 1)
6: **if** $U \leq \underline{f}_J(\theta)/G_J g_J(\theta)$ **then**
7: accept θ
8: **else**
9: **if** $U \leq f(\theta)/G_J g_J(\theta)$ **then**
10: accept θ
11: **end if**
12: $\mathcal{D} \leftarrow \mathcal{D} \cup \{\theta\}$; $J \leftarrow J + 1$; update $\mathcal{L}$.
13: **end if**

The other noteworthy feature of the algorithm is the *adaptive* step in line 12, where the set of evaluation points $\mathcal{D}$ is augmented with the point θ; i.e. since we had to evaluate $f(\theta)$ at line 9, we have another evaluation point and another chord to add to $\mathcal{L}$. Thus, over successive iterations of the algorithm, with successively more evaluation points, the upper and lower envelopes collapse on the target function and the algorithm becomes quite effective. Indeed, as J grows, $\underline{f}_J \to f$ and the ratio $\underline{f}_J(\theta)/G_J g_J(\theta) \to 1$, and the probability of accepting θ at line 6 approaches 1.0. In turn, this means that the algorithm is unlikely to visit the steps at lines 8 through 12 after an initial adaptive phase. These initial iterations may be relatively costly, involving a reasonable amount of computing to generate the set of chords $\mathcal{L}$, the envelopes $\underline{g}_J$ and $\bar{g}_J$, and the normalizing constant G_J each time $\mathcal{D}$ grows and J increments. But after this initial, adaptive phase, ARS is quite effective. In fact, the development of ARS was a critical step in developing a general purpose computer program for simulation-based Bayesian statistical analysis; we return to this point in Chapter 6. Another key development is the extension of ARS to the case of non-log-concave densities (e.g. Gilks, Best and Tan 1995), using methods that we discuss in Section 5.1.

A 'standalone' version of ARS is available in the `R` package `ars` (Rodriguez 2007). This implementation in `R` is essentially a 'wrapper' to a `C++` port of the original `Fortran` code developed by Gilks and Wild (1992), requiring that the user provide an expression for the first derivative of h with respect to θ.

3.5 Further reading

Devroye (1986) is a standard text on random number generation, covering a wide range of densities; see also Ripley (1987). Chapter 2 of Robert and Casella (2004) provides a compact survey of methods for random variable generation. The volumes by Johnson, Kotz and Balakrishnan (1994, 1995) also provide many references to techniques for sampling from the wide array of different probability densities and mass functions they survey.

Liu (2001) is a book length treatment of Monte Carlo methods and their application in statistics and many other fields of science. The book by Krauth (2006) provides an accessible introduction to the uses of Monte Carlo methods in its *locus classicus* of statistical mechanics; Chapter 1 of the book is a particularly helpful introduction to Monte Carlo methods. See also Landau and Binder (2000).

Problems

3.1 Most 'random number generators' that you will encounter are actually 'pseudo-random number generators', a deterministic sequence of numbers that is designed to display extremely (!) low autocorrelation, long periodicity and low levels of dependence on its initial state; e.g. the Mersenne-Twister (Matsumoto and Nishimura 1998), the default pseudo-random number generator in R. Given a starting point in the sequence – the so-called 'seed' – the 'random draws' are actually deterministic. With this as background, consider the following questions.

1. What is the default seed in your preferred statistical computer program? How does one set the seed of the pseudo random number generators in your preferred statistical computer program?
2. What is the default (uniform) pseudo-random number generator in your preferred statistical computer program?
3. Generate 1000 random draws from a uniform density on the unit interval, using the default seed. Note the mean of these realized values.
4. Set the seed to a known value (e.g. an integer such as 1234 or 31459, or perhaps some other form that your software will accept). Generate 1000 random draws from a uniform density on the unit interval. Note the mean of these sampled values. Do they differ from your answer to the previous question? Why? Do both set of answers differ from 0.5? Why? What is the source of this variation around the known, analytic mean of 0.5?
5. Repeat the previous question. That is, set the seed to the *same* value, generate 1000 random draws from the uniform density on the unit interval. Verify that you get exactly the same answer as before.
6. Now generate an additional 1000 random draws from a uniform density on the unit interval. Note the last value drawn. Now repeat the last two steps: (a) set the seed; (b) generate 1000 random draws; (c) generate another 1000 draws. Verify that the last generated value in each instance are the same number.

3.2 Consider computing $\pi \approx 3.141592654\ldots$ by Monte Carlo methods.

1. A crude Monte Carlo method for estimating π can be generated as follows. Let $\mathbb{D}$ be the disc enclosed by the unit circle

$$\mathbb{D} = \{(x, y) \in \mathbb{R}^2 : x^2 + y^2 < 1\}$$

with area π, and let $\mathcal{S} = [0, 1] \times [0, 1]$ be the closed unit square, with area 1. Then $\mathcal{Q} = \mathbb{D} \cap \mathcal{S}$ has area $\pi/4$. This implies that if $U \sim \text{Unif}(0, 1)$ and $V \sim \text{Unif}(0, 1)$ then $\Pr(\sqrt{U^2 + V^2} < 1) = \pi/4$, and hence suggests a Monte Carlo strategy for estimating π: sample U and V many times and simply note the proportion of draws (U, V) that lie in $\mathcal{Q}$. The function `simpi` in the `R` package `pscl` (Jackman 2008b) implements this Monte Carlo estimator of π. How accurate is this Monte Carlo approach to estimating π? Observe that this

Monte Carlo based estimate of π is stochastically converging on the true value as the number of Monte Carlo iterates gets larger.

2. The algorithm sketched in the previous question draws $U \sim \text{Unif}(0, 1)$ and $V \sim \text{Unif}(0, 1)$, i.e. $(U, V) \in \mathcal{S}$. What is $p = \Pr[(U, V) \in \mathcal{Q}]$? Given p, use what you know about the binomial probability mass function (or a large n normal approximation to the binomial) to answer the following question: assuming we possess a computing technology that generates approximately independent, uniformly distributed samples from $\mathcal{S}$, how many independent draws will we require such that this Monte Carlo estimate of π has a 95 % chance of being correct to the j-th decimal place? In a table or a graph, supply answers for $j = 1, 2, \ldots, 8$.

3.3 Refer to Example 3.3. Use Monte Carlo methods to compute the posterior density of the quantities

1. θ_1/θ_0
2. the odds ratio OR (recall Equation 2.9)
3. the log odds ratio.

Use histograms and numerical summaries to communicate interesting features of these posterior densities.

3.4 Repeat the previous problem, but with reference to Example 2.8. Compare your results with Figure 2.5.

3.5 The logistic density is $p(\theta) = \exp(\theta)/(1 + \exp(\theta))^2$:

1. What is the CDF?
2. What is the inverse CDF?
3. Write a computer program to generate samples from the logistic CDF using the inverse-CDF method described in Section 3.4.1.
4. Show that this density is log-concave.

3.6 Polling firms often report results by rounding to the nearest percentage point. Suppose a polling firm reports that 50 % of respondents intend voting Democratic at the next election. The poll is based on a sample size of $n = 1400$.

1. Calculate a 95 % interval for the population proportion of voters intending to vote Democratic (θ), ignoring the rounding (i.e. use the asymptotically-valid normal approximation for the posterior density of θ, with an improper, uninformative prior density for θ).
2. Now consider the marginal posterior of θ obtained by integrating out the uncertainty as to the actual poll result; i.e. if the polling firm reported 50 %, then all we know is $\hat{\theta} \sim \text{Uniform}(.495, .505)$. What is a 95 % interval for θ after we acknowledge this additional uncertainty due to rounding?

3. Does this 'rounding-corrected' interval differ substantially from the interval generated ignoring the rounding?

3.7 Consider $\boldsymbol{\theta} \in \mathbb{R}^p$ with density $p(\boldsymbol{\theta}) \equiv N(\boldsymbol{\mu}, \boldsymbol{\Sigma})$. We are interested in inference over the order statistics of $\boldsymbol{\theta}$. Specifically, we seek $\Pr[r(\theta_j) = k]$, where r is a ranking function, and $j, k \in \{1, \ldots, p\}$.

1. Suggest a Monte Carlo based strategy for computing $\Pr[r(\theta_j) = k]$.
2. Suppose $p = 3$, $\boldsymbol{\mu} = c(0, 1, 1)'$ and

$$\boldsymbol{\Sigma} = \begin{bmatrix} 1.5 & .5 & -.8 \\ .5 & 1 & .2 \\ -.8 & .2 & 1 \end{bmatrix}$$

Use Monte Carlo simulation to induce a probability mass function over

(a) the rank of each θ_j, $j = 1, 2, 3$

(b) which component of $\boldsymbol{\theta}$ has rank k, $k = 1, 2, 3$

(c) over the 3! possible rank orderings

See also Clinton, Jackman and Rivers (2004a, b) and Example 9.2.

3.8 Let f be a proper probability density on $\Theta \subseteq \mathbb{R}$. Suppose g majorizes f. Is g a proper density? (see Definition B.13)

3.9 Show that a posterior density $p(\boldsymbol{\theta}|\mathbf{y}) \propto p(\boldsymbol{\theta}) f(\mathbf{y}|\boldsymbol{\theta})$ is log-concave if both the prior $p(\boldsymbol{\theta})$ and $f(\mathbf{y}|\boldsymbol{\theta})$ are log-concave.

3.10 Refer to the definition of a log-concave function (Definition 3.1) in § 3.4.4.

1. (Robert and Casella 2004, Problem 2.40 (a)). The exponential family of densities is defined as

$$f(\mathbf{x}) = h(\mathbf{x}) \exp[\boldsymbol{\theta}' T(\mathbf{x}) - A(\boldsymbol{\theta})], \quad \boldsymbol{\theta}, \mathbf{x} \in \mathbb{R}^k$$

For simplicity, focus on the univariate case ($k = 1$). Show that this family of (univariate) densities is log-concave with respect to $\mathbf{x}$.

2. Consider the univariate normal density (see Definition B.30). Show that this density is *not* concave (at least not globally); solve for the inflection points where the normal density changes from being a concave function, to a convex function. Then show that the normal density is log-concave.
3. Is the following proposition true or false (and why)?: a concave function is log-concave.
4. Is the following proposition true or false (any why)?: if x_0 is the mode of a function $f(x)$ then it is the mode of $\log f(x)$.

4

Markov chains

In the previous chapter we considered three methods for generating samples from arbitrary densities. These three methods all generate *independent* samples from the densities of interest, or at least in theory (the qualification 'in theory' here is necessary, since in practice, computers usually don't generate truly random numbers, but deterministic sequences of 'pseudo' random numbers). But we will now see that Monte Carlo methods can let us learn about random quantities *even when the samples from the target density are not independent*, but are serially dependent or 'autocorrelated'.

The key mathematical tool here is a Markov chain, a particular kind of random process that we define below. Markov chains are named for the Russian mathematician who invented them, Andrey Markov (1856–1922).

In the preface to his authoritative treatement of Markov chains, Brémaud (1999, ix) writes

> When Markov introduced his famous model in 1906, he was not preoccupied with applications. He just wanted to show that independence is not necessary for the law of large numbers... An example that he considered was the alternation of consonants and vowels in Pushkin's *Eugene Onegin*, which he described as a two-state chain. (This, however, does not say much about the plot!)

Today, Markov chains appear in virtually every branch of science; Bradley and Meek (1986) provide a survey of social science applications for a lay audience. I stress that our interest in Markov chains here is quite focused, and I will not provide a general introduction to the topic. More information can be found by consulting the references cited throughout this chapter or listed in section 4.5.

This chapter is a little more technical than some of the preceeding chapters. For that reason, a brief preview may be helpful. A Markov chain is a stochastic process: a useful, physical analogy is a particle moving randomly in some space. In the context of Bayesian

Bayesian Analysis for the Social Sciences S. Jackman

statistics, the trajectory of the particle is the output of a Monte Carlo algorithm and the 'space' is a parameter space, $\boldsymbol{\Theta}$, the support of a posterior density $p(\boldsymbol{\theta}|\text{data})$.

In the first part of this chapter I introduce notation and definitions that will in turn allow us to state an important theorem about Markov chains, an *ergodic* theorem (Proposition 4.7). To set the stage, remember that we want to characterize the density $p(\boldsymbol{\theta}|\text{data})$ in some sense: in Chapter 3 we saw how Monte Carlo methods let us learn about a random variable $\boldsymbol{\theta}$ via repeatedly *sampling* from the density of $\boldsymbol{\theta}$. What we will see in this chapter is that it is possible to construct Markov chains on parameter spaces such that the chain visits locations in the parameter space with frequencies proportional to the probability of those locations under a density of interest, such as a posterior density $p(\boldsymbol{\theta}|\text{data})$. A Markov chain with this property is said to be *ergodic*: i.e. the time-averaged probability of a state – how *often* the Markov chain visits a point/region in $\boldsymbol{\Theta}$ – is equal to the probability of that point/region under a density of interest.

In turn, this means that after we set up an appropriately constructed Markov chain on the parameter space, we can store the 'path', 'trajectory' or 'iterative history' of the chain, treating these values as a series of samples from the posterior density of interest. We then use the Monte Carlo principle considered in the previous chapter, since the random sequence generated by an arbitrarily large number of iterations of the Markov chain will generally constitute a sample from the density of interest, albeit usually not an independent series of draws. The lack of independence here is not fatal: ergodicity is a form of a law of large numbers for dependent sequences, of the sort generated by a Markov chain. The only real consequence of the dependence is a lack of efficiency – we will need to let the Markov chain run longer and generate more samples than we would if we could generate independent samples – but ergodicity ensures a form of *simulation consistency* holds (again, see Proposition 4.7 later in this chapter).

Much of this chapter prepares the way for seeing the conditions under which ergodicity holds for Markov chains on general state spaces. We will draw on these concepts in Chapter 5 to show that the workhorse MCMC algorithms – the Gibbs sampler and the Metropolis-Hastings algorithm – usually generate ergodic Markov chains with respect to posterior densities. We also briefly consider some results as to the *rate* at which simulation consistency 'kicks in' for these algorithms in various settings, a question of considerable practical importance.

4.1 Notation and definitions

I begin by defining a Markov process or a Markov chain, a random process or sequence with a particular kind of serial dependence.

Definition 4.1 (Markov Property; Markov Process, Markov Chain). *Let* $\{\boldsymbol{\theta}^{(t)}\}$ *be a stochastic process, a collection of random variables indexed by time,* t. *If the stochastic process has the property that*

$$Pr[\boldsymbol{\theta}^{(t+a)} = y|\boldsymbol{\theta}^{(s)} = x_s, s \leq t] = Pr[\boldsymbol{\theta}^{(t+a)} = y|\boldsymbol{\theta}^{(t)} = x_t], \forall\, a > 0.$$

then the process is said to possess the Markov property, and any such process is said to be a Markov process or a Markov chain.

The Markov property is simply a *conditional independence* restriction: beliefs about the state of the process today depend on the past history of the state only through the 'immediate past'. Put differently, conditional on knowing the state of the process today, predictions about future states do not depend on earlier states of the process. Note that we are restricting our attention to *discrete time* Markov chains, rather than the more general class of continuous time Markov processes, since we are interested in the application of Markov chain theory to the process of sampling sequentially from a posterior density (i.e. the iterations of our sampling algorithms and the samples generated at each iteration possess the discrete index $t = 1, 2, \ldots, T$).

4.1.1 State space

Throughout this chapter I reserve the symbol $\boldsymbol{\Theta}$ for the *state space* of a Markov chain, the set of values that $\boldsymbol{\theta}$ might take. In the specific context of using Markov chains to explore posterior densities of parameters, the state space is alternately the *parameter space* (or at least a subset of the parameter space where the posterior density is positive).

One school of thought holds that since computers can only represent a finite subset of the real numbers (typically limited by the amount of floating point precision available on a particular piece of hardware, or the amount of precision implementable in a particular algorithm on that hardware), all implementable Markov chains are on finite state spaces (e.g. Hastings 1970). The treatment I provide here provides definitions and key results for Markov chains on both discrete and continuous state spaces. The definitions and theorems are relatively easy to grasp for the former case, while some measure-theoretic buttressing is required for the latter. When working with Markov chains on continuous state spaces, the probability of a move to any specific value $\boldsymbol{\theta} \in \boldsymbol{\Theta}$ is zero, and so we will consider the probability of a transition to a *region* in the parameter space, $\mathcal{A} \subset \boldsymbol{\Theta}$. Any attempt to cover this material with a semblence of rigor must introduce some definitions from measure theory, although I will do my best to keep this to a minimum. A review of the definitions in Appendix B.1 might be helpful.

4.1.2 Transition kernel

Suppose we are interested in a random quantity $\boldsymbol{\theta} \in \boldsymbol{\Theta} \subseteq \mathbb{R}^p$ with density $p(\boldsymbol{\theta})$. Suppose we possess a sampling technology that generates a sequence of sampled values $\{\boldsymbol{\theta}^{(t)}\}$. Unlike the situation considered in Chapter 3, these sampled values need not be serially independent. The Markov property given in Definition 4.1 can be restated as

$$\Pr(\boldsymbol{\theta}^{(t)} \in \mathcal{A} | \boldsymbol{\theta}^{(t-1)}, \boldsymbol{\theta}^{(t-2)}, \ldots) = \Pr(\boldsymbol{\theta}^{(t)} \in \mathcal{A} | \boldsymbol{\theta}^{(t-1)}), \ \forall \ \mathcal{A} \subset \boldsymbol{\Theta}, \tag{4.1}$$

or, more technically, $\forall \ \mathcal{A} \subset \mathcal{B}(\boldsymbol{\Theta})$, where $\mathcal{B}(\boldsymbol{\Theta})$ is the Borel σ-algebra on $\boldsymbol{\Theta}$ (see Definition B.2). Equation 4.1 defines a *transition function* or, more specifically, a *transition probability*, the conditional probability that at step t, the Markov chain will 'jump' from $\boldsymbol{\theta}^{(t-1)}$ to the set $\mathcal{A}$, given that the Markov chain is at $\boldsymbol{\theta}^{(t-1)}$.

When the state space of the Markov chain is discrete (i.e. $\boldsymbol{\theta}$ is a discrete random variable), the transition probabilities can be represented as a matrix, specifically, a square matrix with non-negative entries known as a *transition matrix*:

Definition 4.2 (Transition Matrix, Discrete Markov chain). *Consider a Markov chain* $\{\boldsymbol{\theta}^{(t)}\}$ *on a discrete state space* $\boldsymbol{\Theta}$. *Then the matrix* $\mathbf{K} = (K_{ij})$ *with* $K_{ij} = \Pr(\boldsymbol{\theta}^{(t)} = j|\boldsymbol{\theta}^{(t-1)} = i)$ *characterizes the transition kernel of the Markov chain, and is referred to as a transition matrix.*

For the case of a Markov chain with J discrete states, the vector $\mathbf{p}^{(t)} = (p_1^{(t)}, \ldots, p_J^{(t)})'$ contains the probabilities that the chain is in state j at time t; i.e. $p_j^{(t)} = \Pr(\theta^{(t)} = j)$, $j = 1, \ldots, J$. The transition matrix defines a transition equation for a discrete Markov chain:

$$\mathbf{p}^{(t)} = \mathbf{p}^{(t-1)}\mathbf{K}. \tag{4.2}$$

Note that in this discrete setting we have the obvious constraint that $\sum_{i \in \boldsymbol{\Theta}} K(i, \mathcal{S}) = 1, \forall\, \mathcal{S} \in \boldsymbol{\Theta}$, implying that each row of the transition matrix $\mathbf{K}$ sums to one.

■ Example 4.1

Two state, 'mover-stayer' Markov chain. Consider the following two state Markov chain, often used to provide a simple, stylized representation of social or political processes. At any given time point t, there are two possible states of the world; generically, we label the states '0' and '1', but in applications these may correspond to labor market status (employed or unemployed), incarceration (in jail, not in jail), martial status (married, not married), military conflict (war or peace), or majority control of the Congress (Democrats or Republicans). These models are sometimes referred to as 'mover-stayer' models, since at any given time point one can either move from or stay in the current state. Transitions between the two states occur stochastically, but subject to the probabilities laid out in the following table:

		$\theta^{(t)}$	
		0	1
$\theta^{(t-1)}$	0	$1-p$	p
	1	q	$1-q$

i.e. $\Pr(\theta^{(t)} = 1|\theta^{(t-1)} = 0) = K(0, 1) = p$ and $\Pr(\theta^{(t)} = 0|\theta^{(t-1)} = 1) = K(1, 0) = q$. Since $\theta^{(t)} \in \{0, 1\}\, \forall\, t$, we can form a 2 by 2 transition probability matrix

$$\mathbf{K}(\theta^{(t-1)}, \theta^{(t)}) = \begin{bmatrix} \Pr(0 \to 0) & \Pr(0 \to 1) \\ \Pr(1 \to 0) & \Pr(1 \to 1) \end{bmatrix} = \begin{bmatrix} 1-p & p \\ q & 1-q \end{bmatrix},$$

where $\Pr(i \to j) \equiv \Pr(\theta^{(t)} = j|\theta^{(t-1)} = i)$.

The transition matrix $\mathbf{K}$ is the key to computing probabilities over future states of the chain, given the current state. Note that by the law of total probability (Proposition 1.2), for the two state chain here we have

$$\begin{aligned}\Pr(\theta^{(t)} = 1) = \Pr(\theta^{(t)} &= 1|\theta^{(t-1)} = 1) \times \Pr(\theta^{(t-1)} = 1) \\ &+ \Pr(\theta^{(t)} = 1|\theta^{(t-1)} = 0) \times \Pr(\theta^{(t-1)} = 0).\end{aligned}$$

This suggests using the following equation to compute probabilities over the state of the Markov chain:

$$\begin{aligned}\mathbf{p}^{(t)} &= \left[\ \Pr(\theta^{(t)}=0) \quad \Pr(\theta^{(t)}=1)\ \right]\\ &= \left[\ \Pr(\theta^{(t-1)}=0) \quad \Pr(\theta^{(t-1)}=1)\ \right]\left[\begin{array}{cc}\Pr(0\to 0) & \Pr(0\to 1)\\ \Pr(1\to 0) & \Pr(1\to 1)\end{array}\right]\\ &= \mathbf{p}^{(t-1)}\mathbf{K}.\end{aligned}$$

Further, r-step ahead probabilities can be computed as $\mathbf{p}^{(t+r)} = \mathbf{p}^{(t)}\mathbf{K}^r$.

For purposes of illustration, suppose we initialize the chain in state '0' (i.e. $\theta^{(0)} = 0$), and we have $p = .9$, $q = .25$, such that

$$\mathbf{K} = \left[\begin{array}{cc} .1 & .9 \\ .25 & .75 \end{array}\right]. \tag{4.3}$$

Then $\Pr(\theta^{(1)} = 1|\theta^{(0)} = 0) = p = .9$, and $\Pr(\theta^{(1)} = 0|\theta^{(0)} = 0) = 1 - p = .1$. After another iteration of the chain, our beliefs about the likely state of the chain are given by

$$\begin{aligned}\Pr(\theta^{(2)} = 1|\theta^{(0)} = 0) &= \Pr(\theta^{(2)} = 1|\theta^{(1)} = 0) \times \Pr(\theta^{(1)} = 0|\theta^{(0)} = 0)\\ &\quad + \Pr(\theta^{(2)} = 1|\theta^{(1)} = 1) \times \Pr(\theta^{(1)} = 1|\theta^{(0)} = 0)\\ &= p(1-p) + (1-q)p\\ &= p(2-p-q) = .9(2-.9-.25) = .765.\end{aligned}$$

and so $\Pr(\theta^{(2)} = 0|\theta^{(0)} = 0) = 1 - .765 = .235$. A more straightforward route to obtaining $\mathbf{p}^{(2)}$ is to multiply the transition matrix $\mathbf{K}$ in Equation 4.3 by itself to yield $\mathbf{p}^{(2)} = \mathbf{p}^{(0)}\mathbf{K}\mathbf{K}$ where since $\theta^{(0)} = 0$, we have $\mathbf{p}^{(0)} = (1, 0)'$. Thus,

$$\mathbf{p}^{(2)} = \left[\ 1 \quad 0\ \right]\left[\begin{array}{cc} .1 & .25 \\ .9 & .75 \end{array}\right]\left[\begin{array}{cc} .1 & .25 \\ .9 & .75 \end{array}\right] = \left[\ .235 \quad .765\ \right],$$

producing the same result as the 'manual' calculation given above.

For the continuous case, we consider a probability measure p on the state space $\mathbf{\Theta}$ (or, actually, a measure on the Borel σ-algebra of $\mathbf{\Theta}$, $\mathcal{B}(\mathbf{\Theta})$, see Definitions B.1 to B.8); throughout the rest of this chapter I will also refer to p as a probability density on $\mathbf{\Theta}$. The *transition kernel* K can be thought of as an *operator* that governs how $p^{(t)}$ evolves, and hence in turn generates the Markov chain $\{\boldsymbol{\theta}^{(t)}\}$:

Definition 4.3 (Transition kernel, Markov chain on a continuous state space). *Consider a Markov chain $\{\boldsymbol{\theta}^{(t)}\}$ on a state space $\mathbf{\Theta}$. The transition kernel of the Markov chain is a function K that takes two arguments, (1) $\boldsymbol{\theta} \in \mathbf{\Theta}$, and (2) $\mathcal{A} \in \mathcal{B}(\mathbf{\Theta})$, such that*

1. *$K(\boldsymbol{\theta}, \cdot)$ is a probability measure (Definition B.8), $\forall\ \boldsymbol{\theta} \in \mathbf{\Theta}$*
2. *$K(\cdot, \mathcal{A})$ is an $\mathcal{B}(\mathbf{\Theta})$-measurable function (Definition B.4), $\forall\ \mathcal{A} \in \mathcal{B}(\mathbf{\Theta})$.*

That is, given that the chain is at $\boldsymbol{\theta}^{(t)}$ at iteration t, $K(\boldsymbol{\theta}^{(t)}, \cdot)$ is a probability measure over where the chain will jump to at the next iteration. Likewise, given $\mathcal{A} \in \mathcal{B}(\boldsymbol{\Theta})$, the kernel K is a measurable function, and some special cases aside, it is often a probability measure also.

With this definition, the transition kernel K lets us generate statements like the following:

$$\Pr(\boldsymbol{\theta}^{(t)} \in \mathcal{A}|\boldsymbol{\theta}^{(t-1)}) = \int_{\mathcal{A}} K(\boldsymbol{\theta}^{(t-1)}, d\boldsymbol{\theta}), \; \forall \, \mathcal{A} \in \mathcal{B}(\boldsymbol{\Theta}). \tag{4.4}$$

Alternatively, where p is understood to be a probability density function over $\boldsymbol{\Theta}$ – i.e. $p \equiv p(\boldsymbol{\theta})$, including the special case where p is a posterior density $p(\boldsymbol{\theta}|\text{data})$ – then the transition kernel generates *functional equations* such as

$$p^{(t)} = \int_{\boldsymbol{\Theta}} K(\boldsymbol{\theta}^{(t-1)}, \cdot) p^{(t-1)} d\boldsymbol{\theta}^{(t-1)}, \tag{4.5}$$

showing how the density of $\boldsymbol{\theta}$, $p^{(t)}(\boldsymbol{\theta})$, evolves over successive iterations of the Markov chain.

■ **Example 4.2**

First-order, autoregressive Gaussian process (a Markov chain on a continuous state space). One of the most widely encountered stochastic processes in the social sciences is the following first-order autoregressive process with Gaussian shocks or 'innovations': $\theta^{(t)} = \rho\theta^{(t-1)} + \epsilon^{(t)}, \theta^{(t)} \in \mathbb{R}$, where $\epsilon^{(t)} \overset{\text{iid}}{\sim} N(0, \sigma^2)$. This defines a Markov chain $\{\theta^{(t)}\}$, $\theta \in \boldsymbol{\Theta} = \mathbb{R}$, where $p^{(t)} \equiv N(\rho\theta^{(t-1)}, \sigma^2)$ and further,

$$p^{(t)} = \int_{\boldsymbol{\Theta}} \frac{1}{\sqrt{2\pi\sigma^2}} \exp\left[\frac{-(\theta^{(t)} - \rho\theta^{(t-1)})^2}{2\sigma^2}\right] p^{(t-1)} d\theta^{(t-1)}.$$

4.2 Properties of Markov chains

The following questions are key to our use of Markov chains:

1. Over successive iterations of the Markov chain, does the chain gravitate towards one state or the other, or to a stable, *equilibrium* distribution over its state space?
2. If such a distribution exists – also known as a *stationary*, *invariant* or *limiting* distribution – is it unique?
3. If a given Markov chain has a unique stationary distribution, *how long* does it take to get there after we start the Markov chain at some arbitrary point in its state space?
4. How do we assess *how close* a Markov chain has come to reaching its stationary distribution?

5. Can summaries of the trajectory of the Markov chain be taken as summaries of the invariant distribution? For instance, if over a long run of the Markov chain $\{\boldsymbol{\theta}^{(t)}\}$ we observe that the chain visits region $\mathcal{A}$ with relative frequency $\hat{p}_{\mathcal{A}}$, then under what conditions is $\hat{p}_{\mathcal{A}}$ a good estimate of the probability of $\Pr(\boldsymbol{\theta} \in \mathcal{A})$ under the chain's invariant distribution, p. This latter property is known as *ergodicity*, which we explore in detail in Section 4.4.

Understanding these features of Markov chains are of great practical importance in many applications: e.g. in Markov models of phenomena such as the progression of a fatal disease, will the disease lead to species extinction (or some steady state above extinction level), and if so, when?

In the specific context of exploring a posterior density, we are interested in the following problem: if we start a Markov chain at an arbitrary point $\boldsymbol{\theta}^{(0)}$, perhaps a sample from an arbitrary density p_0, then after T transitions, each defined by the transition kernel $K(\boldsymbol{\theta}^{(t-1)}, \cdot)$, then what is the density of $\boldsymbol{\theta}^{(T)}$? In particular, does this density, $p(\boldsymbol{\theta}^{(T)})$, converge to a unique, stationary density, say, the posterior density $p(\boldsymbol{\theta}|\text{data})$, as $T \to \infty$? In turn, does ergodicity hold? That is, can numerical summaries of the history of the Markov chain be taken as an estimate of the corresponding feature of $p(\boldsymbol{\theta}|\text{data})$?

We shall now see that if the transition kernel is defined appropriately, then the answers to the last three questions are 'yes'. We investigate these properties in turn, beginning by considering the conditions under which a Markov chain possesses a unique stationary distribution.

4.2.1 Existence of a stationary distribution, discrete case

For a Markov chain on a discrete state space, the existence of a stationary distribution turns on whether we can solve the following system of equations for $\mathbf{p}$:

$$\mathbf{p} = \mathbf{p}\mathbf{K} \Rightarrow \mathbf{p}(\mathbf{I} - \mathbf{K}) = \mathbf{0} \Rightarrow (\mathbf{I} - \mathbf{K})'\mathbf{p}' = \mathbf{0}' \tag{4.6}$$

where (in the case of a discrete state space) $\mathbf{K}$ is the transition matrix of the Markov chain. Clearly, if the transition matrix $\mathbf{K}$ is such that we can solve Equation 4.6 for $\mathbf{p}$, then $\mathbf{p}$ is a stationary distribution of the Markov chain. Finding a solution to Equation 4.6 is straightforward when we recognize that the form of Equation 4.6 implies that $\mathbf{p}'$ is an *eigenvector* of $\mathbf{K}'$, and, specifically, the eigenvector of $\mathbf{K}'$ corresponding to an *eigenvalue* of 1.0. Eigenvectors and eigenvalues are defined in the Appendix at Definition A.19; using that definition we see that since $\mathbf{K}'$ is a square matrix then

$$\lambda\mathbf{x} = \mathbf{K}'\mathbf{x} \tag{4.7}$$

where $\mathbf{x}$ is an eigenvector of $\mathbf{K}'$ and λ is an eigenvalue of $\mathbf{K}'$. That is, we recover the definition of a Markov chain's stationary distribution in Equation 4.6 if we set $\mathbf{p}'$ equal to an eigenvector $\mathbf{x}$ of $\mathbf{K}'$ associated with the eigenvector $\lambda = 1$. This suggests a method for obtaining the stationary distribution of a Markov chain on a discrete state space (and more). We explore this point in more detail presently (see Proposition 4.1, below).

4.2.2 Existence of a stationary distribution, continuous case

For a Markov chain $\{\boldsymbol{\theta}^{(t)}\}$ on a continuous state space, we want to know if we can solve the functional equation in Equation 4.5 for p, i.e. we seek a density p such that $p = \int_{\boldsymbol{\Theta}} p\, K(\boldsymbol{\theta}, \cdot) d\boldsymbol{\theta}$. That is, does there exist a density $p(\boldsymbol{\theta})$ that makes the same probability assignments over the state space $\boldsymbol{\Theta}$ over successive iterates of the chain? If such a density exists then it is called the *invariant density* of the Markov chain.

We provide a formal definition of this term (e.g. Meyn and Tweedie 2009, 229):

Definition 4.4 (Invariant measure, invariant probability measure). *Let* $\{\boldsymbol{\theta}^{(t)}\}$, $\boldsymbol{\theta} \in \boldsymbol{\Theta}$ *be a Markov chain with transition kernel* $K(\cdot, \cdot)$.

- *A σ-finite measure p (Definition B.6) is said to be invariant wrt the transition kernel $K(\cdot, \cdot)$ if*

$$p(\mathcal{A}) = \int_{\boldsymbol{\Theta}} K(\boldsymbol{\theta}, \mathcal{A}) p(d\boldsymbol{\theta}) \;\forall\; \mathcal{A} \in \mathcal{B}(\boldsymbol{\Theta}) \tag{4.8}$$

- *If p is a probability measure (Definition B.8), then it is said to be an invariant probability measure wrt $K(\cdot, \cdot)$ and hence wrt the Markov chain.*

Invariance is an important property for a Markov chain to possess. If a Markov chain is invariant, then it means that averaging over wherever we happen to be in the state space – i.e. integrating with respect to $p(d\boldsymbol{\theta})$ over $\boldsymbol{\Theta}$ – we use the same (invariant) density p to make probability assignments to all regions $\mathcal{A}$ of the state space (on the left-hand side of Equation 4.8). Put differently, the transition kernel K is *measure-preserving* or an *endomorphism* with respect to the measure (or density) p: an additional iteration of the Markov chain by passing through the kernel K leaves the measure/density p unchanged.

This has important implications for the dynamic properties of the Markov chain. If invariance holds (and p is a probability measure), then p is the stationary distribution of the Markov chain, in the sense that if $\boldsymbol{\theta}^{(t)} \sim p$, then it is also the case that $\boldsymbol{\theta}^{(t+a)} \sim p, \forall\, a > 0$. The point of this can be summarized as follows:

1. we may start the Markov chain without knowing the posterior density $p(\boldsymbol{\theta}|\text{data})$,
2. we can nonetheless construct a Markov chain with an appropriately constructed transition kernel K, such that...
3. after a sufficiently large number of iterations of the chain K is generating probability assignments over the parameter space that not only correspond to the posterior density of interest $p(\boldsymbol{\theta}|\text{data})$...
4. but will continue to do so over repeated iterations of the chain (i.e. repeated calls to the operator K).

In turn, the trajectory of the Markov chain $\{\boldsymbol{\theta}^{(T+r)}\}$ for sufficiently large T and $r > 0$ are *samples* from the invariant density, in this case the posterior density of interest $p(\boldsymbol{\theta}|\text{data})$.

We consider the conditions under which such an invariant density exists in Section 4.2.5. And later, we consider the converse problem; that is, how to construct a Markov chain over a state space $\mathbf{\Theta}$ with transition kernel K such that the invariant density of the Markov chain is a posterior density $p(\mathbf{\theta}|\text{data})$. It is this latter problem that is of immense practical relevance. Thus we now consider the properties that a Markov chain must have if it is going to be a useful way to explore a posterior density.

4.2.3 Irreducibility

If we are using a Markov chain to generate a random tour of a state space $\mathbf{\Theta}$, we want the Markov chain to go everywhere in $\mathbf{\Theta}$ that it ought to go, given where the chain currently is: e.g., for a Markov chain on a continuous state space, can we get from $\mathbf{\theta}^{(t-1)}$, the current state of the chain, to any region $\mathcal{A} \in \mathbf{\Theta}$ (or, more technically, any $\mathcal{A} \subset \mathcal{B}(\mathbf{\Theta})$ with positive measure)? This property of a Markov chain – being able to reach any region of the state space given the current state – is called *irreducibility*, and we now work towards a formal definition of the term.

Definition 4.5 (Absorbing state, closed state). *Consider a Markov chain $\{\mathbf{\theta}^{(t)}\}$ with state space $\mathbf{\Theta}$. State $\mathcal{S}$ is said to be an absorbing state or a closed state if $Pr(\mathbf{\theta}^{(t)} \notin \mathcal{S}|\mathbf{\theta}^{(t-1)} \in \mathcal{S}) = 0$ or equivalently $Pr(\mathbf{\theta}^{(t)} \in \mathcal{S}|\mathbf{\theta}^{(t-1)} \in \mathcal{S}) = 1$.*

An absorbing state is a state from which a Markov chain cannot leave, once it enters that state (e.g. extinction of a species, and possibly, a tenured position at Stanford). A Markov chain that has an absorbing state is easily recognized if the Markov chain has a small number of states, since the transition matrix can be inspected directly. If a Markov chain's transition matrix has a '1' in the j-th diagonal element (and hence row j of the transition matrix has zeros everywhere else), then state j is an absorbing state.

Definition 4.6 (Accessible States). *State $\mathcal{P}$ of a Markov chain $\{\mathbf{\theta}^{(t)}\}$ is said to be accessible from state $\mathcal{Q}$ if $Pr(\mathbf{\theta}^{(t+r)} \in \mathcal{P}|\mathbf{\theta}^{(t)} \in \mathcal{Q}) > 0$ for some $r > 0$.*

Note that $\mathcal{P}$ may be accessible from $\mathcal{Q}$ even with $\Pr(\mathbf{\theta}^{(t+1)} \in \mathcal{P}|\mathbf{\theta}^{(t)} \in \mathcal{Q}) = 0$. Note also that if state $\mathcal{P}$ is closed, then while it may be accessible from states $\mathcal{R} = \mathbf{\Theta} \setminus \mathcal{P}$, those states $\mathcal{R}$ are not accessible from it.

Definition 4.7 (Communicating states). *States $\mathcal{P}$ and $\mathcal{Q}$ of a Markov chain are said to communicate with each other if $\mathcal{P}$ is accessible from $\mathcal{Q}$ and $\mathcal{Q}$ is accessible from $\mathcal{P}$.*

Definition 4.8 (Communication class). *A collection $\mathcal{C}$ of states of a Markov chain that communicate with each other are said to form a communication class.*

Note that the states that form a communicating class can all reach one another over iterations of the Markov chain.

The notion of a communicating class plays a key role in the definition of irreducibility, which we first state for the case of a Markov chain on a discrete state space:

Definition 4.9 (Irreducible Markov chain, discrete state space). *Consider a Markov chain on a finite state space $\mathbf{\Theta}$ with transition matrix $\mathbf{K}$. If $\mathbf{\Theta}$ is a communicating class then the Markov chain and its transition matrix are said to be irreducible.*

For a Markov chain on a continuous state space, the definition of irreducibility comes with some additional measure-theoretic baggage (recall the definitions B.1 and B.8):

Definition 4.10 (Irreducibile Markov chain, continuous state space). *For some measure φ, a Markov chain $\{\boldsymbol{\theta}^{(t)}\}$ on state space $\boldsymbol{\Theta}$ with transition kernel $K(\boldsymbol{\theta}, \mathcal{A})$ (see Definition 4.3) is said to be φ-irreducible if $\forall\ \mathcal{A} \in \mathcal{B}(\boldsymbol{\Theta})$ with $\varphi(\mathcal{A}) > 0$, $\exists\ n$ such that $K^n(\boldsymbol{\theta}, \mathcal{A}) > 0 \,\forall\, \boldsymbol{\theta} \in \boldsymbol{\Theta}$.*

If this condition holds with $n = 1 \,\forall\, \mathcal{A} \in \mathcal{B}(\boldsymbol{\Theta})$ with $\varphi(\mathcal{A}) > 0$ then the Markov chain is said to be strongly irreducible.

Irreducibility is thus the property that a Markov chain can wander from any one state to any other state (perhaps not directly, but via another intermediate state). A Markov chain that is not irreducible is said to be *reducible*. The key implication of irreducibility for our purposes is that if a Markov chain is irreducible, it doesn't matter where we start that chain: eventually, the chain will visit all regions of the state space.

■ Example 4.3

Irreducibility of first-order autoregressive Gaussian process (Example 4.2, continued). In this case

$$K(\theta^{(t-1)}, \mathcal{A}) = \int_{\mathcal{A}} \frac{1}{\sqrt{2\pi\sigma^2}} \exp\left[\frac{-(\theta - \rho\theta^{(t-1)})^2}{2\sigma^2}\right] d\theta$$

which is positive for all $\mathcal{A} \in \mathcal{B}(\boldsymbol{\Theta})$ save for sets of measure zero. That is, this chain is φ-irreducible.

For a Markov chain on a discrete state space, irreducibility is sufficient to ensure the existence of a stationary distribution, which is the implication of the following famous result:

Proposition 4.1 (Perron-Frobenius Theorem). *Suppose $\mathbf{K} = (K_{ij})$ is a real n-by-n irreducible matrix with non-negative entries, i.e. $K_{ij} \geq 0$. Then $\mathbf{K}$ has an eigenvalue r with the properties:*

1. *r is real*
2. $r > 0$
3. $r \geq |\lambda|$ *for any eigenvalue of* $\mathbf{K}$, λ
4. $\min_i \sum_j K_{ij} \leq r \leq \max_i \sum_j K_{ij}$
5. *the eigenvector $\mathbf{v}$ associated with r has positive entries.*

Proof. Seneta (1981), Theorems 1.1 and Corollary 1, Theorem 1.5. ◁

Remark. The Perron-Frobenius Theorem is named for Oskar Perron and Ferdinand Georg Froebenius, German mathematicians who derived the results for the theorem that today bears both their names, in separate work published in the period 1907 to 1912.

Part 4 of the Perron-Frobenius theorem says that for an irreducible matrix with non-negative entries, the largest eigenvalue lies between the minimum column sum and maximum column sum. For the transpose of the transition matrix of a Markov chain, all the columns sum to 1, and so the largest eigenvalue of the transition matrix is 1. In addition, the eigenvector associated with that eigenvalue has all positive entries, and as we shall now see, corresponds to the stationary distribution of the Markov chain (up to a factor of proportionality).

Consider the simple, two-state Markov chain presented in Example 4.1, with transition matrix

$$\mathbf{K} = \begin{bmatrix} 1-p & p \\ q & 1-q \end{bmatrix}.$$

Direct inspection shows us that $\mathbf{K}$ is irreducible: the two states of the Markov chain defined by the transition matrix $\mathbf{K}$ clearly communicate with each other provided $p = \Pr(\theta^{(t)} = 1|\theta^{(t-1)} = 0) \neq 0$ and similarly $q = \Pr(\theta^{(t)} = 0|\theta^{(t-1)} = 1) \neq 0$. By Proposition 4.1 (the Perron-Frobenius Theorem) we know that $\mathbf{K}'$ has an eigenvalue equal to 1. In fact, the eigenvalues of $\mathbf{K}'$ are $\boldsymbol{\lambda} = (1, 1-p-q)'$ and the eigenvector corresponding to the unit eigenvalue is $\mathbf{x} = (q/p, 1)$. Normalized to sum to one (so as to correspond to a proper probability mass function), the first eigenvector becomes $\tilde{\mathbf{x}} = [q/(q+p), p/(q+p)]'$ and is sometimes referred to as the *Perron vector* of $\mathbf{K}$ (e.g. Horn and Johnson 1985, 497).

■ Example 4.4

Stationary distribution of a two-state Markov chain (Example 4.1, continued). The transition matrix

$$\mathbf{K} = \begin{bmatrix} .1 & .9 \\ .25 & .75 \end{bmatrix}$$

has eigenvalues $\boldsymbol{\lambda} = (1, -.15)'$ and (according to R) the eigenvector of $\mathbf{K}'$ associated with the (unique) unit eigenvalue is $\mathbf{x}_1 = (-.27, -.96)'$ Normalized to sum to one, the eigenvector is $\tilde{\mathbf{p}} = (.21, .78)'$, which is (to rounding error) equal to the analytic result $\tilde{p} = (q/(p+q), p/(p+q))'$. Note that over a long run of the chain, and *irrespective of the initial state of the Markov chain*, the chain is much more likely to be in state 1 than state 0; this follows from the fact that the probability of the transition to state 1 from state 0 is .9, while the probability of the reverse transition is just .25.

Stationary distributions need not be unique, say, if the Markov chain is reducible; see Exercise 4.4, Exercise 4.5 and the following example.

■ Example 4.5

A three-state Markov chain with a non-unique stationary distribution. Consider a three state Markov chain with transition matrix

$$\mathbf{K} = \begin{bmatrix} 1-p & p & 0 \\ q & 1-q & 0 \\ 0 & 0 & 1 \end{bmatrix}$$

Note immediately that the chain is reducible, having two closed communication classes: one that encompasses states 0 and 1, and the other consisting solely of state 2 (i.e. if the Markov chain starts in state 2 then it stays there forever but otherwise the Markov chain never visits this state). Thus we can expect that the stationary distribution for this Markov chain is not unique, but will depend on where the chain starts (in state 2, or not). The three eigenvalues of $\mathbf{K}$ are $\boldsymbol{\lambda} = (1, 1, 1 - p - q)'$. Suppose $p \neq 0$ and $q \neq 0$. One stationary distribution corresponds to the Markov chain alternating between states 0 and 1, and never visiting state 2; the other stationary distribution corresponds to the chain never leaving state 2. Thus, the behavior of this chain will depend entirely on where we start it.

■ Example 4.6

Existence of stationary distribution, first-order, autoregressive Gaussian process (Example 4.2, continued). Recall that for this example we have $p^{(t)} = p(\theta^{(t)}) \equiv N(\rho\theta^{(t-1)}, \sigma^2)$, for which $E(\theta^{(t)}) = \rho E(\theta^{(t-1)})$ and $V(\theta^{(t)}) = \rho^2 V(\theta^{(t-1)}) + \sigma^2$. If p is a stationary distribution of the Markov chain then the mean and variance given above are not changing over successive iterations of the Markov chain; note also that since p is a normal density, the mean and variance are sufficient statistics. That is, stationarity implies we can drop the indexing by t to obtain

$$E(\theta) = \rho E(\theta),$$
$$V(\theta) = \rho^2 V(\theta) + \sigma^2.$$

These stationarity conditions imply that $V(\theta) = \sigma^2/(1 - \rho^2)$, which is a positive quantity only if $\sigma^2 > 0$ and $\rho \in (-1, 1)$. Given this constraint on ρ, the first stationarity condition can only be satisifed if $E(\theta) = 0$. That is, the stationary distribution of the chain is

$$p(\theta) \equiv N\left(0, \sigma^2/(1 - \rho^2)\right).$$

We can verify this by checking that p given above does in fact solve the functional equation given in Equation 4.5 with $K(\theta^{(t-1)}, \theta^{(t)}) \equiv N(\rho\theta^{(t-1)}, \sigma^2)$. The proof is left as an exercise for the reader (Problem 4.9); see also Lancaster (2004, Example 4.11).

To establish the existence of a unique stationary density for a Markov chain on a continuous state space, we must introduce some additional concepts.

4.2.4 Recurrence

We consider the concept of *recurrence*, and in particular, *positive recurrence* (Definition 4.14). For Markov chains on continuous state spaces we consider a condition known as *Harris recurrence* (Definition 4.16). We begin by defining the return time of a state of a Markov chain, which allows us to categorize states of Markov chains as either *transient* or *recurrent*.

Definition 4.11 (Return time). *Consider a Markov chain in state j at iteration t, and without loss of generality, say $t = 0$. Let T_j be the iteration at which the Markov chain first returns to state j, i.e. $T_j = \min\{r : \theta^{(r)} = j | \theta^{(0)} = j\}$. The quantity T_j is a random variable and is known as the return time.*

Definition 4.12 (Transient state). *If state j of a Markov chain has a return time that is possibly infinite, i.e. $Pr(T_j < \infty) < 1$, then the state is said to be transient.*

Definition 4.13 (Recurrent state). *If state j of a Markov chain is not transient, it is said to be recurrent.*

Definition 4.14 (Positive recurrence; null recurrence).

1. *If the expected waiting time of a recurrent state is finite, i.e. $E(T_j) < \infty$ then the state is said to be positive recurrent.*
2. *If all states of a Markov chain are positive recurrent then the Markov chain is said to be positive recurrent.*
3. *A recurrent state j with $E(T_j) = \infty$ is said to be null recurrent.*

Proposition 4.2 (Irreducibility and recurrence, Markov chains on discrete state spaces).

1. *For an irreducible Markov chain on a discrete state space, either all states are positive recurrent, or none is.*
2. *For an irreducible Markov chain on a finite state space, all states are positive recurrent, and hence a unique stationary distribution exists.*

Proof. 1. Rosenthal (2006), Corollary 8.4.8; Theorem 8.4.9. 2. Rosenthal (2006), Proposition 8.4.10. $\triangleleft$

Note that positive recurrence strengthens the result in the Perron-Frobenius Theorem (Proposition 4.1); irreducibility is sufficient to ensure the existence of a stationary distribution for a Markov chain on a discrete state space, and the additional assumption of positive recurrence ensure the uniqueness of the stationary distribution.

For Markov chains on continuous state spaces, we use a slightly stronger form of recurrence. We begin with yet another definition:

Definition 4.15 (Passages through a state). *For a Markov chain $\{\boldsymbol{\theta}^{(t)}\}$ with state space $\boldsymbol{\Theta}$, consider a subset of the state space $\mathcal{A} \in \mathcal{B}(\boldsymbol{\Theta})$. The number of passages of the Markov chain through $\mathcal{A}$ in a infinitely long run of the chain is $v(\mathcal{A}) = \sum_{t=1}^{\infty} \mathbb{I}(\boldsymbol{\theta}^{(t)} \in \mathcal{A})$, where $\mathbb{I}(\cdot)$ is a binary indicator function.*

Clearly, if a subset of $\boldsymbol{\Theta}$ is recurrent, then as the Markov chain is allowed to run that state is visited infinitely many times; going further, if all states are recurrent, then the Markov chain will visit of all them infinitely many times, and irrespective of which state we happen to start in. We formalize this idea in the next definition (e.g. Meyn and Tweedie 2009, p199):

Definition 4.16 (Harris recurrence). *Consider a Markov chain $\{\boldsymbol{\theta}^{(t)}\}$, $\boldsymbol{\theta} \in \boldsymbol{\Theta}$.*

1. *Suppose the Markov chain is initialized with $\boldsymbol{\theta}^{(0)} \sim p_0$. A set $\mathcal{A} \in \boldsymbol{\Theta}$ is Harris recurrent if $Pr(v(\boldsymbol{\theta}) = \infty) = 1 \forall \boldsymbol{\theta} \in \mathcal{A}$.*

2. *The Markov chain is Harris recurrent if (a) there is a measure φ such that the chain is φ-irreducible (Definition 4.10) and (b)* $\forall\ \mathcal{A} \subset \boldsymbol{\Theta}$ *s.t.* $\varphi(\mathcal{A}) > 0$, *$\mathcal{A}$ is Harris recurrent.*

Alternatively, let $L(\boldsymbol{\theta}, \mathcal{A})$ be the probability that a Markov chain started at $\boldsymbol{\theta}$ ever enters the set $\mathcal{A} \subset \boldsymbol{\Theta}$. We can also define φ-irreducibility as the property that for a measure φ on $(\boldsymbol{\Theta}, \sigma(\boldsymbol{\Theta}))$, $L(\boldsymbol{\theta}, \mathcal{A}) > 0 \,\forall\ \boldsymbol{\theta} \in \boldsymbol{\Theta}$ and $\forall$ φ-positive $\mathcal{A}$. Harris recurrence then is a slightly stronger notion than φ-irreducibility, with $L(\boldsymbol{\theta}, \mathcal{A}) = 1, \forall\ \boldsymbol{\theta} \in \boldsymbol{\Theta}$ and $\forall$ φ-positive $\mathcal{A}$. The sense in which Harris recurrence is stronger than φ-irreducibility is quite technical. φ-irreducibility is almost the same as Harris recurrence, in that the former implies that there is a φ-null set N (i.e. a set of φ-measure zero, or $\varphi(N) = 0$), such that $L(\boldsymbol{\theta}, \mathcal{A}) = 1$ holds $\forall\ \boldsymbol{\theta} \notin N$ and all φ-positive $\mathcal{A}$; Harris recurrence removes this φ-null set such that the condition $L(\boldsymbol{\theta}, \mathcal{A}) = 1$ holds $\forall\ \boldsymbol{\theta} \in \boldsymbol{\Theta}$ (Meyn and Tweedie 2009, Proposition 9.1.1).

Harris recurrence typically holds for the sorts of Markov chains we construct in exploring posterior densities via Markov chain Monte Carlo. In fact, we have the following proposition (which assumes the existence of a unique invariant distribution for the Markov chain, but nonetheless):

Proposition 4.3 (Sufficient conditions for Harris recurrence). *Consider a Markov chain* $\{\boldsymbol{\theta}^{(t)}\}$, $\boldsymbol{\theta} \in \boldsymbol{\Theta}$, *with transition kernel* $K(\cdot, \cdot)$. *If*

1. *the chain is φ-irreducible,*
2. *φ is the unique, invariant distribution of the chain, and*
3. $K(\boldsymbol{\theta}, \cdot)$ *is absolutely continuous wrt* $\varphi\ \forall\ \boldsymbol{\theta}$,

then the chain is Harris recurrent.

Proof. Tierney (1994), Corollary 1. ◁

4.2.5 Invariant measure

The following proposition provides conditions under which a Markov chain will be invariant, underlining the importance of φ-irreducibility (Definition 4.10):

Proposition 4.4 (φ-irreduciblility, recurrence and unique invariant measure). *A φ-irreducible, recurrent Markov chain has an invariant measure that is unique up to a multiplicative constant.*

Proof. Nummelin (1984), Corollary 5.2. ◁

That is, φ-irreducibility and recurrence implies the existence of a unique invariant measure. If the unique invariant measure is a probability measure then we have yet another definition:

Definition 4.17 (Positive Markov chain). *A Markov chain that is φ-irreducible (Definition 4.10) and admits an invariant probability measure (Definition 4.4) is said to be positive.*

■ Example 4.7

Existence of unique invariant measure, first-order autoregressive Gaussian process (Example 4.2, continued). In Example 4.6 we saw that $|\rho| < 1$ is necessary to ensure the existence of a unique, stationary distribution for the Gaussian AR(1) process. A Gaussian AR(1) process is φ-irreducible irrespective of the value of the autoregressive parameter ρ (see Definition 4.10 and Example 4.3). When $|\rho| > 1$, the measure φ on the state space $\mathbf{\Theta} = \mathbb{R}$ is not finite, and every state of the chain (and the Markov chain itself) is null recurrent (see Definition 4.14). On the other hand, if $|\rho| < 1$ then we have a φ-irreducible, positive recurrent Markov chain, with the unique invariant density given in Example 4.6.

4.2.6 Reversibility

Another way to establish the existence of an invariant distribution is to show that the Markov chain is reversible. The idea is relatively simple. We begin with a definition of the *detailed balance* condition:

Definition 4.18 (Detailed Balance Condition). *Consider a Markov chain $\{\boldsymbol{\theta}^{(t)}\}$ with state space $\mathbf{\Theta}$, transition kernel K and stationary distribution p. If*

$$p(\boldsymbol{\theta}^{(t)})K(\boldsymbol{\theta}^{(t)}, \boldsymbol{\theta}^{(t+1)}) = p(\boldsymbol{\theta}^{(t+1)})K(\boldsymbol{\theta}^{(t+1)}, \boldsymbol{\theta}^{(t)}) \tag{4.9}$$

$\forall\, \boldsymbol{\theta} \in \mathbf{\Theta}$ *then the Markov chain is said to possess detailed balance.*

Definition 4.19 (Reversible Markov Chain). *A Markov chain that possesses detailed balance is said to be reversible.*

■ Example 4.8

Detailed balance for a two-state Markov chain. *Consider a two-state Markov chain with transition matrix*

$$\mathbf{K} = \begin{bmatrix} 1-p & p \\ q & 1-q \end{bmatrix}$$

and a unique stationary distribution $\mathbf{p} = (q/(p+q), p/(p+q))'$. For this two-state chain, detailed balance implies that if $\mathbf{p} = (p_0, p_1)'$ is a stationary distribution given the kernel K, then $K_{01}p_0 = K_{10}p_1$. We verify this condition holds in this case since $p \times q/(p+q) = q \times p/(p+q)$.

Reversibility (and detailed balance) is a sufficient condition for determining if a density p over a state space is invariant. To see this, consider integrating both sides of

Equation 4.9 with respect to $\boldsymbol{\theta}^{(t)}$: i.e.

$$\int_{\boldsymbol{\Theta}} p(\boldsymbol{\theta}^{(t)})K(\boldsymbol{\theta}^{(t)}, \boldsymbol{\theta}^{(t+1)})d\boldsymbol{\theta}^{(t)} = \int_{\boldsymbol{\Theta}} p(\boldsymbol{\theta}^{(t+1)})K(\boldsymbol{\theta}^{(t+1)}, \boldsymbol{\theta}^{(t)})d\boldsymbol{\theta}^{(t)}$$

$$= p(\boldsymbol{\theta}^{(t+1)}) \int_{\boldsymbol{\Theta}} K(\boldsymbol{\theta}^{(t+1)}, \boldsymbol{\theta}^{(t)})d\boldsymbol{\theta}^{(t)} = p(\boldsymbol{\theta}^{(t+1)}),$$

which is merely a restatement of the definition of invariance in Equation 4.8. That is, if the density p is such that detailed balance holds (Equation 4.9), then

$$p(\boldsymbol{\theta}^{(t+1)}) = \int_{\boldsymbol{\Theta}} p(\boldsymbol{\theta}^{(t)})K(\boldsymbol{\theta}^{(t)}, \boldsymbol{\theta}^{(t+1)})d\boldsymbol{\theta}^{(t)}$$

which means that p is the invariant density of the Markov chain.

What this means is that if we can come up with transition kernels K that are *reversible*, satisfying the *symmetry* inherent in the detailed balance condition, then the resulting Markov chains will converge on the invariant density, p. Note that this is a stronger result than we require so as to establish the existence of a unique, invariant density for a Markov chain (cf Proposition 4.4), but it does suggest a strategy for generating algorithms for sampling from p. We will observe this in some detail in Chapter 5 when we consider the Metropolis algorithm.

4.2.7 Aperiodicity

We will also require that Markov chains be *aperiodic*. We again begin with the discrete case.

Definition 4.20 (Period of a state, discrete state space). *Let* $\{\boldsymbol{\theta}^{(t)}\}$, $\boldsymbol{\theta}^{(t)} \in \boldsymbol{\Theta}$ *be a Markov chain on a discrete state space with transition function* $K(\cdot, \cdot)$. *The period of a set* $\mathcal{A} \in \boldsymbol{\Theta}$ *is the greatest common divisor of the set* $\{m \geq 1; K^m(\mathcal{A}, \mathcal{A}) > 0\}$.

If the Markov chain is irreducible (i.e. the Markov chain's state space $\boldsymbol{\Theta}$ forms a communicating class, as per Definitions 4.8 and 4.9), then there is only one value for the period of the communicating class, and hence for the period of the whole chain itself.

Definition 4.21 (Aperiodicity). *An irreducible Markov chain with period 1 is said to be aperiodic.*

■ **Example 4.9**

Periodic Markov chain on a discrete state space. Consider a Markov chain on $\boldsymbol{\Theta} = \{0, 1, 2\}$ with transition kernel

$$\mathbf{K} = \begin{bmatrix} 0 & 1 & 0 \\ p & 0 & 1-p \\ 0 & 1 & 0 \end{bmatrix}.$$

The chain is irreducible: all states communicate with one another. But note the zeros on the diagonal of $\mathbf{K}$. In particular, consider state 0: $K(0, 0) = 0$, since $K(0, 1) = 1$ (i.e. if we are in state 0, we move to state 1 with probability 1). Now $K(1, 0) = p$ and so while $K^1(0, 0) = 0$, $K^2(0, 0) = p$, from which we conclude that the period of state 0 is 2. Since the chain is irreducible, the period of the chain can be no smaller than the period of any state (and so can be no smaller than the period of state 0, which is 2) and so the chain is periodic.

As the preceeding example makes clear – at least for the relatively simple case of a Markov chain on a discrete state space – the fact that the Markov chain contains some deterministic elements in its transition kernel means that it can not converge to its stationary distribution. The effect of an initial state never dies away.

■ **Example 4.10**

Periodic Markov chain on a discrete state space (Example 4.9, continued). Suppose we start the chain in state 0, such that $\mathbf{p}_0 = (1, 0, 0)'$. Then,

$$\mathbf{p}_1 = \mathbf{p}_0\mathbf{K} = (0, 1, 0)'$$

$$\mathbf{p}_2 = \mathbf{p}_0\mathbf{K}^2 = (p, 0, 1 - p)'$$

$$\mathbf{p}_3 = \mathbf{p}_0\mathbf{K}^3 = (0, 1, 0)'$$

$$\mathbf{p}_4 = \mathbf{p}_0\mathbf{K}^4 = (p, 0, 1 - p)'$$

i.e. $K^n = K \;\forall$ odd n and $K^n = K^2 \;\forall$ even n. Letting this chain run from $\mathbf{p}_0$ would see the chain in state 1 every odd-numbered step, and in states 0 or 2 with probabilities p and $1 - p$ every even-numbered step. Over the long run then, we would observe the Markov chain visiting states with probabilities $\tilde{\mathbf{p}} = (p/2, 1/2, (1 - p)/2)'$, which is not the stationary distribution of the chain (a fact that you are invited to verify in Problem 4.7). That is, a stationary distribution exists, but the periodicity encoded in $\mathbf{K}$ means that the Markov chain can not converge to it.

4.3 Convergence of Markov chains

We have established the conditions under which Markov chains possess unique stationary densities in the previous sections. Periodicity is one way that a Markov chain cannot converge to a stationary density. We now investigate the question of convergence to a stationary density more thoroughly. To reiterate the key question here: if a Markov chain has an invariant distribution p, under what conditions will the chain reach p from an arbitrary starting density?

We first combine Harris recurrence, positivity and aperiodicity in the following definition:

Definition 4.22 (Ergodic Markov chain). *If a Markov chain* $\{\boldsymbol{\theta}^{(t)}\}$, $\boldsymbol{\theta} \in \boldsymbol{\Theta}$ *is*

1. *positive (Definition 4.17)*

2. Harris recurrent (Definition 4.16)

3. aperiodic (Definition 4.21)

then the Markov chain is said to be ergodic.

A commonly used measure of the 'closeness' of a Markov chain is to its invariant distribution $p(\boldsymbol{\theta})$ is with the total variation norm:

Definition 4.23 (Total variation norm). *Let f and g be measures over $\mathcal{A}$. Then the total variation norm is*

$$||f - g||_{TV} = \sup_{\mathcal{A}} |f(\mathcal{A}) - g(\mathcal{A})|$$

A useful way to think about the total variation norm is the maximum difference between two distributions over some common support (i.e. here, the parameter space $\boldsymbol{\Theta}$). With this definition of 'closeness' we now state the following result:

Proposition 4.5 (Convergence to stationary distribution in total variation norm). *If a Markov chain $\{\boldsymbol{\theta}^{(t)}\}$, $\boldsymbol{\theta} \in \boldsymbol{\Theta}$ is ergodic (Definition 4.22) then*

$$\lim_{t \to \infty} \left\| \int_{\boldsymbol{\Theta}} K^t(\boldsymbol{\theta}, \cdot) h(d\boldsymbol{\theta}) - p \right\|_{TV} = 0,$$

for every initial distribution h, and where (via Proposition 4.4) p is the Markov chain's unique invariant distribution.

Proof. Nummelin (1984), Corollary 6.7 (ii). See also Tierney (1994), Theorem 1 and Robert and Casella (2004), Theorem 6.51. ◁

To exploit this result we need to construct a Markov chain with an appropriate transition kernel K, a topic we consider in Chapter 5, below. We also need some theory and/or practical wisdom for assessing when T is sufficiently large, such that we are in fact exploiting the asymptotic result stated in Proposition 4.5.

To this end, we can also make stronger statements about *how quickly* a Markov chain will converge on its stationary distribution, using other notions of ergodicity:

Definition 4.24 (Geometric ergodicity). *An ergodic Markov chain on $\boldsymbol{\Theta}$ with invariant distribution p is said to be geometrically ergodic if there exists a constant $r > 1$ such that*

$$\sum_{n=1}^{\infty} r^n \left\| K^n(\boldsymbol{\theta}, \cdot) - p(\cdot) \right\|_{TV} < \infty, \; \forall \, \boldsymbol{\theta} \in \boldsymbol{\Theta}.$$

Geometric ergodicity is potentially good news, in that it implies that the convergence result in Proposition 4.5 can be realized 'sooner' rather than 'later', depending on the value of r in the preceeding definitions. Note that values of r arbitrarily close to 1 are of not much comfort, since $r = 1$ implies a linear rate of convergence in total variation norm. Thus, while it may be possible to construct a Markov chain that is geometrically ergodic in a technical, mathematical sense, a given Markov chain may be converging geometrically, but extremely slowly. In short, more work is needed before we can take any great comfort from knowing that a given Markov chain is geometrically ergodic.

If the rate of geometric convergence ergodicity holds over the entire parameter space, then we have a stronger form of ergodicity:

Definition 4.25 (Uniform ergodicity). *A Markov chain* $\{\boldsymbol{\theta}^{(t)}\}$ *on* $\boldsymbol{\Theta}$ *with invariant distribution* p *is said to be uniformly ergodic if*

$$\lim_{n\to\infty} \sup_{\boldsymbol{\theta}\in\boldsymbol{\Theta}} \left\| K^n(\boldsymbol{\theta}, \cdot) - p \right\|_{TV} = 0.$$

Proposition 4.6 (Alternative characterization of uniform ergodicity). *A Markov chain is uniformly ergodic if there exist constants* $R < \infty$ *and* $r > 1$ *such that*

$$\left\| K^n(\boldsymbol{\theta}, \cdot) - p \right\|_{TV} < Rr^{-n}, \forall\, \boldsymbol{\theta} \in \boldsymbol{\Theta}.$$

Proof. Nummelin (1984, Theorem 6.15). ◁

Most of the Markov chains we construct in Bayesian statistics can be shown to satisfy uniform ergodicity.

4.3.1 Speed of convergence

These definitions are only that, and in any practical application of MCMC methods we would prefer some guidance as to *how quickly* the Markov chain is converging on the target, posterior density, since this answers perhaps one of the most pressing questions when using MCMC: how many iterations are sufficient? A review with extensive references appears in Diaconis (2009). In a few cases we have bounds on running time for specific models (e.g. Diaconis, Khare and Saloff-Coste 2008; Jones and Hobert 2001). But usually these results are quite abstract, and of limited utility to practitioners.

For the discrete case we can exploit a nice relationship between the eigenstructure of the transition kernel $\mathbf{K}$ and running time. For a discrete Markov chain we have $\mathbf{p}^{(t)} = \mathbf{p}^{(t-1)}\mathbf{K}$ (Equation 4.2) and more generally $\mathbf{p}^{(t)} = \mathbf{p}^{(0)}\mathbf{K}^t$. If the chain is converging on the stationary mass function $\mathbf{p}$ – from any arbitrary initial mass function $\mathbf{p}^{(0)}$ – then the question of running time is obviously connected to what is happening to the matrix $\mathbf{K}^t$ as t gets large. In turn, the behavior of $\mathbf{K}^t$, $t \to \infty$ is tied to the eigenstructure of $\mathbf{K}$; indeed, recall Proposition 4.1 tells us that the stationary distribution for a Markov chain on a discrete space is the Perron vector of $\mathbf{K}$, obtained by normalizing the first eigenvector of $\mathbf{K}'$ to sum to one. To explore this further, we return to the 2-state Markov chain considered earlier.

■ **Example 4.11**

Rate of convergence for a two-state Markov chain. We seek $\mathbf{K}^t$ where

$$\mathbf{K} = \begin{bmatrix} 1-p & p \\ q & 1-q \end{bmatrix}.$$

The transition kernel $\mathbf{K}$ has eigenvectors that we normalize to $\mathbf{x}_1 = (1, 1)'$ and

$$\mathbf{x}_2 = \left[\frac{p}{1-\lambda_2}, \frac{-q}{1-\lambda_2}\right]'$$

with the assumption that $p > 0, q > 0$ and $p + q \neq 1$ and where the eigenvalues are $\boldsymbol{\lambda} = (\lambda_1, \lambda_2)' = (1, 1 - p - q)'$. We observe that $\mathbf{K}$ admits the diagonalization (Proposition A.5) $\mathbf{K} = \mathbf{X}\boldsymbol{\Lambda}\mathbf{X}^{-1}$ where $\mathbf{X}$ is a matrix of eigenvectors of $\mathbf{K}$ and $\boldsymbol{\Lambda}$ is a diagonal matrix containing the eigenvectors of $\mathbf{K}$. That is,

$$\mathbf{X} = \begin{bmatrix} 1 & \dfrac{p}{1-\lambda_2} \\ 1 & \dfrac{-q}{1-\lambda_2} \end{bmatrix} \quad \text{and} \quad \mathbf{X}^{-1} = \begin{bmatrix} \dfrac{q}{1-\lambda_2} & \dfrac{p}{1-\lambda_2} \\ 1 & -1 \end{bmatrix}.$$

Now, by Proposition A.6, we have $\mathbf{K}^t = \mathbf{X}\boldsymbol{\Lambda}^t\mathbf{X}^{-1}$; observe that this equality holds for the normalized eigenvectors. With these definitions we have (e.g. Graham 1987, §7.4)

$$\mathbf{K}^t = \frac{1}{p+q}\begin{bmatrix} q & p \\ q & p \end{bmatrix} + \frac{(1-p-q)^t}{p+q}\begin{bmatrix} p & -p \\ -q & q \end{bmatrix}.$$

The first term on the right hand side of this equality is the stationary distribution for the Markov chain, obtained in the usual way, normalizing the eigenvector of $\mathbf{K}'$ associated with the unit eigenvector. The second term is a function of $(1 - p - q)^t$, which goes to zero as $t \to \infty$, since $|1 - p - q| < 1$.

That is, as $t \to \infty$, $\mathbf{p}^{(t)} = \mathbf{K}^t\mathbf{p}^{(0)} \to \mathbf{p}$, the stationary distribution of the chain, the difference between $\mathbf{p}^{(t)}$ and $\mathbf{p}$ declining geometrically in $|1 - p - q|$. The perhaps surprising result here is that $|1 - p - q| = |\lambda_2|$, the second eigenvalue of the transition kernel $\mathbf{K}$.

This result – that the rate of convergence of a Markov chain is a function of the size of the second eigenvalue of the transition kernel – is actually quite general. This is a key result in deriving bounds on running times in more general settings than the simple two-state Markov chain considered in Example 4.11. For more general discrete chains we can obtain similar results by investigating properties of the Markov transition graph; there are some lovely mathematical ideas at work in these results, but beyond our scope here. Some summaries appear in Liu (2001, §12.5) and Diaconis (2009). Extensions to general state spaces are not trivial, and practical implications for researchers running MCMC algorithms on specific models and/or data sets remain rare.

4.4 Limit theorems for Markov chains

Finally, we state perhaps the most important result of this chapter, the 'ergodic theorem', sometimes also called a Law of Large Numbers for Markov chains:

Proposition 4.7 (Pointwise ergodic theorem; Law of Large Numbers for Markov chains). *Let* $\{\boldsymbol{\theta}^{(t)}\}$ *be a Harris recurrent Markov chain on* $\boldsymbol{\Theta}$ *with a* σ*-finite invariant measure* p*. Consider a* p*-measurable function* h *s.t.* $\int_{\boldsymbol{\Theta}} |h(\boldsymbol{\theta})| dp(\boldsymbol{\theta}) < \infty$*. Then*

$$\lim_{T\to\infty} T^{-1} \sum_{t=1}^{T} h(\boldsymbol{\theta}^{(t)}) = \int_{\boldsymbol{\Theta}} h(\boldsymbol{\theta}) dp(\boldsymbol{\theta}) \equiv E_p h(\boldsymbol{\theta}).$$

Proof. The result follows from Birkhoff's (1931) pointwise ergodic theorem, which in turn can be shown to be a corollory of more general result due to Chacon and Ornstein (1960); e.g. see Revuz (1975, Chapter 4). ◁

Remark. Breiman (1968, Theorem 6.21) gives a relatively accessible proof of a pointwise ergodic theorem. Norris (1997) is one of many texts giving the result where the Markov chain $\{\boldsymbol{\theta}^{(t)}\}$ is on a discrete state space. It bears pointing out that the Strong Law of Large Numbers itself is a corollory of the ergodic theorem (e.g. Brieman 1968, Corollary 6.25).

The ergodic theorem allows us to use the output of an appropriately constructed Markov chain in a very useful way, as suggested in the introductory paragraphs of this chapter. To reiterate, if we can construct a Markov chain the 'right way' (and more on this in Chapter 5), then

1. the Markov chain will have a unique, limiting distribution, a posterior density that we happen to be interested in, $p \equiv p(\boldsymbol{\theta}|\text{data})$;
2. no matter where we start the Markov chain, if we let it run long enough, it will eventually wind up generating a random tour of the parameter space, visiting sites in the parameter space $\mathcal{A} \in \boldsymbol{\Theta}$ with relative frequency proportional to $\int_{\mathcal{A}} p(\boldsymbol{\theta}|\text{data}) d\boldsymbol{\theta}$;
3. the ergodic theorem means that averages $\bar{h} = T^{-1} \sum_{t=1}^{T} h(\boldsymbol{\theta}^{(t)})$ taken over the Markov chain output are simulation-consistent estimates of

$$E[h(\boldsymbol{\theta})|\text{data}] = \int_{\boldsymbol{\Theta}} h(\boldsymbol{\theta}) p(\boldsymbol{\theta}|\text{data}) d\boldsymbol{\theta}.$$

In short, we can generate valid characterizations of posterior densities with a sampling technology that does not generate independent samples from the posterior density. In turn, this means that the power of the Monte Carlo method we encountered in Chapter 3 extends to situations where we do not have independent random draws from $p(\boldsymbol{\theta}|\text{data})$.

4.4.1 Simulation inefficiency

There is a cost to the fact that the samples produced by a Markov chain are not independent. As the dependence between successive draws of the Markov chain increases, the Markov chain's stochastic exploration of the parameter space $\boldsymbol{\Theta}$ is slower. If $\boldsymbol{\theta}^{(t+1)}$ is closely tied to $\boldsymbol{\theta}^{(t)}$, then we will require more steps of the Markov chain to generate a random tour of the parameter space. Provided the Markov chain's transition kernel K satisfies the conditions given above we still have simulation consistency. So long as we have ergodicity, the lack of independence is more an annoyance than anything, generating what we might call *simulation inefficiency*, in the sense we will require more iterations of the Markov chain to characterize features of the posterior density than if we were exploring the parameter space with an independence sampler.

To get a sense of the problem here, suppose we are interested in some scalar quantity of interest $h(\boldsymbol{\theta})$. We can estimate $E(h(\boldsymbol{\theta})|\mathbf{y})$ – the mean of the posterior density of $h(\boldsymbol{\theta})$ – with the average $\bar{h}_T = T^{-1}\sum_{t=1}^{T} h(\boldsymbol{\theta}^{(t)})$, and for an ergodic Markov chain this is a simulation consistent estimator. But the rate at which $\bar{h}_T$ converges on $E(h(\boldsymbol{\theta})|\mathbf{y})$ – the rate at which the Monte Carlo error of $\bar{h}_T$ approaches zero – is not as fast as the $\sqrt{T}$ rate we get from an independence sampler (recall Proposition 3.2).

To formalize matters, we rely on a concept from the statistical physics literature, known as the *integrated autocorrelation time* (e.g., Landau and Binder 2000):

Definition 4.26 (Integrated auto correlation time). *Let $\boldsymbol{\theta}^{(1)}, \boldsymbol{\theta}^{(2)}, \ldots, \boldsymbol{\theta}^{(T)}$ be realizations from p, the stationary distribution of the Markov chain $\{\boldsymbol{\theta}^{(t)}\}$, and let $h(\boldsymbol{\theta})$ be some (scalar) quantity of interest. If ρ_j is the lag-j autocorrelation of the sequence $\{h(\boldsymbol{\theta}^{(t)})\}$ then*

$$\tau_{int}[h(\boldsymbol{\theta})] = \frac{1}{2} + \sum_{j=1}^{\infty} \rho_j.$$

is the integrated autocorrelation time of the chain.

Following Liu (2001, §5.8) we now show that the 'effective sample size' of the chain $\{h(\boldsymbol{\theta}^{(t)})\}$ is $T/(2\tau_{\text{int}}[h(\boldsymbol{\theta})])$. Let $h_t = h(\boldsymbol{\theta}^{(t)})$, let $\sigma^2 = \text{var}(h_t)$ with respect to the stationary distribution p, let $\gamma_j = \text{cov}(h_t, h_{t-j})$ be the j-th autocovariance of the sequence $\{h_t\}$ and let ρ_j be the corresponding autocorrelation. Then

$$\begin{aligned}
T\,\text{var}(\bar{h}_T) &= T\,\text{var}\left(T^{-1}\sum_{t=1}^{T} h_t\right) = T^{-1}\left(\sum_{t=1}^{T}\text{var}(h_t) + 2\sum_{i<j}\text{cov}\left(h_i, h_j\right)\right) \\
&= T^{-1}\left(T\sigma^2 + 2\sum_{j=1}^{T}(T-j)\gamma_j\right) \\
&= T^{-1}\left(T\sigma^2 + 2T\sigma^2\sum_{j=1}^{T}\frac{T-j}{T}\frac{\gamma_j}{\sigma^2}\right) \\
&= \sigma^2\left(1 + 2\sum_{j=1}^{T}\frac{T-j}{T}\rho_j\right) \approx \sigma^2\left(1 + 2\sum_{j=1}^{\infty}\rho_j\right) \qquad \text{(for large } T\text{)} \\
&= \sigma^2 2\tau_{\text{int}}[h(\boldsymbol{\theta})],
\end{aligned}$$

or $\text{var}(\bar{h}_T) = \sigma^2/T \times 2\tau_{\text{int}}[h(\boldsymbol{\theta})]$. That is, the factor $2\tau_{\text{int}}[h(\boldsymbol{\theta})]$) is a measure of how the dependency inherent in the Markovian exploration of $p(\boldsymbol{\theta}|\mathbf{y})$ is degrading the precision of the summary statistic $\bar{h}$. Obviously, when the autocorrelations are relatively large and decay slowly, $\tau_{\text{int}}[h(\boldsymbol{\theta})]$ is relatively large; on the other hand, in the limiting case of an independence sampler, all the autocorrelations are zero and we obtain the standard result that $\text{var}(\bar{h}_T) = \sigma^2/T$.

An equivalent result follows from noting that variance of the mean $\bar{h}$ is given by the spectral density of the sequence $\{h_t\}$ at frequency zero, divided by T. If $\{h_t\}$ is an independent series, the spectral density at frequency zero is simply the variance of h_t. We apply this result to generate an estimate of the 'effective sample size' of the simulation run. That is, if $\hat{\lambda}_0$ is the estimated spectral density of $\{h_t\}$ at frequency zero, then $\widehat{\text{var}}(\bar{h}_T) = \hat{\lambda}_0/T$ whereas the naïve estimate of this variance (assuming independence) is simply $\text{var}(h_t)/T$. Thus, the quantity $T^* = T\text{var}(h_t)/\lambda_0$ is an estimate of the 'effective' number of simulations, in the sense that the precision in the estimate of $\bar{h}_T$ is the same as that from an independence sampler run for T^* iterations; note that $\text{var}(h_t) \leq \lambda_0 \Rightarrow T^* \leq T$ with equality only in the case of an independence sampler. A function for computing T^* is provided in the R package, coda (Plummer *et al.* 2008); see Section 6.2 for additional discussion.

For the case where the chain is first order Markov, the form of the efficiency loss has a special form. That is, if $E(h_t|h_{t-1}) = \rho h_{t-1}$, $|\rho| < 1$, and $\text{var}(h_t) = \sigma^2$ then

$$\begin{aligned} T\text{var}(\bar{h}_t) &= \sigma^2\left(1 + 2\sum_{j=1}^{T}\frac{T-j}{T}\rho^j\right) = \sigma^2\left(1 + 2\sum_{j=1}^{T}\rho^j - \frac{2}{T}\sum_{j=1}^{T}j\rho^j\right) \\ &\approx \sigma^2\left(1 + \frac{2\rho}{1-\rho}\right), \\ &\quad \text{for large } T\text{, noting that } \sum_{j=1}^{\infty}\rho^j = \sum_{j=0}^{\infty}\rho^j - 1 = \frac{1}{1-\rho} - 1 = \frac{\rho}{1-\rho}, \\ &= \sigma^2\left(\frac{1-\rho+2\rho}{1-\rho}\right) = \sigma^2\frac{1+\rho}{1-\rho}, \end{aligned} \tag{4.10}$$

and so $\text{var}(\bar{h}_T) = \dfrac{\sigma^2}{T}\dfrac{1+\rho}{1-\rho}$.

As $\rho \to 0$, we tend to the independence sampler; as $\rho \to 1$, the dependency increases, and the factor $(1+\rho)/(1-\rho) \to \infty$. For example, suppose we are using an AR(1) process with $\rho = .9$ to generate a Monte Carlo estimate of $E(h(\theta))$, and we seek to drive the standard deviation of the Monte Carlo error to a given level. Then we would require $\sqrt{19} \approx 4.36$ as many iterates as those obtained from the independence sampler.

Note that this result applies to the relatively simple case of estimating the expectation of $h(\theta)$. Markovian random tours can result in even greater inefficiencies if we consider their performance with respect to extreme quantiles (e.g. the 2.5 and 97.5 percentiles of a density). In such a case the difficulty is that the Markov chain will take a relatively long time to find its way out to an extreme quantile and to then wander out to the other tail. Now if the Markov chain satisfies the regularity conditions discussed earlier then it will *eventually* wind up visiting regions of the parameter space with relative frequencies equal to the probability of those regions under the density being explored; it just may take an extremely long time for simulation consistency to 'kick in' for an extreme quantile. We explore this in the next example.

■ Example 4.12

Stationary, first-order, autoregressive normal process (Example 4.2, continued). Consider the first-order autoregressive process $z_t = \rho z_{t-1} + \epsilon_t$, where $\epsilon_t \sim N(0, \omega^2)$, $|\rho| < 1$ and $\omega^2 = 1 - \rho^2$. This process is stationary, with mean zero and variance one. We note immediately that the conditional distribution of z_t given z_{t-1} is

$$z_t | z_{t-1} \sim N(\rho z_{t-1}, 1 - \rho^2). \tag{4.11}$$

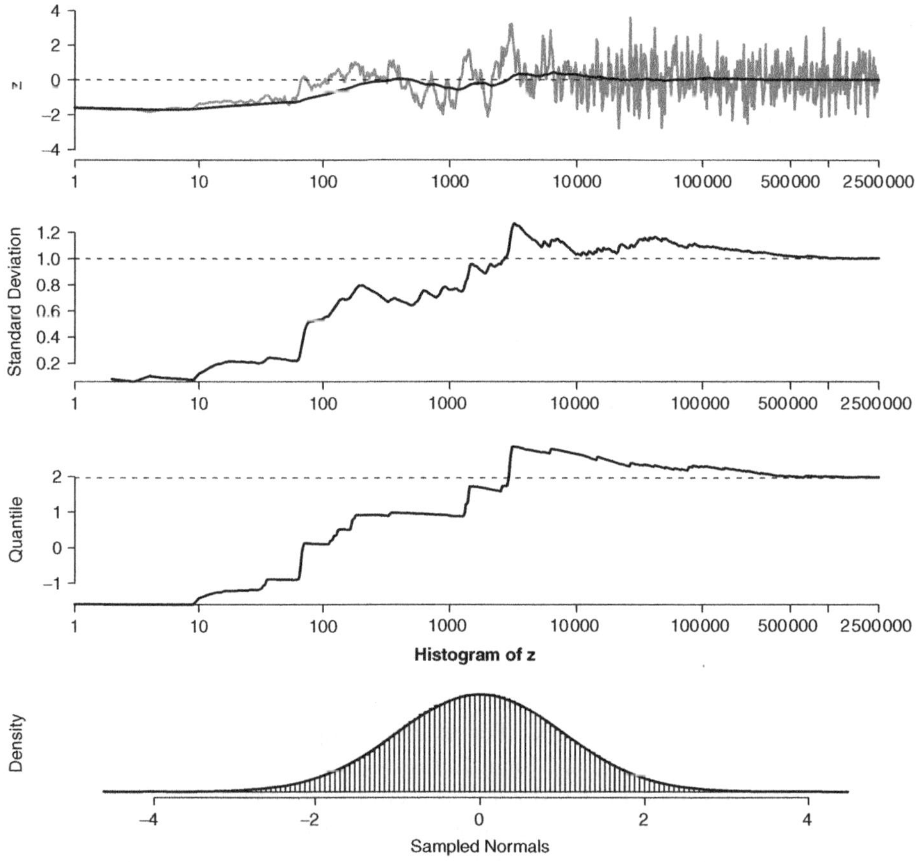

Figure 4.1 History of stationary, first-order normal process, with $\rho = .995$. The top panel displays the series itself, with the cumulative mean overlaid as the darker line. The second panel shows the trajectory of the Monte Carlo estimate of the standard deviation of z, converging on 1.0. The third panel shows the trajectory of the Monte Carlo estimate of the 97.5 percentile of z, converging on 1.96. In the upper three panels, the sample count is plotted on a logarithmic scale. The lower panel displays a histogram of the 2.5 million sampled values of z, with the normal density overlaid.

Now, suppose we are interested in learning about a random variable z, where $z \sim N(0, 1)$. Suppose further we can't sample directly from the $N(0, 1)$ density, but instead explore its density via a first-order Markov process, specifically, the AR(1) process sketched above. Suppose that this stochastic exploration of $p(z)$ is extremely inefficient, in that the AR(1) process is extremely 'sticky' with $\rho = .995$. That is, any sampled value from $p(z)$, say $z^{(t)}$, generated using the conditional distribution in equation (4.11), will be *extremely similar* to the preceding sampled value $z^{(t-1)}$, and so a large number of sampled values are required to generate an accurate characterization of $p(z)$.

The trajectory of the random tour of $p(z)$ provided by this AR(1) process is shown in the top panel of Figure 4.1. The extremely high level of serial correlation is visually apparent in the early stages of the series, but is less visible as the plotted points are closer together (the horizontal axis is plotted on a logarithmic scale). Indeed, when successive iterates are plotted on a highly compressed scale (e.g. iterations 100 000 to 2 500 000), the series appears to be independent 'white noise'. The graphical illusion aside, there is an important point at work here: a highly autocorrelated (but nonetheless stationary) series will display mean reversion, if we let it run long enough. To be sure, as the level of autocorrelation increases, we will need potentially *many* more iterations to generate as many 'mean crossings' as one would observe with an independent series.

4.4.2 Central limit theorems for Markov chains

Finally, we can provide central limit theorems for Markov chains, generalizing the corresponding result for the independence sampler in Proposition 3.2. The details here are not of much direct interest to the applied researcher – the more practical results are simulation consistency and the nature of the simulation inefficiency discussed above – and we will not dwell on this topic. Unsurprisingly, once we leave the independence setting (but nonetheless have a stationary Markov chain), the rate at which asymptotic results 'kick in' depend on the level of dependence in the chain. Central limit theorems require a little more than that required for simulation consistency; in order for a normal law to apply to the Monte Carlo error in a quantity like $\bar{h}_T$, we need not just stationarity, but that the dependence in the Markov chain dies away at a sufficiently fast rate. Geometric ergodicity (Definition 4.24) and (more strongly) uniform ergodicity (Definition 4.25 and Proposition 4.6) are sufficient conditions for CLTs to apply to Markov chains. For example, Robert and Casella (2004, Theorem 6.82) restate a result given by Tierney (1994, Theorem 5), relying on the somewhat strong assumption of uniform ergodicity:

Proposition 4.8 (Central limit theorem for uniformly ergodic Markov chains). *Let* $\{\boldsymbol{\theta}_t\}$ *be a stationary uniformly ergodic Markov chain. For any function* $h(\cdot)$ *s.t.* $\text{var}(h(\boldsymbol{\theta}_t)) = \sigma^2 < \infty$, *there exists a real number* τ_h *s.t.*

$$\sqrt{T}\tau_h^{-1}\left[T^{-1}\sum_{t=1}^{T} h(\boldsymbol{\theta}_t) - E(h(\boldsymbol{\theta}))\right] \xrightarrow{d} N(0, 1)$$

Proof. Tierney (1994, 1717) cites Cogburn (1972, Corollary 4.2(ii)). ◁

CLTs that do not rely on these stronger forms of ergodicity appears in the discussion following Tierney (1994); for instance, see Chan and Geyer (1994) and Robert (1994). In particular, if a Markov chain is reversible (Definition 4.19) and aperiodic and irreducible, then a CLT applies to any estimand $\bar{h}_T$ with finite variance. A useful summary of these and other results appears in Robert and Casella (2004, §6.7). Detailed discussion appears in Meyn and Tweedie (2009, Chapter 17).

4.5 Further reading

The literature on Markov chains is massive. Any serious treatment of probability theory and its applications considers the topic; hence Markov chains appear prominently in classic works such as Feller (1968). Here I list some more recent works that I have found useful sources. Norris (1997) is an accessible treatment for discrete-time Markov chains on discrete state spaces. The book by Brémaud (1999) covers a lot of relevant ground, moving from a treatment of Markov chains to their application in sequential, Monte Carlo explorations of parameter spaces. Rosenthal (2006) also provides a concise introduction to the elements of Markov chain theory relevant to statistics, again focused on the case of Markov chains on discrete state spaces. The recent review article by Diaconis (2009) is especially helpful and contains many references to relevant literature outside of statistics.

Meyn and Tweedie (1993) remains one of the most thorough treatments of Markov chains on general state spaces; a second edition (Meyn and Tweedie 2009) has recently been published. Nummelin (1984) is an older yet very valuable book on the topic, relied on by more recent treatments. Rigorous-yet-accessible treatments in the statistics literature include Robert and Casella (2004, Chapter 6) and Liu (2001). The article by Tierney (1994) is often cited as authority for many of the definitions and propositions given above; see also the accompanying discussion pieces (e.g. Chan and Geyer 1994; Robert 1994) and other summary articles (Tierney 1996, 1997). The issue of 'mixing' or 'running time' – informally, how quickly a Markov chain moves through its state space and in turn, how long we need to run the Markov chain – is the subject of a recent book by Levin, Peres and Wilmer (2008).

Problems

4.1 Consider the two state Markov chain presented in Example 4.1, with the transition matrix given in Equation 4.3. Suppose this Markov chain is initialized with the marginal distribution $\mathbf{p}^{(0)} = (p_0^{(0)}, p_1^{(1)})' = (.5, .5)'$. Show how the marginal distribution progresses towards the Markov chain's stationary distribution over successive iterations, by graphing $p_0^{(t)}$ and $p_1^{(t)}$ over iterations $t = 0, 1, 2, \ldots$. Comment on the way the Markov chain converges on its stationary distribution.

4.2 Repeat the previous exercise, but with the transition matrix

$$\mathbf{K} = \begin{bmatrix} .01 & .99 \\ .99 & .01 \end{bmatrix}$$

and with the initial marginal distribution $\mathbf{p}^{(0)} = (.99, .01)'$.

4.3 Consider a two-state Markov chain with transition matrix

$$\mathbf{K} = \begin{bmatrix} 1 & 0 \\ 0 & 1 \end{bmatrix}$$

i.e. the identity matrix $\mathbf{I}_2$. Comment on the dynamics of this chain and its stationary distribution.

4.4 Consider a two-state Markov chain with transition matrix

$$\mathbf{K} = \begin{bmatrix} 0 & 1 \\ 1 & 0 \end{bmatrix}.$$

Comment on the dynamics of this chain and its stationary distribution.

4.5 Consider the two-state Markov chain with transition matrix

$$\mathbf{K} = \begin{bmatrix} 1-p & p \\ q & 1-q \end{bmatrix},$$

where $p, q \in [0, 1]$.

1. State the conditions under which the Markov chain has a unique stationary distribution.
2. Use the fact that detailed balance (Definition 4.18) holds only for the Markov chain's stationary distribution to derive the stationary distribution of this chain.

4.6 Consider the following three definitions:

Definition 4.1 (Transition graph). *A transition matrix* $\mathbf{K}$ *may be represented by a directed graph (or digraph),* $\mathcal{G}$*, with vertices corresponding to the states of the Markov chain, and with a directed edge from vertex* i *to vertex* j *if* $K_{ij} = Pr(\boldsymbol{\theta}^{(t)} = j | \boldsymbol{\theta}^{(t-1)} = i) > 0$*. Such a graph is called the transition graph of the transition matrix (e.g. Brémaud 1999, 56).*

Definition 4.2 (Strongly connected digraph). *A directed graph is said to be strongly connected if it is possible to reach any vertex starting from any other vertex, proceeding along edges in the direction in which they point.*

Definition 4.3 (Reducible matrix). *A square, n-by-n matrix* $\mathbf{A} = (A_{ij})$ *is reducible if the indices* $1, 2, \ldots, n$ *can be divided into two disjoint, nonempty sets* $i_1, i_2, \ldots, i_p$ *and* $j_1, j_2, \ldots, j_q$*, where* $p + q = n$ *such that* $A_{i_u j_v} = 0, \forall\, u = 1, \ldots, p$ *and* $\forall\, v = 1, \ldots, q$.

1. Show that a Markov chain on a finite state space is irreducible if and only if its transition graph is strongly connected.
2. Consider the directed graph (digraph) displayed in Figure 4.2. Is this graph strongly connected?
3. Construct the transition matrix implied by the transition graph in Figure 4.2.

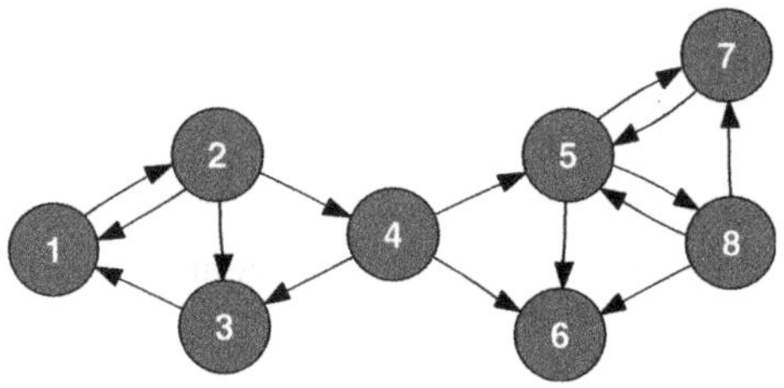

Figure 4.2 Directed graph for Problem 4.6.

4. If the graph in Figure 4.2 is considered a transition graph, then how many communication classes are there in the corresponding Markov chain?

5. If the graph in Figure 4.2 is considered a transition graph, then is corresponding Markov chain irreducible?

4.7 Consider the periodic Markov chain given in Example 4.10. Does a stationary distribution exist? If so, what is it?

4.8 (Robert and Casella 2004, Problem 6.9) Show that an aperiodic Markov chain on a finite state-space with transition matrix $\mathbf{K}$ is irreducible if and only if there exists $N \in \mathbb{N}$ such that $\mathbf{K}^N$ has no zero entries.

4.9 Verify the claim in Example 4.6. That is, suppose $\{\theta^{(t)}\}$ is a Markov chain with $p(\theta^{(t)}) \equiv N(\rho\theta^{(t-1)}, \sigma^2)$, with $|\rho| < 1$, i.e. a stationary, first order autoregressive Gaussian process. Show that

$$p(\theta) \equiv N\left(0, \sigma^2/(1-\rho^2)\right)$$

solves Equation 4.5 and so is the stationary distribution of the chain.

4.10 Show that the AR(1) process defined in the previous problem (and in Example 4.2) is Markovian. That is, show that for such a process $(\theta_t \perp \theta_{t-r})|\theta_{t-1}, \forall\, r > 1$.

4.11 Consider Example 4.12. Replicate Figure 4.1, but with the AR(1) parameter $\rho = .9$. In R, the `arima.sim` function will be helpful.

4.12 Conduct a Monte Carlo experiment to investigate the result in Equation 4.10. That is, repeat the following steps a large number of times, indexed by $m = 1, \ldots, M$, with M set to a large number (e.g. $M = 5000$):

1. Generate a large sample from the stationary, Gaussian, AR(1) process $z_t|z_{t-1} \sim N(\rho z_{t-1}, 1-\rho^2)$, with ρ set to a relatively large value (e.g. $\rho = .95$) and $E(z_t) = 0$. Note that marginally, $\text{var}(z_t) = 1$. Try $T = 10^5$ or so. Again, the `arima.sim` function in R is an easy way to do this. Compute and store the mean of the sampled z_t, $\bar{z}_T^{(m)}$.

2. Use an independence sampler to generate T draws from the marginal distribution $y_t \sim N(0, 1)$; e.g. use the `rnorm` function in R (although since the default

method in `rnorm` uses a pseudo-random number generator it is not an 'independence' sampler, but is more than close enough for all practical purposes). Compute and store the mean of the sampled y_t, $\bar{y}_T^{(m)}$.

Over the M replicates of this Monte Carlo experiment, compute the variance of the means $\bar{z}_T^{(m)}$ and $\bar{y}_T^{(m)}$. Up to Monte Carlo error, you should observe the result in Equation 4.10; the independence sampler generates $\bar{y}_T$ that are less dispersed around zero than the AR(1) sampler, with $\text{var}(\bar{z})/\text{var}(\bar{y}) \approx (1+\rho)/(1-\rho)$.

5

Markov chain Monte Carlo

In Chapter 3 we considered the Monte Carlo principle: anything we want to know about a random variable $\boldsymbol{\theta}$, we can learn by sampling repeatedly from its density, $p(\boldsymbol{\theta})$. In Chapter 4 we used results from Markov chain theory to see that the Monte Carlo principle applies even when the samples are not independent, but form a Markov chain. In this chapter we consider the problem of how to generate Markov chains that have a given target density – a posterior density $p(\boldsymbol{\theta}|\text{data})$ – as the Markov chain's invariant density. This problem is all about constructing the transition kernel of the chain the 'right way'. That is, we will work backwards from the goal of wanting a chain to have a particular stationary density to consider what kind of a transition kernel we need.

This way of sampling from a posterior density – combining the Monte Carlo principle with ideas from Markov chain theory – is known as Markov chain Monte Carlo (MCMC). We will see that MCMC is a very powerful class of techniques. Here we consider the two core MCMC algorithms: the Metropolis algorithm (Section 5.1) and the Gibbs sampler (Section 5.2).

5.1 Metropolis-Hastings algorithm

The Metropolis-Hastings algorithm defines a set of 'jumping rules' that generate a Markov chain on the support of $p(\boldsymbol{\theta}|\text{data})$, $\boldsymbol{\Theta}$. At the start of iteration t, we have $\boldsymbol{\theta}^{(t-1)}$ and we make the transition to $\boldsymbol{\theta}^{(t)}$ as follows:

Algorithm 5.1 Metropolis-Hastings

1: sample $\boldsymbol{\theta}^*$ from a "proposal" or "jumping" distribution $J_t(\boldsymbol{\theta}^{(t-1)}, \boldsymbol{\theta}^*)$.
2:

$$r \leftarrow \frac{p(\boldsymbol{\theta}^*|\mathbf{y})J_t(\boldsymbol{\theta}^*, \boldsymbol{\theta}^{(t-1)})}{p(\boldsymbol{\theta}^{(t-1)}|\mathbf{y})J_t(\boldsymbol{\theta}^{(t-1)}, \boldsymbol{\theta}^*)}, \tag{5.1}$$

Bayesian Analysis for the Social Sciences S. Jackman

3: $\alpha \leftarrow \min(r, 1)$
4: sample $U \sim \text{Unif}(0, 1)$
5: **if** $U \leq \alpha$ **then**
6: $\boldsymbol{\theta}^{(t)} \leftarrow \boldsymbol{\theta}^*$
7: else
8: $\boldsymbol{\theta}^{(t)} \leftarrow \boldsymbol{\theta}^{(t-1)}$
9: **end if**

The quantity r is an acceptance ratio, assessing the plausibility of the candidate point $\boldsymbol{\theta}^*$ relative to the current value $\boldsymbol{\theta}^{(t-1)}$.

This scheme means that if $r > 1$ then the algorithm makes the transition $\boldsymbol{\theta}^{(t)} \leftarrow \boldsymbol{\theta}^*$ with probability 1; otherwise we make that transition with probability r (i.e. if $U \sim \text{Unif}(0, 1)$ then $\Pr(r \leq U) = r$). With probability $1 - r$ the algorithm does not move at iteration t, setting $\boldsymbol{\theta}^{(t)} \leftarrow \boldsymbol{\theta}^{(t-1)}$.

The candidate density J is the key to the algorithm, as we also saw when considering the accept-reject algorithm in Section 3.4.3. Note also that J takes two arguments, indicating that the choice of a candidate point $\boldsymbol{\theta}^*$ may well depend on the current state of the algorithm $\boldsymbol{\theta}^{(t-1)}$. With this in mind, note the reversal of the arguments passed to the proposal density J in the denominator of equation 5.1, a point we explore further, below.

The fact that J has a subscript t indicates that J may change as iterations progress, in an effort to 'tune' the algorithm. For instance, an initial choice of J may generate values of r that are small, generating few acceptances, and hence a very inefficient exploration of the parameter space; on the other hand, J may generate high acceptance ratios, but only in a small neighborhood around $\boldsymbol{\theta}^{(t-1)}$, again resulting in a computationally inefficient exploration of $p(\boldsymbol{\theta}|\mathbf{y})$.

5.1.1 Theory for the Metropolis-Hastings algorithm

The original Metropolis *et al*. (1953) algorithm computed the acceptance ratio as

$$r_M = p(\boldsymbol{\theta}^*|\mathbf{y})/p(\boldsymbol{\theta}^{(t-1)}|\mathbf{y}). \tag{5.2}$$

Hastings (1970) provides the modification of r_M to the r given in Equation 5.1. Liu (2001, 111–112) reviews numerous proposals that have been made for acceptance ratios over the years; a key result (which Liu attributes to a personal communication from Charles Stein) is that we have acceptance ratios of the form

$$r_S(\boldsymbol{\theta}^{(t-1)}, \boldsymbol{\theta}^*) = \frac{\delta(\boldsymbol{\theta}^{(t-1)}, \boldsymbol{\theta}^*)}{p(\boldsymbol{\theta}^{(t-1)}|\mathbf{y}) J_t(\boldsymbol{\theta}^{(t-1)}, \boldsymbol{\theta}^*)}$$

where (a) $\delta(\boldsymbol{\theta}^{(t-1)}, \boldsymbol{\theta}^*)$ is a symmetric function, i.e. $\delta(\boldsymbol{\theta}^{(t-1)}, \boldsymbol{\theta}^*) = \delta(\boldsymbol{\theta}^*, \boldsymbol{\theta}^{(t-1)})$ and (b) δ is such that $r(\boldsymbol{\theta}^{(t-1)}, \boldsymbol{\theta}^*) \leq 1, \forall\ \boldsymbol{\theta}^{(t-1)}, \boldsymbol{\theta}^*$.

We now show that this scheme generates a Markov chain that has the posterior density $p(\boldsymbol{\theta}|\mathbf{y})$ as its invariant distribution. The argument is rather clever. We rely on the definition of reversibility (Definition 4.18). For the generic Metropolis scheme given above with acceptance ratio r_S we have a Markov chain with transition kernel

$$K(\boldsymbol{\theta}^{(t)}, \boldsymbol{\theta}^{(t+1)}) = J_t(\boldsymbol{\theta}^{(t)}, \boldsymbol{\theta}^{(t+1)}) r_S = J_t(\boldsymbol{\theta}^{(t)}, \boldsymbol{\theta}^{(t+1)}) \frac{\delta(\boldsymbol{\theta}^{(t)}, \boldsymbol{\theta}^{(t+1)})}{p(\boldsymbol{\theta}^{(t)}|\mathbf{y}) J_t(\boldsymbol{\theta}^{(t)}, \boldsymbol{\theta}^{(t+1)})}$$
$$= \frac{\delta(\boldsymbol{\theta}^{(t)}, \boldsymbol{\theta}^{(t+1)})}{p(\boldsymbol{\theta}^{(t)}|\mathbf{y})}$$

Now by the symmetry of δ, we have

$$p(\boldsymbol{\theta}^{(t)}) K(\boldsymbol{\theta}^{(t)}, \boldsymbol{\theta}^{(t+1)}) = p(\boldsymbol{\theta}^{(t+1)}) K(\boldsymbol{\theta}^{(t+1)}, \boldsymbol{\theta}^{(t)})$$

which means that the resulting Markov chain is reversible (Definition 4.19) and, critically, that p is the invariant distribution of the chain (recall the argument in Section 4.2.6).

All that remains to be shown is that the particular acceptance ratio used by the Metropolis-Hastings algorithm (Equation 5.1 in Algorithm 5.1) generates a symmetric function that will satisfy the reversibility conditions just given. This is easily done, and here we reproduce an argument that appears in many places in the literature: e.g. Liu (2001, 113) or the expository article by Chib and Greenberg (1995).

The transition kernel for the Markov chain generated by Algorithm 5.1 is for any $\boldsymbol{\theta}^{(t)} \neq \boldsymbol{\theta}^{(t+1)}$

$$K(\boldsymbol{\theta}^{(t)}, \boldsymbol{\theta}^{(t+1)}) = J_t(\boldsymbol{\theta}^{(t)}, \boldsymbol{\theta}^{(t+1)}) \min(r, 1)$$
$$= J_t(\boldsymbol{\theta}^{(t)}, \boldsymbol{\theta}^{(t+1)}) \min\left\{1, \frac{p(\boldsymbol{\theta}^{(t+1)}|\mathbf{y}) J_t(\boldsymbol{\theta}^{(t+1)}, \boldsymbol{\theta}^{(t)})}{p(\boldsymbol{\theta}^{(t)}|\mathbf{y}) J_t(\boldsymbol{\theta}^{(t)}, \boldsymbol{\theta}^{(t+1)})}\right\},$$

and so

$$p(\boldsymbol{\theta}^{(t)}|\mathbf{y}) K(\boldsymbol{\theta}^{(t)}, \boldsymbol{\theta}^{(t+1)}) = p(\boldsymbol{\theta}^{(t)}|\mathbf{y}) J_t(\boldsymbol{\theta}^{(t)}, \boldsymbol{\theta}^{(t+1)}) \min\left\{1, \frac{p(\boldsymbol{\theta}^{(t+1)}|\mathbf{y}) J_t(\boldsymbol{\theta}^{(t+1)}, \boldsymbol{\theta}^{(t)})}{p(\boldsymbol{\theta}^{(t)}|\mathbf{y}) J_t(\boldsymbol{\theta}^{(t)}, \boldsymbol{\theta}^{(t+1)})}\right\}$$
$$= \min\left\{p(\boldsymbol{\theta}^{(t)}|\mathbf{y}) J_t(\boldsymbol{\theta}^{(t)}, \boldsymbol{\theta}^{(t+1)}), p(\boldsymbol{\theta}^{(t+1)}|\mathbf{y}) J_t(\boldsymbol{\theta}^{(t+1)}, \boldsymbol{\theta}^{(t)})\right\}$$

which is a symmetric function in $\boldsymbol{\theta}^{(t)}$ and $\boldsymbol{\theta}^{(t+1)}$. In turn, this gives us reversibility of the resulting Markov chain, and hence p as the chain's invariant distribution.

To apply the 'simulation-consistency' results from Section 4.4 we need some additional theoretical results. Given that we have established the existence of a stationary distribution for the Markov chain generated by Algorithm 5.1, we can establish that the chain is ergodic (Definition 4.22) if we can establish Harris recurrence (Definition 4.16) and aperiodicity (Definition 4.21). Harris recurrence is key, a necessary condition for the Ergodic Theorem we stated in Proposition 4.7.

Irreducibility with respect to the chain's invariant measure p follows if $J_t(\boldsymbol{\theta}^{(t)}, \boldsymbol{\theta}^{(t+1)}) > 0 \, \forall (\boldsymbol{\theta}^{(t)}, \boldsymbol{\theta}^{(t+1)}) \in \boldsymbol{\Theta} \times \boldsymbol{\Theta}$. That is, we need J_t to be such that the algorithm can jump to any region in $\boldsymbol{\Theta}$ where p is positive. This is sufficient for Harris recurrence (Robert and Casella 2004, Lemma 7.3); we can obtain the same result if we add absolute continuity with respect to p to the conditions we impose on J_t (see Proposition 4.3). Aperiodicity follows if J_t (and hence the transition kernel K of the resulting Markov chain) is such that $\Pr(\boldsymbol{\theta}^{(t+1)} = \boldsymbol{\theta}^{(t)}) > 0$. In general, these are not restrictive conditions on J_t. Hence, most Markov chains generated by the Metropolis-Hastings algorithm are ergodic and the limit theorems given in Section 4.4 can be applied.

5.1.2 Choosing the proposal density

Clearly, there are better or worse choices of the proposal density J_t, which will lead to greater or lesser rates of rejection and/or better mixing of the resulting Markov chain through the parameter space. There is a long line of literature on the optimal choices of both the proposal density J_t and the acceptance ratio r. A detailed consideration of this literature is beyond our scope here, although we will pause to consider some highlights; interested readers ought to consult Liu (2001 §13.3).

Random-walk Metropolis

A popular proposal density for Metropolis algorithms is a 'random walk' proposal: i.e. select a candidate point $\boldsymbol{\theta}^*$ by taking a random perturbation around the current point $\boldsymbol{\theta}^{(t)}$, i.e. $\boldsymbol{\theta}^* = \boldsymbol{\theta}^{(t)} + \boldsymbol{\epsilon}$. The possibilities for the density for $\boldsymbol{\epsilon}$ are endless, but include

- $\epsilon_j \sim \text{Unif}(-\delta_j, \delta_j)$, where j indexes the components of $\boldsymbol{\epsilon} = (\epsilon_1, \ldots, \epsilon_J)'$. Tuning this algorithm for better performance involves judicious selection of the δ_j parameters.
- $\boldsymbol{\epsilon} \sim N(0, \boldsymbol{\Omega})$. Here the key parameter is $\boldsymbol{\Omega}$; large values on the diagonal of $\boldsymbol{\Omega}$ will generate candidate points that potentially move the algorithm a long way in the parameter space (but may be rejected because the candidate point lies in a region of low posterior probability); small values on the diagonal of $\boldsymbol{\Omega}$ will generate small moves but may be more likely to be accepted.
- a finite mixture distribution $f(\boldsymbol{\epsilon}) \equiv \sum_{m=1}^{M} \lambda_m f_m(\boldsymbol{\epsilon})$, where $\sum_{m=1}^{M} \lambda_m = 1$ and $0 < \lambda_m < 1 \, \forall \, m$. As we saw in Section 2.1.1 and in Figure 1.9, mixtures are extremely flexible and permit a wide array of unusual densities to be represented in a parametric form.

Clearly, the resulting sequence of sampled values will display serial dependence and simulation inefficiency; how much is a case-by-case proposition, but will depend on the size of the proposed steps generated by sampling from $f(\boldsymbol{\epsilon})$ relative to the target density $p(\boldsymbol{\theta}|\mathbf{y})$.

■ Example 5.1

Sampling from a normal density via random-walk Metropolis. Suppose $\boldsymbol{\theta} \sim N$. Put aside the fact that direct sampling from this density is trivial (see the remark in Example 3.6). If we use a random-walk Metropolis algorithm for this problem, what should we use as $f(\boldsymbol{\epsilon})$? How do we adjudicate among the possibilities?

Variants of this problem appear throughout the literature. Chib and Greenberg (1995) considered a number of proposals for $f(\boldsymbol{\epsilon})$. Here we briefly replicate their analysis with a bivariate normal target density, $\boldsymbol{\theta} \sim N(\boldsymbol{\mu}, \boldsymbol{\Sigma})$, where

$$\boldsymbol{\mu} = \begin{bmatrix} 1 \\ 2 \end{bmatrix} \quad \text{and} \quad \boldsymbol{\Sigma} = \begin{bmatrix} 1 & .9 \\ .9 & 1 \end{bmatrix},$$

and where $\epsilon_j \sim \text{Unif}(-\delta, \delta)$, $j = 1, 2$. We consider the performance of the resulting Markov chain as we vary δ. In every instance the MH algorithm is initialized at $\boldsymbol{\theta}^{(0)} = (0, 0)'$ where $p(\boldsymbol{\theta})$ is small.

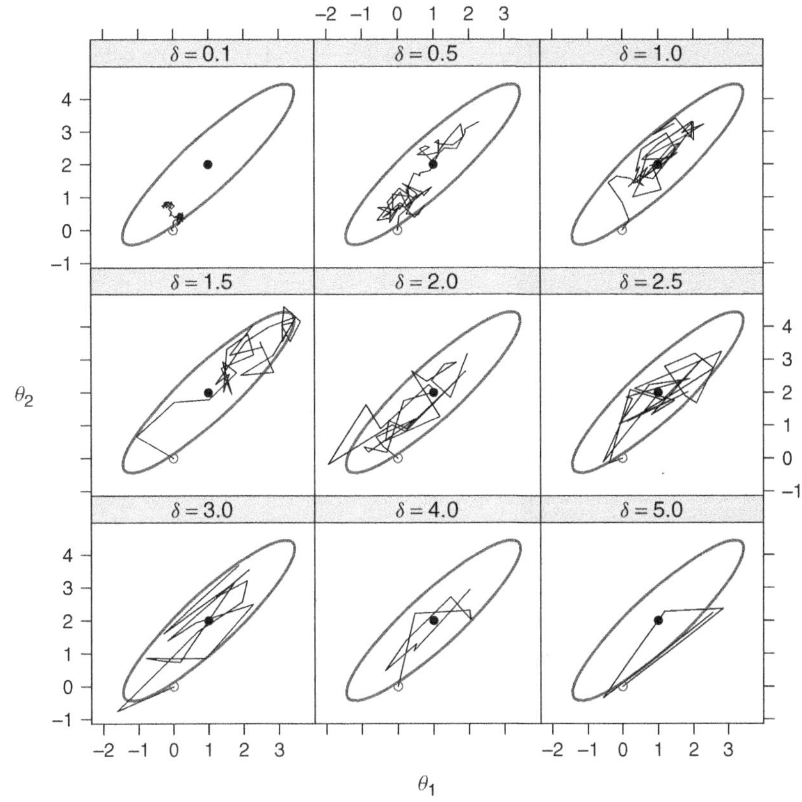

Figure 5.1 Trajectories of random-walk Metropolis-Hastings algorithms over the first 100 iterations, Example 5.1. Each panel shows the trajectory of the first 100 iterations of the algorithm with the indicated level of δ. In each instance the algorithm is initialized at $(0, 0)$. The ellipse encloses a 95 % HDR for the target density.

Figure 5.1 shows the trajectory of the algorithm over its first 100 iterations for a series of values of δ. Small values of δ result in many small steps in the parameter space; overly large values of δ result in very few steps, since the candidate points are almost always proposing steps into regions of low density, and are seldom accepted. Thus, both very small and very large values of δ result in an inefficient exploration of the parameter space, but with much worse results from setting δ too small rather than too large. In this case, visual inspection of Figure 5.1 suggests that values of δ in the range 1.5 to 3.0 generate relatively efficient explorations of the target density.

Further confirmation comes from inspecting the autocorrelations of the trajectories of each parameter (e.g. Figure 5.2 shows the autocorrelations for the sampled θ_1). For very low values of δ the autocorrelations are decaying extremely slowly, consistent with an extremely slow exploration of the target density. The autocorrelation function appears to be decaying most quickly at around $\delta = 2.0$, and slowly degrading again for larger values of δ. Note also that none of the autocorrelation functions suggest a particularly efficient exploration of the target density; even in the best case scenario, the random walk Metropolis algorithms considered here tend to display a moderate amount of autocorrelation and

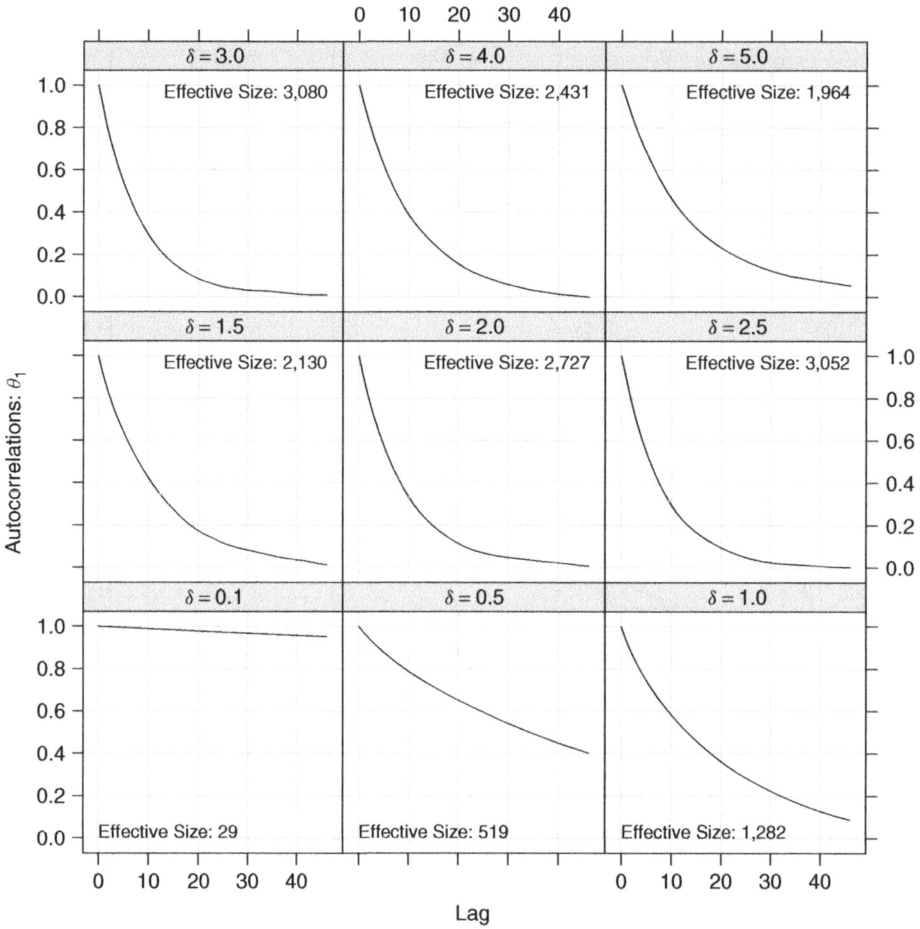

Figure 5.2 Autocorrelations, random-walk Metropolis-Hastings algorithm, Example 5.1. Each panel shows the autocorrelations of the sampled θ_1 for each indicated value of δ. The effective sample size is an estimate of the number of independent draws from the target density that would yield the same precision as that generated by the serially dependent sequence of 50 000 iterates; see § 4.4.1.

hence will require a relatively large number of iterations to generate a good characterization of the target density.

Figure 5.3 considers two performance indicators of the random walk Metropolis algorithm with respect to the θ_1 parameter – the effective sample size (defined in Section 4.4.1) of a 50 000 iteration run of the algorithm, and the acceptance ratio r – as functions of one another, and of the user-set tuning parameter, δ. Recall that the effective sample size is a function of the autocorrelation spectrum of the sampled θ_1. The largest effective sample sizes (the most efficient exploration of the target density) are just over 3000, and occur with relative small acceptance rates, around .16 to .18, putting aside the small amount of Monte Carlo error apparent in Figure 5.3. In turn, these acceptance rates correspond to $\delta \approx 2.5$, which is surprisingly large given that $V(\theta_1) = \mathrm{V}(\theta_2) = 1$. That

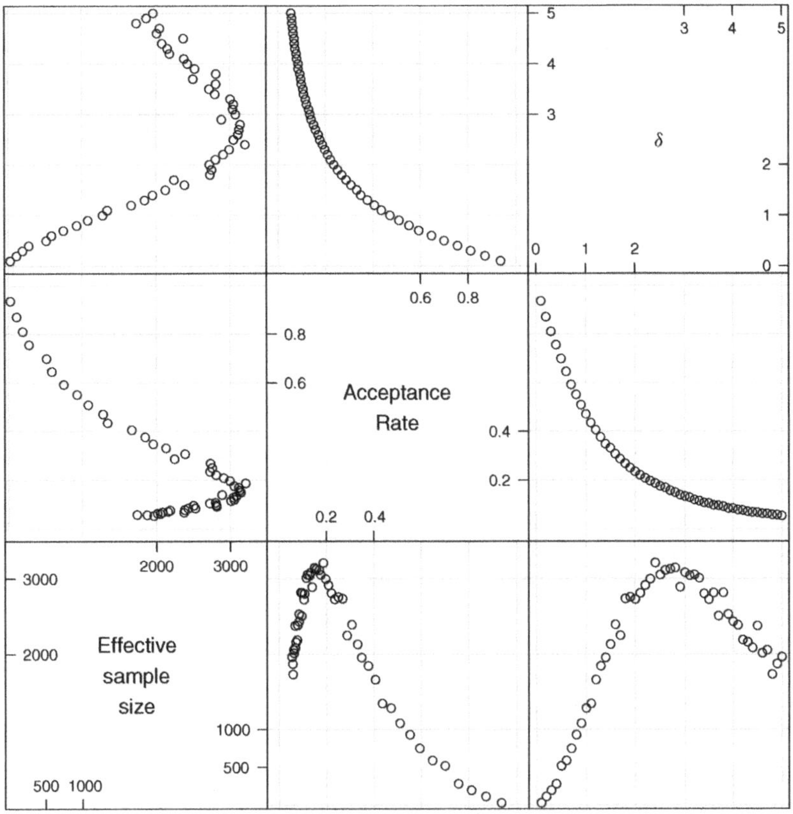

Figure 5.3 Scatterplot matrix, indicators of performance of random-walk Metropolis-Hastings algorithm used in Example 5.1. We assess the performance of the algorithm with respect to the sampled values of θ_1. Each panel is a scatterplot, showing the relationship between the effective sample size of the 50 000 iteration run of the random walk Metropolis-Hastings algorithm (see Section 4.4.1), the acceptance rate of the algorithm, and the tuning parameter δ.

is, with $\delta = 2.5$, 50 % of the proposed $\boldsymbol{\theta}^*$ will lie a distance of more than 2.5 σ away from the current point (in either direction); of these proposed moves, only about 17 % are accepted, but this appears to be optimal using the uniform-in-each-direction proposal density.

The random-walk Metropolis algorithm used in the previous example used an extremely simply 'uniform-in-each-direction' proposal density, centered on the current point. Other proposals are certainly possible, perhaps better suited to the problem at hand. Indeed, since the target density is a normal density, a normal density would make a terrific proposal. One might be tempted to see this as a trite problem, in that we are considering solving the (hypothetical) problem of sampling from a normal by using a

Metropolis algorithm that samples candidate moves from a normal. But this approach is not as superficial as it sounds, in that (a) many posterior densities are well-approximated by a normal and (b) it is trivial to sample from a normal. Thus, understanding how well a normal proposal density performs when the target is a normal is far from a moot exercise, in that understanding the performance of the Metropolis algorithm under these conditions provides considerable insight into how the Metropolis algorithm performs in a reasonably wide set of conditions.

To this end, Gelman, Roberts and Gilks (1995) investigated the properties of a random-walk Metropolis algorithm where the target density is $N(0, 1)$ and the proposals are drawn as $\theta^* = \theta^{(t)} + \epsilon, \epsilon \sim N(0, \sigma^2)$; setting $\sigma = 2.38$ generates a Markov chain with the lowest level of autocorrelation, generating an acceptance rate of about 44 %. This suggests a benchmark of sorts, in that if a target/posterior density is approximately normal, then one should be able to tune a normal proposal density such that a random walk Metropolis algorithm should be able to go close to reaching the 44 % acceptance rate obtained by Gelman, Roberts and Gilks (1995). In Problem 6.1 you are invited to verify this result. When the target density is multidimensional (as we considered in Example 5.1), the optimal acceptance rate may be lower, as we saw with the uniform-in-each-direction proposal density. Roberts, Gelman and Gilks (1997) offer advice on the acceptance rates to aim for when the dimensionality of the target density increases; see Geyer and Thompson (1995) for an interesting counter-example.

The proposal density used in the preceding example takes no account of the correlation structure between θ_1 and θ_2, nor does it remember what the algorithm is revealing about the target density over the course of its iterative history. A proposal density that is better to start with (in that it closely approximates the target density) would generate considerable computational efficiencies for this problem. In the following example we use a random-walk Metropolis algorithm with a 'smarter' proposal density than the independent 'uniform-in-each-direction' proposal used above.

■ Example 5.2

Poisson regression model for count data via an random-walk Metropolis; article production by biochemistry PhD students. Regression models for counts are typically specified as follows: $y_i|(\mathbf{x}_i, \boldsymbol{\beta}) \sim \text{Poisson}(\lambda_i)$, $\lambda_i = \exp(\mathbf{x}_i\boldsymbol{\beta})$ where $y_i \in \{0, 1, 2, \ldots\}$, $\mathbf{x}_i$ is a vector of covariates, $\boldsymbol{\beta}$ is a vector of k unknown coefficients and $i = 1, \ldots, n$ indexes observations. To obtain the likelihood for this model, we assume that the counts are conditionally independent given the $\mathbf{x}_i$ and $\boldsymbol{\beta}$, such that the joint probability mass function of $\mathbf{y}$ factors as the product of the observation-specific Poisson pmfs (see Definition B.23): i.e.

$$f(\mathbf{y}|\mathbf{X}, \boldsymbol{\beta}) = \prod_{i=1}^{n} \frac{\lambda_i^{y_i}}{y_i!} \exp(-\lambda_i) \tag{5.3}$$

where $\lambda_i = \exp(\mathbf{x}_i\boldsymbol{\beta})$.

Long (1990, 1997) uses this model to analyze variation in the number of published articles produced by 915 biochemistry graduate students in the last 3 years of their PhD studies, with a particular interest in differences by gender. These data are available as the data frame `bioChemists` in the R package `pscl` (Jackman 2008b). The modal number of article counts is zero, recorded among 30 % of the graduate students; as is

Table 5.1 Summary statistics, data on article production by biochemistry PhD students. The response variable is Articles, a count of the number of articles produced in the last three years of the student's PhD study; Kids < 5 is the number of children under the age of five; PhD Prestige is an indicator tapping the prestige of the student's department; Mentor is the number of articles produced by the student's faculty mentor over the preceding 3 years. q_{25} indicates the 25th percentile, etc. See (Long 1990) for additional details.

Articles		Gender		Marital status		Kids < 5		PhD prestige		Mentor	
Min.	0	M	494	Single	309	Min.	0.0	Min.	0.8	Min.	0
q_{25}	0	F	421	Married	606	q_{25}	0.0	q_{25}	2.3	q_{25}	3
q_{50}	1					q_{50}	0.0	q_{50}	3.2	q_{50}	6
Mean	2					Mean	0.5	Mean	3.1	Mean	9
q_{75}	2					q_{75}	1.0	q_{75}	3.9	q_{75}	12
Max.	19					Max.	3.0	Max.	4.6	Max.	77

typical of count data in many social science settings, the distribution of the counts are quite long tailed, with a 95-th percentile of 5 articles and a maximum count of 19 articles. A summary of the article counts and covariates appears in Table 5.1. We keep the analysis simple, with each covariate entering the regression model in a linear additive fashion on the scale of log counts; i.e. recall that with the conventional log-link formulation we have $E(y_i|\mathbf{x}_i, \boldsymbol{\beta}) = \lambda_i = \exp(\mathbf{x}_i\boldsymbol{\beta})$.

Bayesian analysis. There is no prior for $\boldsymbol{\beta}$ that is conjugate with respect to the likelihood in Equation 5.3, and so the posterior density for $\boldsymbol{\beta}$ is not available in closed form. The posterior density is usually expressed as proportional to the prior times the likelihood: i.e.

$$p(\boldsymbol{\beta}|\mathbf{y}, \mathbf{X}) \propto p(\boldsymbol{\beta}) f(\mathbf{y}|\mathbf{X}, \boldsymbol{\beta}), \tag{5.4}$$

where the function f is the likelihood in Equation 5.3. Nonetheless, Bayesian analysis of this model is relatively straightforward via a Metropolis type algorithm, with the only real technical challenge being selecting and tuning the proposal density. A random-walk Metropolis algorithm for exploring the posterior density of $\boldsymbol{\beta}$ is implemented in `MCMCpack` (Martin, Quinn and Park 2009) as the function `MCMCpoisson`. The function presumes the user has specified a multivariate normal prior for $\boldsymbol{\beta}$, $\boldsymbol{\beta} \sim N(\mathbf{b}_0, \mathbf{B}_0^{-1})$. The Metropolis proposal density is $\boldsymbol{\beta}^* \sim N(\boldsymbol{\beta}^{(t)}, \mathbf{P})$ where $\mathbf{P} = \mathbf{T}(\mathbf{B}_0 + \mathbf{V}^{-1})^{-1}\mathbf{T}$, with $\mathbf{T}$ a k-by-k diagonal, positive definite matrix containing tuning parameters and $\mathbf{V}$ is the large-sample approximation to the frequentist sampling covariance matrix of the maximum likelihood estimates $\hat{\boldsymbol{\beta}}$, computed as minus the inverse of the Hessian matrix, evaluated at the MLEs: i.e. $\mathbf{V} = -\mathbf{H}^{-1}$ where

$$H_{ij} = \left. \frac{\partial^2 \log f(\mathbf{y}|\boldsymbol{\beta})}{\partial \beta_i \partial \beta_j} \right|_{\boldsymbol{\beta}=\hat{\boldsymbol{\beta}}} \tag{5.5}$$

where $i, j \in \{1, \ldots, k\}$ index elements of the parameter vector $\boldsymbol{\beta}$. The defaults are $\mathbf{T} = 1.1 \cdot \mathbf{I}$ and to initialize the random-walk Metropolis algorithm at the MLEs $\hat{\boldsymbol{\beta}}$. We use vague priors, with $\mathbf{b}_0 = \mathbf{0}$ and $\mathbf{B}_0 = 10^4 \cdot \mathbf{I}$.

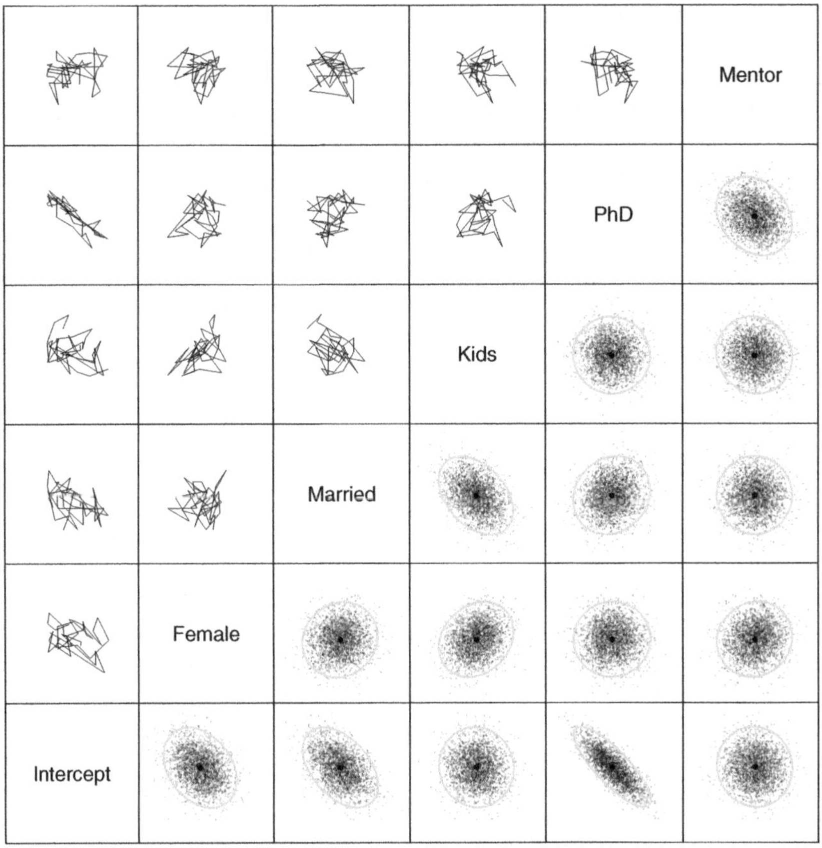

Figure 5.4 Random-walk Metropolis algorithm, Poisson regression model, Example 5.2. Each panel shows the algorithm working in a series of two-dimensional slices through the parameter space. Panels above the diagonal show the algorithm's progress over the first 250 iterations. Panels below the diagonal show 2500 iterations (evenly spaced over the full 50 000 iteration run), with the ellipse indicating a 95 % HDR (based on a normal approximation). The labels on the diagonal identify the particular element of the parameter vector $\boldsymbol{\beta}$.

Figure 5.4 summarizes the output of 50 000 iterations from the random-walk Metropolis algorithm, via a scatterplot matrix layout. The upper panels show a trace plot of the algorithm in two dimensions, for the first 250 iterations of the algorithm. The lower panels summarize the full 50 000 iterations, plotting the algorithm's history at each of 2500 evenly-spaced iterations over the full 50 000 iterations. With the vague prior employed here and a large sample, the posterior density of $\boldsymbol{\beta}$ will be close to the normal approximation to the frequentist sampling distribution of $\hat{\boldsymbol{\beta}}$ with covariance matrix $\mathbf{V}$, which means that the covariance matrix of the proposal density $\mathbf{P}$ is very close to the covariance matrix of the target, posterior density; this should bode well for the performance of the algorithm.

Nonetheless, the random-walk Metropolis algorithm still generates a relatively inefficient exploration of the posterior density, as shown in Figure 5.5. The autocorrelations

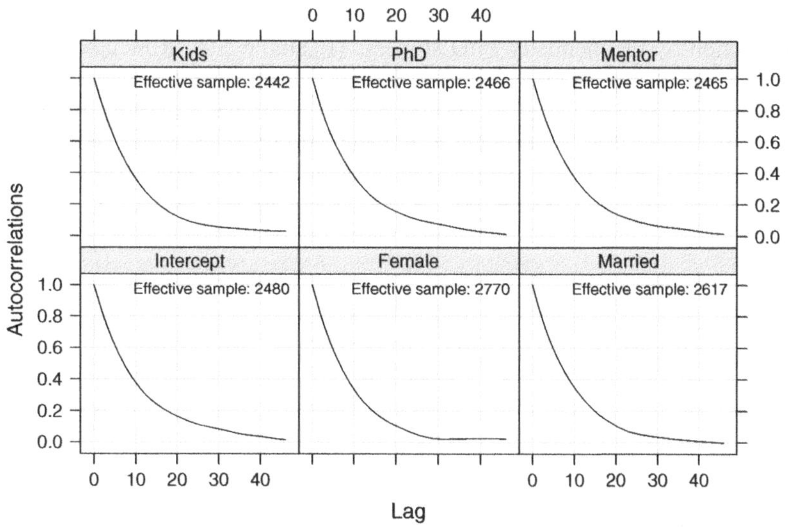

Figure 5.5 Autocorrelations, random-walk Metropolis algorithm, Poisson regression model, Example 5.2. Fifty thousand iterations were produced by the random-walk Metropolis algorithm using the `MCMCpoisson` function in the `R` package `MCMCpack` (Martin, Quinn and Park 2009). Each panel shows the autocorrelations of the sampled values from the marginal posterior density of a given element of β_j, where the labels indicate the name of covariate j. The 'effective sample size' (see Section 4.4.1) is the number of independent draws represented by the highly autocorrelated 50 000 random-walk Metropolis draws.

for each sequence – the sampled $\boldsymbol{\beta}_j^{(t)}$, $j = 1, \ldots, k$ – decay reasonably slowly for most parameters, giving rise to a high level of simulation inefficiency. In fact, the dependence in the sampled $\boldsymbol{\beta}^{(t)}$ is such that the 50 000 iterations generate as much information about the mean of each marginal posterior density as that we would get from about 2500 independent draws (recall the discussion of 'effective sample size' in Section 4.4.1). Yet, the acceptance rate for the sequence of sampled values $\{\boldsymbol{\beta}^{(t)}\}$ is 23 %, which is close to the suggestions of Roberts, Gelman and Gilks (1997). It would seem that these reasonably high levels of serial dependence and the resulting simulation inefficiency are simply 'par for the course' when using random-walk Metropolis algorithms.

With a large number of iterations, the simulation inefficiency here is not particularly consequential. The 50 000 iterations generated here take just a few seconds to generate (so a larger number of iterations could be easily generated), and in any event, the equivalent of 2500 independent draws is sufficient to provide a reasonably precise characterization of the posterior density. Table 5.2 compares Bayes estimates of $\boldsymbol{\beta}$ (the mean of the 50 000 samples from the posterior density of $\boldsymbol{\beta}$ produced by the random-walk Metropolis algorithm) and maximum likelihood estimates (MLEs), the latter produced by the `glm` function in `R`. There is very little to distinguish the two sets of estimates, which is the result to be expected with vague priors and a reasonably large sample.

This might seem an awful lot of work merely to reproduce MLEs, and indeed, there are far more simple ways to compute MLEs for a Poisson regression model than a Metropolis algorithm. But the point of this analysis is to demonstrate that a long run of

Table 5.2 Comparison of Bayes estimates and MLEs, Poisson regression analysis of article production by biochemistry PhD students (Example 5.2). Row labels identify a particular element of the parameter vector, $\boldsymbol{\beta}$. Cell entries are MCMC estimates of the mean of the corresponding marginal posterior density ('Bayes') and maximum likelihood estimates ('MLE'). Intervals in brackets are 95 % marginal HDRs for Bayes estimates and approximate 95 % confidence intervals for MLEs (using the asymptotically-valid normal approximation to the frequentist sampling distribution of $\hat{\boldsymbol{\beta}}$).

	Bayes	MLE
Intercept	0.30 [0.088, 0.50]	0.30 [0.10, 0.51]
Female	−0.22 [−0.33, −0.12]	−0.22 [−0.33, −0.12]
Married	0.16 [0.044, 0.28]	0.16 [0.035, 0.28]
Kids < 5	−0.18 [−0.27, −0.11]	−0.19 [−0.26, −0.11]
PhD prestige	0.013 [−0.040, 0.065]	0.014 [−0.039, 0.065]
Mentor articles	0.026 [0.022, 0.030]	0.026 [0.022, 0.029]

the random walk Metropolis algorithm 'works' in a case where (a) Bayesian computation of the posterior density is non-trivial, but (b) with a vague prior and a reasonable sample size the MLEs will be close to the Bayes estimates (and the normal approximation to the sampling distribution of the MLEs will closely resemble the posterior density).

Independence Metropolis

An independence Metropolis-Hastings algorithm uses a proposal density that does not depend on the current point. In this way the resulting sequence $\{\boldsymbol{\theta}^{(t)}\}$ may display less serial dependence than does a random-walk Metropolis sampler, thereby generating a more efficient exploration of the target density.

Once again, the effectiveness of the independence Metropolis sampler depends on how well the proposal density approximates the target density. Some popular ideas include using the prior density $p(\boldsymbol{\theta})$ as a proposal density: e.g. see Gamerman and Lopes (2006 §6.3.3), West (1995), or Knorr-Held (1997). This will work well when (a) the prior is of a form that is relatively easy to sample from; (b) the prior and the likelihood are not in great conflict, such that the proposal/prior and the target/posterior densities are not markedly different from one another.

Alternatively, one might use an approximation to the likelihood – such as the normal, which will often be reasonable when working with a large sample – which, in turn, will be a good approximation to the target/posterior density. Specifically, we could use a $N(\hat{\boldsymbol{\theta}}, \mathbf{V})$ density as the proposal density, where $\hat{\boldsymbol{\theta}}$ is a vector containing the maximum likelihood estimates of $\boldsymbol{\theta}$ and the covariance matrix $\mathbf{V}$ is the large-sample approximation to the frequentist sampling covariance matrix of $\hat{\boldsymbol{\theta}}$ (see the discussion around Equation 5.5). Scaling $\mathbf{V}$ by some constant term c (and typically, $c > 1$) improves the performance of the resulting exploration of the posterior density, and (at the very least) will help ensure that the proposal density places positive probability on all regions of the parameter space where the target density has non-zero mass. Examples using this proposal density appear in the expository essays by Chib and Greenberg (1995) and Bennett, Racine-Poon and Wakefield (1996, §19.2.4). Independence Metropolis samplers are implemented at various places in the `R` package `bayesm` (Rossi and McCulloch 2008) (e.g. the `rmnlIndepMetrop` function for Bayesian analysis of multinomial logit models).

Going further, one could plausibly use an even better approximation to the target/posterior density by forming a proposal that is a combination of the prior and the likelihood. If normals are good approximations to both the prior and the likelihood, then we can form a proposal density that exploits the precision-weighted averaging form of a normal posterior density that we encountered in Section 2.5.3. For instance, with a prior that is (perhaps only approximately) $\boldsymbol{\theta} \sim N(\boldsymbol{\theta}_0, \mathbf{V}_0)$, and a likelihood that (again, perhaps only approximately) can be represented as $\mathbf{y}|\boldsymbol{\theta} \sim N(\hat{\boldsymbol{\theta}}, \mathbf{V})$, a sensible proposal density would be $\boldsymbol{\theta}^* \sim N(\boldsymbol{\theta}_1, \mathbf{V}_1)$ where

$$\boldsymbol{\theta}_1 = (\mathbf{V}_0^{-1} + \mathbf{V}^{-1})^{-1}(\mathbf{V}_0^{-1}\boldsymbol{\theta}_0 + \mathbf{V}^{-1}\hat{\boldsymbol{\theta}}) \tag{5.6}$$

and $\mathbf{V}_1 = (\mathbf{V}_0^{-1} + \mathbf{V}^{-1})^{-1}$. Again, 'scaling up' the covariance matrix can help improve the performance of the algorithm, or even using a t density in place of the normal, with the longer and/or heavier tailed version of the proposal helping the speed the algorithms exploration of the target density; an example appears in an analysis of a GARCH model in Rachev *et al.* (2008).

Yet another idea in the literature is to use mixtures of normals or (heavier-tailed) t densities (e.g. Tierney 1994) as a proposal density. Potentially even better is an implementation in which the form of the mixture proposal density is updated as iterations progress (Ardia, Hoogerheide and van Dijk 2008). This will work well when the target/posterior density has a decidedly non-elliptical shape, as may be the case with Bayesian regression analysis using instrumental variables; e.g. see Kleibergen and van Dijk (1998), Kleibergen and Zivot (2003), Lancaster (2004 , Ch 8), or Hoogerheide, Kaashoek and van Dijk (2007).

An adaptive Metropolis sampler known as ARMS (adaptive rejection Metropolis sampling) was developed by Gilks, Best and Tan (1995); see also Tierney (1991). The idea is relatively simple: to build a better proposal density by 'remembering' the value of the target density at proposal points $\boldsymbol{\theta}^*$. This algorithm builds on the ARS algorithm (Gilks and Wild 1992; Gilks 1992) discussed in Section 3.4.4, with a Metropolis-Hastings acceptance step at the end of each iteration. While ARS lets us sample from distributions that are log-concave, ARMS permits sampling from non-log-concave distributions. Both algorithms played a key role in making it possible to produce the general purpose MCMC software package `BUGS` (Spiegelhalter *et al.* 2003; Spiegelhalter, Thomas and Best 1996; Thomas, Spiegelhalter and Gilks 1992) and its close cousin `JAGS` (Plummer 2009a), which we will explore in detail in Chapter 6.

5.2 Gibbs sampling

When $\boldsymbol{\theta}$ is high dimensional, as is often the case in many statistical models, sampling from the posterior density $p(\boldsymbol{\theta}|\text{data})$ is simply too hard for any of the sampling methods canvassed in Chapter 3 or the Metropolis-Hastings algorithm. Another complicating factor is that the posterior density may be scaled quite differently or have pronounced skew or even multi-modality in one dimension, such that finding a good, joint proposal density is very difficult. In these cases we will often rely on algorithms that have a 'divide-and-conquer' nature to them, breaking the hard problem of sampling from the high-dimensional posterior density $p(\boldsymbol{\theta}|\text{data})$ into a series of inter-related, easier, lower-dimensional sampling problems. We will see how to do this such that the resulting sequence of sampled values $\{\boldsymbol{\theta}^{(t)}\}$ is a Markov chain with stationary distribution $p(\boldsymbol{\theta}|\text{data})$.

The idea that drives this 'divide-and-conquer' approach is the following:

> *joint probability densities can be completely characterized by their component conditional densities.*

That is, rather than sample from the possibly high-dimensional density $p(\boldsymbol{\theta}|\text{data})$, we will sample from the lower-dimensional *conditional* densities that together characterize the joint density. This idea drives one of the most widely used MCMC algorithms, the Gibbs sampler.

Consider partitioning the parameter vector $\boldsymbol{\theta}$ into d blocks or sub-vectors (possibly scalars), $\boldsymbol{\theta} = (\boldsymbol{\theta}_1, \boldsymbol{\theta}_2, \ldots, \boldsymbol{\theta}_d)'$. Then the Gibbs sampler works as follows, with t indexing iterations:

Algorithm 5.2 Gibbs Sampler

1: **for** $t = 1$ to T **do**
2: sample $\boldsymbol{\theta}_1^{(t+1)}$ from $g_1(\boldsymbol{\theta}_1 \mid \boldsymbol{\theta}_2^{(t)}, \boldsymbol{\theta}_3^{(t)}, \ldots, \boldsymbol{\theta}_d^{(t)}, \mathbf{y})$.
3: sample $\boldsymbol{\theta}_2^{(t+1)}$ from $g_2(\boldsymbol{\theta}_2 \mid \boldsymbol{\theta}_1^{(t+1)}, \boldsymbol{\theta}_3^{(t)}, \ldots, \boldsymbol{\theta}_d^{(t)}, \mathbf{y})$.
4: ...
5: sample $\boldsymbol{\theta}_d^{(t+1)}$ from $g_d(\boldsymbol{\theta}_d \mid \boldsymbol{\theta}_1^{(t+1)}, \boldsymbol{\theta}_2^{(t+1)}, \ldots, \boldsymbol{\theta}_{d-1}^{(t+1)}, \mathbf{y})$.
6: $\boldsymbol{\theta}^{(t+1)} \leftarrow (\boldsymbol{\theta}_1^{(t+1)}, \boldsymbol{\theta}_2^{(t+1)}, \ldots, \boldsymbol{\theta}_d^{(t+1)})'$.
7: **end for**

That is, at the end of these d sampling steps (lines 2 through 5 of the algorithm) we have $\boldsymbol{\theta}^{(t+1)} = (\boldsymbol{\theta}_1^{(t+1)}, \boldsymbol{\theta}_2^{(t+1)}, \ldots, \boldsymbol{\theta}_d^{(t+1)})'$. Note how at every step within an iteration of the Gibbs sampler, the sampled value becomes a conditioning argument in subsequent steps: i.e., after sampling $\boldsymbol{\theta}_j^{(t+1)}$ we condition on it when sampling from the conditional density of $\boldsymbol{\theta}_{j'}, \forall\, j' > j$.

■ **Example 5.3**

Gibbs Sampler for a Bivariate Normal Model. Suppose $\mathbf{y}_i|\boldsymbol{\mu}, \boldsymbol{\Sigma} \sim N(\boldsymbol{\mu}, \boldsymbol{\Sigma})$, $i = 1, \ldots, n$, where $\boldsymbol{\mu} = (\mu_1, \mu_2)'$ and $\boldsymbol{\Sigma}$ is a known 2-by-2 covariance matrix; i.e. the unknown parameters here are $\boldsymbol{\theta} = \boldsymbol{\mu} = (\mu_1, \mu_2)'$. Our goal is to compute the posterior density $p(\boldsymbol{\theta}|\mathbf{y})$. With independent, conjugate prior densities for each element of $\boldsymbol{\mu}$, say, $\boldsymbol{\mu} = (\mu_1, \mu_2)' \sim N(\boldsymbol{\mu}_0, \boldsymbol{\Sigma}_0)'$ with

$$\boldsymbol{\Sigma}_0 = \begin{bmatrix} \sigma_{01}^2 & 0 \\ 0 & \sigma_{02}^2 \end{bmatrix},$$

the posterior density for $\boldsymbol{\mu}$ is a bivariate normal density and is trivial to compute without recourse to simulation-based methods like the Gibbs sampler; indeed, this bivariate normal mean problem (with the covariance matrix known) is simply a generalization of the model we considered in Section 2.4.1 and is the subject of Problem 2.23. In particular, the posterior density for $\boldsymbol{\mu}$ is $N(\boldsymbol{\mu}^*, \boldsymbol{\Sigma}^*)$ where

$$\boldsymbol{\mu}^* \equiv (\mu_1^*, \mu_2^*)' = (\boldsymbol{\Sigma}_0^{-1} + \boldsymbol{\Sigma}^{-1} n)^{-1} (\boldsymbol{\Sigma}_0^{-1}\boldsymbol{\mu}_0 + \boldsymbol{\Sigma}^{-1} n\bar{\mathbf{y}})$$

and

$$\boldsymbol{\Sigma}^* = (\boldsymbol{\Sigma}_0^{-1} + \boldsymbol{\Sigma}^{-1} n)^{-1} = \begin{bmatrix} \sigma_{11}^* & \sigma_{12}^* \\ \sigma_{21}^* & \sigma_{22}^* \end{bmatrix}.$$

Moreover, even if simulation methods are to be used, it is straightforward to sample directly from the bivariate normal posterior density for $\boldsymbol{\mu}$. Nonetheless, the Gibbs sampler for this problem is as follows. At iteration t,

1. sample $\mu_1^{*(t)}$ from its conditional distribution $g_1(\mu_1^*|\mu_2^{*(t-1)}, \boldsymbol{\Sigma}^*, \mathbf{y})$, a normal density with mean $\mu_1^* + \frac{\sigma_{12}^*}{\sigma_{22}^*}(\mu_2^{*(t-1)} - \mu_2^*)$ and variance $\sigma_{11}^* - \sigma_{12}^{*2}/\sigma_{22}^*$ (i.e. see the discussion following Proposition B.2).
2. sample $\mu_2^{*(t)}$ from its conditional distribution $g_2(\mu_2^*|\mu_1^{*(t)}, \boldsymbol{\Sigma}^*, \mathbf{y})$, a normal density with mean $\mu_2^* + \frac{\sigma_{12}^*}{\sigma_{11}^*}(\mu_1^{*(t)} - \mu_1^*)$ and variance $\sigma_{22}^* - \sigma_{12}^{*2}/\sigma_{11}^*$.

Note that we condition on $\mu_2^{*(t-1)}$ when sampling $\mu_1^{*(t)}$; then, given the sampled value $\mu_1^{*(t)}$, we condition on it when sampling $\mu_2^{*(t)}$. The particular order in which we sample components of the parameter vector is not important, at least not in this simple case.

We implement this scheme by returning to the absentee ballot data considered in Example 2.13. Let $\mathbf{y}_1$ be the Democratic margin among absentee ballots and let $\mathbf{y}_2$ be the Democratic margin among ballots cast on election day, both expressed in terms of percentage points. We consider the problem of jointly estimating the means of $\mathbf{y}_1$ and $\mathbf{y}_2$, assuming that the covariance matrix $\boldsymbol{\Sigma} = \text{var}(\mathbf{y}_1, \mathbf{y}_2)$ is known (using the sample covariance matrix). It is worth repeating that this is not a particularly interesting problem, nor one for which we need MCMC; our interest in this inferential problem is purely expository. For the parameters $\boldsymbol{\theta} \equiv \boldsymbol{\mu} = (\mu_1, \mu_2)'$ we use the prior $p(\boldsymbol{\theta}) \equiv N(\mathbf{0}, \boldsymbol{\Sigma}_0)$, $\boldsymbol{\Sigma} = 50^2 \cdot \mathbf{I}_2$; recall that both $(y_{i1}, y_{i2}) \in (-100, 100) \times (-100, 100)$, and so this prior on means $\boldsymbol{\mu}$ is quite vague in terms of the theoretically permissible values for the data. With these priors the posterior density $p(\boldsymbol{\mu}|\mathbf{y})$ is bivariate normal with mean vector $(29.44, 37.72)'$ and covariance matrix

$$\boldsymbol{\Sigma}^* = \begin{bmatrix} 38.41 & 30.59 \\ 30.59 & 43.17 \end{bmatrix},$$

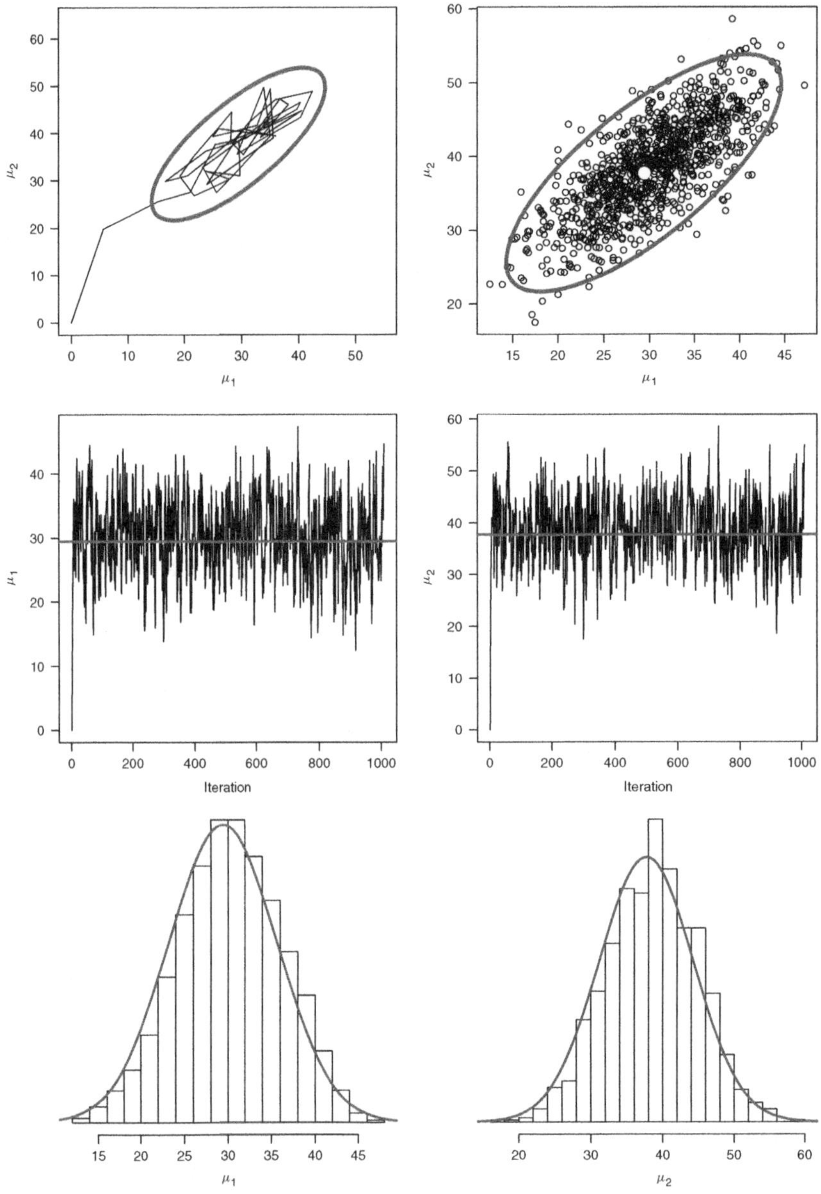

Figure 5.6 Graphical summaries of the Gibbs sampler, bivariate normal means problem. The top left panel shows the trajectory of the sampler for its first 50 iterations through the two-dimensional parameter space; the top right panel shows the points visited by the sampler on the next 1000 iterations (the gray line encloses a 95 % posterior HDR). The middle panels are 'trace plots' (the time series generated by the Gibbs sampler with respect to a single parameter); the horizontal line shows the known mean of the marginal posterior density for the respective parameter. The lower panels are histograms summarizing the Gibbs sampler output for each parameter over this same set of 1000 iterations with the known, analytic marginal posterior densities (univariate normals) superimposed.

to two decimal places of precision. Since we can derive the posterior density for this elementary problem, we know the 'target density' the Gibbs sampler is trying to reach and sample from. In what follows we will compare the performance of the Gibbs sampler with the analytic result.

The Gibbs sampling scheme above can be implemented with just a few lines of code in R; the sampler is initialized at $\boldsymbol{\theta}^{(0)} = (\mu_1^{(0)}, \mu_2^{(0)})' = (0, 0)$ and run for 1,000 iterations. Figure 5.6 presents a series of graphical summaries of the Gibbs sampler output. The top left panel shows the path of the sampler through the parameter space $\boldsymbol{\Omega} \equiv \mathbb{R}^2$ for its first 50 iterations, starting at the initial value $\boldsymbol{\theta}^{(0)} = (0, 0)'$ in the lower left corner. The Gibbs sampler quickly moves away from the initial values to explore the region of the parameter space with substantial posterior probability; note that the grey ellipse encloses the 95 % posterior HDR. The top right panel plots each point visited by the Gibbs sampler over iterations 11 through 1010; note the 'swarm' of points around the posterior mean (the large white point) and the visible decrease in the density of sampled points towards the edge of the 95 % HDR and beyond.

The middle panels show the trajectory of the Gibbs sampler with respect to each of the parameters μ_1 and μ_2 separately. Again we see a rapid transition away from the arbitrary start values. Indeed, with the exception of the outlying initial values, these marginal trajectories resemble stationary, autocorrelated time series, which in fact they are (the gray lines overlaid on each trajectory are the known posterior means for each series).

The bottom panel summarizes the output of the Gibbs sampler for each parameter (iterations 11 to 1010) with a histogram; the known, analytic marginal posterior densities (univariate normals, respectively) are superimposed as gray lines. In each case it appears that the Gibbs sampler is successfully recovering the the posterior density.

Table 5.3 provides numerical summaries of the performance of the Gibbs sampler for both the 1000 iterations summarized in Figure 5.6 and a longer run of an additional 49 000 iterations (for a total of 50 000 iterations used with the numerical summaries). The results from the longer, 50 000 iteration run more closely match the known analytic results than the results based on 1000 iterations. That is, we have some empirical evidence of 'simulation consistency' for estimands computed with the output from the Gibbs sampler. From the theoretical discussion in Chapter 4, we know that simulation consistency obtains not just for the independence samplers considered in Chapter 3, but also when a Monte Carlo exploration of a posterior density exhibits a definite lack of independence (as is almost always the case with MCMC). In this case, the series of sampled μ_1 have an AR(1) parameter of .564 while the sampled μ_2 have an AR(1) parameters of .568; note that these AR(1) parameters are approximately the square of the posterior correlation

Table 5.3 Comparison of Gibbs sampler with known analytic results.

	Analytic	1000 iterations	50 000 iterations
$E(\mu_1\|\mathbf{y})$	29.44	30.34	29.38
$E(\mu_2\|\mathbf{y})$	37.72	38.63	37.65
$V(\mu_1\|\mathbf{y})$	38.41	35.54	38.91
$V(\mu_2\|\mathbf{y})$	43.17	40.12	43.84
$C(\mu_1, \mu_2\|\mathbf{y})$	30.59	28.07	31.12

between μ_1 and μ_2, i.e. $30.59^2/(38.41 \times 43.17) \approx .7512^2 \approx .5642$ (we return to this phenomenon in Section 6.4.2). This lack of independence is really only an annoyance, meaning that in order to drive the Monte Carlo error down to some target level, the Gibbs sampler will need to be run for a comparatively larger number of iterations relative to the number of iterations with the independence sampler; recall the discussion in Section 4.4.1.

5.2.1 Theory for the Gibbs sampler

We now briefly review the theory that makes this algorithm work. A key result – and a surprising one at that – is that by sampling from this series of d, lower-dimensional conditional densities we have 'done enough' to generate a sample from the joint density of $\boldsymbol{\theta}$, or at least to generate a Markov chain on $\boldsymbol{\Theta}$ that has the joint (posterior) density of $\boldsymbol{\theta}$ as its invariant density. The key result here is the one highlighted on page 214: conditional densities are sufficient to characterize the joint density.

In the following proposition we consider the simple case where the parameter vector $\boldsymbol{\theta}$ is partitioned into two components.

Proposition 5.1 (Joint Densities are Characterized by Conditional Densities). *Let* $\boldsymbol{\theta} = (\boldsymbol{\theta}_1, \boldsymbol{\theta}_2)'$ *and let* $g_1 = g_1(\boldsymbol{\theta}_1|\boldsymbol{\theta}_2)$ *and* $g_2 = g_2(\boldsymbol{\theta}_2|\boldsymbol{\theta}_1)$. *Then*

$$g(\boldsymbol{\theta}_1, \boldsymbol{\theta}_2) = \frac{g_2(\boldsymbol{\theta}_2|\boldsymbol{\theta}_1)}{\int g_2(v|\boldsymbol{\theta}_1)/g_1(\boldsymbol{\theta}_1|v)dv}.$$

Proof. The proof appears in numerous places in the literature (e.g. Robert and Casella 2004, Theorem 9.3). Since $g(\boldsymbol{\theta}_1, \boldsymbol{\theta}_2) = g_1(\boldsymbol{\theta}_1|\boldsymbol{\theta}_2)g(\boldsymbol{\theta}_2) = g_2(\boldsymbol{\theta}_2|\boldsymbol{\theta}_1)g(\boldsymbol{\theta}_1)$, we have

$$\int \frac{g_2(v|\boldsymbol{\theta}_1)}{g_1(\boldsymbol{\theta}_1|v)}dv = \int \frac{\frac{g(v,\boldsymbol{\theta}_1)}{g(\boldsymbol{\theta}_1)}}{\frac{g(v,\boldsymbol{\theta}_1)}{g(v)}}dv = \int \frac{g(v)}{g(\boldsymbol{\theta}_1)}dv = \frac{1}{g(\boldsymbol{\theta}_1)}$$

and the result follows, provided the integral $\int \frac{g_2(v|\boldsymbol{\theta}_1)}{g_1(\boldsymbol{\theta}_1|v)}dv$ exists. ◁

A proposition due to Hammersley and Clifford (but never published) generalizes this result to the case of $d > 2$ components in $\boldsymbol{\theta}$; the necessary technical condition is that each of the marginal distributions be positive over their support (Besag 1974), a requirement that was implicit in Proposition 5.1, but which we now formalize:

Definition 5.1 (Positivity; Besag 1974). *Suppose* $\boldsymbol{\theta} = (\boldsymbol{\theta}_1, \ldots, \boldsymbol{\theta}_J)' \sim g$ *and let* g_j *be the marginal density of* $\boldsymbol{\theta}_j$, $j = 1, \ldots, J$. *If* $g_j(\boldsymbol{\theta}_j) > 0 \; \forall \; j \Rightarrow g(\boldsymbol{\theta}) > 0$ *then* g *is said to satisfy the positivity condition.*

Positivity is a sufficient condition to ensure that the Gibbs sampler generates a Markov chain that is irreducible (Definition 4.10): i.e. if positivity holds, then each sample from any of the conditional distributions $g(\boldsymbol{\theta}_j|\boldsymbol{\theta}_{-j})$, $j = 1, \ldots, J$ does not drive the Markov chain into a region where $g(\boldsymbol{\theta})$ has zero support; conversely, and of more interest, positivity means that the Markov chain generated by the Gibbs sampler can jump from $\mathcal{A}$ to $\mathcal{A}'$ in one iteration of the sampler, for any $\mathcal{A}, \mathcal{A}' \in \mathcal{B}(\boldsymbol{\Theta})$ with positive measure.

An alternative set of sufficient conditions

Further, if the transition kernel for the Gibbs sampler – the product of the J conditional distributions $g(\boldsymbol{\theta}_j|\boldsymbol{\theta}_{-j}), j = 1, \ldots, J$ – is absolutely continuous with respect to the chain's invariant density, then the Markov chain generated by the Gibbs sampler is Harris recurrent (Definition 4.16); for restatements and proofs see Tierney 1994, Corollory 1, Roberts and Smith (1994), Robert and Casella (2004, 345) or Geweke (2005, 137–138). This continuity condition – a stronger set of conditions than positivity – usually holds for Gibbs samplers, when we consider that the transition kernel is the product of the J conditional distributions that together characterize the joint, target density $g(\boldsymbol{\theta})$. More to the point, these continuity requirements are usually easier to verify than positivity.

We now verify that the Gibbs sampler does in fact produce a sample from the joint posterior density $p(\boldsymbol{\theta}|\mathbf{y})$. For simplicity, we consider the case of $d = 2$ (e.g. Gamerman and Lopes 2006, 147) and see an elaboration for the general case in Robert and Casella (2004, Theorem 10.6). The Gibbs sampler produces a transition from $\boldsymbol{\theta}^{(t-1)} = (\boldsymbol{\theta}_1^{(t-1)}, \boldsymbol{\theta}_2^{(t-1)})'$ to $\boldsymbol{\theta}^{(t)} = (\boldsymbol{\theta}_1^{(t)}, \boldsymbol{\theta}_2^{(t)})'$ by (a) sampling $\boldsymbol{\theta}_1^{(t)} \sim g_1(\boldsymbol{\theta}_1|\boldsymbol{\theta}_2^{(t-1)}, \mathbf{y})$ and (b) sampling $\boldsymbol{\theta}_2^{(t)} \sim g_2(\boldsymbol{\theta}_2|\boldsymbol{\theta}_1^{(t)}, \mathbf{y})$. This means that the kernel for the resulting Markov chain is

$$K(\boldsymbol{\theta}^{(t-1)}, \boldsymbol{\theta}^{(t)}) = g_1(\boldsymbol{\theta}_1^{(t)}|\boldsymbol{\theta}_2^{(t-1)}, \mathbf{y}) g_2(\boldsymbol{\theta}_2^{(t)}|\boldsymbol{\theta}_1^{(t)}, \mathbf{y}) \tag{5.7}$$

If the Gibbs sampler is in fact sampling $\boldsymbol{\theta}$ from a target density g, say $g \equiv p(\boldsymbol{\theta}|\mathbf{y})$, then g needs to be the invariant distribution of the Markov chain created by the Gibbs sampler (see Definition 4.4). That is, if we Gibbs sample from the conditional distributions g_1 and g_2 that together are sufficient for g, then we should be able to show that $P(\boldsymbol{\theta}^{(t)} \in \mathcal{A}) = \int_{\mathcal{A}} g(\boldsymbol{\theta}^{(t)}) d\boldsymbol{\theta}^{(t)}$. We begin by noting that

$$\Pr(\boldsymbol{\theta}^{(t)} \in \mathcal{A}) = \int_{\boldsymbol{\Theta}} \int_{\boldsymbol{\Theta}} \mathcal{I}(\boldsymbol{\theta}^{(t)} \in \mathcal{A}) \, K(\boldsymbol{\theta}^{(t-1)}, \boldsymbol{\theta}^{(t)}) \, g(\boldsymbol{\theta}^{(t-1)}) \, d\boldsymbol{\theta}^{(t-1)} d\boldsymbol{\theta}^{(t)}$$

where $\mathcal{I}$ is a binary indicator function set to one if its logical argument is true and zero otherwise. Substituting for the kernel of the Gibbs sampler in Equation 5.7 we have

$$\begin{aligned}\Pr(\boldsymbol{\theta}^{(t)} \in \mathcal{A}) = \int_{\boldsymbol{\Theta}_2} \int_{\boldsymbol{\Theta}_1} \int_{\boldsymbol{\Theta}_2} \int_{\boldsymbol{\Theta}_1} & \mathcal{I}(\boldsymbol{\theta}^{(t)} \in \mathcal{A}) g_1(\boldsymbol{\theta}_1^{(t)}|\boldsymbol{\theta}_2^{(t-1)}, \mathbf{y}) g_2(\boldsymbol{\theta}_2^{(t)}|\boldsymbol{\theta}_1^{(t)}, \mathbf{y}) \\ & \times g(\boldsymbol{\theta}_1^{(t-1)}, \boldsymbol{\theta}_2^{(t-1)}|\mathbf{y}) \, d\boldsymbol{\theta}_1^{(t-1)} d\boldsymbol{\theta}_2^{(t-1)} d\boldsymbol{\theta}_1^{(t)} d\boldsymbol{\theta}_2^{(t)},\end{aligned} \tag{5.8}$$

where $\boldsymbol{\Theta} = \boldsymbol{\Theta}_1 \times \boldsymbol{\Theta}_2$. Let $g_{-1} = \int_{\boldsymbol{\Theta}_1} g(\boldsymbol{\theta}_1, \boldsymbol{\theta}_2|\mathbf{y}) d\boldsymbol{\theta}_1 = g_{-1}(\boldsymbol{\theta}_2|\mathbf{y})$ be the marginal posterior density of $\boldsymbol{\theta}_2$. Note that g can always be expressed as the product of a conditional density and a marginal density: i.e. $g = g_1(\boldsymbol{\theta}_1|\boldsymbol{\theta}_2, \mathbf{y}) g_2(\boldsymbol{\theta}_2|\mathbf{y})$. Integrate $\boldsymbol{\theta}_1^{(t-1)}$ out of Equation 5.8 and exploit the identity

$$g_{-1}(\boldsymbol{\theta}_2^{(t-1)}|\mathbf{y}) g_1(\boldsymbol{\theta}_1^{(t)}|\boldsymbol{\theta}_2^{(t-1)}, \mathbf{y}) = g(\boldsymbol{\theta}_1^{(t)}, \boldsymbol{\theta}_2^{(t-1)}|\mathbf{y})$$

to obtain

$$\Pr(\boldsymbol{\theta}^{(t)} \in \mathcal{A}) = \int_{\boldsymbol{\theta}_2}\int_{\boldsymbol{\theta}_1}\int_{\boldsymbol{\theta}_2} \mathcal{I}(\boldsymbol{\theta}^{(t)} \in \mathcal{A}) g(\boldsymbol{\theta}_1^{(t)}, \boldsymbol{\theta}_2^{(t-1)}|\mathbf{y}) \times g_2(\boldsymbol{\theta}_2^{(t)}|\boldsymbol{\theta}_1^{(t)}, \mathbf{y})\, d\boldsymbol{\theta}_2^{(t-1)} d\boldsymbol{\theta}_1^{(t)} d\boldsymbol{\theta}_2^{(t)}. \tag{5.9}$$

Now write $g(\boldsymbol{\theta}_1^{(t)}, \boldsymbol{\theta}_2^{(t-1)}|\mathbf{y}) = g_2(\boldsymbol{\theta}_2^{(t-1)}|\boldsymbol{\theta}_1^{(t)}, \mathbf{y}) g_{-2}(\boldsymbol{\theta}_1^{(t)}|\mathbf{y})$ and integrate $\boldsymbol{\theta}_2^{(t-1)}$ out of Equation 5.9 to obtain

$$\begin{aligned}\Pr(\boldsymbol{\theta}^{(t)} \in \mathcal{A}) &= \int_{\boldsymbol{\theta}_2}\int_{\boldsymbol{\theta}_1} \mathcal{I}(\boldsymbol{\theta}^{(t)} \in \mathcal{A})\, g_2(\boldsymbol{\theta}_2^{(t)}|\boldsymbol{\theta}_1^{(t)}, \mathbf{y}) g_{-2}(\boldsymbol{\theta}_1^{(t)}|\mathbf{y})\, d\boldsymbol{\theta}_1^{(t)} d\boldsymbol{\theta}_2^{(t)} \\ &= \int_{\boldsymbol{\theta}} \mathcal{I}(\boldsymbol{\theta}^{(t)} \in \mathcal{A}) g(\boldsymbol{\theta}^{(t)}|\mathbf{y})\, d\boldsymbol{\theta}^{(t)} = \int_{\mathcal{A}} g(\boldsymbol{\theta}^{(t)}|\mathbf{y})\, d\boldsymbol{\theta}^{(t)}\end{aligned} \tag{5.10}$$

which is the desired result. That is, the Markov chain generated by sampling from the conditional densities g_1 and g_2 (and so on) has the posterior density $g \equiv p(\boldsymbol{\theta}|\mathbf{y})$ as an invariant density.

Now further, since positivity ensures that the Gibbs sampler generates a Markov chain that is irreducible (Definition 4.10) and Harris recurrent (Definition 4.14), the invariant density is unique (Proposition 4.4). Moreover, ergodicity follows, along with the geometric rates of convergence.

Sampling from marginal posterior densities via the Gibbs sampler

Another extremely useful feature of the Gibbs sampler is that it also generates samples from marginal posterior densities as well. That is, as $t \to \infty$, each component of the vector $\boldsymbol{\theta}^{(t)} = (\boldsymbol{\theta}_1^{(t)}, \boldsymbol{\theta}_2^{(t)}, \ldots, \boldsymbol{\theta}_d^{(t)})'$ is a draw from the respective marginal posterior density; i.e. as $t \to \infty$

$$\boldsymbol{\theta}_j^{(t)} \sim p(\boldsymbol{\theta}_j|\mathbf{y}) = \int_{\boldsymbol{\Theta}_{-j}} p(\boldsymbol{\theta}_1, \ldots, \boldsymbol{\theta}_d|\mathbf{y}) d\boldsymbol{\theta}_{-j} \tag{5.11}$$

where the notation $\boldsymbol{\theta}_{-j}$ denotes the parameter vector $\boldsymbol{\theta}$ absent the component $\boldsymbol{\theta}_j \in \boldsymbol{\Theta}_j$, and so analogously $\boldsymbol{\theta}_{-j} \in \boldsymbol{\Theta}_{-j} \subset \boldsymbol{\Theta}$.

This result follows from the nature of the conditional densities that drive the Gibbs sampler. As $t \to \infty$, $\boldsymbol{\theta}^{(t)} \sim p(\boldsymbol{\theta}|\mathbf{y})$. Since these sampled values appear as conditioning arguments in lines 2 through 5 of Algorithm 5.2, each sample from the conditional densities are effectively draws from the corresponding marginal density. That is, the Gibbs sampler is performing the integration in Equation 5.11 by Monte Carlo methods.

The argument here is relatively easy to make for the case of a parameter vector with just two components, $\boldsymbol{\theta} = (\theta_1, \theta_2)'$, $\theta_1 \in \Theta_1$, $\theta_2 \in \Theta_2$, and the following argument largely reproduces the exposition in Casella and George (1992, §4.1). More detailed treatments are provided by Tanner and Wong (1987) and Gelfand and Smith (1990). We adopt the notation of Proposition 5.1 and begin with the identity

$$g_1(\theta_1) = \int_{\Theta_2} g(\theta_1, \theta_2) d\theta_2 = \int_{\Theta_2} g_{1|2}(\theta_1|\theta_2) g_2(\theta_2) d\theta_2.$$

Writing the marginal density $g_2 = \int_{\Theta_1} g_{2|1} g_1(t) dt$ we have

$$g_1(\theta_1) = \int_{\Theta_2} g_{1|2}(\theta_1|\theta_2) \int_{\Theta_1} g_{2|1}(\theta_2|t) g_1(t)\, dt\, d\theta_2.$$

Making whatever assumptions are required so as to change the order of integration,

$$g_1(\theta_1) = \int_{\Theta_1} \left[\int_{\Theta_2} g_{1|2}(\theta_1|\theta_2) g_{2|1}(\theta_2|t) d\theta_2 \right] g_1(t) dt = \int_{\Theta_1} h(\theta_1, t) g_1(t) dt \qquad (5.12)$$

where $h(\theta_1, t)$ is the term in square brackets.

Equation 5.12 is an integral equation of the same sort we considered when examining Markov chains in Chapter 4 (e.g. Equation 4.5). Note that g_1 appears on both sides of Equation 5.12, meaning that we can interpret g_1 as the solution to a fixed point integral equation. What this means is that in performing the integrations in Equation 5.12 by Monte Carlo methods – sequentially sampling from one conditional density, then from the other, following the Gibbs recipe in Algorithm 5.2 – we have a scheme that in equilibrium is producing samples from marginal densities.

In this way the Gibbs sampler is roughly analogous to a stochastic version of an algorithm looking for a fixed point solution to an optimization problem, such as the *EM* algorithm (Dempster, Laird and Rubin 1977) or an alternating, conditional maximum likelihood estimation procedure; but rather than find a fixed point *per se*, the Gibbs sampler finds an equilibrium density (in this case, a posterior density). We return to this feature of the Gibbs sampler in Section 5.2.5, below.

5.2.2 Connection to the Metropolis algorithm

Gibbs sampling can be considered a variant of the Metropolis-Hastings algorithm in the sense that each component of $\boldsymbol{\theta}$ is updated sequentially and the implicit jumping distributions are simply the conditional densities $p(\boldsymbol{\theta}_j|\boldsymbol{\theta}_{-j}^{(t-1)}$, data); this means that in Algorithm 5.1 $r = 1$ and each candidate point is always accepted.

The Metropolis-Hastings algorithm is often used in conjunction with a Gibbs sampler for those components of $\boldsymbol{\theta}$ that have conditional distributions that can be evaluated, but can not be sampled from directly, typically because the distribution is known only up to a scale factor. The Metropolis-Hastings algorithm ensures that MCMC algorithms can still be constructed for these cases. All that is required is that the analyst have some approximating density J from which it is possible to sample, and then be able to evaluate the ratio r with the sampled candidate point. General purpose programs for Bayesian analysis such as BUGS (Thomas, Spiegelhalter and Gilks 1992) make extensive use of this strategy. In fact, there is some theoretical support for the idea that one or more Metropolis steps can actually *improve* the performance of a Gibbs sampler, at least for the case where $\boldsymbol{\theta}$ is a discrete random variable (Besag *et al*. 1995; Liu 1996). For the continuous case, some care needs to be taken so as to ensure that the resulting Markov chain is irreducible; we need the Metropolis proposal density to be such that we are 'mimicking' the positivity condition that is typically satisfied when sampling directly from the conditionals themselves (i.e. the proposal density is positive everywhere the conditional is positive).

We demonstrate a Metropolis-within-Gibbs algorithm in the following example.

■ Example 5.4

Negative binomial regression analysis of overdispersed counts via a Metropolis-within-Gibbs algorithm (Example 5.2, continued). Recall from Definition B.23 that the Poisson probability mass function has the interesting property that its mean is equal to its variance. But count data are often found to be over-dispersed relative to the Poisson, even conditional on any available covariates. This often arises because the conditional model for the data, $E(y_i|\mathbf{x}_i, \boldsymbol{\beta}) = \lambda_i = \exp(\mathbf{x}_i\boldsymbol{\beta})$ is mis-specified, with the true model being $E(y_i|\mathbf{x}_i, \boldsymbol{\beta}) = \exp(\mathbf{x}_i\boldsymbol{\beta} + \epsilon_i) = \exp(\mathbf{x}_i\boldsymbol{\beta})\exp(\epsilon_i) = \lambda_i\delta_i$.

A simple model that incorporates this extra-Poisson variation in the data is to assume that $\delta_i \sim \text{Gamma}(\alpha, \alpha)$, $\alpha > 0$, which gives rise to a negative binomial regression model for the counts, again conditional on the covariates $\mathbf{x}_i$ and the *over-dispersion* parameter α; see Definition B.24 and the discussion in §2.3.1 where we obtained the negative binomial probability mass function as a Gamma mixture of Poissons. Conditional on α, $E(y_i|\mathbf{x}_i, \boldsymbol{\beta}) = \lambda_i = \exp(\mathbf{x}_i\boldsymbol{\beta})$ which is the same as the Poisson model, since $E(\delta_i) = \alpha/\alpha = 1$. However, the conditional variance is $V(y_i|\mathbf{x}_i, \boldsymbol{\beta}, \alpha) = \lambda_i(1 + \lambda_i/\alpha)$ which gives us overdispersion relative to the Poisson $\forall\, 0 < \alpha < \infty$. As $\alpha \to \infty$, the negative binomial tends to the Poisson.

If $y_i|\mathbf{x}_i, \boldsymbol{\beta} \sim \text{NegBin}(\lambda_i, \alpha_i)$ and the y_i are conditionally independent given $\mathbf{x}_i$, $\boldsymbol{\beta}$ and α, then the likelihood for the data is

$$f(\mathbf{y}|\mathbf{X}, \boldsymbol{\beta}, \alpha) = \prod_{i=1}^{n} \frac{\Gamma(y_i + \alpha)}{y_i!\Gamma(\alpha)} \left(\frac{\alpha}{\alpha + \lambda_i}\right)^{\alpha} \left(\frac{\lambda_i}{\alpha + \lambda_i}\right)^{y_i} \tag{5.13}$$

where again $\lambda_i = \exp(\mathbf{x}_i\boldsymbol{\beta})$. Maximum likelihood estimates of $\boldsymbol{\theta} = (\boldsymbol{\beta}, \alpha)'$ can be obtained via the `glm.nb` function available in the `R` package `MASS` (Venables and Ripley 2002).

Bayesian analysis. Like many generalized linear models, there is no conjugate prior for Poisson nor negative binomial regression models. One strategy for Bayesian analysis is to explore the posterior density $p(\boldsymbol{\beta}, \alpha|\mathbf{y}, \mathbf{X})$ via a Metropolis algorithm. This can be a tricky problem, in that the marginal posterior density for the overdispersion parameter α often has a skewed shape and coming up with proposal density covering both $\boldsymbol{\beta}$ and α is not straightforward. Alternatively, we can take a Gibbs sampling approach, or actually, a Metropolis-within-Gibbs approach. At iteration t we sample

1. $\boldsymbol{\beta}^{(t)} \sim g(\boldsymbol{\beta}|\mathbf{y}, \mathbf{X}, \alpha^{(t-1)})$, recognizing that conditional on α, we have a relatively simple Poisson likelihood, placing us back in the situation studied in Example 5.2 where we used a random-walk Metropolis algorithm to sample from the posterior of $\boldsymbol{\beta}$.

2. $\alpha^{(t)} \sim g(\alpha|\mathbf{y}, \mathbf{X}, \boldsymbol{\beta}^{(t)})$, using a random-walk Metropolis algorithm.

Note that both sampling steps in this approach are generated by a Metropolis algorithm.

This algorithm – a Gibbs-like alternating series of steps taken by random-walk Metropolis sampling – is implemented in the `rnegbinRw` function in the `R` package `bayesm` (Rossi and McCulloch 2008). The function presumes priors of the form $p(\boldsymbol{\beta}, \alpha) = p(\boldsymbol{\beta})p(\alpha)$ where $\boldsymbol{\beta} \sim N(\mathbf{b}_0, \mathbf{B}_0)$ and $\alpha \sim \text{Gamma}(a, b)$ with defaults $\mathbf{b}_0 = \mathbf{0}$, $\mathbf{B}_0 = 10 \cdot \mathbf{I}$, $a = .5$ and $b = .1$. Note that the default priors assume *a priori* $E(\alpha) = 5$,

$q_{2.5}(\alpha) = .005$ and $q_{97.5}(\alpha) = 25.12$, with the long right tail of this default prior extending prior probability in the direction of the Poisson assumption of the conditional mean equaling the conditional variance.

We run this algorithm with the default priors for 50 000 iterations, producing the output summarized graphically in Figure 5.7. The top row shows the progress of the algorithm alternating between separate random-walk Metropolis steps for α and the $\boldsymbol{\beta}$; accepted proposals in one dimension, but not the other, give rise to the vertical or horizontal moves shown in these panels. Note that the $\boldsymbol{\beta}$ parameters are drawn from a multivariate proposal density *en bloc*, so we see a the algorithm stepping through the joint parameter space for these parameters.

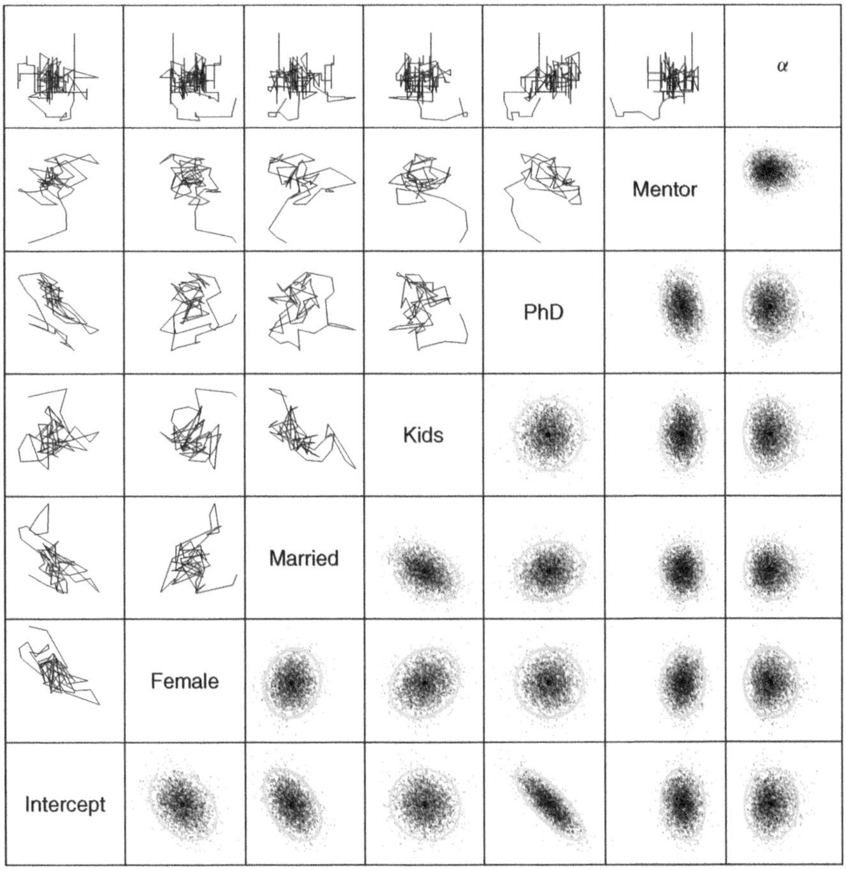

Figure 5.7 Random-walk Metropolis within Gibbs algorithm, negative binomial regression model, Example 5.4. Each panel shows the algorithm working in a series of two-dimensional slices through the parameter space. Panels above the diagonal show the algorithm's progress over the first 250 iterations. Panels below the diagonal show 2500 iterations (evenly spaced over the full 50 000 iteration run), with the ellipse indicating a 95 % HDR (based on a normal approximation). The text labels on the diagonal identify the particular element of the parameter vector $\boldsymbol{\beta}$; α is the negative binomial over-dispersion parameter.

The random-walk Metropolis-within-Gibbs algorithm generates a reasonably slow exploration of its target density, the joint posterior density $p(\boldsymbol{\beta}, \alpha|\mathbf{y}, \mathbf{X})$. The 50 000 iterations generated by the algorithm display high levels of serial dependence, as revealed in Figure 5.8. The autocorrelation functions decay quite slowly for the $\boldsymbol{\beta}$ parameters, but considerably faster for the over-dispersion parameter α. The 50 000 sampled values for the $\boldsymbol{\beta}$ parameters carry the same precision as about 2600 draws from an independence sampler, while the 50 000 draws for α correspond to an effective sample size (see Section 4.4.1) of almost 10 000 iterations.

Finally, Figure 5.9 shows a series of histograms, summarizing the draws of the random-walk, Metropolis-within-Gibbs algorithm for each parameter. These MCMC-based estimates of the marginal posterior density of each parameter closely correspond to the normal approximation to the large-sample frequentist sampling distribution of the MLEs (shown as the solid black lines); there are some very small differences, which may be due to Monte Carlo error or the fact that the prior is not entirely uninformative, at least with respect to the over-dispersion parameter α. The panel for α in Figure 5.9 also superimposes the Gamma(.5, .1) prior for that parameter (dotted line), showing that the Gamma prior is concentrating most of its probability mass on values of α that lie considerably to the left of the MLEs. Since the posterior

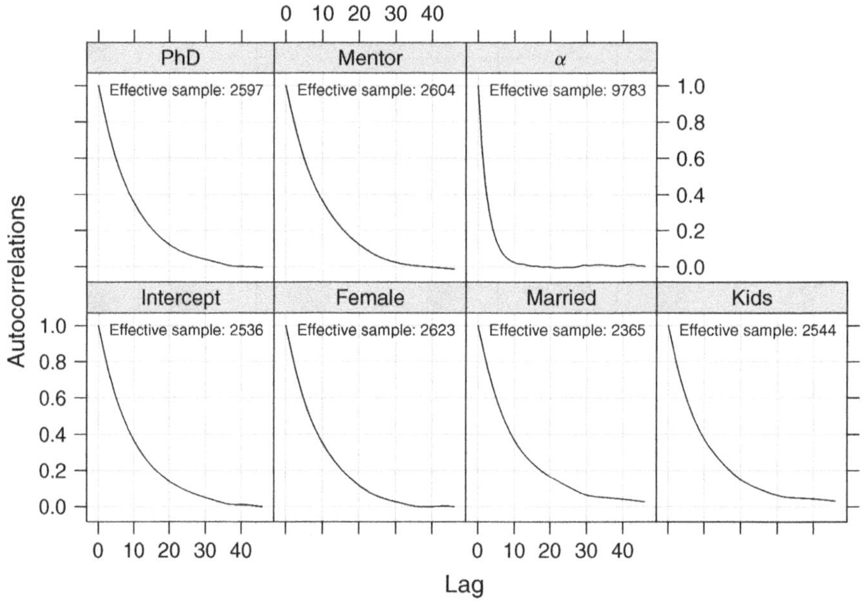

Figure 5.8 Autocorrelations, random-walk Metropolis within Gibbs algorithm, negative binomial regression model, Example 5.4. Fifty thousand iterations were produced by the random-walk Metropolis algorithm using the `rnegbinRw` function in the `R` package `bayesm` (Rossi and McCulloch 2008). Each panel shows the autocorrelations of the sampled values from the marginal posterior density of a given element of β_j and the over-dispersion parameter α. The 'effective sample size' (see Section 4.4.1) is the number of independent draws represented by the highly autocorrelated 50 000 random-walk Metropolis draws.

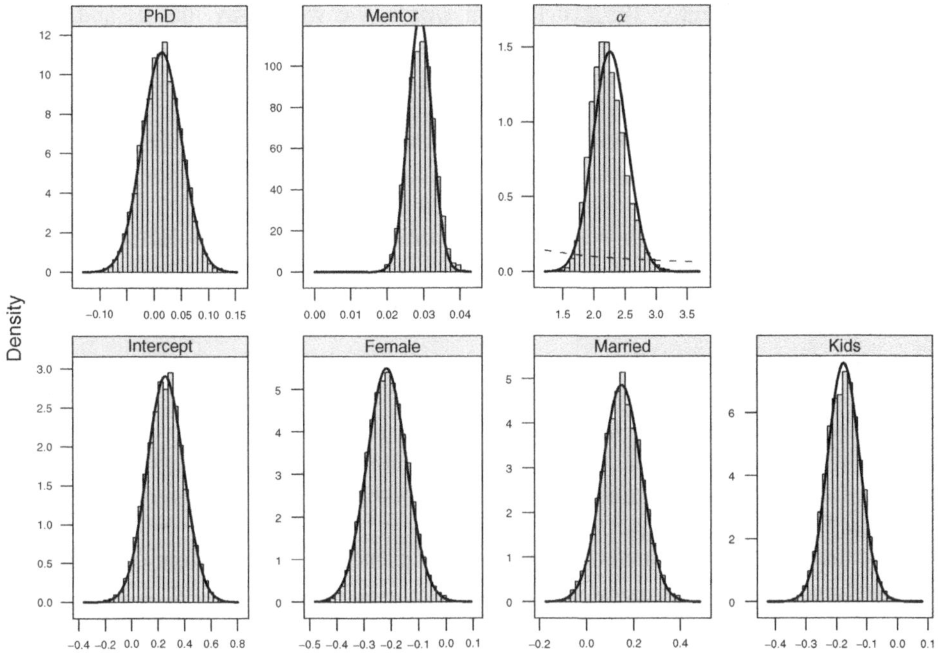

Figure 5.9 Marginal posterior densities, negative binomial regression model produced by the random-walk Metropolis-within-Gibbs algorithm, Example 5.4. Histograms summarize the 50 000 iterations produced by Metropolis-within-Gibbs algorithm. Solid lines are normal densities corresponding to the asymptotic approximation to the frequentist sampling distribution of the MLEs. For the over-dispersion parameter α, the dotted line indicates the prior, $\alpha \sim \text{Gamma}(.5, .1)$, which concentrates most of its probability mass on values of α to the left of the range in the plot.

represents a combination of prior and likelihood, we observe that the posterior for α is shifted slightly to the left of the large-sample, frequentist sampling distribution. In turn, since α and $\boldsymbol{\beta}$ are not independent in the likelihood, prior information about α is not innocuous with respect to the marginal posterior densities we generate for the components of $\boldsymbol{\beta}$, representing another source for the slight divergence between the posterior densities and likelihood functions apparent in Figure 5.9.

5.2.3 Deriving conditional densities for the Gibbs sampler: statistical models as conditional independence graphs

An iteration of the Gibbs sampler cycles over the components of the parameter vector $\boldsymbol{\theta}$, sampling from the conditional distributions for each component given the current values of other elements of $\boldsymbol{\theta}$ and the data. But in most applications, the structure of the model means that the conditional distribution for a given component $\boldsymbol{\theta}_d$ does not depend on *all* of the other elements of $\boldsymbol{\theta}$. That is, most statistical models embody some conditional independence structure, such that $\boldsymbol{\theta}_d$ may only depend on a subset of the remaining $d-1$ components. That is, the form of a given model may imply that $p(\boldsymbol{\theta}_3|\boldsymbol{\theta}_2, \boldsymbol{\theta}_1, \mathbf{y}) =$

$p(\boldsymbol{\theta}_3|\boldsymbol{\theta}_2, \mathbf{y})$ or equivalently, $\{\boldsymbol{\theta}_3 \perp\!\!\!\perp \boldsymbol{\theta}_1\}|\boldsymbol{\theta}_2$ (in words, we would say that $\boldsymbol{\theta}_3$ and $\boldsymbol{\theta}_1$ are conditionally independent given $\boldsymbol{\theta}_2$).

The conditional independence structure of a given model is often easy to deduce when the model is represented as a graph, where the word 'graph' is being used in its formal, mathematical sense. That is, components of a statistical model – data and parameters – are the nodes or vertices of a graph, and the relations among the nodes – stochastic or deterministic – generate the edges of the graph. In particular, many statistical models can be represented as *directed acyclic graphs* (DAGs). Graphs are said to be directed if their edges are directed, and are acyclic if a journey around the edges of the graph does not cycle endlessly, but can terminate somewhere.

Example 5.5

Directed acyclic graph for linear regression. A simple DAG appears in Figure 5.10, showing the directed acyclic graph representation for a linear regression model under normality, $y_i|\mathbf{x}_i \sim N(\mathbf{x}_i\boldsymbol{\beta}, \sigma^2)$, with the conditionally conjugate prior densities $\boldsymbol{\beta} \sim N(\mathbf{b}_0, \mathbf{B}_0)$ and $\sigma^2 \sim$ inverse-Gamma$(\nu_0/2, \nu_0\sigma_0^2/2)$. As is conventional in this literature, quantities represented as circular nodes in the DAG are random quantities (y_i, $\boldsymbol{\beta}$ and σ^2) and square/rectangular nodes are non-stochastic (in this case, $\mathbf{x}_i$). The DAG is not sufficient to completely characterize the model, since it makes no reference to the specific functional form of the model (in this case, normality), nor of the prior densities for $\boldsymbol{\beta}$ or σ^2. The arrows are 'directed edges' running from *parent* nodes to *children* nodes; in this case the y_i are children of the $\mathbf{x}_i$, $\boldsymbol{\beta}$ and σ^2. The graph is acyclic in that the stochastic relations in the model flow in the direction of the terminal node y_i; there is no edge connecting y_i to $\mathbf{x}_i$ or $\boldsymbol{\beta}$.

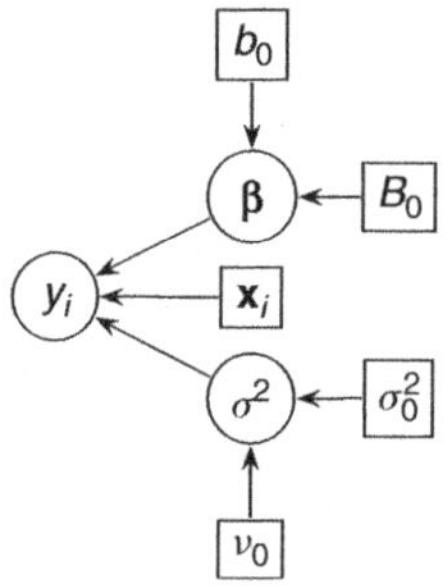

Figure 5.10 Directed acyclic graph for linear regression model. Circles indicate stochastic nodes; squares indicate fixed nodes; directed edges (arrows) indicate parent–child relations.

In particular, nodes in a graph only depend on their immediate set of neighbors, sometimes called the 'Markov cover' of a given node. In the regression model in the preceding example, $\{y_i \perp\!\!\!\perp (\mathbf{b}_0, \mathbf{B}_0)\}|\boldsymbol{\beta}$, since the path from $\mathbf{b}_0$ to $\mathbf{B}_0$ runs through $\boldsymbol{\beta}$. In this way it becomes quite easy to derive the conditional distributions for any given component of a statistical model, once the model has been represented as a DAG:

Proposition 5.2 *If a statistical model can be expressed as a directed acyclic graph $\mathcal{G}$, then the conditional density of node θ_j in the graph is*

$$f(\theta_j|\mathcal{G} \setminus \theta_j) \propto f(\theta_j|parents[\theta_j]) \times \prod_{w \in children[\theta_j]} f(w|parents[w]), \tag{5.14}$$

where $\mathcal{G} \setminus \theta_j$ denotes all nodes in $\mathcal{G}$ other than θ_j.

Proof. See Spiegelhalter and Lauritzen (1990). ◁

One handy consequence of this result is that given a node's parents, that node is conditionally independent of its grandparents (and any other ancestors). As we shall see in Chapter 7 this formulation is especially helpful in the context of hierarchical models, where the hierarchy of the model corresponds to a nested series of parent-child (conditional independence) relations among data and/or parameters. In turn, deriving the conditional densities needed to implement MCMC for a hierarchical model becomes quite straightforward. We provide a brief taste in the following example.

■ Example 5.6

Two-level hierarchical model for one-way analysis of variance; J normal means; 'random effects;' mixed model for variance components; etc. The following model is so simple and ubiquitous that it goes by various names in the literature, as indicated in the long title of this example (see Chapter 7 for more detail on the nomenclature). We have multiple observations $i = 1, \ldots, n$ for each of $j = 1, \ldots, J$ units (e.g. students indexed by i in schools indexed by j) on a real-valued variable y_{ij}. We are interested in the model

$$y_{ij}|\mu, \boldsymbol{\alpha}, \sigma^2 \sim N(\mu + \alpha_j, \sigma^2)$$

where μ is a grand mean and $\boldsymbol{\alpha} = (\alpha_1, \ldots, \alpha_J)'$ are unit-specific offsets around the grand mean. The prior for the α_j depends on hyper-parameters that in turn have priors; this is what makes the model 'hierarchical' (and we have much more to say on this in Chapter 7).

For now, we keep the discussion quite general, focusing on the structural features of the model. We specify priors for the α_j of the form $\alpha_j|\omega^2 \sim p(\alpha_j; \omega^2)$; the hyper-parameter ω^2 is referred to as the *between* unit variance, while σ^2 is the *within* unit variance. Priors on μ, ω^2 and σ^2 complete the specification of the model. We impose *a priori* independence for these parameters, $p(\mu, \omega^2, \sigma^2) = p(\mu)p(\omega^2)p(\sigma^2)$.

The hierarchical model embodies a number of conditional independence restrictions; these are expressed in the 'parent–child' relations represented in the DAG, $\mathcal{G}$, shown in Figure 5.11. Two DAGs are shown in Figure 5.11. The one on the left shows the DAG with the node y_{ij} representing the likelihood contribution of a single observation; the DAG on the right shows the model over two distinct units j and k, with the nodes labeled $\mathbf{y}_j$ and $\mathbf{y}_k$ representing the data from those units. The y_{ij} are children of μ, the α_j parameters and σ^2, while ω^2 is a parent of the α_j. Thus, given the α_j, $\mathbf{Y}$ is conditionally independent of ω^2, as are μ and σ^2. That is, once we have the α_j, there is no additional

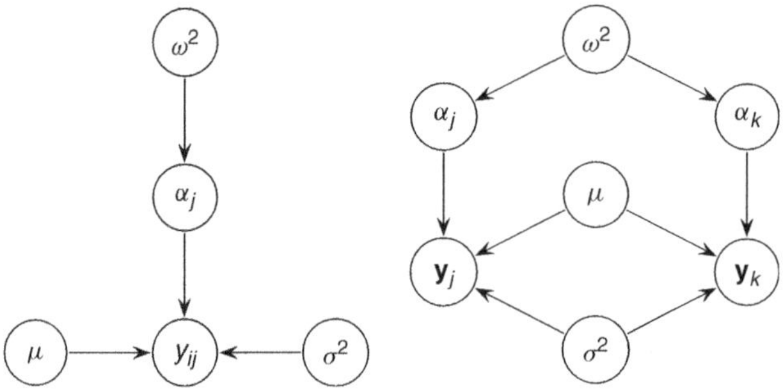

Figure 5.11 Directed acyclic graphs for hierarchical model for one-way analysis of variance in Example 5.6. In the DAG on the left the node y_{ij} represents the likelihood contribution of a single observation; the DAG on the right depicts the hierarchical model's conditional independence structure for two distinct units j and k, with the nodes labeled $\mathbf{y}_j$ and $\mathbf{y}_k$ representing the data from those units. The conditional independence relations among the random quantities in the model (data and parameters) can be 'read off' the graph; e.g., given the α_j, y_{ij}, μ and σ^2 are all conditionally independent of ω^2; $\{\alpha_j \perp\!\!\!\perp \alpha_k\}|\omega^2, \forall\ j \neq k$, and so on.

information about the y_{ij} in ω^2; similarly, the α_j and α_k are conditionally independent given ω^2. To stretch the 'parent–child' metaphors, ω^2 is a 'grandparent' of y_{ij}. These facts will prove helpful when we turn to the form of the conditional densities that drive the Gibbs sampler for this problem.

By Bayes' Rule, the posterior density for this problem is

$$p(\sigma^2, \omega^2, \boldsymbol{\alpha}, \mu|\mathbf{Y}) \propto f(\mathbf{Y}|\mu, \boldsymbol{\alpha}, \sigma^2) f(\boldsymbol{\alpha}|\omega^2) p(\mu) p(\omega^2) p(\sigma^2) \tag{5.15}$$

This density is potentially high dimensional – at least for anything other than a moderate number of units (J) – with $J+3$ components given by the J elements of $\boldsymbol{\alpha}$, the two variances and the 'grand mean' parameter μ. Each iteration of the Gibbs sampler for this problem consists of sampling from the following conditional densities:

1. $g(\sigma^2|\mathcal{G}_{-\sigma^2})$, where $\mathcal{G}_{-\sigma^2}$ is an admittedly awkward notation for "all nodes in $\mathcal{G}$ other than σ^2". Since $(\sigma^2 \perp\!\!\!\perp \omega^2)|\boldsymbol{\alpha}$ we have

   $$g(\sigma^2|\mathcal{G}_{-\sigma^2}) = g(\sigma^2|\mathbf{Y}, \mu, \boldsymbol{\alpha}) \propto f(\mathbf{Y}|\mu, \boldsymbol{\alpha}, \sigma^2) p(\sigma^2).$$

   Note that we obtain this form of the conditional density for σ^2 by using the conditional independence structure of the model to 'pick off' the appropriate elements of $\mathcal{G}_{-\sigma^2}$, from the representation of the posterior density on the right-hand side of Equation 5.15, using the result in Proposition 5.2. In particular: σ^2 appears as a conditioning argument in the density for $\mathbf{Y}$ (i.e. σ^2 is a parent of $\mathbf{Y}$); we condition on μ and $\boldsymbol{\alpha}$ so we can ignore the terms $p(\mu)$ and $f(\boldsymbol{\alpha}|\omega^2)$; by conditional independence, we lose the $p(\omega^2)$ terms; and we retain the prior over σ^2, $p(\sigma^2)$. We will not go into as much detail in the derivation of the other conditional densities; see Chapter 7.

2. $g(\omega^2|\mathcal{G}_{-\omega^2}) = g(\omega^2|\boldsymbol{\alpha})$, since $\{\omega^2 \perp\!\!\!\perp (\mathbf{Y}, \mu, \sigma^2)\}|\boldsymbol{\alpha}$. Thus, $g(\omega^2|\boldsymbol{\alpha}) \propto f(\boldsymbol{\alpha}|\omega^2)$ $p(\omega^2)$.

3. for $j = 1, \ldots, J$, $g(\alpha_j|\mathcal{G}_{-\alpha_j}) = g(\alpha_j|\mathbf{y}_j, \sigma^2, \omega^2, \mu)$, where $\mathbf{y}_j$ is a vector of the observations from unit j, since $\{\alpha_j \perp\!\!\!\perp \alpha_k\}|\omega^2 \, \forall \, j \neq k$; i.e.

$$g(\alpha_j|\mathbf{y}_j, \sigma^2, \omega^2, \mu) \propto f(\mathbf{y}_j|\mu, \alpha_j, \sigma^2)p(\alpha_j|\omega^2).$$

4. $g(\mu|\mathcal{G}_{-\mu}) = g(\mu|\mathbf{Y}, \boldsymbol{\alpha}, \sigma^2)$, since $\{\mu \perp\!\!\!\perp \omega^2\}|\boldsymbol{\alpha}$; i.e.

$$g(\mu|\mathbf{Y}, \boldsymbol{\alpha}, \sigma^2) \propto f(\mathbf{Y}|\mu, \boldsymbol{\alpha}, \sigma^2)p(\mu).$$

In an effort to keep the discussion general, we have not specified the parametric forms of the priors for this example, nor those of the conditional densities; that level of specificity awaits the detailed discussion of hierarchical models in Chapter 7. But given those parametric forms, implementing the Gibbs sampler for this problem would involve coding up samplers for these conditional densities, and wrapping them in a loop, or getting a program like `BUGS` or `JAGS` to do the job for us.

The point here is that we have a simple recipe for generating the conditionals that drive the Gibbs sampler. Moreover, the conditional independence structure of a model often means that these conditional densities have surprisingly simple forms, as we will see in the applications presented in subsequent chapters.

5.2.4 Pathologies

Improper posterior density

The theoretical results for the Gibbs sampler presented in Section 5.2.1 assume that the invariant density $g(\boldsymbol{\theta})$ – typically, a posterior density $p(\boldsymbol{\theta}|\mathbf{y})$ – actually exists. But it is entirely possible for the conditional densities $g_j(\boldsymbol{\theta}_j|\boldsymbol{\theta}_{-j})$ to be well defined even when the joint density $g(\boldsymbol{\theta})$ is improper. That is, although Proposition 5.1 says that a joint distribution is characterized by its conditional densities, the converse is not necessarily true: i.e. it is possible for there to be a set of mutually consistent conditional densities without the existence of a proper joint density (see Problem 6.3). Arnold and Press (1989) refer to a set of conditional densities that generate a joint density as *compatible*; Hobert and Casella (1996) define *functionally compatible* conditional densities as those that generate a joint density, reserving the shorter label 'compatible' for conditionals that generate a proper joint density. That is, it is entirely possible for the analyst to specify a model where the posterior density is improper; in this case one can derive conditional distributions that are 'functionally compatible' and implementable, but the resulting Gibbs sampler is exploring an improper and hence meaningless density.

As Robert and Casella (2004, 403) explain:

> This problem is not a *defect* of the Gibbs sampler, nor even a simulation problem, but rather a problem of carelessly using the Gibbs sampler in a situation for which the underlying assumptions are violated.

This point warrants some elaboration. The Gibbs sampler has not somehow generated an improper posterior. To the extent there is a problem it is because the analyst has specified a model and priors that have generated an improper posterior. The interesting and perhaps surprising point here is that the Gibbs sampler – successively sampling from conditional densities – may not reveal the impropriety.

■ Example 5.7

Hierarchical model with improper priors (Example 5.6, continued). Hobert and Casella (1996) consider the simple one-way model we introduced in Example 5.6.

In this case we specify the prior $\boldsymbol{\alpha} \sim N(\mathbf{0}, \omega^2\mathbf{I})$, or equivalently, $\alpha_j \overset{\text{iid}}{\sim} N(0, \omega^2)\ \forall\ j = 1, \ldots, J$, and specify a prior over the hyper-parameter ω^2. Priors are also needed for the grand mean μ, and the within-group error variance σ^2. A common practice is to specify independent, improper, reference priors: $p(\mu) \propto 1$, $p(\sigma^2) \propto 1/\sigma^2$ and $p(\omega^2) \propto 1/\omega^2$. Note that these priors are directly analogous to the priors we used as improper, reference priors for the linear regression model in Section 2.5.4.

With the improper, reference priors given above, it is known that this posterior density is improper (Hill 1965). Nonetheless, a Gibbs sampler can be implemented for this problem, since the requisite conditional densities are well-defined and easy to sample from. Moreover, the problem here is particularly pernicious in that nothing in the behavior of the Gibbs sampler for this problem would suggest that the posterior density is improper, and the results are meaningless. Indeed, Hobert and Casella (1996) note some examples of published work where Gibbs samplers have been deployed for this one-way hierarchical model where the posterior is improper (e.g. Gelfand *et al*. 1990; Wang, Rutledge and Gianola 1993). In Problem 6.9 you are invited to derive the conditional densities for the Gibbs sampler in this case and implement the sampler.

In Chapter 7 we consider hierarchical models in considerable detail, using proper priors that will ensure the propriety of the resulting posterior densities. Indeed, with computational tools like `BUGS` or `JAGS`, one has to go out of one's way to specify improper priors in a hierarchical model. Of course, many users of these programs specify 'barely proper' priors over variance parameters, but we will generally avoid this practice as well (see Chapter 7 for details).

The point here is that it is certainly possible for the Gibbs sampler to appear to be sampling from a posterior density when in fact no such density exists. Avoiding improper priors is one way to help guard against this. Simulated data examples can help too, particularly as one ventures into a new class of model for the first time. But in general, there is no substitute for bringing analytics to bear, deriving whatever properties of the posterior density of interest one can before deploying simulation-based approaches to inference: as the preceding example demonstrates, sometimes MCMC can appear to be giving us a meaningful inference when analytics would reveal that no such thing can exist.

Nonconnected support

The Gibbs sampler can give misleading results if the support of the posterior density is a series of disconnected regions. That is, suppose $\boldsymbol{\theta} = (\theta_1, \ldots, \theta_d)' \in \boldsymbol{\Theta} \subseteq \mathbb{R}^d$ but that $p(\boldsymbol{\theta}|\mathbf{y}) = 0\ \forall\ \boldsymbol{\theta} \notin \mathcal{D} = \{\mathcal{D}_1, \ldots, \mathcal{D}_J\}$ and where $\cap_{j=1}^{J} \mathcal{D}_j = \emptyset$. Moreover, let

$\mathcal{D}_j = \Theta_{j1} \times \ldots \times \Theta_{jd}$, where $\Theta_{jk} \subset \mathbb{R}, \forall\ j = 1, \ldots, J$ and $k = 1, \ldots, d$. In addition to the disconnectedness condition just given, we will also stipulate that the regions of positive support $\mathcal{D}_j$ are dimension-wise *non-overlapping*: i.e. $\cap_{j=1}^{J} \Theta_{jk} = \emptyset\ \forall\ k = 1, \ldots, d$. Suppose now we start a Gibbs sampler with d components with $\boldsymbol{\theta}^{(0)} \in \mathcal{D}_1$. The sampler will never leave that region, and will never discover the other $\mathcal{D}_j$, $j > 1$. In this case, the resulting characterization of the posterior density generated by Gibbs sampler will be misleading.

The following contrived example appears in numerous places in the literature: e.g. Roberts (1996, 52–53); Robert and Casella (2004, 379); O'Hagan (2004); Gamerman and Lopes (2006, 148).

■ Example 5.8

Gibbs sampler, density with nonconnected support. Suppose $\mathcal{D}^+$ and $\mathcal{D}^-$ are open, unit disks in $\mathbb{R}^2$ with centers (1, 1) and (−1, −1), respectively. Note that $\mathcal{D}^+ \cap \mathcal{D}^- = \emptyset$. The density of interest is uniform on these two disks, but zero everywhere else; i.e.

$$f(\theta_1, \theta_2) = \begin{cases} \frac{1}{2\pi} & \text{if } (\theta_1, \theta_2) \in \mathcal{D} \equiv \{\mathcal{D}^+ \cup \mathcal{D}^-\} \\ 0 & \text{otherwise.} \end{cases}$$

A Gibbs sampler for this problem would consist of sampling from the two conditional densities $g_1(\theta_1|\theta_2)$ and $g_2(\theta_2|\theta_1)$. Suppose we initialize the Gibbs sampler with $\boldsymbol{\theta}^{(0)} \in \mathcal{D}^+$. Now consider $g_1(\theta_1| -2 < \theta_2 < 0)$. This is a uniform density over the interval with endpoints given by solving for θ_1 given θ_2: i.e. since a circle with unit radius centered on (1, 1) can be represented by the equation $(\theta_1 - 1)^2 + (\theta_2 - 1)^2 = 1$, if we are given θ_2 then we can deduce that $\theta_1 \in (1 - d, 1 + d)$ where $d = \sqrt{1 - (\theta_2 - 1)^2}$. Thus the conditional density is

$$g_1(\theta_1| -2 < \theta_2 < 0) = \begin{cases} \frac{1}{2d} & 1 - d < \theta_1 < 1 + d \\ 0 & \text{otherwise} \end{cases}$$

where $d = \sqrt{1 - (\theta_2 - 1)^2}$ and $-2 < \theta_2 < 0$. Note that this conditional density places zero mass on the event $\theta_1 < 0$. In the language of Chapter 4, the Markov chain formed by this Gibbs sampler is not irreducible (Definition 4.9). The sampler defined by these conditional densities can not jump from $\mathcal{D}^+$ to $\mathcal{D}^-$. Moreover, we can verify that this example does not meet the sufficient conditions for convergence of the Gibbs sampler.

Robert and Casella (2004, 379) point out that if we perform a simple change of coordinates to $z_1 = \theta_1 + \theta_2$ and $z_2 = \theta_1 - \theta_2$ then the difficulty is removed. See the discussion on reparameterization in Section 6.4.3 and Problem 6.12.

Multi-modal posterior densities

A less dire problem (though still pressing problem) for the Gibbs sampler can arise when the density being explored has multiple, well-separated modes. In this case sufficient conditions like positivity hold and the Markov chain generated by the Gibbs sampler is ergodic and simulation-consistency results still hold. The problem here is a practical issue. Depending on where the Gibbs sampler is started, it may take an extraordinarily long time for the sampler to discover the presence of all modes of the target density. If the waiting time is too long – longer than an analyst is prepared to wait – then without prior, analytic knowledge as to presence of the multiple modes, the analyst may walk away from the data analysis without discovering them.

Note the delicate distinction between theory and practice here. Sufficient conditions for ergodicity hold, but even a long run of the Gibbs sampler is simply insufficient for the Markov chain to traverse those regions of the parameter space with very low posterior probability separating regions of substantial posterior probability. The practical result is that we wind up in the same situation as that considered in the previous example, where the posterior consists of regions of nonconnected support; i.e. the characterization of the posterior density generated by the Gibbs sampler is misleading and wholly dependent on where we start the sampler. We briefly consider an example using a latent variable model, the subject of Chapter 9.

■ Example 5.9

Multiple modes in latent variable models. Consider a generalized latent variable model, $E[g(y_{ij})] = x_i\beta_j$, where y_{ij} is the observed value for unit i on indicator j ($i = 1, \ldots, n$; $j = 1, \ldots, m$), $x_i \in \mathbb{R}$ is the unobserved score of unit i on a some latent dimension of interest, and $\beta_j \in \mathbb{R}$ is an unknown parameter tapping the reliability of indicator j as a measure of x. Factor analysis and item-response theory (IRT) models are special cases of this general model, as we detail in Chapter 9; higher dimensional versions of the model are also possible, with $\mathbf{x}_i, \boldsymbol{\beta}_j \in \mathbb{R}^d, d > 1$. Our discussion here focuses on the unidimensional case.

The unknown parameters for this model are $\mathbf{x} = (x_1, \ldots, x_n)'$ and $\boldsymbol{\beta} = (\beta_1, \ldots, \beta_m)'$ and are usually not identified (see Definition B.17). Typically, the parameters $\mathbf{x}$ and $\boldsymbol{\beta}$ enter the likelihood $f(\mathbf{Y}|\mathbf{x}, \boldsymbol{\beta})$ linearly, as indicated in the mean function above. In this case the parameter values $\{\tilde{\mathbf{x}}, \tilde{\boldsymbol{\beta}}\}$ generate the same likelihood contributions as $\{h(\tilde{\mathbf{x}}), h^{-1}(\tilde{\boldsymbol{\beta}})\}$, for any invertible, affine transformation $h : \mathbb{R} \mapsto \mathbb{R}$. The lack of identification poses no formal problem for Bayesian analysis *per se*. Indeed, with proper priors over the parameters $\mathbf{x}$ and $\boldsymbol{\beta}$ the posterior density is proper, and the Gibbs sampler (or other MCMC algorithms) can be deployed.

A naïve analyst might be tempted to conclude that that is the end of the story (e.g. Jackman 2001). That is, say with priors $p(\mathbf{x}) = \prod_{i=1}^{n} p(x_i)$, $p(x_i) \equiv N(0, \tau_x^2)$ and $p(\boldsymbol{\beta}) = \prod_{j=1}^{m} p(\beta_j)$, $p(\beta_j) \equiv N(0, \tau_\beta^2)$, $0 < \tau_x^2 < \infty, 0 < \tau_\beta^2 < \infty$, we have a proper posterior density $p(\mathbf{x}, \boldsymbol{\beta}|\mathbf{Y}) \propto f(\mathbf{Y}|\mathbf{x}, \boldsymbol{\beta})p(\mathbf{x})p(\boldsymbol{\beta})$, and the fact that the parameters are unidentified is of no great consequence. This is a misleading conclusion, in that while the posterior density is proper, it is also bimodal given the priors above. In particular, if $\{\tilde{\mathbf{x}}, \tilde{\boldsymbol{\beta}}\}$ is a mode of the posterior density, then so too is $\{c \cdot \tilde{\mathbf{x}}, c \cdot \tilde{\boldsymbol{\beta}}\}$, $c = -1$, a case of *invariance to reflection* about the origin. Less formally, there are two 'mirror image' modes in the posterior density.

This result will be familiar to anyone who has built scale measures and used them in correlation or regression analysis: one can simply reverse the 'polarity' or 'direction' of the scale with the understanding that nothing changes, save for a change in the sign of the coefficient on the scale in any subsequent regression analysis.

Usually, the fact that the posterior is bimodal is relatively innocuous, if the two mirror images modes are well-separated. The Gibbs sampler will explore one mode and never visit the mirror image mode. Of course, strictly speaking, the Gibbs sampler is generating a misleading characterization of the posterior density. A thoughtful researcher will recognize that there are two mirror image posterior modes and that they are exploring just one of them. Alternatively – and this is perhaps the wiser course of action – the researcher could use priors that generate a unimodal posterior density: say, with a sign restriction on at least one of the x_i or β_j. Stronger restrictions are certainly possible, by imposing a normalization on the resulting scale: e.g. two non-collinear point restrictions on any two parameters in $\mathbf{x}$, but these are stronger than necessary to ensure a unimodal posterior density.

To examine this phenomenon, I fit the Gibbs sampler to a small data set: $n = 10$, $m = 5$, and each y_{ij} is a binary observation, and so the model is fit using an IRT model (see Section 9.3). I employ priors of the sort described above, that generate a bimodal posterior density due to invariance to reflection. Because the data set is small and the priors are relatively uninformative ($\tau_x^2 = 1$; $\tau_\beta^2 = 10^2$), the two modes of the posterior density are relatively close to one another and we can expect the Gibbs sampler to find them. The Gibbs sampler is run for just over 5000 iterations. Traceplots are displayed in Figure 5.12. The top two panels show marginal trace plots for x_3 and β_3; it appears that the Gibbs sampler is exploring two modes for each of these parameters, with the flips around zero in each traceplots mirroring each one another. The lower panel shows the Gibbs sampler's trajectory with respect to both x_3 and β_3, with each plotted point $(x_3^{(t)}, \beta_3^{(t)})$ indicating the location of the sampler at iteration t. The gray forty-five degree line highlights the invariance to reflection around the origin. The 'mirror image' nature of the posterior density for this problem is clearly apparent, with the Gibbs sampler switching between the two modes quite easily: the two modes are not particularly well-separated in this example, and the sampler is quite capable of 'jumping' between modes.

If we were to recover traceplots like this in a real world application we ought to immediately suspect some kind of invariance in the posterior density. An obvious solution in this case would be a more restrictive prior on one or more of the parameters; imposing a sign restriction on one or more of the β parameters would be sufficient to remove the posterior bimodality in this case.

With more data, the posterior density is much less dispersed around each mode, and even a long run of the Gibbs sampler will not reveal the bimodality. To see this, I re-run the previous example but with a much larger data set, $n = 102$ and $m = 544$, the same priors as outlined above, and a much longer run of the Gibbs sampler (25 000 iterations, with every 10th iteration saved for analysis, see Section 6.4.1). In these circumstances the Gibbs sampler gravitates towards the mode closest to its starting values, and does not discover the other mirror image mode (at least not over the 25 000 iteration run reported here). A second run of the Gibbs sampler is required to discover the second mode of the posterior density, and so the 2nd sampler is initialized with start values that are the mirror image of those used to start the first sampler.

Figure 5.13 shows trace plots and marginal posterior densities that result for this problem, for two of the latent variables (labelled x_1 and x_2). The samplers quickly converge on the posterior mode closest to their respective start values: one sampler has $x_1^{(0)} = -1$ and $x_2^{(0)} = +1$ and explores the posterior density in the neighborhood of a mode with components $(x_1, x_2) = (-.3, .2)$; the other sampler has 'mirror image' start

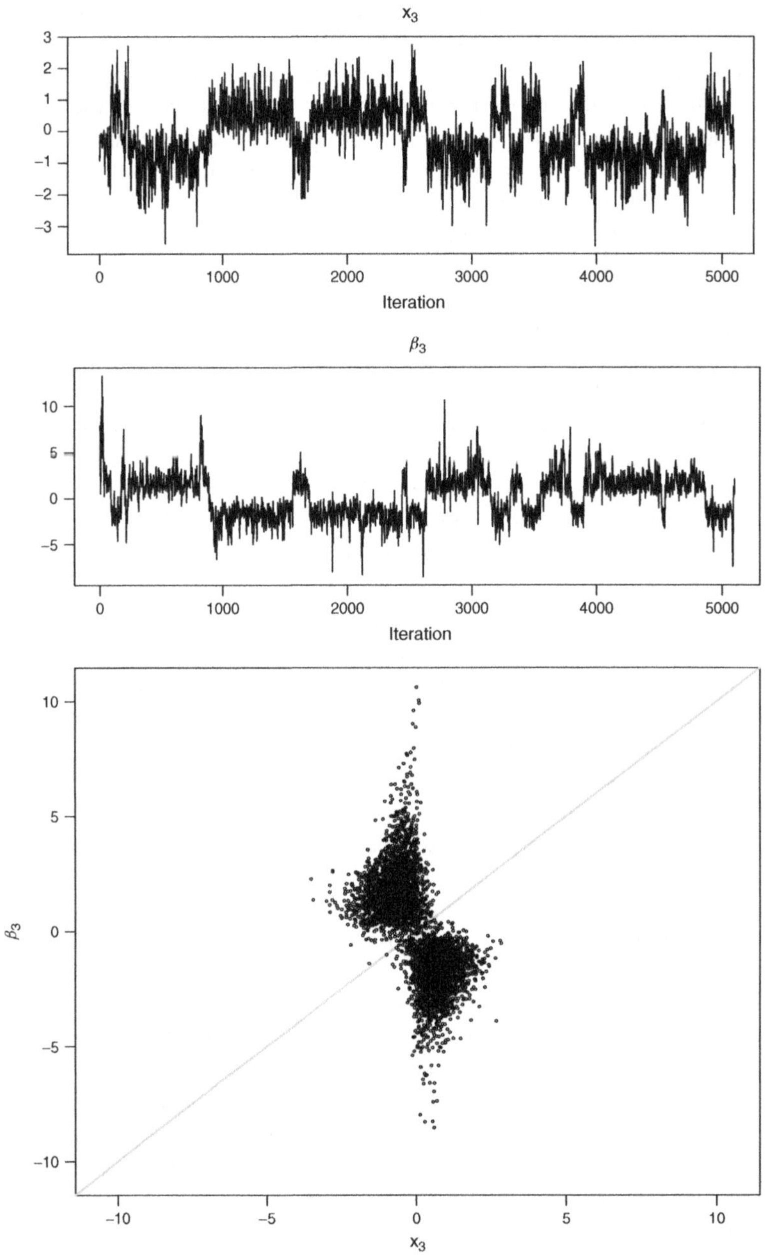

Figure 5.12 Trace plots for Gibbs sampler exploring a bimodal posterior density. The top two panels show time-series style traceplots for two parameters, x_3 and β_3. The lower panel shows the history of the Gibbs sampler with respect to x_3 and β_3 simultaneously. The 'mirror image' nature of the density being explored is clearly evident; the gray line is sloped at 45 degrees, and helps to highlight the reflection around the origin.

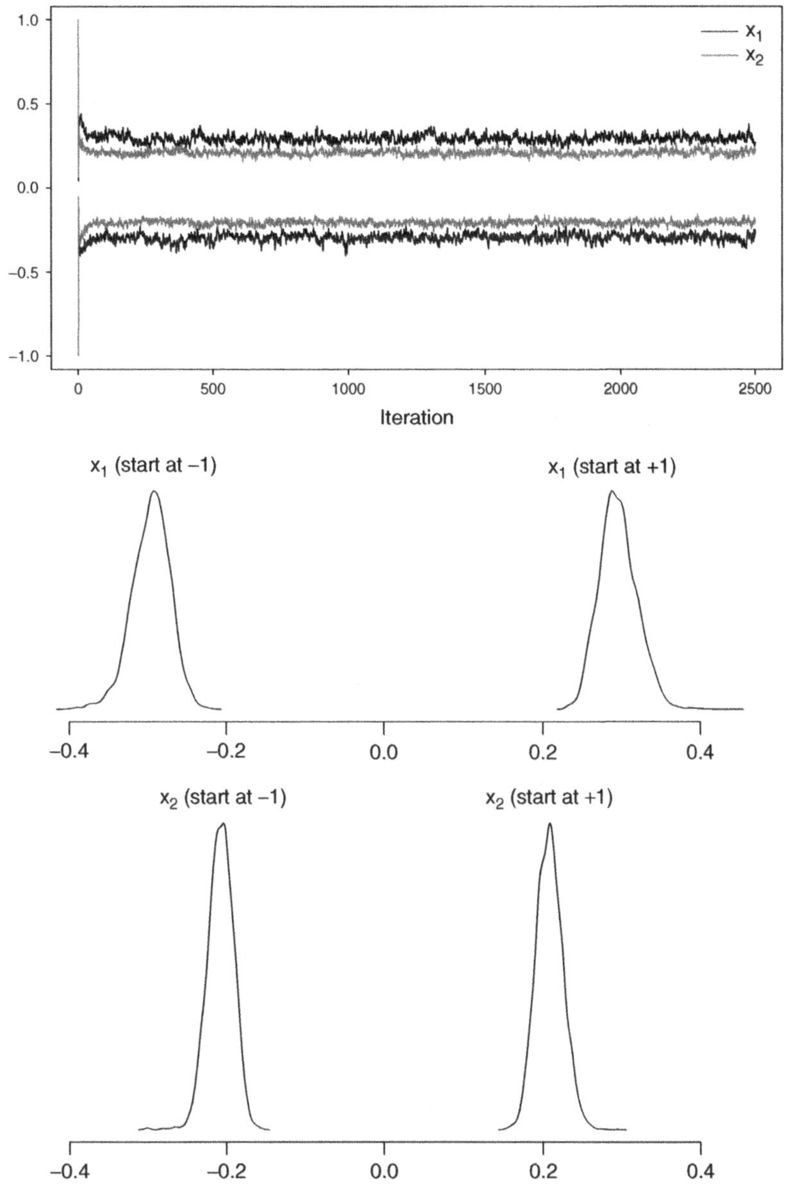

Figure 5.13 Trace plots (upper panel) and marginal posterior densities, for latent variables (labelled x_1 and x_2) whose posterior density is bimodal. Two runs of the Gibbs sampler are required to recover the two posterior modes since they are so well separated.

values and explores the posterior density in the neighborhood of the 'mirror image' mode. Over the course of the 25 000 iterations produced here, there is no hint that the samplers are going to jump the wide gulf of infinitesimal posterior probability separating the two modes.

5.2.5 Data augmentation

The Gibbs sampler is remarkably flexible, in that the parameter vector $\boldsymbol{\theta}$ can be expanded to include auxiliary quantities that are not of direct substantive interest, but make the rest of the analysis 'easier' to conduct. This is tremendously helpful in the case of models with latent variables, where *conditional on the latent variables*, the rest of the analysis is relatively straightforward. That is, computing the posterior density $p(\boldsymbol{\theta}|\mathbf{Y})$ is 'hard' in some sense, but sampling from $f(\boldsymbol{\theta}|\mathbf{Y}, \mathbf{Z})$ is 'easy', suggesting that we might use simulation based methods to compute

$$p(\boldsymbol{\theta}|\mathbf{Y}) = \int_{\mathcal{Z}} f(\boldsymbol{\theta}|\mathbf{Y}, \mathbf{Z}) f(\mathbf{Z}|\mathbf{Y}) d\mathbf{Z},$$

an identity so fundamental that Tanner (1996, 90) refers to it as the *posterior identity*. Exploiting this identity is known as data augmentation, in that the quantities $\mathbf{Z}$ can be thought of as additional data, augmenting the observed data.

■ **Example 5.10**

Data-augmented Gibbs sampler, probit regression model. In an early demonstration of the Gibbs sampler, Albert and Chib (1993) showed how a simple, iterative sampling scheme could be used for Bayesian analysis of a probit model for binary responses. We return to this model in some detail in §8.1. For now we note that the probit model is simply a latent variable regression model with normal errors, where the latent dependent variable y_i^* is censored, observed only in terms of its sign: i.e. $y_i^* = \mathbf{x}_i\boldsymbol{\beta} + \epsilon_i, \epsilon_i \sim N(0, 1) \,\forall\, i = 1, \ldots, n$ with the censoring $y_i = 0 \Rightarrow y_i^* < 0$ and $y_i = 1 \Rightarrow y_i^* \geq 0$. This yields the following restatement of the model:

$$\begin{aligned}\Pr(y_i = 1|\mathbf{x}_i, \boldsymbol{\beta}) = \Pr(y_i^* \geq 0|\mathbf{x}_i, \boldsymbol{\beta}) &= \Pr(\mathbf{x}_i\boldsymbol{\beta} + \epsilon_i \geq 0) = \Pr(\epsilon_i \geq -\mathbf{x}_i\boldsymbol{\beta}) \\ &= \Pr(\epsilon_i < \mathbf{x}_i\boldsymbol{\beta}) = \Phi(\mathbf{x}_i\boldsymbol{\beta}),\end{aligned}$$

with the last equality corresponding to the more conventional presentation of the probit regression model.

The latent data $\mathbf{y}^* = (y_1^*, \ldots y_n^*)'$ augment the inferential problem here in the following way. We wish to compute the posterior density $p(\boldsymbol{\beta}|\mathbf{y}, \mathbf{X}) \propto p(\boldsymbol{\beta})p(\mathbf{y}|\mathbf{X}, \boldsymbol{\beta})$. This is a 'hard problem', in that the product of the prior $p(\boldsymbol{\beta})$ and likelihood $p(\mathbf{y}|\mathbf{X}, \boldsymbol{\beta})$ is mathematically intractable, for any reasonable choice for the form of the prior density (see the remarks in Example II.1). But by augmenting the problem with the latent data $\mathbf{y}^*$, we turn the Bayesian inferential problem into something much easier. That is, instead of trying to compute the posterior $p(\boldsymbol{\beta}|\mathbf{y}, \mathbf{X})$ directly, we work with the data-augmented posterior density $p(\boldsymbol{\beta}, \mathbf{y}^*|\mathbf{y}, \mathbf{X})$, which turns out to be a much easier distribution from which to sample (via the Gibbs sampler) than $p(\boldsymbol{\beta}|\mathbf{y}, \mathbf{X})$.

The data-augmented posterior can be marginalized with respect to either $\boldsymbol{\beta} \in \mathbf{B}$ or $\mathbf{y}^* \in \mathcal{Y}^*$ as follows:

$$p(\boldsymbol{\beta}|\mathbf{y}, \mathbf{X}) = \int_{\mathcal{Y}^*} p(\boldsymbol{\beta}, \mathbf{y}^*|\mathbf{y}, \mathbf{X}) d\mathbf{y}^* = \int_{\mathcal{Y}^*} p(\boldsymbol{\beta}|\mathbf{y}^*, \mathbf{X}) p(\mathbf{y}^*|\mathbf{y}, \mathbf{X}, \boldsymbol{\beta}) d\mathbf{y}^* \tag{5.16}$$

$$p(\mathbf{y}^*|\mathbf{y}, \mathbf{X}) = \int_{\mathbf{B}} p(\boldsymbol{\beta}, \mathbf{y}^*|\mathbf{y}, \mathbf{X}) d\boldsymbol{\beta} = \int_{\mathbf{B}} p(\mathbf{y}^*|\boldsymbol{\beta}, \mathbf{y}, \mathbf{X}) p(\boldsymbol{\beta}|\mathbf{y}, \mathbf{X}) d\boldsymbol{\beta} \tag{5.17}$$

Equation 5.16 is quite suggestive: if it weren't for the fact that $\boldsymbol{\beta}$ appears as a conditioning argument in $p(\mathbf{y}^*|\mathbf{y}, \mathbf{X}, \boldsymbol{\beta})$, we could use the method of composition (Algorithm 3.3 in Section 3.3) to generate samples from $p(\boldsymbol{\beta}|\mathbf{y}, \mathbf{X})$. Iterative methods are typically used to solve integral equations like Equation 5.16; in this case we use a Gibbs sampler over the data-augmented problem to retrieve samples from the (marginal) posterior density for $\boldsymbol{\beta}$. Each iteration of the Gibbs sampler consists of the following two steps: (1) sample $\boldsymbol{\beta}$ from $g(\boldsymbol{\beta}|\mathbf{y}^*, \mathbf{X})$ and (2) sample $\mathbf{y}^*$ from $g(\mathbf{y}^*|\boldsymbol{\beta}, \mathbf{X}, \mathbf{y})$. Further details appear in Section 8.1 with this specific model revisited in Example 8.1 and Algorithm 8.1 in particular.

The class of problems to which data augmentation can apply is massive. As the preceding example suggests, many discrete choice models can be formulated this way, and we will revisit these models in some detail in Chapter 8, which will give us the opportunity to consider an important extension of data augmentation known as *marginal data augmentation* (see Section 8.1.2). Measurement models are fundamentally models for observable indicators conditional on unobserved latent data, and we examine this class of model in Chapter 9. The deep similarities between statistical inference for parameters and imputation for missing data underpins the review of simulation-based Bayesian inference in Jackman (2000). In this vein, we now consider a Bayesian treatment of inference in the presence of missing data.

5.2.6 Missing data problems

The idea behind data augmentation is that there is nothing in the theory of the Gibbs sampler that stops us from working with an augmented state vector $\tilde{\boldsymbol{\theta}} = \{\boldsymbol{\theta}, \mathbf{Z}\}$, where $\mathbf{Z}$ could literally be almost any quantity. A typical use is where $\mathbf{Z}$ is something that makes sampling $\boldsymbol{\theta}$ easier to do (as in the preceding example). An important set of applications in this class are missing data problems, where $\mathbf{Z}$ is missing data. In this respect the data augmented Gibbs sampler is very similar to a stochastic version of the *EM* algorithm for missing data (Dempster, Laird and Rubin 1977), where instead of 'plugging in' the conditional expectation of the missing data given the observed data and the current guess as to $\boldsymbol{\theta}$ (the *E* step of *EM*), we sample from the conditional distribution for $\mathbf{Z}$.

We begin by partitioning data $\mathbf{Y}$ into $\mathbf{Y}_{\text{obs}}$ (observed data) and $\mathbf{Y}_{\text{miss}}$ (missing data). We want to make inferences about $\boldsymbol{\theta}$, a vector of unknown parameters that indexes a probability model for $\mathbf{Y}$, $f(\mathbf{Y}|\boldsymbol{\theta})$. As in any Bayesian analysis, we use the observed data to update prior beliefs about the unknown quantities $\boldsymbol{\theta}$: i.e. via Bayes Rule

$$p(\boldsymbol{\theta}|\mathbf{Y}_{\text{obs}}) \propto p(\boldsymbol{\theta}) f(\mathbf{Y}_{\text{obs}}|\boldsymbol{\theta}).$$

In many cases, this formulation of the posterior density conditional on the *observed data* is difficult to compute. Nonetheless, the *complete data* likelihood and corresponding posterior density may be easy to compute, from which we obtain the *observed data* posterior density by integration: i.e. with $\mathbf{Y}_{\text{miss}} \in \mathcal{Y}_{\text{miss}}$ and $\boldsymbol{\theta} \in \boldsymbol{\Theta}$,

$$\begin{aligned} p(\boldsymbol{\theta}|\mathbf{Y}_{\text{obs}}) &= \int_{\mathcal{Y}_{\text{miss}}} p(\boldsymbol{\theta}|\mathbf{Y}_{\text{obs}}, \mathbf{Y}_{\text{miss}}) p(\mathbf{Y}_{\text{miss}}|\mathbf{Y}_{\text{obs}}) d\mathbf{Y}_{\text{miss}} \\ &\propto \int_{\mathcal{Y}_{\text{miss}}} f(\mathbf{Y}|\boldsymbol{\theta}) p(\boldsymbol{\theta}) p(\mathbf{Y}_{\text{miss}}|\mathbf{Y}_{\text{obs}}) d\mathbf{Y}_{\text{miss}}, \end{aligned} \tag{5.18}$$

where $f(\mathbf{Y}|\boldsymbol{\theta})$ is the complete data likelihood and $p(\boldsymbol{\theta}|\mathbf{Y}_{\text{obs}}, \mathbf{Y}_{\text{miss}}) \propto f(\mathbf{Y}|\boldsymbol{\theta})p(\boldsymbol{\theta})$ is the complete data posterior density for $\boldsymbol{\theta}$. Note however that the predictive density for the missing data will typically involve $\boldsymbol{\theta}$, i.e.

$$p(\mathbf{Y}_{\text{miss}}|\mathbf{Y}_{\text{obs}}) = \int_{\boldsymbol{\theta}} p(\mathbf{Y}_{\text{miss}}|\mathbf{Y}_{\text{obs}}, \boldsymbol{\theta}) p(\boldsymbol{\theta}|\mathbf{Y}_{\text{obs}}) d\boldsymbol{\theta}. \tag{5.19}$$

Note that the observed data posterior $p(\boldsymbol{\theta}|\mathbf{Y}_{\text{obs}})$ appears in this expression; substituting 5.19 into Equation 5.18 we obtain the observed data posterior density as the following integral equation

$$p(\boldsymbol{\theta}|\mathbf{Y}_{\text{obs}}) \propto \int_{\mathcal{Y}_{\text{miss}}} f(\mathbf{Y}|\boldsymbol{\theta}) p(\boldsymbol{\theta}) \left[\int_{\boldsymbol{\Theta}} p(\mathbf{Y}_{\text{miss}}|\mathbf{Y}_{\text{obs}}, \boldsymbol{\theta}) p(\boldsymbol{\theta}|\mathbf{Y}_{\text{obs}}) d\boldsymbol{\theta} \right] d\mathbf{Y}_{\text{miss}}.$$

The form of this integral equation is identical to the integral equation we encountered in the previous discussion on data augmentation (Equation 5.16), Gibbs sampling (Equation 5.12) and Markov chains (Equation 4.5). The Gibbs sampler provides an iterative solution to this equation, as follows: at iteration t,

1. sample $\mathbf{Y}_{\text{miss}}^{(t)}$ from $g(\mathbf{Y}_{\text{miss}}|\mathbf{Y}_{\text{obs}}, \boldsymbol{\theta}^{(t-1)})$
2. sample $\boldsymbol{\theta}^{(t)}$ from $g(\boldsymbol{\theta}|\mathbf{Y}^{(t)})$, where $\mathbf{Y}^{(t)} = \{\mathbf{Y}_{\text{obs}}, \mathbf{Y}_{\text{miss}}^{(t)}\}$ is a complete data set.

This scheme produces a series of imputed values $\{\mathbf{Y}_{\text{miss}}^{(t)}\}$ that are equivalent to what Rubin (1987) calls *multiple imputations*. The sampled values $\{\boldsymbol{\theta}^{(t)}\}$ are each based on analyses of complete data sets, with each complete data set formed by sampling from the predictive density for the missing data. In this way the resulting characterization of the posterior density of $\boldsymbol{\theta}$ reflects uncertainty over the missing data. Over many iterations of the Gibbs sampler – sampling from the posterior density of $\boldsymbol{\theta}$ conditional on a draw from the predictive density of $\mathbf{Y}_{\text{miss}}$ – we are 'averaging over' or (more precisely) integrating $\mathbf{Y}_{\text{miss}}$ out of the complete-data posterior. In this way the sampled $\boldsymbol{\theta}^{(t)}$ are draws from the density of substantive interest, the observed-data posterior in Equation 5.18.

Assumptions that let us ignore the missing data mechanism

A brief word on the assumptions that make this approach valid is warranted. The fact that the data are subject to missingness is considered a mere nuisance in the treatment above, and not a threat to inference for $\boldsymbol{\theta}$ *per se*. This approach is only valid if (a) the missing data are *missing-at-random* or MAR (Little and Rubin 2002; Rubin 1976); (b) the missing data mechanism is *ignorable*. We consider each of these properties.

We begin by assuming that there are two sets of observables in any data analysis: the observed data $\mathbf{Y}_{\text{obs}}$ and a set of indicators $\mathbf{R} = (r_{ij})$ with $r_{ij} = 1$ if y_{ij} is missing and zero otherwise. As before, missing data is denoted $\mathbf{Y}_{\text{miss}}$ and $\mathbf{Y} = \{\mathbf{Y}_{\text{obs}}, \mathbf{Y}_{\text{miss}}\}$. Parameters $\boldsymbol{\theta}$ index a probability model for $\mathbf{Y}$, $f(\mathbf{Y}|\boldsymbol{\theta})$. Parameters $\boldsymbol{\phi}$ index a probability model for the missing data. Even data sets not subject to missingness can be considered this way, in the sense that there may have been more data that was not observed or not collected (possibly in a benign way such as not being randomly selected for inclusion in a sample, etc.).

Bayesian analysis uses observables to update beliefs about unobservables, via Bayes Rule. That is, the posterior density of interest here would seem to be

$$p(\boldsymbol{\theta}, \boldsymbol{\phi}|\text{Y}_{\text{obs}}, \mathbf{R}) \propto p(\mathbf{Y}_{\text{obs}}, \mathbf{R}|\boldsymbol{\theta}, \boldsymbol{\phi})p(\boldsymbol{\theta}, \boldsymbol{\phi}), \tag{5.20}$$

although in most applications the parameters of most substantive importance are $\boldsymbol{\theta}$. We shall now consider assumptions that let us simplify this joint posterior density so we can focus on the substantive problem of inference for $\boldsymbol{\theta}$, with this discussion a slightly expanded restatement of the exposition in Little and Rubin (2002, §6.2).

We write the joint density for the complete data $\mathbf{Y}$ and $\mathbf{R}$ as the product of a marginal density for $\mathbf{Y}$ given $\boldsymbol{\theta}$ and a conditional density for $\mathbf{R}$ given $\mathbf{Y}$ and $\boldsymbol{\phi}$: i.e.

$$p(\mathbf{Y}, \mathbf{R}|\boldsymbol{\theta}, \boldsymbol{\phi}) = p(\mathbf{Y}|\boldsymbol{\theta})\, p(\mathbf{R}|\mathbf{Y}, \boldsymbol{\phi})$$

Notice that in this factorization that we are making the following reasonably weak conditional independence assumptions: (a) $p(\mathbf{Y}|\boldsymbol{\theta}, \boldsymbol{\phi}) = p(\mathbf{Y}|\boldsymbol{\theta})$, i.e. the parameters $\boldsymbol{\phi}$ are not informative with respect to $\mathbf{Y}$, at least given the parameters $\boldsymbol{\theta}$ (and we will consider strengthening this assumption shortly); (b) $p(\mathbf{R}|\mathbf{Y}, \boldsymbol{\phi}, \boldsymbol{\theta}) = p(\mathbf{R}|\mathbf{Y}, \boldsymbol{\phi})$, i.e. the parameters $\boldsymbol{\theta}$ are not informative with respect to the missingness indicators $\mathbf{R}$ after we condition on $\mathbf{Y}$ and $\boldsymbol{\phi}$.

We now turn to the joint density of the observed quantities, $\mathbf{Y}_{\text{obs}}$ and $\mathbf{R}$. As in the derivation of the observed data posterior density (Equation 5.18), we express this density by integrating $\mathbf{Y}_{\text{miss}}$ out of the density over $\mathbf{Y}$ and $\mathbf{R}$: i.e.

$$\begin{aligned} p(\mathbf{Y}_{\text{obs}}, \mathbf{R}|\boldsymbol{\theta}, \boldsymbol{\phi}) &= \int_{\mathcal{Y}_{\text{miss}}} p(\mathbf{Y}_{\text{obs}}, \mathbf{Y}_{\text{miss}}, \mathbf{R}|\boldsymbol{\theta}, \boldsymbol{\phi})d\mathbf{Y}_{\text{miss}} \\ &= \int_{\mathcal{Y}_{\text{miss}}} p(\mathbf{Y}_{\text{obs}}, \mathbf{Y}_{\text{miss}}|\boldsymbol{\theta})\, p(\mathbf{R}|\mathbf{Y}_{\text{obs}}, \mathbf{Y}_{\text{miss}}, \boldsymbol{\phi})d\mathbf{Y}_{\text{miss}}. \end{aligned} \tag{5.21}$$

If the pattern of missingness $\mathbf{R}$ is conditionally independent of the values of the missing data given (a) the observed data $\mathbf{Y}_{\text{obs}}$ and (b) the parameters $\boldsymbol{\phi}$, then the data are said to be *missing at random* (MAR). The conditional independence assumption in MAR can be expressed as

$$p(\mathbf{R}|\mathbf{Y}_{\text{miss}}, \mathbf{Y}_{\text{obs}}, \boldsymbol{\phi}) = p(\mathbf{R}|\mathbf{Y}_{\text{obs}}, \boldsymbol{\phi}), \forall\ \mathbf{Y}_{\text{miss}}, \boldsymbol{\phi}. \tag{5.22}$$

MAR can be contrasted with an even stronger assumption *missing completely at random* (MCAR), in which missingness is conditionally independent of both $\mathbf{Y}_{\text{obs}}$ *and* $\mathbf{Y}_{\text{miss}}$ given $\boldsymbol{\phi}$.

$$p(\mathbf{R}|\mathbf{Y}_{\text{miss}}, \mathbf{Y}_{\text{obs}}, \boldsymbol{\phi}) = p(\mathbf{R}|\boldsymbol{\phi}). \tag{5.23}$$

Simple random sampling can be considered as generating this kind of missingness, where $r_{ij} = 1$ for everyone in a population not randomly sampled. Random subsampling of a large data set will generate data that are MCAR.

The other possibility is *missing not at random* (MNAR), in which the conditional independence assumption in Equation 5.22 does not hold. In such a case we will require a joint model for $\mathbf{Y} = \{\mathbf{Y}_{\text{obs}}, \mathbf{Y}_{\text{miss}}\}$ *and* the pattern of missingness $\mathbf{R}$.

MAR generates a very handy decomposition of the joint density in Equation 5.21: substituting the right-hand side of 5.22 into 5.21 yields

$$\begin{aligned} p(\mathbf{Y}_{\text{obs}}, \mathbf{R}|\boldsymbol{\theta}, \boldsymbol{\phi}) &= \int_{\mathcal{Y}_{\text{miss}}} p(\mathbf{Y}_{\text{obs}}, \mathbf{Y}_{\text{miss}}|\boldsymbol{\theta}) p(\mathbf{R}|\mathbf{Y}_{\text{obs}}, \boldsymbol{\phi}) d\mathbf{Y}_{\text{miss}} \\ &= p(\mathbf{R}|\mathbf{Y}_{\text{obs}}, \boldsymbol{\phi}) \int_{\mathcal{Y}_{\text{miss}}} p(\mathbf{Y}_{\text{obs}}, \mathbf{Y}_{\text{miss}}|\boldsymbol{\theta}) d\mathbf{Y}_{\text{miss}} \\ &= p(\mathbf{R}|\mathbf{Y}_{\text{obs}}, \boldsymbol{\phi}) p(\mathbf{Y}_{\text{obs}}|\boldsymbol{\theta}). \end{aligned} \tag{5.24}$$

This factorization of the joint density is especially helpful, suggesting that inference for $\boldsymbol{\theta}$ can take place conditioning on the observed data $\mathbf{Y}_{\text{obs}}$. In fact, with one more assumption, *ignorability*, we will have completely separated the issues of inference for $\boldsymbol{\theta}$ from modeling the missing data mechanism $p(\mathbf{R}|, \mathbf{Y}_{\text{obs}}, \boldsymbol{\phi})$.

Armed with the decomposition in 5.24, let us return to the joint posterior density in 5.20. Substituting the right-hand side of the last equality in 5.24, we obtain

$$p(\boldsymbol{\theta}, \boldsymbol{\phi}|\mathbf{Y}_{\text{obs}}, \mathbf{R}) \propto p(\mathbf{R}|\mathbf{Y}_{\text{obs}}, \phi) p(\mathbf{Y}_{\text{obs}}|\boldsymbol{\theta}) p(\boldsymbol{\theta}, \boldsymbol{\phi}).$$

If we can assume *a priori* independence of $\boldsymbol{\theta}$ and $\boldsymbol{\phi}$,

$$p(\boldsymbol{\theta}, \boldsymbol{\phi}) = p(\boldsymbol{\theta}) p(\boldsymbol{\phi}), \tag{5.25}$$

then we obtain

$$p(\boldsymbol{\theta}, \boldsymbol{\phi}|\mathbf{Y}_{\text{obs}}, \mathbf{R}) \propto p(\mathbf{R}|\mathbf{Y}_{\text{obs}}, \boldsymbol{\phi}) p(\boldsymbol{\phi}) p(\mathbf{Y}_{\text{obs}}|\boldsymbol{\theta}) p(\boldsymbol{\theta}) \propto p(\boldsymbol{\phi}|\mathbf{Y}_{\text{obs}}, \mathbf{R}) p(\boldsymbol{\theta}|\mathbf{Y}_{\text{obs}}),$$

i.e. the joint inferential problem has been separated into two distinct Bayesian problems. The prior independence assumption in 5.25 is referred to as an *ignorability* assumption, and together with MAR, ensure that $\boldsymbol{\theta}$ and $\boldsymbol{\phi}$ are independent *a posteriori* (Little and Rubin, 2002, 120).

Note that the argument here is very similar to the way we developed the Bayesian approach to the linear regression model in Section 2.5.1. In that context our problem was to separate the posterior for the regression coefficients $p(\boldsymbol{\beta}, \sigma^2|\mathbf{y}, \mathbf{X})$ from the joint posterior $p(\boldsymbol{\beta}, \sigma^2, \boldsymbol{\theta}_x|\mathbf{y}, \mathbf{X})$ where $\boldsymbol{\theta}_x$ are parameters that index the density of the regressors $\mathbf{X}$. The assumptions of weak exogeneity and prior independence of $(\boldsymbol{\beta}, \sigma^2)$ and $\boldsymbol{\theta}_x$ play similar roles to MAR and ignorability in this missing data context.

We now put these arguments to work in examining a famous, manufactured data set subject to a somewhat extreme pattern of missingness.

■ Example 5.11

Extreme missingness in bivariate normal data; Example 1.7, continued. We return to the manufactured data set originally presented in Table 1.1 as part of Example 1.7. The data $\mathbf{y}_i = (y_{i1}, y_{i2})'$, $i = 1, \ldots, n = 12$, are presumed to be exchangeable draws from a bivariate normal density with known mean vector $\boldsymbol{\mu} = \mathbf{0}$ and unknown covariance matrix

$$\boldsymbol{\Sigma} = \begin{bmatrix} \sigma_{11}^2 & \sigma_{12} \\ \sigma_{21} & \sigma_{22}^2 \end{bmatrix}.$$

The data are subject to extreme pattern of missingness (see Table 1.1 in Example 1.7), with just 4 of the 12 data points fully observed, in the sense that both y_{i1} and y_{i2} are observed, and the remaining 8 observations partially observed. If all the data were observed, then given the modeling assumptions above we could compute the complete data likelihood

$$f(\mathbf{y}|\boldsymbol{\Sigma}) \propto |\boldsymbol{\Sigma}|^{-n/2} \prod_{i=1}^{n} \exp\left(\frac{-1}{2}\mathbf{y}_i'\boldsymbol{\Sigma}^{-1}\mathbf{y}_i\right),$$

using the definition of the multivariate normal density in Definition B.31. The parameter of interest is the correlation $\rho = \sigma_{12}/(\sigma_{11}\sigma_{22})$. The conjugate prior for $\boldsymbol{\Sigma}$ in this case is the inverse-Wishart density (Definition B.41), or equivalently, a Wishart density (Definition B.40) for the precision matrix $\boldsymbol{\tau} = \boldsymbol{\Sigma}^{-1}$. That is, if $\boldsymbol{\Sigma} \sim$ inverse-Wishart$(\boldsymbol{\Psi}, n_0)$, then

$$\boldsymbol{\Sigma}|\mathbf{Y} \sim \text{inverse-Wishart}(\mathbf{S} + \boldsymbol{\Psi}, n_0 + n)$$

where $\mathbf{S} = \sum_{i=1}^{n}(\mathbf{y}_i - \bar{\mathbf{y}})(\mathbf{y}_i - \bar{\mathbf{y}})'$ is the matrix of mean-corrected sums of squares and cross-products, or alternatively, $\mathbf{S} = n\hat{\boldsymbol{\Sigma}}$, where $\hat{\boldsymbol{\Sigma}}$ is the maximum likelihood estimate of $\boldsymbol{\Sigma}$. Given this form for the posterior density of $\boldsymbol{\Sigma}$ it is straightforward to use analytical or (easier yet) simulation methods to derive the posterior density over the correlation ρ.

In the presence of missing data, the analysis gets more complicated. In particular, inference for the correlation parameter ρ would seem to be a delicate matter given that 8 of the 12 bivariate observations are partially observed, in the sense that only one of the two data points $(y_{i1}, y_{i2})'$ are observed. We might be tempted to base inference on the 4 fully-observed cases, but this ignores the information about the marginal variances σ_{11}^2 and σ_{22}^2 (and hence the correlation ρ) in the 8 partially observed cases. Note that when using just the 4 fully observed cases we obtain the MLE of ρ as zero.

Given that we are using a bivariate normal model for these data, there is an obvious imputation or predictive model for the missing data here: we simply factor the bivariate normal joint density for any partially observed $\mathbf{y}_i$ into the product of a marginal univariate normal for the observed data point, and a conditional univariate normal for the missing data point given its observed partner and $\boldsymbol{\Sigma}$: i.e.

$$f(\mathbf{y}_i|\boldsymbol{\Sigma}) \equiv f(y_{i,\text{obs}}, y_{i,\text{miss}}|\boldsymbol{\Sigma}) = f(y_{i,\text{obs}}|\boldsymbol{\Sigma})f(y_{i,\text{miss}}|y_{i,\text{obs}}, \boldsymbol{\Sigma}). \tag{5.26}$$

By Proposition B.2, both the marginal and the conditional densities above are normal densities, and easy to compute or to sample from. In fact, the key density here is the conditional density $f(y_{i,\text{miss}}|y_{i,\text{obs}}, \boldsymbol{\Sigma})$, a predictive density for the missing data from which we will sample so as generate complete data sets, 'filling in the blanks' of Table 1.1; equivalently, we can think of this density as a prior density if we are treating the missing data as parameters. In turn, this means that we can easily implement a data-augmented Gibbs sampler for this problem, exploiting the factorization of the posterior density $p(\boldsymbol{\Sigma}|\mathbf{Y}_{\text{obs}})$

given in Equation 5.18. Moreover, we can recover the posterior density $p(\boldsymbol{\Sigma}|\mathbf{Y}_{\text{obs}})$ through the integration/marginalization inherent in the Gibbs sampler.

We use the conjugate prior $\boldsymbol{\Sigma} \sim$ inverse-Wishart$(\boldsymbol{\Psi}, n_0)$, with $\boldsymbol{\Psi} = \mathbf{I}_2$ and $n_0 = 4$, which generates a reasonably vague prior over $\rho \in [-1, 1]$ centered on $E(\rho) = 0$ (a fact you are invited to verify in Problem 6.5). We proceed by implementing a data augmented Gibbs sampler with two steps per iteration: i.e. at iteration t,

1. sample $\boldsymbol{\Sigma}^{(t)}$ from $g(\boldsymbol{\Sigma}|\mathbf{Y}^{(t-1)})$, where $\mathbf{Y}^{(t-1)} = \{\mathbf{Y}_{\text{obs}}, \mathbf{Y}_{\text{miss}}^{(t-1)}\}$. This an inverse-Wishart density with parameters $\mathbf{S}^{(t-1)} + \boldsymbol{\Psi}$ and $n_0 + n$, where $\mathbf{S}^{(t-1)} = (\mathbf{Y}^{(t-1)} - \mathbf{m})'(\mathbf{Y}^{(t-1)} - \mathbf{m})$ where $\mathbf{m}$ is the vector of column means of $\mathbf{Y}^{(t-1)}$ and $\mathbf{Y}^{(t-1)}$ is the complete data matrix formed by replacing missing elements with the corresponding components of $\mathbf{Y}_{\text{miss}}^{(t-1)}$.

2. sample $\mathbf{Y}_{\text{miss}}^{(t)}$ from $g(\mathbf{Y}_{\text{miss}}|\mathbf{Y}_{\text{obs}}, \boldsymbol{\Sigma}^{(t)})$. For the missing y_{i1}, we sample

$$y_{i1}^{(t)}|y_{i2}, \boldsymbol{\Sigma}^{(t)} \sim N(m_{i1|2}^{(t)}, V_{1|2}^{(t)})$$

where $m_{i1|2}^{(t)} = \sigma_{12}^{(t)}/\sigma_{22}^{2(t)} y_{i2}$ and $V_{1|2}^{(t)} = \sigma_{11}^{2(t)} - \sigma_{12}^{2(t)}/\sigma_{22}^{2(t)}$, drawing on results in Proposition B.2. Similarly, for missing y_{i2} we sample

$$y_{i2}^{(t)}|y_{i1}, \boldsymbol{\Sigma}^{(t)} \sim N(m_{i2|1}^{(t)}, V_{2|1}^{(t)})$$

where $m_{i2|1}^{(t)} = \sigma_{12}^{(t)}/\sigma_{11}^{2(t)} y_{i1}$ and $V_{2|1}^{(t)} = \sigma_{22}^{2(t)} - \sigma_{12}^{2(t)}/\sigma_{11}^{2(t)}$.

This scheme is relatively easy to program in R. Iterations are computationally cheap for this tiny data set, and so we examine the output of a 250 000 iteration run of the data-augmented Gibbs sampler.

Figure 5.14 summarizes the implied posterior density for $\rho \equiv \rho(\boldsymbol{\Sigma}) = \sigma_{12}/(\sigma_{11}\sigma_{22})$, formed by computing $\rho^{(t)} \equiv \rho(\boldsymbol{\Sigma}^{(t)})$ with each sampled $\boldsymbol{\Sigma}^{(t)}$, where t indexes iterations

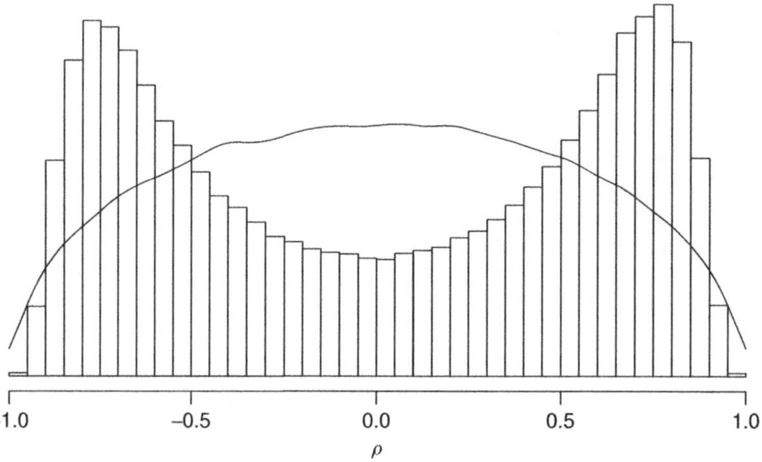

Figure 5.14 Posterior density for ρ, Example 5.11. The histogram summarizes the output of 250 000 iterations of the data-augmented Gibbs sampler. The solid line is a Monte Carlo estimate of the prior for ρ, implied by the inverse-Wishart$(2 \cdot \mathbf{I}_2, 4)$ prior for $\boldsymbol{\Sigma}$.

of the data-augmented Gibbs sampler (see Problem 6.5). We also show the implied prior density for ρ (or, more precisely, a Monte Carlo estimate of this prior density), formed by sampling 250 000 times from the prior $\mathbf{\Sigma} \sim \text{inverse} - \text{Wishart}(2 \cdot \mathbf{I}_2, 4)$, computing $\rho^{(t)} = \rho(\mathbf{\Sigma}^{(t)})$ and estimating the density of the $\rho^{(t)}$ using the `density` function in `R`. The large number of iterations helps drive the Monte Carlo error to very low levels, helping to 'smooth out' any Monte Carlo noise in the histogram in Figure 5.14 (fitted with 51 bins, using the `hist` function in `R`), as well as the estimate of the prior density.

The bimodality of the posterior density for ρ is clearly apparent in Figure 5.14; the fact that the posterior differs so markedly from the unimodal prior confirms the fact that this seemingly unusual result is not driven by the prior. If anything, the prior is slightly informative, tending to concentrate prior probability mass around $\rho = 0$. A less informative prior results with smaller values of n_0 than the $n_0 = 4$ prior used here; with $n_0 = 3$ the inverse-Wishart prior is still proper (though the density is not finite) and becomes improper with $n_0 \leq p$ (with $p = 2$ for this bivariate normal problem). An improper, reference prior is $p(\mathbf{\Sigma}) \propto |\mathbf{\Sigma}|^{-(p+1)/2}$ (see the remarks in Definition B.41) which for these

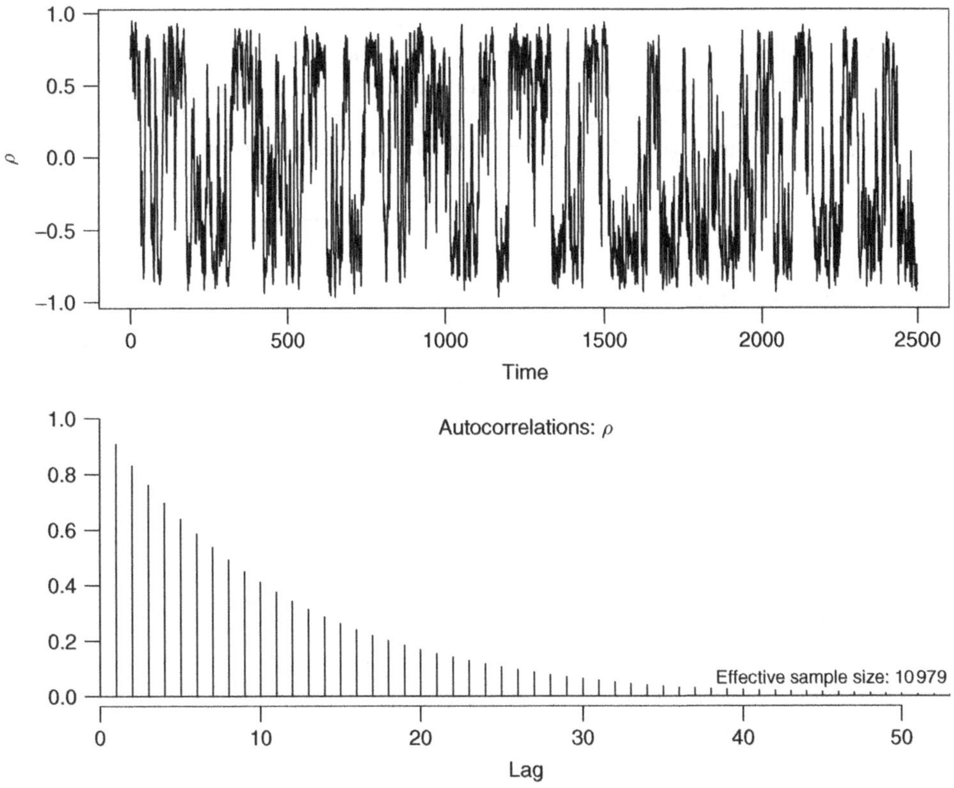

Figure 5.15 Trace plot and autocorrelation function for sampled $\rho^{(t)}$, Example 5.11. Top panel shows the trace plot of $\rho^{(t)}$ for the first 2500 iterations of the 250 000 iteration run; the switching between the two modes of the posterior density is quite apparent. The lower panel shows the slowly decaying autocorrelation function of the sampled $\rho^{(t)}$.

data generates the posterior density $p(\rho|\mathbf{Y}_{\text{obs}}) \propto (1-\rho^2)^{(8+1)/2}/(1.25-\rho^2)^8$ (Tanner, 1996, 96), which has modes at approximately $\pm.824$. These modes are pushed further towards ± 1 than the posterior modes obtained with the proper inverse-Wishart$(2 \cdot \mathbf{I}_2, 4)$ prior used above, which occur at approximately $\pm.75$, since the proper prior concentrates more mass around $\rho = 0$ than the improper, reference prior (in Problem 6.6 you are invited to investigate properties of the improper, reference proper prior for $\mathbf{\Sigma}$).

As we noted when we first encountered this unusual data set in Example 1.7, this example highlights the fact that reducing posterior densities to a single point estimate (a Bayes estimate, see Section 1.6.1) is sometimes misleading. The mean of the posterior density for ρ is zero, which is the same value as the MLE based on a 'complete case' analysis of the data, and the same as the mean of the prior density over ρ. But reporting $\rho = 0$ would be highly misleading in this instance; the posterior mean of ρ lies at the base of a trough of low posterior probability, with modes close to $\pm.9$.

Finally, note that the data-augmented Gibbs sampler employed here does not generate a particularly efficient exploration of the posterior density of ρ. Figure 5.15 shows the trace of the sampled $\rho^{(t)}$ for the first 2500 iterations of the 250 000 iteration run of the Gibbs sampler. The switching between the two modes is easily discerned. The fact that the sampler is trying to explore this bimodal posterior density means that the sampler is relatively inefficient, as revealed in the slow decay of the autocorrelation function of the sampled $\rho^{(t)}$. The high level of dependence in the Gibbs output means that the 250 000 iteration run generates the same level of Monte Carlo error as a 11 000 iteration run of an independence sampler (recall the discussion of effective sample size and simulation inefficiency in Section 4.4.1).

5.2.7 The slice sampler

We conclude this discussion of the Gibbs sampler by focusing on a particular form of the Gibbs sampler known as the *slice sampler*. The slice sampler uses the logic of the Gibbs sampler to generate samples from any arbitrary density. In this way we might well have considered the slice sampler in our discussion of sampling techniques in Section 3.4, and indeed the slice sampler is widely deployed in applications of MCMC. We will make frequent (if not always explicit) use of the slice sampler in the applications presented in Part III.

The term 'slice sampling' is due to Neal (2003, 1997), although some of the underlying ideas can be found through a string of articles stemming from statistical physics: e.g. Neal (2003) cites closely related earlier work by Edwards and Sokal (1988) and Besag and Green (1993), *inter alia*; see also Higdon (1998) and Damien, Wakefield and Walker (1999). The idea underlying the slice sampler is quite elegant and closely related to the data augmentation approach just presented, as well as to the accept-reject sampling method (§ 3.4.3).

We want to sample $\theta \sim p(\theta)$, where $\theta \in \Theta \subseteq \mathbb{R}$, restricting ourselves to the one-dimensional case for the time being. This is equivalent to sampling the pair (θ, U) uniformly from the set $\mathcal{J} = \{(\theta, u) : 0 < u < p(\theta)\}$. That is, consider the expanded parameter space $\tilde{\Theta} = \Theta \times [0, m]$, where $p(\theta) \leq m \, \forall \, \theta \in \Theta$. Now pick a random point (θ^*, U^*) in this space. If $0 < U^* < p(\theta^*)$, then accept the draw. This seemingly simple accept-reject algorithm would yield θ^* that are samples from the target density $p(\theta)$ (see Problem 6.4).

Sampling over the u dimension is equivalent to integrating u out of the joint density $f(\theta, u)$ to recover $p(\theta)$. In this way the U^* can be thought of augmented data of a sort, or 'auxiliary variables', though of a more primitive and fundamental sort than considered previously; that is, by definition, any marginal density $p(\theta)$ can be expressed as

$$p(\theta) = \int_0^{p(\theta)} f(u)du = \int_0^{p(\theta)} du,$$

since $f(u)$ here is a constant. The generality here is compelling. It would appear that we have another extremely general technique for sampling from a density, that involves only evaluating the target density $p(\theta)$; and in fact, $p(\theta)$ need not be a density *per se*, but a function that is proportional to a density (the form in which we typically derive posterior densities).

The interesting practical question is how to implement this strategy. One proposal is to sequentially sample from the conditional densities $g(U|\theta)$ and $g(\theta|U)$. Via the discussion in Chapter 4, it can be shown this algorithm will induce a Markov chain on $\mathcal{J}$ that has a uniform distribution over $\mathcal{J}$ as its stationary distribution. In turn, the marginal density of the sampled θ will be the target density $p(\theta)$. This so-called 'slice sampler' proceeds as follows:

Algorithm 5.3 Slice-sampler

Given $(\theta^{(t-1)}, U^{(t-1)})$:
1: sample $U^{(t)} \sim \text{Unif}(0, p(\theta^{(t-1)}))$
2: sample $\theta^{(t)} \sim \text{Unif}(\mathcal{A}^{(t)})$ where $\mathcal{A}^{(t)} = \{\theta : p(\theta) \geq U^{(t)}\}$.

We will not dwell on the properties of this algorithm, other than to note that it is a special case of the Gibbs sampler, and many of the theoretical results derived for that algorithm carry over to the slice sampler.

Roberts and Rosenthal (1998, 1999, 2002) have derived results specifically on the slice sampler, and even provide bounds on running times for some special cases; a concise review appears in Robert and Casella (2004, §8.3). Tierney and Mira (1999) established uniform ergodicity (Definition 4.25) for the slice sampler; Mira and Tierney (2002) show that slice samplers dominate independence Metropolis-Hastings algorithms (Section 5.1.2) in terms of speed of convergence to the target density.

The potentially difficult part in implementing the slice sampler is evaluating the set $\mathcal{A}^{(t)}$; this is part of the 'art' of slice sampling, and the key challenge in developing efficient, general purpose algorithms for slice sampling. Here we briefly explore an extremely simple idea that has worked well in relatively low-dimensional cases. When the support of $p(\theta)$ is bounded, e.g. $\theta \in \Theta \equiv [a, b]$, then we might implement step two of the algorithm via a naïve rejection step: i.e. replace step 2 in Algorithm 5.3 with (2a) sample $\theta^* \sim \text{Unif}(a, b)$; (2b) set $\theta^{(t)}$ to θ^* if $p(\theta^*) \geq U^{(t)}$, otherwise return to (2a).

■ Example 5.12

Slice sampling from a Beta density. We examine the slice sampler by considering a simple example. Suppose $p(\theta) \equiv \text{Beta}(2, 5)$. There is no need to resort to Monte Carlo methods if $\theta \sim$ Beta (many features of the Beta density can be computed directly, see Definition B.28); our goal here is to see how well the slice sampler performs in a case where the target density is well known.

Note that the support of the Beta density is bounded, i.e. $\theta \in \Theta \equiv [0, 1]$. We initialize the algorithm with $\theta = .5$. Evaluations of the Beta density are easy to compute and so sampling $U^{(t)}$ is straightforward. We sample $\theta^{(t)} \sim \text{Unif}(\mathcal{A}^{(t)})$ via the naïve rejection method discussed above: i.e. in this case step (2a) is to sample $\theta^* \sim \text{Unif}(0, 1)$.

Figure 5.16 shows the progress of the algorithm over the first ten iterations (top left) and the first 250 iterations (top right). The samples from $\theta|U$ move the algorithm in a horizontal direction, while the samples from $U|\theta$ move the algorithm in a vertical direction. Letting the slice sampler run for 500 000 iterations, we recover an excellent approximation to the target density (see the histogram in the lower left panel of Figure 5.16). In addition, the slice sampler is surprisingly efficient, with only a small amount of autocorrelation in the sampled $\theta^{(t)}$; the AR(1) parameter for the 500 000 sampled values is just .11, and the effective sample size (see Section 4.4.1) of the 500 000 sampled values is 382 074.

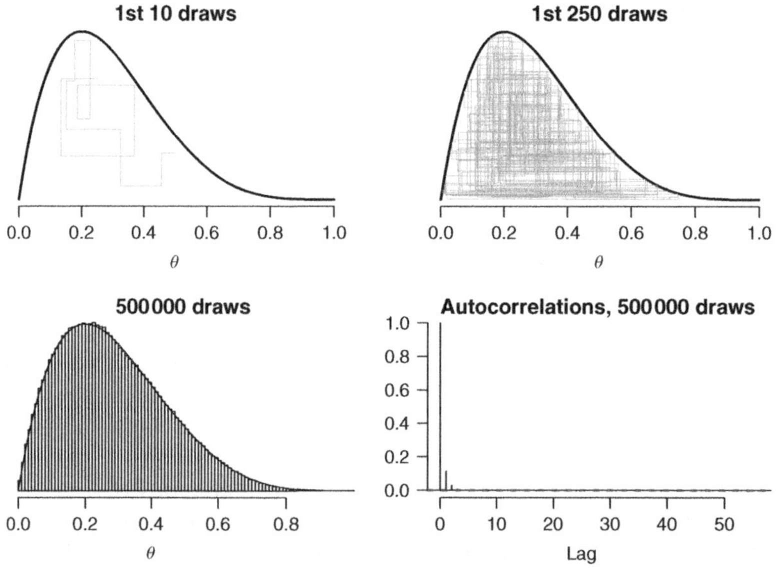

Figure 5.16 Slice sampler output for a Beta(2,5) target density. The top two panels show the algorithm's progress as a series of horizontal (θ) and vertical (U) moves under the target density. The lower left panel shows a histogram summarizing 500 000 draws from the slice sampler, with the target Beta(2,5) density superimposed. The lower right panel shows the autocorrelation function for the 500 000 sampled values; the AR(1) parameter is just .11, higher order autocorrelations are close to zero and the effective sample size (see § 4.4.1) of the 500 000 sampled values is 382 074.

Some intuition into the performance of the slice sampler for this problem follows from noting that when the sampler does venture into the right tail of the target density, it is more likely to step back towards the mode of the density than to linger in the tail. Conditional on having drawn a $\theta^{(t-1)}$ such that $p(\theta)$ is relatively small, consider the constraint that

$$\theta^{(t)} \in \mathcal{A}^{(t)} \equiv \{\theta^* : p(\theta^*) \geq u, u \sim \text{Unif}(0, p(\theta^{(t-1)}))\}$$

A θ^* satisfying this constraint is more likely to lie to the left of $\theta^{(t-1)}$, and this helps push the sampler back in to a region of higher density, and – at least for this example – generates a random tour with a relatively low degree of autocorrelation.

Multivariate slice sampling

The idea underlying the slice sampler extends to the case of multiple parameters, say, where $\boldsymbol{\theta}$ is a vector of parameters, $\boldsymbol{\theta} \in \boldsymbol{\Theta} \subseteq \mathbb{R}^k$, and $p(\boldsymbol{\theta})$ is the target density (again, usually, an unnormalized posterior density). In such a case step 2 of the slice sampling algorithm (Algorithm 5.3) is to sample $\boldsymbol{\theta}^{(t+1)} \sim \text{Unif}(\mathcal{A}^{(t+1)})$ where $\mathcal{A}^{(t+1)}$ is a subset in $\mathbb{R}^k$ (typically, a hyperrectangle). It may be very difficult to actually define this region; Neal (2003) provides a number of possible avenues of attack on this problem.

In these cases it may be easier to break $\boldsymbol{\theta}$ into components, $\boldsymbol{\theta} = (\theta_1, \ldots, \theta_d)'$, and slice sample successively from the corresponding lower-dimensional sets $\mathcal{A}_1^{(t+1)}, \ldots, \mathcal{A}_d^{(t+1)}$; i.e. it is easier to sample from the intervals $\mathcal{A}_j^{(t+1)}$, $j = 1, \ldots, d$ than it is to generate a draw from the hyperrectangle $\mathcal{A}^{(t+1)} = \mathcal{A}_1^{(t+1)} \times \mathcal{A}_2^{(t+1)} \times \ldots \mathcal{A}_d^{(t+1)}$. The potential drawback of this splitting of $\boldsymbol{\theta}$ is that we greatly expand the state vector being carried by the MCMC algorithm, with the introduction of the auxiliary variables u_j, $j = 1, \ldots, d$; note that d is potentially a very large number in latent variable models of the sort considered in Chapters 8 and 9. The trade-off here is a familiar one in MCMC, and one that we will return to in Section 6.4.2. That is, in slice sampling $\boldsymbol{\theta}$ *en bloc* we would generate a reasonably efficient random tour of the target density, but with each iteration being potentially quite expensive in terms of computing time and programming effort. Breaking $\boldsymbol{\theta}$ into components and slice-sampling the components will generally mean that more iterations are required to generate a thorough exploration of the target density, but each iteration will be reasonably cheap to compute.

■ Example 5.13

Extreme missingness in bivariate normal data; Example 5.11, continued. We return to the missing data problem with a different set of priors. Instead of the conditionally conjugate inverse-Wishart prior for the covariance matrix $\boldsymbol{\Sigma}$, we put priors on the two standard deviations σ_1 and σ_2 and the correlation ρ; i.e.

$$\sigma_j \sim \text{Unif}(0, b_j) \quad j = 1, 2$$
$$\rho \sim \text{Unif}(-1, 1)$$

where b_j is a large constant relative to the apparent scale of the data; e.g. in this case, we set $b_j = b = 15$. This formulation has the virtue that the priors are over parameters that are easy to interpret and for which it is easy to specify uninformative priors: e.g. the uniform densities over the standard deviations and an uninformative prior over the $[-1, 1]$ range of the correlation parameter. Letting $\boldsymbol{\theta} = (\sigma_1, \sigma_2, \rho)' \in \boldsymbol{\Theta}$, the priors above induce the constraint $\boldsymbol{\Theta} = [0, b] \times [0, b] \times [-1, 1]$.

Note that these priors do not generate a standard, conditionally conjugate, inverse-Wishart prior over the covariance matrix $\boldsymbol{\Sigma}$. But note that there is nothing special about conjugate priors, beyond their mathematical and computational convenience; armed with modern computing tools, there is no compelling reason for limiting oneself to conjugate priors. It is reasonably straightforward to deduce the prior over each variance σ_j^2, $j = 1, 2$ given the corresponding prior over each standard deviation (the change of variables – see Proposition B.7 – is rather straightforward). In turn, the uniform prior over the correlation ρ induces a prior over the covariance $\sigma_{12} = \rho\sigma_1\sigma_2$. While this particular method of generating priors over the elements of $\boldsymbol{\Sigma}$ ensures that the resulting $\boldsymbol{\Sigma}$ matrix is positive definite – and we will make use of this 'trick' in our consideration of hierarchical models (see Example 7.8) – these priors are not conditionally conjugate with respect to a normal likelihood for the data.

With the parameterization given above, iteration t of the data-augmented Gibbs sampler for this problem consists of

1. sample the elements of $\mathbf{Y}^{(t)}_{\text{miss}}$ from their respective conditional distributions given $\mathbf{Y}_{\text{obs}}$ and $\boldsymbol{\Sigma}^{(t-1)}$, where

$$\boldsymbol{\Sigma}^{(t-1)} \equiv \boldsymbol{\Sigma}(\boldsymbol{\theta}^{(t-1)}) = \begin{bmatrix} [\sigma_1^{(t-1)}]^2 & \rho^{(t-1)}\sigma_1^{(t-1)}\sigma_2^{(t-1)} \\ \rho^{(t-1)}\sigma_1^{(t-1)}\sigma_2^{(t-1)} & [\sigma_2^{(t-1)}]^2 \end{bmatrix},$$

   with the choice of notation here emphasizing that we are working with the standard deviations and the correlation coefficient;

2. sample the elements of $\boldsymbol{\Sigma}^{(t)}$ from their conditional distributions, given $\mathbf{Y}^{(t)}$.

We can perform the sampling in the 2nd step here via the slice sampler, either treating the parameters $\boldsymbol{\theta} = (\sigma_1, \sigma_2, \rho)'$ *en bloc*, or separately. Consider the *en bloc* approach. An iteration of the slice sampler for $\boldsymbol{\theta}$ consists of

1. sample $u^{(t)} \sim \text{Unif}(0, f(\boldsymbol{\theta}^{(t-1)}))$ where $f(\boldsymbol{\theta}^{(t-1)}) \propto \boldsymbol{\phi}(\mathbf{Y}^{(t)}|\boldsymbol{\Sigma}^{(t-1)})$ and where $\boldsymbol{\phi}$ is a bivariate normal density (recall that $\boldsymbol{\mu} = (0, 0)'$ for this problem, and the conditioning on this fixed set of parameters is implicit in the remainder of the discussion).

2. sample $\boldsymbol{\theta}^{(t)} \sim \text{Unif}(\mathcal{A}^{(t)})$, where $\mathcal{A}^{(t)} = \left\{\boldsymbol{\theta}^* \in \boldsymbol{\Theta} : f(\boldsymbol{\theta}^*) \geq u^{(t)}\right\}$.

Delimiting the set $\mathcal{A}^{(t)}$ is the 'hard' part of the algorithm; part of the difficulty here stems from the fact that the marginal posterior density of the correlation coefficient ρ is bimodal (see Figure 5.14), and so the set $\mathcal{A}^{(t)}$ may well comprise two disjoint regions. One approach is to use naïve rejection sampling: i.e.

1. sample $\sigma_j^* \sim \text{Unif}(0, b)$, $j = 1, 2$

2. sample $\rho^* \sim \text{Unif}(-1, 1)$

3. form $\boldsymbol{\theta}^* = (\sigma_1^*, \sigma_2^*, \rho^*)'$
4. set $\boldsymbol{\theta}^{(t)}$ to $\boldsymbol{\theta}^*$ if $f(\boldsymbol{\theta}^*) \geq u^{(t)}$; otherwise return to step 1

This approach can be very slow, in that it may take a large number of attempts to randomly generate a $\boldsymbol{\theta}^*$ that satisfies the constraint. Remember that $\mathcal{A}^{(t)}$ is a slice of $\boldsymbol{\Theta}$, and may be a 'small' slice of $\boldsymbol{\Theta}$ if $\boldsymbol{\theta}^{(t-1)}$ is in the neighborhood of the posterior mode and $u^{(t)}$ happens to be large, which is not an unlikely occurrence; hitting this slice when it is small with three, independent, uniform draws over the dimensions of $\boldsymbol{\Theta}$ is all but guaranteed to be very inefficient.

The other approach is to sample from the conditional densities of σ_1, σ_2 and ρ sequentially. Slice sampling remains an attractive way to proceed in this case, since each of the conditional densities here are non-standard, and indeed, in the case of ρ, not even globally concave. In the case an iteration of the Gibbs sampler consists of the sampling for $\mathbf{Y}_{\text{miss}}^{(t)}$ given $\boldsymbol{\Sigma}^{(t-1)}$, as described above, to generate a complete data set $\mathbf{Y}^{(t)} = \{\mathbf{Y}_{\text{obs}}, \mathbf{Y}_{\text{miss}}^{(t)}\}$. Then,

1. sample $u_1^{(t)} \sim \text{Unif}(0, f_1(\sigma_1^{(t-1)}))$. With the flat priors used here, f_1 is simply the (complete data) likelihood function considered as a function of σ_1; note that the likelihood $\boldsymbol{\phi}(\mathbf{Y}^{(t)}|\boldsymbol{\Sigma}^{(t-1)})$ does not factor cleanly into a set of terms that are solely a function σ_1, and so the $f_1(\sigma_1^{(t-1)})$ function referred to here is (up to a factor of proportionality) the complete data likelihood.

2. sample $\sigma_1^{(t)} \sim \text{Unif}(\mathcal{A}(\sigma_1)^{(t)})$ where $\mathcal{A}(\sigma_1)^{(t)} = \{\sigma_1^* \in [0, b] : f_1(\sigma_1^*) \geq u_1^{(t)}\}$.

3. repeat the previous two steps, but with respect to σ_2 instead of σ_1. Note that when we evaluate the complete data likelihood at this step we replace $\boldsymbol{\Sigma}^{(t-1)}$ with a 'working' version of this matrix with $\sigma_1^{(t)}$ appearing whenever $\sigma_1^{(t-1)}$ appeared.

4. sample $u_3^{(t)} \sim \text{Unif}(0, f_3(\rho^{(t-1)}))$ where again, the $f_3(\rho^{(t-1)})$ function is (up to a factor of proportionality) the complete data likelihood.

5. sample $\rho^{(t)} \sim \text{Unif}(\mathcal{A}(\rho)^{(t)}$ where $\mathcal{A}(\rho)^{(t)} = \{\rho^* \in [-1, 1] : f_3(\rho^*) \geq u_3^{(t)}\}$.

We implement the uniform sampling from the regions $\mathcal{A}(\sigma_1)^{(t)}$, $\mathcal{A}(\sigma_2)^{(t)}$ and $\mathcal{A}(\rho)^{(t)}$ by simple rejection sampling. Since each of σ_1, σ_2 and ρ are given a uniform prior with finite bounds, the parameter space for each parameter is simply the interval supporting the respective uniform prior. Rejection sampling can be then implemented by sampling uniformly over that interval (equivalent sampling from the prior density) until we obtain a draw in the corresponding set $\mathcal{A}(\cdot)^{(t)}$; i.e. a uniform density over $\boldsymbol{\Theta}_j$ is also a uniform density over $\mathcal{A}_j \subseteq \boldsymbol{\Theta}_j$, where j indexes the three parameters in this problem. Checking whether a draw satisfies the constraint this involves evaluations of the corresponding f function, which are not particularly expensive in this case, since the f_j functions are merely the bivariate normal, complete data likelihood.

We run the slice sampler for 50 000 iterations, implementing the scheme described above in R. We can assess the inefficiency of the naïve rejection sampling used when performing the 'horizontal' move for the slice sampler by noting how many times we need to sample σ_1^*, σ_2^* and ρ^* before we find a value that satisfies the respective constraints on each parameters. Figure 5.17 presents graphical summaries of the output from the slice sampler, with marginal posterior densities shown as histograms (left panels)

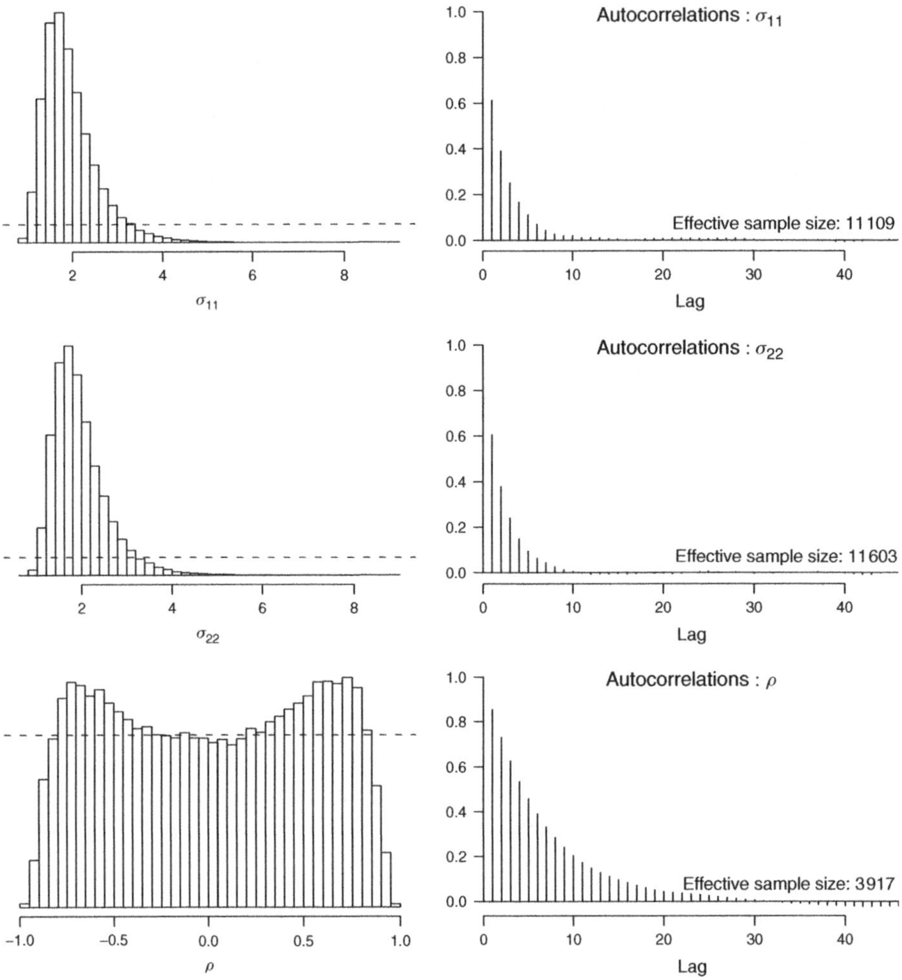

Figure 5.17 Marginal posterior densities and autocorrelation functions, Example 5.13. Left panels shows histograms summarizing 50 000 slice-samples; the dotted horizontal lines indicate the uniform prior for each of the three parameters. Autocorrelation functions are displayed in the panels on the right hand side of the graph.

and autocorrelation functions shown in the right panels. The slice-sampler recovers the bimodality in the posterior density for ρ. Note the two top panels on the left are posterior densities over the standard deviations σ_{11} and σ_{22} and concentrate posterior probability mass well away from their upper bound of $b = 15$. The autocorrelation functions suggest that the slice sampler is generating a reasonably efficient exploration of the posterior density. The 50 000 iterations generated here correspond to an effective sample size (see Section 4.4.1) of over 11 000 for the standard deviations, but just roughly 4000 for the correlation parameter ρ (the bimodality of the posterior density for ρ is partially to blame here).

6

Implementing Markov chain Monte Carlo

6.1 Software for Markov chain Monte Carlo

There are many software packages that implement MCMC algorithms for specific models. Examples include `MLwin` for multi-level models, `AMOS` for structural equation modeling, or `R` packages such as `MCMCpack` (Martin, Quinn and Park 2009), `bayesm` (Rossi and McCulloch 2008) and `pscl` (Jackman 2008b) that support Bayesian analysis of 'standard' problems (e.g. regression, logistic regression, factor analysis, item-response models). A large and growing list of freely-available, open-source `R` packages that implement Bayesian procedures appears as part of the Bayes 'task view' web page on CRAN (the Comprehensive R Archive Network), http://CRAN.R-project.org/.

The closest thing we have to a general purpose, MCMC-based environment for Bayesian modeling remains the `BUGS` program (Thomas, Spiegelhalter and Gilks 1992), the acronym standing for 'Bayesian inference Using Gibbs Sampling'. `BUGS` consists of (1) a parser, which inspects a user-supplied declaration of a statistical model – using a syntax that is very similar to `R` – and user-supplied data, deducing the DAG that corresponds to the model syntax; and (2) an expert system for deducing the form of the resulting conditional distributions, exploiting conjugacy where possible, but otherwise utilizing a suite of sampling algorithms, including the examples we discussed in Chapter 3. `BUGS` then implements the resulting MCMC algorithm, with the user also specifying the desired number of MCMC iterates, the nodes in the DAG to be 'monitored' over the iterations, and so on.

`BUGS` started life on `UNIX`, with a primitive command line interface (e.g. Spiegelhalter *et al.* 1996). `WinBUGS` wraps a GUI around the earlier versions of `BUGS`, but

Bayesian Analysis for the Social Sciences S. Jackman

also represents a considerable improvement in terms of the samplers available and hence the types of models and prior densities that can be deployed (Spiegelhalter *et al*. 2003). There is now an open source development project based on the BUGS core code known as OpenBUGS. There are also various interfaces for calling BUGS from R such as R2WinBUGS (Sturtz, Ligges and Gelman 2005), which greatly speed the work flow of specifying, running and debugging BUGS programs; while WinBUGS has a GUI, it requires a good many mouse clicks in numerous dialog boxes to actually run a model. An R package like R2WinBUGS means the user need not interact with WinBUGS at all, with the output of WinBUGS coming back into R as an object ready for further analysis.

JAGS (Plummer 2009a) is a BUGS 'look-alike'; the acronym stands for 'Just Another Gibbs Sampler'. JAGS does not have a GUI, and is an open-source project developed in C++. Note that OpenBugs, like BUGS and WinBUGS, is written in Component Pascal, a relatively obscure programming language, without compilers for OS/X (to the best of my knowledge) and so neither BUGS, WinBUGS nor OpenBUGS runs natively on the Macintosh platform at this time. Of course, one can run OpenBugs on the Macintosh by running a Windows virtual machine (e.g. via Parallels).

JAGS does run on the Macintosh, and indeed on any platform with C++ compilers. Aside from providing a native, BUGS-like, Bayesian inference program for the Macintosh (my preferred computing environment), JAGS also has a number of subtle improvements over BUGS: I draw attention to some of these in the applications that appear throughout the rest of the book. rjags (Plummer 2009b) is an interface from R to JAGS that is a key component of my Macintosh-based work flow, which I briefly detail below.

6.2 Assessing convergence and run-length

Recall the general results in Chapters 4 and 5: under a wide set of circumstances, MCMC algorithms initialized at arbitrary starting values will eventually generate samples from the desired posterior density of interest. Then, despite the fact that the samples from the posterior density are generally serially dependent, we can nonetheless rely on results such as the ergodic theorem (Proposition 4.7) to be confident that if we let the MCMC algorithm run 'long enough', we can generate simulation-consistent estimates of any feature of the posterior density. Two practical questions arise. First, has the sampler converged on the target, posterior density? Second, have we run the sampler long enough to recover estimates with sufficiently small levels of Monte Carlo error?

General propositions are hard to come by, as we noted in Section 4.3. Tierney (1997, 397) notes that 'universally useful, reliable [convergence] diagnostics do not exist, and cannot exist', given the problem-specific Markov chains generated by MCMC. Accordingly, a more 'empirical' approach is typically adopted, in which we inspect the output of the MCMC algorithm for non-stationarity and for assessments of the rate at which Monte Carlo error is diminishing in simulation length.

A large and growing literature deals with techniques for monitoring convergence of MCMC algorithms; Cowles and Carlin (1996) provide a comprehensive review of 13 diagnostics; other reviews include Brooks and Roberts (1998) and Robert and Casella (2004, Chapter 12). We will not review all of the proposals, focusing instead on some of the more widely used diagnostics. We defer examples to the applications that appear throughout the remainder of the book.

Graphical inspection of the output of an MCMC algorithm is critically important in assessing problems with convergence. In conjunction with formal, analytic diagnostics, graphs of the iterative history of a MCMC algorithms ('trace plots') help researchers identify some common problems with convergence. Non-stationarity is often obvious from looking at these kinds of graphs, as will the less troubling problem of slow mixing. In addition, multi-modal posterior distributions are sometimes obvious from inspecting a trace plot. Slow mixing and multi-modal posteriors are not fatal in and of themselves – the theoretical results guaranteeing convergence to the posterior distribution apply to a wide range of circumstances – but they do mean that the MCMC algorithm may have to be run for a very long period in order to reassure oneself that the algorithm is generating a valid characterization of the posterior density.

The other 'cheap' graphical diagnostics I rely on extensively are autocorrelation functions of the sampled $\theta^{(t)}$; the rate at which the autocorrelation functions decay provide a clear signal as to the efficiency of the algorithm in a given application. Indeed, extremely slow decay in an autocorrelation function is usually a clear sign that the sampled values will fail formal diagnostics of stationarity or fall short of run-length recommendations (see below). Closely related to the autocorrelation function is the computation of the 'effective sample size' which is based on an estimate of the spectral density of the MCMC output at frequency zero (see the discussion of simulation inefficiency in Section 4.4.1). The slower the decay in the autocorrelation function, the greater the integrated autocorrelation time (Definition 4.26) and the smaller the effective sample size. In my applied work I prefer not to base MCMC estimates on anything less than the equivalent of several thousand draws from an independence sampler: anything less tends to give reasonably noisy estimates of extreme quantiles of the marginal posterior density (see the discussion of the Raftery-Lewis run-length diagnostic, below), but sometimes we have to settle for less. An estimate of the effective sample size is provided by the `effectiveSize` function in the `R` package, `coda` (Plummer *et al*. 2008).

Geweke test of non-stationarity

A popular convergence diagnostic due to Geweke (1992) compares the average of the MCMC output for a given element of $\boldsymbol{\theta}$ across two stages of the MCMC run. If these averages are statistically different from one another, then this could be due to non-stationarity. In making the comparison of means with respect to a scalar estimand $g(\boldsymbol{\theta})$, we rely on the fact that the spectral density of the time series $\{g(\boldsymbol{\theta}^{(t)})\}$ can be used to estimate the asymptotic variance of an estimate of the average of the time series. In turn, this permits a comparison of two stages of the Markov chain (say 'early' with T_A iterations and 'late' based on the last T_B iterations), which yield estimates $\bar{g}(\boldsymbol{\theta})_A$ and $\bar{g}(\boldsymbol{\theta})_B$. The difference of these means divided by the asymptotic standard error of the difference tends to a standard normal distribution as $T \to \infty$ (holding T_A/T and T_B/T constant and $T_A + T_B < T$). Cowles and Carlin (1996, 866) discuss the strengths and weaknesses of this diagnostic. In particular, it is unclear how large T_A and T_B should be, relative to T, although Geweke suggested $T_A = .1T$ and $T_B = .5T$ and these are the defaults in the `geweke.diag` function in the `R` package `coda`.

Heidelberger-Welch test of non-stationarity

This diagnostic assesses both stationarity and the accuracy of the MCMC-based estimate of the mean of the posterior density of a scalar estimand. The underlying idea of the stationarity test is to use the Cramér-von-Mises statistic to successively test the null hypothesis that the sampled values come from a stationary distribution; see Heidelberger and Welch (1983) for details. The implementation in the `coda` package in `R` (the `heidel.diag` function) applies the stationarity test first to the entire chain, and then after discarding the first 10 %, then 20 % and so on, up until the null of stationarity is accepted, or 50 % of the sample is discarded. If at this point the null of stationarity has not been accepted, then the test reports that a longer MCMC run is needed; alternatively, if the null is accepted, then the implementation in `R` reports how many iterations can be deemed to be samples from a stationary distribution (i.e. the posterior density).

The 'half width' test of simulation accuracy computes the width of a 95 % confidence interval for the mean of the MCMC output (using the spectral density when estimating the standard error of the mean). The test statistic is the ratio of half the width of the 95 % confidence interval to the estimate of the mean. If this is less than a prescribed value (the default in `heidel.diag` is .1) then the test if deemed to have passed, in the sense that the Monte Carlo error of the MCMC-based estimate of the mean of the posterior density of the scalar estimand is relatively small.

Raftery-Lewis run-length estimate

Raftery and Lewis (1992b) proposed a method of estimating the number of iterations that are required to estimate a given quantile q of a marginal posterior density to a desired level of accuracy. The diagnostic is implemented as `raftery.diag` in the `coda` package in `R`. The defaults are to assess the accuracy of the MCMC-based estimate of the $q = .025$ quantile with a 95 % bound on the estimate no greater than .005 (in quantile terms). The procedure examines the performance of the Markov chain with respect to the target quantile q, forming the binary sequence $\{Z^{(t)}\}$, with $Z_t = \mathcal{I}(\theta^{(t)} \leq u_q)$, where $\mathcal{I}$ is a binary indicator function and u_q is the q-th quantile of $\{\theta^{(t)}\}$. The run-length estimate is based on an estimate of how long a sample would be required so as to make the sequence $\{Z^{(t)}\}$ behave as if it were a Markov chain. $\{Z^{(t)}\}$ can not be a Markov chain, so some thinning is required to make it approximate a Markov chain and a result, the estimates reported by the Raftery-Lewis procedure can be conservative; cf Guihenneuc-Jouyaux and Robert (1998) and Robert and Casella (2004, §12.4.2). With the default values in `raftery.diag`, at least 3746 samples are required to implement the test. Useful summaries also appear in Raftery and Lewis (1992a) and Raftery and Lewis (1996).

For a slow-mixing sampler, estimating extreme quantiles such as $q = .025$ is quite a struggle, and the estimated running time reported by the Raftery-Lewis diagnostic can be surprisingly long; in these cases we often see the sampler slowly meandering from one tail of the marginal posterior density to the other, such that while the estimate of a quantity such as the mean of the marginal posterior is not terribly bad, the performance of the sampler with respect to extreme quantiles is horrible. It is not surprising to see the Raftery-Lewis diagnostic based on a 'short', 5000 iteration run of a MCMC algorithm report that 50 000 iterations are required; then, the 50 000 iteration run reveals that 300 000 iterations might be required. Such is the plight of the researcher stuck with a slow-mixing chain: in moving from 5000 iterations to 50 000 iterations we may be learning more about the woeful performance of the algorithm than about the posterior density! Put differently, not only did the 5000 iteration run provide a poor characterization of the marginal posterior density, but it also underestimated the autocorrelation of the Markov chain.

Multiple chains

Just as practitioners are routinely advised to try different starting values with maximization routines, Gelman and Rubin (1992) recommend starting several MCMC algorithms with overdispersed starting points, where here ‘overdispersed’ means that the variance among the different starting points should be greater than that thought to exist in the target distribution; at the same time, the starting values should not be ‘wildly inaccurate’ (Gelman and Rubin 1992, 458–9). This proposal can be especially helpful working with a posterior distribution reasonably thought to be multi-modal, when there is a risk that the MCMC algorithm will get ‘stuck’ at one of the modes and never discover the presence of the other modes in the posterior, thereby giving rise to a very misleading characterization of the posterior (cf Example 5.9). In practice this ‘multiple chains’ proposal is best done by running several MCMC algorithms samplers in parallel.

Given output from multiple MCMC algorithms, Gelman and Rubin propose a comparison of the between-chain variation and the within-chain variation for each scalar component of $\boldsymbol{\theta}$, ψ. They define

$$\widehat{\text{var}}^{+}(\psi|y) = \frac{T-1}{T}W + \frac{1}{T}B,$$

where W is the (average) within-chain variance and B is the between-chain variance, for some scalar of interest ψ, conditional on observed data y. As $T \rightarrow \infty$ (i.e. the MCMC algorithm is run for longer periods), the contribution of the between-chain variation gets smaller, since it picks up weight $1/T$ in contributing to $\widehat{\text{var}}^{+}(\psi|y)$. Simultaneously, the within-chain variance increasingly dominates this term with additional iterations. Thus this estimate of the marginal posterior variance is an *overestimate* of the true marginal posterior variance for any finite length chain (hence the ‘+’ superscript). Accordingly, Gelman and Rubin propose the following statistic as a convergence diagnostic:

$$\sqrt{\hat{R}} = \sqrt{\frac{\widehat{\text{var}}^{+}(\psi|y)}{W}}.$$

This quantity declines to 1 as $T \rightarrow \infty$, and can be interpreted as the ‘potential scale reduction’ that might result from continuing to run the MCMC algorithm. Given streams of output from multiple MCMC samplers, this statistic can be calculated after a pre-specified number of iterations; Gelman *et al*. (2004, 332) suggest that values of $\sqrt{\hat{R}}$ below 1.2 are ‘acceptable’, but any determination of convergence will vary from data set to data set. This is the original version of the Gelman and Rubin convergence diagnostic; see the modification proposed by Brooks and Gelman (1998) and the `gelman.diag` function in the `R` package, `coda`.

There is an inevitable tradeoff between one long run of a MCMC algorithm versus several shorter runs. Consequently, a consensus position lies somewhere between the ‘one long chain’ and ‘shorter multiple chains’ positions (e.g. Cowles and Carlin 1996, 903). Of course, ‘more is better’ with MCMC, both in terms of the number of sequences run, and the length of each sequence, given the theoretical results that guarantee that

MCMC algorithms will eventually reach the target posterior distribution in most settings. Convergence may be slow in a specific context, and so we might prefer devoting computer resources to one long MCMC run versus several shorter runs. I almost always find myself preferring one longer run to several longer runs, since the practical issue I usually encounter is not convergence *per se*, but the fact that the MCMC algorithm is generating a relatively inefficient exploration of the posterior density. Given that computational power is cheap, and getting cheaper, the 'more is better' advice is increasingly easy to implement, as some of the examples below make clear; in fact, with multiple processor desktop machines now *de rigueur* and becoming easier to use, the 'embarassingly parallel' nature of MCMC simulation can be exploited to make the 'multiple chains' proposal less costly.

But above all, there is no substitute for a clear understanding of the model, priors and data being passed to a MCMC algorithm, and how they might possibly impede convergence or lead to an inefficient exploration of the resulting posterior density.

6.3 Working with BUGS/JAGS from R

To demonstrate the way that R and BUGS/JAGS work together, consider the regression model presented in Example 2.15, analyzing the data on suspected vote fraud in Pennsylvania senate elections. The data for this example are available in the data frame `absentee`, part of the R package `pscl`. We are interested in the regression of y_i, the Democratic lead (expressed as share of the two party vote, in percentage points) among absentee ballots on x_i, the Democratic lead of the two-party vote among votes cast on voting machines as the predictor.

First, some R code that gets the data in shape for BUGS/JAGS:

R code

```
1 library(pscl)
2 data(absentee)
3 attach(absentee)
4 ## create variables for regression analysis
5 y <- (absdem - absrep)/(absdem + absrep)*100
6 x <- (machdem - machrep)/(machdem + machrep)*100
7
8 ## environment for passing to JAGS
9 forJags <- list(y=y[1:21],x=x[1:21],n=21,xstar=x[22])
10
11 ## initial values, one chain
12 inits <- list(list(beta=c(0,0),sigma=5,
13                    .RNG.seed=1234,
14                    .RNG.name="base::Mersenne-Twister"))
```

Lines 1 and 2 load the data from the `pscl` library; lines 5 and 6 create the variables we need for the analysis. Line 9 creates a `list` named `forJags`, which we will then pass to JAGS (or equivalently, BUGS). The suspect data point is the last data point, observation number 22, and so we do not include this observation in the vectors `y` and `x` passed to JAGS. We will generate samples from the posterior predictive density for the suspect data point, and so we pass the predictor for that observation (x_{22}) to JAGS as the scalar xstar. We also pass the number of observations in the regression analysis, 21, as the scalar n.

Finally, line 12 of the code fragment generates some initial values for the MCMC algorithm. Initial values can be supplied for any unobserved or partially observed node in the model, including parameters or unobserved quantities that are functions of parameters, such as missing data or predicted values, and so on. The form of the initial values in R is a list of lists, with each component list corresponding to a separate Markov chain: here we will run just one chain, and so there is only one list in the list of lists that defines inits. Initial values are not required by either BUGS or JAGS, and if BUGS/JAGS is invoked without user-supplied initial values the Markov chains are initialized by sampling from the priors and/or predictive densities. Note that when vague priors are specified this can sometimes result in initializations of the chains in regions of extremely low posterior probability, and can even generate numerical instabilities. Accordingly, for complicated models it is almost always a good idea to pass user-supplied initial values to BUGS/JAGS.

The last two elements of the inits list specify a seed for the random number generator (RNG) to be used by JAGS, and the particular RNG (here we use the popular Mersenne Twister). Specifying the seed means that the results can be replicated, since the (pseudo) random number generators will generate the same sequence. Without explicitly seeding the RNGs, results will vary from run to run, picking up a different amount of Monte Carlo error; by explicitly setting the seed we guarantee that we get the same Monte Carlo error from a MCMC run of a given length, since the RNG is generating the same set of random digits.

In this demonstration we will specify vague normal priors on the intercept β_1 and slope β_2 in the regression model and a non-conjugate prior on σ, the standard deviation of the regression's disturbances. We will also generate samples from the posterior predictive distribution for y for the suspect data point.

The following BUGS/JAGS program specifies the regression model, the priors and the posterior predictive density for y:

JAGS code

```
1 model {
2         for(i in 1:n){    ## regression model for y
3               mu[i] <- beta[1] + beta[2]*x[i]
4               y[i] ~ dnorm(mu[i],tau)
5         }
6         ## priors
7         beta[1] ~ dnorm(0,.0001)
8         beta[2] ~ dnorm(0,.0001)
9         sigma ~ dunif(0,100)
10        tau <- pow(sigma,-2)
11
12        ## out of sample predicton for suspect case
13        mustar <- beta[1] + beta[2]*xstar
14        ystar ~ dnorm(mustar,tau)
15 }
```

Every BUGS/JAGS program begins with a model statement followed by an opening curly brace and ends with an enclosing curly brace (line 15). In this case we specify a loop over the n observations (lines 2 through 5) using a syntax that is very similar to

R. Line 3 generates a node `mu[i]`, the conditional mean of the regression model for y; `beta[1]` is the intercept and `beta[2]` is the slope in the regression.

Line 4 is the actual stochastic specification of the regression, making use of the 'twiddle' or 'tilde' character '~' which denotes a stochastic relation in BUGS/JAGS. Generically, stochastic relations in BUGS/JAGS are of the form `node ~ ddens(arguments)` where `node` is a stochastic quantity and `ddens` is the name of a function implementing one of the densities supported by BUGS/JAGS. These functions all start with the letter `d`, as in `dnorm` for the normal density used here, `dbin` for a binomial mass function, `dbern` for a Bernoulli mass function, `dgamma` for a Gamma density, `dpois` for a Poisson mass function, and so on; see the documentation for BUGS and/or JAGS for a complete list. Note that for both BUGS and JAGS the second argument to the `dnorm` function is a precision, not a variance: line 4 corresponds to the specification $y_i \sim N(\mu_i, \sigma^2)$, where $\sigma^2 = 1/\tau$.

Lines 7 through 10 specify the priors for this model: vague $N(0, 1000)$ priors for `beta[1]` and `beta[2]`, and a non-conjugate, prior on the standard deviation, $\sigma \sim \text{Unif}(0, 100)$. The model for `y[i]` in line 4 refers to the precision parameter `tau`; at line 10 we convert the prior over `sigma` into a prior over the 'operational' parameter `tau`.

It is worth noting that this specification treats the regression coefficients $\boldsymbol{\beta} = (\beta_1, \beta_2)'$ as *a priori* independent of σ, which is a departure from the conventional, conjugate analysis of the normal, linear regression model we encountered in section 2.5, where we specify $p(\boldsymbol{\beta}, \sigma^2) = p(\boldsymbol{\beta}|\sigma^2)p(\sigma^2)$. This alternative specification of the prior with $\boldsymbol{\beta} \sim N$ and independently $\sigma \sim$ inverse-Gamma corresponds to the conditionally conjugate setup we considered in Proposition 2.8; the extension to the regression case was considered in Problem 2.21.

Lines 13 and 14 generate samples from the posterior predictive density for the suspect observation, y_{22}. The quantity `xstar` is x_{22}, but the quantity `ystar` is not given in the data. BUGS/JAGS treats `ystar` as missing at random, and will generate a sample from its predictive density (line 14), given β_1, β_2 and τ. Over the course of many iterations, BUGS/JAGS will sample β_1, β_2 and τ from its posterior density, meaning that the samples from the density in line 14 are draws from the posterior predictive density for y_{22}. This is a general feature of BUGS/JAGS: if a stochastic node in a DAG has no children, then the software generates samples from its predictive density via 'forward sampling', equivalent to treating the node as data that is missing at random.

At compile time, BUGS and JAGS must be able to resolve all references to the variables given in the model syntax. A useful way to think about how the software works is to recall the DAG representation referred to earlier. For instance, in the example given above, the `y[i]` nodes are children of `mu[i]` and `tau`: `y` is supplied as data to the program, and so the compiler will then look for where the parents of the `y[i]` are defined. `mu[i]` is a deterministic function of the `beta` nodes and `x[i]`. `x[i]` is supplied as data, and so the references to `beta` need to be resolved; `beta[1]` and `beta[2]` appear on the left-hand side of the stochastic relations given in lines 7 and 8, with the argument in the respective `dnorm` functions given as fixed constants. Similarly, the reference to `tau` in line 4 can also be resolved up to a known set of parents: line 10 expresses `tau` as a child of `sigma`, and line 9 provides a specification of `sigma` in terms of known quantities (the bounds on the uniform density given in the `dunif` function). For the unobserved node `ystar`, the parent nodes are `mustar` and `tau`; as we just established, `tau` is well-defined, and `mustar` is a deterministic function of the `beta` nodes and the quantity `xstar`, the latter supplied as data.

At compile time, having seen the data supplied in the `forJags` list in R, BUGS and/or JAGS will deduce that `beta` and `sigma` are stochastic nodes, with conditional distributions given by applying Proposition 5.2. Sampling from these conditional distributions in turn generates a Markov chain on the parameter space for $\boldsymbol{\theta} = (\beta_1, \beta_2, \sigma)'$. The theory outlined in Chapter 4 establishes that this Markov chain has the joint posterior density $p(\boldsymbol{\theta}|\mathbf{y}, \mathbf{x})$ as its stationary distribution.

It is also important to note that BUGS and JAGS are *declarative* languages, not *procedural* languages, such as R, C or Fortran. That is, the BUGS/JAGS syntax given above is merely a model declaration, and does not define a set of computational steps to be run sequentially. At compile time the model declaration syntax is turned into a set of instructions that would correspond to a program in the more conventional sense, but this is never seen by the user. Because BUGS and JAGS are declarative, the precise order in which statements are given in the model declaration is not consequential: all that matters is that the compiler can resolve the references to the named quantities in the model syntax as it tries to build the DAG representation of the model.

Calling JAGS from R We can call JAGS from R using functions provided in the `rjags` package, as demonstrated in the following code fragment:

R code

```
1 require(rjags)
2 foo <- jags.model(file="regression.bug",
3                   data=forJags,
4                   inits=inits)
5 out <- coda.samples(model=foo,
6                     variable.names=c("beta","sigma"),
7                     n.iter=50000,thin=5)
```

Line 1 loads the `rjags` package if it is not already loaded. Lines 2-4 invoke the `jags.model` command, sending the file `regression.bug` to JAGS for parsing and compilation; the file `regression.bug` contains the JAGS model syntax given above. The `data` argument (line 3) directs JAGS to use the data in the `forJags` object created earlier. Line 4 directs JAGS to initialize the MCMC algorithm with the initial values created in the earlier R code block. The object `foo` is of class `jags.model` and we can pass it to other functions in the `rjags` package.

The call to `jags.model` will also generate an initial 1000 iterations of the MCMC algorithm, which is sometimes sufficient for 'burn-in', the iterations where the MCMC algorithm is moving away from the initial values towards the posterior density. In this case the 1000 iterations are more than sufficient for this simple regression example, where convergence to the target posterior density is almost instantaneous. If a longer burn-in period is desired, we can request this by either changing the number of initial iterations generated by `jags.model` with the argument `n.adapt`; the name of this argument reflects the fact that some of the samplers used by BUGS/JAGS have an adaptive phase as they learn about the conditional density from which they are trying to sample (e.g. the slice sampler). Alternatively, the user can always request iterations beyond this adaptive phase but before MCMC iterates are monitored and saved to memory, with the command `update`, which takes an object of class `jags.model` and the number of iterations as arguments: e.g. `update(foo,5000)` would generate (but not save) 5000 iterations of the MCMC algorithm associated with the object `foo`.

Line 5 is a call to `coda.samples`, which also takes an object of class `jags.model` as an argument. In this case we are requesting 50 000 iterations of the MCMC algorithm, with every 5th iteration saved for inference; on line 6 we specify the variables we will track over the 50 000 MCMC iterates in the argument `variable.names`.

Inspecting MCMC output with coda The object returned by the `coda.samples` function is of class `mcmc.list`, part of the `coda` library (Plummer *et al*. 2008), a suite of functions and classes for analyzing the output of programs such as `BUGS` and `JAGS`. For the one-chain example considered here, the object `out` is a list with just one component, an array of 10 000 rows and 3 columns tagged to be an object of class `mcmc`, with rows corresponding to each saved MCMC iterate and the columns corresponding to each monitored quantity; multiple chains produce `mcmc.list` objects with multiple components. The result of calling `summary` on the object `out` appears below:

R output

```

Iterations = 1005:51000
Thinning interval = 5
Number of chains = 1
Sample size per chain = 10000

1. Empirical mean and standard deviation for each variable,
   plus standard error of the mean:

            Mean     SD Naive SE Time-series SE
beta[1] -5.0686 5.8350 0.058350       0.062940
beta[2]  0.8618 0.1186 0.001186       0.001305
sigma   15.7706 2.9612 0.029612       0.023823

2. Quantiles for each variable:

           2.5%     25%    50%    75%  97.5%
beta[1] -16.761 -8.8334 -5.143 -1.261  6.553
beta[2]   0.621  0.7852  0.863  0.939  1.093
sigma    11.248 13.6961 15.342 17.337 22.895

```

Objects of class `mcmc.list` or `mcmc` also have `plot` methods defined in `coda`, and can be passed to any of the numerous MCMC convergence diagnostics implemented in `coda`. For instance, the Raftery-Lewis diagnostic (Raftery and Lewis 1992a) is invoked with the `coda` function `raftery.diag`, producing the output shown below for the regression example considered here:

R output

```
[[1]]

Quantile (q) = 0.025
Accuracy (r) = +/- 0.005
Probability (s) = 0.95

```

```
           Burn-in  Total Lower bound  Dependence
           (M)      (N)   (Nmin)       factor (I)
  beta[1]  10       19810 3746         5.29
  beta[2]  10       19325 3746         5.16
  sigma    10       19325 3746         5.16
```

We can also manipulate the MCMC output directly, by extracting the array(s) of MCMC output. Many of the graphs and tabular summaries of posterior densities reported in this book are produced after first extracting the MCMC output from an object of class `mcmc` or `mcmc.list`.

In this section we have looked at a Macintosh-based work flow, linking `R` and `JAGS`. The Windows-based work flow looks very similar, with the `R` package `R2WinBUGS` managing the communication between `R` and `WinBUGS`.

6.4 Tricks of the trade

We now briefly consider some standard 'tricks of the trade' for improving the performance of MCMC algorithms, or for managing the output produced by a MCMC algorithm. These 'tricks' share a common goal: to improve the computational and/or storage burdens imposed by MCMC-based inference.

Recall that in general, MCMC algorithms do not produce a sequence of independent samples from the posterior density of $\boldsymbol{\theta}$, $p(\boldsymbol{\theta}|\text{data})$. Many iterations of the MCMC algorithm may be required to generate Monte Carlo estimates of interesting features of the posterior density up to the level of precision demanded by the researcher. It is not unusual to find that the Markov chain produced by `BUGS/JAGS` is slow mixing: i.e. the trace plots for some or all parameters reveal very long cycles, with slowly decaying autocorrelation spectra and formal diagnostics indicating that an extremely large number of iterations may be required (e.g. the Raftery-Lewis diagnostic).

Waiting for the computer to produce these iterations can be frustrating; simply storing and analyzing the potentially large number of iterations can start to tax the computing resources typically available to social scientists. But there are a few simple 'tricks' that may alleviate these problems.

6.4.1 Thinning

Suppose we implement a MCMC algorithm and find it to be slow-mixing. Convergence diagnostics suggest an extremely long run of the MCMC algorithm is required to obtain precise estimates of interesting features of the posterior density, perhaps tens of thousands or even hundreds of thousands of iterations. In addition, the model has literally thousands of parameters that the researcher cares about, as can arise in many hierarchical models (see Chapter 7) or latent variable models (Chapter 9). Suppose we have 1000 parameters of interest, and convergence diagnostics indicating that we need to run 250 000 iterations. Using something like the `rjags` package (Plummer 2009) in `R`, the object containing this output from the sampler would be a 250 000 by 1000 matrix, an array of some 250 million numbers. `R` would consume about 1.8 GB of memory in merely storing this object, which is large enough to tax the desktop computing resources typically available to social scientists.

Of course, not all of this output need be stored in memory. One might consider writing some of the output to disk as the MCMC algorithm runs, mitigating the need to store massive arrays of MCMC output in RAM; this strategy is an option in the `ideal` function in the `R` package `pscl` (Jackman 2008b). Small amounts of the output can then be read into RAM for inspection by the analyst.

Thinning is another approach to the problems posed by massive amounts of MCMC output. The idea is extremely simple. If the reason we need a massive number of MCMC iterates is that the MCMC algorithm is slow-mixing, then we probably don't lose a lot by *not* storing every iterate. If the algorithm is slow-mixing, then the sampled values $\boldsymbol{\theta}^{(t)}$ generated at iteration t will look a lot like the $\boldsymbol{\theta}^{(t+1)}$ recovered at iteration $t+1$, and $t+2$ and so on. The contribution of each additional MCMC iterate to the quality of inferences about the posterior density is small: indeed, this is precisely why we need so many iterates from a slow-mixing MCMC algorithm. Hence, thinning simply retains every n-th MCMC iterate – where n is known as the 'thinning interval' – such that the 'thinned' series of MCMC iterates displays better mixing than the original series. In short, the original slow-mixing tour of the parameter space is being looked at every n-th iteration, with the shorter, thinned series providing almost as much information about the posterior density as did the original series.

Recall that an independence sampler is the 'gold standard' of Monte Carlo simulation. Indeed, slow-mixing chains need to be run for a long time in order for them to mimic the exploration of a density we would get from an independence sampler: i.e. an independence sampler would generate relatively many mean crossings, rapid traverses from tail to tail, relatively fast convergence of Monte Carlo estimates of extreme quantiles, and so on. Thinning is designed to make a slow-mixing chain approximate an independence sampler, by picking off MCMC iterates far enough apart such that they display low autocorrelations.

Of course, there is a price to be paid with thinning, consistent with one of the key maxims of data analysis: throwing away data is never a good thing. It needs to be stressed that there is nothing in the theory of Markov chain Monte Carlo that says one ought to thin the output of a MCMC algorithm. The considerations here are almost entirely practical. Is storing and summarizing potentially millions of iterations too expensive (in terms of disk, memory and time)? Indeed, there are risks to thinning, if done too aggressively, as we demonstrate in the following example.

■ Example 6.1

Thinning a highly autocorrelated series (Example 4.2, continued). Consider the stationary AR(1) process

$$\theta^{(t)}|\theta^{(t-1)} \sim N(\rho\theta^{(t-1)}, 1-\rho^2), \ |\rho| < 1.$$

Marginally, $E(\theta) = 0$ and $V(\theta) = \rho^2 V(\theta) + 1 - \rho^2 = 1$ and the 2.5 percentile of θ is -1.96. We set $\rho = .98$ and simulate 1 000 000 observations from this process. Table 6.1 shows the results of thinning the resulting series by successively wider intervals for three estimands: the mean $E(\theta) \equiv 0$, the standard deviation of θ, equal to 1.0, and the 2.5 percentile. With no thinning, the 1 000 000 iterations provide reasonably good

Table 6.1 Effect of various thinning intervals on inferences for $\theta \sim N(0, 1)$. One million samples are drawn from the $N(0, 1)$ density for θ, but (by design) are highly autocorrelated ($\rho = .98$). Entries in the column labelled AR(1) are the empirical AR(1) coefficient for the thinned sub-series.

Thinning Interval	Nominal n	Mean	SD	2.5 %	AR(1)
1	1 000 000	0.01	1.00	−1.94	0.98
5	200 000	0.01	1.00	−1.94	0.90
10	100 000	0.01	1.00	−1.94	0.81
25	40 000	0.01	0.99	−1.93	0.60
50	20 000	0.01	0.99	−1.93	0.36
100	10 000	0.01	1.00	−1.93	0.12
250	4000	−0.0002	0.99	−1.90	0.02
500	2000	0.02	0.98	−1.90	−0.02
1000	1000	0.04	0.98	−1.86	−0.02
2500	400	−0.03	0.97	−1.85	0.03
5000	200	0.05	0.97	−1.83	−0.13
10 000	100	0.14	0.99	−1.82	0.06

estimates for all three quantities of interest; the fact that Monte Carlo analysis of a highly-autocorrelated but stationary series can provide good estimates is not in dispute and indeed is precisely the point of Chapter 4 and the ergodic theorem (Proposition 4.7). Identical estimates (to two decimal places) result when the series is thinned by 5 and 10 iterations, reducing the thinned series to 200 000 and 100 000 iterations respectively, with AR(1) coefficients of .90 and .81 respectively. More aggressive thinning sees the autocorrelation of the thinned series fall; n.b., $.98^{100} \approx .13$, $.98^{200} \approx .018$ and $.98^{500} < 10^{-4}$. Thinning the original series of 10^6 iterations beyond a factor of 200 or so does not significantly alter the level of autocorrelation, but reduces the number of the samples available for analysis, resulting in poorer estimates of the quantities of interest. That is, we appear to hit a 'sweet spot' in terms of accuracy and storage demands when the series is thinned by an amount just sufficient to reduce it to approximately an independence sampler, in this case with a thinning interval around about 200 iterations. Note, however the degradation of the estimate of the 2.5 percentile.

To repeat the points here: model fitting with MCMC often involves some short runs, turning to MCMC diagnostics to assess the level of dependence in the resulting Markov chain. We can then determine what a 'safe' run length might be, such that after thinning we have a reasonably large set of MCMC iterates that are 'approximately serially independent.' High levels of autocorrelation in a MCMC algorithm are not fatal in and of themselves, but they will indicate that a very long run of the sampler may be required. Thinning is not a strategy for avoiding these long runs, but it is a strategy for dealing with the otherwise overwhelming amount of MCMC output. Blocking (Section 6.4.2) and/or re-parameterizations (Section 6.4.3) offer ways to perhaps avoid these extremely long runs of the sampler in the first place, and are strategies to be preferred over thinning the output of a very inefficient sampler.

6.4.2 Blocking

Slow mixing MCMC algorithms usually stem from the fact that components of the parameter vector $\boldsymbol{\theta}$ are highly correlated with one another. An MCMC algorithm like the Gibbs sampler will usually generate a very slow random tour of the parameter space in this case. To see why, suppose we have $\boldsymbol{\theta} = (\theta_1, \theta_2)'$ with a high degree of correlation between θ_1 and θ_2, such that $\theta_1 \approx m(\theta_2)$, where m is a linear transformation. Then the sampled value $\theta_1^{(t)} \sim g_1(\theta_1|\theta_2^{(t-1)})$ will lie close to $\theta_2^{(t-1)}$, as will $\theta_2^{(t)} \sim g_2(\theta_2|\theta_1^{(t)})$. As a consequence, $\boldsymbol{\theta}^{(t)} \approx \boldsymbol{\theta}^{(t-1)}$ and the resulting sequence $\{\boldsymbol{\theta}^{(t)}\}$ is a slow-mixing Markov chain.

■ **Example 6.2**

Slow-mixing MCMC algorithm; highly-correlated components of the parameter vector. Suppose

$$\boldsymbol{\theta} = \begin{bmatrix} \theta_1 \\ \theta_2 \end{bmatrix} \sim N(\boldsymbol{\mu}, \boldsymbol{\Sigma}),\ \boldsymbol{\mu} = \begin{bmatrix} \mu_1 \\ \mu_2 \end{bmatrix},\ \boldsymbol{\Sigma} = \begin{bmatrix} \sigma_{11} & \sigma_{12} \\ \sigma_{12} & \sigma_{22} \end{bmatrix},$$

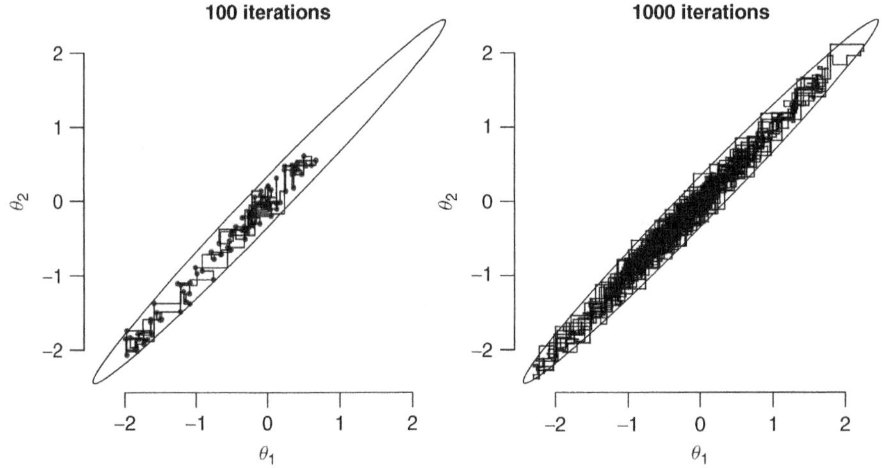

Figure 6.1 Progress of Gibbs sampler with highly correlated components. The target density is a bivariate normal density centered on the origin, with variances one and covariance 0.98 (with unit variances, the covariance is also a correlation). The ellipse (corresponding to a classical 95 % joint confidence region for $\boldsymbol{\mu}$) shows the shape of the target density. The Gibbs sampler is initialized at the origin. The trajectory of the Gibbs sampler is shown with plotted points, and each iteration's move in each dimension is shown as horizontal and vertical lines. The two-component Gibbs sampler makes slow progress through the parameter space because it is exploring the thin, elliptical density with an alternating sequence of small, random, horizontal and vertical moves (i.e. the samples from the two component conditional distributions of the Gibbs sampler). The left panel demonstrates that after 100 iterations the Gibbs sampler is yet to visit the right tail of the target joint density.

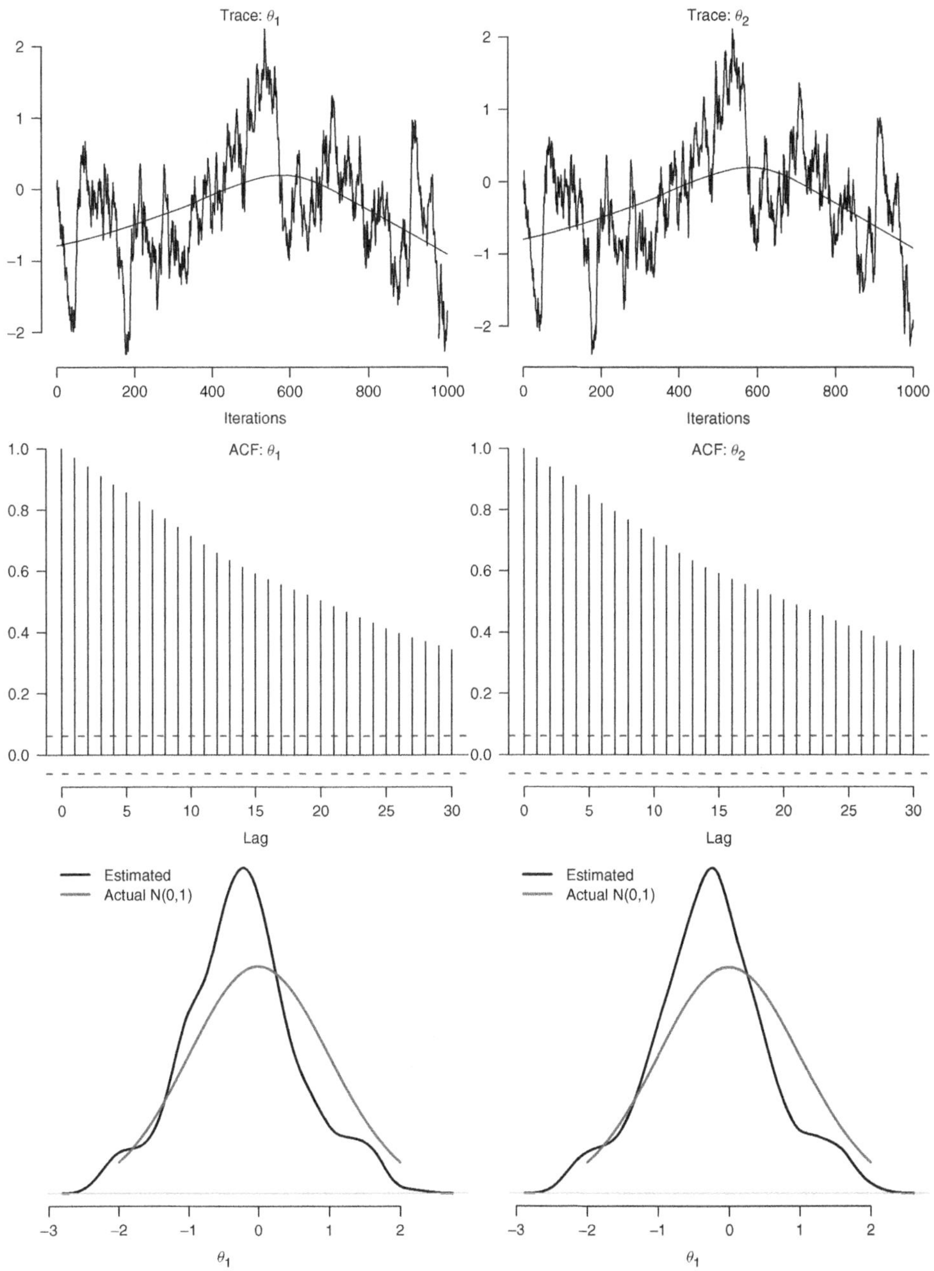

Figure 6.2 Graphical diagnostics, Gibbs sampler with highly correlated components. Top panels show trace plots over 1000 iterations; autocorrelation spectra appear in the middle row of panels; estimated densities based on the 1000 saved iterations appear in the lower panels, with the known $N(0, 1)$ marginal densities for θ_1 and θ_2 overlaid.

with the elements of $\boldsymbol{\mu}$ and $\boldsymbol{\Sigma}$ known, and we use a Gibbs sampler to generate samples from this density. Sampling from a bivariate normal density is no great technical challenge, but suppose we use a two-component Gibbs sampler; i.e. we sample θ_1 conditional on θ_2 and vice-versa. This is trivial to do, in that if the joint density of θ_1 and θ_2 is bivariate normal, then the conditionals of each component given the other are both univariate normals; see Proposition B.2. In fact, iteration t of the Gibbs sampler is

1. sample $\theta_1^{(t)}$ from $N(\mu_{1|2}, \sigma_{1|2})$ where $\mu_{1|2} = \mu_1 + \sigma_{12}(\theta_2^{(t-1)} - \mu_2)/\sigma_{22}$ and $\sigma_{1|2} = \sigma_{11} - \sigma_{12}^2/\sigma_{22}$.
2. sample $\theta_2^{(t)}$ from $N(\mu_{2|1}, \sigma_{2|1})$ where $\mu_{2|1} = \mu_2 + \sigma_{12}(\theta_1^{(t)} - \mu_1)/\sigma_{11}$ and $\sigma_{2|1} = \sigma_{22} - \sigma_{12}^2/\sigma_{11}$.

Take a simple case, where $\mu_1 = \mu_2 = 0$ and $\sigma_{11} = \sigma_{22} = 1$ and so $\sigma_{12} \in [-1, 1]$ is a correlation. Consider the transition from iteration t to $t+1$. Given $\theta_2^{(t)}$, $E(\theta_1^{(t+1)}|\theta_2^{(t)}) = \sigma_{12}^2\theta_2^{(t)}$. Then $E(\theta_2^{(t+1)}|\theta_2^{(t)}) = E(\sigma_{12}^2\theta_1^{(t+1)}) = \sigma_{12}^4\theta_2^{(t)}$. Note also that the variances of the conditional distributions $\sigma_{1|2}$ and $\sigma_{2|1}$ tend to zero as $|\sigma_{12}| \rightarrow 1.0$. The consequence is that as $|\sigma_{12}| \rightarrow 1.0$, the Gibbs sampler will take ever smaller steps in the parameter space, iteration to iteration.

Figure 6.1 demonstrates the slow progress of the two-component Gibbs sampler for this problem, for which $\sigma_{12} = .98$. The Gibbs sampler is initialized at the origin, and each plotted point corresponds to $\boldsymbol{\theta}^{(t)} = (\theta_1^{(t)}, \theta_2^{(t)})'$ produced by iteration t of the sampler. The left panel of Figure 6.1 shows the history of the first 100 iterations of the sampler; 1000 iterations are shown in the right panel. The moves from $\theta_1^{(t-1)}$ to $\theta_1^{(t)}$ are shown with horizontal lines; moves from $\theta_2^{(t-1)}$ to $\theta_2^{(t)}$ are shown with vertical lines. The figure vividly demonstrates the difficulty of exploring a highly elliptical density with a sequence of small, random horizontal and vertical moves (generated by the two stages of each iteration of the Gibbs sampler).

Figure 6.2 presents graphical diagnostics for this slow-mixing, two-component Gibbs sampler. The trace plots (top panels) show the slow progress of the sampler in traversing the marginal densities of θ_1 and θ_2, even over a reasonably long run of 1000 iterations: the sampler clearly does not spend enough time exploring the upper tail of either marginal density (see also the lower panels of Figure 6.2). The autocorrelation functions (middle panels) also indicate slow mixing, with the AR(1) parameters for θ_1 and θ_2 .964 and .960, and the auto correlation functions decaying very slowly.

One strategy for dealing with this problem is to form 'blocks' of parameters that are correlated, and to treat the 'block' as a component in the MCMC updating scheme. That is, suppose we have a parameter vector $\boldsymbol{\theta} = (\theta_1, \theta_2, \theta_3)'$ with θ_1 and θ_2 exhibiting a high degree of posterior correlation. A naïve implementation of the Gibbs sampler will have three steps per iteration: at iteration t

1. sample $\theta_1^{(t)}$ from $g_1(\theta_1|\theta_2^{(t-1)}, \theta_3^{(t-1)}, \text{data})$.
2. sample $\theta_2^{(t)}$ from $g_2(\theta_2|\theta_1^{(t)}, \theta_3^{(t-1)}, \text{data})$.
3. sample $\theta_3^{(t)}$ from $g_3(\theta_3|\theta_1^{(t)}, \theta_2^{(t)}, \text{data})$.

Since θ_1 and θ_2 are highly correlated, this algorithm will tend to display slow mixing of the sort shown in Example 6.2. But if it is possible to sample from the conditional distribution of θ_1 *and* θ_2 considered as a block, then tremendous computational savings may be possible. That is, we would replace the sampling scheme described above with the following: at iteration t

1. sample $(\theta_1^{(t)}, \theta_2^{(t)})'$ from $g_{12}(\theta_1, \theta_2 | \theta_3^{(t-1)}, \text{data})$.
2. sample $\theta_3^{(t)}$ from $g_3(\theta_3 | \theta_1^{(t)}, \theta_2^{(t)}, \text{data})$.

In general, one should always 'bundle' or 'block' parameters if it is possible to sample directly from the conditional distribution of the block, or if a sample can be generated reasonably efficiently via one of the sampling algorithms discussed in Section 3.4. Indeed, the limiting version of this argument would see us bundle all elements of $\boldsymbol{\theta}$ into one block and sample from the joint posterior density directly; of course, if this was feasible we would do it (at which point we would be merely using Monte Carlo methods, not Markov chain Monte Carlo methods). The fact that the joint posterior density is high dimensional and can not be sampled from directly is why we resort to MCMC techniques.

Regression models provide an especially simple example of where blocking can be put to good use. This is easily accomplished via BUGS/JAGS, as we demonstrate in the following example.

■ Example 6.3

'Blocking' parameters to speed up convergence of a Gibbs sampler in a regression analysis. Recall the regression analysis of the data on suspected vote fraud in Pennsylvania senate elections (Example 2.15, and considered above in section 6.3). In the BUGS/JAGS program in Section 6.3 we specified priors over the intercept and slope of the regression model with the following lines of code:

JAGS code

```
## priors
beta[1] ~ dnorm(0,.0001)
beta[2] ~ dnorm(0,.0001)
```

This approach directs BUGS/JAGS to consider β_1 and β_2 as separate nodes in the DAG representation of the Bayesian regression model. The resulting MCMC algorithm generated by BUGS/JAGS will have separate updating steps for each parameter: i.e. iteration t of the MCMC algorithm for this problem will involve

1. sample $\beta_1^{(t)}$ from $g_1(\beta_1 | \beta_2^{(t-1)}, \sigma^{(t-1)}, \mathbf{y}, \mathbf{X})$
2. sample $\beta_2^{(t)}$ from $g_2(\beta_2 | \beta_1^{(t)}, \sigma^{(t-1)}, \mathbf{y}, \mathbf{X})$
3. sample $\sigma^{(t)}$ from $g_3(\sigma | \beta_1^{(t)}, \beta_2^{(t)}, \mathbf{y}, \mathbf{X})$

'Blocking' the intercept and slope parameters into the vector $\boldsymbol{\beta} = (\beta_1, \beta_2)'$ will result in a MCMC algorithm that mixes better than the 'unblocked' setup just described. Indeed, the conventional, Bayesian analysis of the linear regression model treats the regression parameters *en bloc*, as we saw in Section 2.5. Sampling from the conditional distribution of $\boldsymbol{\beta}$ given σ and $\mathbf{y}$ and $\mathbf{X}$ is not at all difficult, at least in the typical case where $\boldsymbol{\beta}$ has a normal prior. In this conditionally conjugate case, the conditional distribution of $\boldsymbol{\beta}$ given $\mathbf{y}, \mathbf{X}$ and σ is multivariate normal with an easily computed mean vector and covariance matrix. Frankly, if one were programming a MCMC algorithm for the normal, linear regression model, handling $\boldsymbol{\beta}$ *en bloc* via a draw from a multivariate conditional distribution is far easier than cycling over the individual, scalar components of $\boldsymbol{\beta}$.

Code for the *en bloc* version of the regression model appears below:

JAGS code

```
 1 model{
 2  for(i in 1:n){   ## regression model for y
 3    mu[i] <- beta[1] + beta[2]*x[i]
 4    y[i] ~ dnorm(mu[i],tau)
 5  }
 6   ## priors
 7   beta[1:2] ~ dmnorm(b0[1:2],B0[1:2,1:2])
 8   sigma ~ dunif(0,100)
 9   tau <- pow(sigma,-2)
10 }
```

The only difference with the 'unblocked' version of the code is the way the normal priors over β_1 and β_2 are specified as a bivariate normal prior for $\boldsymbol{\beta} = (\beta_1, \beta_2)'$ at line 7. The 2-by-1 vector of prior means `b0` and the 2-by-2 prior precision matrix `B0` are passed to BUGS/JAGS as part of the data supplied by the user. In this case, the priors $\beta_j \sim N(0, 1000)$, $j = 1, 2$ are operationalized by setting `b0` to a vector of zeros, and `B0` to $\kappa \cdot \mathbf{I}_2$, where $\kappa = 1/1000$.

Figure 6.3 compares the performance of the two MCMC algorithms for this problem: one that treats the β parameters as separate nodes, and one that uses the 'blocked', bivariate prior specification shown in the code block above. The trace plots and auto-correlation functions for both β parameters reveal that the *en bloc* approach produces a far superior exploration of the posterior density, with no detectable autocorrelation in the sampler's trajectory through the parameter space. The 'unblocked' specification produces a relatively slow-mixing Markov chain, with AR(1) parameters around .64. As a result, many more iterations of the 'unblocked' specification are required to obtain the same amount of precision resulting from a relatively short run of the *en bloc* sampler. In fact, using the simulation inefficiency results from section 4.4.1, a sampler with an AR(1) coefficient of $\rho = .64$ would require $\sqrt{(1 + .64)/(1 - .64)} \approx 2.13$ as many iterations to obtain inferences about the posterior means as precise as those generated by an independence sampler (and in this case the *en bloc sampler* approximates the behavior of an independence sampler). The situation is worse when we consider inferences about extreme quantiles of the marginal posterior densities for β_1 and β_2. The Raftery-Lewis (1992a) diagnostic suggests that the 'unblocked' sampler requires over 10 000 iterations to generate precise estimates of the 2.5 and 97.5 percentiles of the marginal posterior density of β_1, and over 7000 iterations for the slope parameter β_2; the corresponding numbers from the *en bloc* sampler are 3900 and 3800 iterations. This difference in the efficiency of the two samplers stems from the fact that β_1 and β_2 are highly correlated, at about $-.80$ in this case; we return to the issue of correlated parameters in regression analysis in Section 6.4.3, below.

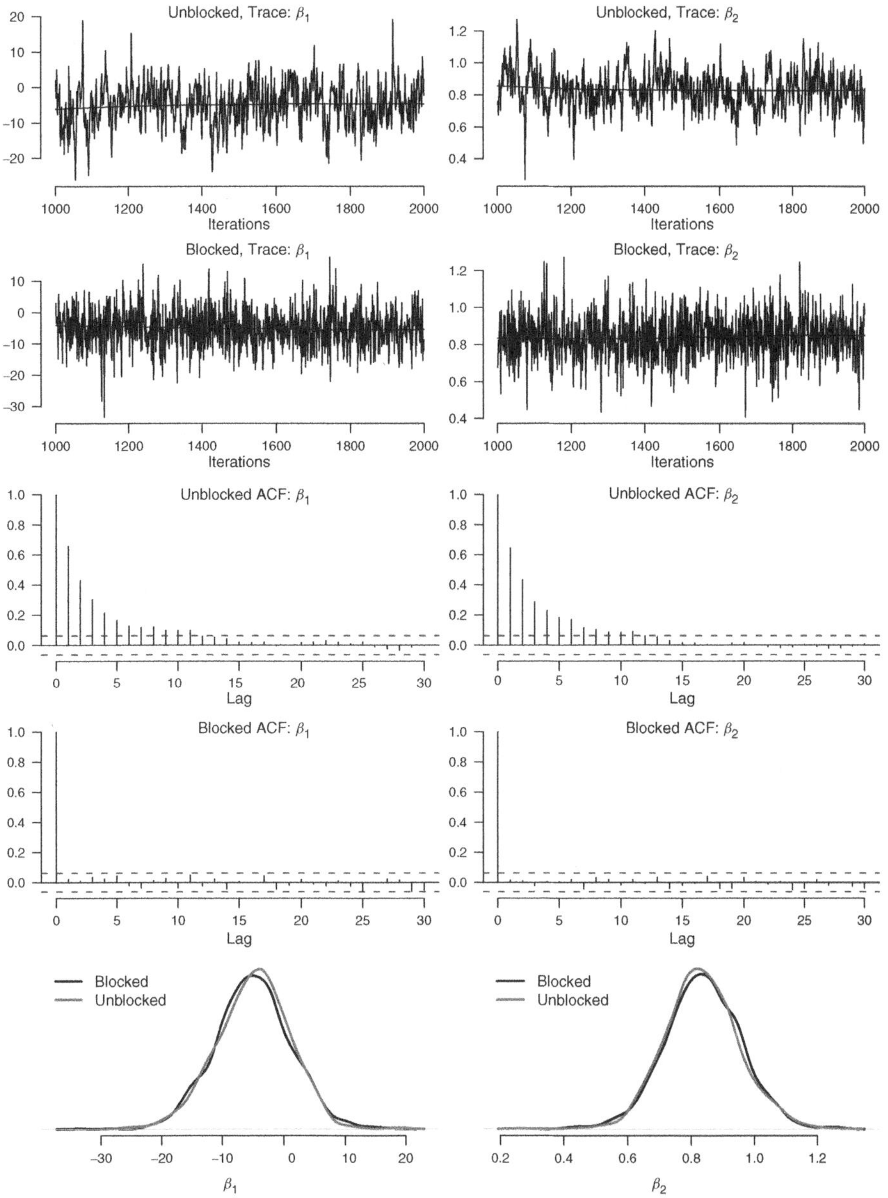

Figure 6.3 Comparison of Gibbs samplers, with and without blocking. Regression models are fit to the data on suspected voter fraud in Pennsylvania state senate elections (Example 2.15). The graphs compare the performance of two different Gibbs samplers, one with the intercept (β_1) and the slope (β_2) coefficients specified as separate nodes in the DAG (the 'unblocked' specification), and the other with the two regression coefficients treated *en bloc* (the 'blocked' specification). Top panels show traces of the 2nd 1000 iterations of a 2000 iteration run of the respective MCMC algorithms. Middle rows show autocorrelation spectra of the 2nd 1000 iterations. The lower row of panels compare the marginal posterior densities of each parameter as estimated by the two different samplers.

Clearly, the *en bloc* sampler outperforms the 'unblocked' version of the sampler. Results like these can be even more striking as the size of the block grows. Here we considered the most simple example of blocking possible: combining two, correlated scalar parameters into a vector of length two. Regression models are seldom so spartan, typically involving more than a single predictor. In cases such as these the advantages of blocking are considerable; blocking can greatly reduce the number of iterations required to explore the posterior density and should be implemented whenever possible.

6.4.3 Reparameterization

It is often the case that a re-parameterized version of a model will generate a Markov chain that mixes more quickly than the Markov chain generated from the original model. In many cases the reparameterization is relatively innocuous, in that key parameters or quantities of substantive interest remain unchanged, or can be easily recovered from the reparameterization.

We noted in Section 6.4.2 that a MCMC algorithm will exhibit slow mixing if the components of the parameter vector $\boldsymbol{\theta}$ are highly correlated with one another. Blocking is one strategy for dealing with this, treating the correlated components as one component and sampling from the conditional distribution of the 'block' of correlated components. Another strategy is to reparameterize the model, coming up with a set of working parameters that display less correlation than that between θ_1 and θ_2.

Perhaps the most common way in which we encounter this problem in practice is with regression type structures, which we elaborate in the following example.

■ **Example 6.4**

Mean-deviating predictors in regression analysis. In Section 2.5 we considered Bayesian analysis of the iid linear regression model under normality, i.e. $\mathbf{y}|\mathbf{X} \sim N(\mathbf{X}\boldsymbol{\beta}, \sigma^2\mathbf{I})$, where $\mathbf{X}$ is a n-by-k matrix of predictors and so $\boldsymbol{\beta}$ is a k-by-1 vector of unknown parameters, usually including a constant term and so a column of $\mathbf{X}$ is a unit vector of length n.

Proposition 2.11 established that if *a priori* $\boldsymbol{\beta}|\sigma^2 \sim N(\mathbf{b}_0, \sigma^2\mathbf{B}_0)$, then *a posteriori* $\boldsymbol{\beta}|\mathbf{y}, \mathbf{X}, \sigma^2 \sim N(\mathbf{b}_1, \sigma^2\mathbf{B}_1)$, with $\mathbf{b}_1 = (\mathbf{B}_0^{-1} + \mathbf{X}'\mathbf{X})^{-1}(\mathbf{B}_0^{-1}\mathbf{b}_0 + \mathbf{X}'\mathbf{X}\hat{\boldsymbol{\beta}})$ and $\mathbf{B}_1 = (\mathbf{B}_0^{-1} + \mathbf{X}'\mathbf{X})^{-1}$. Consider the case where we have a just a single predictor and a constant term in the regression model, such that $\mathbf{X} = [\boldsymbol{\iota}\,\mathbf{x}]$, $\boldsymbol{\iota} = (1, \ldots, 1)'$ and

$$\mathbf{X}'\mathbf{X} = \begin{bmatrix} \boldsymbol{\iota}' \\ \mathbf{x}' \end{bmatrix} \begin{bmatrix} \boldsymbol{\iota} & \mathbf{x} \end{bmatrix} = \begin{bmatrix} \boldsymbol{\iota}'\boldsymbol{\iota} & \boldsymbol{\iota}'\mathbf{x} \\ \mathbf{x}'\boldsymbol{\iota} & \mathbf{x}'\mathbf{x} \end{bmatrix} = \begin{bmatrix} n & \sum_{i=1}^n x_i \\ \sum_{i=1}^n x_i & \sum_{i=1}^n x_i^2 \end{bmatrix}.$$

Suppose we employ a reasonably vague set of priors for $\boldsymbol{\beta} = (\beta_1, \beta_2)'$, say $\boldsymbol{\beta} \sim N(\mathbf{0}, \sigma^2\mathbf{B}_0)$ with $\mathbf{B}_0 = \kappa \cdot \mathbf{I}_2$ where κ is an arbitrarily large, positive constant. Note that with this diagonal form of the variance-covariance matrix for $\boldsymbol{\beta}$ we are specifying that β_1 and β_2 are independent *a priori*, recalling that zero covariance is necessary and sufficient for independence under normality (see Proposition B.2). Then the posterior variance-covariance matrix of $\boldsymbol{\beta}$ is $\sigma^2\mathbf{B}_1$ where

$$\begin{aligned}
\mathbf{B}_1 &= \left(\frac{1}{\kappa}\mathbf{I}_2 + \mathbf{X}'\mathbf{X}\right)^{-1} = \left(\begin{bmatrix} \frac{1}{\kappa} & 0 \\ 0 & \frac{1}{\kappa} \end{bmatrix} + \begin{bmatrix} n & \sum x_i \\ \sum x_i & \sum x_i^2 \end{bmatrix}\right)^{-1} \\
&= \begin{bmatrix} n + \frac{1}{\kappa} & \sum x_i \\ \sum x_i & \sum x_i^2 + \frac{1}{\kappa} \end{bmatrix}^{-1} \\
&= \frac{1}{\left(n + \frac{1}{\kappa}\right)\left(\sum x_i^2 + \frac{1}{\kappa}\right) - (\sum x_i)^2} \begin{bmatrix} \sum x_i^2 + \frac{1}{\kappa} & -\sum x_i \\ -\sum x_i & n + \frac{1}{\kappa} \end{bmatrix},
\end{aligned} \tag{6.1}$$

and so the posterior covariance between the intercept and slope parameters in this regression model (conditional on σ^2) is

$$\text{cov}(\beta_1, \beta_2 | \mathbf{y}, \mathbf{X}, \sigma^2) = \sigma^2 \frac{-\sum x_i}{\left(n + \frac{1}{\kappa}\right)\left(\sum x_i^2 + \frac{1}{\kappa}\right) - (\sum x_i)^2}. \tag{6.2}$$

Note that this quantity is non-zero whenever the numerator $\sum x_i \neq 0$, a point we will return to momentarily.

Consider an example. In the Pennsylvania voting data (Example 2.15, and reconsidered in Section 6.3), we have

$$\mathbf{X}'\mathbf{X} = \begin{bmatrix} 21 & 854.21 \\ 854.21 & 53530.38 \end{bmatrix}$$

and so

$$\mathbf{B}_1 = \frac{1}{\left(21 + \frac{1}{\kappa}\right)\left(53530.38 + \frac{1}{\kappa}\right) - 854.21^2} \begin{bmatrix} 53530.38 + \frac{1}{\kappa} & -854.21 \\ -854.21 & 21 + \frac{1}{\kappa} \end{bmatrix}$$

and

$$\text{cov}(\beta_1, \beta_2, | \mathbf{y}, \mathbf{X}, \sigma^2) = \sigma^2 \frac{-854.21}{\left(21 + \frac{1}{\kappa}\right)\left(53530.38 + \frac{1}{\kappa}\right) - 854.21^2}.$$

With $\kappa = 1000$ and σ^2 set to its posterior mean of 15.78^2, we have $\text{cov}(\beta_1, \beta_2, |\mathbf{y}, \mathbf{X}, \sigma^2) = -.54$ and expressed as a correlation $\text{cor}(\beta_1, \beta_2, |\mathbf{y}, \mathbf{X}, \sigma^2) = -.81$. This high level of dependence between the intercept and slope parameter can cause a MCMC algorithm to run slowly, for the reasons discussed above: the joint posterior density of the regression parameters is highly elliptical, and a 'component-by-component' MCMC algorithm will explore this elliptical density very slowly.

We have already seen that sampling *en bloc* from the bivariate normal posterior density for $\boldsymbol{\beta} = (\beta_1, \beta_2)'$ is one way to avoid the poor performance of a component-by-component Gibbs sampler for this problem. But, if we were stuck with a component-by-component MCMC algorithm for this problem, we can generate a sizeable improvement in performance by deviating the predictors in the regression, $\mathbf{X}$, around their means. By construction, if a predictor $\mathbf{x} = (x_1, \ldots, x_n)'$ has been deviated around its mean, then $\bar{x} = n^{-1}\sum x_i = 0 \Rightarrow \sum x_i = 0$. Under these conditions, and with β_1 and β_2 independent *a priori*, we see that β_1 and β_2 are also independent *a posteriori*; i.e. the term $\sum x_i$ in Equation (6.2) is zero.

Note that the re-parameterization induced by mean-deviating the predictor is rather mild, changing the interpretation of the intercept, reducing the correlation between the intercept and slope parameter to zero, but leaving the slope parameter unchanged.

In the applications in the remainder of the book we will make extensive use of reparameterization in looking at hierarchical models. Specifically, we will use *redundant parameterizations* or *over-parameterizations* to generate MCMC algorithms with better performance than the MCMC algorithms generated by working with the typical parameterization of a given model. We will also make repeated use of the mean-deviating 'trick' introduced above.

6.5 Other examples

We conclude this chapter by examining some additional examples, showcasing the flexibility of `BUGS/JAGS`, and in some cases, how to 'trick' these programs into doing what we want. The books by Congdon (2003; 2005; 2007) are nearly encyclopedic in their coverage of what can be done with `BUGS`. The numerous examples in Part III of this book present `JAGS/BUGS` code as well.

■ Example 6.5

Unidentified/over-parameterized normal model. As a formal matter, the question of identification poses no special problem for a Bayesian analysis. That is, the question of whether the likelihood function $p(y|\boldsymbol{\theta})$ has a unique maximum with respect to $\boldsymbol{\theta}$ does not stop us from applying Bayes Rule. Whatever the shape of the likelihood function, it is still the case that $p(\boldsymbol{\theta}|\mathbf{y}) \propto p(\boldsymbol{\theta})p(\mathbf{y}|\boldsymbol{\theta})$. If the prior density is proper, then so too will be the posterior density and Bayesian analysis is feasible, at least as a technical matter. Moreover, if the posterior density is proper (irrespective of the shape of the likelihood), MCMC methods can be 'safely' deployed; contrast the discussion in Section 5.2.4 on the 'unsafe' use of the Gibbs sampler when the posterior density is not proper.

Now, one might rightly question the practical value of Bayesian analysis when the model parameters are not identified in the likelihood. Nonetheless, in this brief example we demonstrate that technologies like `JAGS` and `BUGS` do not care one way or the other about identification of the model parameters in the likelihood. We use an almost ludicrously simple example due to Carlin and Louis (2000, 174).

We have a data set consisting of one observation, $y = 0$, which we model as arising from the over-parameterized normal model $y \sim N(\beta_1 + \beta_2, 1)$. The identification issue here is obvious: the normal density (the likelihood function for the data) has just one parameter for its mean, μ, which here is parameterized as $\beta_1 + \beta_2$. Whatever μ is, any two numbers β_1, β_2 such that $\beta_1 + \beta_2 = \mu$ will maximize the likelihood function. In this case, $y = 0 \Rightarrow \hat{\mu} = 0$. The likelihood function $\mathcal{L}(\beta_1, \beta_2; y) : \mathbb{R}^2 \mapsto \mathbb{R}$ has a ridge along the locus of points

$$\mathcal{R} = \{(\beta_1, \beta_2) : \beta_1 + \beta_2 = 0\}$$

i.e. the MLEs for this problem are not unique, since any point in the set $\mathcal{R}$ maximizes the likelihood.

In our Bayesian analysis of this model we will put independent, proper, normal priors on β_j, $j = 1, 2$: i.e. $\beta_j \sim N(0, 10^2)$. The model is easily expressed in JAGS or BUGS:

JAGS code

```
model{
        y[1] ~ dnorm(mu,1)
        mu <- beta[1] + beta[2]
        beta[1] ~ dnorm(0,.01)
        beta[2] ~ dnorm(0,.01)
}
```

JAGS happily generates Gibbs samples from the posterior density $p(\beta_1, \beta_2|y = 0)$; the fact that the likelihood has no unique maximum is no impediment. The results of 5000 Gibbs samples are shown in Figure 6.4. The top two rows of panels show the output with respect to each β parameter. The trace plots reveal that the Gibbs sampler generates an inefficient, highly autocorrelated exploration of the posterior density with respect to the unidentified parameters, β_1 and β_2; the AR(1) parameters of both sequences are about .98 and the effective sample size (see Section 4.4.1) of the 5000 Gibbs samples is about 59 observations. This high level of simulation inefficiency is to be expected, recalling that the likelihood is maximized by the linear combination $\beta_1 + \beta_2 = 0$, inducing a high level of posterior correlation between β_1 and β_2 (about $-.99$), which the Gibbs sampler will inherit in sequentially sampling from the conditional distributions $g_1(\beta_1|\beta_2, y)$ and $g_2(\beta_2|\beta_1, y)$.

Perhaps curiously, the marginal posterior densities of the unidentified parameters β_1 and β_2 have 'shrunk' relative to their prior densities (contrast the solid and dotted lines in the top two panels in the right column of Figure 6.4). That is, even though the parameters are not identified in the likelihood, it is not correct to say that the data are 'uninformative' with respect to these parameters; the marginal posterior standard deviations of β_1 and β_2 are approximately $\sqrt{50} \approx 7.08$, considerably smaller than the standard deviations of the marginal prior densities (10.0). The single observation $y = 0$ (with known variance one) generates the likelihood that is quite informative with respect to the linear combination $\mu = \beta_1 + \beta_2$ (see the lower two rows of Figure 6.4); in turn, through the particular dependency between β_1 and β_2 in the likelihood, we actually learn something about the marginal densities of β_1 and β_2 (they are each, marginally, a little more concentrated around zero than we thought, *a priori*).

The fact that Bayesian analysis can proceed when likelihood methods ought to 'fail' or 'break' – i.e. placing proper priors over parameters that are unidentified by the data through the likelihood function – can take some getting used to for the novice, applied Bayesian. Later on, we will actually exploit this feature of Bayesian analysis, to help us 'speed up' MCMC algorithms: a slow, inefficient exploration of the posterior density of a set of identified parameters can be improved by putting the MCMC algorithm to work on a higher dimensional problem in which the identified parameters can be recovered as known, deterministic functions of unidentified parameters. Indeed, in a sense, that is what we just did with the present example: the Gibbs sampler over β_1 and β_2 performs rather poorly with respect to those (unidentified) parameters, but is equivalent to an independence sampler with respect to the output it generates for the identified parameter $\mu = \beta_1 + \beta_2$. We return to this idea briefly in Example 6.9 before exploiting it repeatedly in Chapters 7 and 8.

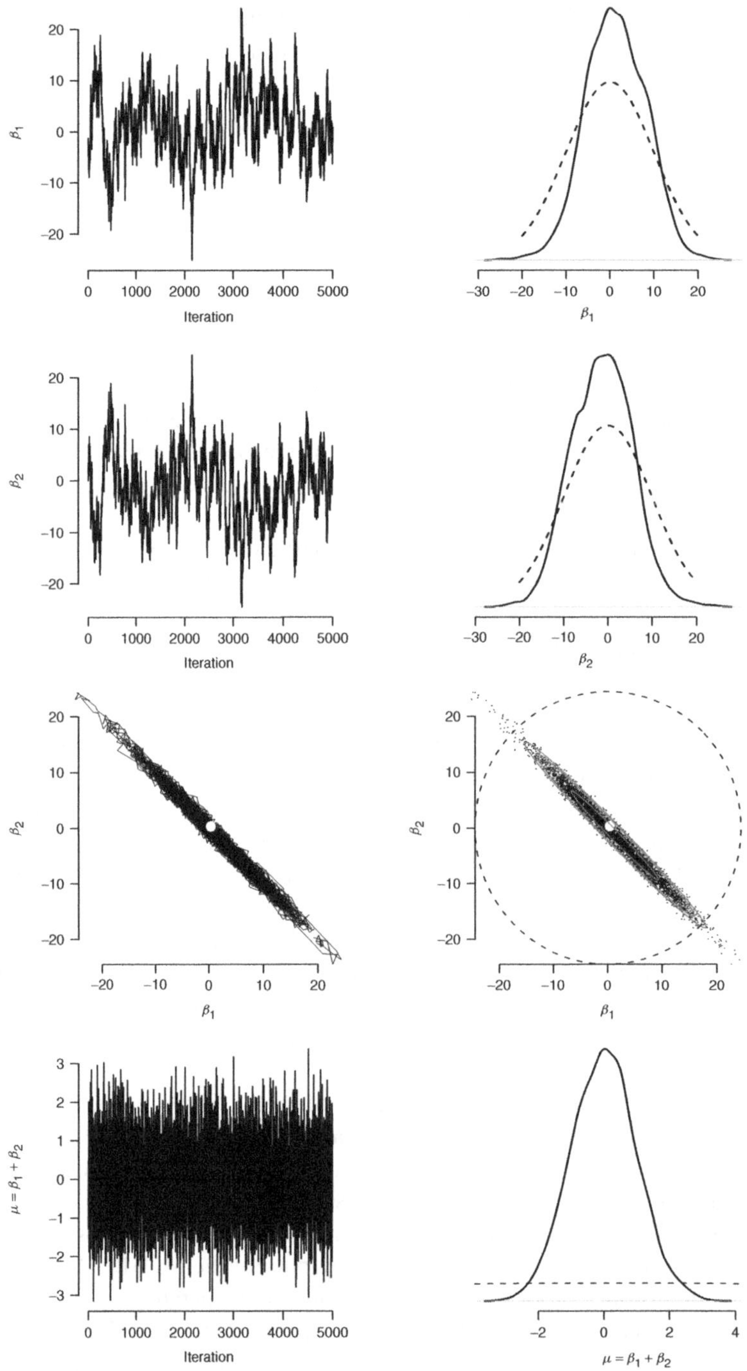

Figure 6.4 Trace plots and posterior densities, Example 6.5. The parameters β_1 and β_2 are not identified in the model $y \sim N(\beta_1 + \beta_2, 1)$, but the linear combination $\mu = \beta_1 + \beta_2$ is; contrast the trace plot and density for μ with those for β_1 and β_2. Dotted lines in the top two left panels indicate the $N(0, 10^2)$ prior densities for β_1 and β_2; the dotted circle is a prior 95 % credible region for $\boldsymbol{\beta} = (\beta_1, \beta_2)'$.

The other lesson here is that one can not confidently rely on the output of a MCMC algorithm to diagnose something as fundamental as a lack of identification. Sometimes the structure of the model will be such that the posterior does in fact equal the prior for a given parameter, from which we can infer that the data are uninformative with respect to that parameter, and that maybe the model is overly elaborate. But as this example reveals, sometimes the interaction between a set of proper priors and the likelihood is more subtle.

In short, with great power (easily implemented, simulation-based Bayesian inference for a very wide class of user-defined models) comes great responsibility. Or even more simply, 'MCMC-for-the-masses' in the form of programs like JAGS and BUGS can sometimes mean 'garbage in, garbage out', with a little Monte Carlo error attached. The trick is for the novice user to be able to recognize 'garbage out', particularly if that user hasn't been able to recognize 'garbage in'.

■ Example 6.6

Inference for a correlation coefficient in a data set subject to extreme missingness; Example 5.11, continued. We return to the missing data example in Example 5.11, showing how the data-augmented Gibbs sampler can be easily implemented in JAGS and/or BUGS. We begin by setting up the data in R:

R code

```
y <- matrix(c(-1, -1, 1, 1, 2, -2, 2, -2, NA, NA, NA, NA,
               -1, 1, -1, 1, NA, NA, NA, NA, 2, -2, 2, -2),
              12, 2)
forJags <- list(y=y)
```

The BUGS code for this example is quite simple, exploiting the fact that recent versions of BUGS can handle partially observed multivariate quantities:

BUGS code

```
## can't do this in JAGS or earlier versions of BUGS
## no imputations for partially observed multivariate nodes
model{
        for (i in 1:12){
             y[i,1:2] ~ dmnorm(mu[1:2],tau[,])
        }
        mu[1] <- 0.0;    ## mean known to be zero
        mu[2] <- 0.0;
        R[1,1] <- 1      ## Wishart prior, scale
        R[1,2] <- 0
        R[2,1] <- 0
        R[2,2] <- 1
        k <- 4           ## Wishart prior, degree of freedom
        tau[1:2,1:2] ~ dwish(R[1:2,1:2],k)

        Sigma <- inverse(tau)  ## convert to covariance matrix

        ## compute correlation
        rho <- Sigma[1,2]/sqrt(Sigma[1,1]*Sigma[2,2])
}
```

That is, the quantity `y[i,1:2]` is partially observed for observations 5 through 12. When recent versions of `BUGS` encounter this partially observed data the program deduces that the likelihood contributions can be made via the factorization of the complete-data likelihood in Equation 5.26 into a marginal density for the observed component of `y[i,]` and a conditional density for the missing component of `y[i,]` given the observed component. This means we have an extremely simple form for the model, given at line 5, where we model the data `y[i,]` as coming from a bivariate normal density, via the `dmnorm` construct.

The remainder of the code is overhead for setting the elements of the mean vector `mu` to zero (lines 7 and 8), setting up the scale matrix `R` (line 9 through 12) and degrees of freedom `k` of the Wishart prior over the precision matrix `tau`. Line 16 converts `tau` to a covariance matrix, `Sigma`. Finally, line 17 computes the correlation `rho` as a function of the appropriate elements of `Sigma`.

There is certainly more than one way to get `JAGS` and/or `BUGS` to do what we want it to do. And in this case, we can 'trick' `JAGS` (and earlier versions of `BUGS`) into handling the missing data problem here. We will also use this example to show how we can depart from the use of a conditionally conjugate inverse-Wishart prior for the covariance matrix $\mathbf{\Sigma}$. Consider the following `JAGS` program:

JAGS code

```
1 model{
2   for (i in 1:12){
3     y[i,1] ~ dnorm(0.0,tau[1,1]);   ## y1 marginal model
4     y[i,2] ~ dnorm(mu[i],tau21);    ## y2 conditional on y1
5     mu[i] <- beta*y[i,1];           ## E(y2|y1)
6 }
7
8   ## priors
9   sigma[1] ~ dunif(0,6)
10  sigma[2] ~ dunif(0,6)
11  rho ~ dunif(-1,1)                    ## correlation
12  Sigma[1,1] <- pow(sigma[1],2)        ## variance
13  Sigma[2,2] <- pow(sigma[2],2)        ## variance
14  Sigma[1,2] <- rho*sigma[1]*sigma[2]  ## covariance
15  Sigma[2,1] <- Sigma[1,2]             ## covariance
16
17  ## convert covariance matrix to precision matrix
18  tau <- inverse(Sigma)
19
20  ## conditional variance
21  Sigma21 <- Sigma[2,2] - pow(Sigma[1,2],2)/Sigma[1,1]
22  tau21 <- 1/Sigma21                   ## conditional precision
23
24  beta <- Sigma[1,2]/Sigma[1,1]        ## regression coefficient
25 }
```

Lines 3 and 4 decompose the bivariate normal problem into a marginal, univariate normal model for `y[i,1]`, and a conditional, univariate normal model for `y[i,2]` given `y[i,1]`. Note that at line 5, `y[i,1]` appears as a 'parent' node of `y[i,2]`, via `mu[i]`. The fact that `y[i,1]` is sometimes missing is handled via the fact that we have defined a sampling model for `y[i,1]` at line 3. Via this decomposition of the problem we have side-stepped the limitation on JAGS not being able to handle unobserved components in a multivariate node.

Now, to accommodate this reformulation of the model we need to define the marginal and conditional precisions (tau11 and tau21, respectively) and a definition of the term beta that appears in line 5. In this case we work with priors over the standard deviations σ_1 and σ_2 and the correlation ρ. I deploy uniform priors on a fairly generous range for the standard deviations, $\sigma_j \sim \text{Unif}(0, 6)$, $j = 1, 2$, and a uniform prior on $\rho \in [-1, 1]$. We can construct a positive definite covariance matrix $\boldsymbol{\Sigma}$ as a deterministic function of these quantities, with the covariance term $\sigma_{12} = \rho\sigma_1\sigma_2$ (line 14). We convert the covariance matrix to a precision matrix tau with the inverse function at line 18, extracting the marginal precision of y_{i1} as tau11 in line 19. We compute the conditional variance and precision of y_{i2} given y_{i1} at lines 22 and 23, respectively. Finally, we compute the quantity beta on line 24, which we use in computing the conditional expectation of y_{i2} given y_{i1}; for the expressions used in computing these conditional expectations and variances/precisions, see Proposition B.2 and the exposition in Example 5.11.

We call this JAGS code from R via the functionality provided by the rjags library:

R code

```
1 inits <- list(list(sigma1=3,sigma2=3,rho=.99),
2               list(sigma1=.25,sigma2=.25,rho=-.99))
3 foo <- jags.model(file="bimodal3.bug",
4                   data=forJags,
5                   n.adapt=5000,
6                   n.chains=2,
7                   inits=inits)
8
9 out <- coda.samples(foo,
10                    n.iter=250000,
11                    thin=50,
12                    variable.names=c("Sigma","rho"))
```

We here run two parallel Markov chains, starting at quite different sets of starting values; note the nchain = 2 option at line 6. With two chains we need two sets of starting values; see the two lists in inits in lines 1 and 2. We also specify a slightly longer than usual adaption phase for the MCMC algorithm here, since a previous run with the defaults generated a warning message from JAGS; this stems from the non-standard uniform priors being used over σ_1, σ_2 and ρ, and the transformations between those parameters and the marginal and conditional precisions that appear in the model for the data. We specify that JAGS generate 250 000 iterations (line 10), saving every 50th iteration for output (line 11).

Figure 6.5 displays traceplots and densities for the output produced by JAGS. In each case the chains appear to mix quite well and produce virtually identical characterizations of the respective marginal posterior densities. The dotted line on the density plots (left panels of Figure 6.5) indicate the uniform priors. For the two standard deviation parameters, the posterior is approximately zero at the from the upper limit of the uniform prior at 15.0, suggesting that this *a priori* bound is not overly restrictive. The bimodality of the marginal posterior density of the correlation parameter ρ mirrors the result we obtained with the 'hand-coded' data-augmented Gibbs sampler in Example 5.11 (see Figure 5.14).

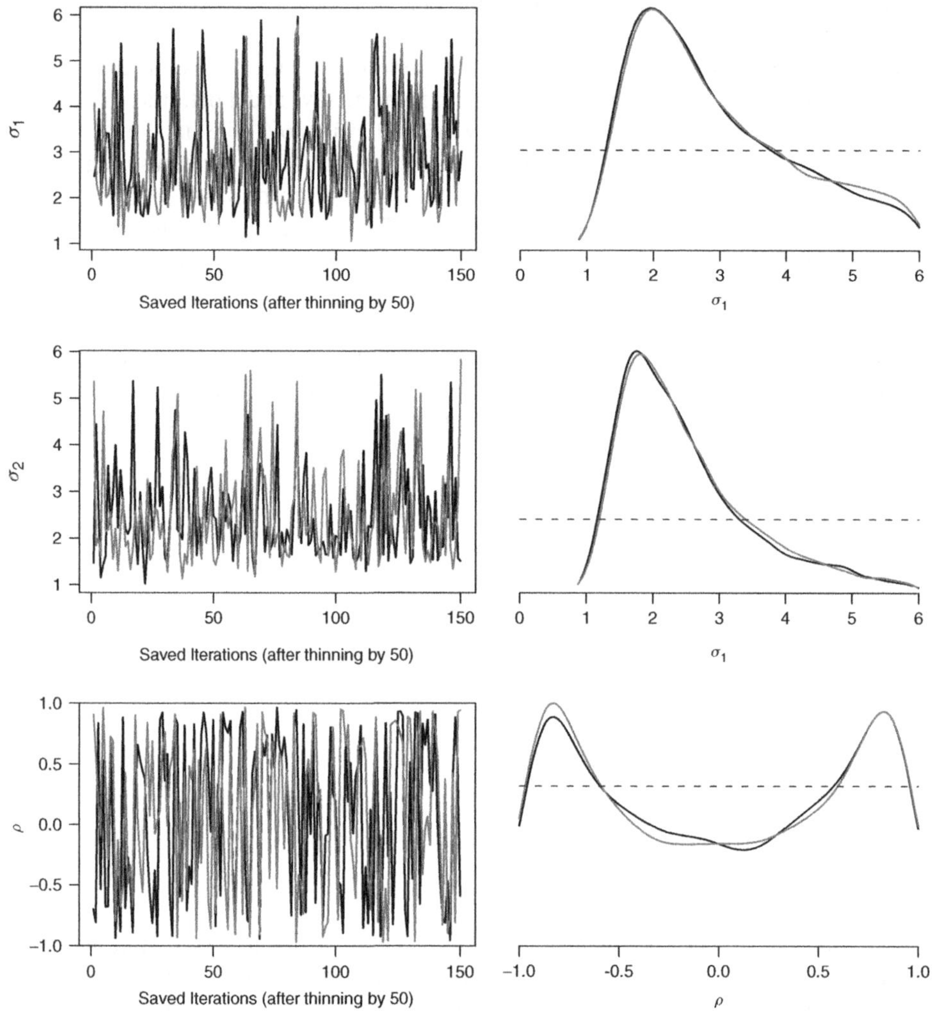

Figure 6.5 Trace plots and densities, JAGS output for Example 6.6. Gray and black lines distinguish the two parallel chains. Trace plots on the left are shown for the first 150 saved iterations of the 250 000 iteration runs (each thinned by 50). Dotted lines on the density plots on the right indicate the uniform priors for each parameter.

■ Example 6.7

Negative binomial regression model; Example 5.2, continued. We return to the data on article production by biochemistry graduate students. In Example 5.2 we used the MCMCpoisson function in the R package MCMCpack (Martin, Quinn and Park 2009) to fit a Poisson regression model to these data; in Example 5.4 we used the rnegbinRw function in the R package bayesm (Rossi and McCulloch 2008) to fit a negative binomial regression model. Here we show how the negative binomial regression model can be fit in JAGS.

The 'setup' R code appears below:

R code

```
1 require(pscl)
2 data(bioChemists)
3
4 forJags <- list(n=dim(bioChemists)[1],  ## sample size
5                 fem=as.numeric(bioChemists$fem=="Women"),    ## covariates
6                 mar=as.numeric(bioChemists$mar=="Married"),
7                 kid5=bioChemists$kid5,
8                 phd=bioChemists$phd,
9                 ment=bioChemists$ment,
10                y=bioChemists$art,     ## response
11                b0 = rep(0,6),         ## prior hyperparameters
12                B0 = diag(.0001,6))
13
14 inits <- list(list(beta=rep(0,6),
15                    r=1))
```

The forJags object, a list, contains not just the article counts and the regressors, but the sample size n and the prior hyperparameters b0 and B0, the prior mean and prior precision matrix of the regression parameters $\boldsymbol{\beta}$. Initial values are supplied in the list inits, containing start values for beta and the over-dispersion parameter r.

The JAGS code for this model appears below:

JAGS code

```
1 model{
2         for(i in 1:n){
3               mu[i] <- beta[1] + beta[2]*fem[i] + beta[3]*mar[i]
4                                + beta[4]*kid5[i] + beta[5]*phd[i]
5                                + beta[6]*ment[i]
6              lambda[i] <- exp(mu[i])
7              p[i] <- r/(r+lambda[i])
8              y[i] ~ dnegbin(p[i],r)
9         }
10
11        beta[1:6] ~ dmnorm(b0[1:6],B0[,])
12        r ~ dunif(0,50)
13 }
```

Lines 3 through 6 generate the regression part of the model, $\lambda_i \equiv E(y_i|\mathbf{x}_i, \boldsymbol{\beta}) = r(1 - p_i)/p_i = \lambda_i = \exp(\mathbf{x}_i\boldsymbol{\beta})$. The JAGS parameterization of the negative binomial probability mass function is the same as that used in Definition B.24, and so at line 7 we define the quantity p[i] $\equiv p_i = r/(r + \lambda_i)$. We then pass p[i] and the over-dispersion parameter r to the dnegbin density at line 8, which generates the likelihood for the article counts y[i]. Line 11 specifies the prior $\boldsymbol{\beta} \sim N(\mathbf{b}_0, \mathbf{B}_0^{-1})$ where the quantities $\mathbf{b}_0 \equiv$ b0 and $\mathbf{B}_0 \equiv$ B0 are defined in the R setup code (above, at lines 11 and 12), with $\mathbf{b}_0 = \mathbf{0}$ and $\mathbf{B}_0 = 10^{-3} \cdot \mathbf{I}_6$. Recall that JAGS/BUGS use precisions, not variances, and so we pass a prior precision matrix as the 2nd argument to the multivariate normal density dmnorm at line 11 of the JAGS code.

Finally, note the uniform prior for the over-dispersion parameter `r` at line 12. This puts an upper bound on `r` at 50, which is not at all restrictive: recall that the negative binomial tends to the Poisson as `r` $\rightarrow \infty$, and the constraint that `r` < 50 might seem restrictive. But for all practical purposes the negative binomial is indistinguishable from the Poisson once the over-dispersion parameter gets anywhere close to being that large. The over-dispersion parameter is constrained to the positive half-line, for which a commonly used prior is a reasonably vague Gamma density; e.g. recall from Example 5.4 that the implementation of the negative binomal regression model in `bayesm` uses a Gamma(.5,.1) prior as the default.

Using the `rjags` library (Plummer 2009b), we can get `R` to put `JAGS` to work on this model, with the MCMC output coming back to `R` as a `coda` object: e.g.,

R code

```
require(rjags)
foo <- jags.model(file="negbin.bug",
                  data=forJags,
                  inits=inits)

update(foo,5000)

out <- coda.samples(foo,
                    variable.names=c("beta","r"),
                    n.iter=50000)
```

In Problem 6.14 you are invited to implement this model. In `JAGS` this resulting Markov chain produces woefully slow-mixing and several hundred thousand iterations are required to generate a thorough exploration of the joint posterior density $p(\boldsymbol{\beta}, r|\mathbf{y}, \mathbf{X})$. `OpenBUGS` may fare better.

■ Example 6.8

Three-component normal mixture; vote shares in US Congressional elections; Example 1.2, continued. Casual inspection of Democratic vote shares in Congressional elections suggests that the data fall into three clumps: Republican seats, where Democratic vote shares are below .5; Democratic seats where the Democratic candidates win vote shares in excess of .5; and a small number of extremely pro-Democratic seats, where Democratic seat shares are in excess of .8. Clearly, these data do not appear to have come from a normal distribution, or, any other standard distribution for that matter. Mixtures of densities are a convenient and flexible way to express the density of data that are multimodal, and in this case we model these vote shares as coming from a three-component mixture of normal densities, i.e. with i indexing n observations,

$$p(y_i) = \sum_{j=1}^{J} \lambda_j \cdot \phi_j[(y_i - \mu_j)/\sigma_j] \tag{6.3}$$

where

- $y_i \in [0, 1]$ is the share of the vote won by Democratic candidates in $n = 371$ contested districts in the 2000 US Congressional elections,

- ϕ_j is a standard normal density (μ_j and σ_j^2 are the mean and variance of the j-th component density, respectively),
- λ_j is a set of mixing parameters that obey the constraints $\lambda_j \in [0, 1]\ \forall\ j = 1, \ldots, J$ and $\sum_{j=1}^{J} \lambda_j = 1$ (i.e. $\boldsymbol{\lambda} = (\lambda_1, \ldots, \lambda_J)' \in \Delta^{J-1}$, where Δ^{J-1} is a standard $J-1$-simplex; see Definition B.25),

and in this case $J = 3$.

The unknown parameters for this problem are $\boldsymbol{\theta} = \{\boldsymbol{\lambda}, \mu_1, \ldots, \mu_J, \sigma_1^2, \ldots, \sigma_J^2\}$. Under an assumption of conditional independence across observations given $\boldsymbol{\theta}$, the likelihood is simply $p(\mathbf{y}|\boldsymbol{\theta}) = \prod_{i=1}^{n} p(y_i|\boldsymbol{\theta})$ where p is given in Equation (6.3). Direct maximization of the likekihood function often runs into difficulties (e.g. see McLachlan and Peel 2000; Titterington, Smith and Makov 1985). The likelihood surface is unbounded on the edge of the parameter space; i.e. a single data point y_i can be fit perfectly by assigning it to its own component density j, in which case $\mu_j = y_i$, $\lambda_j = \frac{1}{n}$, $\sigma_j^2 = 0$ and the likelihood is infinite. This can happen for any i and for $J - 1$ components, and so the model is usually fit with constraints to ensure $\lambda_j > \frac{1}{n}$ and $\sigma^2 > 0$.

Even with these problems solved the likelihood does not globally identify the model parameters, since we can encounter the phenomenon known as 'label switching'. Suppose $\boldsymbol{\theta}_j = (\lambda_j, \mu_j, \sigma^2)$, $j = 1, \ldots, J$ are MLEs in a two-component normal mixture; then the vectors $\boldsymbol{\theta}_{\pi(j)}$ are also MLEs, for all permutations π of the indices $j = 1, \ldots, J$. This problem is more an inconvenience than anything, and similar to the 'mirror image' problem discussed in Section 5.2.4. In a simulation-based approach to inference we have to pay special attention to this problem; over a long run of a MCMC algorithm, it is entirely possible that the algorithm will jump from exploring the posterior density associated with one permutation of the component densities to another, such that final result is something of a mess, a combination of the MCMC sampler jumping from permutation to permutation. We deal with both of these problems – the unbounded likelihood problem of the previous paragraph, and the 'label switching' problem – via restrictions on the parameters introduced in the priors, as we now demonstrate by looking at the Congressional elections data.

An interesting and useful by-product of fitting a mixture model are posterior probabilities of component membership. Let the component memberships be $z_i \in \{1, \ldots, J\}$, a latent, discrete variable. Conditional on z_i the model is extremely straightforward, in that $y_i|z_i = j \sim N(\mu_j, \sigma_j^2)$, and suggests that data-augmentation (§5.2.5) would be a fruitful strategy. Indeed, likelihood methods typically exploit this strategy via the *EM* algorithm, treating the latent z_i as missing data. In the Bayesian approach, we use a discrete analog of the posterior identity $p(\boldsymbol{\theta}|\mathbf{y}) = \int_{\mathcal{Z}} p(\boldsymbol{\theta}|\mathbf{z}, \mathbf{y}) p(\mathbf{z}|\mathbf{y}) d\mathbf{z}$, and so iteration t of the Gibbs sampler consists of

1. sample $\boldsymbol{\theta}^{(t)} \sim g(\boldsymbol{\theta}|\mathbf{z}^{(t-1)}, \mathbf{y})$
2. sample $\mathbf{z}^{(t)} \sim g(\mathbf{z}|\boldsymbol{\theta}^{(t)}, \mathbf{y})$,

where $\mathbf{z} = (z_1, \ldots, z_n)' \in \mathcal{Z}$ is a vector containing the latent component indicators. Note that by Bayes Rule

$$\Pr(z_i = j|\mathbf{y}) = \frac{\Pr(z_i = j) p(y_i|z_i = j)}{\sum_{k=1}^{J} \Pr(z_i = k) p(y_i|z_i = k)} \tag{6.4}$$

where $p(y_i|z_i = j) \equiv \phi([y_i - \mu_j]/\sigma_j)$ and $\Pr(z_i = j) \equiv \lambda_j$. We will not derive the conditional distributions that drive the data-augmented Gibbs sampler for this problem; for the conjugate of a finite mixture of normals, see Robert (1996) and Problem 6.15.

In a Bayesian analysis we put priors on the parameters $\boldsymbol{\theta}$. As mentioned above, the priors provide an opportunity for us to impose constraints that help make the posterior interpretable. In this case, we have strong priors beliefs as to the ordering and location of the three means, $\mu_j, j = 1, 2, 3$. This ordering constraint will also induce a unique labeling of the components. That is, we specify priors that embody the restriction $\mu_1 < \mu_2 < \mu_3$, such that $j = 1$ is the 'safe Republican' component, $j = 2$ is the 'competitive' component and $j = 3$ is the 'safe Democratic' component. We impose this ordering constraint through the priors on μ_j: i.e. $\mu_1 \sim \text{Unif}(0, .5)$, $\mu_2 \sim \text{Unif}(.5, .75)$ and $\mu_3 \sim \text{Unif}(.75, 1)$.

The standard deviations of components '1' and '2' are assigned uniform priors, $\sigma_j \sim \text{Unif}(0.001, 0.25)$, while I fix the standard deviation of the third component at .03; the restriction on σ_3 helps the MCMC algorithm 'find' the relatively small 'extremely Democratic' component and is based a crude, visual inspection of a histogram of the data. I employ a reasonably vague Dirichlet prior (Definition B.29) for $\boldsymbol{\lambda}$ to complete the specification, with the prior operationalizing the belief that components '1' and '2' (Republican and Democratic) are reasonably large relative to the 'extremely Democratic' component: $\boldsymbol{\lambda} \sim \text{Dirichlet}(2, 2, 1)$. With this prior $E(\boldsymbol{\lambda}) = (2/5, 2/5, 1/5)'$ and the marginal priors over each element of $\boldsymbol{\lambda}$ are $\lambda_1 \sim \text{Beta}(2, 3)$, $\lambda_2 \sim \text{Beta}(2, 3)$ and $\lambda_3 \sim \text{Beta}(1, 4)$.

A JAGS program that operationalizes this model and prior set of densities is:

JAGS code

```
1 model{
2         for(i in 1:NOBS){
3               T[i] ~ dcat(lambda[]);
4               y[i] ~ dnorm(mu[T[i]],tau[T[i]]);
5         }
6
7         ## priors
8         lambda[1:3] ~ ddirch(a[1:3]);
9
10        for(j in 1:3){
11         mu[j] ~ dunif(muLims[j,1],muLims[j,2])
12         tau[j] <- 1/pow(sigma[j],2);
13        }
14        sigma[1] ~ dunif(.01,.20)
15        sigma[2] ~ dunif(.01,.20)
16        sigma[3] <- .03
17 }
```

The core of the modeling takes place at lines 3 and 4. At line 3 we sample a latent component indicator `T[i]` $\in \{1, 2, 3\}$ using the `dcat` construct, which takes a vector of probabilities that sum to one as its argument (the JAGS version of `dcat` will normalize the argument vector to sum to one); i.e. if `T[i] ~ dcat(lambda[])` then $\Pr(T_i = j) = \lambda_j$, where $\lambda_j \in [0, 1]\ \forall\ j$ and $\sum_j \lambda_j = 1$. At line 4, we use the sampled `T[i]` to index the appropriate element of the vector of component-specific means `mu[]` and precisions `tau[]`.

The Dirichlet prior over $\boldsymbol{\lambda}$ appears at line 8 of the JAGS program, where we use the `ddirch` construct; this takes a vector of non-negative integers as its argument, each interpretable as a prior number of observations per component of the density. Line 11 generates the prior for the `mu[j]`, with the lower and upper limits of the uniform priors for these parameters being passed to `JAGS` as part of the data from `R` (this allows us to experiment with different priors by changing a few lines of code in `R`, rather than than the `JAGS` job; where we might tinker with these priors is a matter of taste). Lines 14–16 define priors for `sigma[1]` and `sigma[2]`, and the constraint on `sigma[3]`; at line 12 we convert the standard deviations to precisions.

We display the results in a series of graphs; 50 000 MCMC iterations are run, and thinned by a factor of 10. We omit the usual series of trace plots and autocorrelation functions; these suggest that the MCMC algorithm is generating a reasonably efficient exploration of the posterior density for this problem, and that 50 000 iterations is sufficient to generate reasonably accurate estimates of the extreme quantiles of the marginal posterior densities for this problem (e.g. via the Raftery-Lewis diagnostic). We examine the fit of the mixture model to the data in the top panel of Figure 6.6. We compute the mixture density with the model parameters set to their posterior means, and overlay this mixture density on a histogram of the data. The three modes in the data are well recovered by the model, with the means of the three component densities well separated (middle panel of Figure 6.6). The data are quite informative about these parameters, as well as for the two, unconstrained standard deviation parameters (lower panel of Figure 6.6), in that the marginal posterior densities of the means and standard deviations differ markedly from their respective, uniform priors (dotted lines).

Figure 6.7 shows the prior and posterior densities of the mixing parameters $\boldsymbol{\lambda}$ on the standard 2-simplex. The graphs show 5000 samples from the prior (left panel) and posterior (right panel), with the distance of any given point from a vertex indicative of the size of the corresponding element of $\boldsymbol{\lambda} = (\lambda_1, \lambda_2, \lambda_3)'$. The prior mean $E(\boldsymbol{\lambda}) = (.4, .4, .1)'$ is indicated with the white circle on the left panel; with $E(\lambda_3) = 1/10$, the prior mean for $\boldsymbol{\lambda}$ lies close to the bottom of the graph, relatively distant from the λ_3 vertex. The posterior mean of $\boldsymbol{\lambda} = E(\boldsymbol{\lambda}|\mathbf{y}) = (.49, .46, .05)'$ lies reasonably close to the prior mean (both points are shown in the right panel of Figure 6.7). The posterior density over $\boldsymbol{\lambda}$ is far more precise than the prior, but is otherwise not in great conflict with the prior. The relatively small values of λ_3 recovered *a posteriori* indicate that very few of the Congressional seats belong to this component of the mixture density.

Classifying districts. An interesting feature of the mixture model in Equation 6.3 is that it provides a straightforward way to assign data points to components (see Equation 6.4), and in this sense the mixture model is also a simple version of a parametric, classification problem. The component membership indicators for the data are $\mathbf{z} = (z_1, \ldots, z_n)'$ with each $z_i \in \{1, 2, \ldots, J\}$. The posterior mass function of each z_i can be obtained as a posterior predictive mass function, essentially treating the z_i as missing data: i.e. $p(z_i|\mathbf{y}) = \int_{\boldsymbol{\Theta}} p(z_i|y_i, \boldsymbol{\theta})p(\boldsymbol{\theta}|\mathbf{y})d\boldsymbol{\theta}$, and we will perform the integration via the sampling in the data-augmented Gibbs sampler (see our earlier discussion on missing data in Section 5.2.6).

In fact, we are not restricted to assigning the y_i to components, but we can derive the predictive mass function over $\tilde{z}_i$, the component for a hypothetical $\tilde{y}_i$; i.e. $p(\tilde{z}_i|\mathbf{y}, \tilde{y}_i) = \int_{\boldsymbol{\Theta}} p(\tilde{z}_i|\tilde{y}_i, \boldsymbol{\theta})p(\boldsymbol{\theta}|\mathbf{y})d\boldsymbol{\theta}$. We accomplish this by evaluating Equation (6.4) for a given hypothetical $\tilde{y}$ repeatedly over iterations of the Gibbs sampler: i.e. at iteration t of the

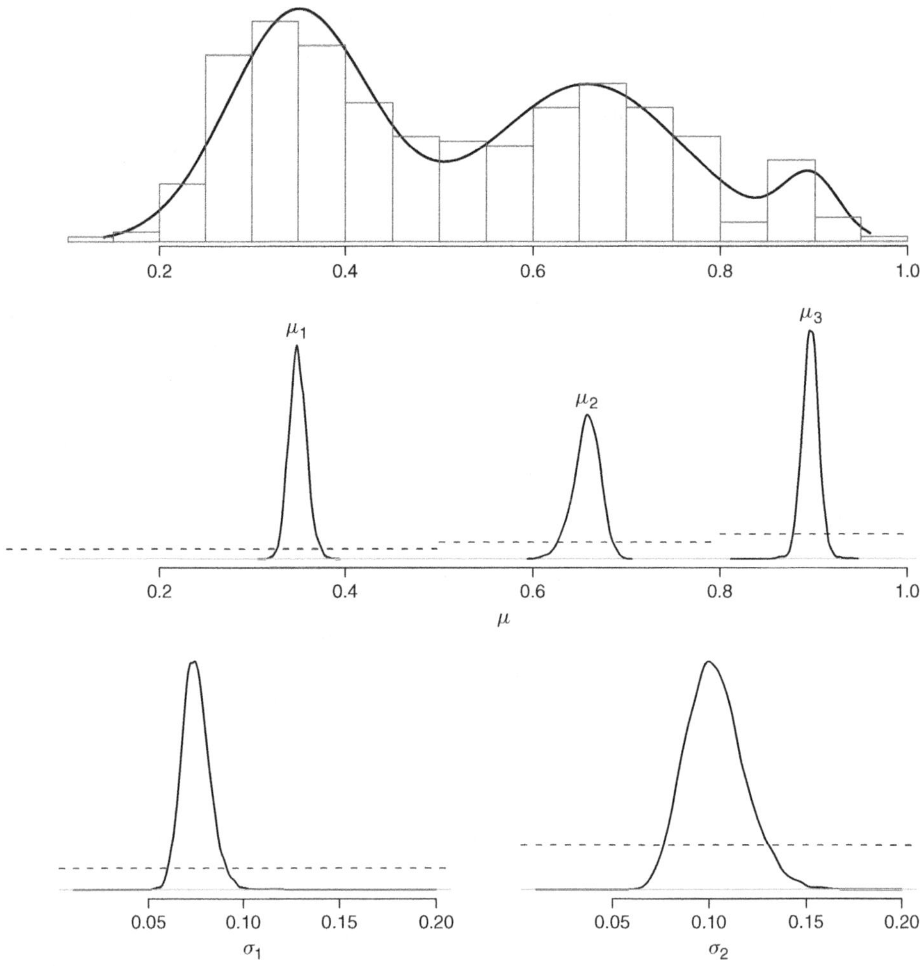

Figure 6.6 Mixture density fit to Congressional vote shares data, Example 6.8. The top panel shows a histogram fit to the vote share data with the fitted mixture density superimposed: the mixture is computed by evaluating the density at a series of grid points with the mixture's parameters held at their posterior means. The middle panel shows three densities, summarizing the marginal posterior densities of the component-specific means, μ_j, $j = 1, 2, 3$; dotted lines indicate the uniform prior densities of each of these parameters. The two graphs in the lower row show the marginal posterior density of the two unconstrained standard deviation parameters, σ_1 and σ_2; again, the dotted lines indicate the prior densities of each of these parameters.

Gibbs sampler, compute

$$\Pr(\tilde{z}_i^{(t)} = j | \tilde{y}_i, \mathbf{y}) = \frac{\lambda_j^{(t)} \cdot \phi\left((\tilde{y} - \mu_j^{(t)})/\sigma_j^{(t)}\right)}{\sum_{k=1}^{J} \lambda_k^{(t)} \cdot \phi\left((\tilde{y} - \mu_k^{(t)})/\sigma_k^{(t)}\right)}$$

where ϕ is the normal density function.

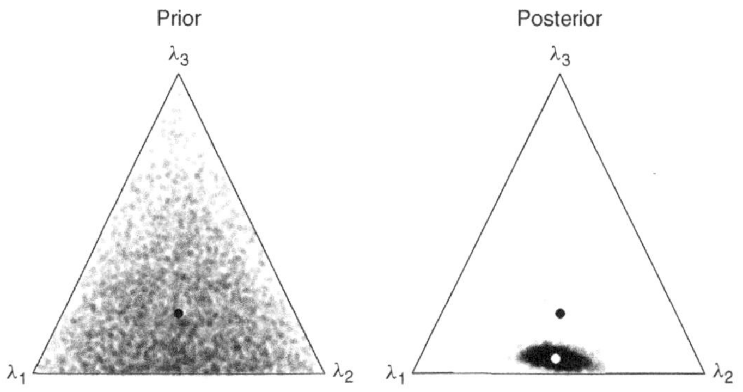

Figure 6.7 Prior and posterior densities of $\boldsymbol{\lambda}$ on the standard 2-simplex, Example 6.8. Each graph is generated by sampling from the prior/posterior density of $\boldsymbol{\lambda}$ five thousand times (50 000 iterations, thinned by a factor of 10) and plotting the results. The standard k-simplex is defined in Definition B.25; $\lambda_1 = (1, 0, 0)$, $\lambda_2 = (0, 1, 0)$ and $\lambda_3 = (0, 0, 1)$ are the vertices of the equilateral triangles that enclose the standard 2-simplex. The solid circles show the mean of the prior and posterior densities (black and white circles, respectively).

We implement this over a grid of values for $\tilde{y}$ that span the Congressional seat share data, with the results displayed in Figure 6.8. At each point g on the grid, and with the output from each iteration of the Gibbs sampler, we compute the component memberships $\Pr(\tilde{z}_g = j | \tilde{y}_g, \mathbf{y})$ for $j \in \{1, 2, 3\}$ and plot the results as a function of $\tilde{y}$. We superimpose the lines over many iterations of the Gibbs sampler to allow us to visualize uncertainty in the classification probabilities. The three panels of Figure 6.8 correspond to the three components of the mixture density, with the component corresponding to Republican strongholds on the left, Democratic seats in the middle panel, and the small component for overwhelmingly Democratic seats shown on the right; the thicker, gray lines indicate

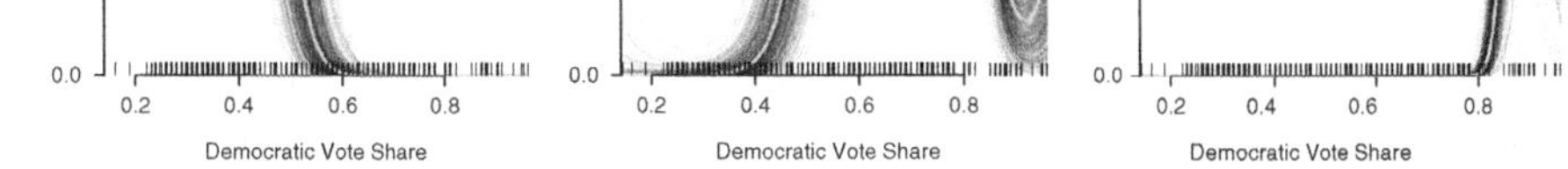

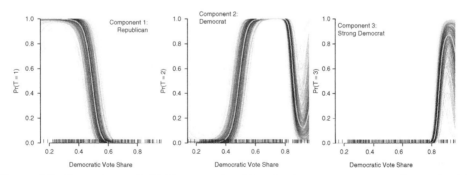

Figure 6.8 Posterior probabilities, membership of mixture components, Example 6.8. Each panel shows the probability of belonging to a specific component (vertical axis), over the range of Democratic vote shares observed in the data (horizontal axis). Each thin line shows the component probabilities given by the parameters as they vary over iterations of the MCMC algorithm; the thicker, gray lines show the component membership probabilities with the parameter values set at the means of their posterior densities.

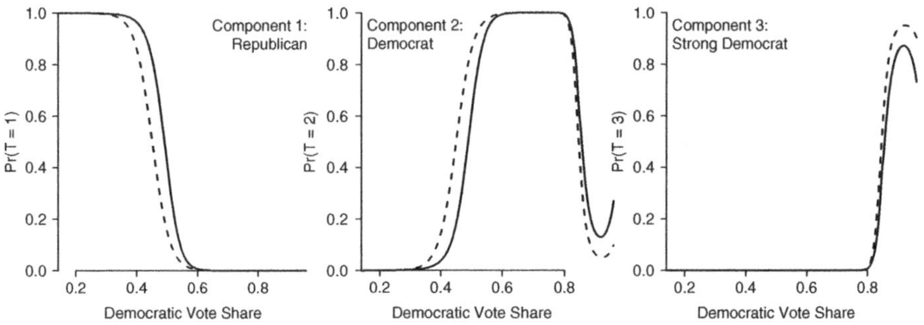

Figure 6.9 Prior and posterior probabilities, membership of mixture components, Example 6.8. Each panel shows the probability of belonging to a specific component (vertical axis), over the range of Democratic vote shares observed in the data (horizontal axis). Dotted lines show the prior classification probabilities, with parameters set at their prior means; solid lines show the posterior probabilities, generated with parameter values set at the means of their posterior densities.

the classification probabilities that arise when the parameters $\boldsymbol{\theta}$ are set at their posterior means.

The curves shown in Figure 6.8 range from zero to one (or to about .8, for component 3 on the right of the figure) and make fairly abrupt transitions between these extreme values. This suggests that the component densities are reasonably distinct, and the assignment of observations to components is relatively unambiguous. Of course, we used relatively informative priors to resolve the three components in this example, and the posterior classification probabilities differ little from those implied by the prior (see Figure 6.9). That is, in this case, the priors reflected little uncertainty as to how to assign seats to components, but rather, uncertainty as to the particular values of the parameters that characterize the components.

■ Example 6.9

Multivariate t regression model for compositional data. Katz and King (1999) examine data from multi-party elections to the United Kingdom's House of Commons. We focus on a subset of the data, covering elections in 1992 in 521 constituencies. The data are *compositional*, in that each observation is a vector of vote proportions that sums to one, by construction: i.e. vote proportions for the Labor Party, the Conservative Party, and the SDP-Liberal-Alliance (Lib-Dems). Votes for all other candidates, spoiled ballots and non-voting are ignored in this analysis, as in the original analysis by Katz and King (1999). In our reanalysis we also ignore seats that were not contested by all three of these parties; we return to this issue below.

A standard approach in the analysis of compositional data is to transform the J-vector of proportions that sum to one,

$$\mathbf{p}_i = (p_{i1}, \ldots, p_{iJ})', \; p_{ij} \in [0, 1], \forall \; j = 1, \ldots, J; \sum_{j=1}^{J} p_{ij} = 1 \, \forall \; i$$

to a $J - 1$ vector of log-odds ratios $\mathbf{y}_i$ where $y_{ij} = \log(p_{ij}/p_{iJ}) \forall\ j = 1, \ldots, J - 1$; see Aitchison (1986). $J = 3$ in the analysis here. Katz and King (1999) note that the log-odds ratios generated by these data appear to be heavy-tailed relative to a $J - 1$-variate normal model, even conditional on covariates, and so use a multivariate t regression model for these data, with unknown degrees of freedom. Katz and King (1999) fit the model via maximum likelihood methods; here we sketch a Bayesian approach, using MCMC methods implemented in JAGS/BUGS.

The two log-odds ratios in each $\mathbf{y}_i$ are formed by dividing the Labor and Conservative vote shares by the Lib-Dem vote shares. There are seven predictors in the analysis: a constant (x_{i1}), the lagged values (from the previous election) of the two log odds ratios (x_{i2} and x_{i3}), an indicator for whether the Conservative candidate is the incumbent (x_{i4}), the incumbency status of the Labor candidate (x_{i5}), the incumbency status of the Lib-Dem candidate (x_{i6}) and an indicator for seats in which no incumbent is running (x_{i7}). Note that $x_{i1} = \sum_{k=4}^{7} x_{ik} = 1 \forall\ i$, and so the model is over-parameterized (the underlying regression model suffers from perfect collinearity) and the parameters attaching to those variables are not identified. We proceed ignoring this lack of identification for the moment.

Let $\mathbf{x}_i$ be the vector of covariates. Then the multivariate t regression model is

$$\mathbf{y}_i = \begin{bmatrix} y_{i1} \\ y_{i2} \end{bmatrix} \sim t_\nu \left(\begin{bmatrix} \mathbf{x}_i \boldsymbol{\beta}_1 \\ \mathbf{x}_i \boldsymbol{\beta}_2 \end{bmatrix}, \boldsymbol{\Sigma} \right) \tag{6.5}$$

where the $\boldsymbol{\beta}_j$ are vectors of regression parameters ($j = 1, 2$), $\boldsymbol{\Sigma}$ is a squared scale matrix (a symmetric, $J - 1$-by-$J - 1$ positive definite matrix) and ν is a degrees of freedom parameter (see Definition B.38 for a definition of the multivariate t density). Under an assumption of conditional independence given the covariates $\mathbf{x}_i$ and the regression model, the likelihood for this problem is simply the n-fold product of these observation-specific multivariate t densities. The model is closely related to the seemingly-unrelated regression (SUR) model (Zellner 1962) save for the use of the multivariate t density in place of the multivariate normal.

A Bayesian analysis requires a prior for the elements $\boldsymbol{\theta}$. I impose the prior independence assumptions:

$$p(\boldsymbol{\theta}) = p(\sigma_{11})p(\sigma_{22})p(\rho)p(\nu) \times \prod_{k=1}^{7} \prod_{j=1}^{2} p(\beta_{kj})$$

where $\beta_{kj} \sim N(0, 1000)$, $p(\sigma_{jj}) \sim \text{Unif}(0, .5)$, $\rho \sim \text{Unif}(-1, 1)$, $\sigma_{12} = \rho\sigma_{11}\sigma_{22}$ and $\nu \sim \text{Unif}(3, 30)$. Note that we do not impose an inverse-Wishart prior on the squared scale matrix $\boldsymbol{\Sigma}$, but put uniform priors on the diagonal elements σ_{11} and σ_{22}, and then recover a positive definite $\boldsymbol{\Sigma}$ by defining $\sigma_{12} = \rho\sigma_{11}\sigma_{22}$ where $\rho \sim \text{Unif}(-1, 1)$.

The uniform prior on the degrees of freedom parameter ν encompasses a very heavy-tailed density at the lower bound of $\nu = 3$; although the t tends to the normal as $\nu \to \infty$, a t density with degrees of freedom at the upper bound of $\nu = 30$ is virtually indistinguishable from the normal. In this sense the prior on ν is quite permissive, allowing the data to adjudicate between the normal or a low degree-of-freedom, heavy-tailed t density. Note also that while we are used to thinking of degrees of freedom as positive integers, in the context of indexing a t density *per se* the restriction to integer values of ν is unnecessary.

Speeding up the MCMC algorithm via over-parameterization. As noted above, the model is over-parameterized. We do this deliberately. A lack of identification poses no formal problem for Bayesian analysis: so long as the prior density is proper, so too is the posterior density. And more pragmatically, a MCMC algorithm that explores a posterior density over a set of unidentified parameters often generates an efficient exploration of the posterior density defined over the (lower-dimensional) space of identified parameters. That is, in many situations, it is to our advantage to specify a model that contains unidentified parameters, let the MCMC algorithm run, and then examine the output of the algorithm projected down into the space of identified parameters. This is a strategy we will explore in greater detail in the context of hierarchical models (e.g. Section 7.2.2) and we defer a detailed discussion until then.

For now, note that the mapping from the set of unidentified parameters to identified parameters is trivial. The unidentified model includes an intercept plus a set of four mutually exclusive and exhaustive indicators of election type. Identification can be obtained by redefining the coefficients on the election type indicators (β_{4j} through β_{7j}, $j = 1, 2$) as offsets around the intercept β_{1j}. This is accomplished by imposing the constraint $\sum_{k=4}^{7} \beta_{kj} = 0$, $j = 1, 2$. We implement this by letting the MCMC algorithm run with the unidentified model. At each iteration t we compute $\beta_{kj}^{*(t)} = \beta_{kj}^{(t)} - \bar{\beta}_j^{(t)}$, $k = 4, \ldots, 7$ and $j = 1, 2$, and where $\bar{\beta}_j^{(t)} = (1/4)\sum_{k=4}^{7} \beta_{kj}^{(t)}$, $j = 1, 2$. In this way we have imposed the constraint $\sum_{k=4}^{7} \beta_{kj}^{*(t)} = 0$, $j = 1, 2, \forall\ t$, with β_{kj}^{*} being the set of identified parameters. We can implement this mapping from unidentified to identified parameters either in the JAGS/BUGS program itself, or 'post-process' the MCMC output in R. In this case we do the latter.

The JAGS code for this problem appears below:

JAGS code

```
model{
        for(i in 1:521){
              for(j in 1:2){
                          mu[i,j] <- inprod(x[i,1:7],beta[1:7,j])
               }
               y[i,1:2] ~ dmt(mu[i,1:2],Prec,nu); ## multivariate t
        }

        ## priors
        for(j in 1:2){
              beta[1:7,j] ~ dmnorm(b0,B0) ## regression parameters
              s[j] ~ dunif(0,.5)          ## sqrt root of diagonal of Sigma
        }
        rho ~ dunif(-1,1)                 ## auxiliary parameter

        ## form Sigma
        Sigma[1,1] <- pow(s[1],2)
        Sigma[2,2] <- pow(s[2],2)
        Sigma[1,2] <- rho*s[1]*s[2]       ## off-diagonal of pos-def Sigma
        Sigma[2,1] <- Sigma[1,2]
        Prec <- inverse(Sigma)            ## convert to precision matrix

        nu ~ dunif(3,30)                  ## degrees of freedom parameter
}
```

This code makes use of the `dmt` construct, denoting a multivariate t density; a univariate t density uses the `dt` construct in `BUGS/JAGS`. Like many densities in `BUGS/JAGS` this function takes the inverse squared scale matrix as an argument, rather than squared scale matrix itself (i.e. recall that the `dnorm` and `dmnorm` functions take precisions as their second arguments). Note also that the unknown degrees of freedom parameter `nu` appears as the third argument to the `dmt` construct.

We also pass the covariates from `R` to `JAGS` as a matrix $\mathbf{X}$, and form the linear predictor $\mu_{ij} = \mathbf{x}_i \boldsymbol{\beta}_j$ via the `inprod` function. As is often the case with `JAGS` and `BUGS`, the model for the data is easy to specify; the bulk of the program is actually devoted to specifying the priors and making transformations. The hyper-parameters `b0` and `B0` are passed to `JAGS` along with the data `y` and `x`.

The `R` code that invokes `JAGS` in this case is:

R code

```
1 forJags <- list(y=y,
2                 x=x,
3                 b0=rep(0,7),
4                 B0=diag(.0001,7))
5
6 inits <- list(list(beta=matrix(0,7,2),        ## two sets of initial values
7                    s=runif(n=2,0,.5),
8                    rho=runif(n=1,-1,1),
9                    nu=4),
10              list(beta=matrix(0,7,2),
11                   s=runif(n=2,0,.5),
12                   rho=runif(n=1,-1,1),
13                   nu=29))
14 require(rjags)
15 foo <- jags.model(file="92plain.bug",
16                   data=forJags,
17                   inits=inits,
18                   nchain=2,
19                   n.adapt=10e3)
20
21 out <- coda.samples(foo,
22                     variable.names=c("beta","Sigma","nu"),
23                     n.iter=50e3)
```

Two sets of initial values are passed to `JAGS` in the object `inits`, a list of lists. Note that we randomly generate start values for `s` and `rho`. This helps ensure that the chains start at different parts of the parameter space, as do the different start values for `nu` (the degrees of freedom parameter, see lines 9 and 13).

We ask `JAGS` to run two multiple chains in parallel with the `nchain` argument (line 18) in the call to `jags.model`. A longer than usual adaption phase is specified with the `n.adapt` argument (line 19), since using the default value of `n.adapt` resulted in `JAGS` generating a warning message. Fifty thousand iterations are requested with the `coda.samples` command at line 21.

After the `JAGS` program has run we map the output of the two chains on the space of unidentified parameters into the space of identified parameters, via the transformations discussed above. We implement this with the following `R` code, with the work being done inside a function, `remap`:

R code

```
## map into space of identified parameters
remap <- function(x){
  foo <- x                            ## copy MCMC output
  int <- x[,"beta[1,1]"]
  k <- cbind(x[,"beta[4,1]"],         ## subset of parameters we need
             x[,"beta[5,1]"],
             x[,"beta[6,1]"],
             x[,"beta[7,1]"])
  kbar <- apply(k,1,mean)             ## mean at each iteration
  k <- sweep(k,1,kbar)                ## subtract out
  foo[,"beta[1,1]"]  <- int + kbar
  foo[,"beta[4,1]"] <- k[,1]          ## write mean-differenced to output
  foo[,"beta[5,1]"] <- k[,2]
  foo[,"beta[6,1]"] <- k[,3]
  foo[,"beta[7,1]"] <- k[,4]

  int <- x[,"beta[1,2]"]
  k <- cbind(x[,"beta[4,2]"],         ## repeat 2nd set of coefficients
             x[,"beta[5,2]"],
             x[,"beta[6,2]"],
             x[,"beta[7,2]"])
  kbar <- apply(k,1,mean)
  k <- sweep(k,1,kbar)
  foo[,"beta[1,2]"] <- int + kbar
  foo[,"beta[4,2]"] <- k[,1]
  foo[,"beta[5,2]"] <- k[,2]
  foo[,"beta[6,2]"] <- k[,3]
  foo[,"beta[7,2]"] <- k[,4]

  return(foo)
}

z <- lapply(out,remap)                ## apply to each chain in out
```

The transformations are not at all difficult to implement (row-wise, mean differencing of a matrix) and the bulk of the work is keeping input and output aligned, and dealing with the fact that we have $J-1$ sets of regression coefficients to remap. Figure 6.10 shows trace plots for the unidentified parameters (left panels) and the transformed, identified parameters in the right panels. The identified parameters mix reasonably well, with Gelman and Rubin R statistics generally no larger than 1.03. On the other hand, Raftery-Lewis diagnostics suggest that the several hundred thousand iterations may be required to generate accurate estimates of the extreme quantiles of some of the marginal posterior densities of $\boldsymbol{\beta}_j$; JAGS uses a Metropolis sampler to update these parameters and may be suffering from a poor choice of a proposal density.

Figure 6.11 presents graphical summaries of the MCMC output for ν, the degrees of freedom parameter in the multivariate t regression model. The two chains produce virtually indistinguishable sets of output for this parameter, as evidenced by the superimposed trace plots (top panel of Figure 6.11), the virtually identical autocorrelation functions (lower left panel) and the density plots (lower right panel); the Gelman-Rubin R statistic for ν is 1.00. The MCMC algorithm is not particularly efficient in its exploration of

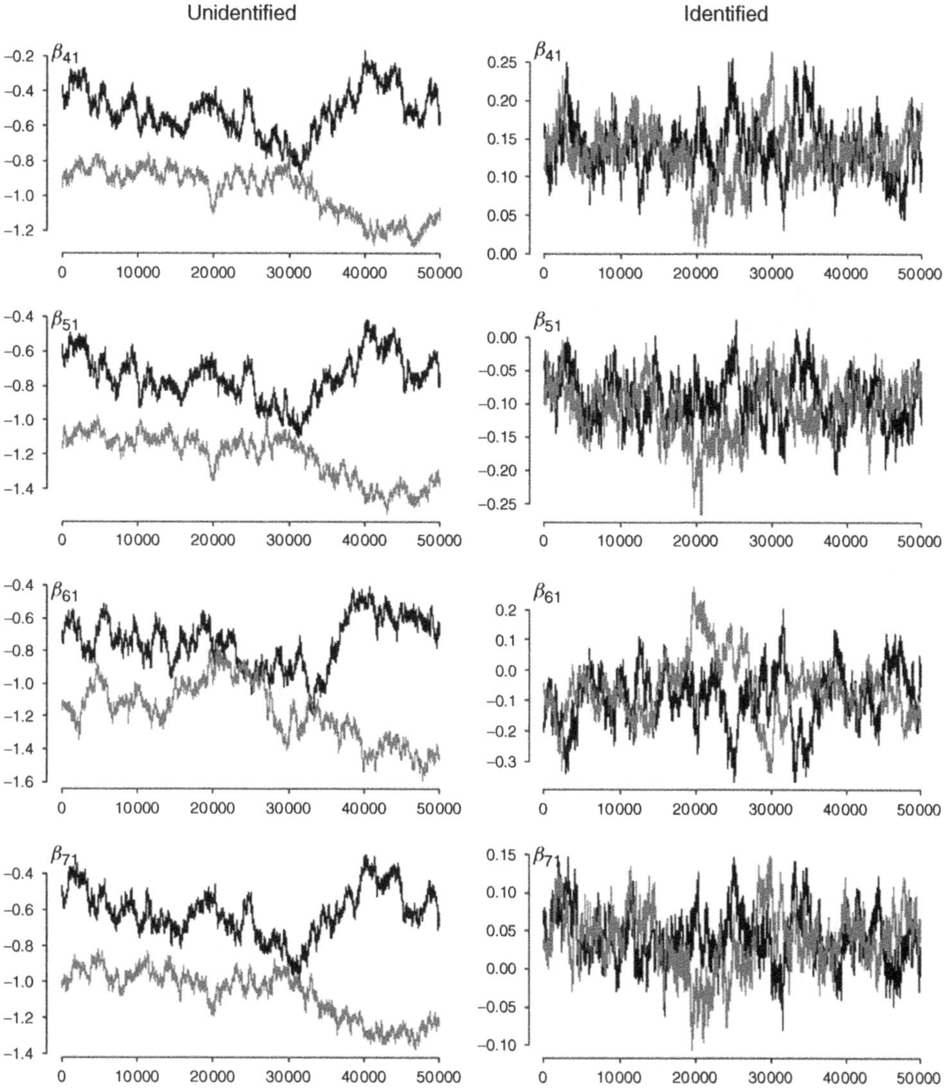

Figure 6.10 Trace plots, unidentified and identified parameters, Example 6.9. Each panels shows trace plots from the two chains; panels on the right are for identified panels. Note the similarity of the trace plots within chains, across parameters, for the unidentified parameters.

the marginal posterior density of ν; the 50 000 iterations per chain used here have the same amount of precision as about 7500 samples from an independence sampler. Finally, note the posterior density for ν is concentrated on values around 3 to 6, indicating that the density of these data is quite heavy-tailed relative to a normal density, even after conditioning on covariates. In Problem 6.16 you are invited to examine these data and assess the consequences of the apparent heavy-tailed nature of the data for the inferences drawn over $\boldsymbol{\beta}$.

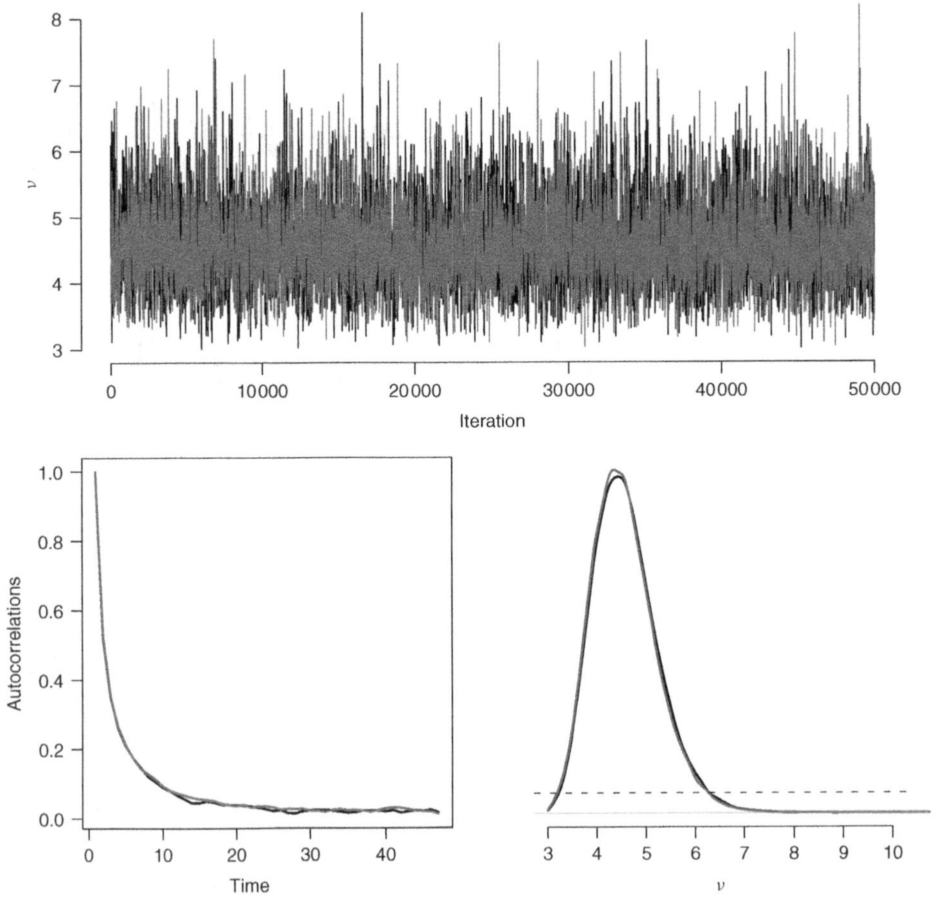

Figure 6.11 MCMC output for degrees of freedom parameter ν, multivariate t regression model, Example 6.9. Top panel shows traceplots for the two chains with respect to ν; lower left panel shows the autocorrelation functions for the two chains; lower right panel displays density plots, with the uniform prior shown as a dotted line.

6.6 Further reading

The 'Gibbs' of 'Gibbs sampling' is Josiah Willard Gibbs (1839–1903), an American theoretical physicist and chemist, and one of the great figures of 19th-century science. Gibbs is credited with founding statistical mechanics via the application of principles of thermodynamics. Several other quantities and functions in statistical mechanics bear his name.

The Gibbs sampler is commonly attributed to Geman and Geman (1984) who used the technique to study image restoration; i.e. the joint distribution of the contents of a field of pixels (pixel values) is usually a complicated high dimensional density, but can be tractably dealt with by using conditional distributions, where the conditioning is on the contents of neighboring groups of pixels. Image reconstruction and other spatial settings remain an active area of development and application of MCMC (Besag 1974; Green 1996; Smith and Roberts 1993b, 18–19) although obviously application of the Gibbs sampler is by no means restricted to Gibbs distributions or statistical inference for spatial processes.

Gelfand and Smith (1990) are generally credited with bringing the Gibbs sampler to the statistical mainstream. The long hiatus between the use of Markov chain Monte Carlo methods in statistical physics and their popularity in statistics *per se* is remarkable, and a great testament to the way technology shapes scientific practice; i.e. until desktop computing became relatively fast and cheap, Markov chain Monte Carlo was something done in large, nationally funded laboratories, not in statistics departments, let alone political science or sociology departments.

The data augmentation algorithm of Tanner and Wong (1987) amounts to a special case of Gibbs sampling, as is Rubin's (1987) work on multiple imputations for missing data.

There are several Metropolis samplers conveniently available in `R` packages. References to `MCMCpack` (Martin, Quinn and Park 2009) appear throughout this chapter and Chapter 8. A general random-walk Metropolis sampler with a normal proposal density is available in the `mcmc` package (Geyer 2005). An implementation of adaptive rejection Metropolis sampling (Gilks, Best and Tan 1995) appears in the `HI` package (Petris and Tardella 2006). User-defined Metropolis samplers are relatively easy to implement in the `Umacs` package (Kerman 2007).

Analyzing MCMC output so as to diagnose problems with convergence and/or stationarity of the sampled sequence is a critical part of a MCMC-based Bayesian analysis. We mentioned the `coda` package (Plummer *et al*. 2008) in this chapter, but see also the `boa` (Smith 2007) and `scapeMCMC` (Magnusson and Stewart 2007) packages.

Problems

6.1 Verify the following result given by Gelman, Roberts and Gilks (1995): i.e. if we are using a random-walk Metropolis algorithm to sample from $\theta \sim N(0, 1)$ and proposals are drawn as $\theta^* = \theta^{(t)} + \epsilon, \epsilon \sim N(0, \sigma^2)$, then setting $\sigma = 2.38$ generates a Markov chain with the lowest level of autocorrelation, generating an acceptance rate of about 44 %. That is, construct a computer simulation exercise to examine this claim, examining the performance of the random-walk Metropolis algorithm with a variety of choices of σ. Does it make any difference to the performance of the algorithm (or the conclusion as to the optimal value of σ) if we use the Metropolis acceptance ratio r_M (Equation 5.2) or the Metropolis-Hastings version of the acceptance ratio r (Equation 5.1)?

6.2 Consider the Poisson regression analysis in Example 5.2, where we used the random-walk Metropolis sampler (implemented in `MCMCpack`). Write a computer program that implements an independence Metropolis sampler for this problem, with the proposal density centered on the precision-weighted average of the prior and the MLEs of $\boldsymbol{\beta}$ as given in Equation 5.6 and where the target, posterior density for the Metropolis algorithm appears in Equation 5.4. Does the independence Metropolis sampler outperform the random-walk version of the Metropolis algorithm?

6.3 (Casella and George 1992). Suppose two random variables θ_1 and θ_2 have the conditional distributions

$$\theta_1|\theta_2 \sim \text{Exponential}(\theta_2), \quad 0 < \theta_1 < \infty,$$
$$\theta_2|\theta_1 \sim \text{Exponential}(\theta_1), \quad 0 < \theta_2 < \infty.$$

1. Consider the application of Proposition 5.1 to this problem. What is $f(\theta_1|\theta_2)/f(\theta_2|\theta_1)$ in this case?
2. What is $\int \frac{f(\theta_1|\theta_2)}{f(\theta_2|}\theta_1)d\theta_1$? Does this integral exist?
3. What does your answer to the previous question imply about the existence of a proper joint density $p(\theta_1, \theta_2)$?

6.4 Prove the following claim that appears in the discussion of the slice sampler (§5.2.7). That is, let p be the density of $\theta \in \Theta \equiv [a, b] \subseteq \mathbb{R}$ and let $m > p(\theta)$. Let (θ^*, U^*) be a uniform draw from the set $\tilde{\Theta} = \Theta \times [0, m]$; only accept the sampled draw if $0 < U^* < f(\theta^*)$. Show that the marginal density of the sampled θ^* is p.

6.5 Consider the inverse-Wishart density (Definition B.41), the conjugate prior for the covariance matrix given a multivariate normal likelihood. A function for sampling from the inverse-Wishart density, `riwish`, is available in the `R` package, `MCMC-pack` (Martin, Quinn and Park 2009); note that the `riwish` function takes a scale matrix as one of its arguments, not a precision matrix.

1. Generate T sampled matrices $\mathbf{\Sigma}^{(t)} \stackrel{\text{iid}}{\sim} \text{inverse-Wishart}(\mathbf{\Psi}, n_0)$ where $p = 2$, $n_0 = 4$ and $\mathbf{\Psi} = \mathbf{I}_p$. Set T to an arbitrarily large number (e.g. $T = 5000$ or larger). Use the T sampled $\mathbf{\Sigma}^{(t)}$ to form a Monte Carlo estimate of $E(\mathbf{\Sigma})$. Verify that up to Monte Carlo error, $E(\mathbf{\Sigma}) = (n_0 - p - 1)^{-1}\mathbf{\Psi}$ (which equals $\mathbf{\Psi} = \mathbf{I}_p$ for the particular values of n_0, p and $\mathbf{\Psi}$ used here).
2. Use the sampled $\mathbf{\Sigma}^{(t)}$ to generate a Monte Carlo estimate of the density of $\rho = \rho(\mathbf{\Sigma}) = \sigma_{12}/(\sigma_{11}\sigma_{22})$, where

$$\mathbf{\Sigma} = \begin{bmatrix} \sigma_{11}^2 & \sigma_{12} \\ \sigma_{21} & \sigma_{22}^2 \end{bmatrix}.$$

That is, for $t = 1, \ldots, T$, compute $\rho^{(t)} = \rho(\mathbf{\Sigma}^{(t)})$; then plot a histogram and compute summary statistics for the $\rho^{(t)}$. What do you observe? Put differently, you are being asked to use Monte Carlo methods to solve the following problem: if $\mathbf{\Sigma}$ is a 2-by-2 covariance matrix and has an inverse-Wishart$(\mathbf{\Psi}, n_0)$ density, then what is the (implied) density over the correlation parameter ρ?
3. Repeat the previous question but with $n_0 = 30$. How has the implied density on ρ changed, if at all?
4. Repeat part 2 of this problem, but with $\mathbf{\Psi} = \text{diag}(4, 2)$. How has the implied density on ρ changed, if at all?

5. Repeat part 2 of this problem, with

$$\mathbf{\Psi} = \frac{1}{n_0}\begin{bmatrix} 1 & .7 \\ .7 & 1 \end{bmatrix}^{-1}$$

and $n_0 = 4$ and then again with $n_0 = 40$. Again, comment on the estimated densities for ρ.

6.6 Let $\mathbf{\Sigma}$ be a 2-by-2 covariance matrix as defined in the previous question. Suppose we assign $\mathbf{\Sigma}$ an improper prior density $p(\mathbf{\Sigma}) \propto |\mathbf{\Sigma}|^{-3/2}$. Derive $p(\rho)$ (at least up to a factor of proportionality) where ρ is the correlation coefficient $\sigma_{12}/(\sigma_{11}\sigma_{22})$. Comment on the shape of this density and its plausibility as a prior density for a correlation coefficient.

6.7 Consider a first order auto-regressive process $\theta_t = \rho\theta_{t-1} + \epsilon_t$ with $\epsilon_t \overset{\text{iid}}{\sim} N(0, \sigma^2)$. Draw samples from this process, for each of $\rho \in \{.5, .8, .9, .95, .99, .995\}$, with sample sizes $T = 10^d, d \in \{2, 3, \ldots, 6\}$. Initialize the simulation with $\theta_0 = 0$ and set $\sigma^2 = 1$. Hint: the `arima.sim` function in `R` is one way to do this.

1. For each (ρ, T) combination above, compute the Geweke (1992) stationarity diagnostic with respect to the series $\{\theta_t\}$. Comment on any pattern you see over the values of ρ and T used in each simulation. The Geweke diagnostic is available via the `geweke.diag` function in the `R` package `coda`.

2. Repeat the previous question, but compute the default Raftery-Lewis (1992a) run-length diagnostic. Report the dependency factor reported by the Raftery-Lewis procedure. Again, comment on any patterns you see over the values of ρ and T used the simulation. The Raftery-Lewis diagnostic is available via the `raftery.diag` function in the `R` package `coda` (Plummer *et al*. 2008).

6.8 Consider the bivariate normal density given in Example 6.2. Sample from this density using a two component Gibbs sampler, as described in Example 6.2; you should be able to code this sampler in a statistical computing environment such as `R`. Run the sampler for at least 5000 iterations. Compute standard convergence diagnostic tests using the ouptut of the sampler. What do these tests report? How many iterations are suggested by the Raftery-Lewis diagnostic? Repeat this exercise with $\sigma_{12} = .999$ and $\sigma_{12} = .8$.

6.9 Refer to Example 5.7. Derive and program the conditional distributions that constitute the Gibbs sampler for this problem. Does the output of the Gibbs sampler reflect the impropriety of the posterior density?

6.10 Consider the normal, linear regression model discussed in Sections 6.3 and 6.4.2: i.e.

$$\mathbf{y}|\mathbf{x} \sim N(\mathbf{X}\boldsymbol{\beta}, \sigma^2\mathbf{I}_n)$$

$$\boldsymbol{\beta} \sim N(\mathbf{0}, \kappa \cdot \mathbf{I}_k),$$

$$\sigma \sim \text{Unif}(0, d)$$

With the (marginal) normal prior over $\boldsymbol{\beta}$ and the normal likelihood for $\mathbf{y}$, this model is said to be conditionally conjugate with respect to $\boldsymbol{\beta}$, in the sense that given σ^2, $\mathbf{y}$ and $\mathbf{x}$, the conditional distribution of $\boldsymbol{\beta}$ is multivariate normal, as is its prior. Derive the conditional distributions for $\boldsymbol{\theta} = (\boldsymbol{\beta}, \sigma)'$ needed to implement a Gibbs sampler for this problem.
Hint: make a change of variables from σ to σ^2; do note that the uniform prior for σ on $[0, d]$ does not induce a uniform prior over σ^2 on $[0, d^2]$.

6.11 Verify the claim of the discussion in §6.4.3. Consider the Ashenfleter absentee ballot data (available as the data frame `absentee` in the `R` package `pscl` and considered in Example 2.15. Perform a Bayesian regression analysis with a Gibbs sampler with the original x variable, and a second analysis with the predictor deviated around its mean. Use priors that specify *a priori* independence between β_1 and β_2. In each case, use a Gibbs sampler that does not sample from the conditional distribution of $\boldsymbol{\beta} = (\beta_1, \beta_2)'$ *en bloc*; rather, sample from the conditional distributions of β_1 and β_2, since the goal of this problem is to investigate the effect of reparameterization (not the effects of 'blocking' parameters). Compare the performance of the two samplers. Verify that (a) the 'unblocked' MCMC algorithm that uses the mean-deviated predictors mixes better than the 'unblocked' MCMC algorithm that uses the raw predictors; (b) the posterior density of the slope parameter is invariant to the reparameterization (i.e. mean-deviating x does not change the slope in a regression, just the intercept); (c) for the mean-deviated version of the model, β_1 and β_2 have zero posterior covariance.

6.12 Refer to Example 5.8 where we considered a Gibbs sampler failing to properly explore a density with regions of nonconnected support. Consider the change of variables suggested by Robert and Casella (2004, 379) to $z_1 = \theta_1 + \theta_2$ and $z_2 = \theta_1 - \theta_2$. What are the conditional densities $g_1(z_1|z_2)$ and $g_2(z_2|z_1)$? Implement this Gibbs sampler. At each iteration t, map the sampled $\mathbf{z}^{(t)} = (z_1^{(t)}, z_2^{(t)})'$ into $\boldsymbol{\theta}^{(t)} = (\theta_1^{(t)}, \theta_2^{(t)})'$. With a long run of the Gibbs sampler, verify that this reparameterized Gibbs sampler is recovering the density of interest over θ_1 and θ_2.

6.13 In Example 2.8 and 2.9 we considered a Bayesian approach to inference for the difference of two proportions. Implement these tests for those examples in `JAGS` or `BUGS`.

6.14 In Examples 5.4 and 6.7 we fit negative binomial regression models to the data on article production by biochemistry graduate students. Replicate these analyses. Compare the performance of the MCMC algorithm used by `BUGS` (or `JAGS`) with that from the `rnegbinRw` function in the `bayesm` package in `R`. Which algorithm seems to generate a more efficient exploration of the posterior density? When you compare the algorithms, consider not just the number of iterations consumed by each algorithm to attain a given level of Monte Carlo error, but also the time taken to produce those iterations by your computer.

6.15 Recall the finite normal mixture model in Example 6.8. Suppose we set up the model with conditionally conjugate priors: i.e.

$$y_i | z_{ij} = 1 \sim N(\mu_j, \sigma_j^2)$$

$$
\begin{aligned}
\mu_j &\sim N(m_j, M_j) \\
\sigma_j^2 &\sim \text{inverse-Gamma}(\nu_j/2, \nu_j s_j^2/2) \\
\mathbf{z}_i = (z_{i1}, \dots, z_{iJ})' &\sim \text{Multinomial}(\boldsymbol{\lambda}, 1) \\
\boldsymbol{\lambda} = (\lambda_1, \dots, \lambda_J)' &\sim \text{Dirichlet}(\mathbf{a})
\end{aligned}
$$

where $y_i, i = 1, \dots, n$ are data and $m_j, M_j, \nu_j, s_j^2, (j = 1, \dots, J)$ and $\mathbf{a} = (a_1, \dots, a_J)'$ are user-supplied hyperparameters. Derive the conditional distributions necessary to implement a data-augmented Gibbs sampler for this problem, where the component indicators z_{ij} are missing data to be augmented.

6.16 Consider the data on 1992 UK House of Commons elections used in Example 6.9. These data are available in the R package `pscl` as the data set `UKHouseofCommons`. Fit the multivariate t regression model to these data; the code presented in Example 6.9 may be helpful. Compare the fit of the multivariate t regression model with that from a multivariate normal regression model, i.e. replacing Equation (6.5) with

$$
\mathbf{y}_i | \mathbf{x}_i \sim N\left(\begin{bmatrix} \mathbf{x}_i \boldsymbol{\beta}_1 \\ \mathbf{x}_i \boldsymbol{\beta}_2 \end{bmatrix}, \boldsymbol{\Sigma} \right).
$$

How do inferences over $\boldsymbol{\beta}$ change as we move from the multivariate t model to the multivariate normal model? The articles by Breusch, Robertson and Welsh (1997) and Tomz, Tucker and Wittenberg (2002) may be of interest as you make these comparisons.

Part III

Advanced Applications in the Social Sciences

7

Hierarchical Statistical Models

7.1 Data and parameters that vary by groups: the case for hierarchical modeling

In many social science settings, the data available for analysis span multiple groups. In these settings it is often plausible that any statistical model we might fit to the data will need to be flexible, so as to capture variation across the groups, typically accomplished by letting some or all of the parameters vary across the groups. Examples include survey data gathered over a set of locations (e.g. states, Congressional districts, countries); experimental studies deployed in multiple locations; and, perhaps the *locus classicus* of hierarchical modeling in the social sciences, studies of educational outcomes where the subjects are students, who are grouped in classes or schools, which nest in school districts, which in turn nest in states.

In studies of data of this type, the researcher is often interested in parameters that vary at the group level. These group level parameters go by different names, in different contexts, in different disciplines, and depending on the estimation method being used. Examples include 'contextual effects', 'fixed effects', 'random effects', and 'varying' or 'stochastic coefficients'. This between-group parameter variation is potentially of great substantive interest, since it speaks to a fundamental issue in empirical social science. Generally, we prefer more parsimonious models to more complicated models. In the search for generalizable propositions about social processes – ideally, perhaps, with the status of a scientific law – we seek simple explanations and simple models, with fewer parameters, rather than more. An empirical regularity worthy of the label ought not to depend on the vagaries of time and place (e.g. King, Keohane and Verba 1994; Przeworski and Teune 1970).

But the social world is not always so accommodating. While a laudable goal, spartan statistical modeling comes at the price of diminished social, political, or historical realism.

As we all quickly discover when fitting models to data, fewer parameters mean a poorer fit to the data. When fitting models to data that span multiple groups, this degradation in goodness of fit could well see the model work well in some groups but poorly in others. The drive towards parsimony often results in models with specifications that do not vary by groups, with global parameters that are weighted averages of the group-specific parameters. In some cases this can result in characterizations of social processes that are reasonable in an 'on average' sense, but ignore considerable heterogeneity between groups. Quite profound policy implications can result. Policy advice given on the basis of 'on average' modeling may well lead to a 'one-size fits all' fallacy: the policy recommendation that would seem to work for the 'average group' could perform poorly or even in a counter-productive way in any given group.

On the other hand, estimating statistical models group-by-group is often just not feasible, even if it was desirable. The amount of data in any given group can sometimes be small, or display very little meaningful variation, such that the within-group analysis yields inferences that are too imprecise to be useful. An example might be breaking down a survey designed to be nationally representative like the General Social Survey or the National Election Studies to generate inferences for states, counties, or congressional districts (Achen 1978; Levendusky, Pope and Jackman 2008; Miller and Stokes 1963).

Now this is not to say that group-by-group analyses should *not* be done: if the groups are sufficiently large then reasonable group-level inferences can be made. Moreover, group-by-group analysis is often an important preliminary step in data analysis – a useful and easily-implemented method for assessing parameter heterogeneity – but one that is often overlooked. Indeed, one of the most vocal proponents of Bayesian hierarchical modeling in the social sciences, Andrew Gelman, facetiously refers to group-by-group analysis as 'secret weapon': a 'weapon' in that group-by-group analysis can be enormously helpful, but 'secret' in that in the rush to implement various panel data estimators or hierarchical models and the like, analysts often neglect to take advantage of the insights available from group-by-group analysis. Many of us teach this advice to our students using other euphemisms: for instance, David Rogosa has long referred to 'unit-by-unit' analysis as 'smart 1st year student' (Rogosa and Saner 1995). But the general point is that breaking a large data set into group-specific pieces will generally result in a better fit to the group-specific data than from a pooled analysis, but at the price of heightened imprecision.

Clearly, there is a tension here, between parsimony and realism, between the high-bias/low-variance results that might be obtained with an analysis that ignores between-group parameter heterogeneity, and the low-bias/high-variance inferences from group-specific analyses. In this chapter we will see that hierarchical statistical models offer a principled way to negotiate this tension, to let us hit the 'sweet spot' between the two extremes of completely pooling across groups, or conducting analyses group-by-group. In fact, we shall see that hitting this 'sweet spot' follows from applying Bayes Rule (e.g. Proposition 7.1).

Hierarchical models deal with the possibility of parameter variation across groups by positing a model for the parameters. The 'hierarchy' arises because the model for the parameters sits 'above' the model for the data. Indeed, in this sense *all* Bayesian models are hierarchical, in that a prior for θ sits above the model for y. Generically, Bayesian hierarchical statistical models have the form

$$
\begin{aligned}
y_j|\theta_j &\sim p(y_j|\theta_j) && \text{(model for the data in group } j = 1, \ldots, J) \\
\theta_j|\nu &\sim p(\theta_j|\nu) && \text{(between-group model or `prior' for the parameters } \theta_j) \\
\nu &\sim p(\nu) && \text{(prior for the \textit{hyperparameter} } \nu),
\end{aligned}
$$

writing the hierarchy from 'bottom' to 'top'. The inferential challenge will be to compute the posterior density of all the parameters, $\boldsymbol{\theta} = (\theta_1, \ldots, \theta_J, \nu)'$ and any marginal posterior densities for specific elements of $\boldsymbol{\theta}$ that may be of interest. We will see that the Gibbs sampler is extremely well-suited to this task.

Perhaps the most simple illustration of a hierarchical model is the following implementation of a one-way ANOVA model:

■ **Example 7.1**

One-way analysis of variance (ANOVA); J normal means. We observe data y_{ij}, where $i = 1, \ldots, n_j$ indexes observations within groups $j = 1, \ldots, J$. Let $n = \sum_j n_j$ be the total number of observations. Within each group the mean of y is α_j; for simplicity, we assume homoskedasticity across groups such that $V(\mathbf{y}_j) = \sigma^2 \, \forall \, j$. The means $\bar{y}_1, \ldots, \bar{y}_J$ and an estimate of the common variance $\hat{\sigma}^2$ are sufficient statistics for the data if we assume a normal model for the data. We are interested in the possibility that the means vary across groups. The following hierarchical model operationalizes this possibility:

$$y_{ij}|\alpha_j, \sigma^2 \sim N(\alpha_j, \sigma^2) \tag{7.1a}$$

$$\alpha_j|\mu, \omega^2 \sim N(\mu, \omega^2). \tag{7.1b}$$

The sense in which this is a hierarchical model is that Equation 7.1a is a model for the data, with parameters α_j and σ^2, while Equation 7.1b is a model for how α_j varies across groups. The parameter μ is the grand mean, the mean of the distribution of the group means, and this group-level distribution has variance ω^2, also known as the *between* variance; σ^2 is known as the *within* variance for group j. The ratio $\omega^2/(\omega^2 + \sigma^2)$ is known as the *intra-class correlation* and is a measure of the relative magnitude of the 'within' variability for clustered or grouped data, in turn interpretable as a measure of the 'relative similarity' of observations in each group (Fisher 1925, Ch7; Koch 1983)

The parameters in the group-level model, μ and ω^2 are known as *hyperparameters*; prior densities for these parameters are necessary to complete the specification of the model, along with a prior for the 'within variance', σ^2. We return to a Bayesian treatment of ANOVA models as hierarchical models in Section 7.2.

In many settings it is often the case that we have some variables that capture the way the groups vary. Covariates of this sort often go by the name 'contextual' predictors, since they are constant for all observations in a given group. These covariates enter the model by structuring the way parameters vary across groups. Hierarchical models of this sort are also often referred to as *multi-level* models:

■ **Example 7.2**

Hierarchical regression model; 'multi-level' model. Perhaps one of the more common applications is the following hierarchical regression model:

$$y_{ij} \sim N(\mathbf{x}_{ij}\boldsymbol{\beta}_j, \sigma_j^2) \tag{7.2a}$$

$$\beta_{jk} \sim N(\mathbf{z}_j\boldsymbol{\gamma}_k, \omega_k^2) \tag{7.2b}$$

where, $i = 1, \ldots, n_j$, indexes observations within groups, $j = 1, \ldots, J$ indexes groups and k indexes the predictors. In this model the group-level covariates $\mathbf{z}_j$ structure the way the regression coefficients β_{jk} vary across groups. In the parlance of multilevel models, the regression model for the y_{ij} in Equation 7.2a is known as the 'level one' model, while the regression structure in Equation 7.2b for the $\boldsymbol{\beta}_j$ is known as the 'level two' model. In a Bayesian treatment of this model, prior densities for the hyperparameters $\boldsymbol{\gamma}_k$ and ω_k^2 complete the model specification, along with priors for the σ_j^2.

We consider a multi-level regression model in Example 7.10, using data from educational testing.

Representation as a 'Mixed Model'. Mixed models are yet another class of models that can be considered in the hierarchical model framework. Mixed models are 'mixed' in that predictors can have coefficients that do not vary over groups ('fixed effects') and coefficients that vary randomly over groups ('random effects'); intercept parameters can also be fixed or random effects in this framework. Incidentally, note that this nomenclature is rather odd to Bayesian ears: since *all* parameters are 'random' variables in the Bayesian approach this distinction between 'fixed' and 'random' parameters is somewhat artificial, but does underline the uncertain status of parameters that vary randomly over groups in the classical approach (these parameters pose no difficulty in a Bayesian setting).

Generically, for a linear, normal model, a mixed model is

$$\mathbf{y} \sim N(\mathbf{X}\boldsymbol{\beta} + \mathbf{Z}\mathbf{b}, \boldsymbol{\Sigma}) \tag{7.3}$$

where $\mathbf{X}$ is a n-by-k matrix of predictors that pick up fixed effects $\boldsymbol{\beta}$ (a k-by-1 vector of coefficients), $\mathbf{Z}$ is a n-by-p matrix of predictors that pick up random effects $\mathbf{b}$, and $\boldsymbol{\Sigma}$ is a n-by-n covariance matrix. The random effects $\mathbf{b}$ have the stochastic structure $\mathbf{b} \sim N(\mathbf{0}, \boldsymbol{\Omega})$, where $\boldsymbol{\Omega}$ is a p-by-p covariance matrix. The zero mean restriction on $\mathbf{b}$ is part of the fixed/random distinction: a covariate $\mathbf{x}$ picks up a coefficient β that is constant over the data set, while the terms in the $\mathbf{Z}$ matrix are interactions of $\mathbf{x}$ with indicators of group membership (i.e. $Z_{ij} = x_i$ if observation i is in group j, and 0 otherwise), such that the effect of $\mathbf{x}$ in group j is $\beta + b_j$.

The mixed model representation also imposes the restriction that the random effects $\mathbf{b}$ are conditionally uncorrelated with the level-one, idiosyncratic disturbances given the fixed effects in the model μ; i.e. if $y_{ij} = \mathbf{x}_i\boldsymbol{\beta} + \mathbf{z}_i b_j + \epsilon_{ij}$, then the mixed model assumes $\text{cov}(b_j, \epsilon_{ij}|\mathbf{x}_i, \boldsymbol{\beta}) = 0 \,\forall\, i, j$. This assumption is standard in the literature, allowing us to treat the random effects b_j and the idiosyncratic disturbances as separate contributions to the variance in $\mathbf{y}$ ('between' and 'within' variance, respectively, see Equation 7.5 in the example below).

■ **Example 7.3**

Hierarchical, one-way analysis of variance model as a mixed model, Example 7.1, continued. The hierarchical version of the one-way ANOVA model is

$$\mathbf{y} \sim N(\boldsymbol{\iota}_n \mu + \mathbf{Z}\mathbf{b}, \boldsymbol{\Sigma}) \tag{7.4}$$

where $\boldsymbol{\iota}_n$ is a n-by-1 unit vector, $\mathbf{Z}$ is a n-by-J *design matrix*, populated with indicators matching observations to groups (i.e. $Z_{ij} = 1$ if observation i is in group j and 0 otherwise), and $\mathbf{b}$ is a J-by-1 vector of parameters with $\mathbf{b} \sim N(\mathbf{0}, \boldsymbol{\Omega})$, where $\boldsymbol{\Omega} = \omega^2 \mathbf{I}_J$, and $\boldsymbol{\Sigma} = \sigma^2 \mathbf{I}_n$. In this parameterization, μ is the grand mean, while the b_j are group-specific offsets around the grand mean, such that $E(y_{\cdot j}) = \mu + b_j$.

Variance components representation. An interesting feature of this model comes from considering it as a *variance components* model. The model decomposes variation in $\mathbf{y}$ around μ into two components: the 'within-group' variance σ^2 and the 'between-group' variance ω^2. In fact, the mixed model representation above implies

$$\begin{aligned}
\text{var}(\mathbf{y}) &= \text{var}(\mu) + \underbrace{\mathbf{Z}\text{var}(\mathbf{b})\mathbf{Z}'}_{\text{'between variance'}} + \underbrace{\text{var}(\boldsymbol{\epsilon})}_{\text{'within variance'}} &(7.5a)\\
&= \omega^2 \mathbf{Z}\mathbf{I}_J\mathbf{Z}' + \sigma^2 \mathbf{I}_n &(7.5b)\\
&= \omega^2 \mathbf{F} + \sigma^2 \mathbf{I}_n, &(7.5c)
\end{aligned}$$

where $\mathbf{F} = \mathbf{Z}\mathbf{Z}'$ is a block diagonal matrix with blocks $\boldsymbol{\iota}_{n_j}\boldsymbol{\iota}_{n_j}'$ (a square n_j-by-n_j matrix of ones), $j = 1, \ldots, J$; in the case of a *balanced* design $n_j = n_k \forall\, j, k$ and $\mathbf{F} = \mathbf{I}_J \otimes \boldsymbol{\iota}_{n_j}\boldsymbol{\iota}_{n_j}'$ (see Definition A.10). For group j, we have

$$\text{var}(\mathbf{y}_j) = \omega^2 \boldsymbol{\iota}_{n_j}\boldsymbol{\iota}_{n_j}' + \sigma^2 \mathbf{I}_{n_j} = \begin{bmatrix} \sigma^2+\omega^2 & \omega^2 & \ldots & \omega^2 \\ \omega^2 & \sigma^2+\omega^2 & \ldots & \omega^2 \\ \vdots & \vdots & \ddots & \vdots \\ \omega^2 & \omega^2 & \ldots & \sigma^2+\omega^2 \end{bmatrix}, \tag{7.6}$$

i.e. the random effects component of the mixed model induces non-zero covariances across observations *within* groups, stemming from the fact that these observations share the group-specific term $b_j \sim N(0, \omega^2)$. Note that in this way, the hierarchical model (aka a 'mixed model' or 'random effects' model) makes this within-cluster covariance explicit. Contrast classical approaches that treat the 'clustered' nature of the data as a nuisance, and perform inference for the fixed effects with 'cluster robust' standard errors.

7.1.1 Exchangeable parameters generate hierarchical models

In the one-way ANOVA example presented above, Equation 7.1b expresses a belief that the group-specific means α_j are *exchangeable*, as we encountered in Section 1.9. That is, we do not possess any knowledge about the groups (or any subset of the groups) that would lead us to assign their corresponding means a prior other than the $N(\mu, \omega^2)$ density assigned to the other groups' means.

In Section 1.9 we introduced exchangeability as a way to generate parametric statistical models given exchangeable probability assignments over *observables*. Here we see that exchangeability is a concept that applies not only to (stochastic) data, but can be a model generating device for parameters as well. In turn, this stems from the fact that parameters are random variables in the Bayesian formulation, and so it is straightforward to extend the concept of exchangeability from observables to parameters. In fact, there are *two* sets of exchangeability assumptions implicit in the model above: (1) exchangeability for the y_{ij} *within* each group, conditional on α_j and σ^2, and (2) exchangeability of the α_j *across* groups, conditional on μ and ω^2.

Put differently, when exchangeability holds for group-specific parameters the implication is that we can and ought to specify hierarchical models such as Equation 7.1. On the other hand, consider the situation where exchangeability does *not* hold. That is, suppose we know that there is something distinct about one or more of the groups such that we would not assign the same probabilistic model to all of the group means. We illustrate this with the following hypothetical example.

■ Example 7.4

Exchangeability and hierarchical models for polling data. Suppose we have data from a survey conducted in J counties in the United States. We ask respondents about their support for border protection, yielding individual level responses $y_{ij} = 1$ if respondent i wants more spending on border protection and 0 otherwise. We possess no individual-level predictors with which to model the responses, and so exchangeability is a reasonable assumption at the micro-level, giving rise to the following binomial model $r_j \sim \text{Binomial}(\theta_j, n_j)$, where $r_j = \sum_{i=1}^{n} y_{ij}$ is the number of survey respondents expressing support for increased spending on border protection, and the θ_j are the unknown county-level proportions in support of increased spending on border protection.

If we lack any prior information with which to distinguish counties from one another an assumption of exchangeability is appropriate for the θ_j, giving rise to the hierarchical model

$$r_j \sim \text{Binomial}(\theta_j, n_j) \tag{7.7a}$$

$$\theta_j \sim \text{Beta}(\alpha, \beta) \tag{7.7b}$$

where α and β are hyperparameters. But when will a hierarchical model be *inappropriate*? The answer to this question turns on the appropriateness of the exchangeability assumption. Suppose we in fact do possess prior information with which to distinguish the θ_j across counties. In particular, we know that some of the counties in the data are from border regions of Arizona and Texas. This suggests that we do have a basis for not assigning all counties the same prior density.

An assumption of *conditional exchangeability* might make sense in this case, given, say, the distance of the geographic centroid of each county from the US-Mexico border, d_j. That is, under this form of conditional exchangeability, we think that the α_j are *a priori* likely to be similar given similar levels of d_j. In this case we might replace the hierarchical model in Equation 7.7 with the *multi-level model*

$$r_j \sim \text{Binomial}(\theta_j, n_j) \tag{7.8a}$$

$$\log\left(\frac{\theta_j}{1-\theta_j}\right) \sim N(\beta_0 + \beta_1 d_j, \omega^2), \tag{7.8b}$$

where prior densities over β_0, β_1 and ω^2 would complete the specification of the model. Note that in this hierarchical model the distance measure d_j appears in a regression structure for the logits of the θ_j, with the θ_j in turn appearing in the binomial model for the county-level survey outcomes.

7.1.2 'Borrowing strength' via exchangeability

As discussed above, hierarchical models are implied when we believe exchangeability holds over group level parameters. An important consequence of exchangeability is a phenomenon known as *borrowing strength* or *combining information*. In the context of Example 7.1, inferences for the group-level parameters α_j reflect not just the information about α_j in group j, but, via the hierarchical model, will also draw on relevant information in the other groups.

Simply put, if the group-specific parameters α_j are exchangeable, then information about the α_j flows 'up' the hierarchy to inform inferences for the other α_j parameters; i.e. via the hyperparameters μ and ω^2 in Equation 7.1b. Bayesian inference for any particular α_j will reflect a combination of information about α_j from the data in group j and the hierarchical component of the model (the 'prior') $\alpha_j \sim N(\mu, \omega^2)$. The interesting thing about hierarchical models is that since μ and ω^2 are unknown parameters, we are using the data to update prior beliefs about those parameters as well. As a result, data from group j helps shape the posterior density over α_k ($\forall\ k \neq j$) via its contribution to inferences for the hyperparameters μ and ω^2. Indeed, this kind of 'sharing' or 'borrowing' information across groups is (a) a consequence of the hierarchical model, and (b) only makes sense if the α_j are exchangeable.

7.1.3 Hierarchical modeling as a 'semi-pooling' estimator

Given a belief of exchangeability among group-level parameters, the Bayesian hierarchical model represents a compromise between two alternative models: a 'no pooling' model and a 'complete pooling' model.

No pooling

In the context of Example 7.1, the no pooling model is simply equation 7.1a, $y_{ij}|\alpha_j, \sigma^2 \sim N(\alpha_j, \sigma^2)$, dropping the hierarchical component of the model in Example 7.1 (Equation 7.1b). In this model, absent any prior information for any given α_j, the only information we have for α_j is the information from group j, and the posterior density for α_j is proportional to the likelihood for y_{ij}. In a large sample, with no other prior information, and under a quadratic loss decision rule (e.g. Section 1.6.1), the Bayes estimate of α_j is simply the maximum likelihood estimate of the mean of y in group j,

$$\hat{\mu}_j^{(\text{NP})} = E(\alpha_j^{(\text{NP})}|\mathbf{y}) = \bar{y}_j = n_j^{-1} \sum_{i=1}^{n_j} y_{ij}.$$

Put differently, it is as if we have J parallel data analyses, none of which inform any other, save for the assumption that σ^2 is constant across groups. This approach to inference for the group-specific means α_j is justifiable if the group means are not exchangeable.

Complete pooling

The complete pooling model arises if we believe the grouping in the data is irrelevant, and we impose the restriction that $\alpha_j = \mu, \forall\ j$, generating the model $y_{ij} \sim N(\mu, \sigma^2)$. The grand mean $\bar{y}$ is the maximum likelihood estimator of μ:

$$\hat{\mu}_j^{(\text{CP})} = E(\alpha_j^{(\text{CP})}|\mathbf{y}) = \bar{y} = J^{-1}\sum_{j=1}^{J}\sum_{i=1}^{n_j} y_{ij} \Big/ \sum_{j=1}^{J} n_j.$$

The Bayesian hierarchical model lies between these two limiting cases, and for this reason Bayes estimates of the group-specific α_j are sometimes referred to as 'semi-pooling' estimators. To see this, again consider the one-way ANOVA model in Example 7.1. Note that we can express both the 'no pooling' and the 'complete pooling' models as special cases of the Bayesian hierarchical model (NP and CP, respectively). Specifically, the two limiting cases imply restrictions on the between-group variance parameter ω^2 in the hierarchical part of the model, $\alpha_j|\mu, \omega^2 \sim N(\mu, \omega^2)$:

1. 'no pooling': $\omega^2 = \infty$. If we regard the hierarchical model as generating a prior for each α_j then as $\omega^2 \to \infty$, this prior becomes less informative, and the information in the likelihood, $y_{ij} \sim N(\alpha_j, \sigma_j^2)$ tends to dominate the posterior density for α_j. In the limiting case of $\omega^2 = \infty$, we are asserting that there is no information in the between-group distribution of the α_j for any particular α_j. As described above, in this case the Bayes estimator of α_j is the group-specific mean $\bar{y}_j$.
2. 'complete pooling': $\omega^2 = 0$. This induces a degenerate between-group density for the α_j, with point mass on the grand mean μ, such that the Bayes estimate of each α_j is simply μ.

Generally, the Bayesian hierarchical model treats the between-group variance ω^2 as an unknown parameter, generating Bayes estimates (posterior means) for the α_j that lie between the grand mean μ and the group-specific means α_j, as we now elaborate.

7.1.4 Hierarchical modeling as a 'shrinkage' estimator

As we saw in Chapter 2, posterior densities are the product of two information sources (prior and likelihood), and are almost always less dispersed than either the prior density or the likelihood function. Moreover, at least for conjugate Bayesian analysis, a Bayes estimate such as the posterior mean is a convex combination of the prior mean and the maximum likelihood estimate. In the specific case of the hierarchical model in Example 7.1, this results in posterior inferences for the group-level α_j that are

1. shifted away from the group-level mean $\bar{y}_j$ in the direction of μ, the mean of the distribution of the group level parameters.
2. are more precise than inferences based on an analysis of group j in isolation from the other groups.

This phenomenon is sometimes referred to as 'shrinkage': the distribution of the group level parameters α_j is 'shrunk' around the grand mean μ, relative to the distribution of group level parameters we obtain with no pooling. We now derive the posterior density of α_j, conditional on the other parameters in the model:

Proposition 7.1 *Assume the model* $y_{ij} \sim N(\alpha_j, \sigma_j^2), \alpha_j \sim N(\mu, \omega^2)$, *where* $i = 1, \ldots, n_j$ *indexes observations within group* j, $j = 1, \ldots, J$. *Then* $\alpha_j | \mathbf{y}_j, \sigma_j^2, \mu, \omega^2 \sim N(\tilde{\mu}_j, V_j)$ *where*

$$\tilde{\mu}_j = \frac{\mu\omega^{-2} + \bar{y}_j \frac{n_j}{\sigma_j^2}}{\omega^{-2} + \frac{n_j}{\sigma_j^2}} \quad \text{and} \quad V_j = \left(\omega^{-2} + \frac{n_j}{\sigma_j^2}\right)^{-1},$$

and where $\bar{y}_j = n_j^{-1} \sum_{i=1}^{n} y_{ij}$ *is the maximum likelihood estimate of* α_j *(i.e. the sample mean). Equivalently,*

$$\tilde{\mu}_j = \lambda_j \mu + (1 - \lambda_j)\bar{y}_j$$

where

$$\lambda_j = \frac{\omega^{-2}}{\omega^{-2} + \frac{n_j}{\sigma_j^2}} = \frac{\frac{\sigma_j^2}{n_j}}{\omega^2 + \frac{\sigma_j^2}{n_j}} = \frac{V(\bar{y}_j)}{V(\bar{y}_j) + \omega^2}$$

is a measure of how much α_j *is 'shrunk' away from* $\bar{y}_j$ *towards* μ.

Proof. See Proposition 2.4. ◁

That is, we obtain the familiar precision-weighted average form of the mean of the posterior density for α_j that we encountered in Section 2.4. When group j provides little information about α_j – say n_j is small, and so $V(\bar{y}_j)$ is large, relative to the between variance ω^2 – then the shrinkage factor λ_j grows, and the Bayes estimate of α_j is pulled towards the grand mean μ. On the other hand, if the between variance ω^2 is relatively large – say, because the groups are quite heterogeneous – then the Bayes estimate of α_j will display less shrinkage, relying more on the information in group j and will reflect less 'borrowing strength' from the other groups.

This estimator of α_j can be justified not only as a Bayesian estimator, but can be shown to dominate the 'no pooling' and 'complete pooling' estimators on frequentist criteria such as mean square error. The 'no pooling' estimator $\bar{y}_j$ is an unbiased for α_j; the 'complete pooling' estimator $\bar{y}$ (the grand mean) is generally a biased estimator of any particular α_j, but because it is based on the data from all groups in the analysis, will have a relatively small standard error. The bias of the 'complete pooling' estimator *increases* to the extent that the groups are *heterogeneous* (i.e. there is large between-group variance, ω^2). But to the extent that there is little between-group variation, then estimating each mean with the grand mean may not be so bad. That is, there is a bias-variance trade off between the two estimators. The Bayesian 'semi-pooled' or 'shrinkage' estimator dominates both the 'no pooling' and 'complete pooling' estimators with respect to mean square error: this result is well known in the frequentist literature and was first noted by Charles Stein (1955). Viewed from a frequentist perspective, the Bayes estimator of the α_j dominates the other alternatives by buying the right amount of bias (shrinkage towards the grand mean), relative to the decrease in variance. The end result is that the total mean square error of the Bayes estimates – i.e. summing the mean square error of each estimate of α_j over the J groups – is smaller than the corresponding quantity for both $\hat{\mu}_j^{(\text{NP})}$ and $\hat{\mu}_j^{(\text{CP})}$; see the vivid demonstration in Example 7.5, below. A discussion of the frequentist properties of Bayes estimators for simple hierarchical models ('random effects' estimators) appears in Carlin and Louis (2000, Chapter 4).

7.1.5 Computation via Markov chain Monte Carlo

Computing the posterior density of the parameters in a hierarchical model is straightforward for the models considered thus far. With normal and inverse-Gamma densities at all levels of the hierarchical model, Markov chain Monte Carlo algorithms can be easily implemented via the Gibbs sampler, since (a) the conditional independence relations among data and parameters that constitute the hierarchical model; (b) the conditional distributions needed to implement the Gibbs sampler are all available in closed-form, as normal or inverse-Gamma densities, drawing on the conjugacy results presented in Chapter 2. With simulation-based Bayesian computing tools like `BUGS` and `JAGS`, we are not restricted to conditionally conjugate forms of hierarchical models. Nonetheless, hierarchical models with conditionally conjugate components are widely used in practice.

Hierarchical models can often involve a large number of parameters, depending on the number of groups in the analysis, and the complexity of the model. The posterior density in these cases can be high dimensional. In this situation, it is not unusual to find Gibbs samplers and other Markov chain Monte Carlo algorithms taking a long time to explore the posterior density. Reparameterizations for speeding up MCMC algorithms are part of the 'tricks of the trade', as we will encounter in various examples, below.

In the meantime, we show that the Gibbs sampler can be used to implement a very simple hierarchical model, via a re-analysis of a 'classic' in the statistics literature.

■ **Example 7.5**

Predicting baseball batting averages via a hierarchical model (Efron and Morris 1975). In a pioneering article, Brad Efron and Carl Morris analyzed the batting averages of 18 major league players over their first 45 at bats in the 1970 season. According to Efron and Morris (1975, 312):

> The problem is to predict each player's batting average over the remainder of the season .... [using only the data from the first 45 at bats]. This sample was chosen because we wanted between 30 and 50 at bats to assure a satisfactory approximation of the binomial by the normal distribution while leaving the bulk of at bats to be estimated.

For the uninitiated, the batting average is simply the number of base hits divided by the number of plate appearences, where a base hit is when the batter safely reaches first base after hitting the ball into fair territory. Efron and Morris used an arc-sine transformation to convert batting averages (proportions) to a quantity that has a normal distribution and unit variance. In a reanalysis of the data Casella (1985) used a normal model for the proportions themselves, using the binomial variance of the average of the observed batting averages $\bar{y}(1-\bar{y})/n$ as the normal variance σ^2; in these data $\bar{y} = .265$ and $n = 45$, so $\sigma^2 = .004332 = .0658^2$. A conjugate, hierarchical normal model for these data is

$$y_i \sim N(\theta_i, \sigma^2) \tag{7.9a}$$

$$\theta_i \sim N(\mu_0, \omega^2) \tag{7.9b}$$

$$\mu_0 \sim N(b_0, B_0) \tag{7.9c}$$

$$\omega^2 \sim \text{inverse-Gamma}(\nu_0/2,\ \nu_0\omega_0^2/2) \tag{7.9d}$$

where i indexes batters, and b_0, B_0, ν_0 and ω_0^2 are user-specified hyperparameters.

Each observed batting average y_i is a draw from a normal density with unknown mean θ_i. The θ_i (the batting average for player i, which, if the model is correct, we would observe after an arbitrarily long sequence of at bats for each player) are drawn from a normal distribution with an unknown mean and an unknown variance. That is we regard the players as exchangeable, lacking any prior information to distinguish any one player from the other. Priors over the unknown mean μ_0 and unknown variance ω^2 of the θ_i distribution complete the model specification; μ_0 is the average batting average, while ω^2 is the variance of the batting averages. We treat these parameters as independent *a priori*, i.e. $p(\mu_0, \omega^2) = p(\mu_0)p(\omega^2)$, with a $N(b_0, B_0)$ marginal prior density for μ_0 and an inverse-Gamma$(\nu_0/2, \nu_0\omega_0^2/2)$ marginal prior density for ω^2.

Priors. Specifying values for the hyper-parameters completes the prior specification. Baseball batting averages computed over a season (for players with at least 45 at bats) lie in a relatively narrow range: certainly no lower than zero, and almost never above .4 (Ted Williams was the last major league hitter to post a season batting average above .4, in 1941). The average batting average μ_0 will lie somewhere in that interval; I guess that the average batting average is unlikely to be greater than .3 and unlikely to be less than .15. The middle of this interval is .225, which I take as the value of $b_0 = E(\mu_0)$; if the interval (.15, .3) corresponds to a 95 % credible prior interval, then $B_0 = (.15/4)^2 = .00140625$. For the variance parameter ω^2, I use an inverse-Gamma prior density with parameters $\nu_0 = 14$ and $\omega_0^2 = .005$. I calibrated this prior by noting that the model in Equation 7.9 implies that the marginal prior density for θ_i is

$$\begin{aligned} p(\theta_i) &= \int_{-\infty}^{\infty}\int_0^{\infty} p(\theta_i, \mu_0, \omega^2)\,d\omega^2 d\mu_0 \\ &= \int_{-\infty}^{\infty}\int_0^{\infty} p(\theta_i|\mu_0, \omega^2)p(\mu_0)p(\omega^2)\,d\omega^2 d\mu_0. \end{aligned}$$

It is straightforward to sample from this density via the method of composition (Section 3.3): for $t = 1, \ldots, T$,

1. sample $\mu_0^{(t)} \sim N(b_0, B_0)$

2. sample $\omega^{2(t)} \sim$ inverse-Gamma$(\nu_0/2, \nu_0\omega_0^2/2)$

3. sample $\theta^{(t)} \sim N(\mu_0^{(t)}, \omega^{2(t)})$

Ten million draws from the marginal prior densities of μ_0 and θ_i are summarized in Figure 7.1. The marginal prior for μ_0 is just a normal density, while the marginal prior for θ has slightly heavier tails than the normal, induced by the mixing over the inverse-Gamma density for the between-variance parameter ω^2. Very little prior probability is given to impossible, negative batting averages (less than 0.5 %); with the normal model for the averages it is impossible to rule out these type of outcomes, although they are extremely unlikely *a priori*. Likewise, ridiculously high batting averages (say, greater than .5) are also given almost zero weight, *a priori*.

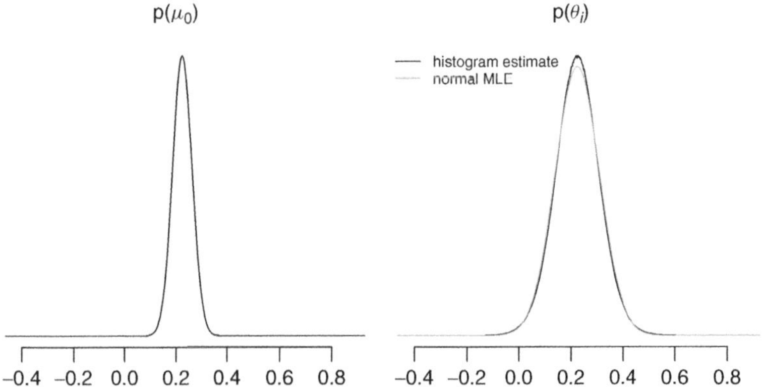

Figure 7.1 Marginal prior densities of μ_0 and θ_i, baseball averages example.

Posterior density. The posterior density for this problem is

$$p(\theta_1, \ldots, \theta_n, \mu_0, \omega^2 | y_1, \ldots, y_n) \propto \underbrace{\prod_{i=1}^{n} p(y_i|\theta_i)}_{\text{likelihood in 7.9a}} \underbrace{p(\theta_i|\mu_0, \omega^2)p(\mu_0)p(\omega^2)}_{\text{prior in 7.9b -- 7.9d}},$$

which looks imposing, but is easy to sample from using the Gibbs sampler. A directed acyclic graph $\mathcal{G}$ corresponding to this model appears in Figure 7.2, making the model's conditional independence relations quite vivid, and highlighting the hierarchical nature of the model. Using Proposition 5.2, we can derive the conditional distributions that drive a Gibbs sampler for this problem:

1. $p(\theta_i|\mathcal{G} \setminus \theta_i)$. The parents of θ_i are the parameters in the hierarchical normal for θ_i, the mean μ_0 and the variance ω^2. The only child node of θ_i is y_i, and with the 'within' variance σ^2 known, the only stochastic parent of y_i is θ_i. Both y_i and θ_i have normal densities, and so the conditions of Proposition 2.4 apply. That is, $\theta_i|\mathcal{G} \setminus \theta_i \sim N(\tilde{\theta}_i, V_\theta)$, where $\tilde{\theta}_i = (\mu_0\omega^{-2} + y_i\sigma^{-2})(\omega^{-2} + \sigma^{-2})^{-1}$ and $V_\theta = (\omega^{-2} + \sigma^{-2})^{-1}$.

2. $p(\mu_0|\mathcal{G} \setminus \mu_0)$. The parents of μ_0 are the prior hyperparameters b_0 and B_0. The children of μ_0 are the θ_i, $i = 1, \ldots, n$, whose parents are μ_0 and ω^2. Thus

$$p(\mu_0|\mathcal{G} \setminus \mu_0) \propto p(\mu_0|b_0, B_0) \times \prod_{i=1}^{n} p(\theta_i|\mu_0, \omega^2),$$

where all of the densities on the right hand side of this equality are normal densities with known variances, and so the conditions of Proposition 2.4 apply. That is,

$$\mu_0|\mathcal{G} \setminus \mu_0 \sim N\left(\frac{b_0 B_0^{-1} + \bar{\theta}\frac{n}{\omega^2}}{B_0^{-1} + \frac{n}{\omega^2}}, \left(B_0^{-1} + \frac{n}{\omega^2}\right)^{-1}\right),$$

where $\bar{\theta} = n^{-1}\sum_{i=1}^{n} \theta_i$.

3. $p(\omega^2|\mathcal{G} \setminus \omega^2)$. The parents of ω^2 are the prior hyperparameters ν_0 and ω_0^2. The children of ω^2 are the θ_i, $i = 1, \ldots, n$, whose parents are μ_0 and ω^2. Thus

$$p(\omega^2|\mathcal{G} \setminus \omega^2) \propto p(\omega^2|\nu_0, \omega_0^2) \times \prod_{i=1}^{n} p(\theta_i|\mu_0, \omega^2).$$

The prior density $p(\omega^2|\nu_0, \sigma_0^2)$ is an inverse-Gamma density, while the θ_i have normal densities, and so the results of Proposition 2.8 apply; i.e. the inverse-Gamma prior over ω^2 is (conditionally) conjugate with respect to the normal densities over the θ_i. Thus

$$\omega^2|\mathcal{G} \setminus \omega^2 \sim \text{inverse-Gamma}\left(\frac{\nu_0 + n}{2}, \frac{\nu_0\omega_0^2 + S_\theta}{2}\right)$$

where $S_\theta = \sum_{i=1}^{n}(\theta_i - \theta_0)^2$.

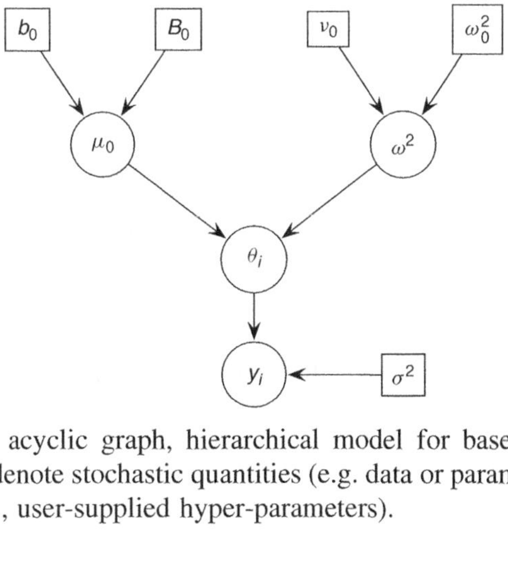

Figure 7.2 Directed acyclic graph, hierarchical model for baseball batting averages, Example 7.5. Circles denote stochastic quantities (e.g. data or parameters); squares denote known quantities (e.g., user-supplied hyper-parameters).

At iteration t the state of the Gibbs sampler is $\boldsymbol{\theta}^{(t)} = (\theta_1^{(t)}, \ldots, \theta_n^{(t)}, \mu_0^{(t)}, \omega^{2(t)})'$. The sampler makes the transition to $\boldsymbol{\theta}^{(t+1)}$ by

1. sampling $\theta_i^{(t+1)}$ from each of the corresponding conditional distributions for the θ_i, given $\mu_0^{(t)}$ and $\omega^{2(t)}$.

2. sampling $\mu_0^{(t+1)}$ from the conditional distribution for μ_0, but given the $\theta_i^{(t+1)}$ sampled in the previous step.

3. sampling $\omega^{2(t+1)}$ from the conditional distribution for ω^2, given the $\theta_i^{(t+1)}$ and $\mu_0^{(t+1)}$ sampled in steps 1 and 2.

I initialized the sampler with a random draw from the prior densities defined above, and then let the sampler run for 51 000 iterations, discarding the first 1000 iterations as burn-in. The following JAGS program implements this scheme:

JAGS code

```
model{
        for(i in 1:18){
              y[i] ~ dnorm(theta[i],prec) ## normal model
              theta[i] ~ dnorm(mu,tau)    ## hierarchical model for theta[i]
        }
        prec <- 1/.004332

        mu ~  dnorm(b0,B0)                ## prior for mu (hyperparameter)
        b0 <- .225
        B0 <- 1/pow(.15/4,2)

        tau ~  dgamma(14/2,.07/2)   ## prior for between-player
                                       inverse-variance
}
```

Results. Figure 7.3 provides a graphical comparison of the Bayes estimates and the MLEs for each θ_i. In each case we can compare the estimates with the season-long batting average for each player (the light gray dot). In almost every instance the Bayes estimate is shrunk away from the data recorded for the first 45 at bats (the MLEs), in the direction of the batting average recorded over the season: the exception is Ron Swoboda (NY-NL), close to the middle of the distribution of batting averages for the 18 players analyzed in this example. For Roberto Clemente, the Bayes estimate shrinks the MLEs too much; Clemente got off to a roaring start in the 1970 season, but fell back to record a nonetheless extremely impressive batting average above .300. At the other end of the data, the Bayes estimate slightly overestimates the extent to which Max Alvis improved on his slow start in the 1970 season. Note also the way that the 95 % intervals for the Bayes estimates are considerably smaller than the corresponding credible intervals for the MLEs.

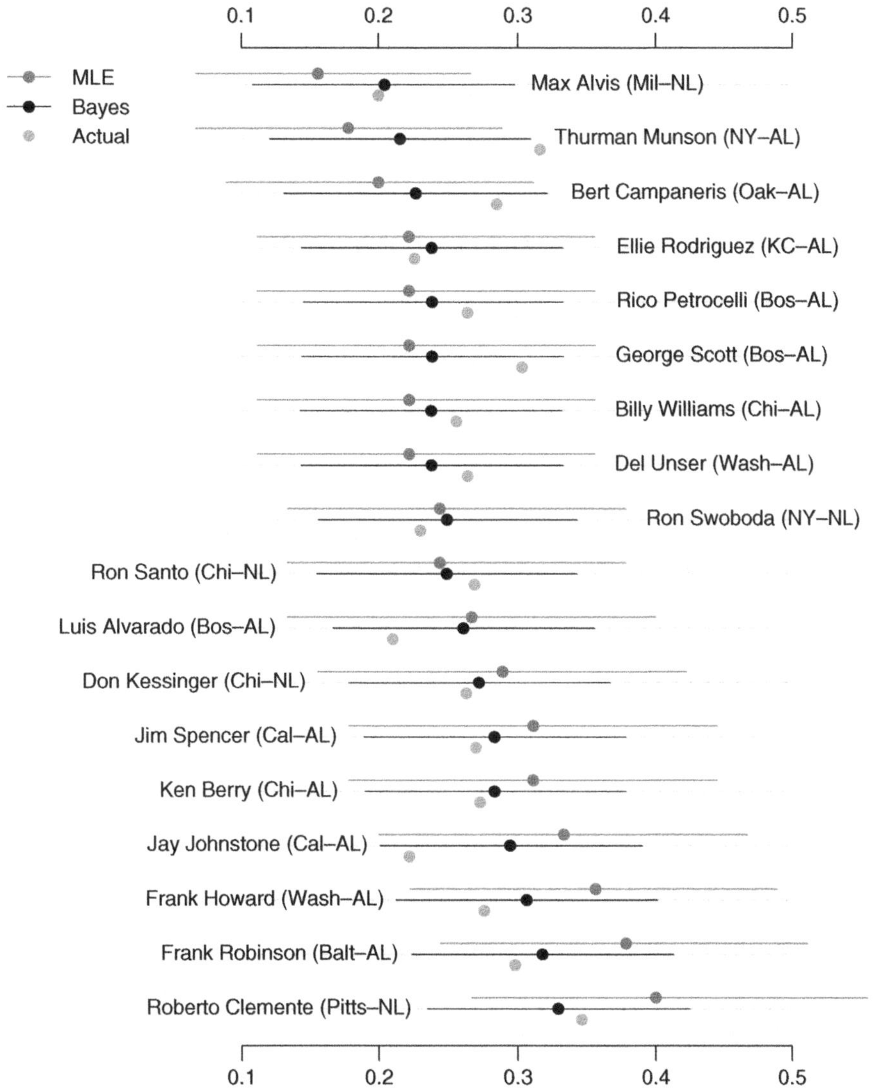

Figure 7.3 Marginal posterior densities, batting averages. For each of the 18 players studied, the black horizontal line extends to cover a 95 % marginal posterior credible interval (estimated with the 2.5 and 97.5 percentiles of the 50 000 iterations produced by the Gibbs sampler); the black dot is the mean of the Gibbs sampler output. The gray dot indicates the batting average recorded in the first 45 at bats of the 1970 season (the MLE of the batting average for the rest of the season), and the horizontal lines around it cover a 95 % interval. The light grey dot is the actual batting average recorded over the remainder of the 1970 season. The data are available as `EfronMorris` in the `R` package, `pscl`.

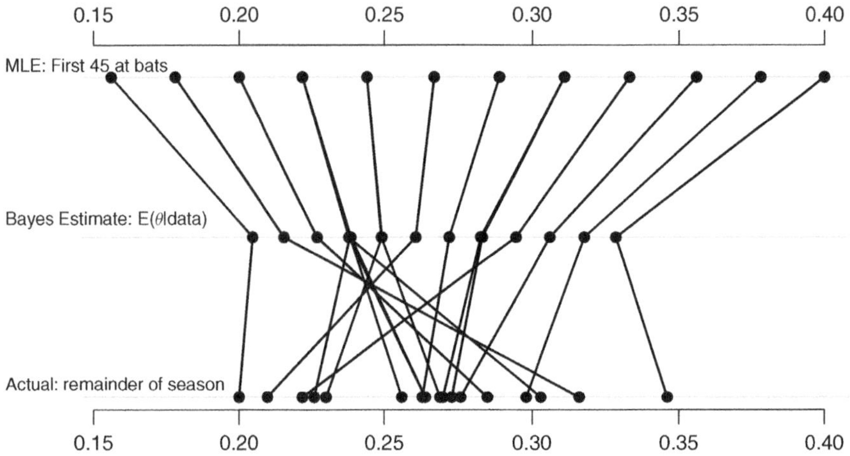

Figure 7.4 Comparison of MLEs, Bayes estimates, and actual batting averages. In almost every case, the Bayes estimates (means of the marginal posterior densities) lie closer to the actual batting averages recorded over the remainder of the season than the MLEs. The MLEs have mean squared error 251 % that of the Bayes estimates.

Figure 7.4 provides a graphical depiction of the shrinkage at work in this example. The Bayes estimates (middle row) display far less dispersion across players than the maximum likelihood estimates (top row). Nonetheless, the Bayes estimates are on average much closer to the rest-of-season batting averages than the MLEs. In fact, the mean squared error of the MLEs is 251 % that of the mean squared error of the Bayes estimates.

Discussion. The hierarchical Bayesian model clearly outperforms the MLEs in this case. But why? What did the assumption of exchangeability do for us in this case? One explanation is that the exchangeable, hierarchical model is an excellent *predictive* model that captures the regression-to-the-mean effect we see over the batting performances in these data. That is, over the remainder of the 1970 season, players off to a strong start regressed back towards the mean of the distribution of batting averages, while players that got off to a poor start improved in the direction of that average. Over the course of a long baseball season, players encounter a vast mix of conditions (pitching, day/night games, home/away, situations, fatigue). After 45 at bats, some players may have seen more 'good situations' than others, and this is in part a cause of the dispersion in the data from the first 45 at bats. Because, via exchangeability, the hierarchical model borrows information *across* batters, the resulting Bayes estimates incorporate some of this information about the mix of conditions that some players may have not yet encountered (but will, over the course of a long season). In short, it is as if the hierarchical model is learning about the regression-to-the-mean effects yet to manifest themselves in these data, by fitting a 'between-player' model (the hierarchical part of the model) at the same time as we fit a model at the player level. We return to this example in the exercises at the end of this chapter.

We now consider hierarchical models in a series of social-science applications, spanning ANOVA models (Section 7.2), panel data (Section 7.3), a cluster randomized field experiment (Section 7.4) and multi-level models (Section 7.5).

7.2 ANOVA as a hierarchical model

ANOVA models are an indispensable, if elementary, statistical tool. Many applications of ANOVA in the social-sciences are designed to answer simple questions: does y vary across groups $j = 1, \ldots, J$, and if so, how much of the variation in $\mathbf{y}$ is due to the grouping of the data by the J groups? For instance, panel data analysis should almost always start with simple ANOVA modeling, so as to understand how much of the variation in $\mathbf{y}$ is cross-sectional or longitudinal. The analysis of experimental data usually starts with an ANOVA decomposition of the variance in $\mathbf{y}$ across treatment and control groups.

Classical implementations typically involve comparing a J-means model for $\mathbf{y}$ against a restricted, null model with just a grand mean, via an F-test. Note that the 'no pooling' model discussed in Section 7.1.3 corresponds to the alternative, J-means model, while the 'complete pooling' model corresponds to the restricted, null model. Two-way and multi-way ANOVA models generalize this idea to the case where the data have multiple grouping factors.

We begin by considering a one-way ANOVA model, showing how the Gibbs sampler can be used for the simple normal/normal model contemplated in Example 7.1. The 'hard work' has been done for us via Proposition 7.1, and the model is essentially identical to the model used for the Efron and Morris baseball data in Example 7.5.

7.2.1 One-way analysis of variance

Adopting the notation of Example 7.1, a full specification of the normal, one-way ANOVA model as a Bayesian hierarchical model is:

$$y_{ij}|\alpha_j, \sigma^2 \sim N(\alpha_j, \sigma^2) \tag{7.10a}$$

$$\alpha_j|\mu, \omega^2 \sim N(\mu, \omega^2) \tag{7.10b}$$

$$\mu \sim N(b_0, B_0) \tag{7.10c}$$

$$\sigma^2 \sim \text{inverse-Gamma}(\nu_0/2, \nu_0\sigma_0^2/2) \tag{7.10d}$$

$$\omega^2 \sim \text{inverse-Gamma}(\kappa_0/2, \kappa_0\omega_0^2/2) \tag{7.10e}$$

A model with unit-wise heteroskedasticity results when we let the 'within-unit' variance parameter σ^2 vary over units (i.e. instead of σ^2 we would have the parameters $\sigma_1^2, \sigma_2^2, \ldots, \sigma_J^2$). The hyperparameters of the normal prior for μ_0 (the mean b_0 and the variance B_0) and the hyperparameters of the priors for σ^2 and ω^2 are user-supplied constants.

Stack the model parameters in the vector $\boldsymbol{\theta} = (\alpha_1, \ldots, \alpha_j, \mu_0, \sigma^2, \omega^2)$. The hierarchical structure of the model implies that the prior density for $\boldsymbol{\theta}$ can be factored as follows:

$$\begin{aligned} p(\boldsymbol{\theta}) &= p(\alpha_1, \ldots, \alpha_j, \mu, \sigma^2, \omega^2) \\ &= p(\alpha_1, \ldots, \alpha_j, |\mu, \omega^2)p(\mu)p(\sigma^2)p(\omega^2) \end{aligned}$$

$$= \prod_{j=1}^{J} p(\alpha_j|\mu, \omega^2)p(\mu)p(\sigma^2)p(\omega^2)$$

Note that exchangeability of the unit-specific α_j implies the equality in the last line, where the conditional density $p(\alpha_1, \ldots, \alpha_j|\mu_0, \omega^2)$ is factored as the J-fold product of the conditional densities specific to each unit, given the hyperparameters μ and ω^2.

Likelihood methods Hierarchical models can pose headaches for likelihood based methods. For one thing, once we specify a stochastic form for the unit-specific effects α_i, treating them as 'random effects' (a curious nomenclature, from the Bayesian perspective, in which all parameters are random quantities), their status for likelihood based inference becomes unclear. Are the 'random effects' parameters or not? There are a variety of approaches in the literature. One approach is to treat the random effects as nuisance parameters and to maximize a *marginal likelihood* in which these parameters have been integrated out: e.g. for the one-way 'random effects' model considered here, this marginal likelihood function is

$$\mathcal{L}_M(\mu, \omega^2, \sigma^2|\mathbf{y}) = \int_{\boldsymbol{\alpha}} f(\mathbf{y}|\mu, \boldsymbol{\alpha}, \omega^2, \sigma^2) f(\boldsymbol{\alpha}|\omega^2, \sigma^2) d\boldsymbol{\alpha}$$

where $\boldsymbol{\alpha} = (\alpha_1, \ldots, \alpha_J)$. Another approach treats the α_j as missing data, but uses the EM algorithm (Dempster, Laird and Rubin 1977): the E step of the algorithm generates imputations for the α_j conditional on the fixed effects and variance parameters, while the M step maximizes the complete data likelihood, updating the fixed effects and variance parameters.

The popular *restricted maximum likelihood* (REML) estimator (Patterson and Thompson 1971; Thompson 1962) takes a different tack, maximizing the likelihood with respect to the variance terms in the model after integrating with respect to the fixed effects in the model, i.e. maximizing

$$\mathcal{L}_R(\omega^2, \sigma^2|\mathbf{y}) = \int f(\mathbf{y}|\mu, \omega^2, \sigma^2) f(\mu) d\mu.$$

The REML likelihood is equivalent to the marginal posterior density over the variance parameters ω^2 and σ^2, given an uninformative prior over μ. The implementation in `R` via the `lmer` function in the `lme4` package (Bates, Maechler and Dai 2008) comes with a `mcmcsamp` option, generating samples from the posterior density of the model parameters assuming improper, uninformative priors.

Bayesian inference The difficulty for Bayesian inference is that the posterior density for this problem $p(\boldsymbol{\theta}|\mathbf{Y}) \propto p(\boldsymbol{\theta})p(\mathbf{Y}|\boldsymbol{\theta})$ may be high dimensional: $\boldsymbol{\theta}$ contains J α_j parameters, plus μ_0, and the two variances σ^2 and ω^2 for a total of $J + 3$ parameters. This where the Gibbs sampler becomes especially useful. Characterizing the $J + 3$ dimensional posterior density $p(\boldsymbol{\theta}|\mathbf{Y})$ can be done by sampling sequentially from the corresponding $J + 3$ conditional densities.

The DAG (directed acyclic graph) in Figure 7.5 provides a graphical representation of the hierarchical model $\mathcal{G}$, from which it will be straightforward to deduce the forms

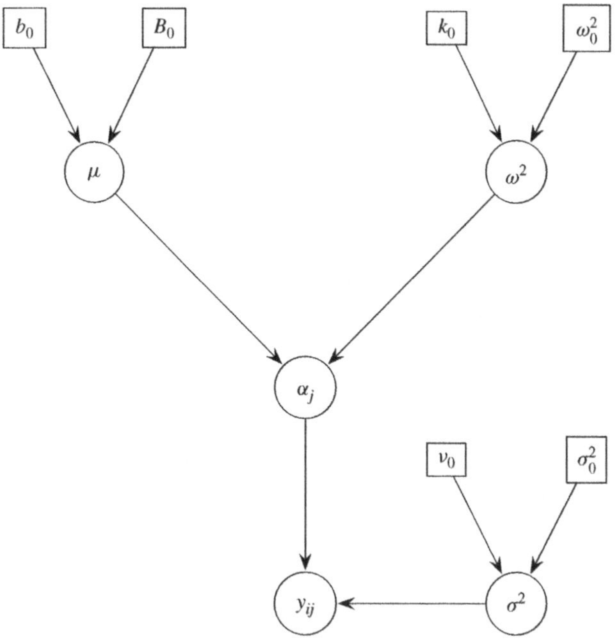

Figure 7.5 Directed acyclic graph, hierarchical model for the one-way ANOVA model (J normal means), Example 7.6.

of the conditional distributions required to implement a Gibbs sampler. The conditional independence relations among the random quantities in the model are made especially vivid by the DAG: e.g. given α_j, the data y_{ij} are conditionally independent of the hyperparameters μ_0 and the between-unit variance ω^2. Moreover, given other components (data and/or parameters) of the model, α_j and α_k are conditionally independent of one another, $\forall\ j \neq k$, meaning that these group-specific parameters can be updated one at a time, in a series of very simple updating steps (e.g. Robert 2001, 467).

We now turn to the specific conditional distributions needed to implement the Gibbs sampler. In each case we apply the simple rule given by Equation 5.14 in Proposition 5.2.

1. $p(\alpha_j|\mathcal{G} \setminus \alpha_j)$, $j = 1, \ldots, J$. The parents of each α_j are μ and ω^2; the children of α_j are the data in unit j, $\mathbf{y}_j = (y_1, \ldots, y_{n_j})'$ and the parents of $\mathbf{y}_j$ are α_j and σ^2. By equation 5.14, $p(\alpha_j|\mathcal{G} \setminus \alpha_j) \propto p(\alpha_j|\mu, \omega^2) \prod_{i=1}^{n_j} p(y_{ij}|\alpha_j, \sigma^2)$. Since $p(\alpha_j|\mu, \omega^2) \equiv N(\mu, \omega^2)$, and $p(y_{ij}|\alpha_j, \sigma^2) \equiv N(\alpha_j, \sigma^2)$, the results of Proposition 7.1 apply: i.e.

$$\alpha_j|\mathcal{G} \setminus \alpha_j \sim N\left(\frac{\mu\,\omega^{-2} + \bar{y}_j \frac{n_j}{\sigma^2}}{\omega^{-2} + \frac{n_j}{\sigma^2}},\ \left(\omega^{-2} + \frac{n_j}{\sigma^2}\right)^{-1}\right), \tag{7.11}$$

where $\bar{y}_j = n_j^{-1} \sum_{i=1}^{n_j} y_{ij}$ (the mean of the data in unit j). That is, we obtain the familiar result from conjugate normal analysis (Section 2.4) that the expected *a posteriori* value of α_j is the precision weighted average of its prior μ, and the maximum likelihood estimate $\bar{y}_j$.

2. $p(\mu|\mathcal{G}\setminus\mu)$. The parents of μ are just its prior hyperparameters, the prior mean and variance b_0 and B_0 respectively. The children of μ_0 are $\boldsymbol{\alpha} = (\alpha_1, \alpha_2, \ldots, \alpha_j)'$. The $\boldsymbol{\alpha}$ have two parents, μ and ω^2. Thus, $p(\mu|\mathcal{G}\setminus\mu) \propto p(\mu|b_0, B_0)\prod_{j=1}^{J} p(\alpha_j|\mu, \omega^2)$. These are all normal densities, and so again we apply the 'precision-weighted averaging' results of Proposition 2.4:

$$\mu|\mathcal{G}\setminus\mu \sim N\left(\frac{b_0 B_0^{-1} + \bar{\alpha}\frac{J}{\omega^2}}{B_0^{-1} + \frac{J}{\omega^2}}, \left(B_0^{-1} + \frac{J}{\omega^2}\right)^{-1}\right), \tag{7.12}$$

where $\bar{\alpha} = J^{-1}\sum_{j=1}^{J}\alpha_j$.

3. $p(\omega^2|\mathcal{G}\setminus\omega^2)$. The parents of ω^2 are just its prior hyperparameters, κ_0 and ω_0^2. The children of ω^2 are the α_j; the parents of α_j are ω^2 and μ. Thus,

$$p(\omega^2|\mathcal{G}\setminus\omega^2) \propto p(\omega^2|\kappa_0, \omega_0^2)\prod_{j=1}^{J} p(\alpha_j|\mu, \omega^2).$$

The prior density $p(\omega^2|\kappa_0, \omega_0^2)$ is an inverse-Gamma density, while the α_j have normal densities, and so the results of Proposition C.5 apply. That is, the inverse-Gamma prior over ω^2 is conjugate with respect to the normal 'likelihood' over the α_j, and so

$$\omega^2|\mathcal{G}\setminus\omega^2 \sim \text{inverse-Gamma}\left(\frac{\kappa_0 + J}{2}, \frac{\kappa_0\omega_0^2 + S_\alpha}{2}\right) \tag{7.13}$$

where $S_\alpha = \sum_{j=1}^{J}(\alpha_j - \mu)^2$.

4. $p(\sigma^2|\mathcal{G}\setminus\sigma^2)$. The parents of σ^2 are just its prior hyperparameters, ν_0 and σ_0^2. The children of σ^2 are the y_{ij}; the parents of the y_{ij} are the α_j and σ^2. Thus,

$$p(\sigma^2|\mathcal{G}\setminus\sigma^2) = p(\sigma^2|\nu_0, \sigma_0^2)\prod_{j=1}^{J}\prod_{i=1}^{n_j} p(y_{ij}|\alpha_j, \sigma^2).$$

The prior density $p(\sigma^2|\nu_0, \sigma_0^2)$ is an inverse-Gamma density, while the y_{ij} have normal densities, and again, the results of Proposition C.5 apply. That is,

$$\sigma^2|\mathcal{G}\setminus\sigma^2 \sim \text{inverse-Gamma}\left(\frac{\nu_0 + n}{2}, \frac{\nu_0\sigma_0^2 + S_{\mathbf{Y}}}{2}\right) \tag{7.14}$$

where $n = \sum_{j=1}^{j} n_j$ is the total number of observations and $S_{\mathbf{Y}} = \sum_{j=1}^{J}\sum_{i=1}^{n_j}(y_{ij} - \alpha_j)^2$ is the total sum-of-squares of $\mathbf{Y}$.

An iteration of the Gibbs sampler consists of sampling from each of these conditional distributions, with the sampled values stored, and available as conditioning arguments in subsequent steps. Formally, the Gibbs sampler makes the transition from

$$\boldsymbol{\theta}^{(t)} = (\alpha_1^{(t)}, \ldots, \alpha_J^{(t)}, \mu^{(t)}, \omega^{2(t)}, \sigma^{2(t)})$$

to $\boldsymbol{\theta}^{(t+1)}$ as follows:

1. sample $\alpha_j^{(t+1)}$ from the normal density given in Equation 7.11, with the conditioning arguments μ, ω^2 and σ^2 set to $\mu^{(t)}$, $\omega^{2(t)}$ and $\sigma^{2(t)}$, respectively, with $j = 1, \ldots, J$.
2. sample $\mu^{(t+1)}$ from the normal density given in Equation 7.12, with the conditioning arguments $\bar{\mu}$ and ω^2 set to $\bar{\mu}^{(t+1)}$ and $\omega^{2(t)}$ respectively (i.e. the α_j were 'updated' in step 1).
3. sample $\omega^{2(t+1)}$ from the inverse-Gamma density in Equation 7.13, with the conditioning argument S_α set to $S_\alpha^{(t+1)}$. That is, S_α is a function of both the α_j and μ; the α_j were updated to $\alpha_j^{(t+1)}$ in step 1 and μ was updated to $\mu^{(t+1)}$ in step 2.
4. sample $\sigma^{2(t+1)}$ from the inverse-Gamma density in Equation 7.14, with the conditioning argument $S_\mathbf{Y}$ updated to $S_\mathbf{Y}^{(t+1)}$. That is, $S_\mathbf{y}$ is a function of the α_j, which were updated to $\alpha_j^{(t+1)}$ in step 1.

After these four steps, a complete $\boldsymbol{\theta}^{(t+1)}$ has been sampled, and we condition on it in the next iteration. For the conjugate case considered here, these conditional densities are reasonably easy to program, meaning that a Gibbs sampler can be constructed 'from scratch'. However, it is much easier to implement this model with just a few commands in a general-purpose MCMC computer program such as `BUGS` or `JAGS`.

Non-conjugate priors for variance parameters The normal/inverse-Gamma form of the hierarchical model in Equation 7.10 is something of a legacy from the pre-MCMC era. To be sure, the normal/inverse-Gamma form of the hierarchical model makes it easy to derive and program the conditional distributions that drive a Gibbs sampler for this problem (as the previous discussion makes clear). Throughout this book we've seen that it takes a good deal of work to elicit inverse-Gamma priors over variance parameters; see, for instance, the 'calibration' exercises reported in Examples 2.13, 2.16 and the hierarchical model for baseball averages considered in Example 7.5.

The availability of modern computational tools for MCMC mean that we no longer have to adhere to conjugacy as a guide when specifying Bayesian models. This has led some scholars to investigate other priors for variance parameters in hierarchical models (e.g. Natarajan and Kass 2000). Proper-but-vague, uniform priors over standard deviations are one of the classes of priors recommended by Gelman (2006) and are used extensively in the hierarchical models presented in Gelman and Hill (2007). These priors require that the analyst supply an upper bound κ on the feasible range of a standard deviation σ; then the prior deployed for σ is simply $\sigma \sim \text{Unif}(0, \kappa)$. When in doubt, setting κ to a large, positive number will usually suffice (where 'large' is relative to the scale of the problem one is working on; see the examples below); recalling the implications of Cromwell's rule (Section 1.4.3) – that if we assign zero prior probability to particular values of a parameter then those values will have zero probability *a posteriori* – we don't want to set κ to too small a value without good reason. In a BUGS/JAGS program, one can convert the standard deviation to a precision as follows: e.g. with $\kappa = 10$,

JAGS code

```
sigma ~ dunif(0,10)
tau <- pow(sigma,-2)
```

An attractive feature of this proper, uniform prior is that it is uninformative over a feasible range of values for σ. Attempts to specify uninformative, inverse-Gamma priors often result in priors that have close to point mass at zero, and an extremely long right tail. For instance, many BUGS applications often employ the prior `tau ~ dgamma(.01,.01)` for precision parameters, a density that is 'barely proper'. This density is actually not uninformative with respect to the variance (the inverse precision) in a neighborhood close to zero (see Problem 7.3) and poses some interesting issues in the context of hierarchical models.

In hierarchical models, when level two variance parameters get close to zero, then the distribution of the level two parameters is becoming degenerate. For instance, consider the one-way ANOVA model under consideration. From Equation 7.10, if $\omega^2 \to 0$, then the group-specific parameters α_j are all approximately tightly clustered around the grand mean μ, at least *a priori*. This can see the Gibbs sampler described above running into problems: if $\omega^2 \approx 0$ then the updates of α_j (equation 7.11) leave these parameters close to μ, which in turn sees ω^2 unlikely to change much when it is updated (the S_α term in Equation 7.13 will be quite small). Accordingly, the Gibbs sampler described above can get 'stuck' in the corner of the parameter space with $\omega^2 \approx 0$ and $\alpha_j \approx \mu$. Left to run long enough, the Gibbs sampler will break out of this region of the parameter space, but there is always the danger that a naïve analyst might look at the output of the 'stuck' Gibbs sampler and think that it has converged. On the other hand, for data sets in which there is *actually* very small between-group variation, then $\omega^2 \approx 0$ and $\alpha_j \approx \mu$ is a correct characterization of the posterior density of these parameters. We need to be able to distinguish these two situations from one another, underscoring the value of descriptive analysis prior to modeling: i.e. is there much between-group variation in the data to be modeled?

A proper, uniform prior on the standard deviation can mitigate some of these problems. A proper, uniform prior on ω doesn't heap prior probability mass on values of ω^2 extremely close to zero, or at least not to the same extent as the 'uninformative' inverse-Gamma prior on ω^2. The proper uniform prior doesn't rule out values of ω^2 close to zero, and so the Gibbs sampler can still run into the problems described above; we consider some other 'tricks of the trade' for dealing with this issue, below. In fact, for the one-way ANOVA model in Equation 7.10, with a proper, uniform prior on ω, $\omega \sim \text{Unif}(0, \kappa)$, $\kappa < \infty$, the updating step for ω^2 is virtually identical to that given in Equation 7.13, except that (a) the prior hyperparameters κ_0 and ω_0^2 are set to zero, and (b) we would reject any draw from the resulting inverse-Gamma density that is greater than κ.

Moreover, standard deviations are easier quantities to grapple with than variances, having the same scale as the variable being modeled. Thus, it is usually easier to elicit an upper bound on a standard deviation (κ, in the discussion above), then it is to elicit the hyper-parameters of an inverse-Gamma prior density for a variance. For these reasons we will use a proper, uniform prior over standard deviation parameters in the applications below.

■ Example 7.6

Math achievement scores across 160 schools; hierarchical model for one-way ANOVA. The 1982 High School and Beyond Survey is a nationally representative sample of US public and Catholic schools, covering 7185 students in 160 schools. The chief outcome of interest is a standardized measure of math ability, with a mean of 12.75 and IQR of [7.28, 18.32]. These data figure prominently in the standard reference on hierarchical linear models (Raudenbush and Bryk 2002), and so are well suited for our purposes here.

We fit a one-way ANOVA model (Equation 7.1 in Example 7.1), recognizing that students are grouped by school. We momentarily ignore school type (public vs Catholic) and other school-level characteristics; these 'level-two' covariates will play a role in the sequel, below.

Classical, preliminary analysis. A naïve estimate of the population mean math score that wrongly ignores the clustered design of this study would yield the sample grand mean of 12.75 with a standard error of .081. This standard error is based on the assumption that the data are a simple random sample; a 'clustered' standard error, based on a Eicker-Huber-White sandwich estimator (Eicker 1963; Huber 1967; White 1980) of the sampling variance of the mean – computed using the `survey` package (Lumley 2004) in `R` – is .24, almost 3 times as large as the naïve standard error, implying a design effect of $(.24/.081)^2 \approx 8.75$. That is, the clustered-by-schools design and a reasonably high level of within-school correlation in math scores (the estimated intraclass correlation is .17) means that the sample of 7185 students is as informative about the grand mean as a simple random sample of $7185/8.75 \approx 821$ students.

Other simple classical procedures also strongly suggest substantial effects from the clustering by school. The F-test from a one-way ANOVA, using school as a grouping factor, yields a test statistic of 10.43 on 159 and 7025 degrees of freedom ($p < .01$). The r^2 from a regression of math score on a set of dummy variables for school – another way of assessing the relative size of the within-class variability to the between-class variability – is .19; i.e. 19 % of the variance in math scores is between-class variability. suggesting a moderate degree of within-class homogeneity.

Hierarchical model. We estimate a hierarchical model for these data using three approaches: two based on the likelihood function, and the Bayesian procedure described above.

We depart from the one-way ANOVA model considered in Equation 7.10, substituting the conditionally conjugate inverse-Gamma priors for the variance parameters with Unif(0, 10) priors on both σ (the 'within' standard deviation) and ω (the 'between' standard deviation). `BUGS/JAGS` code for this model is as follows:

JAGS code

```
model{
        for(i in 1:N){
              mu.y[i] <- alpha[j[i]]
              math[i] ~ dnorm(mu.y[i],tau[1])
        }

        for(p in 1:J){
              alpha[p] ~ dnorm(mu,tau[2])
        }

        mu ~ dnorm(0,.0001)
        for(p in 1:2){
              tau[p] <- pow(sigma[p],-2)
              sigma[p] ~ dunif(0,10)
        }
}
```

The first loop in the program is the 'level one' model. Note the use of the 'double subscript' or 'nested indexing' to assign the right α_j to student i; i.e. we set `j` to be a variable, such that student `i` is in school `j[i]` and `alpha[j[i]]` is the correct α_j for student `i`, where `j` indexes the $J = 160$ schools. With `j` playing the role of a variable, the second loop uses the index `p` to define the level two part of the hierarchical model. A prior for the grand mean `mu` and the uniform priors on the two standard deviations completes the model. Note that for convenience I refer to the standard deviations as `sigma`; this way we simply direct BUGS/JAGS to monitor the node `sigma`, recovering MCMC output for the within standard deviation as `sigma[1]` and the between standard deviation as `sigma[2]`.

After a 1000 iteration burn-in, 10 000 iterations of a MCMC algorithm were run and stored for inference. Trace plots indicate that the algorithm converged quickly on the target posterior density; various formal convergence diagnostics are also consistent with the algorithm having converged on the posterior density. For instance, the Raftery-Lewis diagnostic indicates that 6349 MCMC iterations are required to accurately learn about the 2.5 % and 97.5 % quantiles of the marginal posterior density of ω, which by this metric makes ω the 'slowest mixing' parameter in this model; unsurprisingly, ω also has the highest auto-correlation parameter (.39) of all the model parameters. The 10 000 iterations saved here are more than sufficient to generate accurate characterizations of the posterior density of the model parameters.

For purposes of comparison, we also estimate this model using REML via the `lme4` package (Bates, Maechler and Dai 2008) in R. This package also provides a MCMC method for sampling from the posterior density of the parameters of a mixed model assuming improper, uniform priors, `mcmcsamp`. With the data in a data frame called `data1`, we invoke the mixed model fitting function `lmer` and then generate 10 000 MCMC samples:

R code

```
require(lme4)
l1 <- lmer(math ~ (1 | school),
```

```
          data=data1)

set.seed(1001)            ## retain RNG seed across R sessions
z <- mcmcsamp(l1,n=10000,saveb=TRUE)

```

Numerical summaries of the model fitting appear in Table 7.1, for the grand mean μ, the 'between' standard deviation ω and the 'within' standard deviation σ. The left column summarizes the results of the JAGS run, showing the mean of the MCMC output for each of the three parameters, the standard deviation, and an estimate of the 95 % HDR of the marginal posterior density of each parameter. The substantive import of the findings accord with the classical analysis provided earlier, with about 18 % of the total variation being 'between' variation, and 82 % of the variation in math achievement scores being within-class or individual-level variation. A nice feature of the Bayesian approach is that we can induce a posterior density over the intra-class correlation, since it is merely a deterministic function of ω and σ; this exercise reveals that the ICC is estimated reasonably precisely, in turn stemming from the fact that the two standard deviations are themselves subject to relatively little posterior uncertainty.

The REML estimates almost exactly coincide with the fully Bayesian, MCMC results (unsurprisingly, given the vague, proper priors used in the Bayesian setup), except when we use the `mcmcsamp` to produce uncertainty assessments for the model parameters. The `lmer` + `mcmcsamp` estimates of the 'between' standard deviation ω produces a

Table 7.1 Posterior summaries, one-way ANOVA model for math achievement scores. μ is the grand mean, ω is the 'between' or 'level-2' standard deviation, and σ is the 'within' or 'level 1' standard deviation. ICC is the intraclass correlation coefficient. The MCMC estimates are based on 10 000 iterations produced by JAGS; cell entries are means of the MCMC output, quantities in parentheses are standard deviations of the MCMC output; ranges in square brackets are estimated 95 % HPD intervals, again based on the MCMC output. The REML estimates are produced by the `lmer` function in the R package `lme4` (Bates, Maechler and Dai 2008), version 0.999375-22, with MCMC methods (`lme4::mcmcsamp`) generating 10,000 iterates so as to form standard error estimates and 95 % confidence intervals.

	MCMC	REML	REML/MCMC
μ	12.64	12.64	12.64
	(.25)	(.24)	(.22)
	[12.17, 13.12]		[12.21, 13.07]
ω	2.96	2.93	2.62
	(.19)		(.15)
	[2.61, 3.33]		[2.31, 2.91]
σ	6.23	6.26	6.27
	(.05)		(.05)
	[6.15, 6.36]		[6.17, 6.38]
$ICC = \omega^2/(\omega^2 + \sigma^2)$	.18	.18	.15
	(.02)		(.02)
	[.15, .22]		[.12, .18]

marginal posterior density that is centered a considerable distance below that recovered by JAGS, and a considerable distance from the REML point estimate of ω; in fact, the disparity of the REML/MCMC estimate of ω and the REML estimate is $2.93-2.62 = .31$ or about 2 posterior standard deviations. Note that the REML/MCMC 95 % HPD interval for ω does not cover the MCMC estimate of the posterior mean, 2.96. In turn, this leads to an underestimate of the ICC by the REML/MCMC combination, with the 95 % HPD interval for this quantity barely covering the value recovered by MCMC in JAGS. Figure 7.6 shows a comparison of the marginal posterior densities as histograms, for μ, σ and ω, across the two MCMC-based technologies used here. The underestimate of ω by the `lmer` + `mcmcsamp` approach is especially clear (middle row of Figure 7.6). At the time of writing, the authors of the `lme4` package are aware of this issue and future iterations of `lme4` will do a better job of recovering uncertainty estimates in these parameters.

Estimates of school-specific average levels of math achievement. It is also interesting to examine estimates of the school-specific average levels of math achievement, α_j. Note that in the mixed model parameterization (Equation 7.4), we recover each α_j as the sum of the fixed effect μ and the group-specific offset b_j. Figure 7.7 shows the shrinkage at work for this application, with the hierarchical model's estimates of α_j (posterior means) pulled towards the grand mean μ, relative to the observed school-specific $\bar{y}_j$. Not all schools are shrunk as much as others (some of the lines in Figure 7.7 cross one another), stemming from the unbalanced design here (some schools contribute more data than others, and so their respective $\bar{y}_j$ is estimated more precisely than for other schools).

Figure 7.8 displays comparisons of point estimates of α_j (top panels) from the various methods considered here. The top left panel of Figure 7.8 provides an alternative representation of the shrinkage shown in Figure 7.7; the hierarchical model 'pulls in' extremely low and high $\bar{y}_j$. The top right panel of Figure 7.8 shows the comparison of the two MCMC methodologies used here, JAGS and `lmer` + `mcmcsamp`, displaying no difference in their point estimates of the α_j.

The lower panels of Figure 7.8 compare estimates of the uncertainty in the estimates of α_j. The lower left panel compares the standard errors of the group-specific means (the estimates one would obtain from a conventional, 'no pooling' classical analysis) with the standard deviations of the marginal posterior densities produced by the MCMC-based, Bayesian analysis in JAGS. Here we see the same result we obtained with the baseball analysis in Example 7.5: the hierarchical model produces 'shrunken' estimates of group-specific parameters that are generally estimated more precisely (or at least no more imprecisely) than are their classical counterparts. The lower right panel of Figure 7.8 shows that the estimated marginal posterior standard deviations from the two MCMC-based approaches almost coincide, with a slight suggestion that the `lmer` + `mcmcsamp` estimates are a tad too small. This would not be surprising given the gross underestimate of ω we saw from `lmer` + `mcmcsamp` earlier in Table 7.1 and Figure 7.6. In the other hierarchical modeling examples to come we will use JAGS.

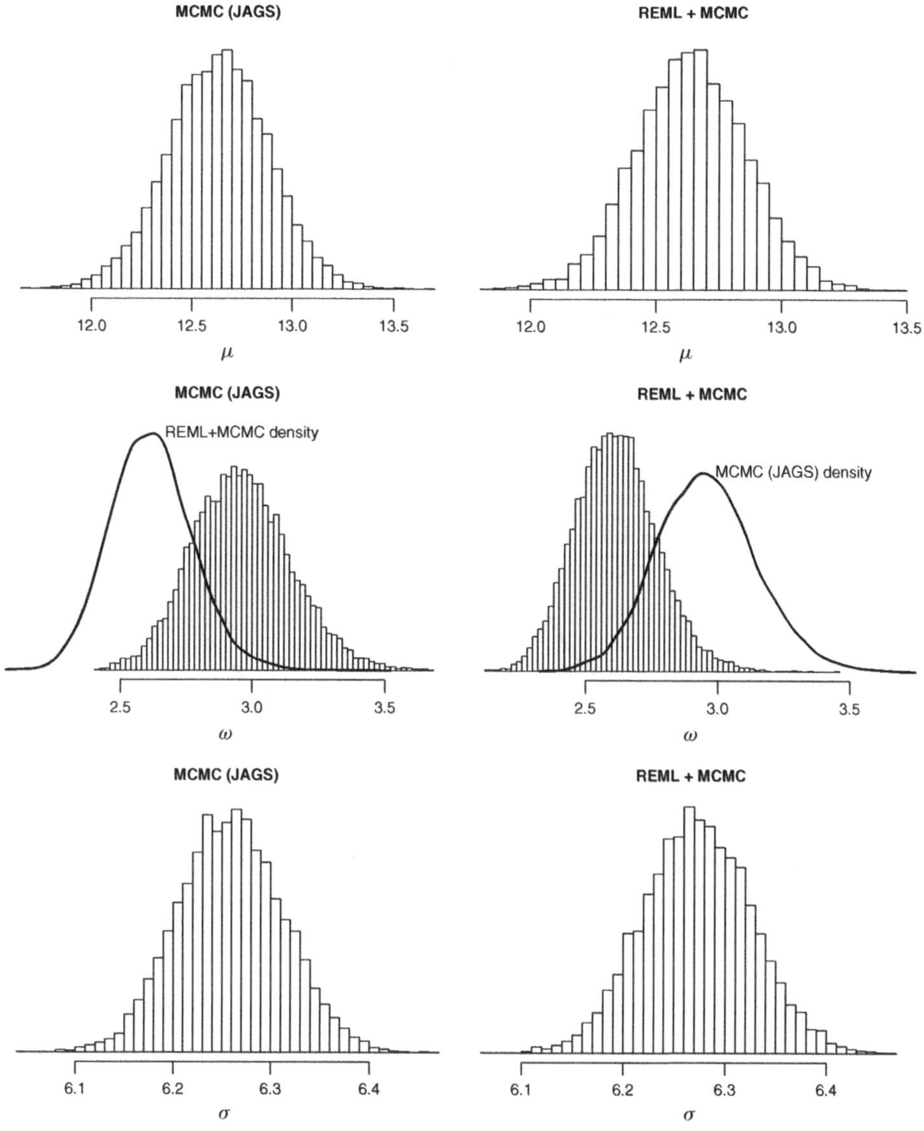

Figure 7.6 Comparisons of marginal posterior densities from MCMC via `JAGS` and the `mcmcsamp` function in `lme4` (version 0.999375-22), hierarchical one-way ANOVA model for mathematics achievement across 160 schools (Example 7.6). Histograms on the left panels summarize 10 000 MCMC iterates generated using `JAGS`; histograms on the right summarize the output of `lme4::mcmcsamp`. The two MCMC-based approaches generate identical results for the 'fixed effect' parameter μ and the 'within' standard deviation σ, but diverge for the 'between' standard deviation ω, with `lme4::mcmcsamp` generating an underestimate of ω. At the time of writing, the authors of `lme4` are aware of the issue.

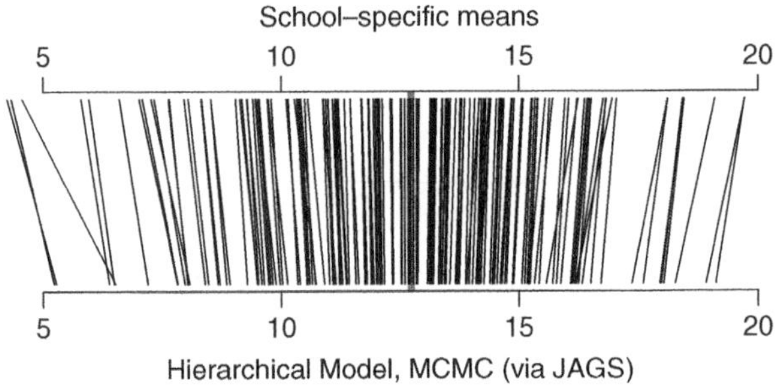

Figure 7.7 Shrinkage of school-specific means via hierarchical modeling, math achievement scores. Top axis shows location of the $J = 160$ school-specific mean achievement scores; the lower axis indicates the 'shrunken' hierarchical estimates, pulled in towards the grand mean $\mu = 12.64$. The thick gray line (barely visible behind the vertical lines) shows the grand mean. Contrast this graph with the similar result obtained with the baseball batting averages data (Figure 7.4, Example 7.5).

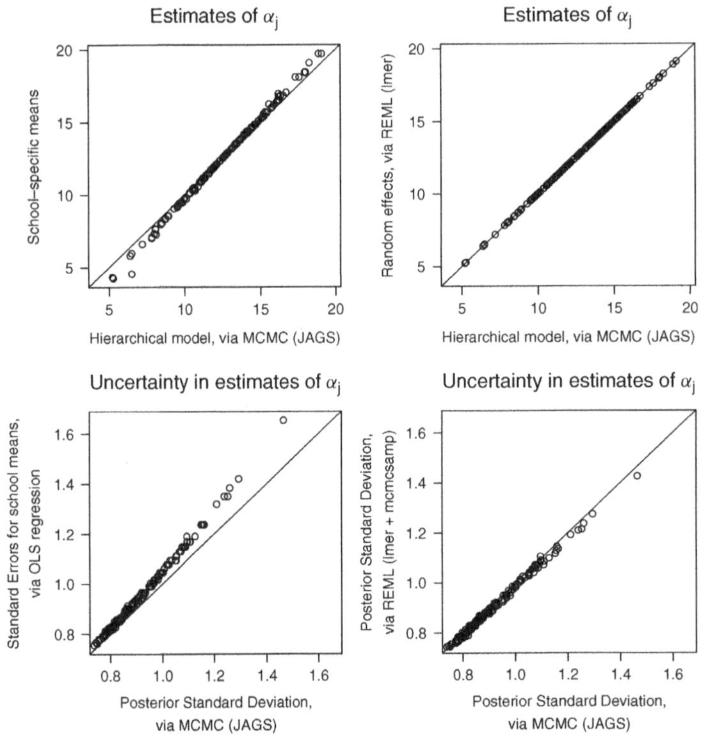

Figure 7.8 Comparison of estimates of school-specific average math achievement scores. Top panels compare three sets of estimates of the $J = 160$ school-specific α_j: the sample means, MCMC via `JAGS`, and REML/MCMC using the `lmer` + `mcmcsamp` approach in `lme4`. Bottom panels compare standard deviations of the marginal posterior densities of each α_j.

7.2.2 Two-way ANOVA

The hierarchical, one-way ANOVA model is easily generalized to handle two grouping factors, as we demonstrate with the next example. This example will also provide us with the opportunity to consider one of the 'tricks of the trade' for MCMC based approaches to computing posterior densities: we will specify an over-parameterized, unidentified version of a model so as to improve the efficiency of a MCMC-based random tour of the posterior density of the identified parameters, with hierarchical models being particularly well-suited to this approach (e.g. Gelman *et al.* 2007).

■ **Example 7.7**

Presidential election returns, by state, by year. In Example 2.5 we saw how prior information could be combined with polling data to improve forecasts of election outcomes. The prior information in that case consisted of state level election returns from previous years and across multiple states. Here we show how that historical information from multiple states can be combined in a simple hierarchical model to yield predictions for future elections.

Let $y_{it} \in [0, 100]$ be the percentage of the vote won by the Democratic candidate in state i at election t. Here we have an unbalanced data set of 945 observations (available as the data frame `presidentialElections` in the `pscl` package in `R`), spanning all 50 states plus the District of Columbia, from presidential elections between 1932 and 2004, inclusive; the imbalance in the data arises because (a) Hawaii and Alaska contribute data from 1960 onwards, (b) the District of Columbia contributes data from 1964 onward, and (c) Alabama has missing data for 1948 and 1964.

Figure 7.9 shows the state-specific time series superimposed on one another, with white lines indicating Southern states (the 11 states of the former Confederacy). The District of Columbia is the outlier at the top of the graph, reliably generating Democratic vote shares in the neighborhood of 80 %, and we drop it from the analysis which follows.

Hierarchical Model. We use the following two-way ANOVA normal model to capture variation across both cross-sectional and longitudinal dimensions of the data:

$$y_{it} \sim N(\mu + \alpha_i + \delta_t, \sigma^2) \tag{7.15}$$

The parameters α_i are state-specific terms, while the parameters δ_t are election-specific terms. With the presence of μ in the model, we will set the α_i and the δ_t to have mean zero across states and across elections, respectively, such that μ is interpretable as the grand mean (and we revisit this point, below).

Note also that this normal model assigns positive probability to the impossible outcomes $y_{it} \notin [0, 100]$. None of the y_{it} get particularly close to 0 % or 100 % and σ^2 can be expected to be reasonably small, so as a practical matter the model predictions lie comfortably to the interior of the feasible [0, 100] interval. An alternative and easily implemented approach would be to model some transformation of the y_{it}, such as the logits of the vote shares.

Inspection of Figure 7.9 reveals considerable within-state volatility, at least for some states, such that the model in Equation 7.15 is probably unrealistic. The period 1932–2004

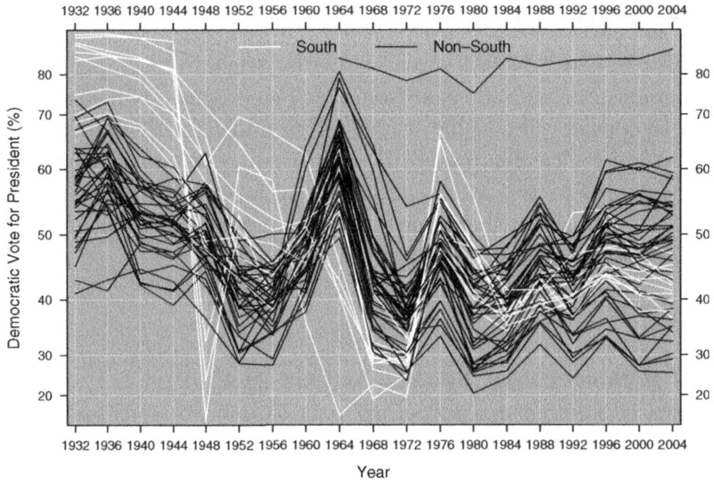

Figure 7.9 Democratic vote for President (%), by state, 1932–2004. States are grouped by South/non-South (white and dark lines, respectively), where Southern states are defined as the eleven states of the former Confederacy. The vertical axis is scaled non-linearly so as to better visualize political interesting variation in the neighborhood of 50 %. The District of Columbia is the outlier at the top of the graph. The analysis in Example 7.7 uses data from the period 1984–2000 to generate 'step ahead' predictions for the 2004 data.

spans the demise of the one-party, Democratic South, and sees considerable change in the partisan composition of many non-Southern states. The simple two-way ANOVA model with iid normal disturbances seems unrealistic for the entire data set: some combination of state-specific or region-specific time-trends might fare better, as might an explicitly dynamic model, perhaps with state-specific error variances. We defer a consideration of these richer specifications for later. The goal of the analysis is to generate predictions for recent elections, and so the issue here is whether the estimates of α_i based on analysis of the entire historical data set would generate useful predictions for any particular state. The model in Equation 7.15 seems more appropriate if fit to a shorter time period, with more within-state homogeneity: i.e. over a shorter, recent time period, σ^2 is likely to be relatively small, and we can expect less variation around the state-specific mean $\mu + \alpha_i$. To this end, we fit the model in Equation 7.15 to the period 1984–2000; the 2004 results will be 'held back' from the analysis so as to assess the out-of-sample performance of the model. Even over this shorter period, the two-way ANOVA model is not particularly realistic, but is nonetheless a useful tool for highlighting the details of Bayesian hierarchical modeling via Markov chain Monte Carlo.

Prior densities. Priors are required for all model parameters. For the grand mean μ we use a $N(50, 15^2)$ prior. A non-conjugate prior is used for the error standard deviation σ, $\sigma \sim \text{Unif}(0, 20)$, which amounts to a reasonably uninformative prior given that the vote shares generally lie well to the interior of the [0, 100] interval.

The state-specific parameters tap time-invariant facets of each state's support for Democratic candidates, while the year-specific parameters tap features of the election that impact vote shares in all states. Since the model includes a parameter for the grand mean, μ, the α_i and δ_t parameters are deviations around this average: for this reason we center their densities at zero. Each set of parameters – the state-specific parameters α_i and the year-specific δ_t – is considered exchangeable and each are given a common prior densities, with mean zero:

$$\alpha_i \sim N(0, \sigma_\alpha^2) \tag{7.16a}$$

$$\delta_t \sim N(0, \sigma_\delta^2) \tag{7.16b}$$

Priors on the hyper-parameters σ_α^2 and σ_δ^2 complete the specification of the model. Here I employ the non-conjugate priors

$$\sigma_\alpha \sim \text{Unif}(0, 15) \tag{7.17a}$$

$$\sigma_\delta \sim \text{Unif}(0, 15) \tag{7.17b}$$

Implementing a MCMC algorithm. The following simple JAGS program implements the model given above:

JAGS code

```
model{
        for(i in 1:n){
                mu.y[i] <- mu + alpha[s[i]] + delta[j[i]]
                demVote[i] ~ dnorm(mu.y[i],tau[1])
        }

        mu ~ dnorm(50,tau.mu)
        tau.mu <- pow(15,-2)

        sigma[1] ~ dunif(0,20)
        sigma[2] ~ dunif(0,15)
        sigma[3] ~ dunif(0,15)

        for(i in 1:50){
                alpha[i] ~ dnorm(0,tau[2])
        }

        for(i in 1:nyear){
                delta[i] ~ dnorm(0,tau[3])
        }

        for(i in 1:3){
                tau[i] <- pow(sigma[i],-2)
        }
}
```

The resulting MCMC algorithm is run for 55 000 iterations, with the initial 5000 iterations discarded as burn-in, and the subsequent 50 000 iterations thinned by 5 iterations each. Figure 7.10 presents traceplots and autocorrelation functions for the grand mean μ and the three standard deviation parameters in the model: σ, the standard deviation

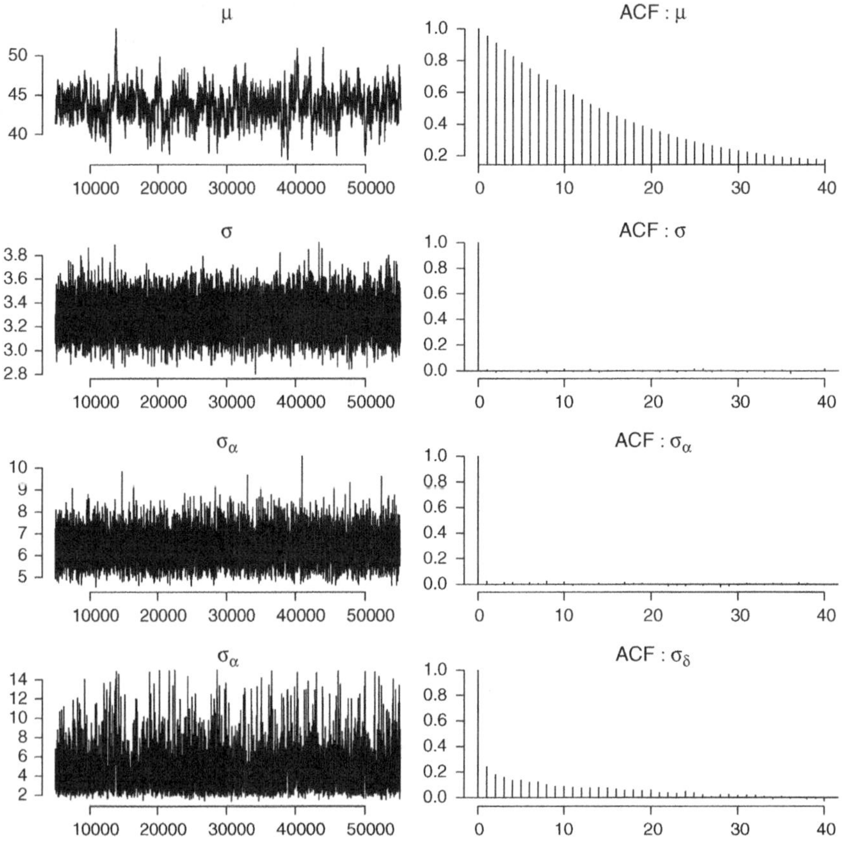

Figure 7.10 Trace plots and autocorrelation functions for grand mean and standard deviation parameters, two-way ANOVA model for presidential election outcomes in Equations 7.15 and 7.16.

of the idiosyncratic disturbance term in Equation 7.15; σ_α, the standard deviation of the state-level offsets α_i; and σ_δ, the standard deviation of the election-specific offsets δ_t. The trace plots and autocorrelation functions indicate some slow mixing for the grand mean μ, but otherwise suggest that the MCMC algorithm has converged on the posterior density.

Formal diagnostics largely confirm this, as presented in Table 7.2. The Geweke diagnostic for the σ_δ parameter suggests non-stationarity for that parameter, but all other convergence diagnostics suggest acceptable performance by the MCMC algorithm; it is interesting that the Geweke and Heidelberger-Welch tests point in difference directions as to the stationarity of the sampled values of σ_δ. The Raftery-Lewis diagnostics also point to the slow-mixing for the grand mean μ, suggesting that over 250 000 iterations might be needed to accurately estimate the 2.5 and 97.5 % percentiles of the marginal posterior density for μ; nonetheless, the other diagnostics for this parameter suggest convergence,

Table 7.2 Convergence diagnostics, MCMC algorithm for two-way ANOVA model. All convergence diagnostics tests are implemented using the `coda` package in `R` (Plummer *et al.* 2008), see Section 6.2 for details. Diagnostics computed on 50 000 iterations, thinned by five iterations, after a 5000 iteration burn-in. For the Heidelberger-Welch diagnostic, the p value is from the Cramér-von-Mises test of stationarity. The Raftery-Lewis N is the estimated number of MCMC iterates required so as to obtain an accurate estimate of the .025 quantile of the corresponding marginal posterior density; I is the 'dependence factor', an estimate of the dependence in the sampled draws (higher values, more dependence).

	Geweke	Heidelberger-Welch	Raftery-Lewis	
Parameter	z	p	N	I
μ	0.74	0.82	256665	68.50
σ	−0.53	0.82	18705	4.99
σ_α	−0.61	0.99	18550	4.95
σ_δ	−1.83	0.89	19170	5.12

despite the slow mixing. This suggests a re-parameterization might be in order, so as to help the MCMC algorithm better deal with the 'sticky' grand mean parameter, μ.

Efficient exploration of the posterior density via over-parameterization. We again exploit a well-known feature of MCMC algorithms: a random tour through the space of *unidentified* parameters $\mathcal{V}$ can be mapped into the space of *identified* parameters $\mathcal{W}$, with the random tour in $\mathcal{W}$ being very efficient. That is, we specify a model we know to be unidentified, but we only perform inference with respect to the identified parameters, which can be recovered as simple, deterministic functions of the unidentified parameters.

We exploit this idea for our two-way ANOVA model as follows. In the specification of the model in Equations 7.16 the priors over the state-specific terms α_i and election-specific terms δ_t have mean zero, respectively. This is because of the presence of the parameter μ in the model: with both $\bar{\alpha} = \mu_\alpha = 0$ and $\bar{\delta} = \mu_\delta = 0$, the parameter μ has the interpretation as the grand mean, and the α_i and δ_t are offsets around the grand mean; indeed, this is the same parameterization used for mixed models (recall Equation 7.4). Note that with either μ_α or μ_δ left as free parameters, the model is unidentified. To see this, consider the likelihood contributions generated by the model in Equation 7.15 but where we replace the hierarchical structure in 7.16 with

$$\alpha_i \sim N(\mu_\alpha, \sigma_\alpha^2) \tag{7.18}$$

$$\delta_t \sim N(\mu_\delta, \sigma_\delta^2). \tag{7.19}$$

Let $\boldsymbol{\theta} = (\mu, \mu_\alpha, \mu_\delta)$. Note that we get the same likelihood contributions from $\boldsymbol{\theta}$ as we get from $\tilde{\boldsymbol{\theta}} = (\tilde{\mu}, \tilde{\mu}_\alpha, \tilde{\mu}_\delta)'$ where

$$\tilde{\mu} = \mu + c \tag{7.20a}$$

$$\tilde{\mu}_\alpha = \mu_\alpha - \lambda c \tag{7.20b}$$

$$\tilde{\mu}_\delta = \mu_\delta - (1 - \lambda)c \tag{7.20c}$$

and $c \neq 0$ and $\lambda \in [0, 1]$. That is, the addition of c to μ can be exactly offset by subtracting c from μ_α and μ_δ, in proportions λ and $1 - \lambda$ respectively. Note that Equations 7.20b and 7.20c imply

$$\tilde{\alpha}_i = \alpha_i - \lambda c, \tag{7.21a}$$

$$\tilde{\delta}_t = \delta_t - (1 - \lambda)c, \tag{7.21b}$$

$\forall\, i, t$, and so

$$\begin{aligned}
\mu + \alpha_i + \delta_t &= \tilde{\mu} + \tilde{\alpha}_i + \tilde{\delta}_t \\
&= \mu + c + \alpha_i - \lambda c + \delta_t - (1 - \lambda)c \\
&= \mu + \alpha_i + \delta_t + c - (\lambda + 1 - \lambda)c \\
&= \mu + \alpha_i + \delta_t.
\end{aligned}$$

That is, the model in which μ, μ_α and μ_δ are unconstrained generates the same likelihood contributions as the model in which just μ is a free parameter. This is nothing other than a *perfect collinearity* problem of the sort sometimes encountered in linear regression models: the design matrix $\mathbf{X}$ implied by the model in 7.15 without constraints on $\boldsymbol{\theta} = (\mu, \mu_\alpha, \mu_\delta)'$ is rank deficient (see Problem 7.5). In short, μ, μ_α and μ_δ are not jointly identified, but a linear combination of these three parameters is.

Carrying all three location parameters μ, μ_α and μ_δ is an example of *over-parameterization* or *redundant* parameterization, usually considered a problem for modeling, but here to be used as a means to an end. The redundancy can be removed by the restrictions $\mu_\alpha = 0$ and $\mu_\delta = 0$ in 7.16, which impose linear restrictions on the α_i and δ_t, specifically

$$\mu_\alpha = 0 \Rightarrow \sum_{i=1}^{n} \alpha_i = 0 \tag{7.22a}$$

$$\mu_\delta = 0 \Rightarrow \sum_{t=1}^{T} \delta_t = 0, \tag{7.22b}$$

that are sufficient for identification.

Our interest in the identification issue here is to exploit the computationally efficient *unidentified* parameterization so as to better analyze the posterior density of the *identified* parameters. We use the unidentified parameterization of the hierarchical, two-way ANOVA model,

$$y_{it} \sim N(\mu + \alpha_i + \delta_t, \sigma^2) \tag{7.23a}$$

$$\mu \sim N(0, 100^2) \tag{7.23b}$$

$$\alpha_i \sim N(\mu_\alpha, \sigma_\alpha^2) \tag{7.23c}$$

$$\delta_t \sim N(\mu_\delta, \sigma_\delta^2) \tag{7.23d}$$

$$\mu_\alpha \sim N(0, 100^2) \tag{7.23e}$$

$$\mu_\delta \sim N(0, 100^2) \tag{7.23f}$$

letting $\boldsymbol{\theta} = (\mu, \mu_\alpha, \mu_\delta)'$ be unconstrained parameters, but subject to the proper (if vague) priors above. We let a MCMC algorithm explore the parameter space of the unidentified model, but map the MCMC output into the space of identified parameters. In this case, the transformation from the unidentified parameters to identified parameters is simple: we impose the constraints $\bar{\alpha} = 0$ and $\bar{\delta} = 0$ by subtracting the sampled values of these quantities out of the model. For instance, at iteration m of the MCMC algorithm, define

$$\alpha_i^{*(m)} = \alpha_i^{(m)} - \bar{\alpha}^{(m)}, \quad i = 1, \ldots, n \tag{7.24a}$$

$$\delta_t^{*(m)} = \delta_t^{(m)} - \bar{\delta}^{(m)}, \quad t = 1, \ldots, T. \tag{7.24b}$$

and

$$\mu^{*(m)} = \mu^{(m)} + \bar{\alpha}^{(m)} + \bar{\delta}^{(m)}. \tag{7.25}$$

Note that the identified model is merely a re-parameterization of the unidentified model: in both cases we generate the same likelihood contributions, since

$$\begin{aligned}\mu^* + \alpha_i^* + \delta_i^* &= \mu + \bar{\alpha} + \bar{\delta} + \alpha_i - \bar{\alpha} + \delta_i - \bar{\delta} \\ &= \mu + \alpha_i + \delta_t.\end{aligned}$$

That is, that both the identified and unidentified model make the same predictions for the data. The distinction between the two models is only consequential for inferences with respect to the location parameters and leaves the variance parameters and predicted values unaffected.

In implementing this model we can 'hardwire' these transformations into a BUGS/JAGS program, or apply the transformations after running the MCMC algorithm ('post-processing' the output of the MCMC algorithm). I use the following JAGS program:

JAGS code

```
model{
        for(i in 1:n){
              mu.y[i] <- mu[1] + alpha[s[i]] + delta[j[i]]
              demVote[i] ~ dnorm(mu.y[i],tau[1])
        }

        sigma[1] ~ dunif(0,20)
        sigma[2] ~ dunif(0,20)
        sigma[3] ~ dunif(0,20)

        for(i in 1:50){
              alpha[i] ~ dnorm(mu[2],tau[2])
        }

        for(i in 1:nyear){
```

```
            delta[i] ~ dnorm(mu[3],tau[3])
    }

    for(i in 1:3){
            tau[i] <- pow(sigma[i],-2)
    }

    for(i in 1:3){
            mu[i] ~ dnorm(0,1E-4)
    }

    ## transformations for identified parameters
    mustar <- mu[1] + mean(alpha[]) + mean(delta[])
    for(i in 1:50){
            alphastar[i] <- alpha[i] - mean(alpha[])
    }
    for(i in 1:nyear){
            deltastar[i] <- delta[i] - mean(delta[])
    }
}
```

As with the identified version of the model, we run the MCMC algorithm for 55 000 iterations, discarding the first 5000 iterations as burnin, and thinning the MCMC output down to every 5th iteration. Convergence diagnostics for the identified parameter μ^* and the three standard deviations σ, σ_α and σ_δ appear in Table 7.3, indicating that 50 000 iterations is more than sufficient for these four parameters.

Trace plots, marginal posterior densities and autocorrelation functions for selected parameters appear in Figure 7.11. The different levels of performance of the MCMC algorithm with respect to the identified and unidentified parameters is clear. The trace plots for the unidentified parameters μ, μ_α and μ_δ display terribly slow mixing, and the autocorrelation functions are nearly flat, even out to 500 lags. On the other hand, we

Table 7.3 Convergence diagnostics, MCMC algorithm for re-parameterized two-way ANOVA model. All convergence diagnostics are implemented using the `coda` package in `R` (Plummer *et al.* 2008), see Section 6.2 for details. Diagnostics computed on 50 000 iterations, thinned by 5 iterations, after a 5000 iteration burn-in. Note that μ, μ_α and μ_δ are unidentified parameters and of no substantive interest, but that $\mu^* = \mu + \bar{\alpha} + \bar{\delta}$ is identified. For the Heidelberger-Welch diagnostic, the p value is from the Cramér-von-Mises test of stationarity. The Raftery-Lewis N is the estimated number of MCMC iterates required so as to obtain an accurate estimate of the .025 quantile of the corresponding marginal posterior density; I is the 'dependence factor', an estimate of the dependence in the sampled draws (higher values, more dependence).

	Geweke	Heidelberger-Welch	Raftery-Lewis	
Parameter	z	p	N	I
$\mu^* = \mu + \bar{\alpha} + \bar{\delta}$	−0.58	0.91	19645	5.24
σ	−0.34	0.53	18855	5.03
σ_α	−0.22	0.21	19010	5.07
σ_δ	1.25	0.75	18705	4.99

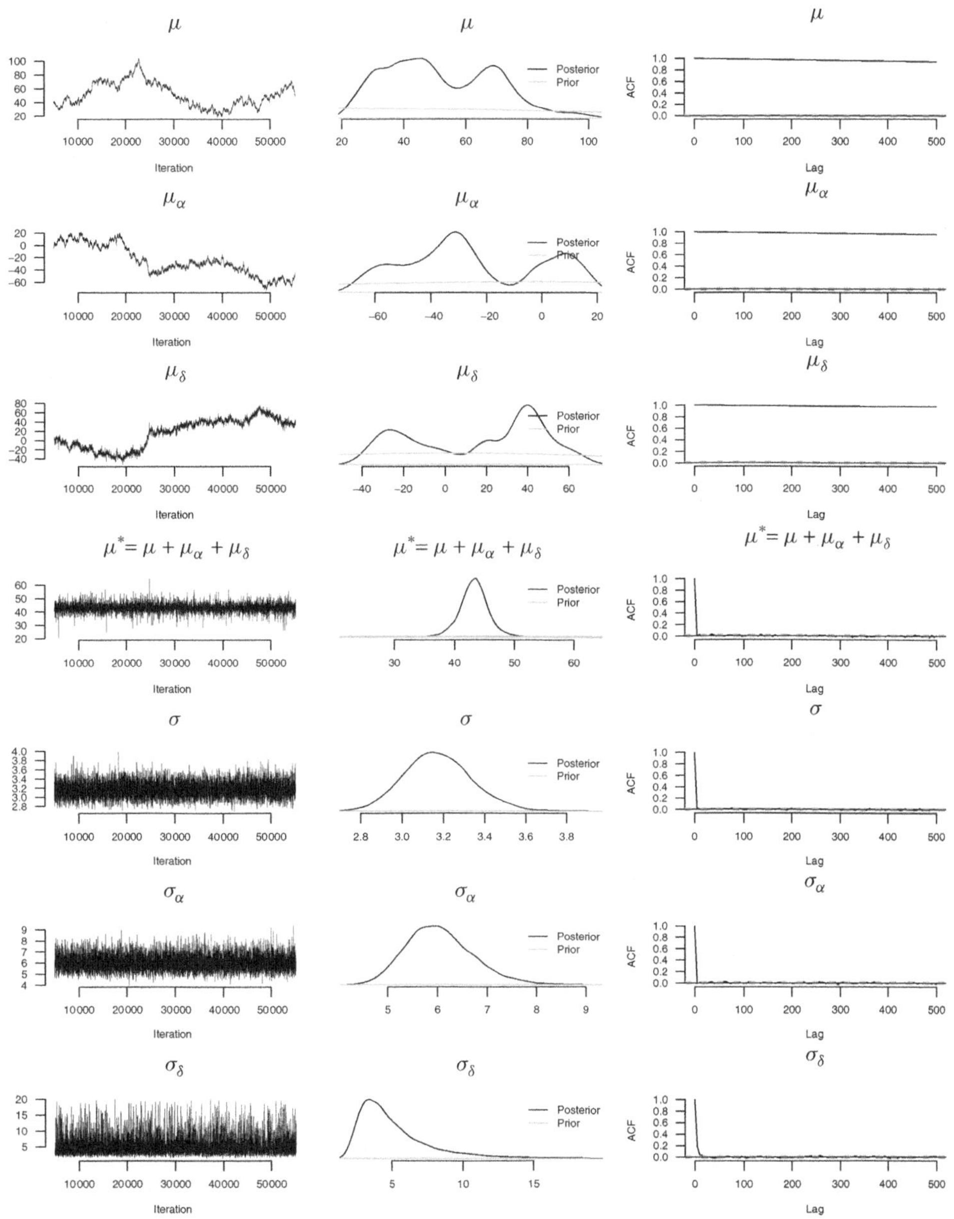

Figure 7.11 Trace plots (left column), marginal posterior densities (middle) and autocorrelation functions (right), unidentified version of the two-way ANOVA model for presidential elections outcomes. Note that μ, μ_α and μ_δ are unidentified parameters and of no substantive interest, but that $\mu^* = \mu + \mu_\alpha + \mu_\delta$ is identified, along with the three standard deviation parameters.

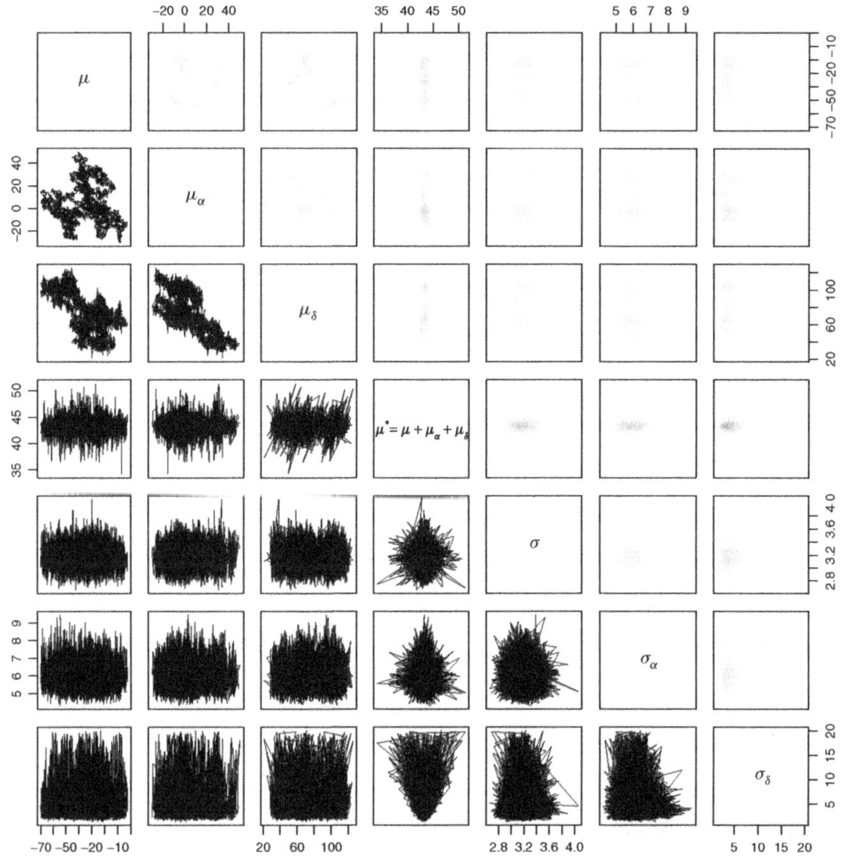

Figure 7.12 Bivariate trace plots, unidentified version of the two-way ANOVA model for presidential election outcomes. Panels below the diagonal display bivariate trace plots for 50 000 iterations of the MCMC algorithm; panels above the diagonal are scatterplots. Observe the poor performance of the MCMC algorithm for the unidentified parameters (μ, μ_α and μ_δ), but contrast the more acceptable performance for the identified combination of those parameters, μ^*.

appear to be generating a very efficient random tour of the posterior density for the the identified parameter μ^* and the three standard deviations. Further evidence on this score appears in Figure 7.12, which shows a number of bivariate trace plots and scatterplots of the MCMC output. Again, the performance of the MCMC algorithm for the unidentified parameters is unsurprisingly poor, while the picture is just the opposite for the identified parameters. For instance, the scatterplots of the identified parameters in the lower right of Figure 7.12 correspond to the 'teardrop' normal/inverse-Gamma contours typical of conjugate Bayesian analysis of normal data.

Results. Numerical summaries of the marginal posterior densities of the four key model parameters appear in Table 7.4. Recall that the parameter σ is the standard deviation of the error term in the model, that part of y_{it} that isn't accounted for by the two-way

Table 7.4 Numerical summaries, marginal posterior densities, two-way ANOVA model for presidential election outcomes, based on 50 000 MCMC iterations, thinned to every 5th iteration, after a 5000 iteration burn-in.

		95 % HDR	
	Mean	lower	upper
μ^*	43.30	37.84	48.93
σ	3.17	2.86	3.48
σ_α	6.03	4.83	7.38
σ_δ	5.19	1.74	10.85

ANOVA model in 7.15. We estimate the mean of the marginal posterior density of σ to be 3.17 percentage points. This is reasonably large, implying that even conditioning on the state-level effects α_i and the election-specific terms δ_t, a 95 % credible interval for a prediction from the two-way ANOVA model spans about $\pm 2 \times 3.17 = \pm 6.34$ percentage points. These are large margins in substantive terms, and suggest that the model will not be particularly useful as a forecasting tool. The mean of the marginal posterior for the residual variance σ^2 is 10.11, and a 95 % highest posterior density interval ranges from 8.10 to 12.10. Conditional on the posterior mean of σ^2, the r^2 for the model is .81, and this quantity has a 95 % highest posterior density interval ranging from .77 to .85. That is, the state-specific α_i and the election-specific δ_t parameters are accounting for the bulk of the variation in the election outcomes. The between-state standard deviation σ_α has a posterior mean of 6.03, and the between-election standard deviation σ_δ has a posterior mean of 5.19 (although this parameter is estimated imprecisely, with a 95 % HPD ranging from 1.74 to 10.85); the posterior means of these parameters are large relative to the posterior mean of the residuals standard deviation σ (6.03 and 5.19 versus 3.17).

Figure 7.13 provides a graphical summary of the marginal posterior densities of the state-specific offsets (α_i). There is considerable variability across states in the α_i, with this between-state variability tapped by the σ_α parameter: in Table 7.4 I report the mean of the posterior density of σ_α as about 6 percentage points. The Bayes estimates (posterior means) of the α_i range from a low of -14.3 in Utah to 10.4 in Rhode Island, spanning almost 25 percentage points of vote share. Note that by construction, the α_i have mean zero across states; Ohio and Kentucky are the states with α_i closest to zero (0.35 and -0.29 percentage points, respectively), meaning that these states tend to produce election results closest to the national average in any given election, $\bar{y}_t = \mu + \delta_t$. Maine has the 3rd smallest estimated α_i, just 1.06 percentage points, suggesting that there is perhaps still some truth to the old election night cliché, 'as Maine goes, so goes the nation.'

But Figure 7.13 also makes clear that there is considerable uncertainty over any given state's α_i; the 95 % HPD intervals are all quite wide, with the posterior standard deviations of the α_i all roughly 1.37 percentage points. That is, while the between-state variation is large with $\sigma_\alpha \approx 6$ percentage points, the uncertainty attaching to any particular α_i is about one-quarter of this, representing an additional, non-trivial source of uncertainty. This uncertainty arises because of the relatively short longitudinal dimension in the data used in this analysis: each α_i is based on just $T = 5$ elections per state. The hierarchical model

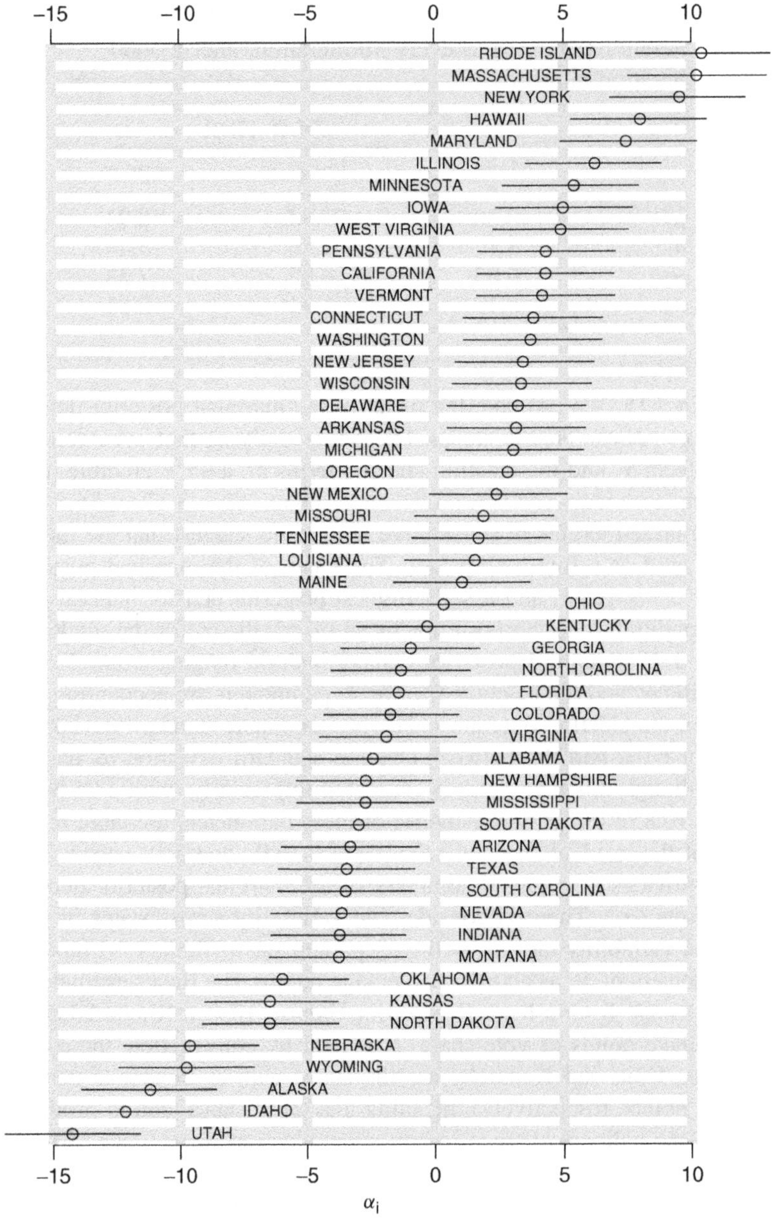

Figure 7.13 Marginal posterior densities for state-specific terms (α_i) and 95 % HPD intervals, two-way ANOVA model for presidential election outcomes. States are ordered vertically by the means of the marginal posterior densities (open plotting symbol); horizontal lines cover 95 % HPD intervals.

generates more precise estimates of α_i than we would obtain via a non-hierarchical model (see Problem 7.7) – consistent with the results in Example 7.5 – but are nonetheless reasonably large.

Figure 7.14 shows the marginal posterior densities for the election-specific δ_t parameters. Like the α_i, the δ_t are centered on zero by construction. Ronald Reagan's re-election in 1984 corresponds to the worst, average Democratic performance in these data, while Bill Clinton's 1996 re-election generates the maximum δ_t in these data. Ross Perot's 1992 candidacy depressed vote shares for the major party candidates in 1992, resulting in a relatively low level of δ_t for that election, despite the Democratic victory in that year.

Note that we have no data for 2004 in this analysis, and so the posterior density for δ_t is actually a posterior *predictive* density, given by the hierarchical model in Equation 7.23d, with μ_δ set to zero by construction, and with σ_δ drawn from its posterior density. Accordingly, the marginal posterior predictive density for δ_{2004} is centered on $\mu_\delta = 0$, and has quite a large standard deviation: recall from Table 7.4 that the posterior mean of σ_δ is over 5 percentage points, which sees the 95 % HPD interval for δ_{2004} extend beyond ± 10 percentage points.

Predictions for the 2004 result. To generate 'out-of-sample' predictions for the 2004 result, we exploit a very convenient feature of the BUGS/JAGS software. If a stochastic node – a quantity appearing on the left hand side of a 'twiddle', the symbol '$\sim$' – is

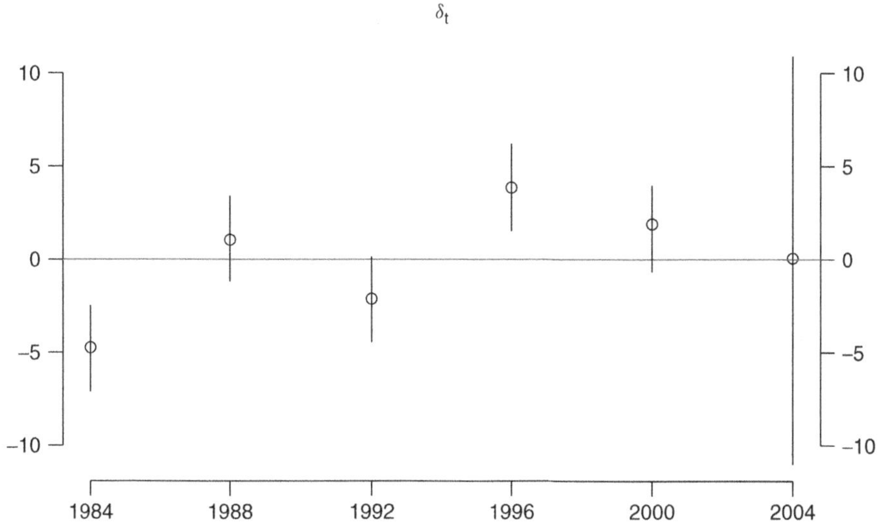

Figure 7.14 Marginal posterior densities for election-specific terms (δ_t) and 95 % HPD intervals, two-way ANOVA model for presidential election outcomes. Note that no data from 2004 is used in the model fitting, and so the marginal posterior (predictive) density for the δ_t term in that year is considerably more dispersed than for the 'in-sample' years, reflecting the posterior uncertainty in the second level parameter σ_δ.

unobserved, then the software will sample from the predictive distribution for that node, conditional on the rest of the model. In the case where the stochastic node is partially observed data, the software will sample from the predictive density for the unobserved elements, functionally equivalent to treating the data as missing at random (MAR) and the sampled values treated as multiple imputations, in the sense of Rubin (1987).

In this case, we treat the 2004 data as missing data while fitting the model and examine the predictive density (or summaries of many draws from the predictive density) as an assessment of model fit. We accomplish this by making a copy of the data, with the 2004 values of y set to `NA` (missing data), and then passing this 'working' copy of the data to `JAGS`. We then add y to the set of nodes to be monitored over iterations of the MCMC algorithm; `JAGS` stores the imputed (sampled) values of the missing y_{it} for 2004 in the corresponding element of the $\mathbf{y}$ vector. Like any other quantity output by a MCMC algorithm, these can be summarized for the purposes of making inferences, or, in this case, for assessing the 'out-of-sample' predictive performance of the model.

The predictive density from which we generate imputations for the missing values of the dependent variables is just the model in Equation 7.15; for a future election r, this predictive density is just $y_{ir} \sim N(\mu + \alpha_i + \delta_r, \sigma^2)$. Since we impose the identifying restriction that $\bar{\delta} = \mu_\delta = 0$ (where the averaging is across elections), it follows that for a given state i, the mean of the posterior predictive density for a future election (election r) is

$$E(y_{ir}|\text{data}) = E(\mu|\text{data}) + E(\alpha_i|\text{data}), \tag{7.26}$$

where the notation $E(\cdot|\text{data})$ signifies the mean of a posterior density. The posterior means of μ^* and α_i are provided by the MCMC algorithm. For a hypothetical election no particular election-specific offset δ_t can be observed, so we sample a δ_t from its predictive density given in Equation 7.23d. Note that posterior uncertainty in σ_δ generates uncertainty as to the value of δ_r that ought to apply to hypothetical election r; while $E(\delta_r) = 0$ posterior uncertainty over σ_δ does influence the variance and hence the width of any credible interval around that best guess.

Because the MCMC algorithm is sampling from the joint posterior density of all model parameters, posterior uncertainty over these parameters is propagated forward when we sample from the predictive density for the sampling for the y_{ir}. At each iteration m of the MCMC algorithm we generate sampled values $\mu^{(m)}$, $\alpha_i^{(m)}$, $\delta_r^{(m)} \sim N(0, \sigma_\delta^{2(m)})$ and $\sigma^{2(m)}$. These sampled values of the parameters then used to form the moments of the predictive density from which we then sample $y_{ir}^{(m)}$, i.e.

$$y_{ir}^{(m)} \sim N(\mu^{(m)} + \alpha_i^{(m)} + \delta_r^{(m)}, \sigma^{2(m)}). \tag{7.27}$$

Predictive performance. Figure 7.15 shows summaries of the posterior predictive densities of Democratic presidential vote shares for each state. The open plotting symbols indicate the Bayes estimates of y_{ir} (the mean of the respective marginal posterior predictive density) and the horizontal lines cover 95 % HPD intervals. Each state's actual level of vote share in 2004 is also plotted (solid, square plotting symbol). The Bayes

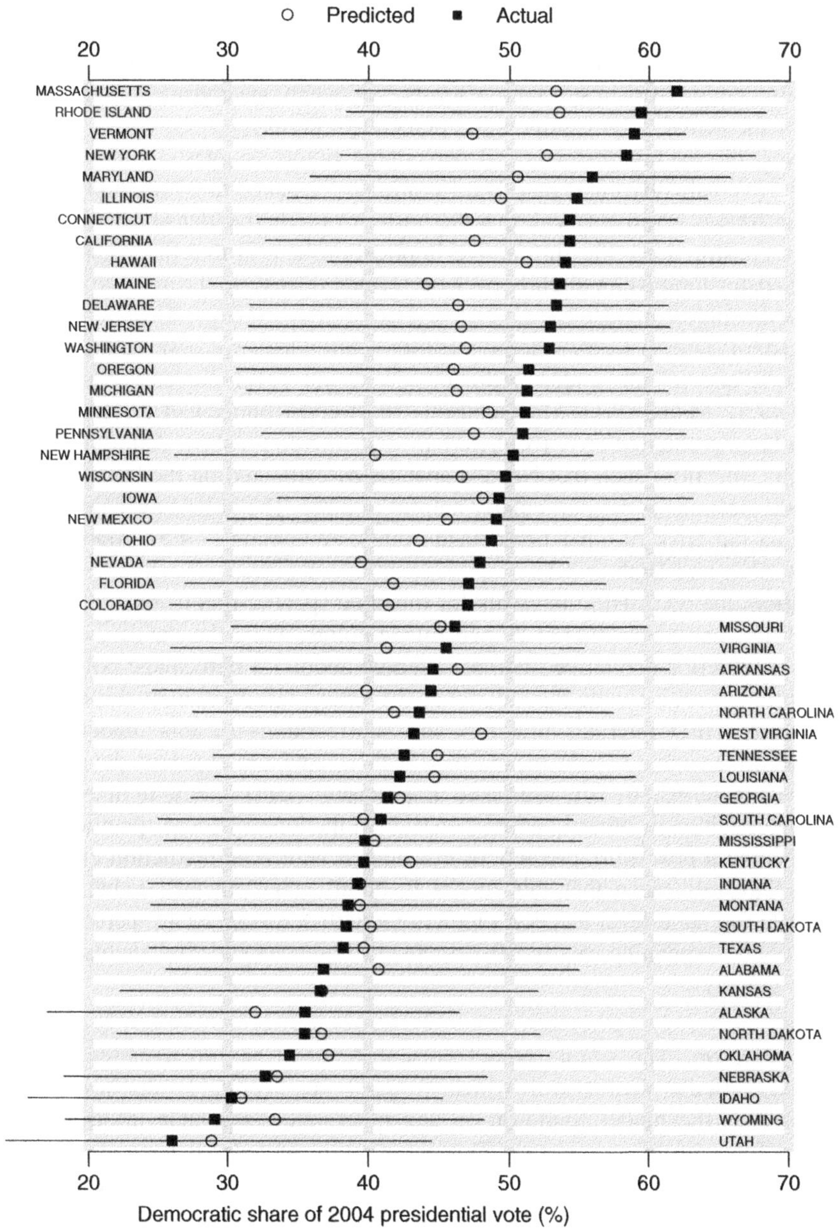

Figure 7.15 Out of sample forecasts and 2004 results, two-way ANOVA model for presidential election outcomes. States ordered vertically by 2004 outcomes (Democratic percentage share of the presidential vote); horizontal lines cover 95 % HPD intervals for each state's marginal posterior predictive density over y_{ir}

estimates of y_{ir} follow the actual 2004 results reasonably closely ($r = .90$). But the prediction errors – defined as the actual 2004 result minus the Bayes estimates of y_{ir} – are actually quite large in substantive terms. The root mean squared error (RMSE) is 4.7 percentage points, and the median absolute prediction error is 3.3 percentage points. These are reasonably large errors in substantive terms, say, when one considers that the standard deviation of the actual 2004 results is 8.5 percentage points.

Moreover, the average error is 2.3 percentage points, suggesting that the model's predictions systematically understate Democratic performance in 2004, implying that the actual δ_t for 2004 is positive (recall that the posterior predictive density for δ_t is centered on zero, by construction); the substantive implication here is that the 2004 Democratic candidate Senator John Kerry slightly over-performed relative to the average of the $T = 5$ previous Democratic candidacies.

Northeastern states – Kerry's home territory – are where we observe the largest prediction errors, and all are under-predictions: Vermont (11.4 percentage points), New Hampshire (9.5), Maine (9.1), Massachusetts (8.4) and Connecticut (7.1). Nevada also generates a large prediction error (8.1), perhaps indicative of a slow, demographic shift in a Democratic direction, such that Nevada-specific α_i estimated on the basis of Nevada's 1984 to 2000 elections is not particularly helpful in 2004. The largest over-predictions are West Virginia (an error of −5.1 percentage points), Wyoming (−4.5), Alabama (−4.2), Kentucky (−3.3) and Utah (−3.1).

In addition, the 95 % HPD intervals for the state-specific predictions are quite wide: the 95 % HPDs span an average of 29 percentage points, while the standard deviations of the marginal posterior predictive intervals average about 7.3 percentage points. Because the HPD intervals are so wide they all cover the actual results in their respective states.

These wide HPD intervals highlight the shortcomings of this model. The short longitudinal dimension ($T = 5$) means that each α_i and the standard deviation of the election-specific terms, σ_δ are estimated imprecisely. Uncertainty in these quantities generates uncertainty in other quantities that are functions of those quantities, such as the state-specific *a posteriori* predictions considered here. The prediction errors and the wide HPD intervals mean that this model is of no great practical value. A political strategist offering state-level election predictions with a RMSE of 4.7 percentage points and 95 % bounds spanning nearly 30 percentage points would find themselves out of work quite quickly.

Model criticism. The simple model presented above has no dynamic component, ignoring the particular sequence in which the previous election results have been generated. Nor does the model condition on any other freely available information about states beyond that in the vote shares. This information includes the region of the country in which state resides, demographic characteristics of states, home-state effects (e.g. Lewis-Beck and Rice 1983) or national-level characteristics that might help us predict each election's δ_t such as the state of the macro-economy, or the incumbent president's approval ratings, variables which appear in the aggregate-level election forecasting literature (e.g. Fair 1990; Hibbs 1987; Lewis-Beck and Rice 1992). If available, these covariates could enter a multilevel model as 'level 2' predictors.

7.3 Hierarchical models for longitudinal data

Grouped data often have a longitudinal dimension, as in the example just considered. That is, the 'group' is a single unit (an individual, a school, a country, etc.), and we have multiple, *sequential* observations on each unit. Such data go by various names in different disciplines: e.g. longitudinal data, repeated measures, or panel data. In these cases we usually need the modeling to address the sequential nature of the data. Precisely because the data form a sequence, it is unlikely that the within-unit conditional independence assumptions of the one-way and two-way ANOVA models are correct.

To make this concrete, recall the the election forecasting example in the previous section: if we believe the particular *order* in which the data were generated would help us predict the 'out-of-sample' 2004 election, then we need a different model than the simple two-way ANOVA model considered above. On the other hand, if we think a simple, 'static' model is sufficient, then we'd be willing to either (a) randomly shuffle the data along the longitudinal dimension, or (b) pick any of the years in the 1984–2004 interval as a candidate election to be held back from the model fitting and used to generate 'out-of-sample' predictions. These assessments are nothing more than judgements as to exchangeability: literally, if we would fit the same 'static' model to the data after randomly permuting the t (time) subscripts on the data, then, by definition, exchangeability holds, or at least a form of conditional exchangeability given the model. If instead we think that, say, data from 2000 are more informative about the 2004 results than data from 1996 or earlier, then clearly our modeling and predictions are not invariant to a permutation of the t subscripts, and (again, by definition), conditional exchangeability does not hold.

So what to do? The field of longitudinal modeling is massive, and there are many modeling strategies available to us that might induce conditional independence across the observations within any given unit. Indeed, the introduction of a longitudinal dimension in the two-way ANOVA model above is a common, easily implemented approach: the δ_t parameters soak up a common trend and the 'fixed effects', classical approach to estimating δ_t figures prominently in many analyses of panel data in the social sciences. Other longitudinal models include auto-regressive models, that condition on lags of y_{it}, or models with time trends. Here we will consider hierarchical models with these explicit, longitudinal or dynamic components in the model.

A 'time trend' is simply some function of time, $g(t)$. Examples include linear trends or low order polynomials, or sine or cosine functions of time capturing cyclic trends, perhaps at different frequencies. Non-parametric and semi-parametric approaches are also often used to model trends: examples include cubic spline functions and local polynomial fitting.

The precision with which we estimate these longitudinal features of the data is a function of the length of the longitudinal dimension: one can't expect to precisely estimate, say, a regression function that includes a cubic polynomial in time with $T = 5$ or thereabouts. Indeed, the cubic polynomial will consume 4 degrees of freedom, which may be a substantial proportion of T for data sets with a short longitudinal dimension.

It is not at all uncommon for panel data in the social sciences to have a short longitudinal dimension. The limiting case of $T = 2$ often means that we can collapse the two waves into a cross-sectional regression, *a la* a 'differences-in-differences' specification, i.e. regressing changes in y on changes in X. For short panels, a hierarchical model with unit-specific effects can actually do a good job of soaking up the longitudinal, within-unit dependence while stopping short of providing an explicit, dynamic model: recall that in Equation 7.6, the covariance between all observations within a given group is ω^2. This covariance does not decay with temporal distance between observations (i.e. $\text{cov}(y_{it}, y_{ir}) = \omega^2, \forall\ i, \forall\ t \neq r$), while covariances do decay in temporal distance for most stationary processes. But for panel data sets with short T, there is very little scope for longitudinal modeling per se, and so the simple way that the hierarchical models considered above capture within-unit covariance (Equation 7.6) may be a useful approach.

Panel data sets with longer longitudinal dimensions are an interesting case. In these cases we might consider a model in which y has a time trend specific to unit i. In any given unit we will estimate this trend imprecisely, say, if T is not particularly long. But if we believe the time trends are exchangeable across units – or more precisely, the parameters characterising the time trends are exchangeable – then hierarchical modeling can be used to estimate the parameters of the unit-specific time trends, 'borrowing strength' across units in thc manner we've encountered in the last few sections. We consider a linear time trend for the presidential elections data in the next example.

■ Example 7.8

Time trends in panel data/repeated measures. We return to the presidential elections data used in Example 7.7, where we considered the problem of forecasting the 2004 result, given the sequences of votes shares won by Democratic candidates for president in each state. Here we consider fitting linear time trends for each state so as to improve the forecasting performance of the hierarchical, two-way ANOVA model considered in Example 7.7. Fitting a linear time trend to each state's data may well generate better state level forecasts for the 2004 outcome than the 'static' model considered in Example 7.7. On the other hand, with just $T = 5$ 'in-sample' observations per state (the presidential elections spanning 1984 to 2000), we will generally estimate the state-specific time trends with consideable imprecision. In turn, this will lead to reasonably imprecise credible intervals in the predictions for the 2004 outcomes.

Hierarchical modeling offers an alternative way forward. We treat the state-specific intercept and slope parameters as exchangeable realizations from a bivariate normal density with unknown hyperparameters:

$$y_{it} \sim N(\beta_{i1} + \beta_{i2}t, \sigma^2) \tag{7.28a}$$

$$\sigma \sim \text{Unif}(0, 20) \tag{7.28b}$$

$$\begin{bmatrix} \beta_{i1} \\ \beta_{i2} \end{bmatrix} = \boldsymbol{\beta}_i \sim N(\boldsymbol{\mu}, \boldsymbol{\Omega}), \tag{7.28c}$$

where $t = 1, \ldots, 5$ indexes the presidential elections 1984 to 2000, $\boldsymbol{\mu} = (\mu_1, \mu_2)'$ and

$$\boldsymbol{\Omega} = \begin{bmatrix} \omega_1^2 & \omega_{12} \\ \omega_{12} & \omega_2^2 \end{bmatrix} = \begin{bmatrix} \omega_1^2 & \rho\omega_1\omega_2 \\ \rho\omega_1\omega_2 & \omega_2^2 \end{bmatrix},$$

with the last equality exploiting the identity $\omega_{12} = \rho\omega_1\omega_2$ where ρ is the correlation between β_1 and β_2 across the $n = 50$ states.

Priors over the 'level 2' parameters complete the model specification. We adopt independent, vague $N(0, 25^2)$ priors for μ_1 and μ_2, and uniform priors over the two standard deviations in $\mathbf{\Omega}$. For the standard deviaton of the intercept parameters I adopt the prior $\omega_1 \sim \text{Unif}(0, 20)$, while the standard deviation of the linear trend parameters of the time trends is given the prior $\omega_2 \sim \text{Unif}(0, 10)$. Given priors over ω_1 and ω_2, a prior over ρ is sufficient to induce a prior over ω_{12} and complete the specification. We adopt the uninformative prior $\rho \sim \text{Unif}(-1, 1)$. Note the convenience of this parameterization of the covariance matrix $\mathbf{\Omega}$; it is considerably easier to specify priors over standard deviations and correlations than over the full covariance matrix via an inverse-Wishart density. Note also that any matrix $\mathbf{\Omega}$ we assemble from these priors over ω_1, ω_2 and ρ is guaranteed to be positive definite.

MCMC algorithm. We exploit both (a) hierarchical centering and (b) a redundant parameterization to speed up the convergence of the algorithm. The hierarchical centering simply centers the time covariate t, via the model

$$E(y_{it}) = \beta_{i1} + \beta_{i2}(t - \bar{t}).$$

Note that this changes the interpretation of the intercepts: now $\beta_{i1} = E(y_{it}|t = \bar{t})$, whereas in the original parameterization $\beta_{i1} = E(y_{it}|t = 0)$. The actual values attaching to the intercepts are not particularly interesting: what it is interesting is variation in the intercepts across the states, and variation in the linear trend parameters β_{i2}, which are left unchanged by the centering. Note also that in the body of the program below we refer to t as `j`. The time counter `j` is passed to the program as an observed variable, with nested indexing or 'double subscripting' selecting the appropriate state-specific intercepts and slopes for a given observation (e.g. `s[i]`) and the correct value of the time counter (e.g. `j[i]`).

The redundant parameterization uses the model $E(y_{it}) = \alpha + \beta_{i1} + \beta_{i2}(t - \bar{t})$; in this parameterization α and μ_1 are not jointly identified, but we recover the identified parameters as $\mu_1^* = \mu_1 + \alpha$ and $\beta_{i1}^* = \beta_{i1} + \alpha, i = 1, \ldots, 50$.

JAGS code

```
model{
        ## loop over data for likelihood
        for(i in 1:n){
             mu.y[i] <- alpha + beta[s[i],1] + beta[s[i],2]*(j[i]-jbar)
             demVote[i] ~ dnorm(mu.y[i],tau)
        }
        sigma ~ dunif(0,20)   ## prior on standard deviation
        tau <- pow(sigma,-2) ## convert to precision

        ## hierarchical model for each state's intercept & slope
        for(p in 1:50){
             beta[p,1:2] ~ dmnorm(mu[1:2],Tau[,]) ## bivariate normal
        }

        ## means, hyper-parameters
        for(q in 1:2){
             mu[q] ~ dnorm(0,.0016)
```

```
        }

        ## priors for standard deviations
        omega[1] ~ dunif(0,20)
        Omega[1,1] <- pow(omega[1],2)  ## convert to variance

        omega[2] ~ dunif(0,10)
        Omega[2,2] <- pow(omega[2],2)  ## convert to variance

        ## uniform prior on correlation
        rho ~ dunif(-1,1)
        Omega[1,2] <- rho*omega[1]*omega[2] ## covariance
        Omega[2,1] <- Omega[1,2]

        ## convert covariance matrix to precision
        Tau[1:2,1:2] <- inverse(Omega[,])

        ## redundant parameter, grand mean
        alpha ~ dnorm(0,.0001)

        ## identified parameters
        mustar[1] <- mu[1] + alpha
        mustar[2] <- mu[2]

        ## identified parameters
        for(p in 1:50){
                betastar[p,1] <- beta[p,1] + alpha
                betastar[p,2] <- beta[p,2]
        }

        ## out-of-sample predictions for 2004 where j = 6
        for(i in 1:50){
                mu.yhat[i] <- betastar[s[i],1] + betastar[s[i],2]*(6-jbar)
                yhat[i] ~ dnorm(mu.yhat[i],tau)
        }

}
```

This program is rather long, since we need to set up the priors over the standard deviations ω_α and ω_β and the correlation parameter ρ, convert these to variances and covariances and then invert the covariance matrix $\boldsymbol{\Omega}$ to form a precision matrix $\boldsymbol{\tau}$. We also have some overhead mapping from unidentified to identified parameters. Note also the lines at the end of the program generating predictions for the out-of-sample 2004 outcomes, for which `j` $=$ `6`.

The MCMC algorithm is run for 250 000 iterations after a burn-in of 5000 iterations; this run length was chosen after inspecting convergence diagnostics which indicate that ω_2 is the slowest mixing parameter, and that more than 100 000 iterations would be required to accurately learn about the extreme quantiles of this parameter (i.e. the Raftery-Lewis run length diagnostic). Every 25th iterate is saved for summaries and inferences. Trace plots, estimated marginal posterior densities and autocorrelation function for selected parameters appear in Figure 7.16. The autocorrelation function confirm that ω_2 is relatively slow mixing in this case, but this aside, all other parameters appear to have converged on their marginal posterior densities. The correlation parameter, ρ, is confined to the $[-1,1]$ interval and in fact a good deal of the posterior density for ρ is close to

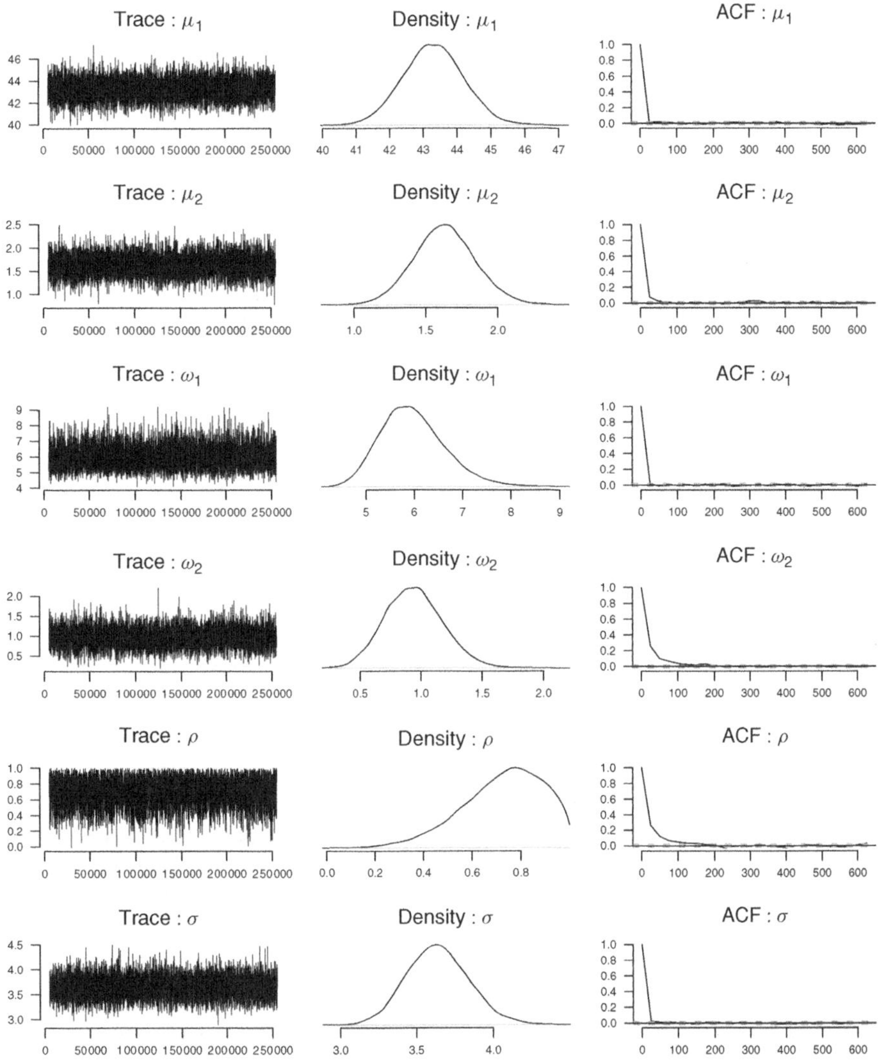

Figure 7.16 Trace plots (left column), marginal posterior densities (middle) and autocorrelation functions (right), for selected parameters from the hierarchical, linear time trend model for state-level presidential elections outcomes.

1.0, with a reasonably long left tail; accordingly, this parameter produces the odd trace plot shown in Figure 7.16 and the reasonably slow autocorrelation function. Nonetheless, convergence diagnostics confirm that the reasonably long run of the MCMC algorithm has generated a valid characterisation of the posterior density of this and other parameters.

Results. Since the specification in Equation 7.28 includes a hierarchical model over two parameters – the intercept β_{i1} and the linear time trend β_{i2} for each state – we can expect to observe 'shrinkage' with respect to both of these parameters. Figure 7.17 displays this two-dimensional shrinkage, with the arrows joining the 'no pooling' estimates

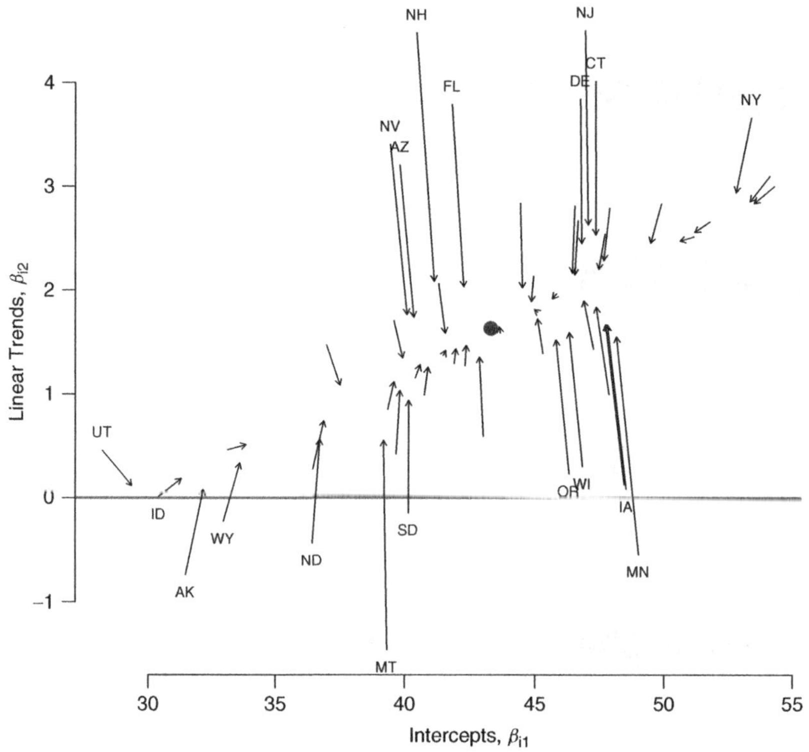

Figure 7.17 Shrinkage in two dimensions, intercept and trend parameters for hierarchical linear time trend model for state-level presidential election outcomes. The arrows indicate the shrinkage from the 'no pooling' least squares estimates to the posterior means from the hierarchical model in Equation 7.28. The solid gray dot indicates the 'complete pooling' estimates. Note that there appears to be more shrinkage for the slopes than the means.

of $\boldsymbol{\beta}_i$ with their counterparts from the hierarchical model (the means of the samples from the posterior densities of $\boldsymbol{\beta}_i$ produced by the MCMC algorithm). The shrinkage here is more pronounced with respect to the linear trend parameters than the intercepts, since the former are estimated with relatively less precision than the latter. Figure 7.17 also shows that the estimates of β_{i1} and β_{i2} produced by the hierarchical model are quite strongly and positively correlated: note that the posterior means for β_{i1} and β_{i2} almost all lie on a line, and indeed *would* lie on a line if $\rho = 1$. As Figure 7.16 shows, the posterior mode of ρ is about .8, and the posterior mean of ρ is .72, and so it is unsurprising that the Bayes estimate of β_{i1} and β_{i2} are almost linear functions of one another.

Figure 7.18 shows the two-dimensional shrinkage induced by the hierarchical model for 16 randomly chosen states. The ellipses are a joint 95 % credible region for each state's $\boldsymbol{\beta}_i$, for both 'no pooling' least squares regression and the hierarchical model. Observe again that 'no pooling' estimates of the trend parameters β_{i2} are estimated relatively imprecisely (the gray ellipses are 'long' in the vertical direction), and so the hierarchical model induces a reasonable amount of shrinkage for these parameters, towards the 'complete pooling' estimate displayed with the cross-hatch. In every instance the 95 %

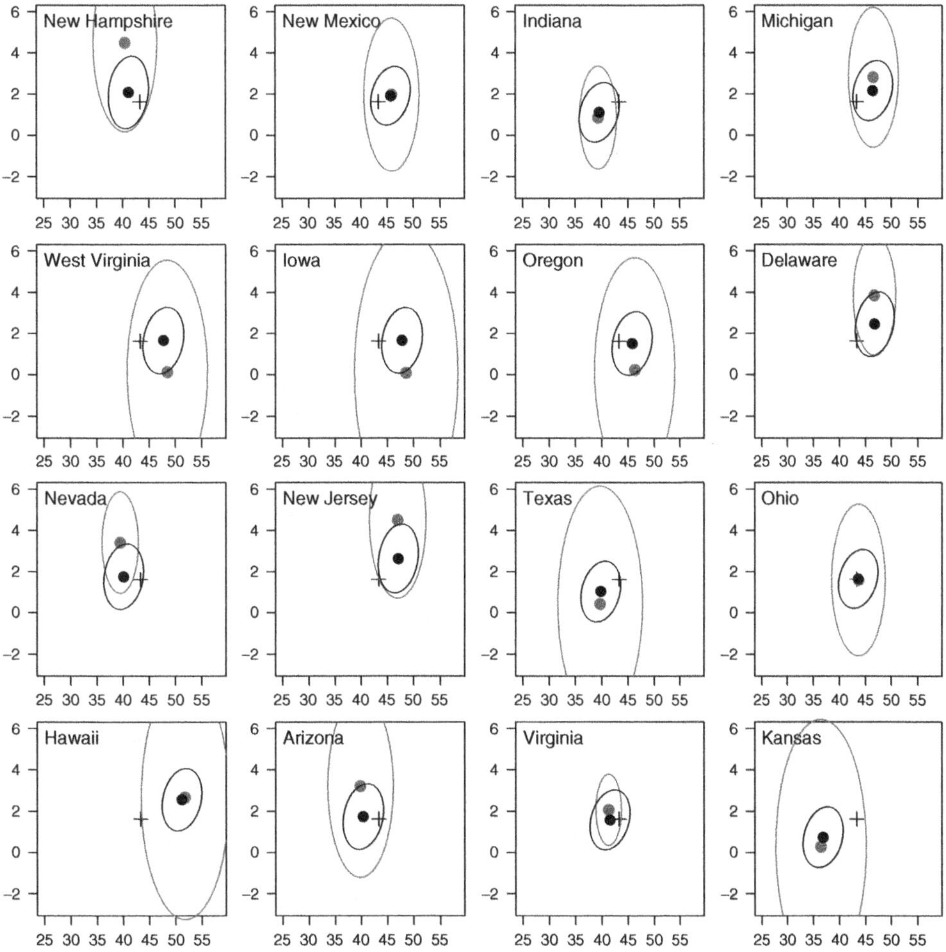

Figure 7.18 Credible regions for hierarchical and 'no pooling' estimates, linear time trend model for state-level presidential election outcomes, for 16 randomly selected states. The vertical axis is the 'trend' or β_{i2} dimension; the horizontal axis is the intercept or β_{i1} dimension. The darker ellipses indicate the 95 % credible regions for the estimates of $\boldsymbol{\beta}_i = (\beta_{i1}, \beta_{i2})'$ from the hierarchical model; the lighter colored ellipses indicate a 95 % confidence ellipse for the 'no pooling' least squares estimates. The plotted points indicate the least squares estimates of $\boldsymbol{\beta}_i$ from the 'no pooling' model (in grey) or the mean of the posterior density of $\boldsymbol{\beta}_i$ from the hierarchical model. The estimates from the hierarchical model are 'shrunk' in the direction of the 'complete pooling' estimates, displayed with cross hairs. Note that there appears to be more shrinkage for the slopes than the intercepts.

confidence ellipse for the hierarchical model (darker color) is considerably smaller and more compact than the corresponding credible region for the least squares estimates.

Likewise, the top two panels of Figure 7.19 show the complete set of 50 state-specific linear time trends for both the hierarchical model and the 'no pooling' model. The state-by-state linear trends generated by the 'no pooling' model display a considerable amount of heterogeneity: some states show decreasing levels of support for Democratic presidential candidates over time, while in other states we estimate positive trends. The state-specific linear trends from the hierarchical model display less variability across states, and are all non-negative. The estimated trends for two states are shown along with the trend lines implied by a random 250 draws from the MCMC output for their corresponding $\boldsymbol{\beta}_i$ parameters, so as to convey a sense of the uncertainty attaching to the estimate of the respective state-specific linear trend. This uncertainty is reasonably large: the standard deviations of the posterior densities for the slope terms average .62, while the between state variation in the posterior means of the slope terms is only slightly larger, with a standard deviation of .73. In addition, 12 of the 50 slope parameters have 95 % posterior HPD intervals that overlap zero. In sum, while the hierarchical model generates a considerable increase in the precision for the slope parameters relative to the 'no pooling' model, the hierarchical model is not in itself a magic bullet: five observations per unit is simply not a lot of data with which to estimate a linear regression, and even a hierarchical model can only do so much in these circumstances.

Predictions for the 2004 outcome. Finally, Figure 7.20 presents a series of comparisons of the various models fit to these data, with respect to the 'out-of-sample' 2004 outcome. The top two panels summarize the predictive performance of the 'no pooling' model (left panel) and the hierarchical linear trend model considered here. The lower right panel shows the predictions for the two-way ANOVA model considered in Example 7.7. The actual outcomes are on the vertical axis; a model with perfect predictive performance would generate predictions that lie on the 45 degree line. The hierarchical time trend model produces predictions that lie closer to the 45 degree line than those from the other two models: the mean squared prediction error is 810.46 for this model (using the parameter values evaluated at their posterior means), but is 940.9 for the 'no pooling' linear time trend model, and 1046.24 for the hierarchical, two-way ANOVA model considered earlier.

Moreover, not only does the hierarchical time trend model produce point predictions that are closer to the actual 2004 outcomes than the other models, but the standard deviations of the predictions – the standard deviation of the posterior predictive density for the 2004 outcomes – is smaller than that from the other two models. The lower right panel of Figure 7.20 makes this especially clear, comparing the standard deviations of the posterior predictive densities for the 2004 outcomes from two models: the hierarchical, linear trend model considered here, and the hierarchical, two-way ANOVA model considered in Example 7.7. In every state, the standard deviations of the 2004 predictions produced by the two-way ANOVA model are considerably larger than those from the linear time trend; the former roughly 6 percentage points of vote share, while the latter are about 4.5 percentage points, representing a large gain in precision, as well as a sizeable improvement with respect to bias. The corresponding calculations and comparisons for the 'no pooling' version of the linear trend model are left as an exercise (Problem 7.10).

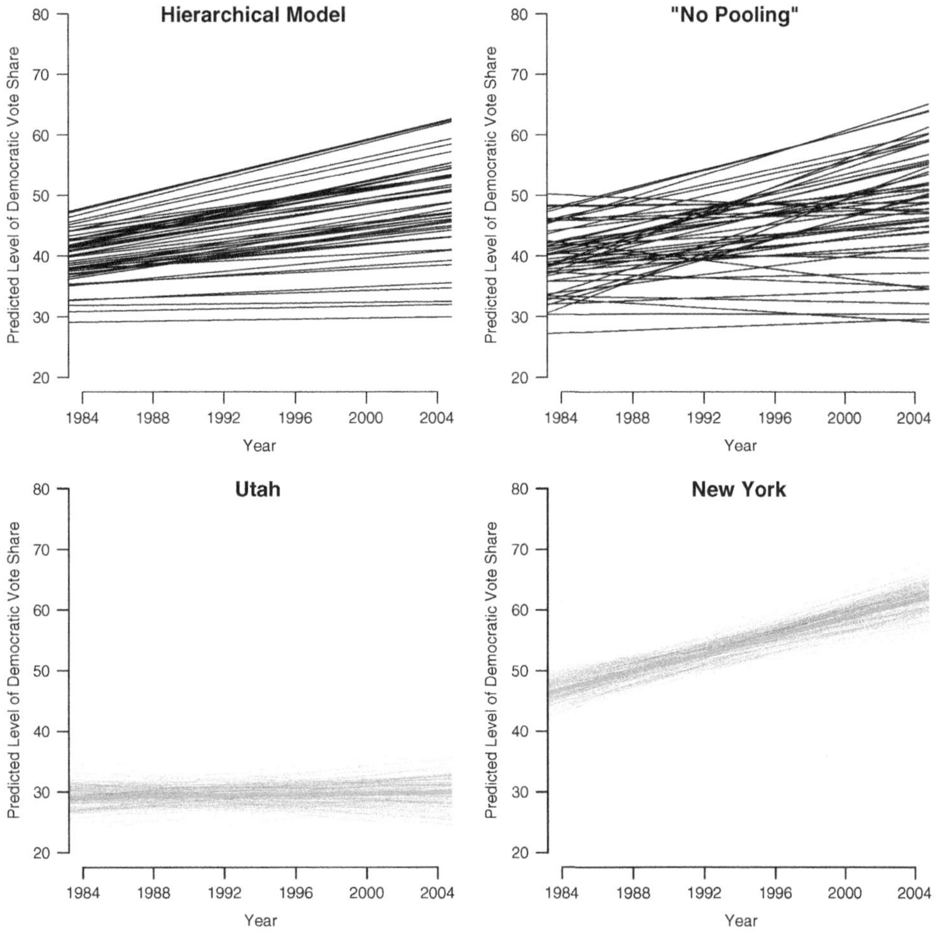

Figure 7.19 Linear time trends for state-level presidential election outcomes. The top left panel plots the 50 state-specific time trends using the posterior means for each state's $\boldsymbol{\beta}_i = (\beta_{i1}, \beta_{i2})'$; the top right panel shows the state-specific linear trends that result from fitting a linear model each state separately (the 'no pooling' model). The lower panels summarize the uncertainty associated with the linear time trends for two states (Utah and New York), by plotting the linear trends implied by 250 random draws from their respective posterior densities for $\boldsymbol{\beta}_i$; the solid gray lines in the lower panels show the linear trends implied by the posterior mean value of $\boldsymbol{\beta}_i$.

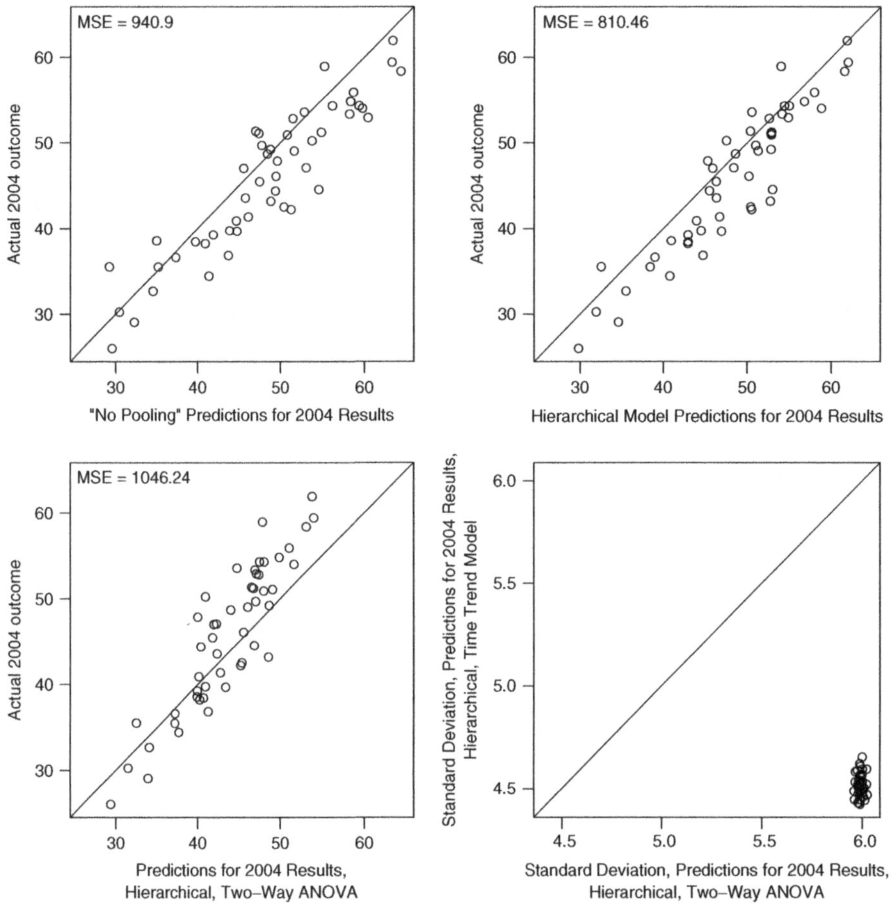

Figure 7.20 Predictions for 2004 state-level presidential election outcomes, plotted against actual outcomes. The predictive performance of the hierarchical, linear time trend model is compared against that of 'no pooling' OLS and the hierarchical, two-way ANOVA model considered in Example 7.7. For the hierarchical models the plotted points are the means of the posterior predictive densities. The lower right panel plots the standard deviations of the marginal posterior predictive densities for each states' outcome from the hierarchical, time-trend model against the standard deviations of the predictions from the two-way ANOVA model.

7.4 Hierarchical models for non-normal data

The preceeding examples have used normal models at both levels of two-level, hierarchical models. The normal model may be an appropriate model for continuous responses, like the math achievement scores considered in Example 7.6 or the vote shares considered in Example 7.7 and Example 7.8. This is well and good, but discrete data abound in the social sciences, and Chapter 8 considers Bayesian analysis of models for discrete data in some detail. The next example presents a hierarchical generalized linear model, with a discrete, binomial response at level one, but with a normal model at level two, from data generated by a cluster-randomized field experiment on voter turnout.

This will not be the only time that we combine generalized linear models for discrete data and hierarchical modeling in this book: the ideas that motivate hierarchical modeling – e.g. parameters that are exchangeable over groups or clusters – are hardly restricted to the case of continuous responses. For instance, in Example 8.6 we will encounter a hierarchical model for ordinal data.

■ Example 7.9

Hierarchical Model for a Cluster-Randomized Experiment. Green and Vavreck (2008) implemented a cluster-randomized experimental design in assessing the effects of a voter mobilization treatment in the 2004 U.S. presidential election. The clusters in this design are geographic areas served by a single cable television system. So as to facilitate analysis, the researchers restricted their attention to small cable systems whose reach is limited to a single zip code. Further, since the experiment was fielded during the last week of the election campaign, the researchers restricted their search to cable systems that were not in the 16 hotly-contested 'battleground' states (as designated by the *Los Angeles Times*). Further details on the selection of cable systems for this experiment appear in Green and Vavreck (2008). Eighty-five cable systems were available for randomization. Cable systems were assigned to treatment after stratification on previous turnout levels in presidential elections (as determined from analysis of the corresponding states' voter registration files). Each cable system was matched with one or sometimes two other cable systems in the same state, yielding 40 strata. Then within each strata, cable systems were randomly assigned to treatment and control conditions. Strata 3, 8 and 25 have two control cable systems and 1 treated system each, while strata 6 and 20 have two treated cable systems and one control system. The remaining 35 strata have one treated cable system and one control system. In this way there are $38 + 4 = 42$ treated systems, spanning 40 experiment strata.

The treatment involved researchers purchasing prime-time advertising spots on four channels in the respective cable system in which the researchers aired voter mobilization ads. The ads were produced by *Rock the Vote*, targeted at younger voters, and aired four times per night, per channel, over the last eight days of the election campaign. After the election, public records were consulted to assemble data on turnout levels in the treated and control cable systems. In the analysis reported in Green and Vavreck (2008), the researchers focused on turnout among registered voters aged 18 and 19 years old.

We analyze these data with a hierarchical binomial model. Within cable system $i = 1, \ldots, 85$, we observe the number of 18 and 19 year old registered voters, n_i, and the number of those voters who turned out to vote, r_i. Other information available from the voter file such as gender is ignored, since '...these independent variables have almost no predictive power within this narrow band of the voting-age population' (Green and Vavreck 2008, 149). Likewise, past voting behavior, 'which is ordinarily a strong predictor in models of this type, is inapplicable to first-time voters' (Green and Vavreck 2008, 149). Without covariates to distinguish among the n_i registered young voters within cable system i, we can consider the voting decisions as exchangeable Bernoulli trials given a success parameter θ_i, and so a binomial model is appropriate at the level of the cable system: i.e. conditional on θ_i, we assume that turnout decisions are independent across cable systems.

We also lack any covariates at the level of cable systems, aside from the strata identifiers, $s_i = 1, \ldots, 40$, and the assignment status of each cable system (i.e. $T_i = 1$ if treated, or $T_i = 0$ if in the control condition). Presumably, the researchers possess information about the demographic and political characteristics of the cable systems, but this information is not available here. Hence, exchangeability conditional on strata and treatment status is a plausible assumption for the θ_i.

Recall that the goal of the analysis is to assess if the cable-specific turnout rates θ_i vary systematically as a function of treatment status T_i. Within each experimental strata this amounts to a fairly simple comparison of binomial proportions. Under exchangeability, we will embed this comparison in a hierarchical model for the θ_i. We consider the hierarchical model

$$r_i \sim \text{Binomial}(\theta_i, n_i) \tag{7.29a}$$

$$\log\left(\frac{\theta_i}{1-\theta_i}\right) = \alpha_{s(i)} + \delta_i T_i \tag{7.29b}$$

$$\alpha_s \sim N(\mu_\alpha, \sigma_\alpha^2) \tag{7.29c}$$

$$\delta_i \sim N(\mu_\delta, \sigma_\delta^2) \tag{7.29d}$$

$$\mu_\alpha \sim N(0, 2^2) \tag{7.29e}$$

$$\mu_\delta \sim N(0, 2^2) \tag{7.29f}$$

$$\sigma_\alpha \sim \text{Unif}(0, 2) \tag{7.29g}$$

$$\sigma_\delta \sim \text{Unif}(0, 2) \tag{7.29h}$$

where s indexes experimental strata such that $s(i)$ is the strata of cable system i. On the logit scale, δ_i is the *intent-to-treat* effect for cable system i. Note the assumption that the intent-to-treat effect is additive on the logit scale. Note also that we refer to δ_i as an intent-to-treat effect because we lack any knowledge as to who, if anyone, in the treated cable systems actually viewed the advertisements. Both the strata-specific baselines α_s and the treatment effects δ_i are considered exchangeable, with the corresponding hierarchical specification given in Equations 7.29c through 7.29h. Note the non-conjugate priors for the standard deviations σ_α and σ_δ; these non-conjugate priors pose no special problems for programs like BUGS or JAGS.

Note also that this model is close to being *saturated*, in the sense that a saturated model has as many parameters as there are data points. Recall that in strata with just two cable systems (one treated, one control), the model is saturated, with two parameters $\alpha_{s(i)}$ and δ_i perfectly fitting the two binomial observations. But the model here is not quite saturated in that (1) α and δ are not 'fixed' effects but are considered exchangeable with the (hierarchical) prior given in Equation 7.29, and (2) 3 strata each have 2 control cable systems, but just a single baseline parameter α_s.

Capturing over-dispersion via hierarchical modeling. Another way to think about the model is that the variance terms σ_α^2 and σ_δ^2 pick up *extra-binomial* variation in the data. That is, if the strata-specific baselines rates of turnout α_s were constant across strata, say, $\alpha_s = \alpha \, \forall \, s$, then $\sigma_\alpha^2 = 0$. Likewise, if the intent-to-treat effects were constant across treated cable systems, say, $\delta_i = \delta \, \forall i$, then $\sigma_\delta^2 = 0$. If both these restrictive conditions hold, then the model reduces to a simple difference in binomial proportions: $r_i \sim \text{Binomial}(\theta_i, n_i)$, where $\log[\theta_i/(1-\theta_i)] = \alpha + \delta T_i$, with δ the key 'difference' parameter on the logit scale. In this very simple model, variation in the turnout proportions across cable systems is due to treatment status and binomial variation.

Regarding the latter, recall that if $r_i \sim \text{Binomial}(\theta_i, n_i)$ even if the success probability θ_i and number of trials n_i are fixed, the number of successes r_i varies around its expected value $E(r_i) = \theta_i n_i$ with variance $V(r_i) = n_i\theta_i(1-\theta_i)$; see Definition B.22. Additional variation in the r_i can not be captured by this simple, restrictive model, say if the θ_i are varying across cable systems in ways not captured by the simple $\alpha + \delta T_i$ model given above. From the perspective of this simple, restrictive model, this kind of additional variation in the θ_i can be considered extra-binomial variation; put differently, we would see that the r_i are *over-dispersed* relative to dispersion we'd expect under the simple binomial model. And in fact, the possibility of extra-binomial variation is what the hierarchical model lets us pick up: the 'level two' variance parameters σ_α^2 and σ_δ^2 pick up additional variation in the r_i beyond that due to merely knowing the treatment status of cable system i and the size of the target population in cable system i, n_i.

Implementation. A BUGS/JAGS program for the hierarchical model is as follows:

JAGS code

```
model{
        for(i in 1:85){ ## loop over cable systems
                logit(theta[i]) <- v[i]
                v[i] <- alpha[strata[i]] + delta[treatedIndex[i]]*treated[i]
                r[i] ~ dbin(theta[i],n[i])
        }

        for(j in 1:40){ ## loop over baseline effects
                alpha[j] ~ dnorm(mu[1],tau[1])
        }

        for(j in 1:42){ ## loop over treatment effects
                delta[j] ~ dnorm(mu[2],tau[2])
        }

        for(k in 1:2){  ## hyper-parameters
                mu[k] ~ dnorm(0,.25)
                tau[k] <- 1/pow(sigma[k],2)
                sigma[k] ~ dunif(0,2)
        }

        ## out of sample prediction for avg system and hypothetical system
        delta.new ~ dnorm(mu[2],tau[2])
        logit(p.new) <- delta.new
        pvalue[1] <- step(mu[2])      ## Pr(mu[2] > 0)
        pvalue[2] <- step(delta.new)  ## Pr(delta.new > 0)
}
```

Line 4 is the heart of the model, operationalizing Equation 7.29b. The counter `i` indexes cable systems. The indexing variable `strata[i]` indicates the experimental strata to which system `i` belongs; the variable `treatedIndex` is a counter indexing the 42 treated cable systems, incrementing if and only if cable system `i` is a treated cable system. The variable `treated` is a dummy variable, coded 1 if cable system `i` is a treated unit, and 0 otherwise, corresponding to the variable T_i in Equation 7.29b. This dummy variable formulation means that the treatment effects δ only make likelihood contributions for treated cable systems. The quantity `v[i]` puts the baseline levels `alpha` and treatment effects `delta` together on the logit scale; line 3 provides the mapping from the logit scale to a probability `theta[i]`, which then appears in the binomial model on line 5. We pass `strata`, `treatedIndex`, `treated`, `r` and `n` as data from `R` to `BUGS/JAGS`; the quantities `alpha` and `delta` are unobserved parameters, along with the hyperparameters `mu[1]` (μ_α in Equation 7.29), `mu[2]` (μ_δ) and the variances `sigma[1]` (σ_α) and `sigma[2]` (σ_δ).

The last few lines of the JAGS code warrant elaboration. These lines of code generate samples from the posterior predictive density for the treatment effect in a hypothetical cable system, $\delta_{\text{new}} \sim N(\mu_\delta, \sigma_\delta^2)$, where of course the hyperparameters μ_δ and σ_δ^2 are also subject to posterior uncertainty. Line 25 of the program uses the `step` function in `BUGS/JAGS` to keep track of the posterior probability that $\mu_\delta > 0$: the `step` function returns a 1 if its argument is positive, and 0 otherwise. Line 26 computes the corresponding quantity for δ_{new}, the treatment effect for a hypothetical cable system. The average of these binary quantities over many iterations of the MCMC algorithm is a simulation-consistent, Monte Carlo based estimate of the posterior probability that the respective argument passed to the `step` function is positive.

Note an important distinction here: $\mu_\delta = E(\delta_i)$ is the average treatment effect across strata, while δ_{new} is a draw from the distribution of possible treatment effects. Of course, $E(\delta_{\text{new}}) = \mu_\delta$ and so point summaries like the posterior means will coincide for these two quantities. But δ_{new} is perhaps the more interesting quantity, incorporating uncertainty not only as to the average treatment effect μ_δ, but additional uncertainty given that cable systems may well vary in the way they respond to treatment, and that we are uncertain about the magnitude of this heterogeneity in the treatment effects. Any future implementation of this treatment in an 'out-of-sample' cable system can't be guaranteed to yield the average treatment effect μ_δ, and so δ_{new} is perhaps the more relevant quantity to understand. Note that given the additional sources of uncertainty picked up by δ_{new} we can expect the posterior density of δ_{new} to be more dispersed than that of μ_δ, and so `pvalue[2]` (line 26) will be a smaller quantity than `pvalue[1]` (line 25).

MCMC algorithm. We initialize a single-chain MCMC algorithm with all α and δ parameters set to zero. After a 1000 iteration burn-in, we use the `JAGS` program given above to generate 50 000 samples from the posterior density of the model parameters, with every 5th iteration saved for inferences. Standard convergence diagnostics indicate that the MCMC algorithm quickly converges on the joint posterior density and is reasonably efficient, moving well through the parameter space. For instance, the Raftery-Lewis diagnostic indicates that approximately 20 000 iterations are sufficient to generate good estimates of the extreme quantiles of the marginal posterior densities of the α and δ parameters, and a similar number of iterations appear sufficient to give accurate characterizations of the marginal posterior densities of the level 2 parameters μ_α, μ_δ, σ_α and σ_δ.

Results. Figure 7.21 displays graphical summaries of the marginal posterior densities of the intent-to-treat effects δ_i for the 42 treated cable systems. Also shown are the maximum likelihood estimates of δ_i, obtained by running a binomial GLM in R. The dots are point estimates (means of 10 000 MCMC-generated samples from the marginal posterior densities for the hierarchical model) and the lines cover 95% credible intervals (highest posterior density intervals for the hierarchical model).

The estimates of δ_i are quite dispersed across the treated cable systems and relatively wide credible intervals attach to each δ_i. The familiar shrinkage pattern we obtain from hierarchical modeling is also apparent, with the hierarchical model's point estimates (posterior means) pulled towards the estimated average intent-to-treat effect shown at the top of the graph, relative to the system-specific MLEs of δ_i. Note also that the hierarchical model produces slightly more precise estimates of the δ_i than does MLE. Nonetheless, few of the δ_i have 95% HDRs than do not overlap zero: just 10 of the 42 treated units have δ_i that are unambiguously positive (in the sense that the 95% HDRs of their marginal posterior densities lie above zero).

Figure 7.21 also shows considerable variability in the widths of the 95% HPDs and confidence intervals across treated cable systems. This cross-system variability stems from the fact that there are differing numbers of individuals aged 18–19 in each cable system, the quantity denoted n_i above. These numbers range from 30 to 990, with an average of 281. In smaller cable systems we will estimate the corresponding δ_i with less precision, other things being equal.

Table 7.5 presents numerical summaries of the marginal posterior densities of some key parameters in the hierarchical model. The average intent-to-treat effect, μ_δ has a posterior mean of .12, with a 95% HDR interval that just overlaps zero; the posterior probability that $\mu_\delta > 0$ is .97. This effect is also reasonably small in substantive terms. The average baseline rate of voter turnout is given by the inverse-logit transformation of the posterior mean of μ_α, $1/(1 + \exp(-.06)) = .51$. Adding the average intent-to-treat effect boosts the baseline rate of voter turnout from 51% to $1/(1 + \exp(-.06 - .12)) = .54$, or about 3 percentage points. Posterior uncertainty in this effect is reasonably small, such that the posterior probability that this average effect is positive is .97.

Moreover, the variation in treatment effects across treated cable systems is quite large; the posterior mean of σ_δ is .33, again recalling that the δ intent-to-treat effects are on the

Table 7.5 Summaries of Marginal Posterior Densities, selected parameters, voter turnout experiment. Both the α and δ parameters are on the logit scale; see Equation 7.29.

		Mean	95 % HDR interval	
Baseline levels:	μ_α	0.06	−0.07	0.20
$\alpha_{s(i)} \sim N(\mu_\alpha, \sigma_\alpha^2)$	σ_α	0.40	0.30	0.50
Intent-to-treat effects:	μ_δ	0.12	−0.01	0.24
$\delta_i \sim N(\mu_\delta, \sigma_\delta^2)$	$\Pr(\mu_\delta > 0)$	0.97		
	σ_δ	0.33	0.25	0.44
Predictive density:	δ_{new}	0.12	−0.55	0.78
$\delta_{\text{new}} \sim N(\mu_\delta, \sigma_\delta^2)$	$\Pr(\delta_{\text{new}} > 0)$	0.64		

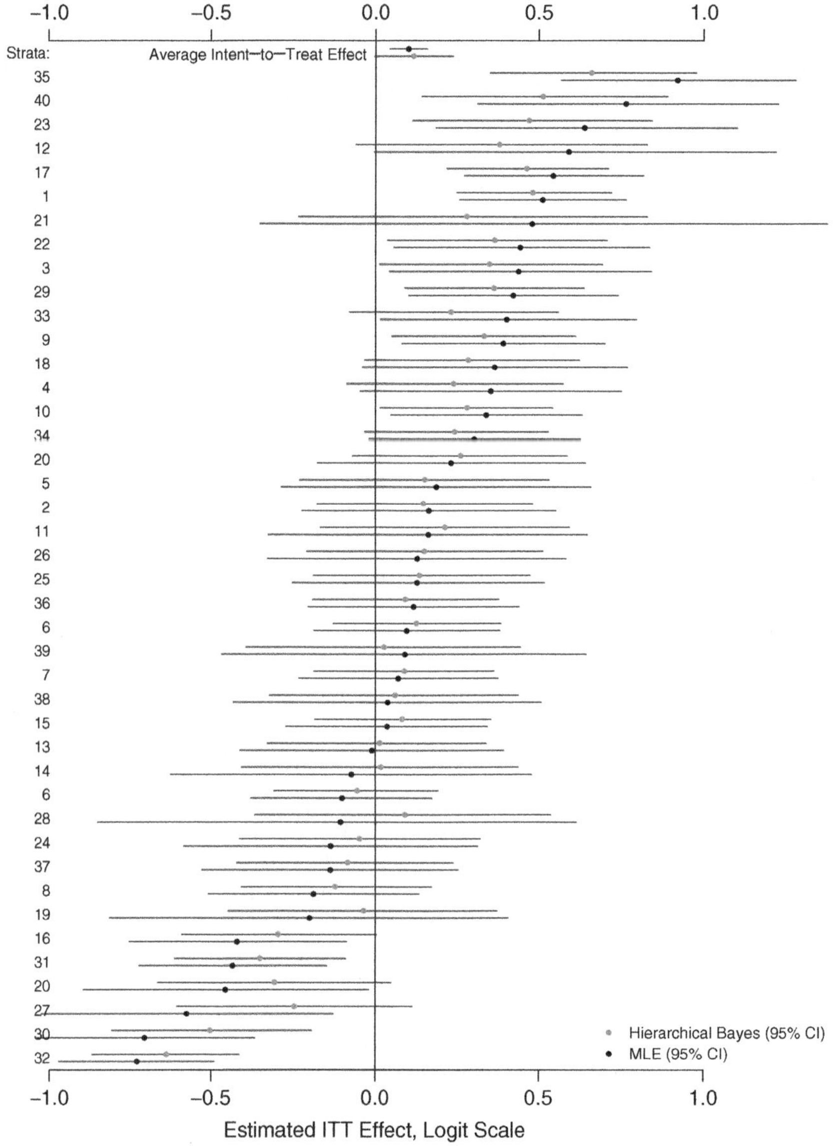

Figure 7.21 Marginal Posterior Densities and Maximum Likelihood Estimates for 42 Intent-to-Treat Effects for Voter Turnout Experiment. The horizontal lines cover 95 % HPDs. Lighter dots indicate the posterior means for δ from the hierarchical model given in equation 7.29; darker dots indicate the MLEs. The two estimates of the average intent-to-treat effects, μ_δ, are summarized at the top of the graph. Note how the hierarchical model's estimates of δ_i have been shrunk towards the average effect, relative to the MLEs of δ_i. The numbers on the left of the graph indicate the experimental strata to which each system belongs.

logit scale. Conditional on setting σ_δ to its posterior mean, this implies that many of the δ_i intent-to-treat effects can be expected to be negative, as we observed in Figure 7.21: note that the posterior mean of σ_δ (.33) is almost 3 times the size of the posterior mean of the average δ (.12). But there is also considerable posterior uncertainty as to σ_δ, with a 95% HDR ranging from .25 to .44. We use simulation to assess what this implies for the intent-to-treat effects δ. Our 'best guess' as to the magnitude of the intent-to-treat effect in a randomly chosen cable system is simply the average treatment effect, μ_δ, and note in Table 7.5 that the posterior mean of δ_{new} is the same as that of μ_δ (.12 on the logit scale). But the posterior density of δ_{new} is quite dispersed: the last row of Table 7.5 shows that the marginal posterior probability that $\delta_i > 0$ is just .64, since the 95% HPD interval for δ_{new} ranges from -.55 to .78.

That is, although we are reasonably sure that 'on average' the television advertisements do generate a small increase in voter turnout – equivalent to about 3 percentage points of voter turnout – there is a large amount of variability in the intent-to-treat effects across cable systems, such that in any particular cable system we can not be particularly confident that the television ads will actually increase voter turnout. Indeed, there is a 36% chance that the ads will actually *reduce* voter turnout. More work is required so as to better understand the sources of the variability in the intent-to-treat effects δ_i across cable systems. To this end, a hierarchical model with covariates at the level of cable systems would be a useful way to proceed, an easily implemented extension to the simple hierarchical model presented here.

Alternative specification of the hierarchical model. There is certainly more than one model we might fit to these data, even without covariates at the level of cable systems. Suppose we knew nothing about the cable systems: i.e. we don't know which strata the cable systems belong to, nor do we know the treatment status. If this were the case then we might well start with the exchangeable model

$$\log\left(\frac{\theta_i}{1-\theta_i}\right) \sim N(\mu, \sigma^2) \tag{7.30}$$

where θ_i is the proportion of 18-19 year olds turning out to vote in cable system i, and with priors over μ and σ^2 completing the specification of this extremely simple hierarchical model. Thinking of this simple model as a 'baseline' model, we then might add the information we do have about cable systems: their strata and treatment status. That is, we replace the simple mean μ in Equation 7.30 with the richer hierarchical structure given in Equation 7.29: i.e.

$$r_i \sim \text{Binomial}(\theta_i, n_i) \tag{7.31a}$$

$$\log\left(\frac{\theta_i}{1-\theta_i}\right) \sim N(\mu_i, \sigma^2) \tag{7.31b}$$

$$\mu_i = \alpha_{s(i)} + \delta_i T_i \tag{7.31c}$$

and so on, repeating Equations 7.29d through 7.29h.

In this alternative version of the model there are *three* variance terms: (1) σ^2, picking up variability in the log-odds of θ_i that is not accounted for by the additive combination of the strata-specific term $\alpha_{s(i)}$ and the intent-to-treat effect δ_i; (2) σ_α^2, the between strata variability in baseline levels of turnout, $\alpha_{s(i)}$; (3) σ_δ^2, the between cable system variability in intent-to-treat effects δ_i. Clearly, this hierarchical model asks more of the data then the 'nearly-saturated' model considered above, and is close to being unidentified. All but 3 strata have just two cable systems (one treated, one control); within each of these strata the model is saturated with two parameters $\alpha_{s(i)}$ and δ_i which would perfectly fit the two binomial observations. In turn, this would imply that $\sigma^2 = 0$, save for the facts that (1) α and δ are not 'fixed' effects but are considered exchangeable with the (hierarchical) prior given in Equation 7.29, and (2) 3 strata have 2 control cable systems. It would seem that there is not a lot of information in these data regarding the *three* variance parameters in this augmented version of the model; see Problem 7.11.

7.5 Multi-level models

We now consider a hierarchical model where the hierarchical component of the model involves covariates. Models of this sort allow us to introduce group-level covariates $\mathbf{z}_j$ into the analysis in the hierarchical component of the model, where typically these 'group-level', 'contextual' or 'level two' variables structure the way that 'individual-level' or 'level one' variables $\mathbf{x}_{ij}$ effect the response variable y_{ij}. We introduced these models in Example 7.2 with the multi-level regression model in Equation 7.2; as we did there, let j index groups and let k index the level one predictors. In that case – and is typical in the literature – the group-specific, level one regression coefficients β_{jk} are themselves considered response variables in level two, where the hierarchical model for β_{jk} is itself a regression model with predictors $\mathbf{z}_j$ and coefficients $\boldsymbol{\gamma}_k$ (e.g. Equation 7.2b).

Note that the presence of the covariates $\mathbf{z}_j$ in the hierarchical model means we are assuming a form of *conditional exchangeability* given the level two covariates $\mathbf{z}_j$; i.e. among groups with similar values of the group-level covariates $\mathbf{z}_j$ we expect similar level one coefficients β_{jk}. Contrast the multi-level/hierarchical model $\beta_{jk} \sim N(\mathbf{z}_j\boldsymbol{\gamma}_k, \omega_k^2)$ with the unconditional hierarchical model $\beta_{jk} \sim N(\mu_k, \omega_k^2)$. In the multi-level regression model any shrinkage in the β_{jk} induced by the hierarchical structure of the model is towards the point $\mathbf{z}_j\boldsymbol{\gamma}_k$, which varies over groups depending on each group's $\mathbf{z}_j$, while the unconditional model shrinks towards μ_k.

Regression with interactions. An alternative to the multi-level/hierarchical model is a classical regression model with interactions between level one covariates and level two covariates. It is straightforward to see that such a regression model amounts to a restricted version of the multi-level model. For instance, the regression model

$$\begin{aligned} E(y_{ij}|x_{ij}, z_j) &= \beta_{j1} + \beta_{j2}x_{ij} \\ &= \gamma_{10} + \gamma_{11}z_j + (\gamma_{20} + \gamma_{21}z_j)x_{ij} \\ &= \gamma_{10} + \gamma_{11}z_j + \gamma_{20}x_{ij} + \gamma_{21}z_jx_{ij} \end{aligned}$$

is a familiar way of making the group-specific intercepts (β_{j1}) and slopes (β_{j2}) in the regression of y_{ij} on x_{ij} vary depending on the group-level covariate z_j. We will compare this model with the multi-level, hierarchical model in the example, below.

■ Example 7.10

Multi-level regression model for mathematics achievement (Example 7.6, continued). We use a multi-level model for math achievement scores, in which the effects of each student's socio-economic status vary across schools (groups), but in a way that depends on school-level characteristics $\mathbf{z}_j$. Following previous examinations of these data (e.g. Raudenbush and Bryk 2002) we focus on two school-level covariates: (1) a binary indicator `cath` indicating whether the school is a Catholic school; (2) the average socio-economic status of students at the school, `meanses`. The idea is that there may be differences in the ways that a student's socio-economic status maps into math achievement across different schools, depending on school type (Catholic versus public) and the school-averaged SES level. While it is plausible to expect that math achievement scores y increase with student level SES x, do Catholic schools amplify or moderate this relationship, changing both the intercept and the slope of linear mapping from x to y? Similarly, in a school where SES levels are higher than average, is the individual level mapping from SES to math achievement scores steeper than in schools with lower than average SES levels?

Data. There are 160 unique schools in the data available for analysis, of which 70 (43.8 %) are Catholic schools. School-average SES levels range from -1.2 to 0.8, with a mean of 0; at the individual level (level one), SES ranges from -3.75 to 2.69. About 28 % of the individual level (level one) variation in SES is between-school variation. Catholic schools score .30 units higher than non-Catholic school on `meanses`. Since Catholic schools score higher on `meanses` than non-Catholic schools we will be concerned with distinguishing any 'Catholic school' effect from a mean-SES effect in the level-2 part of the analysis.

Preliminary analysis. Ignoring the clustering of the students into schools, the correlation between math achievement and student-level SES is .36. Regression analysis at the student level ($n = 7185$) – a 'complete pooling' regression of `math` on `ses` – finds that a one unit increase in student-level SES is associated with a 3.2 unit increase in math achievement score ($t = 33$). A 'no pooling' analysis – 160 separate, school-specific, linear regressions of `math` on `ses` – generates the estimates intercepts and slopes summarized graphically in Figure 7.22. Running these 160 regressions is extremely simple using the `lmList` command in the `lme4` package (Bates, Maechler and Dai 2008) in R: e.g.

R code

```
secretWeapon <- lmList(formula = math  ~ ses | school,
                       data = data1)
```

The object produced by `lmList` can be passed on to various methods such as `summary`, `plot`, `pairs`, `coef` and so on. We rely on these methods to produce Figures 7.22 and 7.23, below.

There appears to be a reasonable amount of variation in the fitted regressions across school type, with Catholic schools tending to generate fitted regressions with higher intercepts and smaller slopes than those in non-Catholic schools. Figure 7.23 shows the relationship between the 160 school-specific intercept and slope estimates and school-average

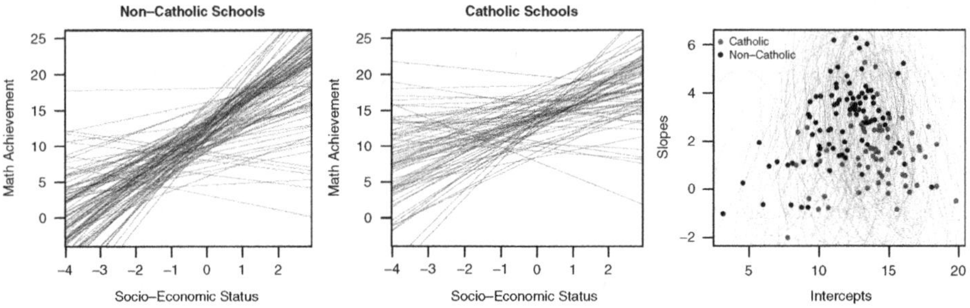

Figure 7.22 Results from 160 separate, school-specific regressions of math achievement score on student level SES. The left and middle panels show the estimated linear regression lines for non-Catholic and Catholic schools, respectively. The right panel plots the estimated intercepts and slopes as a scatterplot, using color to distinguish Catholic and non-Catholic schools; the light lines are 95 % confidence ellipses around the OLS point estimates, highlighting that each school-specific regression is estimated with considerable imprecision. Nonetheless, it is clear that Catholic schools appear to have higher intercepts and lower slopes than non-Catholic schools.

SES levels (`meanses`). It is clear that the intercepts and school-average SES are positively associated (top-left or bottom-right panels of Figure 7.23); the slope coefficients also appear to be increasing as a function of `meanses` after a rough, visual adjustment for the differences between Catholic and non-Catholic schools.

Classical analysis with interaction terms. We begin with some simple regression models, regressing `math` on individual level `ses`, fit by ordinary least squares under standard assumptions. Table 7.6 summarizes three simple regression models. Classical testing procedures indicate that a complete pooling model that fits just a single slope and a single intercept (left column) can be improved by adding school-specific intercepts ('fixed effects'; $F = 6.07$, $p < .01$). In turn, a model that also lets each school pick up a separate slope coefficient generates a further improvement in fit, and is preferred to the model that only has school-specific intercepts ($F = 1.34$, $p \approx .003$). That is, this classical analysis adds formal confirmation to the visual message of Figure 7.22: there is between-school variation in the slope and intercept parameters of the regression model.

We also use classical methods to provide an initial exploration of this variation, adding interaction terms to the regression model: i.e. we estimate the regression

$$E(y_{ij}|x_{ij}, \mathbf{z}_j) = \underbrace{\gamma_{10} + \gamma_{11} z_{j1} + \gamma_{12} z_{j2}}_{\text{intercept } (\beta_{j1})} + \underbrace{(\gamma_{20} + \gamma_{21} z_{j1} + \gamma_{22} z_{j2})}_{\text{slope } (\beta_{j2})} x_{ij} \tag{7.33}$$

where y_{ij} is the math achievement score of student i in school j (`math`), x_{ij} is the SES of student i (`ses`), z_{j1} is a binary indicator (`cath`) coded 1 if school j is a Catholic school and zero otherwise, and z_{j2} is the average SES level in school j (`meanses`). As indicated in Equation 7.33, the school j intercept of the level-one regression is $\gamma_{10} + \gamma_{11} z_{j1} + \gamma_{12} z_{j2}$ and the slope is $\gamma_{20} + \gamma_{21} z_{j1} + \gamma_{22} z_{j2}$, both deterministic, linear,

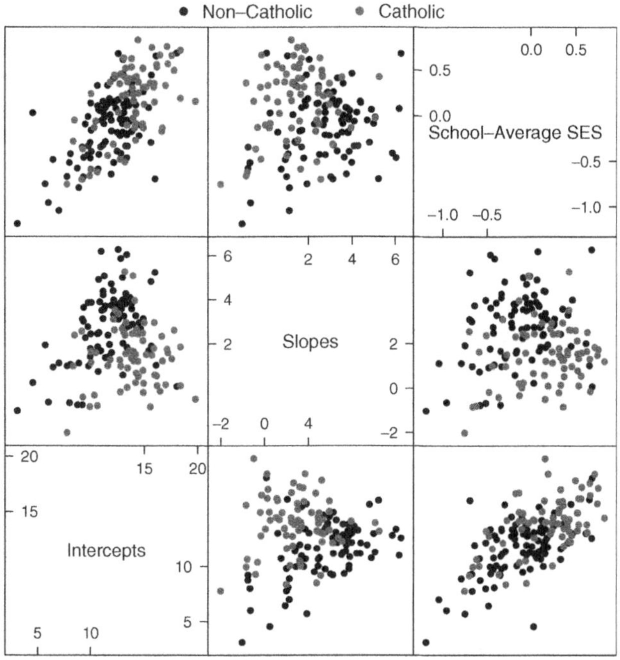

Figure 7.23 Scatterplot matrix, relationships between intercepts, slopes, and school-average SES, 160 separate, school-specific regressions of math achievement on SES. Lighter points indicate Catholic schools.

Table 7.6 Classical regression analysis, math achievement scores and SES in 160 schools. Table entries are least squares slope estimates with standard errors in parentheses. The 3rd SES estimate is the average of the 160 school-specific SES coefficients. All reported F tests are accompanied by p-values less than .01.

	Pooled	School intercepts	School intercepts and slopes
SES	3.18	2.19	2.20
	(0.097)	(0.11)	(0.12)
Model df	2	161	320
r^2	.13	.24	.26
$\hat{\sigma}$	6.42	6.08	6.06
F vs Pooled		6.07	3.73
F vs School Intercepts			1.34

additive functions of $\mathbf{z}_j$. The parameters γ_{11} and γ_{12} are often referred to as the 'main effects' of the school-level Catholic and mean SES variables; likewise the parameter γ_{20} is the main effect of student-level SES.

Classical estimates (OLS) of this model appear in Table 7.7, in the column labelled 'raw'. The parameter estimates suggest a large role for school-level average SES, with a one-unit increase in the school's average level of SES lifting the intercept by 3.15 points. This is a larger point estimate than the estimated effect of individual-level SES; a one-unit increase in SES in a non-Catholic school with an average school-level of SES generates a 2.84 improvement in math achievement score. Likewise, the 'contextual' effects of being in a Catholic school are reasonably large too, lifting the intercept of the school-specific level-one regression of `math` on `ses` by 1.24 units, while depressing the slope by -1.46 units.

Re-parameterization by hierarchical centering. We also consider a version of the model where all covariates are centered. For the level-one predictor `ses`, the centering is done *hierarchically* or within schools; for level-two predictors, there is no within-school variation and the centering is across schools. Using these centered versions of the predictors amounts to a reparameterization that does not and can not generate a superior fit to the data, but does changes the interpretation of the parameters in the model. For instance, the school-specific intercepts are now $\beta_{j1} = E(y_{ij}|x_{ij} = \bar{x}_j) = \gamma_{10} + \gamma_{11}z_{j1} + \gamma_{12}z_{j2}$, the expected level of math achievement for a student with the average SES level in school j, $\bar{x}_j$. This changes the interpretation of the γ_{11} and γ_{12} parameters; for instance, γ_{12} now taps how a unit change in mean SES levels across schools is associated with change in the conditional mean β_{j1} defined above, the expected value of `math` evaluated at the average value of the predictor `ses`.

A key consequence of the hierarchical centering is that since the within-school distributions of SES are now all centered at zero, the level-two variables `meanSES` will now pick up more of the between-school variation in the response variable y (`math`). In the uncentered version of the model the level-one variable x_{ij} (`ses`) is picking up both within-school and between-school variation; we saw earlier that 28 % of the variation in `ses` is between-school variation. After hierarchically centering x_{ij}, all of its variation is within-school variation, meaning that variables with between-school variation – the level-two variables $\mathbf{z}_j$ – explain this between-school variation in y. Accordingly, we can expect the γ_{12} and γ_{22} parameters to be larger in the hierarchically-centered verison of the model relative to the raw, uncentered parameterization.

Another way to think about hierarchical centering in this particular case is to note that the correlation between the uncentered level-one predictor `ses` and the level-two contextual variable `meanses` is .53; creating an interaction between `ses` and `meanses` introduces some multicollinearity to the resulting design matrix. Hierarchical centering reduces this correlation to zero and improves the condition of the design matrix. In short, hierarchical centering helps us 'unconfound' variables that covary across levels in the hierarchy: after hierarchical centering, level-one `ses` only has variation across students, within schools; and conversely, level-two `meanses` taps between-school variation.

Classical estimates of the hierarchically centered version of the model appear in the column of Table 7.7 labelled 'mean centered'. The hierarchically centered model yields exactly the same predicted values and goodness-of-fit statistics as produced by the uncentered parameterization, as it should: to repeat, hierarchical centering is merely a re-parameterization of the model, and brings no new information to bear on the problem of predicting y. The hierarchically centered parameterization does result in larger

Table 7.7 Classical regression analysis, math achievement score regressed on SES, with level-one/level-two interactions.

	Raw		Mean centered	
	Estimate	Std Err	Estimate	Std Err
Intercepts β_{j1}				
Intercept γ_{10}	12.16	0.11	12.64	0.07
Catholic γ_{11}	1.24	0.16	1.28	0.16
Mean SES γ_{12}	3.15	0.22	5.16	0.19
Slopes β_{j2}				
Intercept γ_{20}	2.84	0.15	2.22	0.11
Catholic γ_{21}	−1.46	0.21	−1.64	0.24
Mean SES γ_{22}	0.39	0.24	1.04	0.30
σ	6.25		6.25	
r^2	.18		.18	

estimates of the effects of the level-two/between-school covariate `meanses` (z_{j2}). A one-unit change in `meanses` is associated with more than a 5 unit boost in `math`, holding student-specific levels of `ses` constant; this is considerably larger than the typical within-school effects of `ses` estimated here, i.e. a 2.22 unit boost in `math` for a one unit increase in `ses`. Note also the large contextual effects of `meanses` on the slope coefficients: a one-unit boost in `meanses` would see the level-one effects of `ses` in an otherwise typical school jump from 2.22 to about 3.26. As with the uncentered version of the model, Catholic schools appear to (a) provide a modest boost in math achievement for all students (about 1.3 points) while (b) substantially dampening the impact of student-level SES on math achievement.

Estimation and inference for this type of a hierarchical model is extremely simple: we simply create (or get our software to create) interactions between the level-one predictor `ses` and the level-two covariates `cath` and `meanses`, and estimate the resulting regression model via OLS (subject to the usual assumptions about conditional independence and conditional homoskedasticity). While simple, the cost is that we have a quite restrictive model at level two: in this case, the school-specific regressions have intercepts and slopes that are deterministic functions of their levels of the level two $\mathbf{z}_j$ covariates.

Harking back to Figure 7.23, we see considerable variation in the school-specific intercepts and slopes produced by running regressions in each of the 160 schools. The bottom left panel of Figure 7.23 suggests that the variation in the 160 school-specific intercepts will be fit moderately well by the linear, additive combination of `cath` and `meanses`; the middle panel in the right column of Figure 7.23 suggests that 160 slope coefficients will be fit less well by the linear, additive combination of `cath` and `meanses`. In sum, the simple, interactive specification is constraining the points in Figure 7.23 – the school-specific intercepts (β_{j1}) and slopes (β_{j2}) – to lie on lines defined by the linear, additive, deterministic models for both β_{j1} and β_{j2} in Equation 7.33. This approach does have the virtue of parsimony, but at some cost in terms of lack of fit (as Figure 7.23 makes clear) and scientific realism: intercepts and slopes simply don't vary between schools in the deterministic way contemplated by the relatively spartan regression model in Equation 7.33.

Model comparisons. Classical model comparison tests strongly suggest that the restrictions imposed by this OLS-with-interactions model relative to the 'fixed effects' model (2nd column, Table 7.6) and the 'varying slopes, varying intercepts' models (3rd column, Table 7.6) can be rejected, with F test statistics of 3.54 and 2.44, respectively (both with accompanying p-values below .01). Other classical model comparison tools also suggest that the degradation in fit relative to the models in Table 7.6 is not offset by the simplicity of the OLS-with-interactions model: using the AIC criterion, the 'fixed effects' model is preferred, with an AIC of 46 496, versus an AIC of 46 595 for the 'varying intercepts, varying slopes' model and 46 727 for the OLS-with-interactions model. That is, the OLS-with-interactions model consumes just 6 degrees of freedom, and can generate some heterogeneity in the implied school-specific intercepts and slopes. The model with interactions doesn't fit the data as well as the regression models with school-specific intercepts ('fixed effects'), which consumes 161 degrees of freedom, or the model that fits school-specific intercepts and slopes, consuming 320 degrees of freedom. The data seem to indicate that there is more heterogeneity in the school-specific intercepts and slopes than can be generated by the deterministic, linear additive combinations of `cath` and `meanses` provided by the regression model with interactions. Nonetheless, for substantive reasons, we want to keep these variables as potential sources of between-school variation in the intercepts and slope parameters. This suggests a model in which the school-specific intercepts and slopes are permitted to vary across schools, but in a way not wholly determined by the school-specific covariates (`cath` and `meanses`). This is precisely what the multi-level/hierarchical accomplishes.

Bayesian hierarchical model. The Bayesian hierarchical model differs from the regression model with interactions considered above in that the school-specific intercepts and slopes are now considered random variables; in particular, we consider the intercepts β_{j1} and the slopes β_{j2} to be each *conditionally exchangeable* given a linear, additive combination of the level two covariates $\mathbf{z}_j$, `cath` and `meanses`. The Bayesian hierarchical model for this problem is

$$y_{ij} \sim N(\beta_{j1} + \beta_{j2}x_{ij}, \sigma^2) \tag{7.34a}$$

$$\begin{bmatrix} \beta_{j1} \\ \beta_{j2} \end{bmatrix} \sim N\left(\begin{bmatrix} \gamma_{10} + \gamma_{11}z_{j1} + \gamma_{12}z_{j2} \\ \gamma_{20} + \gamma_{21}z_{j1} + \gamma_{22}z_{j2} \end{bmatrix}, \mathbf{\Omega}\right) \tag{7.34b}$$

$$\mathbf{\Omega} = \begin{bmatrix} \omega_{11}^2 & \omega_{12} \\ \omega_{21} & \omega_{22}^2 \end{bmatrix} \tag{7.34c}$$

with the following priors over the level two parameters: each γ parameter is assigned a vague $N(0, 10^2)$ prior (each independent of the others), and $\mathbf{\Omega} \sim$ inverse-Wishart$(\mathbf{W}, 2)$ (see Definition B.40), where $\mathbf{W} = \kappa\mathbf{I}_2$ and $\kappa = .1$. The level-one error standard deviation σ is assigned a Unif(0, 10) prior; this prior places no probability mass on $\sigma > 10$, but this upper bound is comfortably above the empirical standard deviation of the mathematics achievement scores and so is uninformative in this instance.

MCMC algorithm. The unknown parameters for this problem are (a) the level one intercept and slope parameters $\boldsymbol{\beta}_j = (\beta_{j1}, \beta_{j2})'$, where $j = 1, \ldots, J = 160$ indexes the schools; (b) the level one error standard deviation σ; (c) the level two regression parameters $\boldsymbol{\gamma} = (\gamma_{10}, \ldots, \gamma_{22})'$, and (d) the level two variance-covariance matrix $\boldsymbol{\Omega}$. The posterior density is $p(\boldsymbol{\beta}_1, \ldots, \boldsymbol{\beta}_J, \sigma, \boldsymbol{\gamma}, \boldsymbol{\Omega}|\mathbf{y}, \mathbf{x}, \mathbf{z})$, where $\mathbf{y}$ and $\mathbf{x}$ are level one observables and $\mathbf{z}$ are level two covariates. A MCMC algorithm is reasonably straightforward to implement, with the conditional distributions required to implement a Gibbs sampler for this problem relatively easy to derive and implement. The general form of the MCMC algorithm for this problem is very similar to that for the one-way ANOVA problem in Example 7.6, except now we have an intercept and a slope specific to each group in the analysis rather than just a mean. This said, this problem is extremely well-suited to implementation in BUGS or JAGS, as shown in the following code block:

JAGS code

```
model{
        ## level 1 model
        for(i in 1:N){              ## loop over students
              mu[i] <- b[s[i],1] + b[s[i],2]*sesStar[i]
              math[i] ~ dnorm(mu[i],tau.y)
        }
        sigma ~ dunif(0,10)         ## prior on standard deviation
        tau.y <- pow(sigma,-2)      ## convert to precision

        ## level 2 model
        for(j in 1:J){              ## loop over schools
              for(k in 1:K){        ## loop over level 1 coefficients
                  mu.b[j,k] <- g[k,1] + g[k,2]*(cath[j]-mean(cath[])) +
                                        g[k,3]*(meanses[j]-mean(meanses[]))
              }
              b[j,1:K] ~ dmnorm(mu.b[j,1:K],Tau)   ## mv normal
        }

        ## priors for level 2 model
        for(k in 1:K){
              g[k,1:3] ~ dmnorm(mu.g[1:3],Tau.G)
        }

        Tau[1:K,1:K] ~ dwish(W,2)      ## Wishart prior for precision matrix
        Omega[1:K,1:K] <- inverse(Tau)    ## convert to var-covar
        omega[1] <- sqrt(Omega[1,1])      ## standard deviations
        omega[2] <- sqrt(Omega[2,2])
        rho <- Omega[1,2]/(omega[1]*omega[2]) ## correlation
}
```

Some comments on the implementation in BUGS/JAGS are in order. There are two key loops in the model syntax, one for each level in the model. The level one loop makes use of the nested indexing trick to assign the correct intercept and slope coefficients to each student: the quantity `s[i]` indicates the school of student `i`, such that `beta[s[i],1]` is the appropriate school-specific intercept for student `i`. The level two loop simply implements the multivariate normal model in Equation 7.34b, building elements of the mean vector as functions of the γ parameters (referred to as `g` in the JAGS program), and the level two covariates `cath` and `meanses`; note also the centering of these covariates 'on-the-fly'. There is some extra code specifying the Wishart prior over the precision matrix `Tau` via the `dwish` density, inverting that matrix to obtain $\boldsymbol{\Omega}$, and converting variances to standard deviations and the covariance `Omega[1,2]` to a more easily interpreted correlation `rho`. Note also the way we specify the priors for the γ parameters via the multivariate normal `dmnorm` density; the vector `mu.g` is simply a vector of zeros and the prior precision matrix `Tau.G` $= .01 \cdot \mathbf{I}_3$, both passed to `JAGS` as data from `R`, implementing the priors $\gamma \stackrel{\text{iid}}{\sim} N(0, 10^2)$.

We begin with a short run of the MCMC algorithm: just 5000 iterations, which on inspection reveals slow mixing and a lack of convergence for many parameters. A longer run of 100 000 iterations produces a better looking set of results, according to the standard convergence diagnostics, but with some slow mixing for the level-two parameters for the slope parameters (the $\gamma_{2\cdot}$ parameters and the standard deviation ω_{22}); e.g., see the AR(50) column in Table 7.8. Posterior inferences are based a longer run of 500 000 iterations, thinned by 50. Convergence diagnostics for this long run appear in Table 7.8, with the Raftery-Lewis diagnostic indicating that the 500 000 iteration run used here is considerably longer than the 250 000 or so iterations required to generate an accurate estimate of the extreme quantiles of the marginal posterior densities for the $\gamma_{2\cdot}$ and ω_{22} parameters.

Results. Numerical summaries of the level two parameters appear in Table 7.9. The posterior means are extremely similar to the point estimates from the classical analysis using interaction terms between level one and level two, reported in Table 7.7. School-level average SES elevates the school-specific intercept by a large amount, holding student-level SES constant: a one standard deviation change in school-level average SES (.41) sees average math achievement scores rise by $5.33 \times .41 = 2.19$ units, larger than the corresponding difference as one moves between Catholic and non-Catholic schools (1.22 units). Changes in school-level average SES also increase the level one SES slopes

Table 7.8 Convergence diagnostics for selected parameters, MCMC output for multi-level regression model for math achievement, using the parameterization and priors given in Equation 7.34. All convergence diagnostics are implemented using the `coda` package (Plummer *et al.* 2008) in `R`. The AR(50) diagnostic is based on the fact that the MCMC output is every 50th iteration from a run of 500 000 iterations. The row labelled ρ is the correlation between the level one slopes and intercepts, given by $\omega_{12}/(\omega_{11}\omega_{22})$.

Parameter	Geweke	Heidelberger-Welch	AR(50)	Raftery-Lewis	
	z	p	ρ	N	I
γ_{10}	−1.02	0.26	0.01	188550	50.30
γ_{20}	0.41	0.84	0.38	256150	68.40
γ_{11}	−1.15	0.29	0.02	184000	49.10
γ_{21}	−0.94	0.90	0.40	251900	67.20
γ_{12}	0.01	0.47	−0.02	185500	49.50
γ_{22}	−0.01	0.35	0.42	258300	69.00
ω_{11}	0.71	0.92	0.01	191700	51.20
ω_{22}	0.67	0.23	0.40	213350	57.00
ρ	−0.34	0.83	0.49	288500	77.00
σ	0.17	0.12	0.02	185550	49.50

Table 7.9 Numerical summaries, marginal posterior densities, multilevel regression model for math achievement scores. Cell entries are posterior means; 95 % HPD intervals in brackets. Estimates based on 500 000 MCMC iterations, thinned by 50 iterations, after a 5000 iteration burn-in.

	Intercept	Catholic	Mean SES	ω
Intercepts β_{j1}	12.63	1.22	5.33	1.52
	[12.37, 12.92]	[0.63, 1.81]	[4.62, 6.05]	[1.30, 1.76]
SES β_{j2}	2.22	−1.65	1.04	0.26
	[2.00, 2.43]	[−2.11, −1.17]	[0.48, 1.65]	[0.10, 0.45]
$\rho = \omega_{12}/(\omega_{11}\omega_{22})$	.41			
	[−.48, .97]			
σ (level one)	6.06			
	[5.96, 6.16]			

by a sizable amount: a one standard deviation change in mean SES (.41) would see the coefficient on SES for a typical non-Catholic school increase by $1.04 \times .41 = .43$, or about a third of the 1.65 difference between the SES slope in a Catholic and a non-Catholic school.

The final row of Table 7.9 summarizes the marginal posterior density of the standard deviation of the level one disturbances, σ. The posterior mean for this parameter is 6.06, smaller than the OLS estimate $\hat{\sigma} = 6.25$ reported in Table 7.7, indicative of the superior fit of the hierarchical model relative to the OLS-with-interactions model. This estimate of 6.09 for σ compares favorably with the least squares estimates of σ from the classical regression models in Table 7.6, with $\hat{\sigma} = 6.08$ and 6.06, but consuming 161 and 320 degrees of freedom, respectively.

Graphical summaries of the hierarchical component of the model appear in Figure 7.24, comparing the school-specific intercepts and slopes from the hierarchical model (top panels) with those from the OLS-with-interactions model (lower panels). In each panel we plot either estimated school-specific intercepts (left panels) or school-specific slopes (right panels) against school-average SES on the horizontal axis. The lines correspond to the level two relationships between school-average SES and the school-specific intercepts and slopes, with the gray lines and black lines distinguishing Catholic and non-Catholic schools, respectively. For the OLS-with-interactions model, all the estimated intercepts and slopes lie on the displayed lines, since for this model the level two model is a deterministic linear model.

In the upper panels of Figure 7.2 we see that the multi-level model produces school-specific intercepts (β_{j1}) and slopes (β_{j2}) that are stochastic functions of the level-two covariates `cath` and `meanses`. The hierarchical model produces estimates of the school-specific slopes that are tightly clustered around the fitted values from the systematic component of the level two model (top right panel); this is because there is relatively little information about the slope parameters *within* each school, and the 'prior information' embodied in the level two model tends to dominate. Table 7.9 reports the posterior mean of ω_{22} (the standard deviation of the stochastic component of the level two model for the school-specific slopes) to be just 0.26; as in conventional regression analysis, this is the standard deviation of the 'residuals' around the fitted level-two regression shown in the top-right panel of Figure 7.24.

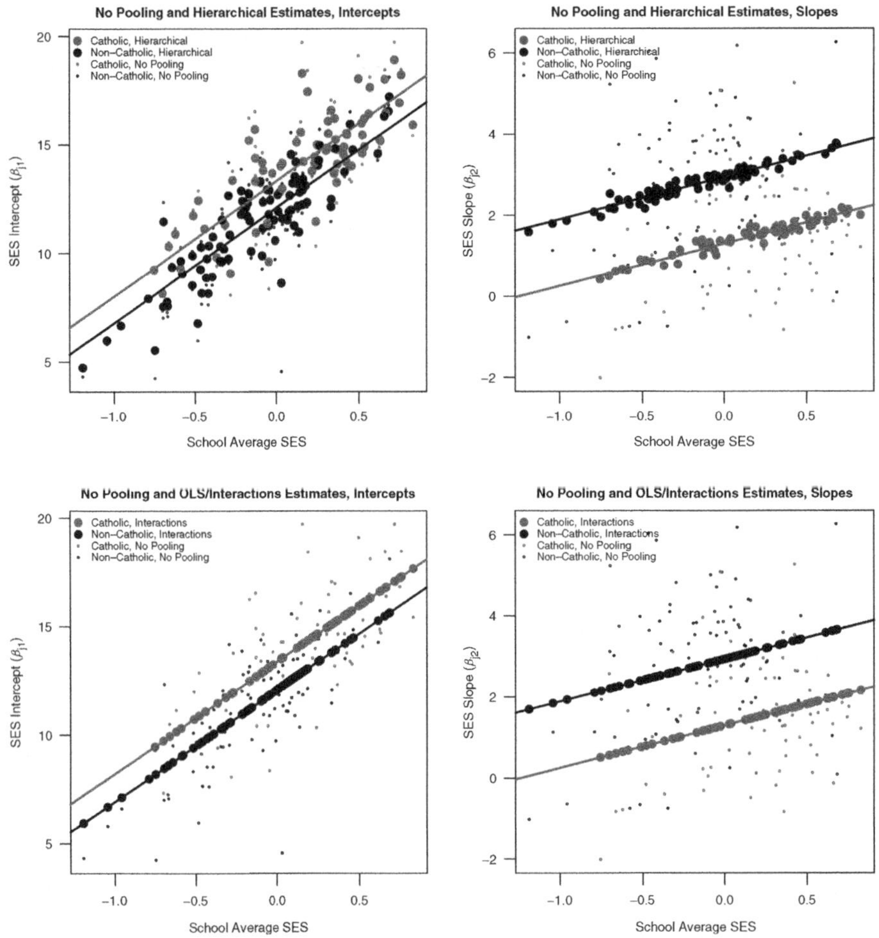

Figure 7.24 Comparison of estimates of school-specific intercept and slope parameters, regression analysis of math achievement scores. Grey dots indicate Catholic schools; black dots indicate non-Catholic schools. Larger dots are posterior means from the multi-level/hierarchical model (top panels) or OLS-with-interactions (lower panels). The lines on each panel correspond to the level-two relationship between `cath` and school-average SES `meanses`; note that for the lower two panels (comparing the no pooling and OLS-with-interactions models), the estimates all lie on the lines, reflecting the fact that the OLS-with-interactions model is a deterministic linear model at level two.

The story is somewhat reversed for the school-specific intercepts. The level-two model fits less well, with the posterior mean of ω_{11} estimated to be 1.52, and the posterior means of the school-specific intercepts (β_{j1}) are less well predicted by the linear, additive combination of `cath` and `meanses`. In this case, there is a reasonable amount of within-school information about the intercept parameters, and so the prior information in the level-two model for the β_{j1} contributes less to the posterior estimate of these parameters.

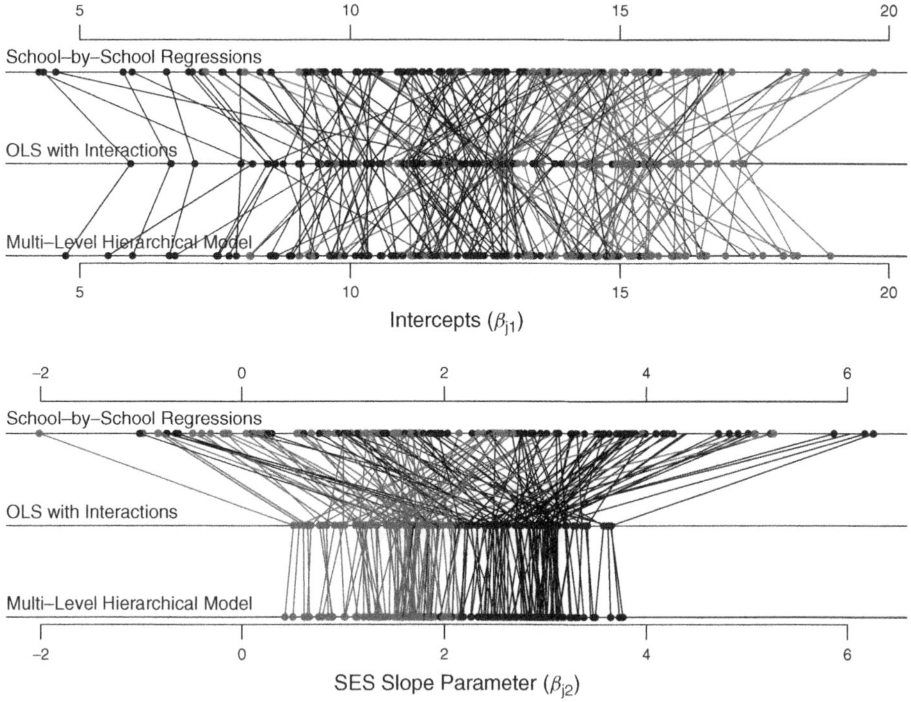

Figure 7.25 Shrinkage of OLS-with-interactions and multi-level/hierarchical model, math achievement scores. The lines connect three different estimates of the school-specific intercepts (top panel) and slopes (lower panels). Grey dots indicates Catholic schools; dark dots indicate non-Catholic schools.

Figure 7.25 conveys a similar message, using the 'shrinkage plots' used in previous examples in this chapter (e.g. see Figures 7.4, 7.7 and 7.17). Each panel of Figure 7.25 presents three sets of estimates of the school-specific intercepts (β_{j1}, top panel) and slopes (β_{j2}, lower panel): from school-by-school regressions (top line), the OLS-with-interactions model (middle line) and the multi-level/hierarchical model (posterior means plotted on the lower line). Color is used to distinguish Catholic (grey) and non-Catholic schools (black). For both the intercepts and slopes we see that the OLS-with-interactions model 'overshrinks', by constraining each school-specific estimate to be a linear, deterministic function of the level two covariates. The hierarchical model generates school-specific estimates that are a combination of the information in the level-two covariates and the school-specific regressions. For the intercepts, the 'over-shrinking' of OLS-with-interactions is quite pronounced, with the hierarchical model's estimates lying relatively close to the estimates generated by school-by-school regressions. For the slope coefficients, (a) the school-by-school estimates are reasonably imprecise and (b) the hierarchical, level two model fits reasonably well (see the top left panel of Figure 7.24), so that the $\hat{\beta}_{j2}$ generated by the OLS-with-interactions model are quite close to the posterior means for the β_{j2} generated by the hierarchical model.

Finally, Figure 7.26 compares the three approaches to modeling these data considered here (school-specific regressions, OLS-with-interactions, and the Bayesian hierarchical

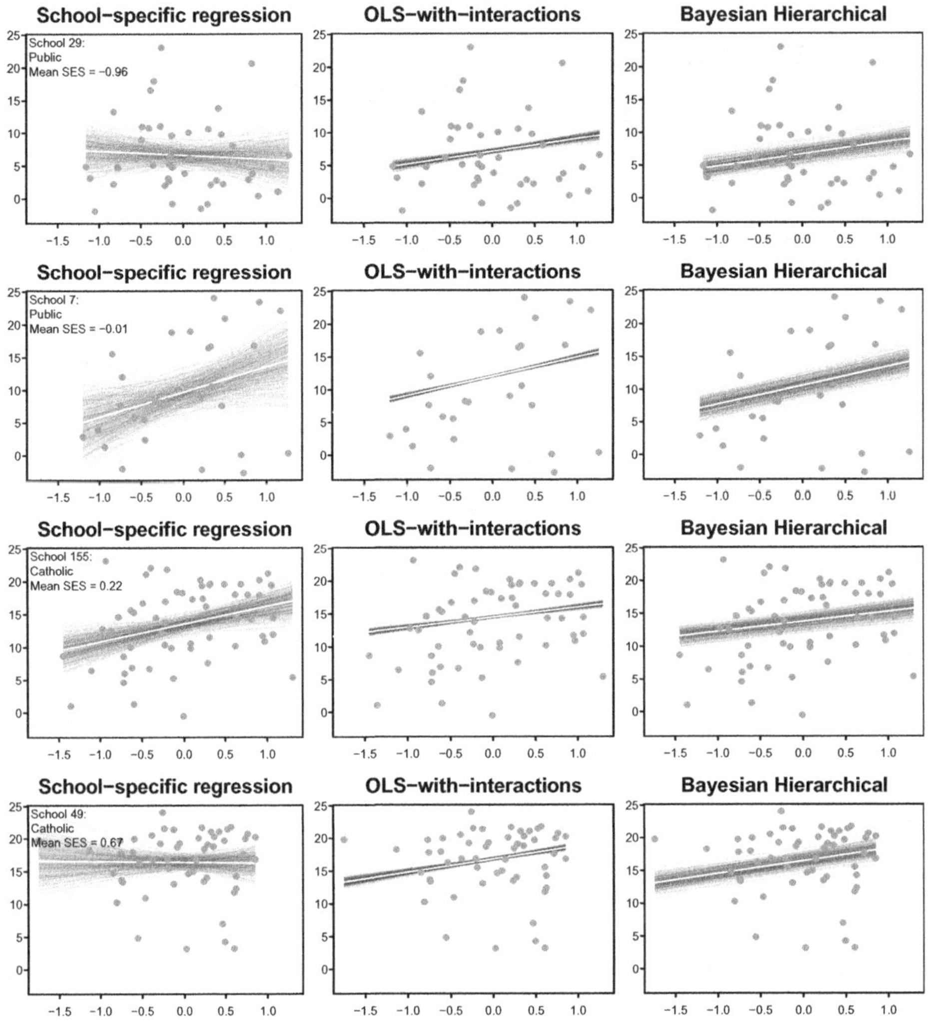

Figure 7.26 Comparison of model fits, four randomly selected schools, math achievement data. Each row shows the results of fitting three different estimators of the regression of math achievement score (vertical axis) on student-level SES (horizontal axis), for the indicated school: a school-specific or 'no pooling' linear regression (left panel), the OLS-with-interactions model reported in Table 7.7 (middle panel), and the Bayesian hierarchical model (right panel). Results from two randomly chosen public schools (top two rows) and two randomly chosen Catholic schools (two lower rows) are displayed. Plotted points show the data from the indicated school. For each estimator, 500 simulated regression lines are superimposed on the plot to graphically convey the uncertainty around the respective regression fit; for the two classical models (school-specific regressions and OLS-with-interactions) these draws are from the posterior density of the school-specific intercept and slope parameters under an improper, reference prior (see Proposition 2.15), while for the Bayesian hierarchical model these 500 draws are from the posterior density produced by the MCMC algorithm.

model), for four selected schools: two public, and two Catholic schools. Uncertainty in the fit to the data is shown via simulation methods, sampling from the posterior density of each school's $\boldsymbol{\beta}_j = (\beta_{j1}, \beta_{j2})'$, for each model. Five hundred draws generate 500 lines which are superimposed on the respective panel of Figure 7.26. In every instance, the school-specific regressions produce the most uncertainty; this is unsurprising, since this 'no pooling' approach relies exclusively on the information within each school when forming an estimate of $\boldsymbol{\beta}_j$. On the other hand, note that this school-by-school approach produces the best linear regression fit to each school's data: by definition, school-specific regressions will produce the smallest sum of squared errors, of the three approaches considered here. OLS-with-interactions (middle column of Figure 7.26) produces the most precise estimates of the school-specific regression parameters $\boldsymbol{\beta}_j$, since it fits a deterministic model at level two, but at some cost in terms of fit to the data at level one. Graphical inspection suggests that the regression fits implied by the OLS-with-interactions model can be quite different from those produced by school-by-school regressions (e.g. school 29 in the top row of Figure 7.26, and school 49 in the bottom row of the Figure).

The third column of Figure 7.26 shows the results for the Bayesian hierarchical model. We see a by-now familiar result: the Bayesian hierarchical model is negotiating a bias-variance tradeoff, accepting some bias in departing from the school-by-school regression fit, but more than making up for this with less variability. The results shown in Figures 7.24 and 7.25 are also apparent in Figure 7.26, with the school-specific regression fits produced by the hierarchical model having slopes that are very close to those from the OLS-with-interactions model, but with intercepts that lie relatively close to those from the school-by-school regressions. One interpretation of what the hierarchical model does is to combine level-two and level-one information, producing a better fit to the data than the OLS-with-interactions model, moving in the direction of the low-bias/high-variance school-by-school regression model, providing a better fit to the data without suffering from the levels of imprecision we see from the school-by-school regressions.

Model comparison. We conclude by comparing the three models used here with respect to total mean square error (TMSE). That is, for each model, we compute the quantity

$$\text{TMSE} = \sum_{j=1}^{J} \sum_{i=1}^{n_j} \left[(y_{ij} - \hat{y}_{ij})^2 + \text{var}(\hat{y}_{ij}) \right]$$

where $\hat{y}_{ij}$ is the predicted value of y_{ij} from a particular model and y_{ij} is the `math` score for student i in school j. TMSE is simply MSE summed over the entire data set, recalling that the MSE criterion combines both the lack of fit to the data (the squared error term) and imprecision (variance).

For the Bayesian model, we set $\hat{y}_{ij}$ to be the posterior mean of the quantity $E(y_{ij}) = \beta_{j1} + \beta_{j2}x_{ij}$ where the quantities β_{j1} and β_{j2} are produced by the level-two model in Equation 7.34b. We compute this posterior mean by computing $E(y_{ij})$ with the values of β_{j1} and β_{j2} produced by each iteration of the MCMC algorithm and averaging over iterations. We compute the posterior variance of $\hat{y}_{ij}$ similarly, by computing the variance in $E(y_{ij})$ over many MCMC iterations. For the classical models (school-by-school regressions and OLS-with-interactions), the predicted values and variances are computed using

Table 7.10 Total mean square error for three models fit to math achievement data.

Model	Sum of squared errors	Sum of variances	TMSE
School-by-school	252,084.5	11,876.7	263.961.2
OLS-with-interactions	280,225.7	234.2	280,459.9
Hierarchical model	259,475.4	3,781.8	263,257.3

standard formulas (corresponding to a Bayesian analysis with an improper, reference prior as in Proposition 2.15).

Table 7.10 compares the three models with respect to TMSE, with the columns of the table showing the decomposition of TMSE into bias and variance components. School-by-school clearly 'wins' in terms of fit to the data, as expected, but suffers from high variance, since we do not 'borrow strength' or share information across schools: recall the graphical depictions of the imprecision of the school-by-school regressions presented in Figures 7.22 and 7.26. The OLS-with-interactions model performs extremely well on variance, since it consumes just 6 degrees of freedom in fitting a deterministic model at level two; however, this model suffers in terms of bias, with the school-specific regressions it produces driven by the deterministic level-two model. The hierarchical model is the best performing model in terms of TMSE, with a little more bias than that produced by school-by-school regressions (a result of the shrinkage we see in Figure 7.25), but with considerably less variance. Again, this is a by now familar result, underscoring the usefulness of hierarchical models as an approach to dealing with parameters that vary across groups.

Problems

7.1 Fit the Efron and Morris (1975) baseball data using a hierarchical binomial model. That is, using a computer program like `BUGS` or `JAGS`, fit the model

$$\begin{aligned} r_i &\sim \text{Binomial}(\theta_i, n_i) \\ \theta_i &\sim \text{Beta}(\alpha, \beta) \\ \alpha &\sim \text{Exp}(2) \\ \beta &\sim \text{Exp}(2) \end{aligned}$$

where r_i is the number of hits recorded by player i in $n_i = 45$ at bats. The data are available in the data frame `EfronMorris`, part of the `R` package `pscl`.

7.2 Again consider the Efron and Morris (1975) baseball data. Suppose you possessed information as to the player's averages in the previous season (this would seem relevant to the problem of trying to predict the current season's batting averages). How would you incorporate this extra information in a model for the current season's averages? Specifically, what kind of model would you specify for this problem?

7.3 Suppose $\tau \sim \text{Gamma}(.01, .01)$, a density often used as an 'uninformative' prior for precision parameters. Using either analytic or Monte Carlo methods, characterize the density of $\sigma^2 = 1/\tau$. What is the mean of σ^2? What is the median, and other critical quantiles? Plot this density of σ^2 in the neighborhood of zero, say, over the interval zero to .25. Comment on what you see. Is this density 'uninformative' in this neighborhood? The functions `densigamma` and `pigamma` and friends in the `R` package `pscl` may be helpful.

7.4 Consider the model $y|\sigma^2 \sim N(0, \sigma^2)$, $\sigma \sim \text{Unif}(0, 1)$. Note that the 2nd density is with respect to the standard deviation σ, not the variance σ^2.

1. Using analytic or Monte Carlo methods, characterize the density of σ^2. Compare this density to $p(\sigma^2) \equiv \text{inverse-Gamma}(.01, .01)$, often used as an 'uninformative' prior for a variance parameter in Bayesian modeling.
2. Using analytic or Monte Carlo methods, characterize the marginal density of y. Compare this marginal density to that obtained where $\sigma^2 \sim \text{inverse-Gamma}(.01, .01)$.
3. Given data $\mathbf{y} = (y_1, \ldots, y_n)'$, what is the posterior density of (a) the variance σ^2, and (b) the precision σ^{-2}?

7.5 Consider the two-way ANOVA model given in Equation 7.15 in Example 7.7, without constraints on the α_i or δ_t parameters.

1. If we were to estimate the parameters $\boldsymbol{\theta} = (\mu, \alpha_1, \ldots, \alpha_n, \delta_1, \ldots, \delta_T)'$ by least squares regression methods, what does the corresponding design matrix look like? That is, if we were to estimate $\boldsymbol{\theta}$ with $\hat{\boldsymbol{\theta}} = (\mathbf{X}'\mathbf{X})^{-1}\mathbf{X}'\mathbf{y}$, what is the content of the $\mathbf{X}$ matrix?
2. Show that without constraints on the α_i and δ_t parameters there exist linear dependencies among the columns of $\mathbf{X}$.
3. Suppose we drop the time-specific terms from the model such that we have the one-way ANOVA model $y_{it} \sim N(\mu + \alpha_i, \sigma^2)$. Repeat the previous two questions assuming no restrictions on the α_i.

7.6 In Example 7.7 we used an over-parameterized version of the two-way ANOVA model, in which the identified parameter μ^*, the grand mean of y, could be recovered as a deterministic function of three unidentified parameters μ, μ_α and μ_δ. To implement this model, we used proper priors on the unidentified parameters. But what implied prior did we use for the identified parameter μ^*?

7.7 Verify that the hierarchical model used in Example 7.7 generates more precise estimates of the state-specific α_i parameters than we would obtain with a non-hierarchical model.

7.8 The hierarchical, two-way ANOVA model in Example 7.7 generates quite large prediction errors. Consider an extremely simple model in which the 2000 presidential election results serve as the predictions for 2004. What is the root mean square error and median absolute error of these predictions? How do these performance measures compare to those from the hierarchical model?

7.9 For the two-way hierarchical model in Example 7.7, consider the joint posterior density over the α_i, the state-specific offsets in average levels of Democratic vote share. Re-estimate the model, but rank the α_i. Let uncertainty over the α_i propagate into uncertainty over the order statistics. That is, if t indexes samples from the joint posterior density of $\boldsymbol{\alpha} = (\alpha_1, \ldots, \alpha_n)'$, then compute ranks $\mathbf{r}^{(t)} = (r_1^{(t)}, \ldots, r_n^{(t)})' = r(\boldsymbol{\alpha}^{(t)})$, where r returns order statistics; i.e. absent any ties, $\mathbf{r}^{(t)}$ will be a permutation of the integers $1, \ldots, n$. Performing this simple computation over many samples from the posterior density for $\boldsymbol{\alpha}$ induces a posterior probability mass function over the rank of each α_i; i.e. the proportion of times we see $r_i^{(t)} = j$ is a Monte Carlo based estimate of the posterior probability that α_i has rank j, $i, j \in \{1, \ldots, n\}$.

7.10 Consider the linear time trend model for the presidential elections data presented in Example 7.8.

1. Estimate both the 'no pooling' and hierarchical versions of this model. Recall that the 'no pooling' model amounts to simply running a regression within each state.

2. Compute and inspect the posterior predictive densities of both models for the 'out-of-sample' 2004 outcomes. Compare the predictive performance of both models by plotting the means of the posterior predictive densities against the actual 2004 outcomes.

3. Also examine the standard deviations of the posterior predictive densities. How much smaller are the standard deviations for the predictions from the hierarchical linear trend model than those from the 'no pooling' version of this model?

4. For each state, and for each model, compute the quantity $\text{MSE}_i = \text{error}_i^2 + \text{variance}_i$, where 'error' is the difference between the actual 2004 outcome and the point forecast (i.e. the mean of the posterior predictive density), 'variance' is the variance of the forecast (the variance of the posterior predictive density), and $i = 1, \ldots, 50$ indexes the states. Compare the models by computing $\text{TMSE} = \sum_{i=1}^{50} \text{MSE}_i$.

7.11 Implement the augmented version of the hierarchical model discussed at the end of Example 7.9. The data are available as the data frame `RockTheVote` in the `R` package `pscl`.

7.12 (Suggested by Don Green.) Reconsider Example 7.9 with a hierarchical model that places zero probability on negative intent-to-treat effects. That is, it might be argued that there is no possible way that 'get-out-the-vote' television advertisements could depress voter turnout and the only scientific uncertainty concerns the magnitude of the positive intent-to-treat effects. What kind of hierarchical model operationalizes this set of prior beliefs over the δ_i in Example 7.9? Implement your model in `BUGS` or `JAGS` and compare the results with those reported above in Example 7.9.
Hint: One possible specification might be a log-normal model, i.e. $\log \delta_i \sim N(\mu_\delta^*, \sigma_\delta^2)$, with priors on the hyperparameters μ_δ^* and σ_δ^2; note that with this prior places zero probability on negative values of δ_i. When implementing this kind of model, some kind of experimentation with the priors on the hyperparameters μ_δ^* and σ_δ^2 might be necessary so as to generate reasonable priors over the δ_i.

8

Bayesian analysis of choice making

8.1 Regression models for binary responses

We encountered binary data in Section 2.1, where we saw that the Beta density is the conjugate prior density for the success parameter in a series of binary trials. We again encountered binary data in the introduction to Part II, where we noted that Bayesian analysis of regression models for binary data can get surprisingly complicated. Markov chain Monte Carlo has changed this state of affairs, and somewhat dramatically. Today, Bayesian analysis of discrete data models is standard fare, and a close look at Bayesian computation for these models provides a useful stepping stone to more complicated models.

Regression models for binary data take the following form. We have data $y_i \in \{0, 1\}$ along with k covariates $\mathbf{x}_i = (x_{i1}, \ldots, x_{ik})'$ where i indexes the n observations. We relate the covariates to the binary response via a vector of k unknown parameters $\boldsymbol{\beta} \in \mathcal{B} \subseteq \mathbb{R}^k$ and a function $F : \mathbb{R} \mapsto [0, 1]$, as follows:

$$
\begin{aligned}
y_i | \mathbf{x}_i &\sim \text{Bernoulli}(\pi_i) \\
\pi_i &= F(\mathbf{x}_i \boldsymbol{\beta})
\end{aligned}
$$

In the parlance of generalized linear models (McCullagh and Nelder 1989), the function F is referred to as the inverse-link function, i.e. $F^{-1}(\pi_i) = \mathbf{x}_i\boldsymbol{\beta}$. A logistic regression or 'logit' model results with $F(t) = [1 + \exp(-t)]^{-1}$, the distribution function of the logistic density (see Definition B.32). A probit model results with $F(t) = \Phi(t) = \int_{-\infty}^{t} \phi(z)dz$ where ϕ and Φ are the standard normal density and its cumulative distribution function, respectively. The logit model tends to be more commonly encountered in practice than the probit model, mainly for historical reasons (i.e. the CDF of the logistic is easily computed, while the normal CDF is not available in closed form). Under conditional independence of the data given the predictors $\mathbf{X}$ and parameters $\boldsymbol{\beta}$, the joint density of the data (the likelihood for $\boldsymbol{\beta}$) is simply the n-fold product of the observation specific probabilities,

$$\mathcal{L}(\boldsymbol{\beta}; \mathbf{y}, \mathbf{X}) = p(\mathbf{y}|\mathbf{X}, \boldsymbol{\beta}) = \prod_{i=1}^{n} F(\mathbf{x}_i\boldsymbol{\beta})^{y_i}[1 - F(\mathbf{x}_i\boldsymbol{\beta})]^{1-y_i} \tag{8.1}$$

i.e. the likelihood contributions of observations with $y_i = 1$ are simply $\pi_i = F(\mathbf{x}_i\boldsymbol{\beta})$ while observations with $y_i = 0$ make likelihood contributions $1 - \pi_i$.

We have already noted that Bayesian analysis of this model is not trivial. There is no conjugate prior for $\boldsymbol{\beta}$ given this likelihood. Nonetheless, MCMC based approaches make short work of this model. First, we can use *data augmentation* to turn the probit model into a linear regression conditional on latent data, for which Bayesian analysis via the Gibbs sampler is straightforward. Then, for the logit model, we will see how a Metropolis sampler can tackle the problem of sampling from the posterior 'head-on', without resorting to the latent data formulation. We begin with a consideration of the probit model, so as to showcase the data augmentation algorithm, and a closely related idea, *marginal data augmentation*.

8.1.1 Probit model via data augmentation

We introduced the concept of a data augmented Gibbs sampler in Section 5.2.5, and for the specific case of a probit model in Example 5.10. Recall that the probit model is simply a latent variable regression model with normal errors, where the latent dependent variable y_i^* is censored, observed only in terms of its sign: i.e.

$$y_i^* = \mathbf{x}_i\boldsymbol{\beta} + \epsilon_i, \quad \epsilon_i \sim N(0, 1) \; \forall \; i = 1, \ldots, n, \tag{8.2}$$

with the censoring rule $y_i = 0 \Rightarrow y_i^* < \tau$ and $y_i = 1 \Rightarrow y_i^* \geq \tau$.

Identification

Setting the censoring threshold τ at zero is an arbitrary restriction necessary so as to identify the intercept parameter in $\boldsymbol{\beta}$. That is, since

$$\begin{aligned}\Pr(y_i = 1|\mathbf{x}_i, \boldsymbol{\beta}) &= \Pr(y_i^* \geq \tau|\mathbf{x}_i, \boldsymbol{\beta}) = \Pr(\beta_0 + \beta_1 x_i > \tau) \\ &= \Pr(\beta_0 + c + \beta_1 x_i > \tau + c) \; \forall \; c \neq 0\end{aligned}$$

it is obvious that the censoring threshold τ and the intercept parameter β_0 are not jointly identified; setting τ to a constant is sufficient to identify β_0 and $\tau = 0$ is simple way to implement this restriction. The restriction that y_i^* has its scale parameter σ^2 set to 1.0 is another arbitrary identification restriction: i.e. consider multiplying both sides of Equation (8.2) by any $m > 0$; $\boldsymbol{\theta} \equiv (\boldsymbol{\beta}, \sigma^2) = (\boldsymbol{\beta}, 1)$ generates the same value of the likelihood function as $\boldsymbol{\theta}^* = (m\boldsymbol{\beta}, m^2), \forall \; m > 0$. Setting σ^2 to any positive constant will pin down the scale of the latent variable (and hence identify $\boldsymbol{\beta}$), and the particular choice of $\sigma^2 = 1.0$ is simply chosen for convenience.

Data augmentation

The latent data $\mathbf{y}^* = (y_1^*, \ldots y_n^*)'$ augment the original inferential problem, as foreshadowed in Section 5.2.5. Instead of trying to compute the posterior $p(\boldsymbol{\beta}|\mathbf{y}, \mathbf{X})$ directly, we work with the data-augmented posterior density $p(\boldsymbol{\beta}|\mathbf{y}^*, \mathbf{y}, \mathbf{X})$, which (as we will see) is much easier to sample from (via the Gibbs sampler) than $p(\boldsymbol{\beta}|\mathbf{y}, \mathbf{X})$.

We use the following iterative scheme (a data augmented Gibbs sampler) for sampling from the posterior for $\boldsymbol{\beta}$:

1. generate augmented data $\mathbf{y}^*$ from the predictive density $p(\mathbf{y}^*|\boldsymbol{\beta}, \mathbf{y}, \mathbf{X})$
2. sample $\boldsymbol{\beta}$ from its conditional density given the augmented data and observed data $p(\boldsymbol{\beta}|\mathbf{y}^*, \mathbf{y}, \mathbf{X})$; and actually, for the probit model, given $\mathbf{y}^*$, conditioning on $\mathbf{y}$ is redundant and so the conditional density for $\boldsymbol{\beta}$ is $p(\boldsymbol{\beta}|\mathbf{y}^*, \mathbf{X})$.

Implementing this scheme is not at all difficult, since the normal density underlying the probit model (Equation 8.2) implies that

$$y_i^*|\mathbf{x}_i, \tilde{\boldsymbol{\beta}}, y_i = 0 \sim N(\mathbf{x}_i\tilde{\boldsymbol{\beta}}, 1)\mathcal{I}(y_i^* < 0)$$
$$y_i^*|\mathbf{x}_i, \tilde{\boldsymbol{\beta}}, y_i = 1 \sim N(\mathbf{x}_i\tilde{\boldsymbol{\beta}}, 1)\mathcal{I}(y_i^* > 0)$$

where $\mathcal{I}(\cdot)$ is an indicator function, evaluating to one if its argument is true and zero otherwise, so as to induce truncated normal sampling. The untruncated predictive density for y_i^* has variance $\sigma^2 = 1$, implementing the identification constraint discussed earlier. We perform the integration over $\mathbf{y}^*$ by Monte Carlo methods, giving rise to the following algorithm:

Algorithm 8.1 Data augmented Gibbs sampler for probit regression (Albert and Chib 1993).

```
1: for t = 1 to T do
2:   sample β^(t) ~ p(β|y*^(t−1), X)
3:   sample y*^(t) ~ p(y*|β^(t), y, X)
4: end for
```

Further, if we use normal priors for $\boldsymbol{\beta}$, $\boldsymbol{\beta} \sim N(\mathbf{b}_0, \mathbf{B}_0)$, then this algorithm becomes extremely easy to implement, since the normal prior for $\boldsymbol{\beta}$ is conjugate with respect to the normal model for the latent y_i^*, and we can deploy standard results on the Bayesian analysis of a linear regression problem (see Section 2.5.3). In this case the algorithm sketched above becomes the following data-augmented Gibbs sampler (e.g. Albert and Chib 1993): at iteration t,

1. sample $\boldsymbol{\beta}^{(t)}|\mathbf{y}^{(t-1)}, \boldsymbol{\beta} \sim N(\mathbf{b}, \mathbf{B})$ where

$$\mathbf{b} = (\mathbf{B}_0^{-1} + \mathbf{X}'\mathbf{X})^{-1}(\mathbf{B}_0^{-1}\mathbf{b}_0 + \mathbf{X}'\mathbf{y}^{*(t-1)})$$
$$\mathbf{B} = (\mathbf{B}_0^{-1} + \mathbf{X}'\mathbf{X})^{-1}$$

   using results on congujate analysis of the linear regression model (see Proposition 2.11).

2. We sample each $y_i^{*(t)}$, $i = 1, \ldots, n$, conditional on $\boldsymbol{\beta}^{(t)}$, $\mathbf{x}_i$ and $y_i \in \{0, 1\}$, as follows:

$$y_i^{*(t)}|y_i = 0, \mathbf{x}_i, \boldsymbol{\beta}^{(t)} \sim N(\mathbf{x}_i\boldsymbol{\beta}^{(t)}, 1)\,\mathcal{I}(y_i^{*(t)} < 0)$$
$$y_i^{*(t)}|y_i = 1, \mathbf{x}_i, \boldsymbol{\beta}^{(t)} \sim N(\mathbf{x}_i\boldsymbol{\beta}^{(t)}, 1)\,\mathcal{I}(y_i^{*(t)} \geq 0)$$

   Sampling from these truncated distributions can be done by rejection sampling (i.e. sampling from the untruncated distribution until we obtain a draw that satisfies the truncation constraint implied by the observed value of y_i). Alternatively, we can use a more sophisticated accept-reject algorithm, as discussed in Section 3.4.3 and Example 3.8, or a 'one-shot' inverse-CDF method as discussed in Section 3.4.1 and Example 3.6.

Reasonable starting values can usually be obtained by fitting the probit model via least squares regression.

■ Example 8.1

Probit model of voter turnout. Nagler (1991) analyzes voter turnout with individual level data from the 1984 Current Population Survey; in presidential election years the CPS administers a 'voter supplement' in which respondents are asked to report whether they voted in the most recent presidential election, and these data have become a near canonical data source for students of voter turnout. Covariates include each survey respondent's age and a categorical measure of their level of education. In addition, measures of each respondent's political context are available: the number of days voter registration closed before the election in the respondent's state, whether or not a gubernatorial election took place in the respondent's state, and whether the respondent lives in the South. This re-analysis uses a random 3000 observation subset of Nagler's original 98 000 observation data set.

We adopt a specification similar to that employed in Nagler's 1991 article. Maximum likelihood estimates are easily obtained for this model, say by fitting a generalized linear model (McCullagh and Nelder 1989) to the binary turnout data with a probit link, i.e.

$$y_i|\mathbf{x}_i \sim \text{Bernoulli}(\pi_i)$$
$$\pi_i = \Phi(\mathbf{x}_i\boldsymbol{\beta})$$

with the likelihood given in equation 8.1, with $F \equiv \Phi$, the normal CDF. In a large sample, the maximum likelihood estimates $\hat{\boldsymbol{\beta}}$ will correspond to the mean of the posterior density given an improper, uniform prior density over $\boldsymbol{\beta}$, say $p(\boldsymbol{\beta}) \propto 1$ (see Definitions B.12 and B.13). We adopt this improper prior here for sake of exposition, so as to assess the performance of the MCMC-based approach: i.e. up to Monte Carlo error, the average of the MCMC draws should correspond to the maximum likelihood estimates.

I use the `R` package `MCMCpack` (Martin, Quinn and Park 2009) to generate the data-augmented Gibbs sampler for this problem; the function `MCMCprobit` implements a Bayesian probit regression. One thousand iterations of the algorithm are discarded as burn-in, and a subsequent 5000 iterations are saved for inference. Graphical inspection of the Gibbs sampler output can be extremely helpful in assessing how effectively the sampler is exploring the posterior density. Figure 8.1 provides a graphical summary of the

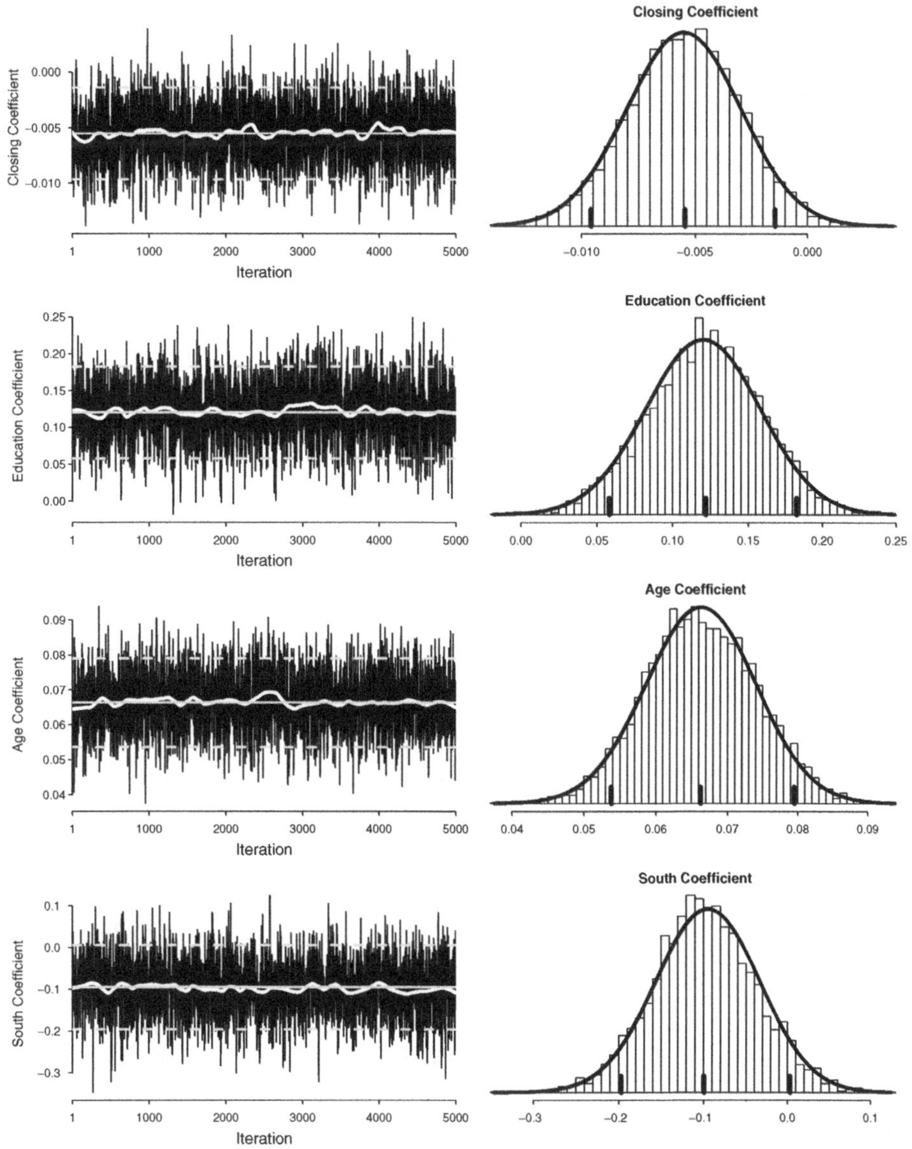

Figure 8.1 Traceplots and densities, data-augmented Gibbs sampler, probit model for voter turnout. In the left panels, each undulating gray line is a moving average superimposed on the Gibbs sampler output (presented as a time series or 'traceplot'); the solid and dotted lines indicate the MLEs and the limits of a 90 % confidence interval. In the right panel, the histograms summarize the samples generated by the data-augmented Gibbs sampler, while the smooth curve is a normal density centered on the MLEs with standard deviation equal to the standard error of the MLEs (i.e. given the improper, uniform prior, and a large sample, the marginal posterior densities of the model parameters will correspond to these densities); the ticks on the horizontal axis indicate the 5th, 50th and 95th percentiles of the Gibbs samples for the respective parameter.

Table 8.1 Comparison of MLEs and data-augmented Gibbs sampler, probit model of voter turnout. For the data-augmented Gibbs sampler (MCMC), the averages of 5000 samples are reported as the point estimates, standard deviations are reported in parentheses and the 2.5 and 97.5 percentiles are reported in brackets; the frequentist 95 % confidence intervals around the MLEs (assuming asymptotic normality) are also reported in brackets.

	MLE	MCMC
Intercept	−2.69	−2.70
	(0.277)	(0.279)
	[−3.23, −2.15]	[−3.24, −2.14]
Closing Day	−0.00555	−0.00549
	(0.00251)	(0.0025)
	[−0.0105, −0.000629]	[−0.0103, −6E − 04]
Education	0.251	0.254
	(0.0833)	(0.0838)
	[0.0877, 0.414]	[0.0891, 0.415]
Education^2	0.00563	0.0053
	(0.00825)	(0.00828)
	[−0.0105, 0.0218]	[−0.0107, 0.0217]
Age	0.0669	0.0669
	(0.00772)	(0.00763)
	[0.0518, 0.082]	[0.0518, 0.082]
Age^2	−0.000475	−0.000474
	(8.05E-05)	(7.94E-05)
	[−0.000633, −0.000317]	[−0.000632, −0.000315]
South	−0.0947	−0.0947
	(0.0615)	(0.062)
	[−0.215, 0.0258]	[−0.221, 0.0251]
Gubernatorial Election	0.0659	0.0681
	(0.066)	(0.0656)
	[−0.0635, 0.195]	[−0.0585, 0.199]

performance of the data-augmented Gibbs sampler for selected parameters; the traceplots are visually indistinguishable from white noise, strongly suggesting that the sampler has converged on the posterior density of the model parameters. Table 8.1 presents a comparison of the MLEs obtained using the `glm` procedure in `R` and the output of the data-augmented Gibbs sampler. The two sets of estimates are virtually identical for this model, as they should be (since we employ an improper, uniform prior).

Multiple chains. Further evidence that the Gibbs sampler works well for this problem comes from considering the behavior of multiple Gibbs samplers, started from widely dispersed points in the parameter space. We accomplish this by running four versions of the Gibbs sampler, started at points that straddle the maximum likelihood estimates by plus or minus 3 standard errors. The results are depicted graphically in Figure 8.2, which

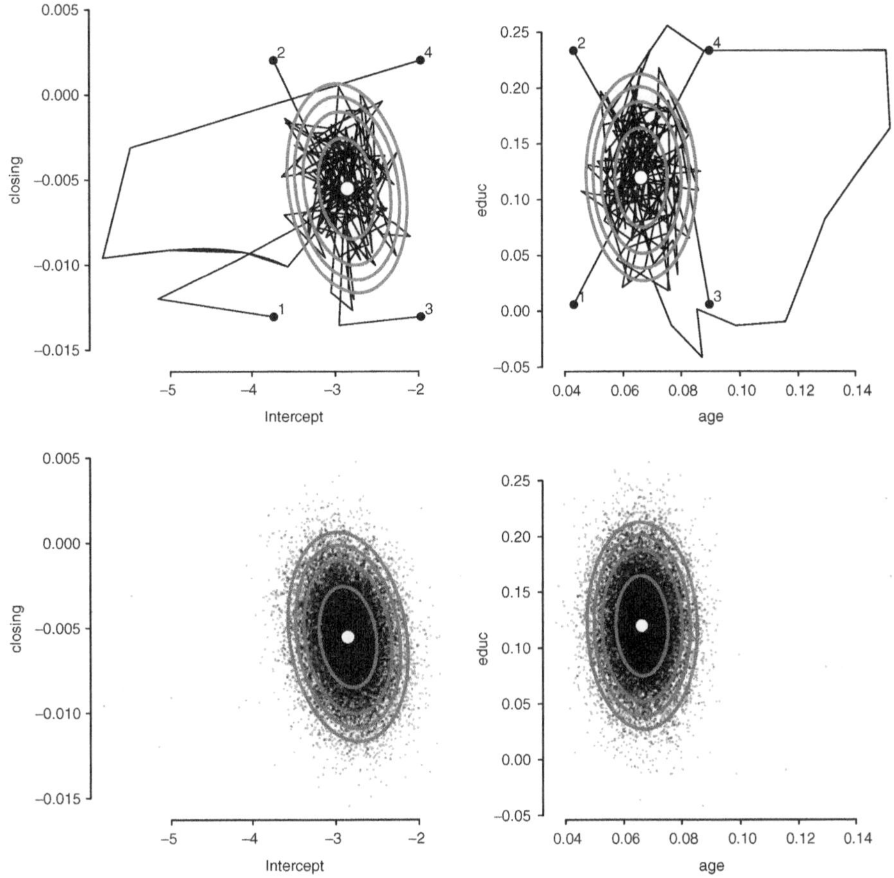

Figure 8.2 Multiple Gibbs samplers, voter turnout example. Four Gibbs samplers were run for 5000 iterations each, starting from points plus or minus 3 standard errors from the maximum likelihood estimates, with an improper uniform prior. The top panels show the trajectories of the samplers in two dimensions, for the first 50 iterations, indicating that the samplers quickly move away from the highly dispersed starting values to regions of high posterior density. The lower panels show the entire output of the samplers as a series of transparent points (i.e. the darker colors indicate regions of high posterior density). The ellipses correspond to contours of the likelihood function, with the circle indicating the location of the maximum likelihood estimates.

shows that the four samplers quickly move away from their starting values to regions of the parameter space with relatively high posterior probability.

Auxiliary quantities of interest

As in many modeling settings, it is straightforward to induce posterior densities over quantities that are functions of the model parameters. In the specific context of binary response models, the 'auxiliary quantities of interest' we will often include goodness of fit measures, predicted probabilities and marginal effects. Generically, any of these auxiliary quantities can be written as $h(\boldsymbol{\beta})$ and so we seek the posterior density $p(h(\boldsymbol{\beta})|\mathbf{y}, \mathbf{X})$. In general we will not generate analytic characterizations of these posterior densities, but rather, use the Monte Carlo methods described in Section 3.2 to estimate features of $p(h(\boldsymbol{\beta})|\mathbf{y}, \mathbf{X})$, e.g. the mean, the standard deviation, quantiles, etc. That is, the data-augmented Gibbs sampler produces sampled values $\boldsymbol{\beta}^{(t)}, t = 1, \ldots, T$, from the posterior density $p(\boldsymbol{\beta}|\mathbf{y}, \mathbf{X})$, and for each t we simply compute the auxiliary quantity $h(\boldsymbol{\beta}^{(t)})$, which constitute a sample from $p(h(\boldsymbol{\beta})|\mathbf{y}, \mathbf{X})$; the mean of the $h(\boldsymbol{\beta}^{(t)})$ is a simulation-consistent estimate of $E(h(\boldsymbol{\beta})|\mathbf{y}, \mathbf{X})$ and so on, for other summaries of $p(h(\boldsymbol{\beta})|\mathbf{y}, \mathbf{X})$.

■ Example 8.2

Posterior densities for auxiliary quantities of interest, probit model of voter turnout (Example 8.1, continued). Figure 8.3 shows histograms summarizing two goodness of fit indicators, percent correctly predicted (PCP) and the area under the receiver operating characteristic (ROC) curve, using the last 4500 draws from each of the 4 MCMC algorithms used to explore the posterior density of $\boldsymbol{\beta}$ (i.e. the histograms in Figure 8.3 are based on $4 \times 4500 = 18\,000$ sampled values).

PCP is the percentage of observations with predicted probabilities π_i that lie on the correct side of a pre-specified threshold, $\kappa \in [0, 1]$; i.e. if $y_i = 1$ and $\pi_i > \kappa$ a correct prediction is recorded; $\pi_i < \kappa$ is considered a correct prediction if $y_i = 0$. PCP is simply

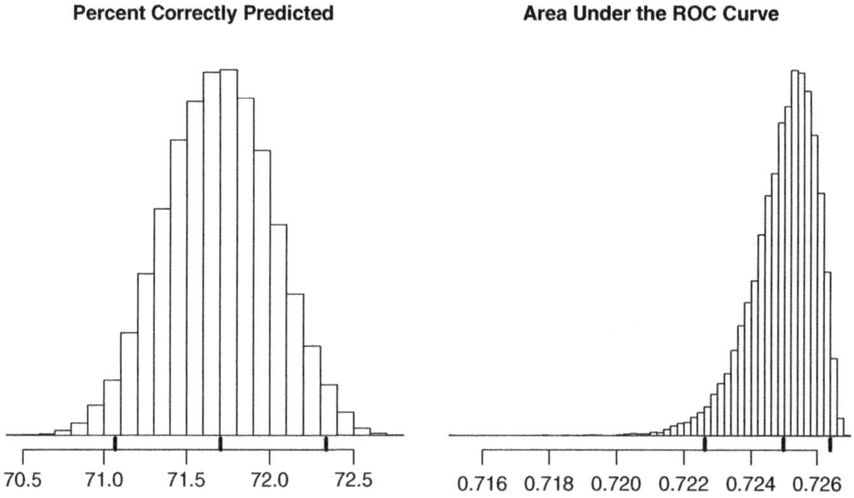

Figure 8.3 Posterior densities for goodness-of-fit indicators, probit model of voter turnout. Percent-correctly-predicted is defined as the percentage of observations where $\pi_i \equiv \Phi(\mathbf{x}_i\boldsymbol{\beta}) > .5$ and $y_i = 1$, or $\pi_i < .5$ and $y_i = 0$. The area under the receiver operatoring characteristic (ROC) curve is a measure of predictive success that is less sensitive to the choice of the classification threshold. Tick marks on the horizontal axes indicate the mean of the respective posterior density (estimated with the mean of the sampled values from the data-augmented Gibbs sampler) and a 95 % credible interval (estimated with the 2.5 and 97.5 percentiles of the output of the Gibbs sampler).

the percentage of cases with correct predicted probabilities for a given level of κ, typically set to .5. In the probit model considered here $\pi_i = \Phi(\mathbf{x}_i\boldsymbol{\beta})$ and so given the data $\mathbf{y}$ and $\mathbf{X}$, PCP is a function of the unknown parameters $\boldsymbol{\beta}$. Thus, uncertainty in $\boldsymbol{\beta}$ induces uncertainty in π_i and hence in a summary measure such as PCP, which is important when considering the predictive performance of one model against another. In this case, the posterior mean of the PCP measure is 71.7 %, with a 95 % credible interval ranging from 71.1 % to 72.3 %.

The ROC curve provides a more general assessment of model fit for a binary classifier (including, as a special case, a regression model for binary responses such as the probit model considered here). The ROC curve is the line formed by plotting the TPF (true positive fraction, or predictions of y_i that are correct) against the FPF (false positive fraction) as the classification threshold κ ranges between 0 and 1. The resulting function is defined on the unit square, and the area under the ROC curve A is a measure of the classificaton success of the model. A value of $A = .5$ indicates random predictions and is a lower bound; $A = 1$ indicates perfect classification. An accessible introduction to the ROC curve appears in Hosmer and Lemeshow (2000, 160); Sing *et al.* (2005) provide a powerful `R` package for visualizing the performance of classifiers, with ROC curves as a special case. In the Bayesian approach here, we again note that conditional on the data, A is a function of $\boldsymbol{\beta}$, and we seek to characterize its posterior density. The histogram in the lower panel of Figure 8.3 summarizes the area under the ROC curve A for the probit model of voter turnout: the mean of the posterior density for A is .725 with a small 95 % credible interval, ranging from .722 to .726. Curiously, there is a long left tail to the posterior density for A, but it is nonetheless obvious that A is bounded well away from its lower bound of .5.

Assessing the marginal effect of a key predictor via posterior predictive densities. Students of voter turnout are particularly interested in the effects of legal requirements concerning voter registration on turnout. To assess the effect of closing day requirement (the number of days before the election that a state's voter registration rolls close), I considered a hypothetical subject with 12 years of education, 40 years of age (the median value in the data set), living in a non-Southern state without a gubernatorial election on their state's ballot. I then considered increasingly onerous closing day requirements, setting the closing day requirement to each of its 15 unique values observed in the data, ranging from same-day registration (0 days) to a fifty day registration requirement. Let $\tilde{\mathbf{X}}$ be a 15-by-k matrix formed by stacking the vectors containing fixed values of the other covariates and the 15 values of the closing day variable. Let $\tilde{\pi}_i = \Phi(\tilde{\mathbf{x}}_i\boldsymbol{\beta})$ be the estimated probability of turning out given $\tilde{\mathbf{x}}_i$; we seek the posterior density of $\tilde{\pi}_i$, which is the posterior predictive density

$$p(\tilde{\pi}_i|\mathbf{y}, \mathbf{X}, \tilde{\mathbf{x}}_i) = \int_{\mathcal{B}} p(\tilde{\pi}_i|\boldsymbol{\beta}, \tilde{\mathbf{x}}_i)p(\boldsymbol{\beta}|\mathbf{y}, \mathbf{X})d\boldsymbol{\beta}.$$

Sampling from this density is trivial, via the method of composition introduced in Section 3.3 and as described above: i.e. the data-augmented Gibbs sampler provides us with samples $\boldsymbol{\beta}^{(t)}, t = 1, \ldots, T$, from $p(\boldsymbol{\beta}|\mathbf{y}, \mathbf{X})$ which we then use in forming $\tilde{\pi}_i^{(t)} = \Phi(\tilde{\mathbf{x}}_i\boldsymbol{\beta}^{(t)})$. Note that the predictive density $p(\tilde{\pi}_i|\boldsymbol{\beta}, \tilde{\mathbf{x}}_i)$ is degenerate, in the sense that $\tilde{\pi}_i$ is a deterministic function of $\boldsymbol{\beta}$ and $\tilde{\mathbf{x}}_i$.

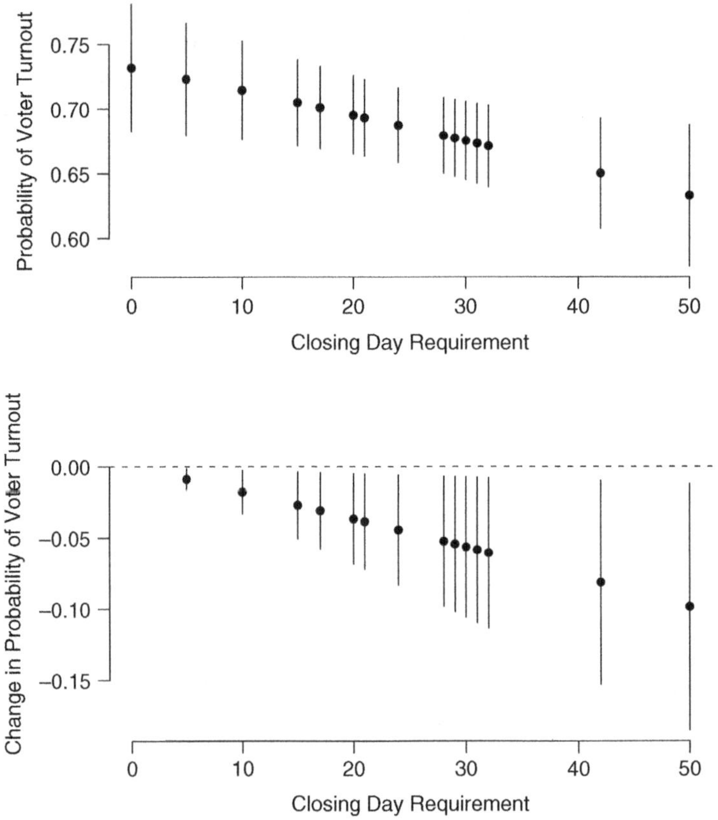

Figure 8.4 Marginal effect of closing day requirement, probit model of voter turnout. The graph shows predicted probabilities of voter turnout for a respondent in a non-Southern state, without a gubernatorial election, with 12 years of education and 40 years of age, as a function of the closing day requirement. Plotted points are the mean of the posterior density of the probability (estimated with the mean of the probabilities computed from the Gibbs sampler output) and the vertical lines cover a 95 % credible interval (estimated with the 2.5 and 97.5 percentiles of the probabilities computed with the Gibbs sampler output). The bottom panel shows the change in the predicted probability (with credible intervals) relative to the baseline condition of same-day voter registration (a closing day requirement of zero days).

Figure 8.4 shows summaries of the posterior densities for $\tilde{\pi}_i$ for the 15 values of the closing day requirement observed in the data, and conditional on the fixed values of the other covariates. The bottom panel of Figure 8.4 summarizes the posterior density of the *change* in the predicted probability of turnout, relative to a baseline condition of a zero day closing day requirement (same day registration); note that in every instance the 95 % credible interval on the change in probability does not overlap zero. This type of graph makes the closing day effect more vivid than the tabular summary of the marginal posterior density for the closing day coefficient provided in Table 8.1. The top panel of Figure 8.4 shows that with same day registration, the hypothetical respondent under consideration has a .73 probability of turning out; with a 30 day closing day requirement this predicted probability has dropped to .68, and under a 50 day registration requirement (the most onerous observed in these data) the predicted probability has dropped to .63. These are not massive effects (10 percentage points over the entire range of closing day requirements observed in the data), but are reasonably large relative to the variation in state-level turnout rates around the United States, and more than large enough to be potentially consequential in close races.

8.1.2 Probit model via marginal data augmentation

The data augmentation (DA) approach sketched above can be extended using an idea due to Meng and van Dyk (1999), known as *marginal data augmentation* (MDA). MDA is closely related to earlier proposals by Meng and van Dyk (1997) and Liu and Wu (1999). Generically, DA introduces latent data $\mathbf{y}^*$ so that we can obtain the posterior density as

$$p(\boldsymbol{\beta}|\mathbf{y}, \mathbf{X}) = \int_{y^*} p(\boldsymbol{\beta}|\mathbf{y}^*, \mathbf{X}) p(\mathbf{y}^*|\mathbf{Y}, \mathbf{X}, \boldsymbol{\beta}) d\mathbf{y}^*$$

with the integration that marginalizes out the latent data performed via Monte Carlo methods, as described in Section 3.3. MDA takes this idea a step further, by introducing a 'working parameter', say α, that appears in the predictive density for the augmented data. It is important to stress that the working parameter α plays no role in the model of substantive interest, and indeed, is unidentified in that model, but is identified conditional on the augmented data (and only appears in the predictive density for the augmented data). That is, while data augmentation generates augmented data by sampling from $p(\mathbf{y}^*|\mathbf{y}, \mathbf{X}, \boldsymbol{\beta})$, marginal data augmentation involves sampling from $p(\mathbf{y}^*|\mathbf{y}, \mathbf{X}, \boldsymbol{\beta}, \alpha)$. This scheme is referred to as *marginal* data augmentation because we obtain the usual data augmentation predictive density by marginalizing with respect to the working parameter α, i.e.

$$p(\mathbf{y}^*|\mathbf{y}, \mathbf{X}, \boldsymbol{\beta}) = \int p(\mathbf{y}^*|\mathbf{y}, \mathbf{X}, \boldsymbol{\beta}, \alpha) p(\alpha) d\alpha.$$

The advantage of MDA over DA is that the resulting MCMC algorithm over $\boldsymbol{\beta}$ displays better mixing than the DA algorithm. That is, it takes fewer iterations to provide a thorough exploration of the posterior density of $\boldsymbol{\beta}$, because the algorithm is more efficiently traversing the parameter space under the posterior density. Although each MDA iteration will involve extra computational effort than a DA iteration, this is usually more than made up for by the better mixing of the MDA algorithm.

At first glance this seems paradoxical: how can introducing an extra working parameter α actually make for a more efficient algorithm? The trick here is to recall that the working parameter α is unidentified in, and auxilary to, the model of substantive interest; α can be thought of as 'partially decoupling' the augmented data from the observed data, making the Markov chain for the identified parameters $\boldsymbol{\beta}$ move further in its parameter space than we would obtain under DA, and so in turn generating a faster exploration of the posterior density for $\boldsymbol{\beta}$. MDA has been successfully deployed in numerous settings (for a discussion, see van Dyk and Meng 2001) and indeed, probit models provide a vivid and simple demonstration of the power of this technique; MDA will resurface in the context of the multinomial probit model (Section 8.4).

MDA for the probit model

The key to making MDA work well – or the 'art', as van Dyk and Meng (2001) put it – is the choice of the working parameter α. For the case of a probit model, there is a fairly obvious choice as to what the working parameter should be. Recall that the probit model is only identified to up to scale: that is, in the unidentified probit model $y_i^*|\mathbf{x}_i \sim N(\mathbf{x}_i\tilde{\boldsymbol{\beta}}, \sigma^2)$ the parameter $\boldsymbol{\beta} = \tilde{\boldsymbol{\beta}}/\sigma$ is identified. This suggests that choosing σ as the working parameter will be a fruitful way to implement a marginal data augmentation algorithm for this model. Given that we introduce the unidentified variance parameter σ^2 into the model, we require it be given a prior density, and we adopt a conditionally conjugate inverse-Gamma prior, i.e. $\sigma^2 \sim$ inverse-Gamma$(\nu_0/2, \nu_0\sigma_0^2/2)$. Different choices of the prior hyper-parameter ν_0 and σ_0^2 induce different MDA algorithms, and van Dyk and Meng (2001) prove that the most efficient choice (in terms of the performance of the resulting algorithm) is the improper prior for σ^2 given when $\nu_0 = 0$, i.e. $p(\sigma^2) \propto \sigma^{-2}$. We retain the normal prior for $\tilde{\boldsymbol{\beta}}$, i.e. $\tilde{\boldsymbol{\beta}} \sim N(\mathbf{b}_0, \mathbf{B}_0)$, noting that $\tilde{\boldsymbol{\beta}}$ and σ^2 are independent *a priori*. Special care needs to be taken in the case where one wishes to incorporate meaningful prior information over the identified parameter vector $\boldsymbol{\beta} = \tilde{\boldsymbol{\beta}}/\sigma$; note that the $N(\mathbf{b}_0, \mathbf{B}_0)$ prior over $\tilde{\boldsymbol{\beta}}$ is not a prior over $\boldsymbol{\beta}$.

With σ^2 as the working parameter, a marginal data augmentation/MCMC algorithm is as follows, corresponding to 'Scheme 1' in van Dyk and Meng (2001) and Imai and van Dyk (2005a, 317–318): i.e. at iteration t,

1. sample $\tilde{y}_i^{*(t)}$ from $p(\tilde{y}_i^*|\boldsymbol{\beta}^{(t-1)}, \mathbf{x}_i, y_i) = \int p(\tilde{y}_i^*|\boldsymbol{\beta}, \sigma^2, \mathbf{x}_i, y_i)p(\sigma^2|\boldsymbol{\beta})d\sigma^2$; the notation $\tilde{y}^*$ indicates that we are working with the version of the model with unidentified parameters. We accomplish this sampling by first sampling σ^2 from $p(\sigma^2|\boldsymbol{\beta}) = p(\sigma^2)$, i.e. sampling σ^2 from its prior, which here is a diffuse-yet-proper inverse-Gamma density. Note that the draw of σ^2 will be discarded at the end of this particular step in the MDA algorithm, consistent with the idea that we are generating latent data after marginalizing with respect to the working parameter σ^2. That is, after sampling $\sigma^2 \sim$ inverse-Gamma$(\nu_0/2, \nu_0\sigma_0^2/2)$, we then sample

$$\tilde{y}_i^{*(t)}|\left(y_i = 0, \mathbf{x}_i, \boldsymbol{\beta}^{(t-1)}, \sigma^2\right) \sim N\left(\sigma\mathbf{x}_i\boldsymbol{\beta}^{(t-1)}, \sigma^2\right) \mathcal{I}(\tilde{y}_i^{*(t)} < 0)$$
$$\tilde{y}_i^{*(t)}|\left(y_i = 1, \mathbf{x}_i, \boldsymbol{\beta}^{(t-1)}, \sigma^2\right) \sim N\left(\sigma\mathbf{x}_i\boldsymbol{\beta}^{(t-1)}, \sigma^2\right) \mathcal{I}(\tilde{y}_i^{*(t)} \geq 0)$$

2. sample $\boldsymbol{\beta}^{(t)}$ from $p(\boldsymbol{\beta}, \sigma^2|\mathbf{y}^{*(t)}, \mathbf{X})$, again discarding the sampled value of the working parameter σ^2. We momentarily work in terms of the unidentified parameter $\tilde{\boldsymbol{\beta}}$, making use of the identity $p(\tilde{\boldsymbol{\beta}}, \sigma^2|\mathbf{y}^*, \mathbf{X}) = p(\tilde{\boldsymbol{\beta}}|\sigma^2, \mathbf{y}^*, \mathbf{X})p(\sigma^2|\mathbf{y}^*, \mathbf{X})$. Thus, to generate a sample from $p(\tilde{\boldsymbol{\beta}}|\sigma^2, \mathbf{y}^*, \mathbf{X})$ we first sample σ^2 from $p(\sigma^2|\mathbf{y}^{*(t)}, \mathbf{X})$ and then sample $\tilde{\boldsymbol{\beta}}$ from $p(\tilde{\boldsymbol{\beta}}|\sigma^2, \mathbf{y}^{*(t)}, \mathbf{X})$, where $\tilde{\boldsymbol{\beta}}$ is the unidentified parameter. At the end of this step in the algorithm we will map back to the identified parameter $\boldsymbol{\beta}^{(t)} = \tilde{\boldsymbol{\beta}}/\sigma$.
   In 'Scheme 1' of Imai and van Dyk (2005a) we use the prior $\boldsymbol{\beta} \sim N(\mathbf{0}, \mathbf{B}_0)$; see Problem 8.2. Then compute $\mathbf{B}_1 = (\mathbf{B}_0^{-1} + \mathbf{X}'\mathbf{X})^{-1}$, $\mathbf{b}_1 = \mathbf{B}_1\mathbf{X}'\mathbf{y}^{*(t)}$ and $S = \sum_{i=1}^n(\mathbf{y}_i^{*(t)} - \mathbf{x}_i\mathbf{b}_1)^2$. We then sample $\sigma^2 \sim$ inverse-Gamma$(\nu_1/2, \nu_1\sigma_1^2/2)$ where $\nu_1 = \nu_0 + n$ and $\nu_1\sigma_1^2 = \nu_0\sigma_0^2 + S + \mathbf{b}_1'\mathbf{B}_0^{-1}\mathbf{b}_1$; again, see Problem 8.2. Then sample $\tilde{\boldsymbol{\beta}} \sim N(\mathbf{b}_1, \sigma^2\mathbf{B}_1)$ and finally set $\boldsymbol{\beta}^{(t)} = \tilde{\boldsymbol{\beta}}/\sigma$.

■ Example 8.3

Probit model, voting on Iraq War use of force resolution, Bayesian analysis via marginal data augmentation. On October 11, 2002, the United States Senate voted 77–23 to authorize the use of military force against Iraq. The only Republican to vote against the resolution was Lincoln Chafee (Rhode Island); Democrats split 29–22 in favor of the resolution. In addition to senators' party affiliations, constituents' preferences are another plausible determinant of the votes cast on this resolution. A widely used proxy for constituent preferences is the state-wide share of the presidential vote won by one of the major party candidates. We use Al Gore's share of the presidential vote in 2000 (the presidential election immediately preceding the 2002 'use of force' vote) as a proxy for the ideological disposition of each state; this variable ranges from a low of 26.34 % in Utah, to a high of 60.99 % in Rhode Island (the District of Columbia is excluded since it does not have representation in the Senate), with a median of 46.46 % recorded in Ohio. Our expectation is that as Gore's vote share increases, the likelihood that the state's senators vote for the war diminishes, net of the effects associated with party affiliation.

We estimate a standard probit model to recover maximum likelihood estimates (e.g. the `glm` function in `R`), as well as the two Bayesian analyses (DA and MDA) using vague $N(0, 1000)$ priors over each element of $\boldsymbol{\beta}$. For MDA, the working parameter σ^2 is given a diffuse prior ($\nu_0 = 1, \sigma_0^2 = 1$). For purposes of comparison, the data augmentation algorithm of Albert and Chib (1993) is also used, again with improper, uninformative priors on $\boldsymbol{\beta}$. Table 8.2 summarizes the output of the samplers, using the last 9000 iterations of a 10 000 iteration run, with cell entries indicating MLEs and their standard errors, and means and standard deviations of the sampled values for the columns for the two Bayesian algorithms.

Figure 8.5 compares the performance of the two algorithms for these data, showing the autocorrelation functions for each element of $\boldsymbol{\beta}$ (rows), and for both algorithms (columns). For each of the three parameters it is apparent that the MDA algorithm is generating a more efficient exploration of the parameter space than the DA algorithm, as evidenced by the respective autocorrelation functions. The consequence is that if we wish to generate a summary of the posterior density at a given level of precision (e.g. a quantile estimate), then the MDA algorithm will require fewer iterations than the DA algorithm. Of course, each iteration of the MDA algorithm requires a little more computing than an iteration

Table 8.2 Summary statistics, probit analysis of senate voting for the use of force in Iraq, Example 8.3. For the two Bayesian analyses, the table entries are the means and standard deviations of the respective marginal posterior densities, as recovered by the indicated data-augmented Gibbs samplers.

	MLE		Marginal DA		DA	
	Estimate	Std Error	Mean	Std Dev	Mean	Std Dev
Intercept	3.51	1.29	3.68	1.27	3.67	1.25
Gore Vote	−.068	.026	−.071	.025	−.071	.025
Republican Senator	1.68	.52	1.85	.56	1.90	.60

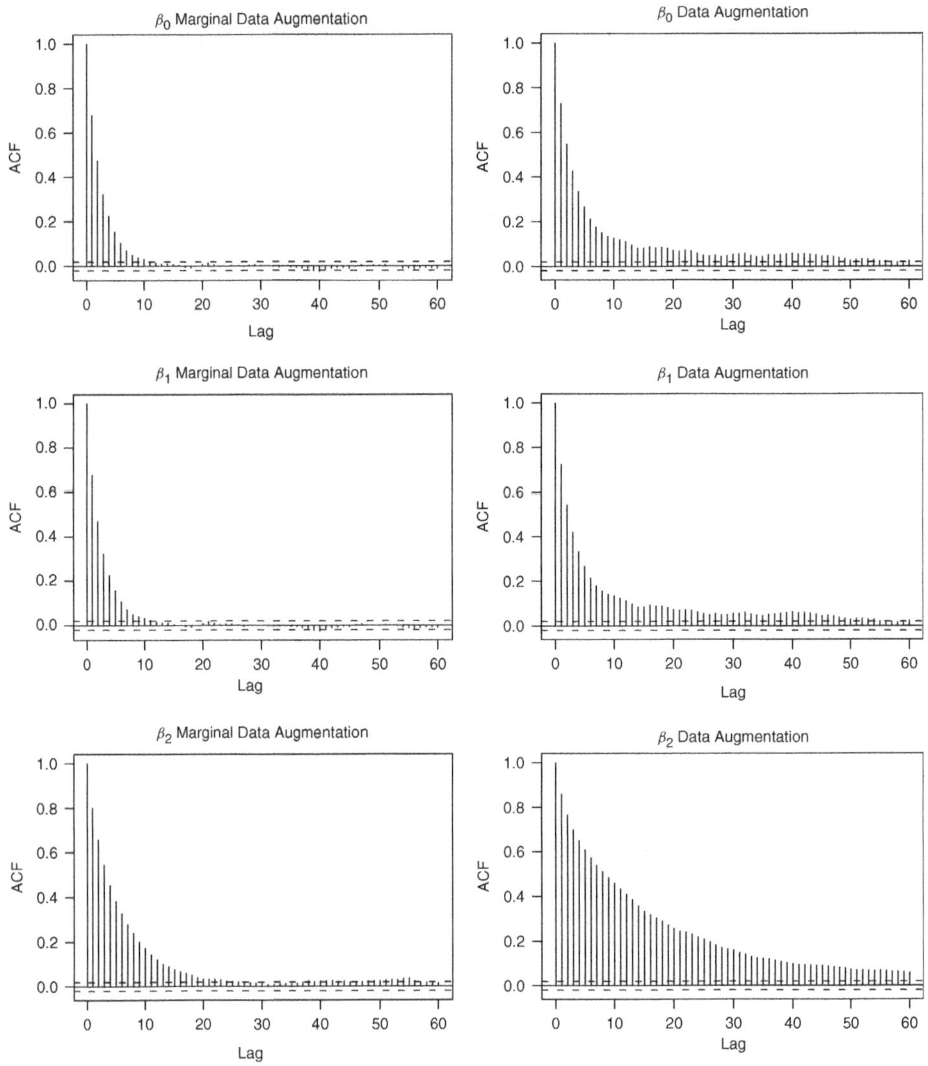

Figure 8.5 Comparison of MCMC output geneated under marginal data augmentation and data augmentation, autocorrelation functions for $\boldsymbol{\beta}$ (probit regression coefficients) from Example 8.3.

of data augmentation, but only a little; and so, at least for this example, marginal data augmentation can be said to outperform data augmentation.

As the preceeding example makes clear, the gains from marginal data augmentation are not always large, and will vary from parameter to parameter; van Dyk and Meng (2001) provide more details and references to the literature. A helpful explanation of the benefits of MDA in the probit case also appears in Liu (2001). We will encounter MDA again when considering the multinomial probit model (Section 8.3).

8.1.3 Logit model

Data augmentation (and marginal data augmentation) makes a Bayesian treatment of the probit model exceptionally easy. Matters are not so straightforward with a logit model. If we have the logistic binary response model $y_i|\mathbf{x}_i \sim \text{Bernoulli}(\pi_i)$, $\pi_i = F(\mathbf{x}_i\boldsymbol{\beta}) = (1 + \exp(-\mathbf{x}_i\boldsymbol{\beta}))^{-1}$, then the latent linear regression form of the model is as in Equation 8.2, except that the latent variables y_i^* have logistic densities (see Definition B.32), with mean $\mathbf{x}_i\boldsymbol{\beta}$, and appropriately truncated given the observed y_i. The difficulty is that with this model for the y_i^*, there is no corresponding conjugate prior for $\boldsymbol{\beta}$. That is, if the y_i^* follow a logistic density, and, say, we use a normal prior for the $\boldsymbol{\beta}$, then the conditional distribution for the $\boldsymbol{\beta}$, $p(\boldsymbol{\beta}|\mathbf{y}^*, \mathbf{X})$, is not a standard distribution that can be easily sampled from. There are two alternative approaches in the literature:

1. Approximate the logistic density for the y_i^* with a finite mixture of normals. With a normal prior on $\boldsymbol{\beta}$ (or an uninformative, improper reference prior), and given a particular normal component of this mixture density, then the conditional distribution of $\boldsymbol{\beta}$ is normal. Averaging over these normals produces a draw from the conditional distribution of $\boldsymbol{\beta}$ given $\mathbf{y}^*$ and $\mathbf{X}$. This idea of approximating a heavy-tailed distribution through a mixture of normals appears in numerous places in the literature – for instance, in the context of stochastic volatility models for financial time series (Chib, Nardari and Shephard 2002; Kim, Shephard and Chib 1998) – and has been recently deployed in the context of logistic regression models (Frühwirth-Schnatter and Frühwirth 2007; Holmes and Held 2006). This idea is also closely related to work by Albert and Chib (1993), who considered Bayesian inference for binary response models with a t_ν distribution as the link function, with the degrees of freedom parameter ν an unknown parameter: $\nu \approx$ 8 or 9 corresponds to a logistic model.

2. Rejection-sampling or Metropolis methods. In this case we dispense with the latent data representation of the model and data-augmentation, and consider a 'direct assault' on the posterior density for $\boldsymbol{\beta}$, $p(\boldsymbol{\beta}|\mathbf{y}, \mathbf{X}) \propto p(\mathbf{y}|\mathbf{X}, \boldsymbol{\beta})p(\boldsymbol{\beta})$. This is how Bayesian analysis software such as `BUGS` or `JAGS` deals with logistic regression models. A random-walk Metropolis algorithm approach is implemented in `R` in the `MCMCpack` package which we investigate in the following example.

■ **Example 8.4**

Voter turnout, logit link via a Metropolis algorithm (Example 8.1, continued). We revisit the voter turnout example with a logit link function. As in Example 8.1 we use an improper, uniform prior so as to compare the results of the MCMC-based approach with the easily computed maximum likelihood estimates. The `MCMClogit` function in the `R` package `MCMCpack` (Martin, Quinn and Park 2009) implements Bayesian analysis of this model via a random-walk Metropolis algorithm.

Figure 8.6 shows the Metropolis alogrithm at work on this problem, for an initial 1000 iterations. The apparent 'step functions' in the traceplots result from the fact that the Metropolis algorithm does not always move to a new location in the parameter space at every iteration (recall the discussion in Section 5.1). The right panels of Figure 8.6 show autocorrelation functions for several parameters; the slow mixing of the Metropolis

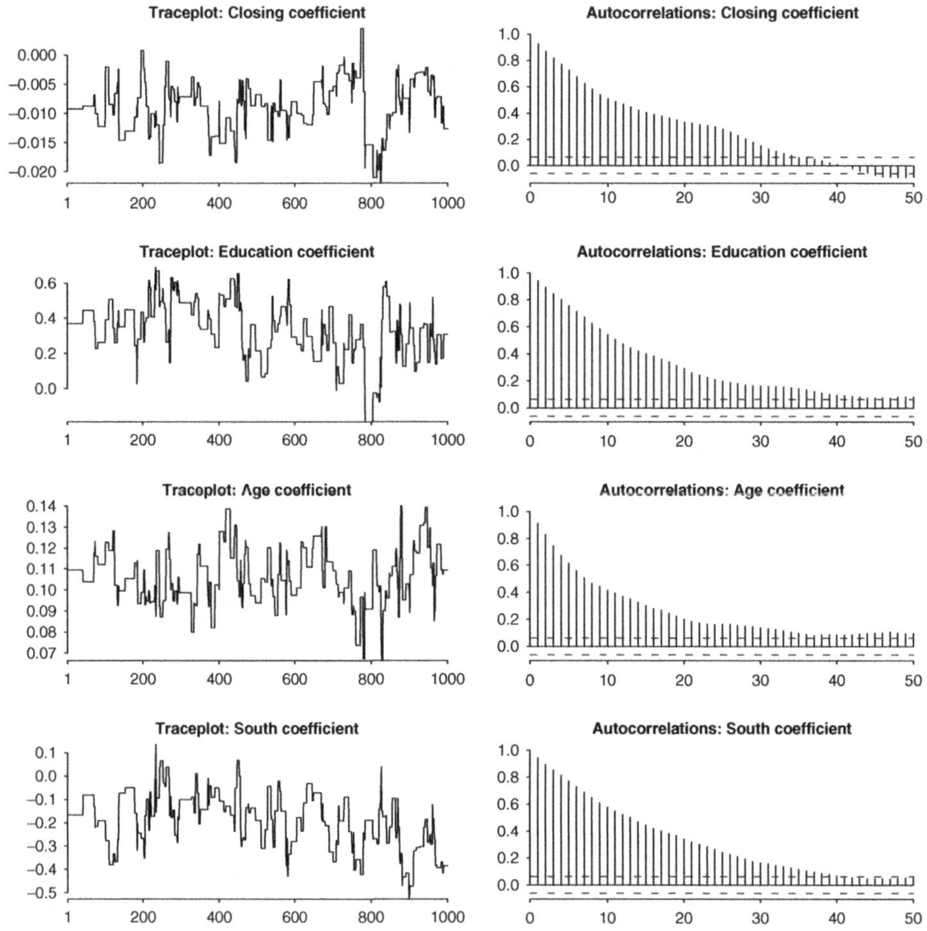

Figure 8.6 Graphical inspection of random-walk Metropolis algorithm, logit model of voter turnout (Example 8.4). Recall that the Metropolis algorithm does not necessarily jump at every iteration, giving rise to the 'steps' in the trace plots (left panels) and slow mixing (the slow decay in the autocorrelation functions, right panels). A substantially longer run than the 1000 iterations shown here will be necessary so as to generate a valid characterization of the posterior density of the model parameters.

algorithm is clearly obvious from the slow decay of the autocorrelations. Raftery-Lewis diagnostics (Raftery and Lewis 1992a) suggest that over 100 000 iterations may be required to get good estimates of relatively extreme quantiles of the marginal posterior densities of $\boldsymbol{\beta}$. Happily, the random-walk Metropolis algorithm for the logistic regression model in `MCMCpack` is not computationally burdensome, and letting the MCMC algorithm generate this many iterations is not particularly time consuming. Table 8.3 shows a comparison of the maximum likelihood estimates and the results of running the Metropolis sampler for 150 000 iterations (after a 1000 iteration burn-in), retaining every 25th sample for inference. The two sets of results are virtually indistinguishable, as they should be: with an improper, uniform prior, the posterior density being explored by the

Table 8.3 Comparison of MLEs and output of the Metropolis sampler, logit model of voter turnout. For the Metropolis sampler (MCMC), the averages of 6000 samples (150 000 iterations, thinned by 25) are reported as point estimates, standard deviations are reported in parentheses and the 2.5 and 97.5 percentiles are reported in brackets; the frequentist 95 % confidence intervals around the MLEs (assuming asymptotic normality) are also reported in brackets.

	MLE	MCMC
Intercept	−4.37	−4.39
	(0.464)	(0.472)
	[−5.28, −3.46]	[−5.34, −3.49]
Closing	−0.00922	−0.00932
	(0.00425)	(0.00422)
	[−0.0176, −0.000891]	[−0.0176, −0.00116]
Education	0.37	0.372
	(0.142)	(0.143)
	[0.0908, 0.648]	[0.0927, 0.654]
Education^2	0.0156	0.0156
	(0.0144)	(0.0143)
	[−0.0126, 0.0438]	[−0.0125, 0.0439]
Age	0.110	0.110
	(0.0129)	(0.0129)
	[0.0843, 0.135]	[0.0852, 0.136]
Age^2	−0.000767	−0.000771
	(0.000135)	(0.000134)
	[−0.00103, −0.000503]	[−0.00104, −0.00051]
South	−0.166	−0.168
	(0.102)	(0.102)
	[−0.367, 0.0347]	[−0.372, 0.0299]
Gubernatorial Election	0.113	0.113
	(0.111)	(0.111)
	[−0.105, 0.331]	[−0.104, 0.33]

Metropolis sampler is (up to a constant of proportionality) the likelihood function; in a large sample, the likelihood will have a normal shape, meaning that the maximum likelihood estimates (the location of the mode of the likelihood function) will correspond to the mean of the posterior density being explored by the random-walk Metropolis sampler.

8.1.4 Binomial model for grouped binary data

Micro-level, binary data can often be aggregated to form group-level binomial data. This can often result in tremendous gains in computational efficiency. For instance, suppose we have the model $y_i|\mathbf{x}_i \sim \text{Bernoulli}(F[\mathbf{x}_i\boldsymbol{\beta}])$, where $\mathbf{x}_i$ is a vector of covariates. Consider a set $\mathcal{C} = \{i : \mathbf{x}_i = \mathbf{x}_\mathcal{C}\}$ i.e. the set of respondents who have covariate vector $\mathbf{x}_\mathcal{C}$. The model does not distinguish between respondents in this set; in fact, our probability assignments over y_i for these respondents are identical and conditionally exchangeable given their common $\mathbf{x}_i$ and $\boldsymbol{\beta}$. Accordingly, an equivalent model for these respondents is the binomial model $r_\mathcal{C} \sim \text{Binomial}(p_\mathcal{C}; n_\mathcal{C})$, where $p_\mathcal{C} = F(\mathbf{x}_\mathcal{C}\boldsymbol{\beta})$, $r_\mathcal{C} = \sum_{i \in \mathcal{C}} y_i$ is the number of 'successes' in $\mathcal{C}$ and $n_\mathcal{C}$ is the cardinality of $\mathcal{C}$. The implication is that we can replace the $n_\mathcal{C}$ binary observations with one binomial observation, greatly easing the computational burden needed to store the data and to compute the likelihood or posterior densities over $\boldsymbol{\beta}$.

The set $\mathcal{C}$ is known as a *covariate class* and we can form a covariate class $\mathcal{C}_c$ for each unique combination $c = 1, \ldots, C$ of the covariates. If there is a lot of redundancy in the covariates and/or there is a small number of covariates, then the resulting representation of the data and model can be extremely efficient, since the number of covariate classes C may be much smaller than n. Note there is no loss of information in representing the data this way: independent Bernoulli data *are* binomial data.

■ Example 8.5

Binomial model for grouped binary data, Example 8.1, continued. The full data set analyzed by Nagler (1991) has almost 99 000 observations. We consider a simple model for these data which has just two covariates: age and education, and quadratic terms for both of these predictors. Even though there are almost 99 000 individual-level observations, there are just 636 unique combinations of age and education. Programs like `BUGS` or `JAGS` can sample from the posterior density for $\boldsymbol{\beta}$ given a likelihood over 636 binomial observations much more quickly than if the likelihood calculations require dealing with almost 99 000 individual-level binary observations. Collapsing the predictors into covariate classes is reasonably easy in `R`; the following code chunk creates a data frame `groupedData` from the unique combinations of `age` and `educYrs` in the data frame `nagler`. The vectors `r` and `n` are the numbers of 'successes' and 'trials' in each of covariate class.

R code

```
## collapse by covariate classes
X <- cbind(nagler$age,nagler$educYrs)
X <- apply(X,1,paste,collapse=":")
covClasses <- match(X,unique(X))
covX <- matrix(unlist(strsplit(unique(X),":")),ncol=2,byrow=TRUE)
r <- tapply(nagler$turnout,covClasses,sum)
n <- tapply(nagler$turnout,covClasses,length)
groupedData <- list(n=n,r=r,
                    age=as.numeric(covX[,1]),
                    educYrs=as.numeric(covX[,2]),
                    NOBS=length(n))
```

We can then pass the `groupedData` data frame to `JAGS`. We specify the binomial model $r_i \sim \text{Binomial}(p_i; n_i)$ with $p_i = F(\mathbf{x}_i\boldsymbol{\beta})$ and vague normal priors on $\boldsymbol{\beta}$ with the following code:

JAGS code

```
model{
  for (i in 1:NOBS){
      logit(p[i]) <- beta[1] + age[i]*beta[2]
                             + pow(age[i],2)*beta[3]
```

```
                    + educYrs[i]*beta[4]
                    + pow(educYrs[i],2)*beta[5]
      r[i] ~ dbin(p[i],n[i]) ## binomial model for each covariate class
   }


         beta[1:5] ~ dmnorm(b0[],B0[,])
}

```

Note that the hyperparameters b0 and B0 are defined in R before we invoke BUGS/JAGS. In this case `b0 <- rep(0,5)` and `B0 <- diag(.001,5)`.

8.2 Ordered outcomes

We now consider dependent variables that are ordinal. For instance, it is customary to employ a 7-point scale when measuring party identification in the US, assigning the numerals $\{0, \ldots, 6\}$ to the categories {'Strong Republican', 'Weak Republican', ..., 'Strong Democrat'}. Note that the difference between 0 and 2 on the coded party identification scale (moving from 'Strong Republican' to 'Republican Leaner') may be quite different from the difference between 2 and 4 ('Republican Leaner' to 'Democrat Leaner'), or 4 and 6 ('Democrat Leaner' to 'Strong Democrat'). These ordinal variables are sometimes also called 'polychotomous' (as opposed to 'dichotomous'). Survey data is a common source of ordinal responses, where respondents are asked to choose one from a number of ordered options designed to tap the respondent's degree of agreement for particular propositions, or strength of a particular attitudinal state.

It should be obvious that ordinary regression analysis is typically inappropriate for ordinal response variables. For one thing, regression is an incoherent model for discrete variables, making continuous predictions that are impossible given the discrete data. An extension of the generalized linear models we encountered for binary data is the typical approach for modeling ordinal data with predictors. By the same token, once the number of ordinal categories exceeds seven, it is customary to treat the variable as continuous.

Here we will use a cumulative link GLM, the most frequently used ordinal response model in political science and economics; see Agresti (2002) or McCullagh and Nelder (1989) for alternative ways of specifying the link function for ordinal data models. The popularity of the cumulative link model dates back to work by McKelvey and Zavoina (1975) on a probit version of the cumulative link model, although an identical model had been proposed earlier by Aitchison and Silvey (1957).

The layout of the model will be familiar given the discussion of data augmentation for binary responses (Section 8.1.1). We begin with the latent linear regression function

$$y_i^* = \mathbf{x}_i \boldsymbol{\beta} + \epsilon_i, \quad \epsilon_i \sim N(0, \sigma^2), \quad i = 1, \ldots, n. \tag{8.3}$$

The assumption of normality for ϵ_i generates the probit version of the model; a logistic density generates the ordinal logistic model. We map from this latent regression to the observed ordinal responses y_i with the following censoring scheme:

$$
\begin{aligned}
y_i = 0 &\iff y_i^* \leq \tau_1 \\
y_i = j &\iff \tau_j < y_i^* \leq \tau_{j+1}, \quad j = 1, \ldots, J-1 \\
y_i = J &\iff y_i^* > \tau_J
\end{aligned}
$$

where the threshold parameters obey the ordering constraint $\tau_1 < \tau_2 < \ldots < \tau_J$. The probabilities of the outcomes are:

$$
\begin{aligned}
\Pr[y_i = 0] &= \Pr[y_i^* \leq \tau_1], \quad \text{and substituting from Equation 8.3,} \\
&= \Pr[\mathbf{x}_i\boldsymbol{\beta} + \epsilon_i \leq \tau_1] = \Pr[\epsilon_i \leq \tau_1 - \mathbf{x}_i\boldsymbol{\beta}] = \Phi[(\tau_1 - \mathbf{x}_i\boldsymbol{\beta})/\sigma] \qquad (8.4a) \\
\Pr[y_i = 1] &= \Pr[\tau_1 < y_i^* \leq \tau_2], \\
&= \Pr[\tau_1 < \mathbf{x}_i\boldsymbol{\beta} + \epsilon_i \leq \tau_2], = \Pr[\tau_1 - \mathbf{x}_i\boldsymbol{\beta} < \epsilon_i \leq \tau_2 - \mathbf{x}_i\boldsymbol{\beta}], \\
&= \Phi[(\tau_2 - \mathbf{x}_i\boldsymbol{\beta})/\sigma] - \Phi[(\tau_1 - \mathbf{x}_i\boldsymbol{\beta})/\sigma]. \qquad (8.4b)
\end{aligned}
$$

It is straightforward to see that $\Pr[y_i = 2] = \Phi[(\tau_3 - \mathbf{x}_i\boldsymbol{\beta})/\sigma] - \Phi[(\tau_2 - \mathbf{x}_i\boldsymbol{\beta})/\sigma]$, and that generically $\Pr[y_i = j] = \Phi[(\tau_{j+1} - \mathbf{x}_i\boldsymbol{\beta})/\sigma] - \Phi[(\tau_j - \mathbf{x}_i\boldsymbol{\beta})/\sigma]$. For $j = J$ (the 'highest' category) we have

$$
\Pr[y_i = J] = 1 - \Phi[(\tau_J - \mathbf{x}_i\boldsymbol{\beta})/\sigma]. \qquad (8.5)
$$

Figure 8.7 provides a graphical depiction of how the ordinal probit model works: i.e. the normal density for y_i^* is being cut into a set of mutually exclusive and exhaustive regions, each region corresponding to the probability of observing the corresponding ordinal response.

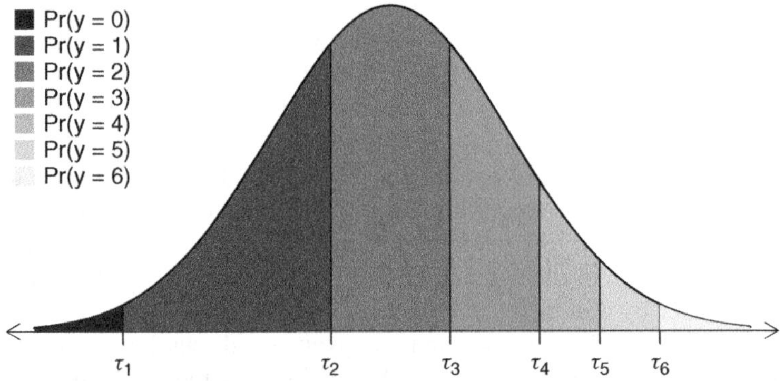

Figure 8.7 Graphical depiction, cumulative link ordinal probit model. The normal curve in the graph is the density of $y_i^*|\mathbf{x}_i, \boldsymbol{\beta}$, centered on $\mathbf{x}_i\boldsymbol{\beta}$ and with variance σ^2 (usually set to 1.0 so as to identify the model). The threshold parameters τ partition the support of y_i^* such that the shaded regions under the normal curve correspond to the probability of a particular outcome, e.g. $\Pr(y_i = 1|\mathbf{x}_i, \boldsymbol{\beta}) = \Pr(\tau_1 < y_i^* \leq \tau_2|\mathbf{x}_i, \boldsymbol{\beta})$, where $y_i^*|\mathbf{x}_i, \boldsymbol{\beta} \sim N(\mathbf{x}_i\boldsymbol{\beta}, \sigma^2)$.

As a notational convenience, we define an indicator variable Z_{ij}, which equals 1 if $y_i = j$ and 0 otherwise. Then the likelihood for this model is simply

$$\mathcal{L} = \prod_{i=1}^{n} \prod_{j=0}^{J} p_{ij}^{Z_{ij}}$$

where $p_{ij} = \Pr(y_i = j)$, as defined above.

8.2.1 Identification

As written, the model's parameters are not identified. Any change in the scale parameter σ can be offset by changes in τ and $\boldsymbol{\beta}$. In addition, the thresholds $\boldsymbol{\tau}$ and any intercept parameter in $\boldsymbol{\beta}$ are not jointly identified: i.e. consider shifting all the threshold parameters up or down, with an offsetting shift in the intercept component of $\boldsymbol{\beta}$. For these reasons the model is typically identified with one of the sets of normalizing constraints outlined in Table 8.4. Perhaps the most common identifying normalization is the second listed option, setting the scale of the error density to a fixed constant (e.g. $\sigma = 1$) and dropping the intercept from the model.

It is worth noting that it is sometimes possible to re-define the latent variable y_i^* as a substantively meaningful quantity, such as money, as in the analyses of congressional roll call voting by Krehbiel and Rivers (1988) and Bartels (1991). In these applications the thresholds are defined as the midpoints between alternatives being voted on, with dollar amounts attaching to each threshold. With two thresholds defined, we are supplying a basis for the latent $y_i^* \in \mathbb{R}$, and the scale parameter σ is identified. Moreover, since the latent scale is interpretable in terms of dollars, the parameters $\boldsymbol{\beta}$ are interpreted as the effects of the predictors on a dollar scale.

Priors

Bayesian inference for the cumulative link ordered response model is only a little more complicated than inference for the binary response models considered earlier. There are two sets of parameters in the model: i.e. $\boldsymbol{\theta} = (\boldsymbol{\beta}, \boldsymbol{\tau})'$ where $\boldsymbol{\tau} = (\tau_1, \ldots, \tau_J)'$. Priors are required for each set of parameters, and the usual approach is to treat $\boldsymbol{\beta}$ and $\boldsymbol{\tau}$ as independent *a priori*; i.e. $p(\boldsymbol{\theta}) = p(\boldsymbol{\beta}, \boldsymbol{\tau}) = p(\boldsymbol{\beta})p(\boldsymbol{\tau})$. Normal priors are typically used for $\boldsymbol{\beta}$, e.g. $\boldsymbol{\beta} \sim N(\mathbf{b}_0, \mathbf{B}_0)$.

Any prior for the $\boldsymbol{\tau}$ has to be coherent, in the sense that the prior must respect the ordering constraint implied by the model, i.e. $\tau_j > \tau_{j-1}, \forall j = 2, \ldots, J$. A simple way to

Table 8.4 Ordered probit model, identification constraints.

Normalization	$\boldsymbol{\beta}$	σ	$\boldsymbol{\tau}$
1	unconstrained	fixed e.g., $\sigma = 1$	one τ_j fixed e.g., $\tau_1 = 0$
2	drop intercept	fixed e.g., $\sigma = 1$	unconstrained
3	unconstrained	unconstrained	two τ_j fixed
	e.g. Krehbiel and Rivers (1988), Bartels (1991).		

accomplish this is via the following recursive structure over the priors for $\boldsymbol{\tau}$:

$$\begin{aligned} \tau_1 &\sim N(t_1, T_1) \\ \delta_j &\sim \text{Exponential}(d), \qquad j = 2, \ldots, J, \\ \tau_j &= \tau_{j-1} + \delta_j, \qquad j = 2, \ldots, J, \end{aligned} \tag{8.6}$$

with the analyst supplying values for the hyper-parameters t_1, T_1, and d. Note that the precise choice of prior density here is somewhat arbitrary; the normal density for $p(\tau_1)$ could well be some other density (e.g. a uniform on a specific interval). The 'trick' here is to impose an ordering constraint on the elements of $\boldsymbol{\tau}$, and this is what the exponentially-distributed quantities δ_j accomplish, since the exponential density only has support on the non-negative half of the real line (see Definition B.33). A vague prior over the thresholds can be obtained, say, by setting $t_1 = 0$, T_1 to a reasonably large scalar (e.g. 5^2), and d to a small positive quantity (e.g. 1). The use of the exponential density as part of the prior is somewhat arbitrary; any density with support restricted to the non-negative half line will suffice (e.g. a uniform density on an interval that does not overlap zero, a log-normal density, etc.). The following code block provides an example of this set of priors, for the case of a four category model:

BUGS code

```
    tau[1] ~ dnorm(0,.01)
    for(j in 1:3){
         delta[j] ~ dexp(2)
         tau[j+1] <- tau[j] + delta[j]
    }
```

An improper-yet-coherent prior for the $\boldsymbol{\tau}$ is also possible, and indeed, this was the prior considered by Albert and Chib (1993, 673) in one of the first Bayesian analyses of this model. An improper-yet-coherent prior for $\boldsymbol{\tau}$ is uniform over the polytope $\mathcal{T} \subset \mathbb{R}^J$, the feasible region of the parameter space for $\boldsymbol{\tau}$; i.e.

$$\mathcal{T} = \{\boldsymbol{\tau} = (\tau_1, \ldots, \tau_J)' \in \mathbb{R}^J : \tau_j > \tau_{j-1}, \forall\, j = 2, \ldots, J\} \tag{8.7}$$

The following code fragment approximates this prior in BUGS for the case of a four category model:

BUGS code

```
  tau[1] ~  dnorm(0,.01)I(,tau[2])
  tau[2] ~  dnorm(0,.01)I(tau[1],tau[3])
  tau[3] ~  dnorm(0,.01)I(tau[2],tau[4])
  tau[4] ~  dnorm(0,.01)I(tau[3],)
```

In JAGS we can make use of the `sort` function:

JAGS code

```
for(j in 1:4){
      tau0[j] ~ dnorm(0,.01)
}
tau[1:4] <- sort(tau0)  ## JAGS only, not in WinBUGS!
```

See Example 8.6, below.

Data augmentation for the ordered probit model. The posterior density for $\boldsymbol{\theta} = (\boldsymbol{\beta}, \boldsymbol{\tau})$ generated by the ordinal response model is difficult to compute. As we saw for the case of binary responses, there is no conjugate prior to exploit, and the introduction of the $\boldsymbol{\tau}$ parameters is an additional complication. The Albert and Chib (1993) data-augmented Gibbs sampler we considered for binary response model data is an attractive computational strategy. In this case, each iteration of the algorithm consists of

1. sample y_i^*, $i = 1, \ldots, n$, given $\boldsymbol{\beta}$, σ^2, $\boldsymbol{\tau}$, y_i and $\mathbf{x}_i$ from a truncated normal density, with mean $\mathbf{x}_i\boldsymbol{\beta}$, variance σ^2 (usually set to 1 for identification), and truncated to the interval implied by the observed y_i; i.e.

$$y_i^* | \mathbf{x}_i, \boldsymbol{\beta}, \sigma^2, y_i = j \sim N(\mathbf{x}_i\boldsymbol{\beta}, \sigma^2)\,\mathcal{I}(\tau_{y_i} < y_i^* \leq \tau_{y_i+1})$$

   where we define $\tau_0 = -\infty$ for the case of $y_i = 0$ and $\tau_{J+1} = \infty$ for the case of $y_i = J$.

2. sample $\boldsymbol{\beta}$ given the latent $\mathbf{y}^*$ and $\mathbf{X}$ from a normal density with mean $\mathbf{b} = (\mathbf{B}_0^{-1} + \mathbf{X}'\mathbf{X})^{-1}(\mathbf{B}_0^{-1}\mathbf{b}_0 + \mathbf{X}'\mathbf{y}^*)$ and variance $\mathbf{B} = (\mathbf{B}_0^{-1} + \mathbf{X}'\mathbf{X})^{-1}$.

3. sample the $\boldsymbol{\tau}$ from their conditional densities, given the latent data $\mathbf{y}^*$. For the improper prior proposed by Albert and Chib (1993), these conditional densities are uniform densities on intervals given by the sampled $\mathbf{y}^*$ and/or the neighboring τ. For instance, for τ_j, sample uniformly from the interval

$$\left[\max\left\{\max(y_i^* | y_i = j-1), \tau_{j-1}\right\}, \quad \min\left\{\min(y_i^* | y_i = j), \tau_{j+1}\right\}\right]$$

   again defining $\tau_0 = -\infty$ for the case of $j = 0$ and $\tau_{J+1} = \infty$ for the case of $j = J$. Cowles (1996) pointed out that this updating scheme for $\boldsymbol{\tau}$ results in extremely slow mixing and proposed a Metropolis scheme for sampling the vector of threshold parameters $\boldsymbol{\tau}$ *en bloc* which is now widely used in implementations of the ordinal probit model; e.g. the `MCMCoprobit` function in the `MCMCpack` `R` package (Martin, Quinn and Park 2009). See also Johnson and Albert (1999). If we are using a proper prior for $\boldsymbol{\tau}$ such as that given in 8.6, slice sampling or Metropolis steps will be needed to sample from the resulting conditional density for $\boldsymbol{\tau}$.

■ Example 8.6

Interviewer ratings of respondents' levels of political information. An important variable in the study of political behavior and public opinion is political information, a person's level of general knowledge of politics, the political parties, political institutions and political processes. A useful indicator of political information comes from the American National Election Studies election-year surveys. At the end of the pre-election interview, the interviewers record their impressions of each respondent's 'general level of information and public affairs' using a 5 point, ordinal scale with the response categories 'Very High', 'Fairly High', 'Average', 'Fairly Low' and 'Very Low'. In 2000, these interviewer rating scores were assigned to exactly 1800 respondents, resulting in the following marginal distribution:

Label	y	n	%
Very Low	0	105	6
Fairly Low	1	334	19
Average	2	586	33
Fairly High	3	450	25
Very High	4	325	18

What explains variation in the interviewer's ratings? For the purposes of exposition, we consider a reasonably simple analysis using the following respondent-specific covariates:

- respondent education; a dummy variable coded 1 if the respondent has a college degree (true for 40 % of the respondents)
- respondent gender; a dummy variable coded 1 if the respondent is female (true for 56 % of the respondents)
- log of the respondent's age (mean age is 47.2 years, with an IQR of 34 years to 58 years)
- is the respondent a home owner?; another dummy variable (true for 67 % of the respondents)
- is the respondent employed in the public sector?; i.e. an employee of the federal government, or of a state or local government (true for 12 % of the respondent).

We also have several variables about the interview and the interviewers which may be helpful in predicting the interviewers' scores. In addition to having a unique identifier for each interviewer (which we will use below), we also know the length of each interview in minutes. A respondent who is skipping many items or is uninterested in the subject matter of the survey (politics) will generate a short interview, while a respondent who is engaged and thoughtful will tend to generate a longer interview. The interview times range from 19 minutes to 234.5 minutes, with a median time of 63.35 minutes; we use log interview time as a predictor of the interviewer ratings. After listwise deletion of cases due to missing data, we lose 10 cases (.55 %), taking the n available for analysis down to 1790.

Both ordinal probit and logistic models are fit to the data. For the probit case, the Bayesian analysis uses improper, uniform priors, and so up to Monte Carlo error, the MCMC results should correspond with the MLEs. The MCMC results for the ordinal probit model are produced using the `MCMCoprobit` function in the `MCMCpack` package (Martin, Quinn and Park 2009) in `R`; note that this model uses the identifying restrictions given in row 1 of Table 8.4, while the MLEs are produced using the normalization in row 2 of Table 8.4. The call to `MCMCoprobit` function is relatively straightforward:

R code

```
require(MCMCpack)
mcmc <- MCMCoprobit(y ~  collegeDegree + female + log(age) +
                    homeOwn + govt + log(length),
                    data=politicalInformation,
```

```
                    burnin=1e4,
                    mcmc=1e5,
                    verbose=1000)
```

Ordinal logistic model via *JAGS/BUGS.* We explore the posterior density $p(\boldsymbol{\beta}, \boldsymbol{\tau}|\mathbf{y}, \mathbf{X})$ for the ordinal logit model with the following JAGS program:

JAGS code

```
1  model{
2          for(i in 1:N){ ## loop over observations
3                ## form the linear predictor (no intercept)
4                mu[i] <- x[i,1]*beta[1] +
5                         x[i,2]*beta[2] +
6                         x[i,3]*beta[3] +
7                         x[i,4]*beta[4] +
8                         x[i,5]*beta[5] +
9                         x[i,6]*beta[6]
10
11               ## cumulative logistic probabilities
12               logit(Q[i,1]) <- tau[1]-mu[i]
13               p[i,1] <- Q[i,1]
14               for(j in 2:4){
15                    logit(Q[i,j]) <- tau[j]-mu[i]
16                    ## trick to get slice of the cdf we need
17                    p[i,j] <- Q[i,j] - Q[i,j-1]
18               }
19               p[i,5] <- 1 - Q[i,4]
20               y[i] ~ dcat(p[i,1:5]) ## p[i,] sums to 1 for each i
21         }
22
23         ## priors over betas
24         beta[1:6] ~ dmnorm(b0[],B0[,])
25
26         ## thresholds
27         for(j in 1:4){
28               tau0[j] ~ dnorm(0, .01)
29          }
30          tau[1:4] <- sort(tau0) ## JAGS only not in BUGS!
31  }
```

Some comments on the JAGS program are warranted. Lines 4 through 9 simply generate the linear predictor `mu[i]`, $\mu_i = \mathbf{x}_i\boldsymbol{\beta}$. Lines 12 through 19 map from `mu[i]` to the conditional probability of observing a given level of y_i given `mu[i]`, implementing the ordinal model in Equations 8.4 to 8.5, albeit now for the logistic case. The `logit` function in JAGS/BUGS sets its argument to the logit of the quantity on the right-hand side of the assignment operator: e.g. at line 12 we have `logit(Q[i,1]) <- tau[1] - mu[i]`, which sets

$$\texttt{Q[i,1]} = F(\texttt{tau[1]} - \texttt{mu[i]}) = \frac{1}{1 + \exp(-[\texttt{tau[1]} - \texttt{mu[i]}])}.$$

For the 'interior' categories $j = 2, 3, 4$, $P(y_i = j) = F(\tau_j - \mu_i) - F(\tau_{j-1} - \mu_i)$, which we operationalize with the loop at lines 14 through 18. We obtain $P(y_i = 5)$ as the complement of $F(\tau_4 - \mu_i)$ at line 19. The result is to form a vector p[i,1:5] which contains the appropriate conditional probabilities of each of the five ordinal outcomes, for each respondent. Line 20 finally links the model to the observed responses, via the dcat density, indicating to JAGS/BUGS that we have a categorical response y[i] with the vector of conditional probabilities p[i,1:5] for each possible response. Although the theoretical discussion of the ordinal response model in the preceding pages presumes that the minimum of $y_i = 0$, both JAGS and BUGS must have $y_i = 1$ as the smallest value of the ordinal response variable; the dcat probability mass function is defined only for positive, integer values. Priors for β and τ complete the program. We specify a multivariate normal prior for beta.

Also note the use of the sort function in the JAGS code above; i.e. we specify priors over a set of unconstrained parameters tau0 (lines 27 to 29), but the sort function imposes an ordering such that tau[1] < tau[2] < ... < tau[4]. In BUGS we would use one of the following methods of specifying ordered priors over the threshold parameters. The first method builds up the tau sequentially, with $\delta_j = (\tau_j - \tau_{j-1}) \sim$ Exponential($j > 1$):

```
                     _____BUGS code_____
1                tau[1] ~ dnorm(0,.01)
2                for(j in 1:3){
3                        delta[j] ~ dexp(2)
4                        tau[j+1] <- tau[j] + delta[j]
5                }
```

Alternatively, we can specify a vague set of priors over the τ parameters that nonetheless obey the ordering restriction:

```
                     _____BUGS code_____
1    tau[1] ~  dnorm(0,.01)I(,tau[2])
2    tau[2] ~  dnorm(0,.01)I(tau[1],tau[3])
3    tau[3] ~  dnorm(0,.01)I(tau[2],tau[4])
4    tau[4] ~  dnorm(0,.01)I(tau[3],)
```

For completeness, I also demonstrate the call to JAGS from R for this problem:

```
                     _____R code_____
1  require(pscl)
2  data(politicalInformation)
3
4  ols <- lm(as.numeric(y) ~ collegeDegree + female + log(age)
5             + homeOwn + govt + log(length),
6             data=politicalInformation,
7             x=TRUE,
8             y=TRUE)
9
10  forJags <- list(y=ols$y,
11                  x=apply(ols$x[,-1],2,function(x)x-mean(x)),
12                  N=length(ols$y),
13                  b0=rep(0,6),
14                  B0=diag(1E-08,6))
15
16  inits <- list(list(beta=coef(ols)[-1],
17                  tau0=2:5))
18
19  require(rjags)
```

```
20
21  foo <- jags.model(file="oLogit.bug",
22                    data=forJags,
23                    inits=inits)
24
25  out <- coda.samples(foo,
26                      variable.names=c("beta","tau"),
27                      n.iter=50000,
28                      thin=10)
```

The code is reasonably straightforward. We load the `pscl` package and access the `politicalInformation` data. A linear regression (line 4) accomplishes several tasks: (1) we remove the small amount of missing data via the default use of listwise deletion in the `lm` function; (2) we store the matrix of regressors `x` and the response vector `y` in the `lm` object, `ols`; (3) the OLS regression coefficients give us plausible start values for `beta` in the MCMC algorithm. We mean-deviate the regressors (line 11) and store values for the hyperparameters of the vague multivariate normal prior over $\boldsymbol{\beta}$, `b0` and `B0`. Line 16 uses the regression coefficients to assign initial values for the `beta` parameters, dropping the intercept parameter which is unidentified in the ordinal model and does not appear in the `JAGS` program. At line 17 we also specify initial values for the threshold parameters `tau0`; note that in the `JAGS` program it is `tau0` that are unknown parameters and so these parameters are assigned initial values rather than the `tau` parameters, which are deterministic functions of the random `tau0`. Line 21 asks `JAGS` to compile the program (stored as `oLogit.bug`) using the data and initial values defined earlier; line 25 requests 50 000 MCMC iterations be generated and that we store every 10th sampled value for `beta` and `tau`. These run lengths and thinning values are generated after experimenting with shorter runs; various convergence diagnostics suggested that runs of this length would be necessary to generate good estimates of the extreme quantiles of the marginal posterior densities of the $\boldsymbol{\beta}$ parameters.

Results. The results for the ordinal logit models appear in columns three and four of Table 8.5. Again there is very little difference between the MLEs and the MCMC estimates of the mean of posterior densities; note that the MCMC implementation of the ordinal logistic model works with centered predictors, which has the effect of shifting the location of the threshold parameters relative to the parameterization used by the MLEs that does not use centered predictors. Note in this case, proper priors are employed for all parameters, with the priors in Equation 8.6 used for the threshold parameters $\boldsymbol{\tau}$, which accounts for some of the differences between the MLEs and the MCMC results. As for the results themselves, all predictors have reasonably large effects that are distinguishable from zero at conventional levels of statistical significance, with the exception of the government employee dummy variable. The effects of college education seem especially large, moving a respondent a distance of .84 on the latent scale, which is about the typical distance between the estimated category thresholds.

It is interesting to examine the performance of the MCMC algorithms for these models. As noted above, the threshold parameters $\boldsymbol{\tau}$ are notorious for being highly correlated with one another and for demonstrating slow mixing. This is the case here, as evidenced by Figure 8.8 for the ordinal probit model and by Figure 8.9 for the logit model. For the ordered probit model implemented in `MCMCpack` we see that the MCMC algorithms are quite efficient at exploring the marginal posterior densities of the slope parameters,

Table 8.5 Analysis of ordinal ratings of political information, ordinal probit and logit models, Example 8.6. MCMC analyses are based on 100 000 iterations discarding the first 10 000 iterations as burn-in for the probit model, and 50 000 iterations for the logit model. MLEs are produced with the `polr` function in the `MASS` package (Venables and Ripley 2002) in `R`; MCMC results for the ordinal probit model are produced with the `MCMCoprobit` function in the `MCMCpack` package (Martin, Quinn and Park 2009) in `R`; MCMC results for the ordinal logit model are produced with `JAGS` and use proper priors over the threshold parameters, see text for details. For the MLEs, quantities in parentheses are estimated asymptotic standard errors; for MCMC output, cell entries are the averages of the iterations retained for inference (MCMC estimates of the means of the marginal posterior densities for each parameters), with standard deviations in parentheses (MCMC estimates of the standard deviation of the marginal posterior density for each parameter). Note that the MCMC implementation of the ordinal logit model uses mean-deviated predictors, shifting the threshold parameters relative to the MLEs.

	Ordinal probit		Ordinal logit	
	MLE	MCMC	MLE	MCMC
Intercept	0	−2.20	0	0
		(.39)		
College degree	.84	.84	1.50	1.46
	(.05)	(.05)	(.10)	(.10)
Female	−.37	−.37	−.66	−.66
	(.05)	(.05)	(.09)	(.09)
log(Age)	.25	.26	.46	.46
	(.07)	(.07)	(.12)	(.12)
Home owner	.27	.26	.45	.45
	(.06)	(.06)	(.10)	(.10)
Government employee	.11	.11	.17	.17
	(.08)	(.08)	(.14)	(.13)
log(Interview Length)	.65	.65	1.10	1.12
	(.09)	(.09)	(.15)	(.15)
Threshold parameters:				
τ_0	2.20	0	3.80	−3.17
	(.39)		(.68)	(.11)
τ_1	3.19	.99	5.60	−1.35
	(.39)	(.05)	(.68)	(.06)
τ_2	4.21	2.01	7.30	0.36
	(.40)	(.06)	(.69)	(.05)
τ_3	5.06	2.86	8.70	1.82
	(.40)	(.06)	(.70)	(.07)

$\boldsymbol{\beta}$, but generate quite slow traversals of the marginal posterior densities for the threshold parameters (n.b., the extremely long decay in the autocorrelation functions for the $\boldsymbol{\tau}$ parameters displayed in Figure 8.8). The ordinal logistic model with no intercept and centered predictors implemented in `JAGS` fares better on this score; see Figure 8.9.

This said, focusing on the MCMC output for ordinal probit model – the model with the slow-mixing threshold parameters – it does appear that the chain is mean stationary,

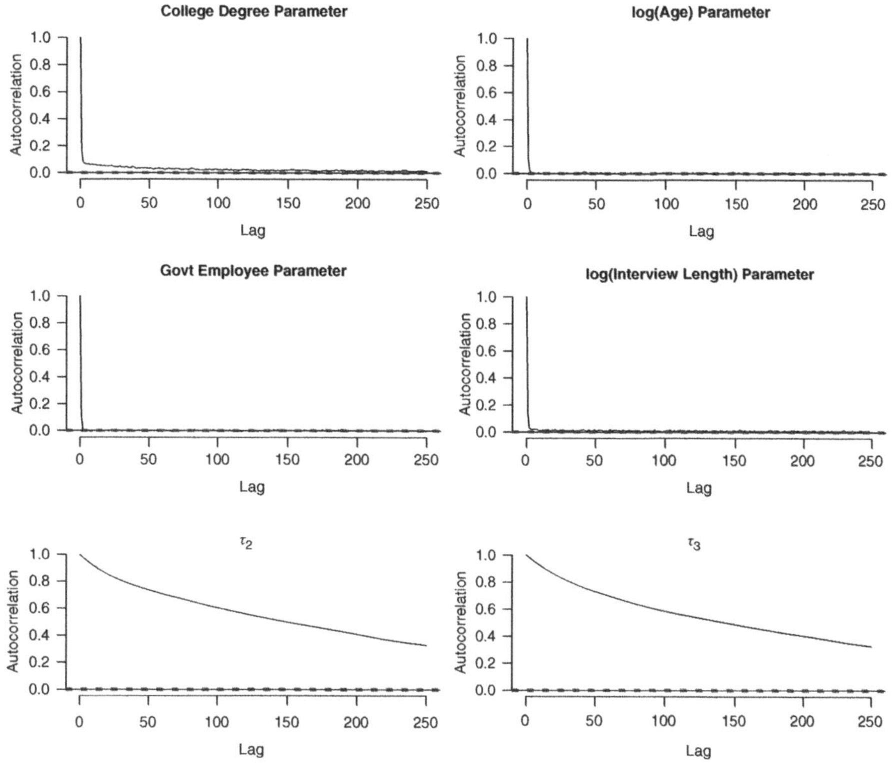

Figure 8.8 Autocorrelation functions, selected parameters, ordinal probit analysis of interviewer ratings of survey respondents' level of political information (Example 8.6). Note the slow mixing for the threshold parameters, τ_2 and τ_3.

at least as indicated by the Geweke z tests, (see Table 8.6). The problem created by a slow mixing MCMC algorithm is with respect to estimating the dispersion of the posterior density, and, in particular, quantiles of the posterior density. This is clearly apparent in Table 8.6, where the Raftery-Lewis diagnostics make it clear that a large number of iterations are required so as to provide good estimates of the 2.5 and 97.5 percentiles of the marginal posterior densities, at least for the threshold parameters.

Re-parameterizing for a more efficient MCMC algorithm. This example would seem to be a case where some of the 'tricks of the trade' from Section 6.4 could be usefully deployed. For a start, we ought to consider centering the predictors around their means, as we did in the logistic version of the ordinal model implemented in `JAGS`. Moreover, a *redundant parameterization* might be worth exploring, say in a modeling environment where that is possible; i.e. this is a case where an implementation in `JAGS/BUGS` might offer some advantages over the problem-specific solutions in the `MCMCpack` package, since the former easily lets us implement over-parameterized models while the latter uses the identified version of the ordinal response model.

We noted in Section 8.2.1 that the model parameters are not identified without restrictions on the threshold parameters and the intercept term. In Problem 8.4 the reader is

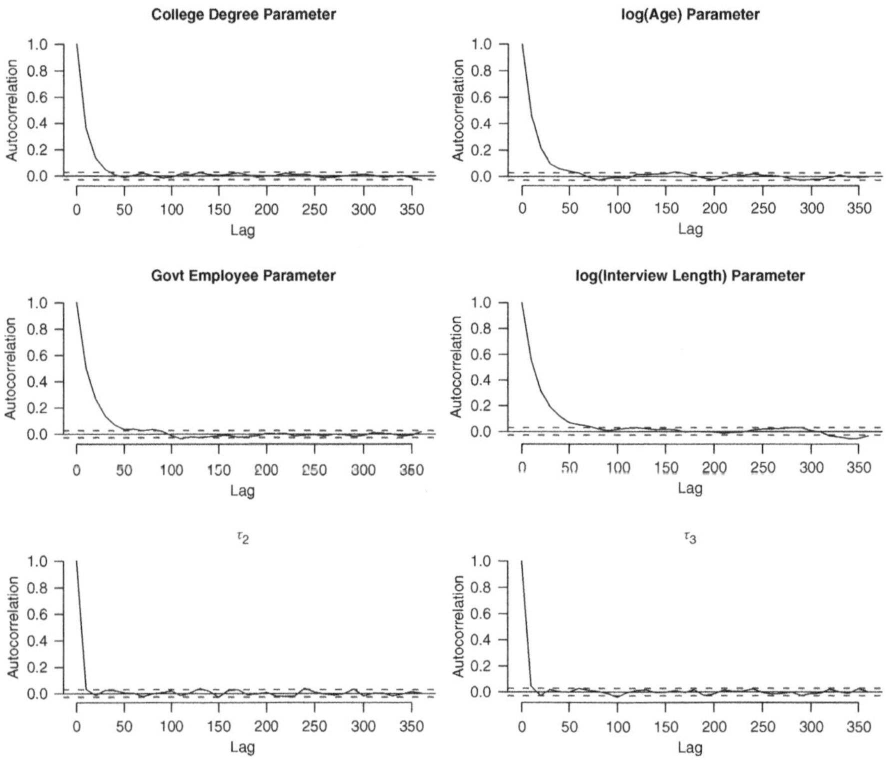

Figure 8.9 Autocorrelation functions, selected parameters, ordinal logit analysis of interviewer ratings of survey respondents' level of political information (Example 8.6). The threshold parameters recovered from this implementation of the ordinal logistic model display much better mixing than those from the probit version of the model (see Figure 8.8).

Table 8.6 Convergence diagnostics, MCMC output, ordinal probit model, Example 8.6.

Parameter	Geweke	AR(1)	Raftery-Lewis N	I
Intercept	−0.31	.12	3913	1.04
College degree	0.16	.17	4021	1.07
Female	−0.33	.13	3935	1.05
log(Age)	−0.21	.11	3821	1.02
Home owner	1.30	.11	3881	1.04
Government employee	0.92	.12	3903	1.04
log(Interview Length)	0.28	.13	3932	1.05
τ_1	−0.11	.98	199 080	53.10
τ_2	0.003	.99	381 810	102.00
τ_3	0.24	.99	354 429	94.60

invited to specify an over-parameterized model in which the intercept and the full set of threshold parameters appear in the model; in Section 7.2.1 we saw that this type of over-parameterized model generates a MCMC algorithm that mixes well in the space of identified parameters. This approach also seems to work well here.

Interviewer effects via scale-use heterogeneity. Thus far, the models fitted to these data ignore any effects that may be specific to each interviewer, relying solely on respondent-specific covariates. Since it is interviewers that are generating the ordinal ratings, it would seem sensible to augment the model with information about the interviewers. The interviewers are presented with the five, ordinal response categories, which are open to considerable subjective interpretation, e.g. interviewer j may have a different standard as to what constitutes a 'Fairly High' rating than interviewer j', and so on, inducing scale-use heterogeneity across interviewers. This argument implies that each interviewer may be using a unique set of thresholds in mapping from the levels of political information they observe in respondent i, y_i^*, into the discrete response categories y_i. We could certainly specify a model with thresholds specific to each interviewer, but this comes with a considerable loss of parsimony, requiring that we estimate four τ parameters for each interviewer. Thus while it is plausible that each interviewer is using a unique set of thresholds, estimating such a model is asking a lot of these data. The 1790 ratings available for analysis here were generated by 115 interviewers and the distribution of interviews across interviewers is quite uneven. Thirty-eight interviewers contribute less than 10 interviews each, while 8 interviewers contribute more than 30 interviews each; 41 is the maximum number of interviews per interviewer; 6 interviewers have 1 interview each and another five interviewers have 2 interviews each. Given this paucity of data, estimating a set of thresholds per interviewer seems overly ambitious.

This is a situation well-suited for Bayesian hierarchical modeling. Without any additional information about the interviewers, we can treat the interviewers as exchangeable, and specify that interview-specific effects are drawn from a common distribution, *a priori*. We initially consider a simple operationalization of 'interviewer effects'. That is, we consider a model in which there is a baseline set of thresholds, but offset on a per interviewer basis via an interviewer-specific parameter η_k, where k indexes the set of interviewers. This constitutes an extremely simple elaboration of the model outlined in Equation 8.3, as follows:

$$\begin{aligned}
\Pr(y_i \geq j) &= F(\tau_j - \mu_i), \quad j = 0, \ldots, J-1 \\
\Pr(y_i = J) &= 1 - F(\tau_{J-1} - \mu_i) \\
\mu_i &= \mathbf{x}_i\boldsymbol{\beta} + \eta_k \\
\eta_k &\sim N(0, \sigma^2)
\end{aligned} \tag{8.8}$$

where k indexes the 115 interviewers, i indexes respondents, j indexes the 5 ordinal response categories, and F is the cumulative distribution function of the normal density for a probit version of the model, or the CDF of the logistic density for a logit version of the model (in the analysis below I use the logistic version of the model). The hierarchical model for the interviewer-specific effects is extremely simple, a normal model with mean zero and unknown variance σ^2; the mean zero restriction is required so as to identify the model parameters, for precisely the same reason that we set the intercept of the model to zero when all the threshold parameters are free parameters (e.g. normalizing restriction 2 in Table 8.4).

Priors over $\boldsymbol{\tau} = (\tau_0, \ldots, \tau_3)'$, $\boldsymbol{\beta}$ and σ^2 complete the model specification. Letting $\boldsymbol{\theta} = \{\boldsymbol{\tau}, \boldsymbol{\beta}, \sigma^2\}$, I assume prior independence among these sets of parameters, $p(\boldsymbol{\theta}) = p(\boldsymbol{\tau})p(\boldsymbol{\beta})p(\sigma^2)$ where

$$\begin{aligned}
\boldsymbol{\beta} &\sim N(\mathbf{0}, 10^2 \cdot \mathbf{I}) \\
\tau_0 &\sim N(0, 10^2)\mathcal{I}(\tau_0 < \tau_1) \\
\tau_j &\sim N(0, 10^2)\mathcal{I}(\tau_{j-1} < \tau_j < \tau_{j+1}), \quad j = 1, 2 \\
\tau_3 &\sim N(0, 10^2)\mathcal{I}(\tau_2 < \tau_3) \\
\sigma &\sim \text{Unif}(0, 2)
\end{aligned}$$

The prior on the standard deviation of the density of the interviewer offsets constrains $\sigma < 2$, both *a priori* and *a posteriori*. At the upper bound of $\sigma = 2$, 95 % of the η_k will lie in the interval $[-4, 4]$, which is not at all restrictive when we recall that the η_k are offsets that operate on the logistic scale, and so this prior encompasses an extremely broad range of plausible interviewer offsets. Notice also the vague normal priors on $\boldsymbol{\beta}$ and the threshold parameters $\boldsymbol{\tau}$, subject to the ordering constraint. The following JAGS program implements this model:

JAGS code

```
model{
        for(i in 1:N){    ## loop over observations
              ## form the linear predictor
              mu[i] <- x[i,1]*beta[1] +
                       x[i,2]*beta[2] +
                       x[i,3]*beta[3] +
                       x[i,4]*beta[4] +
                       x[i,5]*beta[5] +
                       x[i,6]*beta[6] + eta[id[i]]

              ## cumulative logistic probabilities
              logit(Q[i,1]) <- tau[1]-mu[i]
              p[i,1] <- Q[i,1]
              for(j in 2:4){
                   logit(Q[i,j]) <- tau[j]-mu[i]
                   p[i,j] <- Q[i,j] - Q[i,j-1]
              }
              p[i,5] <- 1 - Q[i,4]
              y[i] ~  dcat(p[i,1:5]) ## p[i,] sums to 1 for each i
        }

        ## priors over betas
        beta[1:6] ~ dmnorm(b0[],B0[,])

        ## hierarchical model over etas, note zero mean restriction
        for(k in 1:NID){
             eta[k] ~ dnorm(0.0,eta.tau)
        }
        eta.tau <- 1/pow(sigma,2) ## convert stddev to precision
```

```
30          sigma ~  dunif(0,2)
31
32          ## priors over thresholds
33          for(j in 1:4){
34                 tau0[j] ~  dnorm(0,.01)
35          }
36          tau[1:4] <- sort(tau0)  ## JAGS only, not in WinBUGS!
37 }
```

Note the loop over the `eta` parameters at line 26, inducing the hierarchical prior for the interviewer-specific offsets; `NID` is the number of unique interviewers in the data. The density for the η parameters is constrained to have mean zero, recalling that the threshold parameters $\boldsymbol{\tau}$ and the intercept are not identified, and that the mean of the η density plays the role of the intercept in this version of the model. We could (and perhaps should) relax this restriction here, letting the mean of the η distribution be an unknown parameter with a proper prior. This over-parameterized version of the model is not identified, but (and as we have seen before), letting MCMC algorithms go to work in the space of the unidentified parameters can often generate Markov chains that display excellent mixing in the subspace of identified parameters.

Also note the use of the `sort` function in the `JAGS` code above. As noted earlier, this form of inducing an ordering constraint is specific to `JAGS`; see the code fragments given earlier that generate ordering constraints in `BUGS`.

`JAGS` generated 500 000 MCMC iterations, and we discard the first 10 000 iterations as burn-in. Standard convergence diagnostics indicate the same pattern as for the non-hierarchical model: the MCMC algorithm exhibits good mixing with respect to most parameters, with the exception of two of the elements of $\boldsymbol{\beta}$ and the threshold parameters $\boldsymbol{\tau}$. The standard deviation of the density of the interviewer specific parameters, σ, displays good mixing as well.

Table 8.7 presents estimates of the hierarchical model. Results of the non-hierarchical ordinal logistic model are presented as well, so as to facilitate a comparison. The evidence for scale-use heterogeneity is quite compelling: the marginal posterior density for σ (the standard deviation of the density of the interviewer effects, η) has an estimated mean of .77 and an estimated standard deviation of .08, and so σ is bounded well away from zero (the implied value of σ for the non-hierarchical model). Figure 8.10 displays a histogram of the output of the MCMC algorithm for σ, confirming that the posterior density for σ is tightly concentrated around its mean of .77 and is dramatically different to the uniform prior density; i.e. the data are quite informative with respect to this parameter.

Given the normal model for the interviewer effects, $\eta_k \sim N(0, \sigma^2)$, if we set σ to its posterior mean of .77, then half of the interviewer effects will lie more than 1.35 $\sigma \approx 1.04$ away from zero. That is, conditional on the posterior mean value of σ, half of the survey respondents are subject to shifts in their implied value of y_i^* that are greater than one unit (on the logit scale). This is reasonably large, say, relative to the differences between the thresholds reported in Table 8.7 and clearly will be sufficient to alter the information scores likely to be assigned to respondents who look otherwise identical.

Goodness of fit. Table 8.7 shows the cross-tabulation of predicted outcomes against actual outcomes for the hierarchical model. The classification rule employed is to assign cases to the ordinal response with the greatest posterior probability. The model almost

Table 8.7 Analysis of ordinal ratings of political information, ordinal logit model with interviewer effects, Example 8.6. Cell entries are estimated posterior means, based on 100 000 iterations for the non-hierarchical model, discarding the first 10 000 iterations as burn-in (500 000 iterations for the hierarchical model). Quanities in parentheses are estimates of the standard deviations of the marginal posterior densities, again using the MCMC output.

	Non-hierarchical	Hierarchical
College degree	1.46	1.61
	(.10)	(.10)
Female	−.66	−.76
	(.09)	(.09)
log(Age)	.47	.42
	(.12)	(.13)
Home owner	.45	.48
	(.10)	(.10)
Government employee	.17	.16
	(.14)	(.14)
log(Interview Length)	1.13	1.45
	(.15)	(.18)
σ	0	.77
		(.08)
Threshold parameters:		
τ_0	3.85	4.69
	(.67)	(.75)
τ_1	5.66	6.60
	(.67)	(.75)
τ_2	7.37	8.46
	(.68)	(.76)
τ_3	8.83	10.08
	(.69)	(.77)

never assigns high probability to the lowest ordinal rating; 93 % of the cases actually receiving the lowest interview rating ($y_i = 0$) are predicted to receive a 1 or a 2 rating. Overall, the hierarchical model correctly classifies 41.2 % of cases and the 95 % bound on this quantity ranges from 39.7 % to 42.7 %. This compares quite favorably with the corresponding quantity from the non-hierarchical ordinal logit model, which generates a correct classification rate of 38.3 %, further evidence that the hierarchical model with the interview effects provides a better fit to these data than a model that ignores the interviewer scale-use heterogeneity. Note that a null model which simply assigns each case to the modal ordinal rating attains a classification success rate of 32.6 %.

Assessing the consequences of scale-use heterogeneity among interviewers via posterior predictive mass functions. To better gauge the consequences of the scale-use heterogeneity among interviewers, we engage in some auxiliary computation, examining the posterior predictive mass functions for the ordinal scores y_i by first ignoring the

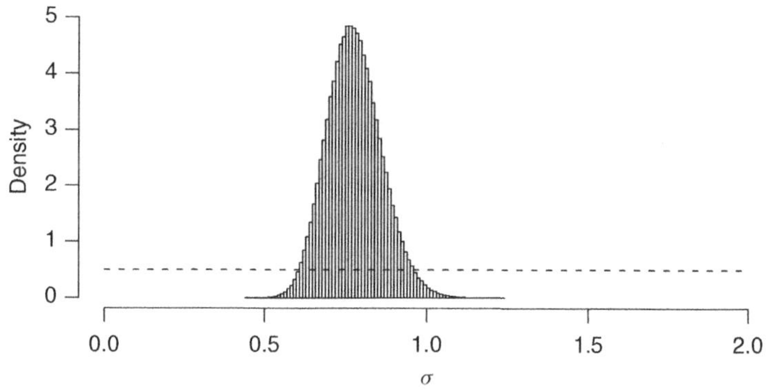

Figure 8.10 Posterior and prior densities for σ, hierarchical model for interviewer ratings, Example 8.6. The histogram summarizes the output of the MCMC algorithm (100,000 iterations, discarding the first 10,000 as burn-in). The dotted horizontal line shows the prior density for σ, a uniform density on [0,2]. The difference between the prior and the posterior indicates that the data are quite informative about this parameter.

Table 8.8 Classification table, ordinal logistic model with interviewer scale-use offsets, Example 8.6. Cell entries are row percentages; 95 % credible intervals appear in brackets. The classification rule is to assign cases to the ordinal response with the greatest posterior probability.

	Predicted (highest probability outcome)				
Actual	0	1	2	3	4
0	1	45	48	5	1
	[0 - 5]	[33 - 56]	[37 - 60]	[2 - 9]	[0 - 2]
1	1	26	62	9	2
	[0 - 3]	[18 - 35]	[52 - 70]	[5 - 13]	[1 - 4]
2	0	13	64	18	5
	[0 - 1]	[8 - 18]	[56 - 72]	[12 - 24]	[3 - 8]
3	0	4	45	33	18
	[0 - 0]	[2 - 7]	[37 - 53]	[22 - 43]	[12 - 25]
4	0	1	24	34	41
	[0 - 0]	[0 - 2]	[19 - 31]	[23 - 45]	[32 - 50]

scale-use heterogeneity, and then adding in the scale-use heterogeneity. First, consider a hypothetical survey respondent, with attributes $\tilde{\mathbf{x}}_i$ that we can multiply by $\boldsymbol{\beta}$ to generate a latent political information score $\tilde{y}_i^* = \tilde{\mathbf{x}}_i\boldsymbol{\beta}$. Note that conditional on the covariates $\tilde{\mathbf{x}}_i$, uncertainty in $\boldsymbol{\beta}$ generates uncertainty in $\tilde{y}_i^*$. The average interviewer has $\eta_k = 0$, by construction. In turn, uncertainty in the location of the thresholds $\boldsymbol{\tau}$ generates uncertainty in the mapping from the latent variable $\tilde{y}_i^*$ to the observed ordinal rating $\tilde{y}_i$. For this average interviewer, the posterior predictive mass function over the ordinal rating $\tilde{y}_i$ is a function of both the fixed set of respondent attributes $\tilde{\mathbf{x}}_i$, and the random quantities $\boldsymbol{\beta}$ and the thresholds $\boldsymbol{\tau}$. *A posteriori*, $\boldsymbol{\beta}$ and $\boldsymbol{\tau}$ are known up to their joint posterior density

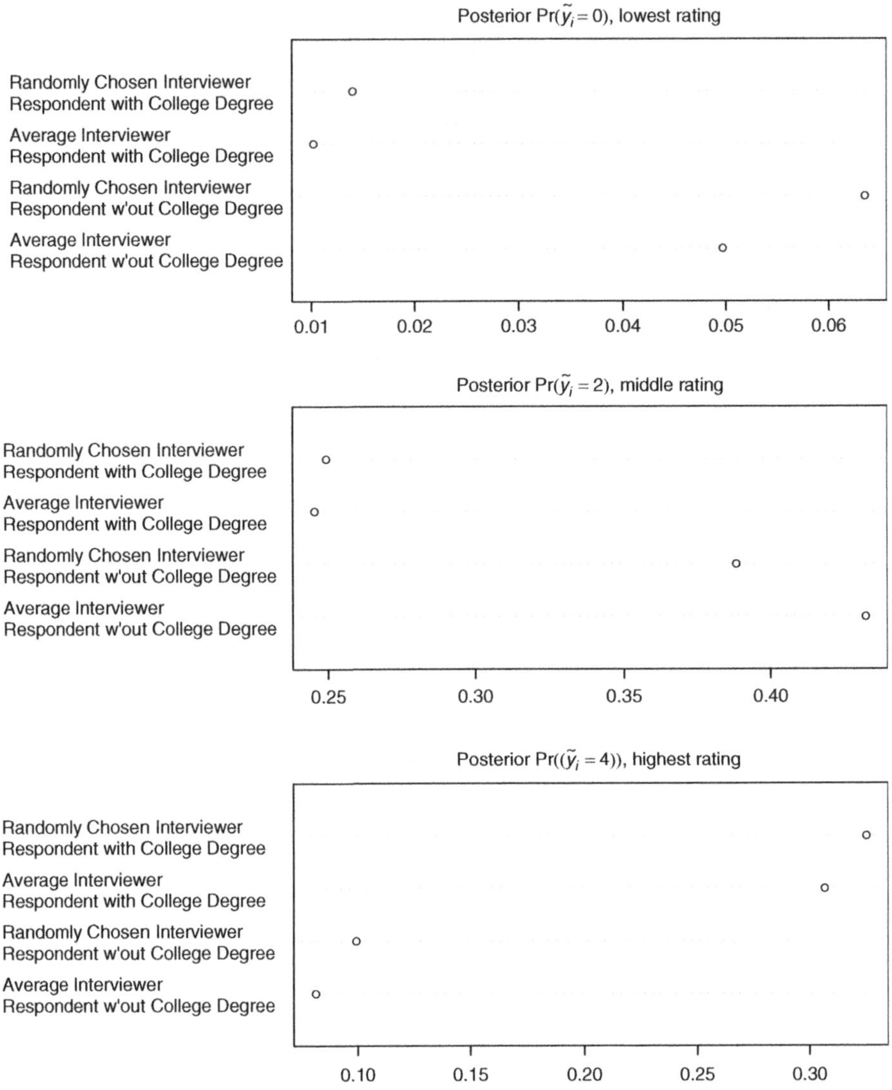

Figure 8.11 Posterior predictive densities, ordinal logit models of interviewer ratings. The three panels show the predicted probability of a respondent receiving the lowest political information score ($y_i = 0$, top panel), the middle score ($y_i = 2$, middle panel), and the highest score ($y_i = 4$, lower panel). When interviewer scale-use heterogeneity is taken into account, the probability of the respondent being given an extreme score is slightly higher than when we ignore the scale-use heterogeneity, indicative of the extra variation (or 'measurement error') in the reported scores that is due to between-interviewer variation in use of the ordinal categories.

$p(\boldsymbol{\beta}, \boldsymbol{\tau}|\mathbf{y}, \mathbf{X})$. That is, we compute

$$p(\tilde{y}_i|\tilde{\mathbf{x}}_i) = \int_{\Theta} p(\tilde{y}_i|\tilde{\mathbf{x}}_i, \boldsymbol{\theta}) p(\boldsymbol{\theta}|\mathbf{y}, \mathbf{X}) d\boldsymbol{\theta}$$

where $(\boldsymbol{\beta}, \boldsymbol{\tau})' = \boldsymbol{\theta} \in \Theta \subset \mathbb{R}^p$. We perform the integration via the method of composition, using the output of the MCMC algorithm. That is, iteration t of the MCMC algorithm produces $\boldsymbol{\theta}^{(t)}$, which we use to sample $\tilde{y}_i^{(t)}$ from $p(\tilde{y}_i|\tilde{\mathbf{x}}_i, \boldsymbol{\theta}^{(t)})$. Storing the results over the iterations of the MCMC algorithm provides a sample from the posterior predictive mass function for $\tilde{y}_i$. I repeat this exercise with the interviewer effects included, in which case we have $\boldsymbol{\theta} = (\boldsymbol{\beta}, \boldsymbol{\tau}, \sigma)$, where $\eta_k \sim N(0, \sigma^2)$ and $\tilde{y}_i^* = \tilde{\mathbf{x}}_i\boldsymbol{\beta} + \eta_k$.

Figure 8.11 presents the results of this exercise under $2 \times 2 = 4$ different scenarios: (1a) ignoring the scale-use heterogeneity by setting $\eta_k = 0$ (the average interviewer), or (1b) incorporating the scale-use heterogeneity by sampling η_k from the predictive density $N(0, \sigma^2)$, equivalent to randomly sampling from the population of interviewers; and, at the same time, considering two hypothetical respondents who differ only in that (2a) one has a college degree while the other does not (2b). The resulting posterior predictive densities in these four configurations are displayed in three panels in Figure 8.11: the top panel displays the probability of being assigned the lowest score, $\Pr(y_i \tilde{=} 0|\tilde{\mathbf{x}}_i)$, the middle panel shows the posterior probability of the middle rating ($\tilde{y}_i = 2$), and the lower panel shows the posterior probability of the maximum rating ($\tilde{y}_i = 4$). Perhaps the most notable feature of Figure 8.11 is the large difference in the probabilities associated with the difference in educational attainment. But note also that the probabilities of being in the extreme categories are slightly higher when we take the between-interviewer scale-use heterogeneity into consideration. That is, the scale-use heterogeneity introduces an additional source of variation into the data, making us slightly less confident about the score that might be assigned to a given respondent.

8.3 Multinomial outcomes

We now consider models for dependent variables that take on values that can not be ordered. Variables of this type are sometimes referred to as nominal variables, with any numeral we assign to them being purely arbitrary and meaningful only for the purposes of labeling or indexing. Classic examples come from labor economics or transport economics, where consumer choices of interest to the analyst take on values such as {'car', 'bus', 'train'} or, say in a voting example {'Conservative', 'Labor', 'Alliance'} or {'Bush', 'Clinton', 'Perot', 'Abstain'}.

In economics and political science, models for unordered responses are usually motivated via a random utility model. That is, we assume that decision-maker i faces a choice over J outcomes. The utility to decision-maker i of choice j is linear in some predictors, plus a random component, e.g. $U_{ij} = \mathbf{x}_i\boldsymbol{\beta}_j + \epsilon_{ij}$, $j = 0, \ldots, J$.

8.3.1 Multinomial logit (MNL)

A common, mathematically convenient assumption is that the ϵ_{ij} follow a Type-1 extreme-value distribution with $F(\epsilon_{ij}) = \exp[-\exp(-\epsilon_{ij})]$ and hence

ϵ_{ij} has the density

$$p(\epsilon_{ij}) = \exp(-\epsilon_{ij}) \exp[-\exp(-\epsilon_{ij})]. \tag{8.9}$$

It is common to further assume that this assumption holds for all decision-makers i and for all alternatives j, and critically, that the ϵ_{ij} are independent across the alternatives in the choice set. These assumptions are chosen more for tractability rather than as specifying something interesting about the utilities; in particular, the assumptions lead to a convenient 'log-odds' or 'logit' interpretation of the linear combination $\mathbf{x}_i\boldsymbol{\beta}_j$. The multinomial probit model (Section 8.4) results if we assume that the $\boldsymbol{\epsilon}_i = (\epsilon_0, \ldots, \epsilon_J)'$ follow a multivariate normal density, allowing for covariation among the stochastic components of the utilities.

Under the postulate of utility maximization, decision-maker i chooses j if $U_{ij} > U_{ik}$, $\forall\, k \neq j$. Because the utilities contain a stochastic component ϵ_{ij}, choice here is probabilistic. Decision-maker i chooses option j with probability

$$\pi_{ij} = \Pr(y_i = j) = \Pr[U_{ij} > U_{ik}], \quad \forall\, k \neq j.$$

Consider a choice set with 3 elements, {'0', '1', '2'}. Suppose we observe $y_i = 2$. Under the multinomial logit assumptions given above the probability of this event is (e.g. Amemiya 1985, 297)

$$\begin{aligned}
&\Pr(y_i = 2) = \Pr(U_{i2} > U_{i1}, U_{i2} > U_{i0}) \\
&= \Pr[\mathbf{x}_i\boldsymbol{\beta}_2 + \epsilon_{i2} > \mathbf{x}_i\boldsymbol{\beta}_1 + \epsilon_{i1},\ \mathbf{x}_i\boldsymbol{\beta}_2 + \epsilon_{i2} > \mathbf{x}_i\boldsymbol{\beta}_0 + \epsilon_{i0}], \\
&= \Pr[\epsilon_{i2} + \mathbf{x}_i\boldsymbol{\beta}_2 - \mathbf{x}_i\boldsymbol{\beta}_1 > \epsilon_{i1},\ \epsilon_{i2} + \mathbf{x}_i\boldsymbol{\beta}_2 - \mathbf{x}_i\boldsymbol{\beta}_0 > \epsilon_{i0}], \\
&= \int_{-\infty}^{\infty} f(\epsilon_2) \left[\int_{-\infty}^{\epsilon_{i2}+\mathbf{x}_i\boldsymbol{\beta}_2-\mathbf{x}_i\boldsymbol{\beta}_1} f(\epsilon_1)\, d\epsilon_1 \cdot \int_{-\infty}^{\epsilon_{i2}+\mathbf{x}_i\boldsymbol{\beta}_2-\mathbf{x}_i\boldsymbol{\beta}_0} f(\epsilon_0)\, d\epsilon_0 \right] d\epsilon_2, \\
&= \int_{-\infty}^{\infty} f(\epsilon_2) \times \exp[-\exp(-\epsilon_{i2} - \mathbf{x}_i\boldsymbol{\beta}_2 + \mathbf{x}_i\boldsymbol{\beta}_1)] \\
&\qquad \times \exp[-\exp(-\epsilon_{i2} - \mathbf{x}_i\boldsymbol{\beta}_2 + \mathbf{x}_i\boldsymbol{\beta}_0)]\, d\epsilon_2, \\
&= \frac{\exp(\mathbf{x}_i\boldsymbol{\beta}_2)}{\exp(\mathbf{x}_i\boldsymbol{\beta}_0) + \exp(\mathbf{x}_i\boldsymbol{\beta}_1) + \exp(\mathbf{x}_i\boldsymbol{\beta}_2)}.
\end{aligned}$$

And so in general,

$$\pi_{ij} = \Pr(y_i = j) = \frac{\exp(\mathbf{x}_i\boldsymbol{\beta}_j)}{\sum_{k=0}^{J} \exp(\mathbf{x}_i\boldsymbol{\beta}_k)}. \tag{8.10}$$

Given conditional independence over decision-makers, the likelihood for the multinomial logit model is simply the product over the probabilities of the outcomes actually observed: i.e.

$$\mathcal{L}(\boldsymbol{\beta}; \mathbf{y}, \mathbf{X}) = p(\mathbf{y}|\boldsymbol{\beta}, \mathbf{X}) = \prod_{i=1}^{n} \prod_{j=0}^{J} \pi_{ij}^{Z_{ij}} \tag{8.11}$$

where $Z_{ij} = 1 \iff y_i = j$ and 0 otherwise.

Identification via normalizing on baseline outcome

Without some restrictions, the parameters of the multinomial logit model are not identified. For instance, since $\boldsymbol{\beta}_j$ appears in both the numerator and denominator of Equation 8.10 there is no unique $\boldsymbol{\beta}_j$ that maximizes the resulting likelihood function; see Problem 8.5. The standard identifying restriction is to set one of the $\boldsymbol{\beta}_j$ to a null vector, thereby normalizing the model with respect to a 'baseline' category. It is conventional to make this baseline outcome the 'zero' outcome, in which case so the identifying restriction is $\boldsymbol{\beta}_j = \mathbf{0}$.

This identifying restriction makes the model reasonably easy to interpret, just as in the case of a logistic regression model. In particular, the model is said to be linear in the parameters with respect the log-odds ratio of an outcome and the baseline category. To see this, first consider the effect of setting $\boldsymbol{\beta}_0 = 0$:

$$\frac{\pi_{ij}}{\pi_{i0}} = \frac{\exp(\mathbf{x}_i\boldsymbol{\beta}_j)}{\exp(\mathbf{x}_i\boldsymbol{\beta}_0)} = \frac{\exp(\mathbf{x}_i\boldsymbol{\beta}_j)}{\exp(0)} = \exp(\mathbf{x}_i\boldsymbol{\beta}_j).$$

Accordingly, $\ln(\pi_{ij})/(\pi_{i0}) = \mathbf{x}_i\boldsymbol{\beta}_j$. After this normalization the probabilities of observing the respective choices are now

$$\Pr(y_i = j) = \frac{\exp(\mathbf{x}_i\boldsymbol{\beta}_j)}{1 + \sum_{k=1}^{J}\exp(\mathbf{x}_i\boldsymbol{\beta}_k)} \quad \forall\ j = 1, 2, \ldots, J \tag{8.12}$$

$$\Pr(y_i = 0) = \frac{1}{1 + \sum_{k=1}^{J}\exp(\mathbf{x}_i\boldsymbol{\beta}_k)} \tag{8.13}$$

The parallels between MNL and the logit model for binary outcomes is no mere coincidence. The MNL model amounts to a series of binary choices, considered simultaneously, but independently.

Conditional logit vs multinomial logit

It is also conventional to distinguish between two types of predictors: covariates that vary over decision makers *and* over choices (x_{ij}), and possibly *only* over choices (x_j), versus covariates that vary only over decision makers but constant over choices (x_i, such as demographic characteristics of the decision makers). A model which uses just the former type of regressors (choice-specific predictors) is called a 'conditional logit' model (McFadden 1974). Conditional logit models have the feature that the predictors vary over choices (and possibly decision makers too), but the parameters do not, and so the conditional model is

$$\pi_{ij} = \Pr(y_i = j) = \frac{\exp(\mathbf{x}_{ij}\boldsymbol{\beta})}{\sum_{j=0}^{J}\exp(\mathbf{x}_{ij}\boldsymbol{\beta})} \tag{8.14}$$

where $\mathbf{x}_{ij}$ is the vector of regressors associated with the ith decision maker on the jth choice. The constraint that the coefficients are constant over choices means that it is differences on the $\mathbf{x}_{ij}$ (or $\mathbf{x}_j$) that account for the different choices observed. That is, for the conditional logit model, the log-odds of choosing option j over k is

$$\ln\left(\frac{\pi_{ij}}{\pi_{ik}}\right) = (\mathbf{x}_{ij} - \mathbf{x}_{ik})\boldsymbol{\beta} \qquad \text{or} \qquad \ln\left(\frac{\pi_j}{\pi_k}\right) = (\mathbf{x}_j - \mathbf{x}_k)\boldsymbol{\beta},$$

which constrasts with the multinomial logit log-odds $\ln(\pi_{ij}/\pi_{ik}) = \mathbf{x}_i(\boldsymbol{\beta}_j - \boldsymbol{\beta}_k)$ with $\mathbf{x}_i$ constant across choices and differences in the $\boldsymbol{\beta}_j$ accounting for the differences in observed behavior. It should be noted that any multinomial logit model can be written as a conditional logit model, and vice-versa; see Long (1997, 180–1) for a simple example.

A hybrid model can be obtained by combining the multinomial and conditional logit models, say, where we have covariates that are constant over choices (e.g. characteristics of the decision-makers) and covariates that vary over choices (and possibly individuals as well), such as costs/benefits of choices or individual-specific perceptions of those costs/benefits. For notational convenience, the more general model can be written as

$$U_{ij} = \mathbf{x}_i\boldsymbol{\beta}_j + \mathbf{z}_{ij}\boldsymbol{\gamma} + \epsilon_{ij} \tag{8.15}$$

where $\mathbf{x}_i$ are characteristics of individuals that are constant across choices, and $\mathbf{z}_{ij}$ are characteristics that vary across individuals and choices, and these covariates pick up parameters $\boldsymbol{\beta}_j$ $(j = 0, \ldots, J)$ and $\boldsymbol{\gamma}$, respectively.

Bayesian analysis of the multinomial logit model

Bayesian inference for the multinomial logit model proceeds as usual, with the posterior density for the model parameters proportional to their prior density times the likelihood in Equation 8.11. As we saw for the binary logistic model (Section 8.1.3), there is no conjugate prior for the multinomial model, and so the posterior density is not available in an easily recognized form; thus, a Monte Carlo based exploration of the posterior density of the model parameters is required. However, the conditional densities for various elements of $\boldsymbol{\theta} = (\boldsymbol{\beta}_0, \ldots, \boldsymbol{\beta}_J, \boldsymbol{\gamma})'$ do not correspond to standard densities that are easy to sample from. As a consequence, most Bayesian algorithms for the multinomial logit problem deploy Metropolis samplers (see Section 5.1), as we will see in the following example.

■ Example 8.7

Vote choice in the 1992 US presidential election. The 1992 US presidential election provided voters with a choice between the Republican incumbent, President George H.W. Bush, the Democratic challenger (and victor) Bill Clinton, and Reform Party candidate Ross Perot. We omit abstentions from the analysis. Data from the 1992 American National Election Studies (Miller *et al*. 1999) provides observations from $n = 909$ respondents who reported voting for one of these candidates, with 34 % reporting having voted for Bush, 46 % for Clinton, and 20 % for Perot.

We use a small set of the available covariates in the data set so as to estimate a relatively simple model for voters' choices, closely following the analysis in Alvarez and Nagler (1995) and utilizing a replication data set available from the Inter-University Consortium for Political and Social Research (Study Number 1112; http://dx.doi.org/10.3886/ICPSR01112).

The utility to voter i from voting for candidate j is presumed to decline in the squared distance between the voter's self-reported ideological location and the average location assigned to candidate j; both sets of locations are measured as points on a unidimensional scale. That is, $U_{ij} = \mathbf{x}_i\boldsymbol{\beta}_j + z_{ij}\gamma + \epsilon_{ij}$, where i indexes decision-makers (survey respondents), $j \in$ {'Bush', 'Clinton', 'Perot'}, $\mathbf{x}_i$ is a set of controls that includes party identification, respondent gender, and retrospective assessments of the national

Table 8.9 Summary statistics, covariates used in multinomial logit analysis, Example 8.7. The quantities in the top four rows are proportions, summing to 1 across each row, showing the distribution of vote choices among Democratic identifiers, Republican identifiers and among women and men. Cell entries in the lower half of the table show means of the indicated covariate (row) by vote choices (columns).

	Bush	Clinton	Perot	Min	Max
Dem	0.05	0.80	0.15	0.00	1.00
Rep	0.70	0.08	0.22	0.00	1.00
Female	0.32	0.52	0.17	0.00	1.00
Male	0.36	0.40	0.23	0.00	1.00
National Economy	−0.45	−0.84	−0.66	−1.00	1.00
Bush Sq Distance	1.19	5.30	2.71	0.10	18.66
Clinton Sq Distance	5.74	1.97	3.21	0.00	16.16
Perot Sq Distance	1.55	2.87	1.63	0.24	12.18

economy ('worse' = −1, 'same' = 1, 'better' = +1) and z_{ij} is the squared ideological distance of respondent i from candidate j. Summary statistics for the covariates appear in Table 8.9. Note that with this model the probability of choosing candidate j over k is

$$\begin{aligned}\Pr(U_{ij} > U_{ik}) &= \Pr(\mathbf{x}_i\boldsymbol{\beta}_j + z_{ij}\gamma + \epsilon_{ij} - \mathbf{x}_i\boldsymbol{\beta}_k - z_{ik}\gamma - \epsilon_{ik} > 0)\\ &= \Pr(\mathbf{x}_i[\boldsymbol{\beta}_j - \boldsymbol{\beta}_k] + [z_{ij} - z_{ik}]\gamma > \epsilon_{ik} - \epsilon_{ij}).\end{aligned}$$

Since the ϵ_{ij} have Type-1 extreme-value densities (equation 8.9), their difference follows a logistic density and we obtain the log-odds ratio

$$\ln(\pi_{ij}/\pi_{ik}) = \mathbf{x}_i(\boldsymbol{\beta}_j - \boldsymbol{\beta}_k) + (z_{ij} - z_{ik})\gamma.$$

Note that while the ideological distance terms z_{ij} appear in the utility functions, it is their pairwise differences among alternatives that determine the log-odds ratios.

We define the 'Bush' outcome as the 'baseline' category and normalize the parameter vector $\boldsymbol{\beta}_0$ to $\mathbf{0}$. Outcome $j = 1$ refers to the 'Clinton' outcome and $j = 2$ to the 'Perot' outcome.

Over-parameterization. To improve the efficiency of the MCMC algorithm used in this case we use an over-parameterized version of the model, similiar to the over-parameterizations we used when considering hierarchical models in Section 7.2. Here we include an intercept (β_{1j}) plus three, mutually exclusive and exhaustive indicators for the party identification categories (Democratic, Republican and Independents/Apoliticals), with parameters β_{2j}, β_{3j} and β_{4j}, respectively. These four parameters are not identified, but we recover

$$\beta^*_{j1} = \beta_{j1} + \frac{1}{3}\sum_{k=2}^{4}\beta_{jk}$$

$$\beta^*_{jl} = \beta_{jl} - \frac{1}{3}\sum_{k=2}^{4}\beta_{jk}, \quad l = 2, 3, 4$$

such that $\sum_{k=2}^{4} \beta^*_{jk} = 0$ with this latter restriction identifying the β^*_{jk} as offsets around the intercept β^*_{j1}. This over-parameterization does not effect the other parameters in the model, nor does it effect the $\boldsymbol{\beta}_1$ parameters set to zero for identification (i.e. we only employ this overparameterization with $j = 2, 3$).

We implement this model – with vague normal priors over the unconstrained coefficients – in the following JAGS program:

JAGS code

```
1  model{
2          for(i in 1:NOBS){
3                  for(j in 1:3){
4                                  mu[i,j] <- beta[j,1]
5                                      + beta[j,2]*dem[i]
6                                      + beta[j,3]*ind[i]
7                                      + beta[j,4]*rep[i]
8                                      + beta[j,5]*female[i]
9                                      + beta[j,6]*natlecon[i]
10                                     + gamma*dist[i,j]
11                         emu[i,j] <- exp(mu[i,j])
12                         p[i,j] <- emu[i,j]/sum(emu[i,1:3])
13                 }
14                 y[i] ~ dcat(p[i,1:3])
15         }
16
17         ## priors
18         for(k in 1:6){
19                 beta[1,k] <- 0 ## identifying restriction
20                 betastar[1,k] <- 0
21         }
22         for(j in 2:3){
23                 beta[j,1:6] ~ dmnorm(b0,B0)
24         }
25         gamma ~ dnorm(0,.01)
26
27         ## map to identified parameters
28         for(j in 2:3){
29                 bmean[j] <- mean(beta[j,2:4])
30                 betastar[j,1] <- beta[j,1] + bmean[j]
31                 for(k in 2:4){
32                                 betastar[j,k] <- beta[j,k] - bmean[j]
33                         }
34                 for(k in 5:6){
35                                 betastar[j,k] <- beta[j,k]
36                         }
37                 }
38  }
```

The program is reasonably straightforward: a loop over the observations, in which we nest a loop over the choices. Line 10 introduces the squared ideological distances, with the covariate `dist[i,j]` varying over individuals and choices, but picking up a constant parameter `gamma`. Line 12 generates the probabilities of each choice, operationalizing Equation 8.10 in the discussion above. We use the `dcat` construct at line 14 to actually model the data, encountered earlier in Example 6.8 when fitting a mixture model; recall that if `y[i] ~ dcat(p[i,1:3])` then `p[i,j]` is simply the probability that $y_i = j, j = 0, 1, 2$.

Vague normal priors are specified at line 23 for $\boldsymbol{\beta}_j, j = 2, 3$, by setting the prior precision matrix `B0` to $k \cdot \mathbf{I}_6$, where k is an arbitrarily small, positive quantity (we set $k = .01$ in the `R` code fragment given below). A $N(0, 10^2)$ prior for γ appears at line 25.

The Metropolis algorithm employed by `JAGS` for this problem requires a reasonably large number of iterations to explore the posterior density. Here we use 150 000 iterations after a 10 000 iteration burn-in, after shorter runs indicated that the Markov chain is quite slow mixing, even after the over-parameterization 'trick' employed here. The `R` code below makes the call to `JAGS`:

R code

```
1  forJags <- list(y=4-as.numeric(vote92$vote),
2                  ind=as.numeric(!vote92$dem & !vote92$rep),
3                  dem=vote92$dem,
4                  rep=vote92$rep,
5                  female=vote92$female,
6                  natlecon=vote92$natlecon,
7                  dist=cbind(vote92$bushdis,
8                    vote92$clintondis,
9                    vote92$perotdis),
10                 NOBS=dim(vote92)[1],
11                 b0=rep(0,6),
12                 B0=.01*diag(6))
13 inits <- list(list(beta=matrix(0,3,6),
14                    gamma=0,
15                    .RNG.name="base::Mersenne-Twister",
16                    .RNG.seed=1234))
17 require(rjags)
18 foo <- jags.model(file="vote92.bug",
19                   inits=inits,
20                   data=forJags)
21 update(foo,n.iter=10e3)              ## 10,000 iterations
22 out <- coda.samples(foo,             ## 150,000 iterations
23                     n.iter=150e3,
24                     variable.names=c("betastar","gamma"))
```

The initial values (`inits`) is a 3-by-6 matrix of zeros for `beta` and a scalar zero for `gamma`. Note also the use of the `.RNG.name` and `.RNG.seed` options in the definition of the initial values at lines 15 and 16. The 10 000 burn-in iterations are generated at line 21, with the 'production' run of 150 000 iterations generated by the `coda.samples` command at line 22, along with the directive to monitor the identified parameters `betastar` and `gamma` (line 24).

Selected convergence diagnostics appear in Table 8.10. These indicators are generally consistent with the Metropolis sampler attaining stationarity. The sampler is not particularly efficient, with the effective sample sizes (see Section 4.4.1) of the 150 000 draws as low as 1360 for β^*_{33}.

Table 8.10 Convergence diagnostics, MCMC analysis of multinomial logit model, Example 8.7. 150 000 Metropolis iterations were run in JAGS, after a 10 000 iteration burn-in. 'Effective sample' refers to the equivalent sample size from an independence sampler, such that we would obtain the same level of Monte Carlo error in the estimate of the mean of the marginal posterior density (see Section 4.4.1).

	Geweke	Heidelberger	Autocorrelation	Raftery-Lewis		Effective
	z	p	ρ, AR(1)	N	I	Sample
β^*_{21}	0.19	0.59	0.96	89 102	23.80	2423
β^*_{31}	0.11	0.14	0.95	74 828	20.00	2290
β^*_{22}	0.04	0.88	0.96	93 300	24.90	1565
β^*_{32}	−0.06	0.83	0.97	124 583	33.30	1512
β^*_{23}	−0.15	0.51	0.97	135 520	36.20	1494
β^*_{33}	−0.05	0.31	0.98	138 495	37.00	1360
β^*_{24}	0.17	0.40	0.96	84 172	22.50	2510
β^*_{34}	0.18	0.17	0.96	91 155	24.30	2249
β^*_{25}	1.06	0.46	0.95	82 956	22.10	2690
β^*_{35}	0.69	0.81	0.96	101 004	27.00	2686
β^*_{26}	0.75	0.44	0.96	86 128	23.00	2880
β^*_{36}	0.77	0.08	0.95	78 767	21.00	3636
γ	−0.57	0.82	0.26	9022	2.41	61 963

Table 8.11 Summaries of marginal posterior densities, multinomial logit model, Example 8.7. The first column labelled 'Clinton' shows results with respect to the log-odds of the probability of the 'Clinton' response relative to the (baseline) 'Bush' response. Cell entries are posterior means, as recovered from the 150 000 iteration run of the MCMC algorithm described in the text; intervals in brackets are 95 % highest density regions. Note the identifying restriction $\sum_{k=2}^{4} \beta^*_{jk} = 0$, $j \in$ {'Clinton', 'Perot'}.

	Clinton	Perot
Intercept, β^*_{j1}	−0.07	0.14
	[−.55, .40]	[−.23, .54]
Democratic Party Id Offset, β^*_{j2}	2.09	0.77
	[1.70, 2.49]	[0.38, 1.18]
Independent/Apolitical Offset, β^*_{j3}	0.26	0.54
	[−.23, 0.75]	[.06, 1.00]
Republican Party Id Offset, β^*_{j4}	−2.35	−1.31
	[−2.73, −1.97]	[−1.63, −1.00]
Female, β^*_{j5}	−0.21	−0.51
	[−.67, .25]	[−.93, −.10]
National Economy, β^*_{j6}	−0.97	−0.43
	[−1.42, −.54]	[−.77, −.10]
Relative Ideological Distance, γ	−0.13	
	[−.17, −.09]	

With the vague priors used here, and a reasonably large sample, the marginal posterior densities are all approximately normal densities centered on the maximum likelihood estimates. Summaries of the marginal posterior densities of the parameters – as recovered from the 150 000 iteration run of the Metropolis sampler implemented by `JAGS` – are reported in Table 8.11. The parameter tapping the effect of relative ideological distance, γ, is unambiguously negative, consistent with ideological distance decreasing the attractiveness of candidate j to voter i. Large offsets are observed for Democratic and Republican partisan identifiers, strongly preferring the candidate of their own party, net of the effects of ideological distance. Retrospective evaluations of the national economy pick up almost twice the weight in the major-party, binary choice (Clinton vs Bush), where $E(\beta^*_{j6}|\mathbf{y}, \mathbf{X}) = -.97$, versus a corresponding posterior mean of $-.43$ for the same parameter in the Perot vs Bush arm of the model.

In Problem 8.6 you are invited to replicate these results and to verify that the Bayes estimates do in fact correspond with the maximum likelihood estimates.

8.3.2 Independence of irrelevant alternatives

The multinomial and/or conditional logit forms of the discrete choice model assume that the stochastic component of the utilities for each alternative are independent. This has the virtue of leading to a relatively simple likelihood function for the model and the interpretation of the parameters estimates in terms of log-odds. Nonetheless, these features of the model rest on what is often a dubious assumption, that the stochastic component of the utility associated with each outcome is uncorrelated with stochastic components of utility for the other outcomes. In practice this is often not the case.

The assumption about the choice probabilities that rationalizes the multinomial logit model is known as the 'independence of irrelevant alternatives' (IIA) The assumption is simply that when choosing alternatives from a set $\mathcal{J}$, with observed attributes $\mathcal{S}$ (attributes of decision-makers and/or choices), the probabilities of choosing x or y obey the restriction

$$\frac{\Pr(y|\mathcal{S}, \{x, y\})}{\Pr(x|\mathcal{S}, \{x, y\})} = \frac{\Pr(y|\mathcal{S}, \mathcal{B})}{\Pr(x|\mathcal{S}, \mathcal{B})}$$

for all possible alternative sets $\mathcal{B} \subseteq \mathcal{J}$; see Axiom 1 in McFadden (1974).

A useful thought experiment to see if this property is valid in a given setting is to ask whether the relative probabilities (i.e. the log-odds ratio) for alternatives j and k remains the same whether or not another alternative l is in the choice-set, even when l is a close substitute for one of j or k. This is sometimes illustrated with the 'red bus – blue bus' problem. As McFadden (1974, 113) introduced the issue:

> Suppose a population faces the alternatives of travel by auto and by bus, and two-thirds choose to use auto. Suppose now a second 'brand' of bus travel is introduced that is in all essential respects the same as the first [i.e. 'red bus' versus 'blue bus']. Intuitively, two-thirds of the population will still choose auto, and the remainder will split between the bus alternatives. However, if the selection probabilities satisfy Axiom 1 [IIA], only half the population will use auto when the second bus is introduced.

In terms of the notation introduced earlier, the issue here turns on the conditional independence assumptions across the choice-specific disturbance terms. Amemiya (1985, 298) puts the problem this way:

> Using McFadden's famous example, suppose that the three alternatives... consist of car, red bus, and blue bus.... In such a case, the independence between ϵ_1 and ϵ_2 is a clearly unreasonable assumption because a high (low) utility for red bus should generally imply a high (low) utility for a blue bus. The probability $\pi_0 = P(U_0 > U_1, U_0 > U_2)$ calculated under the independence assumption would underestimate the true probability in this case because the assumption ignores the fact that the event $U_0 > U_1$ make the event $U_0 > U_2$ more likely.

That is, the model specifies the relative probabilities between a pair of alternatives without regard for the the third alternative. For instance, the probability of choosing j or k is specified without any regard at all for the probability of choosing any other outcome $r \neq j, k$. We can see this by simply noting that given three outcomes j, k and r, the multinomial logit model gives us

$$\begin{aligned}\Pr(y_i = j | y_i = j \text{ or } y_i = k) &= \frac{\Pr(y_i = j)}{\Pr(y_i = j) + \Pr(y_i = k)} \\ &= \frac{\exp(\mathbf{x}_i\boldsymbol{\beta}_j)[\exp(\mathbf{x}_i\boldsymbol{\beta}_j) + \exp(\mathbf{x}_i\boldsymbol{\beta}_k) + \exp(\mathbf{x}_i\boldsymbol{\beta}_r)]^{-1}}{[\exp(\mathbf{x}_i\boldsymbol{\beta}_j) + \exp(\mathbf{x}_i\boldsymbol{\beta}_k)][\exp(\mathbf{x}_i\boldsymbol{\beta}_j) + \exp(\mathbf{x}_i\boldsymbol{\beta}_k) + \exp(\mathbf{x}_i\boldsymbol{\beta}_r)]^{-1}} \\ &= \frac{\exp(\mathbf{x}_i\boldsymbol{\beta}_j)}{\exp(\mathbf{x}_i\boldsymbol{\beta}_j) + \exp(\mathbf{x}_i\boldsymbol{\beta}_k)}\end{aligned}$$

and so the relative probability of j vis-à-vis k is independent of the probability of choosing r, and invariant to the presence or absence of r in the choice set!

We now consider a more general model – the multinomial probit model – that overcomes the IIA restriction, albeit at considerable computational cost.

8.4 Multinomial probit

Multinomial probit (MNP) is a more general model than the MNL model in that the disturbances in the utility functions can be correlated across choices. We begin with a random utility model,

$$U_{ij} = \mathbf{r}_{ij}\boldsymbol{\beta} + v_{ij}, \quad j = 0, 1, \ldots, J; i = 1, \ldots, n, \tag{8.16}$$

momentarily focusing on the case of a k-by-1 vector of covariates $\mathbf{r}_{ij}$ that varies over individuals and choices (but again noting that this includes as a special case covariates that vary over only individuals). We obtain the MNP model by assuming a multivariate normal density for the v_{ij}, $\mathbf{v}_i = (v_{i1}, \ldots, v_{iJ})' \stackrel{\text{iid}}{\sim} N(\mathbf{0}, \mathbf{V})$, where $\mathbf{V}$ is a $(J+1)$-by-$(J+1)$ covariance matrix. Non-zero covariance terms – the off-diagonal elements of $\mathbf{V}$ – explicitly operationalize the possibility that utilities are correlated across choices in ways not tapped by the available covariates.

With a multivariate normal density over the $\boldsymbol{\nu}_i$ the probability of observing outcome j is

$$\begin{aligned}\pi_{ij} &= \Pr(y_i = j) = \Pr(U_{ij} > U_{ik}), \ \forall\, k \neq j \\ &= \int_{-\infty}^{\infty}\int_{-\infty}^{U_{ij}} \cdots \int_{-\infty}^{U_{ij}} f(U_0, U_1, \ldots, U_J)\, dU_0\, dU_1 \ldots dU_j \end{aligned} \tag{8.17}$$

where f is the $J+1$-variate normal density defined above. Evaluating this high-dimensional integral has long been a formidable barrier to researchers interested in deploying the multinomial probit model. Over the last few decades simulation-based approaches have been become the preferred method of tackling the MNP model (e.g. Geweke, Keane and Runkle 1994). We describe a Bayesian, simulation-based approach due to McCulloch and Rossi (1994) and McCulloch, Polson and Rossi (1998). To foreshadow the key result here, we will tackle the high-dimensional integral underlying the MNP model via Monte Carlo methods.

The contemporary, simulation-based, Bayesian approach to the MNP model generalizes the data-augmented, latent variable approach we used for the binary probit model (Section 8.1.1). Utilities are unobserved, but under our assumption of utility maximization, if choice j is observed for person i, we know that $U_{ij} - U_{ik} > 0 \,\forall\, j \neq k$. Without loss of generality choose a 'baseline' outcome, $j = 0$, and define the utility differences $\mathbf{w}_i = (w_{i1}, \ldots, w_{iJ})'$ with $w_{ij} = U_{ij} - U_{i0}$, $j = 1, \ldots, J$; applying this differencing to Equation 8.16 we obtain

$$w_{ij} = (\mathbf{r}_{ij} - \mathbf{r}_{i0})\boldsymbol{\beta} + \nu_{ij} - \nu_{i0} = \mathbf{x}_{ij}\boldsymbol{\beta} + \epsilon_{ij}. \tag{8.18}$$

Note that since the difference of two zero mean normal random variables is also a zero mean normal, we have $\boldsymbol{\epsilon}_i \overset{\text{iid}}{\sim} N(\mathbf{0}, \boldsymbol{\Sigma})$, where $\boldsymbol{\epsilon}$ is a J-by-1 vector. With this differencing we have the following mapping from latent variables to observed choices:

$$y_i = h(\mathbf{w}_i) \equiv \begin{cases} 0 & \text{if} \max(\mathbf{w}_i) < 0 \\ j & \text{if} \max(\mathbf{w}_i) = w_{ij} > 0 \end{cases} \tag{8.19}$$

That is, if all the w_{ij} are negative then we know that $U_{ij} < U_{i0} \,\forall\, j = 1, \ldots, J$ and so individual i chose outcome $y_i = 0$. If not, then the index of the largest w_{ij} corresponds to the choice made by individual i.

Working with these differences in utilities goes some way towards identifying the model, just as it does in the binary case; i.e. the latent variable having a threshold at zero is sufficient to identify an intercept in the binary probit model (see Section 8.1.1). But the model remains invariant to arbitrary changes in scale. That is, for any constant $c > 0$, $\tilde{y}_i = h(c\,\mathbf{w}_i)$ is observationally indistinguishable from $y_i = h(\mathbf{w}_i)$, where $h()$ is the assignment rule in Equation (8.19). Put differently, the distribution of $\mathbf{y}|\mathbf{X}, \boldsymbol{\beta}, \boldsymbol{\Sigma}$ is the same as the distribution of $\mathbf{y}|\mathbf{X}, c\boldsymbol{\beta}, c^2\boldsymbol{\Sigma}$ (e.g. McCulloch and Rossi 1994, 209). Defining σ_{ij} as the ij-th element of $\boldsymbol{\Sigma}$, an easily-implemented normalization sufficient to identify the model is to set $\sigma_{11} = 1$.

Note that these identifying normalizations do not solve the computational challenge posed by the MNP model. In working with the differenced form of the model we have dropped one integral out of the nasty expression for the likelihood contributions in Equation 8.17. Nonetheless, we still face a formidable computing problem for which a simulation-based approach will prove especially helpful.

8.4.1 Bayesian analysis via MCMC

The inferential problem here is to find the posterior density $p(\boldsymbol{\beta}, \boldsymbol{\Sigma}|\mathbf{y}, \mathbf{X})$. Data-augmented Gibbs sampling is an extremely attractive way to proceed, exploiting the fact that conditional on the latent random utilities the problem reduces to a routine multivariate normal regression model, for which Bayesian analysis is straightforward (McCulloch and Rossi 1994; Chib and Greenberg 1997; McCulloch, Polson and Rossi 1998). In this way the troublesome (if not impossible) integrations involved in computing the likelihood contributions π_{ij} are avoided. Iteration t of the data-augmented Gibbs sampler consists of sampling from the following conditional distributions:

1. sample $\mathbf{w}_i^{(t)}$ from $p(\mathbf{w}_i|\boldsymbol{\beta}^{(t-1)}, \boldsymbol{\Sigma}^{(t-1)}, \mathbf{y}, \mathbf{X})$, $i = 1, \ldots, n$, the data-augmentation step
2. sample $\boldsymbol{\beta}^{(t)}$ from $p(\boldsymbol{\beta}|\boldsymbol{\Sigma}^{(t-1)}, \mathbf{W}^{(t)}, \mathbf{y}, \mathbf{X})$.
3. sample $\boldsymbol{\Sigma}^{(t)}$ from $p(\boldsymbol{\Sigma}|\boldsymbol{\beta}^{(t)}, \mathbf{W}^{(t)}, \mathbf{y}, \mathbf{X})$.

The data-augmentation step here is quite straightforward, since ignoring the censoring induced by the observed choice y_i, we have $\mathbf{w}_i \sim N(\mathbf{R}_i\boldsymbol{\beta}, \boldsymbol{\Sigma})$. The censoring rule in Equation 8.16 restricts the sampled $\mathbf{w}_i$ to lie in a particular subset of $\mathbb{R}^J$, suggesting that naïve rejection sampling can be used; contrast the implementation in Imai and van Dyk (2005b).

For step 2, let the prior for $\boldsymbol{\beta}$ be $N(\mathbf{b}_0, \mathbf{B}_0)$. Then the conditional distribution for $\boldsymbol{\beta}$ is $N(\mathbf{b}_1, \mathbf{B}_1)$, where $\mathbf{B}_1 = (\mathbf{X}'\mathbf{G}\mathbf{X} + \mathbf{B}_0^{-1})^{-1}$, $\mathbf{b}_1 = \mathbf{B}_1(\mathbf{X}'\mathbf{G}\mathbf{W} + \mathbf{B}_0^{-1}\mathbf{b}_0)$, and $\mathbf{G} = \boldsymbol{\Sigma}^{-1} \otimes \mathbf{I}_n$, where $\otimes$ is the Kronecker product operator (see Definition A.10). That is, with a diffuse prior we are essentially estimating a system of seemingly unrelated regressions (Zellner 1962).

For step 3, the prior and the conditional distribution for $\boldsymbol{\Sigma}$ is complicated by the identifying constraint $\sigma_{11} = 1$. McCulloch, Polson and Rossi (1998) partition $\boldsymbol{\Sigma}$ around the fixed parameter σ_{11} and sample from the conditional distributions for the remaining elements of $\boldsymbol{\Sigma}$ in turn. Imai and van Dyk (2005a) makes the observation that the MNP model is a prime candidate for marginal data augmentation, which we introduced so as to improve the performance of the data-augmented Gibbs sampler in the binary probit case (Section 8.1.2). The idea is to work with an unidentified version of the covariance matrix, $\tilde{\boldsymbol{\Sigma}}$, assigning it an (unrestricted) proper, inverse-Wishart prior (see Definition B.41), treating $\tilde{\sigma}_{11}$ as the working parameter for MDA, and recovering the identified covariance matrix $\boldsymbol{\Sigma} = \tilde{\boldsymbol{\Sigma}}/\tilde{\sigma}_{11}$. Imai and van Dyk (2005a) demonstrate that their algorithm generates better mixing (in the space of identified parameters) than earlier data-augmented Gibbs samplers for the MNP problem; their algorithm is implemented in the `R` package `MNP` (Imai and van Dyk 2005b), which we will employ in the following example.

■ **Example 8.8**

Vote choice in the 1992 US Presidential election, Example 8.7, continued. We repeat the analysis of vote choice in the 1992 election considered in Example 8.7. There we used a multinomial logit model to model the choices $y_i \in$ {'Bush', 'Clinton', 'Perot'}. The covariates include indicators of each respondent's party identification and gender, and retrospective evaluations of the economy. Squared ideological distances from the candidates also enter the model as choice-specific covariates. That is, in terms of the notation in Equation 8.18 we have

$$w_{ij} = \mathbf{x}_i \boldsymbol{\beta}_j + (r_{ij} - r_{i0})\gamma + \epsilon_{ij}$$

where $j \in \{$'Clinton', 'Perot'$\}$, with the 'Bush' outcome serving as the $j = 0$ 'baseline' outcome. The predictors $\mathbf{x}_i$ vary over individuals (but not over choices) and r_{ij} is the squared ideological distance between respondent i and candidate j. The parameters $\boldsymbol{\beta}_1, \boldsymbol{\beta}_2$ and γ are unknown, and in our Bayesian analysis, are assigned an improper, uninformative prior, which is the default in the implementation by Imai and van Dyk (2005a) in their R package. The disturbance vector $\boldsymbol{\epsilon}_i = (\epsilon_{i1}, \epsilon_{i2})' \sim N(\mathbf{0}, \boldsymbol{\Sigma})$ where $\boldsymbol{\Sigma}$ is a 2-by-2 covariance matrix; we assign $\boldsymbol{\Sigma}$ an inverse-Wishart prior density, with scale term $\mathbf{V}$ and degrees of freedom v, subject to the restriction that $\sigma_{11} = 1$. The defaults in the Imai and van Dyk (2005a) package are to set $\mathbf{V} = \mathbf{I}_{J-1}$ and $v = J$, generating a (barely) proper prior over the unconstrained elements of $\boldsymbol{\Sigma}$.

We request 1.5 million iterations from the `mnp` function, with the large number of iterations stemming from the fact that the resulting Markov chain mixes slowly. Figure 8.12 displays histograms and autocorrelation functions for the unconstrained elements of $\boldsymbol{\Sigma}$, the slowest mixing parameters for this problem. The AR(200) parameters for the sampled σ_{22} and σ_{12} parameters are around .5 to .6, which make them the slowest mixing parameters we will encounter in the many MCMC examples in this book. The 1.5 million iterations used here are equivalent to about 2500 draws from an independence

Table 8.12 Convergence diagnostics, MCMC analysis of multinomial probit model, Example 8.8. 1.5 million iterations were run, using the marginal data augmentation algorithm implemented in the `mnp` function in the R package `MNP` (Imai and van Dyk 2005b). z is from Geweke's test of stationarity; p is the p-value from Heidelberger's test; ρ is the AR(25) parameter of the MCMC output for each parameter; N is an estimate of the number of iterations required to accurately estimate extreme quantiles of the marginal posterior densities using the Raftery-Lewis procedure, and I is an estimate of the "dependence factor"; "EffSamp" refers to the equivalent sample size from an independence sampler, such that we would obtain the same level of Monte Carlo error in the estimate of the mean of the marginal posterior density (see Section 4.4.1).

	z	p	ρ	N	I	EffSamp
β_{11}, Intercept, Perot	−0.82	0.62	0.37	303 375	81.00	5908
β_{21}, Intercept, Clinton	−0.36	0.36	0.46	547 875	146.00	6211
β_{12}, Dem Id, Perot	−1.12	0.49	0.48	320 025	85.40	4076
β_{22}, Dem Id, Clinton	−0.27	0.69	0.73	860 850	230.00	3175
β_{13}, Repub Id, Perot	0.66	0.72	0.34	659 850	176.00	6599
β_{23}, Repub Id, Clinton	0.42	0.72	0.75	958 400	256.00	2763
β_{14}, Female, Perot	−0.07	0.32	0.01	94 025	25.10	58 544
β_{24}, Female, Clinton	1.04	0.51	0.02	99 850	26.70	50 304
β_{15}, Econ Retro, Perot	0.51	0.97	0.18	198 950	53.10	11 272
β_{25}, Econ Retro, Clinton	0.15	0.57	0.59	607 750	162.00	3709
γ, Ideological Distance	0.21	0.57	0.72	967 950	258.00	2778
σ_{12}	−1.46	0.09	0.90	975 375	260.00	2540
σ_{22}	−0.78	0.98	0.88	902 500	241.00	2412
ρ	−1.16	0.33	0.89	1 159 350	309.00	2575

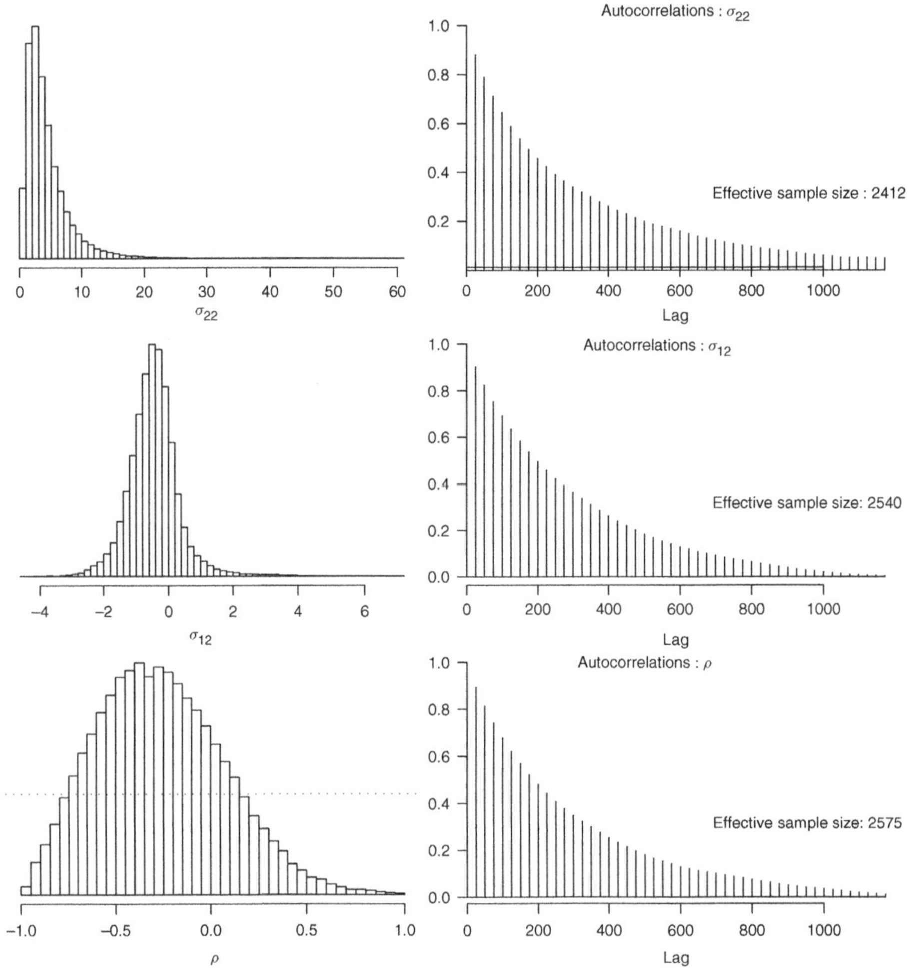

Figure 8.12 Marginal posterior densities and autocorrelation functions, unconstrained elements of $\mathbf{\Sigma}$ in multinomial probit model, Example 8.8. The parameter ρ is computed as $\sigma_{12}/\sqrt{\sigma_{22}}$. Histograms in the left panels show the marginal posterior density of the indicated parameter, summarizing the 60 000 MCMC iterates saved for analysis (1.5 million iterations, thinned by 25). The dotted line in the histogram for ρ shows the Unif$(-1, 1)$ prior density for this parameter. Panels on the right show the autocorrelation functions of the sequence of draws from the posterior density for the indicated parameter; these autocorrelation functions decay extremely slowly, indicating that the MCMC algorithm mixes very slowly with respect to these parameters; see also the relatively small effective sample sizes (Section 4.4.1).

sampler, in terms of the Monte Carlo error inherent in the resulting estimates of the posterior means of σ_{22} and σ_{12}. The marginal data augmentation algorithm performs better than other extant data augmentation strategies for the multinomial probit model, but nonetheless an extremely large number of iterations are required so as to generate a reasonably accurate characterization of the posterior density.

Convergence diagnostics reported in Table 8.12 suggest that the MCMC output is a stationary series, but that a million iterations would be a lower bound on the running time in this case. These kinds of results – suggesting that very long runs of the MCMC algorithm are required – are quite common for the MNP model. This model asks a lot of the data; for one thing, we are trying to estimate covariances among latent variables that are linked to each decision-maker's observed choice through the somewhat stringent censoring given in Equation 8.19. Note the marginal posterior densities of both the covariance parameter σ_{12} and the resulting correlation ρ given in Figure 8.12; the posterior densities for these parameters are centered close to zero and in the case of the correlation parameter the 95 % posterior HDR ranges from −.89 to .37. The posterior probability that $\rho > 0$ is .22 and the corresponding *a priori* probability is .5; i.e. even with 1000 observations the data are not particularly informative with respect to this parameter. The fact that these off-diagonal elements of $\mathbf{\Sigma}$ need not be constrained to zero is one of the features of the MNP model that distinguishes it from the MNL model; nonetheless, we need to be mindful that absent a lot of data, we are probably not going to estimate these covariance parameters with much precision.

Table 8.13 presents numerical summaries of the marginal posterior densities for the parameters of the MNP model deployed here. The indicators for partisanship have predictably large effects with 95 % HPDs that do no overlap zero; the largest effects are in

Table 8.13 Summaries of marginal posterior densities, multinomial probit model, Example 8.8. The first column labelled "Perot" shows results with respect to the Perot minus Bush utility difference, while the "Clinton" results show results with respect to the Clinton minus Bush utility difference. Cell entries are posterior means, as recovered from the 1.5 million iteration run of the MCMC algorithm; intervals in brackets are 95 % highest density regions. The Perot-Bush error variance σ_{11} is constrained to 1.00 so as to identify the model.

	Perot		Clinton
Intercept, β_{j1}	0.08		−0.72
	[−.40, .60]		[−1.83, .23]
Democratic Party Id Offset, β_{j2}	−0.21		2.08
	[−.77, .39]		[.67, 3.75]
Republican Party Id Offset, β_{j3}	−0.77		−2.20
	[−1.25, −.29]		[−3.99, −0.75]
Female, β_{j4}	−0.31		−0.07
	[−.55, −.08]		[−.55, .41]
National Economy, β_{j5}	−0.15		−0.87
	[−.36, .06]		[−1.63, −.23]
Relative Ideological Distance, γ		−0.12	
		[−.19, −.05]	
Error Variances, σ_{jj}	1.00		4.33
			[0.30, 11.11]
Covariance, σ_{12}		−.46	
		[−2.07, 1.08]	
Correlation, $\rho = \sigma_{12}/\sqrt{\sigma_{22}}$		−.27	
		[−.89, .37]	

the Clinton/Bush comparison (shown in the 2nd column of Table 8.13); Republican party identification is associated with a moderate net disutility for the Perot outcome relative to the Bush choice. In the presence of the other covariates used in this model, gender has a small impact on the Perot/Bush vote choice, with women being more likely to support Bush, in the usual *ceteris paribus* sense. The effect of national economic retrospections is distinguishable from zero in the Clinton/Bush arm of the model, with negative economic evaluations leading to higher utility for the Clinton choice relative to the Bush choice. Ideological distance diminishes the utility associated with a particular choice, with the posterior mean for γ roughly the same magnitude as that recovered with the multinomial logit version of the model.

In Problem 8.7 you are invited to reexamine this model for these data, comparing the predictive performance of the MNL model with that of the MNP model.

Problems

8.1 Recall the analysis of votes to authorize the use of force in Iraq (Example 8.3); the data for this problem are part of the `R` package `pscl` (Jackman 2008b). Note that the two Bayes estimates of the probit coefficient on the Republican indicator variables largely agree with one another, but differ from the MLEs for this problem, despite the use of largely uninformative priors. What might be going on here? Specifically:

1. Generate MLEs of the parameters (probit coefficients) using the `glm` command in `R`.

2. Run a long data-augmented Gibbs sampler to generate a large number of samples from the posterior density of the probit coefficients; e.g. use the `MCMCprobit` function in `MCMCpack` (Martin, Quinn and Park 2009). Construct a histogram summarizing the draws from the marginal posterior density of the coefficient on the Republican indicator. Overlay the large n, normal approximation to the sampling distribution of the MLE (i.e. a normal density centered on the MLE, with standard deviation equal to the estimated asymptotic standard error of the MLE). What do you see?

3. Why does the Bayesian posterior density differ from the sampling distribution of the MLE? Hint 1: cross-tabulate the dependent variable against the Republican indicator. Hint 2: try repeating the steps above with Lincoln Chafee (Republican, Rhode Island) excluded from the analysis.

8.2 Consider the marginal data augmentation scheme for the binary probit model presented in Section 8.1.2. That is, we have the latent regression model

$$y_i^* | y_i = 1 \sim N(\mathbf{x}_i\boldsymbol{\beta}, 1)\mathcal{I}(y_i^* \geq 0)$$
$$y_i^* | y_i = 0 \sim N(\mathbf{x}_i\boldsymbol{\beta}, 1)\mathcal{I}(y_i^* < 0)$$

with the prior $\boldsymbol{\beta} \sim N(\mathbf{0}, \mathbf{B}_0)$. We introduce σ^2 as a 'working parameter', assigning it the prior $\sigma^2 \sim$ inverse-Gamma$(\nu_0/2, \nu_0\sigma_0^2/2)$, and considering a version of the model in terms of unidentified parameters: e.g.

$$\tilde{y}_i{}^* | y_i = 1 \sim N(\mathbf{x}_i \tilde{\boldsymbol{\beta}}, \sigma^2) \mathcal{I}(\tilde{y}_i^* \geq 0)$$
$$\tilde{y}_i{}^* | y_i = 0 \sim N(\mathbf{x}_i \tilde{\boldsymbol{\beta}}, \sigma^2) \mathcal{I}(\tilde{y}_i^* < 0)$$

where $\boldsymbol{\beta} = \tilde{\boldsymbol{\beta}}/\sigma$. Step 2 of 'Scheme 1' in Imai and van Dyk (2005a) involves sampling from $p(\tilde{\boldsymbol{\beta}}, \sigma^2 | \tilde{\mathbf{y}}^*, \mathbf{X})$ by first sampling σ^2 from $p(\sigma^2 | \tilde{\mathbf{y}}^*, \mathbf{X})$ and then sampling $\tilde{\boldsymbol{\beta}}$ from $p(\tilde{\boldsymbol{\beta}} | \sigma^2, \tilde{\mathbf{y}}^*, \mathbf{X})$.

1. If the prior for $\boldsymbol{\beta}$ is $N(\mathbf{0}, \mathbf{B}_0)$ then what is the prior for $\tilde{\boldsymbol{\beta}}$?
2. Derive $p(\tilde{\boldsymbol{\beta}}, \sigma^2 | \tilde{\mathbf{y}}^*, \mathbf{X})$.
3. Show that $p(\sigma^2 | \tilde{\mathbf{y}}^*, \mathbf{X})$ is an inverse-Gamma density with parameters $\nu_1/2$ and $\nu_1\sigma_1^2/2$ where $\nu_1 = \nu_0 = n$ and $\nu_1\sigma_1^2 = \nu_0\sigma_0^2 + S + \mathbf{b}_1' \mathbf{B}_0^{-1} \mathbf{b}_1$ and where $\mathbf{B}_1 = (\mathbf{B}_0^{-1} + \mathbf{X}'\mathbf{X})^{-1}$, $\mathbf{b}_1 = \mathbf{B}_1 \mathbf{X}' \tilde{\mathbf{y}}^{*(t)}$ and $S = \sum_{i=1}^n (\mathbf{y}_i^{*(t)} - \mathbf{x}_i \mathbf{b}_1)^2$.
4. Verify that the result in the previous question only holds if the multivariate normal prior density for $\boldsymbol{\beta}$ has mean vector $\mathbf{0}$.

8.3 For the ordinal response model in Section 8.2 there are various ways one might specify the prior over the threshold parameters that respect the ordering restriction implied by the model. Use Monte Carlo methods to explore the properties of these priors; e.g. the priors given in Equation 8.6 or in the code fragments immediately following Equation 8.7. That is, sample many times from a given prior over $\boldsymbol{\tau} = (\tau_1, \ldots, \tau_J)'$, say, with $J = 4$, and use numerical summaries, histograms and scatterplots to characterize the various marginal densities over each element of $\boldsymbol{\tau}$, and the prior bivariate density over each pair of elements in $\boldsymbol{\tau}$ etc.

8.4 Reconsider the ordinal logistic model for levels of political information we examined in Example 8.6. The following `JAGS` code implements an over-parametrized version of the model, containing both an intercept `beta[7]` and all four thresholds `tau`:

JAGS code

```
model{
        for(i in 1:N){  ## loop over observations
              ## form the linear predictor
              mu[i] <- x[i,1]*beta[1] +
                     x[i,2]*beta[2] +
                     x[i,3]*beta[3] +
                     x[i,4]*beta[4] +
                     x[i,5]*beta[5] +
                     x[i,6]*beta[6] + beta[7]

              ## cumulative logistic probabilities
              logit(Q[i,1]) <- tau[1]-mu[i]
              p[i,1] <- Q[i,1]
              for(j in 2:4){
                         logit(Q[i,j]) <- tau[j]-mu[i]
                   ## trick to get slice of the cdf we need
                   p[i,j] <- Q[i,j] - Q[i,j-1]
              }
```

```
            p[i,5] <- 1 - Q[i,4]

            y[i] ~ dcat(p[i,1:5]) ## p[i,] sums to 1 for each i
        }

        ## priors over betas
        beta[1:7] ~ dmnorm(b0[],B0[,])

        ## priors over thresholds
        for(j in 1:4){
            tau0[j] ~ dnorm(0, .01)
        }
         tau[1:4] <- sort(tau0)
}
```

Note that in WinBUGS or OpenBUGS the prior imposing an ordering constraint over the tau parameters would need to be specified using one of the code fragments discussed in Section 8.2.1. Note also the use of the multivariate normal prior over the beta parameters (i.e. a "blocking" strategy, as discussed in Section 6.4.2). Set the hyperparameter B0 to an arbitrarily small constant, so as to induce a vague-yet-proper prior on beta.

Implement this model using the data frame politicalInformation supplied as part of the pscl package (Jackman 2008b) in R. After using BUGS/JAGS to generate samples from the posterior density of $\boldsymbol{\theta} = (\boldsymbol{\beta}, \boldsymbol{\tau})'$, what transformations are required so as to map the sampled $\boldsymbol{\theta}$ into the space of identified parameters? Apply this transformation to the MCMC output. Is the MCMC algorithm mixing well in the space of identified parameters? Compare your results with those obtained in Example 8.6.

8.5 Consider the multinomial logit model

$$\mathbf{y}_i \sim \text{Multinomial}(\boldsymbol{\pi}_i, 1)$$
$$\pi_{ij} = \frac{\exp(\mathbf{x}_i \boldsymbol{\beta}_j)}{\sum_{k=0}^{J} \exp(\mathbf{x}_i \boldsymbol{\beta}_k)}$$

where $i = 1, \ldots, n$ indexes decision-makers and $j = 0, \ldots, J$ indexes choices.

1. Verify that $\sum_{j=0}^{J} \pi_{ij} = 1$.
2. Verify the claim in Section 8.3 that the parameters $\boldsymbol{\beta} = (\boldsymbol{\beta}_0, \ldots, \boldsymbol{\beta}_J)'$ are unidentified. Specifically, show that we obtain the same likelihood contributions π_{ij} with $\boldsymbol{\beta}_j = \boldsymbol{\beta}_j^*$ as we do from $\boldsymbol{\beta}_j = \tilde{\boldsymbol{\beta}}_j = \boldsymbol{\beta}_j^* + c$, $c \neq 0$; i.e. if $\boldsymbol{\beta}_j^*$ is a MLE of $\boldsymbol{\beta}_j$, then so too is $\tilde{\boldsymbol{\beta}}_j$.

8.6 Consider the analysis of vote choices using the multinomial logit model, presented in Example 8.7.

1. Replicate the Bayesian analysis with the vague normal priors given in Example 8.7 using either BUGS or JAGS. The data are available as vote92 in the R package pscl (Jackman 2008b). Up to Monte Carlo error, you should be able to replicate the results in Table 8.11.

2. Estimate the binary logit model that looks at the Clinton versus Bush choice, excluding Perot voters from the analysis entirely, but otherwise using the same choice model as presented in Example 8.7. Compare the posterior density for the parameters in the binary choice model with the posterior density for the corresponding parameters in the multinomial analysis. Comment on what this says about the multinomial logit model.

3. The `R` packages `MCMCpack` (Martin, Quinn and Park 2009) and `bayesm` (Rossi and McCulloch 2008) both provide functions for MCMC-based analysis of the multinomial logit model; the relevant function in the `MCMCpack` package is `MCMCmnl` while the relevant function in the `bayesm` package is `rmnlIndep-Metrop`. Take care passing the covariates to these packages; recall that z_{ij} varies over choices and survey respondents, while variables like party identification and gender are constant over choices. Compare the performance of the algorithms in either package with that of the implementation in `BUGS` and/or `JAGS`. Verify that up to Monte Carlo error your answers are the same. Which implementation do you prefer, and why?

4. Verify that the means of the posterior density of $(\boldsymbol{\beta}, \gamma)'$ reported in Table 8.11 are very close to the maximum likelihood estimates of those parameters.

5. Consider a hypothetical female, Democratic partisan, with negative retrospective views of the economy. What happens to the predicted probabilities of voting for a particular candidate as we vary this hypothetical respondent's ideological self-placement ξ_i from 'extremely liberal' ($\xi_i = 1$) to 'extremely conservative' ($\xi_i = 7$), in integer steps, with the average candidate placements $\bar{\xi}_{\text{Bush}} = 5.32$, $\bar{\xi}_{\text{Clinton}} = 2.98$ and $\bar{\xi}_{\text{Perot}} = 4.49$; recall that the variable that enters the analysis is $z_{ij} = (\xi_i - \bar{\xi}_j)^2$, $j \in \{\text{'Bush'}, \text{'Clinton'}, \text{'Perot'}\}$. For the purposes of answering this question set the model parameters $\boldsymbol{\beta}$ and γ to their posterior means.

6. Repeat the previous question, but this time do not ignore the posterior uncertainty in the model parameters $\boldsymbol{\beta}$ and γ. The easiest way to do this might be to augment the `BUGS/JAGS` code presented in Example 8.7, sampling from the posterior predictive mass function for the vote choice $\tilde{y}_i$ of the hypothetical voter. In a table or graph, summarize the sampled $\tilde{y}_i$ as a function of the ideological self-placements ξ_i we hypothetically ascribe to voter i.

8.7 Replicate the MCMC-based analysis of the multinomial probit model presented in Example 8.8.

1. Verify that up to Monte Carlo error, you obtain the same set of posterior means and 95 % highest posterior density intervals for the parameters as those reported in Table 8.13.

2. Generate predicted probabilities of each choice, for each individual, conditional on the parameters set to their respective posterior means. This is straightforward to do using the `predict` function in the `MNP` `R` package (Imai and van Dyk 2005b).

3. Assign each respondent to a predicted outcome, using the rule that the outcome with the highest predicted probability for a given respondent is the predicted

outcome for that respondent. Compare these predicted outcomes with the actual outcomes in a cross-tabulation. Comment on how well the model fits the data using this criterion.

4. Compare the fit of the multinomial probit model with the fit of the multinomial logit model to these data considered in Example 8.7 and in Problem 8.6. Is there any evidence that the MNP model generates a superior fit to the data than the MNL model?

5. Repeat part 5 of Problem 8.6, but with the fitted MNP model of generating the predicted values. How do the answers generated here compare with the predictions from the MNL model?

9

Bayesian approaches to measurement

Many problems in the social sciences involve making inferences about attributes of observational units that are not directly observable. Here we refer to these quantities as latent states or latent variables, $\boldsymbol{\xi}$. Examples include the ideological dispositions of survey respondents (e.g. Erikson 1990), legislators (Clinton, Jackman and Rivers 2004a), judges (Martin and Quinn 2002), or political parties (Huber and Inglehart 1995); the quantitative, verbal and analytic abilities of applicants to graduate school (e.g. as measured by the Graduate Record Examination, or GREs); levels of democracy in the world's countries (Gurr and Jaggers 1996); locations in an abstract, latent space used to represent actors in a social network (Hoff, Raftery and Handcock 2002); or levels of support for political candidates over the course of an election campaign (e.g. Green, Gerber and Boef 1999). In each instance, the available data are manifestations of the latent quantity (*indicators*) and the inferential problem can be stated as follows: conditional on observable data $\mathbf{y}$, what should we believe about latent quantities $\boldsymbol{\xi}$?

9.1 Bayesian inference for latent states

Bayesian inference is particularly attractive for latent variables. One of the most important inferential questions underlying latent variable models – what should we believe about latent variables $\boldsymbol{\xi}$ given observed indicators $\mathbf{y}$? – has a straightforward interpretation in the Bayesian approach. As we have seen throughout this book, Bayesian inference for $\boldsymbol{\xi}$ is based on the *posterior density* of $\boldsymbol{\xi}$, $p(\boldsymbol{\xi}|\mathbf{y})$, a probability density characterizing beliefs about $\boldsymbol{\xi}$ in light of data $\mathbf{y}$. Thus, to say we want to know about $\boldsymbol{\xi}$, a random variable, is to say that we seek some useful characterization of its posterior density $p(\boldsymbol{\xi}|\mathbf{y})$.

Bayesian Analysis for the Social Sciences S. Jackman

In many applications we are also interested in *indicator quality*, and this problem is usually tackled via a parametric model: e.g. a latent variable $\boldsymbol{\xi}$ maps into a observable response $\mathbf{y}$ via a function parameterized by $\boldsymbol{\beta}$, where $\boldsymbol{\beta}$ taps the quality of the indicator, *inter alia*. Learning about the $\boldsymbol{\beta}$ parameters is critical too, since it lets us refine our methods, say, directing possibly scarce resources for measurement towards better indicators and away from poor indicators. Thus, the inferential problem of latent variable modeling can be re-stated as a *joint* problem, that of learning about $\boldsymbol{\theta} = (\boldsymbol{\xi}, \boldsymbol{\beta})$ from data $\mathbf{y}$.

9.1.1 A formal role for prior information

Via Bayes Rule, the posterior density for $\boldsymbol{\theta}$ is proportional to the prior density times the likelihood of the data, $\mathbf{y}$: i.e. $p(\boldsymbol{\theta}|\mathbf{y}) \propto p(\boldsymbol{\theta})p(\mathbf{y}|\boldsymbol{\theta})$. The prior density $p(\boldsymbol{\theta})$ can play an interesting role in the latent variable context. Suppose that prior beliefs about the latent indicators, $\boldsymbol{\xi}$, are independent of prior beliefs about the veracity of the indicators, such that $p(\boldsymbol{\theta}) = p(\boldsymbol{\xi}, \boldsymbol{\beta}) = p(\boldsymbol{\xi})p(\boldsymbol{\beta})$. Prior information about $\boldsymbol{\xi}$ will vary from application to application, but could be helpful if there is limited information about $\boldsymbol{\xi}$ in the indicators $\mathbf{y}$. For example, prior information about $\boldsymbol{\xi}$ could well include historical sources or expert evaluations, say, if we have only recently started to systematically collect indicators in a form suitable for quantitative, statistical analysis. In analyses with a longitudinal dimension, prior information about the latent state may come from its past values. In spatial models, the prior for $\boldsymbol{\xi}$ might draw on the values of the latent variable in neighboring locations.

In addition, hierarchical modeling (Chapter 7) provides a way for additional information to enter the analysis by writing the prior density for $\boldsymbol{\xi}$ as a function of covariates, say $\mathbf{z}$. These covariates are not indicators of $\boldsymbol{\xi}$ *per se*, but may nonetheless convey important information about $\boldsymbol{\xi}$. Consider the following:

- in standardized testing, responses to the test items are the indicators $\mathbf{y}$, and the subjects' latent ability constitute $\boldsymbol{\xi}$. But data on the subjects' socio-economic backgrounds, the schools they attend, and so on might well constitute important information $\mathbf{z}$, relevant to $\boldsymbol{\xi}$. Indeed, consider the thought experiment of learning about latent ability in the absence of the test responses $\mathbf{y}$ or literally *prior* to obtaining $\mathbf{y}$; in this circumstance, the covariates $\mathbf{z}$ might well be a useful source of information.
- in studies of legislative politics, $\mathbf{y}$ might be the recorded votes of legislators on roll calls and $\boldsymbol{\xi}$ the ideological dispositions of legislators. In this case information such as the legislators' party affiliations and characteristics of their districts might be useful information $\mathbf{z}$, relevant to the problem of learning about $\boldsymbol{\xi}$.

Prior information about the veracity of the indicators can also enter the analysis in a formal way in a Bayesian analysis. For instance, previous research may supply more or less information about the measurement properties of particular indicators: indicator j may be known *a priori* to be a more reliable indicator of the concept under study than item k, and the prior densities for β_j and β_k may be specified accordingly.

9.1.2 Inference for many parameters

Latent variable modeling can give rise to a prodigious number of parameters, and quickly. To take just one example, consider the problem of modeling Congressional roll calls as a function of legislators' latent ideological locations (we return to this problem in Section 9.3). In this example, the votes are the indicators, **y**, and the latent variables $\boldsymbol{\xi}$ are the legislators' latent preferences, represented as a point in a low-dimensional, abstract space (e.g., for a one dimensional model, the space is a line, usually interpreted as a 'left-to-right' ideological continuum). An identical statistical model is used to analyze data from standardized tests, where the latent variable is the test-taker's ability, and each test item has two parameters associated with it, tapping item difficulty and item discrimination, respectively. Thus, with two parameters per item (or vote), and one parameter per test-taker (or legislator), the number of parameters in the model can grow quickly. More generally, with data from n legislators voting on m roll calls, a d-dimensional model has $p = nd + m(d + 1)$ parameters.

Table 9.1 presents values of p for five different data sets. A moderately sized roll call data set (say the 105th U.S. Senate) with $n = 100$, $m = 534$ non-unanimous roll calls and $d = 1$ yields $p = 1168$ unknown parameters, while a two dimensional model has $p = 1802$ parameters. A typical House of Representatives (e.g. the 93rd House) has $n = 442$ and $m = 917$, and so a one dimensional model has $p = 2276$ parameters, while a two dimensional model has $p = 3635$ parameters. Pooling across years dramatically increases the number of parameters: for instance, Poole and Rosenthal (1997) report that fitting a two dimensional model to roughly two hundred years of US House of Representatives roll call data gave rise to an optimization problem with $p > 150\,000$ parameters.

The proliferation of parameters causes several problems. The usual optimality properties of conventional frequentist estimators, such as maximum likelihood, may not hold when, as in this case, the number of parameters is a function of the sample size. This phenomenon is known as the 'incidental parameters problem' (Neyman and Scott 1948); see Lancaster (2000) for a recent survey. Bayesian inference is not immune from this problem, in the sense that a Bayesian point estimate such as the mean of the posterior density also tends to the wrong limit as the sample size grows. Diaconis and Freedman (1986a,b) provide details. Of course, in the Bayesian case, the posterior density is a combination of the prior and the likelihood, and so it is always possible to use a prior that can mitigate the inconsistency due to the incidental parameters problem. This suggests that Bayesian approaches may have superior large sample properties than analysis based

Table 9.1 Parameter proliferation in the analysis of legislative roll call data. Table entries under the 'Dimensions' columns show the number of parameters appearing in an item-response theory model typically fit to roll call data, given the number of number of legislators n, the number of roll calls m and the number of latent dimensions d being fit to the data.

	Legislators	Roll calls	Dimensions (d)		
Legislature	n	m	1	2	3
US. Supreme Court, 1994–97	9	213	435	657	879
105th US Senate	100	534	1168	1802	2436
93rd US House	442	917	2276	3635	4994
US Senate, 1789–1985	1714	37 281	76 276	115 271	154 266
US House, 1789–1985	9759	32 953	75 485	118 017	160 549

solely on the likelihood function, but specifying a prior so as to alleviate the inconsistency due to incidental parameters would seem to be delicate and problem-specific.

The other difficulty presented in this context is computational. With thousands of parameters, if not tens or hundreds of thousands of parameters, computing standard errors for maximum likelihood estimates is often impractical, since the information matrix is too large for direct inversion. Of course, with a simulation-based, Bayesian approach such as the Gibbs sampler we avoid computing and trying to invert massive matrices, building up a Monte Carlo based approximation to the posterior density by sequentially sampling from low dimensional conditional densities.

9.2 Factor analysis

We consider Bayesian approaches to some standard models with latent variables, beginning with a simple factor analytic model. Factor analysis is often presented as a model for the covariance matrix of p indicators, $\mathbf{\Sigma}$; e.g., the orthonormal factor model with k factors, is usually given as $\mathbf{\Sigma} = \mathbf{\Lambda}\mathbf{\Phi}\mathbf{\Lambda}' + \mathbf{\Psi}$ where $\mathbf{\Lambda}$ is a p-by-k matrix of factor loadings, $\mathbf{\Phi} = \mathbf{I}_k$ and $\mathbf{\Psi}$ is a diagonal p-by-p matrix with 'uniquenesses' on the diagonal. This presentation obscures the fact that the factor analytical model is a model for the observed indicators $\mathbf{y}$ in which latent variables $\boldsymbol{\xi}$ and factor loadings $\mathbf{\Lambda}$ appear as parameters, *inter alia*; see, for instance, Mardia, Kent and Bibby (1979, §9.2.1) or Anderson (2003, §14.2.1).

In our fully Bayesian approach, we will work with the latent variable representation of the factor analytic model, the goal being to compute the posterior density over all unknown parameters in the model conditional on the observable indicators and any prior information. For instance, Figure 9.1 presents a directed acyclic graph representing a factor analytic model. The single latent factor, ξ_i, has $p = 4$ observed indicators, y_{ij}, $j = 1, \ldots, p$, all conditionally independent from one another given ξ_i, where $i = 1, \ldots, n$ indexes the observations. The relationship between each indicator and the single latent factor is parameterized in terms of $\boldsymbol{\gamma}_j$ (factor loadings and/or intercepts), and measurement error variances ω_j^2, $j = 1, \ldots, p$.

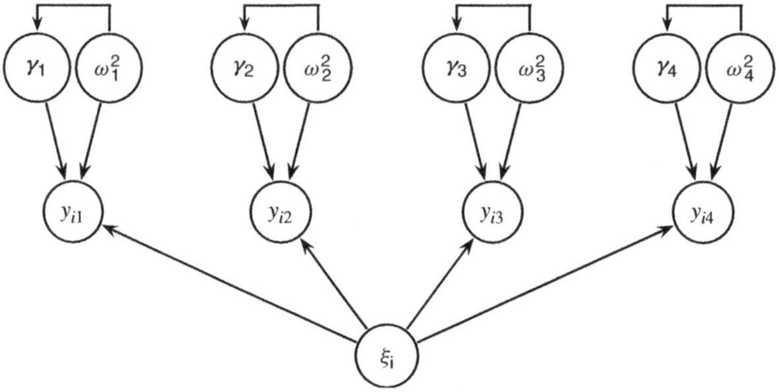

Figure 9.1 Directed acyclic graph, factor analysis model with one single latent factor with four indicators.

9.2.1 Likelihood and prior densities

The DAG in Figure 9.1 represents a factor analysis model at a reasonably high level of abstraction, and does not specify any distributions over the random quantities in the model. We now specify these distributions to complete the description of the model. For the case of continuous indicators, and for the purposes of exposition, normal distributions will suffice, along with inverse-Gamma prior densities over the variance parameters: i.e. for the observed indicators we have

$$y_{ij} \sim N(\gamma_{j0} + \gamma_{j1}\xi_i, \omega_j^2), \tag{9.1}$$

with likelihood function

$$\mathcal{L} \equiv p(\mathbf{Y}|\boldsymbol{\theta}) \propto \prod_{i=1}^{n}\prod_{j=1}^{p} \phi\left(\frac{y_{ij} - \gamma_{j0} - \gamma_{j1}\xi_i}{\omega_j}\right), \tag{9.2}$$

where $\mathbf{Y}$ is the n-by-p matrix of the observed indicators, $\boldsymbol{\theta} = \{\boldsymbol{\Gamma}, \boldsymbol{\psi}, \boldsymbol{\xi}\}$ where $\boldsymbol{\Gamma} = (\gamma_{10}, \ldots, \gamma_{p1})'$, $\boldsymbol{\psi} = (\omega_1^2, \ldots, \omega_p^2)'$ and $\boldsymbol{\xi} = (\xi_1, \ldots, \xi_n)'$ are the unknown parameters for this model and ϕ is the standard normal density. Prior distributions over the components of $\boldsymbol{\theta}$ complete the specification of the model. Assumptions of prior independence let us factor the prior density as

$$\begin{aligned} p(\boldsymbol{\theta}) = p(\boldsymbol{\Gamma}, \boldsymbol{\psi}, \boldsymbol{\xi}) &= p(\xi_1, \ldots, \xi_n)\, p(\boldsymbol{\gamma}_1|\omega_1^2)p(\omega_1^2) \ldots p(\boldsymbol{\gamma}_p|\omega_p^2)p(\omega_p^2) \\ &= \prod_{i=1}^{n} p(\xi_i) \prod_{j=1}^{p} p(\boldsymbol{\gamma}_j|\omega_j^2)p(\omega_j^2) \end{aligned} \tag{9.3}$$

i.e. we assume that the ξ_i are *a priori* independent of one another and of the $\boldsymbol{\gamma}_j$ and ω_j^2, but use a 'conditional-times-marginal' factorization to represent the joint prior density over $(\boldsymbol{\gamma}_j, \omega_j^2)$, subject to the assumptions that $(\boldsymbol{\gamma}_j, \omega_j^2) \perp\!\!\!\perp (\boldsymbol{\gamma}_k, \omega_k^2)\, \forall j, k = 1, \ldots, p, j \neq k$. We operationalize this factorization of the prior with the following conjugate prior densities:

$$\xi_i \stackrel{\text{iid}}{\sim} N(\mu_\xi, \sigma^2), \quad i = 1, \ldots, n, \tag{9.4}$$

$$\boldsymbol{\gamma}_j|\omega_j^2 \sim N(\mathbf{g}_{j0}, \omega_j^2\mathbf{G}_{j0}), \quad j = 1, \ldots, p, \tag{9.5}$$

$$\omega_j^2 \sim \text{inverse-Gamma}(\nu_{j0}/2, \nu_{j0}\omega_{j0}^2/2), \quad j = 1, \ldots, p, \tag{9.6}$$

where $\mu_\xi, \sigma^2, \mathbf{g}_{j0}, \mathbf{G}_{j0}, \nu_{j0}$ and $\omega_{j0}^2, j = 1, \ldots, p$ are user-specified hyper-parameters. In any specific application, the analyst would choose values for these hyperparameters so as to reflect her prior information about the corresponding quantities $\xi_i, \boldsymbol{\gamma}_j$ and ω_j^2, respectively.

An absence of prior information about the factor loadings can be expressed in the usual way, by setting large values for the elements of the prior sum-of-squares matrices $\mathbf{G}_{j0}$, such that the prior is proper but approximately uniform in the region of the parameter space supporting non-negligible likelihood. Similarly, a relative absence of prior information about the measurement-error variances ω_j^2 can be expressed in the choices of the hyper-parameters ν_{j0} and ω_{j0}^2. Note also that absent any prior information distinguishing

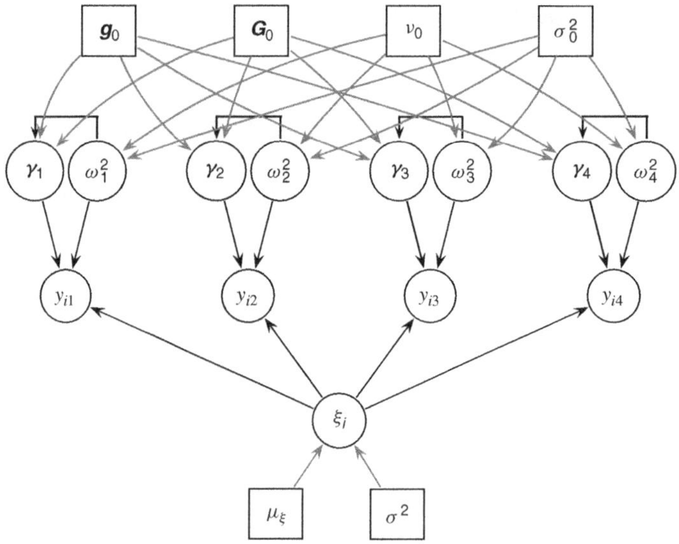

Figure 9.2 Directed acyclic graph for single latent variable model with four indicators, with hyperparameters. The hyperparameters $\mathbf{g}_0$ and $\mathbf{G}_0$ parameterize the (common) prior normal densities for the $\boldsymbol{\gamma}_j$ parameters, while ν_0 and σ_0^2 parameterize the (common) prior inverse-Gamma densities for the ω_j^2 parameters.

the items, the analyst could use the same prior over all $\boldsymbol{\gamma}_j$ and ω_j. As written, the model specifies a common prior over the latent factor ξ_i; but if the researcher had prior information that distinguished the observations by the expected value of the latent variable, then this could be encoded in prior densities specific to each ξ_i, or groups of ξ_i, say via a *hierarchical* model for the ξ_i.

The DAG for the model including the priors appears in Figure 9.2. In this rendering of the model, a common $N(\mathbf{g}_0, \omega_j^2\mathbf{G}_0)$ prior is assumed for all the $\boldsymbol{\gamma}_j$ parameters, along with a common inverse-Gamma$(\nu_0/2, \nu_0 2\sigma_0^2/2)$ prior density for the ω_j^2 parameters. The fixed hyperparameters appear in square boxes since they are deterministic; the edges connecting fixed hyperparameters to their respective child nodes are shown with a lighter color.

9.2.2 Identification

The parameters of the model in Equation 9.1 (generating the likelihood in Equation 9.2) are not identified. For one thing, since the scale of ξ_i is not fixed, any arbitrary scaling of the ξ_i, say, $\tilde{\xi}_i = c\xi_i, c \neq 0$, can be offset by simply dividing each of the γ_{j1} factor loadings by c to produce $\tilde{\gamma}_{j1} = \gamma_{j1}/c$. That is, an infinite number of combinations of $\tilde{\xi}_i$ and $\tilde{\gamma}_{j1}$ provide the same fit to the data; put differently, ξ_i and γ_{j1} are not identified. Identification also fails in that any rescaling of ω_j can be offset by an inverse re-scaling of γ_{j0}, γ_{j1} and the ξ_i, to again produce identical values of the likelihood function. And further, the ξ_i can be all translated (shifted) by some non-zero quantity, and an offsetting shift applied to the intercept parameters γ_{j0}.

As we have noted previously, the fact that the model parameters are not identified poses no formal problem in a Bayesian analysis. Identification (or the absence of it) is a property of the likelihood function, whereas Bayesian inference simply uses the likelihood function to map through the data from prior beliefs to posterior beliefs. This said, the very point of data analysis is to *learn* from data, and so working with unidentified likelihood functions is usually unsatisfactory from a practical standpoint.

In the specific case of factor analysis, the lack of identification is relatively easy to remedy, by imposing *normalizations* on the latent variable ξ_i and/or the item parameters $\boldsymbol{\gamma}_j = (\gamma_{j0}, \gamma_{j1})'$. For instance (and these identifying restrictions will be familiar to experienced users of factor analysis), if we impose the restriction that the latent variable ξ_i has a fixed mean and variance (location and scale), then we induce *local identification*; the parameters are identified up to a 180 degree rotation of the latent scale (i.e. setting $c = -1$ in discussion, above). A sign restriction on one of the γ_j parameters is then sufficient to rule out invariance to rotation, providing *global identification*. Normalizing the ξ_i to have a fixed location rules out translations and ensures that the ξ_i and the intercept parameters γ_{j0} are jointly identified; a similar result follows by simply setting two of the ξ_i to known (but different) constants. Note also that fixing the scale of the ξ_i is also sufficient to identify the measurement error variance parameters ω_j^2. Equivalently, note that we could impose a point restriction on one factor loading (one of the γ_{j1} parameters) to rule out invariance to scale, and another point restriction on an intercept parameter (one of the γ_{j0}) to rule out invariance to translation. Combinations are also possible: a location restriction on the ξ_i and a scale restriction via setting one of the γ_{j1} factor loadings or one of the error variance parameters ω_j^2; alternatively, we could impose a scale restriction on the ξ_i and then rule out invariance to translation via a restriction on one of the intercept parameters γ_{j0}. In short, there are lots of ways to impose the simple identifying normalization.

I conclude this section on identification by noting how a proper prior distribution provides 'near-identification'. I use this term with some hesitation, recalling the earlier observation that identification is a property of the likelihood function, and not of the prior nor the posterior. Nonetheless, if parameter restrictions can be considered limiting forms of prior densities, then there is at least a functional equivalence between introducing prior information about parameters, and imposing identifying restrictions. Thus, given the prior density for the ξ_i in Equation 9.4, $\xi_i \stackrel{\text{iid}}{\sim} N(\mu_\xi, \sigma^2)$, then the mean of the ξ_i, $\bar{\xi} = n^{-1} \sum_{i=1}^{n} \xi_i$, follows a $N(\mu_\xi, \sigma^2/n)$ density. Hence, as n grows large relative to σ^2, the (implied) prior density for the mean of the ξ_i is collapsing to a degenerate density concentrated on μ_ξ. For instance, if we choose $\mu_\xi = 0$ and $\sigma^2 = 1$ then with $n = 100$, we have an implied prior 95 % credible interval on $\bar{\xi}$ of $\pm 1.96\sqrt{1/100} = \pm .196$, and with $n = 1000$ we have an implied prior 95 % credible interval of $\pm 1.96\sqrt{1/1000} = \pm .062$, and so on. Similarly, if $\xi_i \stackrel{\text{iid}}{\sim} N(\mu_\xi, \sigma^2)$, then the prior variance of the ξ_i is just σ^2, which *a priori* imposes a unique scale on the ξ_i. Thus, a proper common iid prior on the ξ_i pushes in the direction of identification, with the implied location restriction on the ξ_i more closely approximating a normalizing point restriction on $\bar{\xi}$ as $n \to \infty$, converging towards a point restriction at rate $1/\sqrt{n}$. All this is to say that although as a technical matter, the prior $\xi_i \stackrel{\text{iid}}{\sim} N(0, 1)$ does not (and can not) identify the likelihood function, as a practical matter there is almost no difference between (a) estimating a latent variable model with normalizing restrictions imposed and (b) performing a Bayesian analysis with the prior $\xi_i \stackrel{\text{iid}}{\sim} N(\mu_\xi, \sigma^2)$ at least when $n \gg \sigma^2$.

9.2.3 Posterior density

Via Bayes Rule, the posterior density is proportional to the prior times the likelihood, and so for the factor analysis problem we have $p(\boldsymbol{\theta}|\mathbf{Y}) \propto p(\boldsymbol{\theta})p(\mathbf{Y}|\boldsymbol{\theta})$ where the prior $p(\boldsymbol{\theta})$ is given in Equation 9.3 and the likelihood $p(\mathbf{Y}|\boldsymbol{\theta})$ is given in Equation 9.2. Note that the posterior density is high-dimensional, with n of the ξ_i parameters, $2p$ of the γ parameters and another p measurement error variance parameters ω_j^2, for a total of $n + 3p$ parameters, absent any identifying restrictions. To compute this high dimensional density we rely on the Gibbs sampler.

The posterior density is the $n + 3p$ dimensional distribution $p(\boldsymbol{\theta}|\mathbf{Y}) = p(\boldsymbol{\Gamma}, \boldsymbol{\psi}, \boldsymbol{\xi}|\mathbf{Y})$. The Gibbs sampler generates samples from this joint posterior density by iterating over the following scheme. At iteration t,

1. sample $\xi_i^{(t)}$ from its conditional density given the current values of other components of the model's DAG. To deduce the form of this conditional density, note that y_{ij} and $y_{i'j}$ are conditionally independent given $\boldsymbol{\Gamma}$, $\boldsymbol{\psi}$, ξ_i and $\xi_l \, \forall i' \neq i$. Letting $\mathcal{G}$ denote the DAG for this model, we apply the result in Proposition 5.2, and obtain the conditional density

$$p(\xi_i|\mathcal{G}_{\backslash \xi_i}) = p(\xi_i|\text{parents}[\xi_i]) \times \prod_{w \in \text{ children } [\xi_i]} p(w|\text{parents}[w]) \qquad (9.7)$$

The parents of ξ_i are its hyperparameters μ_ξ and σ^2. We observe that ξ_i is a parent node of $y_{i1}, \ldots, y_{ip}$, but not of $y_{i'1}, \ldots, y_{i'p}$, $\forall i' \neq i$. In addition, the children of ξ_i are the $y_{i1}, \ldots, y_{ip}$, and their parents are ξ_i, $\boldsymbol{\Gamma}$ and $\boldsymbol{\psi}$. Let g denote the distribution formed by taking the product of the distributions on the right-hand side of equation 9.7. Then, at this stage of the Gibbs sampler we sample

$$\xi_i^{(t)} \sim g(\xi_i \,|\, \mu_\xi, \sigma^2, \boldsymbol{\Gamma}^{(t-1)}, \boldsymbol{\omega}^{2\,(t-1)}, \mathbf{y}_i), \quad i = 1, \ldots, n, \qquad (9.8)$$

where $\mathbf{y}_i$ is the i-th row of the n-by-p matrix of the indicators $\mathbf{Y}$.

2. Sample $\boldsymbol{\gamma}_j^{(t)}$ from its conditional distribution, $j = 1, \ldots, p$. The parents of $\boldsymbol{\gamma}_j$ are the prior hyperparameters $\mathbf{g}_{j0}$ and $\mathbf{G}_{j0}$ and ω_j^2. The only children of $\boldsymbol{\gamma}_j$ are the data on the jth indicator, $\mathbf{y}_j = (y_{1j}, \ldots y_{nj})'$. But the other parents of $\mathbf{y}_j$ are $\boldsymbol{\xi}$, and ω_j^2. Thus at this stage of the Gibbs sampler we sample

$$\boldsymbol{\gamma}_j^{(t)} \sim g(\boldsymbol{\gamma}_j \,|\, \mathbf{g}_{j0}, \mathbf{G}_{j0}, \boldsymbol{\xi}^{(t)}, \omega_j^{2\,(t-1)}, \mathbf{y}_j), \quad j = 1, \ldots, p. \qquad (9.9)$$

3. Sample $\omega_j^{2\,(t)}$ from its conditional distribution, again with $j = 1, \ldots, p$. The parents of ω_j^2 are its prior hyperparameters ν_{j0} and σ_{j0}^2. The children of ω_j^2 are $\boldsymbol{\gamma}_j$ and the data $\mathbf{y}_j = (y_{1j}, \ldots y_{nj})'$, and the other parents of $\mathbf{y}_j$ are $\boldsymbol{\gamma}_j$ and $\boldsymbol{\xi}$. Thus we sample

$$\omega_j^{2,(t)} \sim g(\omega_j^2 | \, \nu_{j0}, \sigma_{j0}^2, \boldsymbol{\gamma}_j^{(t)}, \boldsymbol{\xi}^{(t)}, \mathbf{y}_j), \quad j = 1, \ldots, p. \qquad (9.10)$$

Iterating over these three steps (and their component sub-steps) updates all elements of $\boldsymbol{\theta}^{(t-1)}$ to $\boldsymbol{\theta}^{(t)}$ and constitutes a complete iteration of the Gibbs sampler.

To determine the specific forms of these conditional distributions, we now rely on standard results from Bayesian theory on the analysis of linear regression models with conjugate priors (see Section 2.5.3). Consider first the conditional distribution from which we sample $\xi_i^{(t)}, i = 1, \ldots, n$, in Equation 9.8. Note that we have the 'regression-like' model in Equation 9.1, except that we treat the ξ_i as unknown parameters and the $\boldsymbol{\gamma}_j$ as data. Re-arranging the regression model, we have $w_{ij} = y_{ij} - \gamma_{j0} \sim N(\xi_i \gamma_{j1}, \omega^2 j)$. Applying Proposition 2.11 to this transformed model, we obtain the following expression for the conditional distribution of ξ_i, $i = 1, \ldots, n$:

$$\xi_i | \mu_\xi, \sigma^2, \boldsymbol{\Gamma}, \boldsymbol{\psi}, \mathbf{y}_i \sim N(\mu_\xi^*, \sigma^{2*}) \tag{9.11}$$

where

$$\mu_\xi^* = \frac{\dfrac{\mu_\xi}{\sigma^2} + \dfrac{\hat{\xi}_i}{V(\hat{\xi}_i)}}{\dfrac{1}{\sigma^2} + \dfrac{1}{V(\hat{\xi}_i)}} \quad \text{and} \quad \sigma^{2*} = \frac{\omega_j^2}{\dfrac{1}{\sigma^2} + \dfrac{1}{V(\hat{\xi}_i)}}$$

and where

$$\hat{\xi}_i = (\boldsymbol{\gamma}_1' \boldsymbol{\Omega}^1 \boldsymbol{\gamma})^{-1} \mathbf{w}_i' \boldsymbol{\gamma}_1$$
$$V(\hat{\xi}_i) = \frac{1}{p} (\boldsymbol{\gamma}_1' \boldsymbol{\Omega}^1 \boldsymbol{\gamma})^{-1}$$

where $\boldsymbol{\Omega} = \text{diag}(\omega_1^2, \ldots, \omega_p^2)$, i.e. $\mathbf{w}_i = (w_{i1}, \ldots, w_{ip})' = (y_{i1} - \gamma_{10}, \ldots, y_{ip} - \gamma_{p0})'$ and $\boldsymbol{\gamma}_1 = (\gamma_{11}, \ldots \gamma_{1p})'$.

For the $\boldsymbol{\gamma}_j$ parameters, the conditional distributions are obtained by Bayesian analysis of the regression model in Equation 9.1 in the usual way, with ξ_i as a predictor and $\boldsymbol{\gamma}_j$ as the unknown parameters. That is, for $j = 1, \ldots, p$,

$$\boldsymbol{\gamma}_j | \mathbf{g}_{j0}, \mathbf{G}_{j0}, \boldsymbol{\xi}, \omega_j^2, \mathbf{y}_j \sim N(\mathbf{g}_{j1}, \omega_j^2 \mathbf{G}_{j1}), \tag{9.12}$$

where

$$\begin{aligned}
\mathbf{g}_{j1} &= (\mathbf{G}_{j0}^{-1} + \mathbf{Z}'\mathbf{Z})^{-1} (\mathbf{G}_{j0}^{-1} \mathbf{g}_{j0} + \mathbf{Z}'\mathbf{Z} \hat{\boldsymbol{\gamma}}_j), \\
\mathbf{G}_{j1} &= (\mathbf{G}_{j0}^{-1} + \mathbf{Z}'\mathbf{Z})^{-1}, \\
\mathbf{Z} &= [\boldsymbol{\iota} \ \boldsymbol{\xi}] \quad \text{and} \\
\hat{\boldsymbol{\gamma}}_j &= (\mathbf{Z}'\mathbf{Z})^{-1} \mathbf{Z}'\mathbf{y}.
\end{aligned}$$

That is, $\mathbf{Z}$ is the n-by-2 matrix formed with a unit vector $\boldsymbol{\iota}$ in the first column and the $\boldsymbol{\xi}$ in the second column (a regressor matrix for the purposes of inference for $\boldsymbol{\gamma}_j$).

Finally, for the measurement error variance parameters ω_j^2, we use Proposition 2.11 to deduce that for $j = 1, \ldots, p$,

$$\omega_j^2 | \nu_{j0}, \sigma_{j0}^2, \boldsymbol{\gamma}_j, \boldsymbol{\xi}, \mathbf{y}_j \sim \text{inverse-Gamma}(\nu_1/2, \nu_1 \sigma_1^2/2), \tag{9.13}$$

where

$$\begin{aligned}
\nu_1 &= \nu_0 + n, \\
\nu_1\sigma_1^2 &= \nu_0\sigma_0^2 + S_j + r_j, \\
S_j &= (\mathbf{y} - \mathbf{Z}\hat{\boldsymbol{\gamma}}_j)'(\mathbf{y} - \mathbf{Z}\hat{\boldsymbol{\gamma}}_j) \quad \text{and} \\
r_j &= (\mathbf{g}_{j0} - \hat{\boldsymbol{\gamma}}_j)'(\mathbf{G}_{j0} + (\mathbf{Z}'\mathbf{Z})^{-1})^{-1}(\mathbf{g}_{j0} - \hat{\boldsymbol{\gamma}}_j).
\end{aligned}$$

■ **Example 9.1**

Social attributes of fifty American states. To demonstrate this implementation of the factor analytic model, we consider the following data on the 50 states of the United States, provided as part of the `base` package in `R` (R Development Core Team 2009):

1. `Income`, per capita income in 1974, thousands of dollars
2. `Illiteracy`, 1970, percent of population
3. `Life Exp`, life expectancy in years (1969–71)
4. `Murder`, murder and non-negligent manslaughter rate per 100 000 population in 1976
5. `HS Grad`, percent high-school graduates (1970)

The unique elements of the correlation matrix of the indicators are presented in Table 9.2, along with the five eigenvalues of the correlation matrix. There are few small correlations in the correlation matrix (the −.23 correlation between the income and murder indicators is the smallest). The first eigenvalue is 3.2, with the second eigenvalue just under one (.94), a pattern that is typically interpreted to mean that a one dimensional model will be sufficient to model the data (the 'eigenvalues greater than one' rule of thumb).

Table 9.2 Lower triangle of correlation matrix and eigenvalues, five social attributes of fifty american states.

	Income	Illiteracy	Life Exp	Murder	
Illiteracy	−0.44				
Life Exp	0.34	−0.59			
Murder	−0.23	0.70	−0.78		
HS Grad	0.62	−0.66	0.58	−0.49	
Eigenvalues	3.20	0.94	0.40	0.31	0.15

We fit a one-dimensional latent variable model to this set of $p = 5$ indicators, using the Bayesian model and Gibbs sampler described in the previous section. I use common, diffuse priors on the slopes and intercepts, setting $\mathbf{g}_{j0} = \mathbf{0}$ and $\mathbf{G}_{j0} = 100 \cdot \mathbf{I}_2, \forall\ j = 1, \ldots, p$. Barely proper inverse-Gamma priors are also used for the measurement error variance parameters, $\omega^2 \sim$ inverse-Gamma(.01, .01). I specify a $N(0, 1)$ prior for the ξ_i, but then also impose the normalizing restriction that the ξ_i have mean zero and standard deviation 1. To recap, this model has 50 latent variables ($\xi_i, i = 1, \ldots, n = 50$), and $p = 5$ intercept parameters, $p = 5$ slope parameters and $p = 5$ measurement error variances, for a total of 65 parameters.

The following JAGS program implements this model:

JAGS code

```
1 model{
2         for(i in 1:n){          ## loop over observations
3               for(j in 1:5){ ## loop over indicators
4                     mu[i,j] <- gamma[j,1] + gamma[j,2]*xi[i]
5                     y[i,j] ~ dnorm(mu[i,j],tau[j])
6                      }
7         }
8
9         ## prior for latent variable
10        for(i in 1:n){
11              xistar[i] ~ dnorm(0,1)
12              xi[i] <- (xistar[i]-mean(xistar[]))/sd(xistar[])
13        }
14
15        ## priors for the measurement parameters
16        for(j in 1:5){
17              ## intercepts and slopes
18              gamma[j,1:2] ~ dmnorm(g0[1:2],G0[1:2,1:2])
19              ## measurement error variances
20              tau[j] ~ dgamma(.01,.01)
21              omega[j] <- 1/sqrt(tau[j])
22        }
23 }
```

The code sets up a loop over the data, in which we nest a loop over the indicators. Lines 4 and 5 are the heart of the model, generating a linear regression of `y[i,j]` on `xi[i]`. The remainder of the code generates priors for the unknown parameters; note that we utilize a conditionally conjugate prior for $\boldsymbol{\gamma}_j$, while equation 9.12 in the text specifies a conjugate, normal prior for $\boldsymbol{\gamma}_j$ given ω_j^2.

The Gibbs sampler is run for a burn-in period of 1000 iterations, and then for another 5000 iterations. The resulting Gibbs sampler generates an efficient exploration of the posterior density, with convergence diagnostics (Section 6.2) consistent with stationarity and generally low to moderate autocorrelations for all 65 parameters. Figure 9.3 summarizes the output of the Gibbs sampler for four selected parameters: (1) γ_{21}, the slope parameter for the income indicator; (2) σ_4, the standard deviation of the disturbances in the measurement equation for the murder rate indicator; (3) ξ_5, the latent variable for California; (4) ξ_{18}, the latent variable for Louisiana. The traceplots and autocorrelation plots indicate that the Gibbs sampler has converged on the posterior density of the model parameters, and is generating a reasonably efficient tour of the parameter space. Formal diagnostics support this conclusion, with σ_4 being the slowest mixing parameter here; the Raftery-Lewis diagnostic for this parameter – based on the 5000 saved iterations – suggests that we will require about 30 000 iterations to precisely estimate the extreme quantiles of the marginal posterior density. The autocorrelation function for the sequence of sampled values for this parameter decays slowly, and the effective sample size (see Section 4.4.1) of the 5000 Gibbs samples is less than 400; contrast the relatively fast decay in the autocorrelation sequences for the three other parameters displayed in Figure 9.3.

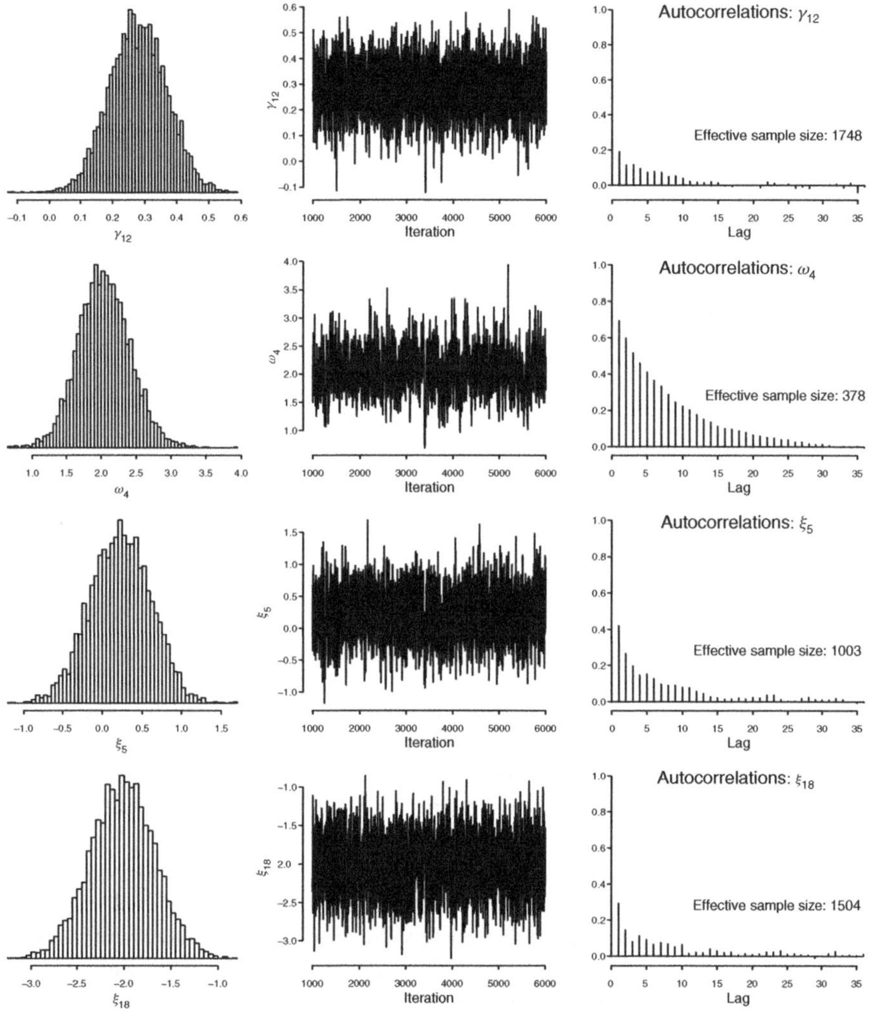

Figure 9.3 Graphical summaries of the Gibbs sampler, selected parameters of the one-factor latent variable model, Example 9.1. The panels in the left column are histograms summarizing the marginal posterior density (as estimated with 5000 draws from the Gibbs sampler). Panels in the middle column display trace plots; the right column shows autocorrelation functions. The four parameters (one per row) are (1) γ_{21}, the slope parameter for the income indicator, (2) ω_4, the standard deviation of the disturbances in the life expectancy equation, (3) ξ_5, the latent variable for California and (4) ξ_{18}, the latent variable for Louisiana. ω_4 is the slowest mixing of the 65 parameters.

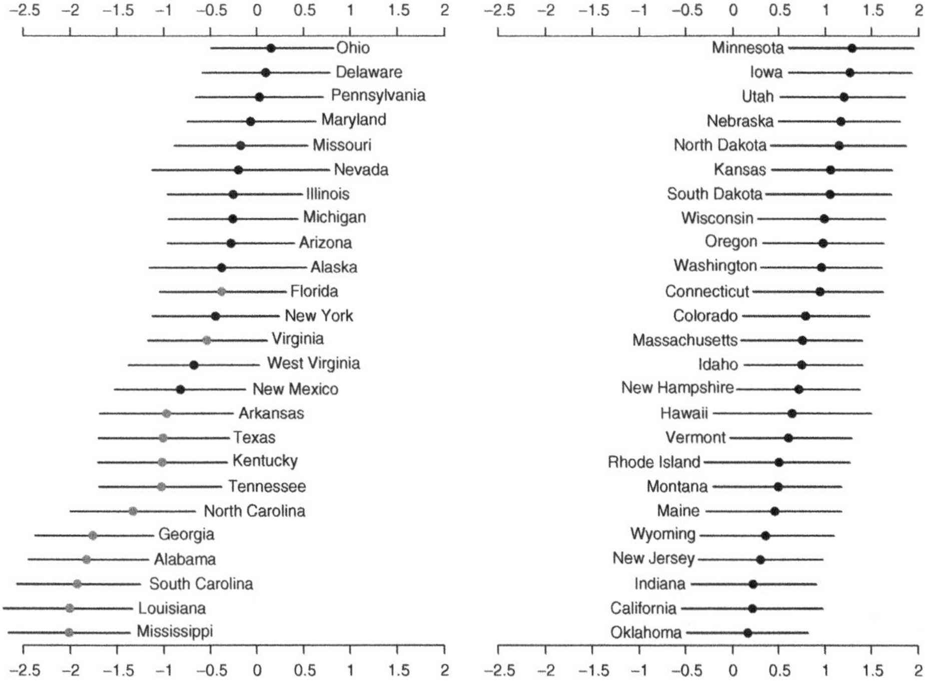

Figure 9.4 Posterior means for latent variable and 95 % credible intervals, one dimensional latent variable model for social attributes of the American states. Gray points indicate Southern states.

Figure 9.4 presents a graphical summary of the marginal posterior densities of the 50 latent variables. Each marginal posterior density is summarized with its mean (the plotted circle) and by an estimated 95 % credible interval (the horizontal bars); the states are ordered from lowest to highest in terms of their posterior means. States from the southern United States (plotted in gray) generally have low values on the recovered latent variable, with Louisiana and Mississippi the two lowest scoring states, while Minnesota and Iowa are the two highest scoring states. Note also the pronounced left tail in the distribution of the latent variable: i.e. the latent variable is normalized to have a mean of zero and standard deviation one, but the lowest scoring states have scores of approximately −2, while the highest scoring states are at about 1.3.

More similarities (and some differences) between maximum likelihood factor analysis and the Bayesian latent variable model are presented in Table 9.3. For the Bayesian model, the table entries consist of point estimates (the estimate of the mean of the posterior density) and 95 % credible intervals for the γ_j parameters, the measurement error standard deviations, and the implied r^2 for each indicator. Analogous quantities from factor analysis – factor loadings and indicator uniquenesses, generated with the maximum likelihood factor analysis function in `R`, `factanal` – are presented in the two right columns of the table. Note none of the 95 % credible intervals for the slope parameters overlaps zero, indicating that all indicators are contributing information to inferences for the latent variable. However, it is apparent that the income indicator is the

Table 9.3 Bayes estimates (posterior means) for indicator-specific parameters, factor analysis model, Example 9.1. 95 % credible intervals shown in brackets. The two columns on the right present analogous quantities from maximum likelihood factor analysis model fit using the `factanal` command in `R`.

	Bayesian latent variable model				Factor analysis (ML)	
Indicator	Intercept γ_{j0}	Slope γ_{j1}	Error SD ω_j	r^2	Factor loading	Indicator uniqueness
Income	4.43	0.279	.561	.177	.457	.791
	[4.28, 4.59]	[0.0965, 0.452]	[0.454, 0.696]	[0.0344, 0.322]		
Illiteracy	1.17	−0.489	.371	.629	−.806	.351
	[1.06, 1.28]	[−0.599, −0.361]	[0.262, 0.488]	[0.473, 0.793]		
Life Exp	70.9	1.11	.773	.667	.833	.306
	[70.6, 71.1]	[0.829, 1.34]	[0.540, 1.04]	[0.490, 0.821]		
Murder	7.37	−3.10	2.03	.694	−.847	.283
	[6.78, 7.96]	[−3.71, −2.35]	[1.34, 2.78]	[0.511, 0.860]		
HS Grad	52.70	5.69	5.83	.480	.710	.495
	[51.1, 54.4]	[3.68, 7.47]	[4.39, 7.48]	[0.304, 0.657]		

odd one out, with just 18 % of the variation in this indicator accounted for by the latent variable, where as for the other indicators the r^2 range from .48 to .69.

9.2.4 Inference over rank orderings of the latent variable

Armed with many samples from the joint posterior density of the parameters $\boldsymbol{\theta}$, we can then induce posterior densities over any quantity of interest that is a function of the $\boldsymbol{\theta}$. Returning to the social attributes data considered in Example 9.1, suppose we are interested in ranking the states by their score on the latent dimension; we considered this problem in Problem 7.9. Uncertainty in the latent variable induces uncertainty in rank orderings on the latent dimension and any inferences as to a given state's rank order ought to reflect this uncertainty. With the output from the Gibbs sampler, it is straightforward to induce a posterior density over each state's rank as follows. At iteration t, the Gibbs sampler produces $\boldsymbol{\xi}^{(t)}$, a sample from the joint posterior distribution of the latent variables, which can can be sorted to produce a vector of ranks $\mathbf{r}^{(t)} = \mathbf{r}(\boldsymbol{\xi}^{(t)}) = (r_1^{(t)}, \ldots, r_n^{(t)})'$. Absent any ties, each element of $\mathbf{r}^{(t)}$ is an integer $r_i^{(t)} \in \{1, \ldots, n\}$, $\forall t$. Thus, by computing and storing ranks over many iterations of the Gibbs sampler, we build a posterior probability mass function over the possible ranks. Thus, our estimate of the posterior probability that a given subject occupies rank r is just the proportion of times we see that event in many samples from the joint posterior density of the latent variables. In the particular context of Example 9.1, we might focus on the probability of occupying the lowest rank, interpretable as the probability of being the 'worst off' state in the United States on the latent dimension tapped by this ensemble of social indicators.

Figure 9.5 shows the posterior probability mass functions over ranks for the nine states with the highest probability of being the 'worst off' state. Louisiana has the highest posterior probability of being the 'worst off' state, .30, followed closely by Mississippi

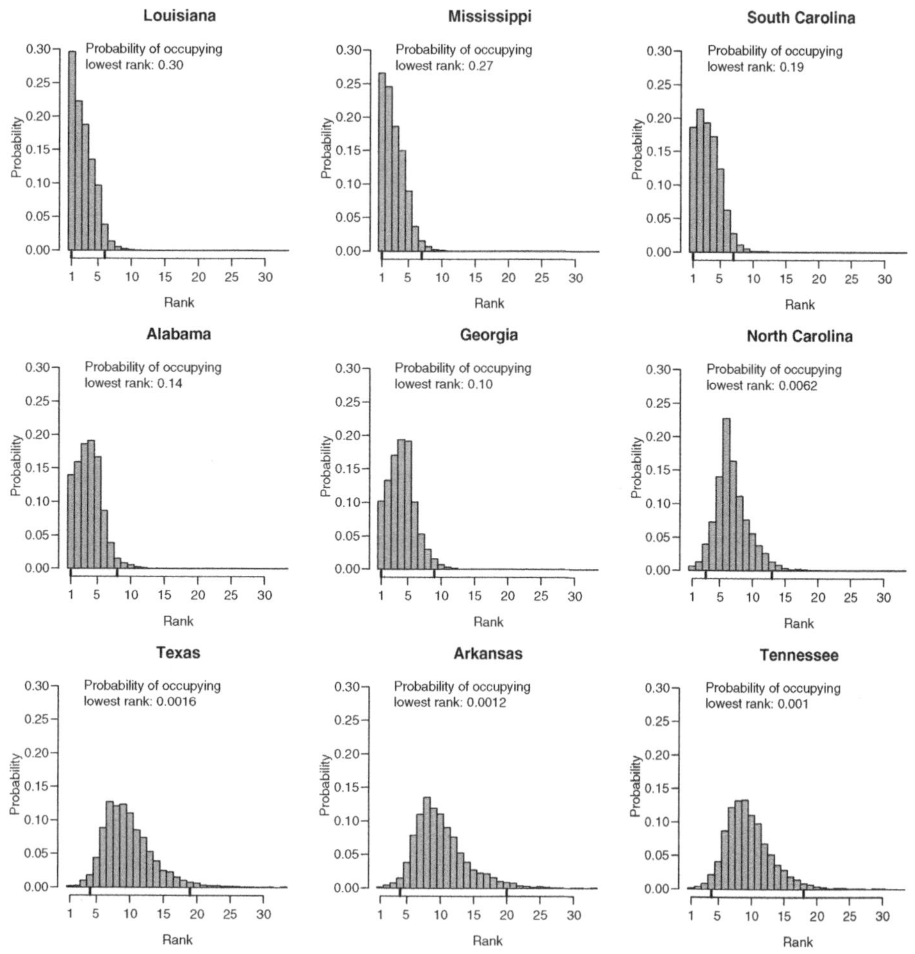

Figure 9.5 Histograms, posterior probability mass functions over ranks of the latent variable for nine states, Example 9.1. The ticks on the horizontal axis correspond to the 2.5th and 97.5th percentiles of the sampled ranks (i.e. a 95 % credible interval).

with .27 and South Carolina with .19. No single state can be assigned the 'worst off' rank unambiguously; more information is needed to more clearly distinguish the relative positions of the states with respect to the latent variable, either in the form of (a) more indicators that discriminate with respect to the recovered dimension, (b) more information as to the way the indicators are structured by the latent variable (i.e. less uncertainty over the $\boldsymbol{\gamma}_j$ and ω_j^2 parameters), or (c) prior information over the latent variable that distinguishes among the states (contrast the common $N(\mu_\xi, \sigma^2)$ prior used here).

9.2.5 Incorporating additional information via hierarchical modeling

Continuing with Example 9.1, the analysis reported above (e.g. Figure 9.4) strongly suggests that the twelve Southern states have lower values on the latent variable than the other 38 states. We do not consider the fact that a state is from the South to be an indicator of the latent trait, on a par with the Income, Illiteracy, Life Expectancy and other indicators. Still, there are good reasons to believe that because of their shared historical circumstances and geographical clustering, states from the South will have similar values on the latent variable.

We allow for the possible distinctiveness of the Southern states by replacing the prior density for ξ_i in Equation 9.4 with the following simple hierarchical model,

$$\begin{aligned} \xi_i|S(i) &\sim N(\alpha_{S(i)}, 1), \\ S(i) &= \begin{cases} 1 & \text{if state } i \text{ is in the South} \\ 0 & \text{otherwise,} \end{cases} \\ \alpha_r &\sim N(0, 1), \quad r \in \{0, 1\}. \end{aligned} \tag{9.14}$$

That is, the two groups of states, Southern and non-Southern, are assigned their own prior densities, with means α_r that vary across the two groups. Since the α parameters are assigned the same $N(0, 1)$ prior density, our prior is agnostic as to whether Southern states have lower scores on the latent variable than non-Southern states; posterior inferences will be based solely on the information in the indicators, $\mathbf{Y}$. All that the prior asserts is that Southern states are similar to one another, and likewise for non-Southern states.

Formally, the hierarchical model here amounts to a stipulation of *conditional exchangeability*. That is, given the label 'Southern' or 'non-Southern', we do not further distinguish between the states, at least *a priori*. Before seeing the data on the indicators, the hierarchical model does not make any *a priori* distinction between Alabama and Mississippi, nor between New York and Massachusetts, but the labeling operationalized in the hierarchical model means that we do not consider Georgia and California as unconditionally 'similar' (again, at least *a priori*). In contrast, in the non-hierarchical model we did not distinguish between the states *a priori*: each state is assigned the same prior for its latent variable (Equation 9.4), as if the data were unaccompanied by any labels at all and exchangeable. In the hierarchical model we stipulate that the states are exchangeable within the classes 'Southern' and 'non-Southern'.

This hierarchical version of the latent variable introduces two new parameters, α_0 and α_1, but poses no great computational challenge. The parameters α_0 and α_1 are added to $\boldsymbol{\theta}$, the set of parameters to be updated at each iteration of the Gibbs sampler. For instance, the conditional distribution for α_0 is simply

$$\begin{aligned} p(\alpha_0|\mathcal{G}_{\backslash\alpha_0}) &= p(\alpha_0|\text{parents}[\alpha_0]) \times \prod_{w \in \text{ children } [\alpha_0]} p(w|\text{parents[w]}) \\ &= \phi(\alpha_0) \times \prod_{i:S(i)=0} \phi(\xi_i - \alpha_0), \end{aligned}$$

where ϕ is the standard normal probability density function; i.e. the $\xi_i : S(i) = 0$ are the only children of α_0, and α_0 is the only parent of the $\xi_i : S(i) = 0$. Since $\phi(\alpha_0) \equiv N(0, 1)$ and $\phi(\xi_i|\alpha_0) \equiv N(\alpha_0, 1)$, then

$$\phi(\alpha_0) \times \prod_{i:S(i)=0} \phi(\xi_i - \alpha_0), \equiv N(\alpha_0^*, \sigma_{\alpha_0}^2)$$

where, letting n_0 be the cardinality of the set $\{i : S(i) = 0\}$,

$$\alpha_0^* = \frac{\bar{x}_0 n_0}{1 + n_0},$$

$$\bar{x}_0 = n_0^{-1} \sum_{i:S(i)=0} \xi_i, \quad \text{and}$$

$$\sigma_{\alpha_0}^2 = 1/(1 + n_0),$$

where $\bar{x}_0$ is the mean of the ξ_i among states where $S_i = 0$. Note that the extra '1' in the denominators of α_0^* and $\sigma_{\alpha_0}^2$ stem from the $N(0, 1)$ prior for α_0. We obtain the conditional distribution of α_1 similarly.

The conditional distribution for ξ_i remains relatively unchanged from that in the model without the hierarchical structure as given in Equation 9.11, except that the μ_ξ term is replaced by $\alpha_{S(i)}$; the σ^2 term in Equation 9.11 can be ignored since we set $\sigma^2 = 1$. As before, the Gibbs sampler is initially run for a burn-in period, with a longer set of iterations then saved for inference. Inspection of the Gibbs sampler output *a la* Figure 9.3 strongly suggests the algorithm again converges quickly on the joint posterior density of the model parameters.

Identification

Identification of the model parameters remains an issue. We will work with an unidentified version of the model when we run the MCMC algorithm, but map the sampled values for the unidentified parameters into the space of identified parameters. The required normalizations here are rather trivial. That is, we have the model

$$y_{ij} \sim N(\mu_{ij}, \omega_j^2)$$

$$\mu_{ij} = \gamma_{j1} + \gamma_{j2}\xi_i$$

$$\xi_i \sim N(\alpha_{S(i)}, 1)$$

which is not identified due to the invariance of the likelihood to linear transformations of the ξ_i. We impose the identifying restriction that $\tilde{\xi}_i = (\xi_i - c)/m$, where $c = \bar{\xi}$ and $m = \text{sd}(\xi)$; i.e. we normalize the ξ_i to have mean zero and standard deviation one, denoting the identified version of the latent traits as $\tilde{\xi}_i$. The question then is what are the transformations of the other model parameters that generate the same likelihood contributions we would get from the parameterization in terms of the unidentified parameters ξ_i. If we define

$$\tilde{\gamma}_{j1} = \gamma_{j1} + \gamma_{j2}\, c/m$$

$$\tilde{\gamma}_{j2} = \gamma_{j2}\, m$$

then $\tilde{\gamma}_{j1} + \tilde{\gamma}_{j2}\tilde{\xi}_i = \gamma_{j1} + \gamma_{j2}\xi_i$, and so we get the same likelihood contributions from both sets of parameters. The transformation of the ξ_i to $\tilde{\xi}_i$ also induces a transformation of the α parameters to $\tilde{\alpha}_{S(i)} = (\alpha_{S(i)} - c)/m$, $S(i) \in \{0, 1\}$. We make these transformations 'iteration-by-iteration', remapping the MCMC exploration of the posterior density of the unidentified parameters into space of identified parameters.

The following JAGS program implements this model, including a block of code that implements the mapping to identified parameters:

JAGS code

```
1 model{
2         for(i in 1:n){               ## loop over observations
3               for(j in 1:5){     ## loop over indicators
4                     mu[i,j] <- gammaTmp[j,1] + gammaTmp[j,2]*xiTmp[i]
5                     y[i,j] ~ dnorm(mu[i,j],tau[j])
6                     }
7         }
8
9         for(i in 1:n){               ## hierarchical model for x
10              muxi[i] <- alphaTmp[south[i]+1]  ## select alpha for this obs
11              xiTmp[i] ~ dnorm(muxi[i],1)
12        }
13
14        for(j in 1:5){              ## priors for measurement parameters
15              gammaTmp[j,1:2] ~ dmnorm(g0[1:2],G0[1:2,1:2])
16              tau[j] ~ dgamma(.01,.01)
17              omega[j] <- 1/sqrt(tau[j])
18        }
19
20        for(j in 1:2){   ## priors over hierarchical model
21              alphaTmp[j] ~ dnorm(0,1)
22        }
23
24        ## transformations to identified parameters
25        m <- sd(xiTmp[1:n])
26        c <- mean(xiTmp[1:n])
27        for(i in 1:n){
28              xi[i] <- (xiTmp[i]-c)/m
29        }
30        for(j in 1:5){
31              gamma[j,1] <- gammaTmp[j,1] + gammaTmp[j,2]*c/m
32              gamma[j,2] <- gammaTmp[j,2]*m
33        }
34        for(j in 1:2){
35              alpha[j] <- (alphaTmp[j]-c)/m
36        }
37        ## difference in average xi, between south and non-south
38        delta <- alpha[1] - alpha[2]
39 }
```

We use 'working' versions of the parameters in the body of the program (i.e. gammaTmp, alphaTmp, xiTmp) which we then map into the identified parameters with the code block starting at line 24. The resulting MCMC output mixes slowly for the 'intercept' parameters γ_{j1} and so we run the algorithm for 250 000 iterations with a thinning interval of 5.

Results

Summaries of the marginal posterior densities of the α parameters for this model appear in Table 9.4. As foreshadowed by the non-hierarchical analysis in the previous section,

Table 9.4 Summaries of marginal posterior densities, hierarchical latent variable model for social attributes of American states. $\delta = \alpha_1 - \alpha_0$ is the difference of the average value of the latent variable in the 12 Southern states versus the 38 non-Southern states.

	Mean	2.5 %	97.5 %
α_0	0.457	0.282	0.630
α_1	−1.420	−1.740	−1.07
$\delta = \alpha_0 - \alpha_1$	1.87	1.47	2.23

the states do cluster along the Southern/non-Southern dichotomy. The difference in the α parameters is large in substantive terms: the mean of the posterior density of $\delta = \alpha_0 - \alpha_1$ is 1.87 (recalling that the latent variable is normalized to have standard deviation one) and the posterior 95 % credible region for δ lies well away zero. That is, on average, Southern states score 187 % of a standard deviation below non-Southern states on the latent variable.

The hierarchical model also leads to increased precision in the estimates of the latent variables. Figure 9.6 shows the 'shrinkage' that is typical of Bayesian hierarchical modeling, relative to non-hierarchical modeling. The top panel of Figure 9.6 shows that the estimates of the latent variable (the means of their respective marginal posterior densities) are more tightly clustered into two distinct groupings, Southern and non-Southern states. In the hierarchical analysis, the distinction between Southern and non-Southern states is much more stark than in the non-hierarchical model, with all 12 Southern states estimated to have lower values on the latent variable than all 38 non-Southern states; in contrast, the non-hierarchical model found some evidence of overlap between Southern and non-Southern states. This result is standard in hierarchical models, with the unit-specific estimates drawn towards one another conditional on the 'grouping' implied by the hierarchical part of the model (the solid gray lines indicate the Southern and non-Southern means of the latent variable estimates, and 'pull apart' in the hierarchical model, corresponding to the estimates of α_r reported in Table 9.4); the consequence is the greater separation of the cases along the Southern/non-Southern dichotomy.

The lower panel of Figure 9.6 shows the two sets of standard deviations of the posterior densities for the latent variables recovered from the two models (hierarchical and non-hierarchical). The data lie below the forty-five degree line in this case, indicating that the standard deviations produced by the hierarchical model are smaller than those produced by the non-hierarchical model. Again, this is a standard result in Bayesian hierarchical modeling: we have more information *a posterior* (i.e. marginal posterior densities with smaller standard deviations) as a result of supplying more information *a priori*. In this case the information supplied in the hierarchical model is quite modest, merely the labeling of a state as 'Southern' or 'non-Southern'. But in this application, the Southern/non-Southern distinction is quite informative and accounts for a large component of the variation in the latent variable, producing the substantial shrinkage apparent in both panels of Figure 9.6.

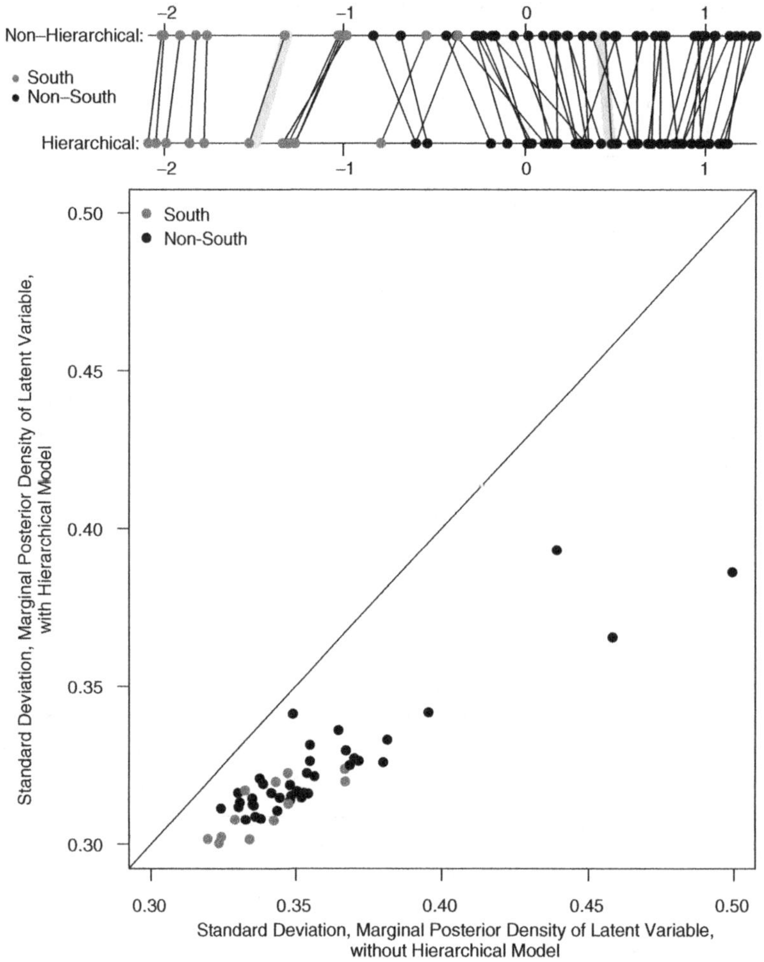

Figure 9.6 Comparison of means (upper panel) and standard deviations (lower panel), marginal posterior densities of latent variables, hierarchical latent variable model (clustering by South/non-South as in Equation 9.14) versus latent variable model without hierarchical structure (previously given in Equation 9.4). The thick light gray lines in the upper panel show how the Southern and non-Southern means of the latent variable estimates differ in the hierarchical and non-hierarchical versions of the model.

9.3 Item-response models

Binary indicators are particularly prevalent in the social sciences: examples include data generated from standardized testing (scored as correct/incorrect), legislative roll call data (recorded as 'Yeas' and 'Nays'), and the recorded votes of committees and judicial bodies. The inferential task is the same as that encountered in factor analysis: what do the observed binary data tell us about the latent scores of the subjects (test-takers, legislators, or judges, as the case may be). Additionally, what do the binary data tell us about the indicators? The model typically used to analyze data of this type comes from a subfield of psychometrics known as item-response theory (IRT). A widely used model in IRT is the following "two-parameter" model

$$\pi_{ij} = \Pr(y_{ij} = 1|\xi_i, \beta_j, \alpha_j) = F(\xi_i\beta_j - \alpha_j) \tag{9.15}$$

where

- $y_{ij} \in \{0, 1\}$ is the ith subject's answer to the jth item (e.g. $y_{ij} = 1$ if correct, $y_{ij} = 0$ if incorrect), where $i = 1, \ldots, n$ indexes respondents and $j = 1, \ldots, m$ indexes items;
- $\xi_i \in \mathbb{R}$ is an unobserved attribute of subject i (typically considered ability in the test-taking context, or revealed preferences ideology in the analysis of legislative data)
- β_j is an unknown parameter, tapping the *item discrimination* of the jth item, the extent to which the probability of a correct answer responds to change in the latent trait ξ_i
- α_j is an unknown *item difficulty* parameter, tapping the probability of a correct answer irrespective of levels of political information
- $F(\cdot)$ is a monotone function mapping from the real line to the unit probability interval, typically the logistic or normal CDF.

A one parameter version of the model results from setting $\beta_j = 1, \forall j$; i.e. items vary in difficulty, but not in terms of their discrimination, and is often called a Rasch model. A three parameter version of the model is available, designed to help correct for guessing on standardized tests. van der Linden and Hambleton (1997) provide a comprehensive survey of IRT models.

Contrast the two-parameter IRT model in equation 9.15 with the factor analysis model we considered in Section 9.2: the discrimination parameter β_j is similar to a factor loading; there is no analog to the difficulty parameter α_j in the factor analysis model (which is typically implemented as a model for covariances, in which mean levels of the indicators and intercept parameters are absent), although the version of the factor analysis model we presented did include intercepts. In addition, in most political science settings we would like to make inferences for the latent variables $\boldsymbol{\xi}$ at the same time as we make inferences for the item parameters $\boldsymbol{\beta}$ and $\boldsymbol{\alpha}$; contrast factor analysis in which we typically generate measures of $\boldsymbol{\xi}$ conditional on estimates of the item parameters $\boldsymbol{\beta}$ (and $\boldsymbol{\alpha}$). For more on the similarities between the IRT models and factor analysis see Takane and de Leeuw (1987) and Reckase (1997).

The statistical problem here is inference for $\boldsymbol{\xi} = (\xi_1, \ldots, \xi_n)'$, $\boldsymbol{\beta} = (\beta_1, \ldots, \beta_m)$ and $\boldsymbol{\alpha} = (\alpha_1, \ldots, \alpha_m)'$. We form a likelihood for the binary data by assuming that given these parameter, the binary responses are conditionally independent across subjects and items; this assumption is called 'local independence' in the argot of IRT. Thus the likelihood has the same form as the likelihood for the binary responses considered in Section 8.1, i.e.

$$\mathcal{L} = \prod_{i=1}^{n}\prod_{j=1}^{m} \pi_{ij}^{y_{ij}}(1 - \pi_{ij})^{1-y_{ij}} \tag{9.16}$$

where π_{ij} is defined in Equation 9.15.

Identification

The model parameters are unidentified, just as we saw with the factor analytic model (§9.2). For instance, any linear transformation of the ξ_i can be offset by appropriate linear transformations for the β_j and α_j; an obvious case is scale invariance, in which $\pi_{ij} = F(\xi_i\beta_j - \alpha_j)$ indistinguishable from the model with $\pi^*_{ij} = F(\xi^*_i\beta^*_j - \alpha_j)$ where $\xi^*_i = c\xi_i$ and $\beta^*_j = \beta_j/c$, $c \neq 0$. A special type of rotational invariance arises in one dimension with $c = -1$. Any two linearly independent restrictions on the latent traits are sufficient for at least local identification, in the sense of Rothenberg (1971); a typical example is a mean-zero, unit variance restriction on the ξ_i, while setting at least one pair of the (β_j, α_j) item parameters to fixed values is one way of obtaining global identification.

The chief political science use of IRT model is in the analysis of legislative roll call data. Operationalizing the Euclidean spatial voting model (Davis, Hinich and Ordeshook 1970; Enelow and Hinich 1984) gives rise to an IRT model, in which the latent trait corresponds to legislators' "ideal points", and the discrimination and difficulty parameters are functions of the locations of the 'Yea' and 'Nay' locations (e.g. Clinton, Jackman and Rivers 2004a; Ladha 1991). In the roll call context it is often interesting to consider a multidimensional version of the model, where $\boldsymbol{\xi} \in \mathbb{R}^d$, $d > 1$; in this case the item-discrimination parameter becomes $\boldsymbol{\beta}_j \in \mathbb{R}^d$. Outside of the roll call context, IRT has been used elsewhere in political science to assess the measurement properties of survey responses by Delli Capini and Keeter (1996) and Jackman (2000).

Estimation and inference for the two-parameter IRT model is not trivial, due to the large number of parameters involved in many settings; recall Table 9.1. Data sets in educational testing can involve many thousands of test-takers and hundreds of items. Baker and Kim (2004) is a book length treatment of estimation strategies for these models. Popular approaches include: marginal maximum likelihood, effectively treating the ξ_i as nuisance parameters and known only up to a distribution, which is then integrated out of the likelihood over the item parameters (Bock and Aitkin 1981); alternating conditional maximum likelihood, used in the NOMINATE algorithms (e.g. Poole *et al.* 2007; Poole 2005; Poole and Rosenthal 1997) for analyzing roll call data; or Bayesian methods, exploring the posterior density of the parameters via Markov chain Monte Carlo methods. We consider the Bayesian approach, which has been applied to the analysis of Congressional roll calls (Clinton, Jackman and Rivers 2004a; Jackman 2000, 2001) and decisions of the Supreme Court (Martin and Quinn 2002).

Bayesian analysis via a data-augmented Gibbs sampler

We consider a multidimesional version of the two-parameter IRT model, with $\boldsymbol{\xi}_i, \boldsymbol{\beta}_j \in \mathbb{R}^d$. Let Ξ denote the n-by-d matrix formed by stacking the $\boldsymbol{\xi}_i, i = 1, \ldots, n$ and let $\mathbf{B}$ be the m-by-$(d+1)$ matrix formed by stacking the vectors $(\boldsymbol{\beta}'_j, \alpha_j)$, $j = 1, \ldots, m$. The inferential problem here is to compute the posterior density $p(\boldsymbol{\theta}|\mathbf{Y})$ where $\boldsymbol{\theta} = \{\Xi, \mathbf{B}\}$. By Bayes Rule, this posterior density is proportional to the prior over $\boldsymbol{\theta}$ times the likelihood in Equation 9.16. We assume prior independence between Ξ and $\mathbf{B}$. Further, we assume that $p(\Xi) = \prod_{i=1}^n p(\boldsymbol{\xi}_i)$, where $p(\boldsymbol{\xi}) \equiv N(\mathbf{v}, \mathbf{V})\ \forall i$, where typically $\mathbf{v} = \mathbf{0}$ and $\mathbf{V} = \mathbf{I}_d$. For $\mathbf{B}$ we make a similar assumption of prior independence across items, such that $p(\mathbf{B}) = \prod_{j=1}^m p(\boldsymbol{\beta}_j, \alpha_j)$ where $p(\boldsymbol{\beta}_j, \alpha_j) \equiv N(\mathbf{d}, \mathbf{D})\ \forall j$, where typically $\mathbf{d} = \mathbf{0}$ and $\mathbf{D} = \kappa \cdot \mathbf{I}_{d+1}$, with κ an arbitrarily large constant.

The resulting posterior density is high-dimensional and analytically intractable, and so we characterize it using MCMC methods. There is no new methodology to introduce here, since the model is essentially a series of probit models, save for the complication that the regressors $\boldsymbol{\xi}$ are unobserved. In Section 8.1.1 we used a data-augmented Gibbs sampler to explore the posterior density of a probit model; in Section 9.2 we derived the conditional densities of the latent variables $\boldsymbol{\xi}$ given observed indicators $\mathbf{Y}$ and item-specific parameters. We now combine these ideas to outline a data-augmented Gibbs sampler for the two-parameter IRT model (Albert 1992; Johnson and Albert 1999; Patz and Junker 1999).

We can formulate the probit model as a latent linear regression; in the IRT context we have m such latent linear regressions, indexed by j. That is, $y^*_{ij} = \boldsymbol{\beta}'_j\boldsymbol{\xi}_i - \alpha_j + \epsilon_{ij}$ with $V(\epsilon_{ij}) = \sigma_j = 1\ \forall j$ so as to identify the model. With the assumption of normality for the ϵ_{ij}, we have a simple, data-augmentation step to obtain the y^*_{ij} as we saw in §8.1.1. With $\boldsymbol{\beta}_j$ and α_j given, $\boldsymbol{\xi}_i$ is a vector of coefficients in the regression of $y^*_{ij} + \alpha_j$ on $\boldsymbol{\beta}_j$ with observations indexed by $j = 1, \ldots, m$; we then sample from the conditional density for $\boldsymbol{\xi}_i$ as per usual for the Gibbs sampler. Then, conditional on $\mathbf{Y}^*$ and $\boldsymbol{\xi}$, we sample from the conditional densities for $\boldsymbol{\alpha}$ and $\boldsymbol{\beta}$, which are given by standard results on the Bayesian analysis of regression models under normality (see Section 2.5).

In detail, letting t index iterations of the data-augmented Gibbs sampler:

1. sample $y^{*(t)}_{ij}$ from $g(y^*_{ij}|y_{ij}, \boldsymbol{\xi}_i, \boldsymbol{\beta}_j, \alpha_j)$. At the start of iteration t, we have $\boldsymbol{\beta}^{(t-1)}_j$, $\alpha^{(t-1)}_j$ and $\boldsymbol{\xi}^{(t-1)}_i$. We sample $y^{*(t)}_{ij}$ from one of the two following densities, depending on whether we observed $y_{ij} = 1$ or $y_{ij} = 0$:

$$y^*_{ij}|(y_{ij} = 0, x^{(t-1)}_i, \boldsymbol{\beta}^{(t-1)}_j, \alpha^{(t-1)}_j) \sim N(\mu^{(t-1)}_{ij}, 1)\mathcal{I}(y^*_{ij} < 0)$$

$$y^*_{ij}|(y_{ij} = 1, x^{(t-1)}_i, \boldsymbol{\beta}^{(t-1)}_j, \alpha^{(t-1)}_j) \sim N(\mu^{(t-1)}_{ij}, 1)\mathcal{I}(y^*_{ij} \geq 0)$$

   where $\mu^{(t-1)}_{ij} = \boldsymbol{\xi}^{(t-1)}_i\boldsymbol{\beta}^{(t-1)}_j - \alpha^{(t-1)}_j$ and $\mathcal{I}(\cdot)$ is an indicator function. For any y_{ij} missing-at-random we sample $y^{*(t)}_{ij}$ from the untruncated normal density $N(\mu^{(t-1)}_{ij}, 1)$.

2. sample $(\boldsymbol{\beta}^{(t)}_j, \alpha^{(t)}_j)$ from $g(\boldsymbol{\beta}_j, \alpha_j|\Xi, y^*_{ij})$. For $j = 1, \ldots, m$, sample $\boldsymbol{\beta}^{(t)}_j$ and $\alpha^{(t)}_j$ from the multivariate normal density with mean vector $(\mathbf{X}'\mathbf{X} + \mathbf{D}^{-1})^{-1}(\mathbf{X}'\mathbf{y}^{*(t)}_{\cdot j} + \mathbf{D}^{-1}\mathbf{d})$ and variance-covariance matrix $(\mathbf{X}'\mathbf{X} + \mathbf{D}^{-1})^{-1}$, where $\mathbf{X}$ is a n-by-$(d+1)$ matrix with typical row $\mathbf{x}_i = (\boldsymbol{\xi}^{(t-1)}_i, -1)$, $\mathbf{y}^{*(t)}_{\cdot j}$ is a n-by-1 vector, and $N(\mathbf{d}, \mathbf{D})$ is the prior for $(\boldsymbol{\beta}'_j, \alpha_j)$. This amounts to computing a 'Bayesian regressions' of $\mathbf{y}^{*(t)}_{\cdot j}$ on $\boldsymbol{\xi}^{(t-1)}_i$ and a negative constant team, and then sampling from the posterior density for the coefficients $\boldsymbol{\beta}_j$ and α_j, for $j = 1, \ldots, m$.

3. sample $\boldsymbol{\xi}_i$ from $g(\boldsymbol{\xi}_i|y^*_{ij}, \boldsymbol{\beta}_j, \alpha_j)$. Re-arranging the latent linear regression yields $w_{ij} = y^*_{ij} + \alpha_j = \boldsymbol{\xi}'_i\boldsymbol{\beta}_j + \epsilon_{ij}$. Collapse these equations over the j subscript, to yield the n regressions $\mathbf{w}_i = \mathbf{B}\boldsymbol{\xi}_i + \boldsymbol{\epsilon}_i$, recalling that $\mathbf{B}$ is the m by d matrix with the j-th row given by $\boldsymbol{\beta}'_j$. That is, the latent variables $\boldsymbol{\xi}_i$ are vectors of d unknown parameters in each of n regressions. The conditional density of each $\boldsymbol{\xi}_i$ is a d-dimensional normal density with mean vector $(\mathbf{B}'\mathbf{B} + \mathbf{V}^{-1}_i)^{-1}(\mathbf{B}'\mathbf{w}_i + \mathbf{V}^{-1}_i\mathbf{v}_i)$ and variance-covariance matrix $(\mathbf{B}'\mathbf{B} + \mathbf{V}^{-1}_i)^{-1}$, noting that $V(\boldsymbol{\epsilon}_i) = 1\ \forall i$.

This data-augmented Gibbs sampler scheme is implemented in the `ideal` function in the `R` package `pscl` (Jackman 2008b); another implementation is available in `MCMCpack` (Martin, Quinn and Park 2009).

■ Example 9.2

Was Barack Obama the 'Most Liberal Senator?'. In the midst of the 2008 presidential campaign, *National Journal* announced that Senator Barack Obama (D, IL) was the 'Most Liberal Senator' in 2008. Interestingly, *National Journal* had also pronounced Senator John Kerry (D, MA) as the 'Most Liberal Senator' in 2004, a claim we examined in considerable detail at that time (Clinton, Jackman and Rivers 2004b). We analyze the validity of this claim by analyzing the votes on the 99 'key votes' used by *National Journal* in devising their 2007 ratings of senators' voting records, fitting the IRT model described above.

The model of roll call voting we employ is quite standard: the Euclidean spatial voting model due to Enelow and Hinich (1984). Roll call $j \in \{1, \ldots, m\}$ presents legislator $i \in \{1, \ldots, n\}$ with a choice between a 'Yea' position $\boldsymbol{\zeta}_j$ and a 'Nay' position $\boldsymbol{\psi}_j$, locations in $\mathbb{R}^d$, where d denotes the dimension of the policy space. Let $y_{ij} = 1$ if legislator i votes 'Yea' on the jth roll call and $y_{ij} = 0$ otherwise. Legislators are assumed to have quadratic utility functions over the proposals and status quos; i.e. $U_i(\boldsymbol{\zeta}_j) = -||\boldsymbol{\xi}_i - \boldsymbol{\zeta}_j||^2 + \eta_{ij}$ and $U_i(\boldsymbol{\psi}_j) = -||\boldsymbol{\xi}_i - \boldsymbol{\psi}_j||^2 + \nu_{ij}$, where $\boldsymbol{\xi}_i \in \mathbb{R}^d$ is the *ideal point* of legislator i, η_{ij} and ν_{ij} are the errors or stochastic elements of utility, and $|| \cdot ||$ is the Euclidean norm. Utility maximization implies $y_{ij} = 1$ if $U_i(\boldsymbol{\zeta}_j) > U_i(\boldsymbol{\psi}_j)$ and $y_{ij} = 0$ otherwise. The voting model is completed by assigning a distribution to the errors. We assume that the errors η_{ij} and ν_{ij} have a joint normal distribution with $E(\eta_{ij}) = E(\nu_{ij})$, $\text{var}(\eta_{ij} - \nu_{ij}) = \sigma_j^2$ and the errors are independent across both legislators and roll calls. It follows that

$$\begin{aligned}
\pi_{ij} &= \Pr(y_{ij} = 1) = \Pr\left(U_i(\boldsymbol{\zeta}_j) > U_i(\boldsymbol{\psi}_j)\right) \\
&= \Pr\left(\nu_{ij} - \eta_{ij} < ||\boldsymbol{\xi}_i - \boldsymbol{\psi}_j||^2 - ||\boldsymbol{\xi}_i - \boldsymbol{\zeta}_j||^2\right) \\
&= \Pr\left(\nu_{ij} - \eta_{ij} < 2(\boldsymbol{\zeta}_j - \boldsymbol{\psi}_j)'\boldsymbol{\xi}_i + \boldsymbol{\psi}_j'\boldsymbol{\psi}_j - \boldsymbol{\zeta}_j'\boldsymbol{\zeta}_j\right) \\
&= \Phi(\boldsymbol{\beta}_j'\boldsymbol{\xi}_i - \alpha_j)
\end{aligned} \tag{9.17}$$

where $\boldsymbol{\beta}_j = 2(\boldsymbol{\zeta}_j - \boldsymbol{\psi}_j)/\sigma_j$, $\alpha_j = (\boldsymbol{\zeta}_j'\boldsymbol{\zeta}_j - \boldsymbol{\psi}_j'\boldsymbol{\psi}_j)/\sigma_j$, and $\Phi(\cdot)$ denotes the standard normal distribution function. This corresponds to a probit model (Section 8.1) but with an unobserved regressor $\boldsymbol{\xi}_i$ corresponding to the legislator's ideal point (a logit model results if the errors have extreme value distributions). The coefficient vector $\boldsymbol{\beta}_j$ is the direction of the jth proposal in the policy space relative to the 'Nay' position. Notice that in one dimension the ratio $\tau_j = \alpha_j/\beta_j = (\zeta_j^2 - \psi_j^2)/2(\zeta_j - \psi_j) = (\zeta_j + \psi_j)/2$ is the cutpoint between the proposals, the point at which a legislator would find themselves indifferent between the two proposals (absent stochastic sources of utility).

It is important to note the highly stylized nature of the model. A legislator's voting record reflects a number of different influences including personal ideology, the ideology of the legislator's constituency, lobbying by interest groups and pressure from party leaders. Without considerably more data than is available here and/or assumptions, the effects of each of these plausible sources of influence can not be ascertained. Accordingly, the estimates of ξ_i produced by this exercise should not be literally treated as a measure of a senators' personal ideology, but rather as a mix of these possible influences on roll call voting, and, in any event, a useful summary of the ideological content of a senators' voting record.

In our reanalysis of the *National Journal* key votes, we assume that proposals status quos and senators' ideal points are points on an ideological continuum (i.e. we assume a unidimensional policy space, setting $d = 1$ in the context of the discussion in the previous paragraph). Given the assumptions of conditional independence of the votes across legislators and roll calls given ideal points and item parameters, the likelihood is the same as that in Equation 9.16.

Identification. When fitting a unidimensional model we typically impose the identifying restriction that the ideal points ξ_i have mean zero and standard deviation one across legislators (the `normalize` option in the implementation of this model is `ideal` in `R`, see Jackman 2008); this is sufficient for local identification, ruling out any observationally-equivalent translations and rescalings of the parameters, save for a reflection of the parameters about the origin. This invariance to reflection is the sense in which identification is only local. We typically find that the lack of global identification is not a practical concern when fitting unidimensional models to data sets from recent US Congresses; e.g. if we initialize the Gibbs sampler with all Democrats 'on the left' and all Republicans 'on the right' (with the corresponding $\xi_i = \pm 1$, respectively), then the algorithm is quickly drawn to the posterior mode that has most (if not all) Democratic ideal points lying to the left of the Republican ideal points; since the posterior densities are typically estimated with reasonable levels of precision (the roll call matrices are quite large, and the unidimensional model provides an excellent fit to data from recent US Congresses) the Gibbs sampler never visits the neighborhood around the 'mirror image' mode of the posterior density; see Example 5.1 Identification is more delicate when fitting higher dimensional models; a working paper by Rivers (2003) provides some details.

Specifying priors for identified parameters. With this identifying restriction on the ideal points ξ_i, it is interesting to then consider the effects of specifying priors on the bill parameters β_j and α_j. Typically we assume priors of the sort $(\alpha_j, \beta_j)' \sim N(\mathbf{0}, \sigma_\beta^2 \cdot \mathbf{I}_2)\ \forall j$, with σ_β^2 a user-supplied hyperparameter. We begin with the unidentified model

$$y_{ij} \sim \text{Bernoulli}(\pi_{ij}), \tag{9.18a}$$

$$\Phi^{-1}(\pi_{ij}) = \xi_i \beta_j - \alpha_j, \tag{9.18b}$$

$$\xi_i \sim N(0, \sigma_\xi^2), \quad i = 1, \ldots, n, \tag{9.18c}$$

$$(\beta_j, \alpha_j)' \sim N(\mathbf{0}, \sigma_\beta^2 \mathbf{I}_2), \quad j = 1, \ldots m. \tag{9.18d}$$

We then impose the normalization $\tilde{\xi}_i = (\xi_i - c)/m$ where $c = \bar{\xi}$ and $m = \text{sd}(\xi)$, i.e. given c and m there is a simple linear mapping from the space of unidentified parameters ξ_i to identified parameters $\tilde{\xi}_i$. Further, given that $\tilde{\xi}_i = (\xi_i - c)/m$ we seek $\tilde{\beta}_j$ and $\tilde{\alpha}_j$ such that $\xi_i \beta_j - \alpha_j = \tilde{\xi}_i \tilde{\beta}_j - \tilde{\alpha}_j$: i.e. the identified bill parameters are $\tilde{\beta}_j = \beta_j m$ and $\tilde{\alpha}_j = \alpha_j - \tilde{\beta}_j c/m = \alpha_j - \beta_j c$. Given that $\xi_i \sim N(0, \sigma_\xi^2)$, *a priori* we have $c = \bar{\xi} \approx 0$ and (at least for a reasonably large n), $m = \text{sd}(\xi) \approx \sigma_\xi$ and so for the identified parameters we have $\tilde{\xi}_i \sim N(0, 1)$, $\tilde{\beta}_j \sim N(0, \sigma_\xi^2 \sigma_\beta^2)$ and $\tilde{\alpha}_j \sim N(0, \sigma_\beta^2)$. The interesting part of these results is that the prior variance for $\tilde{\beta}$ is the product $\sigma_\xi^2 \sigma_\beta^2$. If both σ_ξ^2 and σ_β^2 are set to large quantities then the resulting prior variance for $\tilde{\beta}$ can be potentially massive, or at least much larger than what the user may have intended when specifying the variance of the prior of the unidentified parameter β_j. An interesting case is where the user specifies $\sigma_\xi^2 = 1$, in which case $V(\tilde{\beta}_j) = V(\beta_j) = \sigma_\beta^2$. The normalization from ξ_i to $\tilde{\xi}_i$ means that σ_ξ^2 is largely a redundant parameter; this suggests that in the absence of any prior information over the ξ_i, we may as well set $\sigma_\xi^2 = 1$ such that $V(\tilde{\beta}_j) = \sigma_\beta^2$ and the user supplied value for this quantity actually corresponds to the variance over the identified parameter $\tilde{\beta}_j$.

Missing data. A potentially serious problem is that the *National Journal* ratings ignore the fact that candidates for president miss a substantial fraction of votes. Also, changes in the content of the legislative agenda from year to year changes the nature of the issues being voted on, threatening the validity of comparisons of year-specific voting scores.

In Table 9.5 we report the proportion of missed votes on *National Journal* key votes for Senators who were running from president, in each of 2005, 2006 and 2007. For comparison, we include Senator John Kerry who of course was a presidential candidate in 2004, but not thereafter. We emphasize two features of the data reported in Table 9.5. First, the fraction of votes analyzed by the *National Journal* in 2005, 2006 and 2007 is rather small relative to the total number of recorded votes that year: 19 %, 29 % and 22 %, respectively. Second, the pattern of missingness in the votes selected by the *National Journal* is highly correlated with the presence of a presidential election. Senator John McCain (R, AZ) missed 4.2 % and 4.9 % of the votes scored by the *National Journal* in 2005 and 2006 respectively, but missed 55.6 % in 2007. Likewise, Senator Barack Obama (D, IL) missed only 1.4 % and none of the *National Journal* key votes in 2005 and 2006, respectively, but missed 33.3 % while campaigning during 2007. In the case of Obama in 2007 we have just 66 recorded 'key votes' for analysis; for McCain we have only 44 key votes. For these legislators – and others with similarly short voting histories – we can expect the results to be sensitive to the priors we specify over the bill parameters. We will also use these data to assess the plausibility of extending the basic model so as to incorporate additional sources of information via a hierarchical model.

Table 9.5 Percentage of votes missed by presidential candidates: all votes and *National Journal* 'key votes', by year.

	2005		2006		2007	
Name	All	NJ	All	NJ	All	NJ
Bush (R)	87.7	64.3	74.9	64.6	–	–
Biden (D DE)	8.2	4.3	12.2	2.4	39.1	26.3
Brownback (R KS)	0.3	0	4.3	3.7	30.5	27.2
Clinton (D NY)	3.3	1.4	1.4	0	23.3	17.2
Dodd (D CT)	5.2	4.3	3.6	1.2	37.6	33.3
Kerry (D MA)	1.9	1.4	6.8	2.4	4.8	3.0
McCain (R AZ)	8.7	4.2	9.3	4.9	55.9	55.6
Obama (D IL)	2.2	1.4	1.1	0.0	37.6	33.3
Number of Votes	366	70	279	82	442	99

Bayesian analysis. Initially, we set the hyperparameter $\sigma_\beta^2 = 5^2$ and sample from the resulting posterior density for the ideal points and bill parameters using the data-augmented Gibbs sampler described above. The sampler is initialized with all Democratic senators at -1 and all Republicans at 1; start values for the bill parameters are generating by running a probit of each roll call on these initial values for the ideal points. We generate 1.1 million samples, discarding the first 100 000 samples and thinning the remaining one million samples by a factor of 100, to yield a set of 10 000 samples. The data-augmented Gibbs sampler is not particularly efficient and this large number of samples is needed to generate a reasonable exploration of the posterior densities of the ideal points of senators with short and/or relatively one-sided voting histories (e.g. Sanders, the Independent senator from Vermont, and Obama, among others).

Figure 9.7 shows summaries of the marginal posterior densities of the ideal point of 101 senators voting on the *National Journal* 99 'key votes'. Negative scores are associated with more liberal preferred positions (ideal points) and positive scores represent more conservative preferred positions. There is almost no partisan overlap in the distribution of recovered ideal points, with the ideal points of Joe Lieberman (the most conservative Democrat) and Olympia Snowe (the most liberal Republican) being virtually identical.

The 95 % credible intervals for legislators with more extreme voting histories are relatively wide, since the 99 *National Journal* key votes are generally 'easy' votes for these senators, and do not do much to help us discriminate among legislators in the tails of the distribution of ideal points. The most lop-sided vote among the 99 *National Journal* key votes has 10 senators voting against the proposal; this means that the recorded votes do not do a particularly good job of discriminating among legislators with relatively extreme preferences. The analogy from an educational setting is the student who gets every question asked of them correct: until the student is asked questions that they answer incorrectly, all we know is that the student is relatively smart, but we lack a precise estimate of how smart the student is in an absolute sense. Indeed, this is one of the motivations behind adaptive testing in standardized tests; i.e. tailoring the difficulty of items to the ability of test-takers, as revealed by their responses to items answered earlier in the test. Both features appear to influence the posterior density we recover for Obama's ideal point; our best guess is that Obama is quite liberal, but it is difficult to precisely state 'how liberal' given the combination of a one-sided voting record and the prevalence of missing data.

Figure 9.8 shows the ranks of the estimated voting scores (and 95 % credible intervals) for the nine 'most liberal' senators; we induce a posterior probability mass function over the order statistics of the ideal points using the procedure described in Section 9.2.4. Given that the voting scores graphed in Figure 9.7 are estimated with uncertainty, any rank-ordering based on these scores will also be subject to uncertainty. No single senator can be unambiguously classified as 'the most liberal'. The legislator most likely to occupy the 'most liberal' rank is Sanders, whose ideal point occupies rank 1 in 30 % of the 10 000 samples retained from the Gibbs sampler; for Obama the corresponding figure drops to 19 % and to 15 % for Biden. Again, recall the high rates of absenteeism for these latter two Senators, both of whom were actively campaigning for the presidency in 2007.

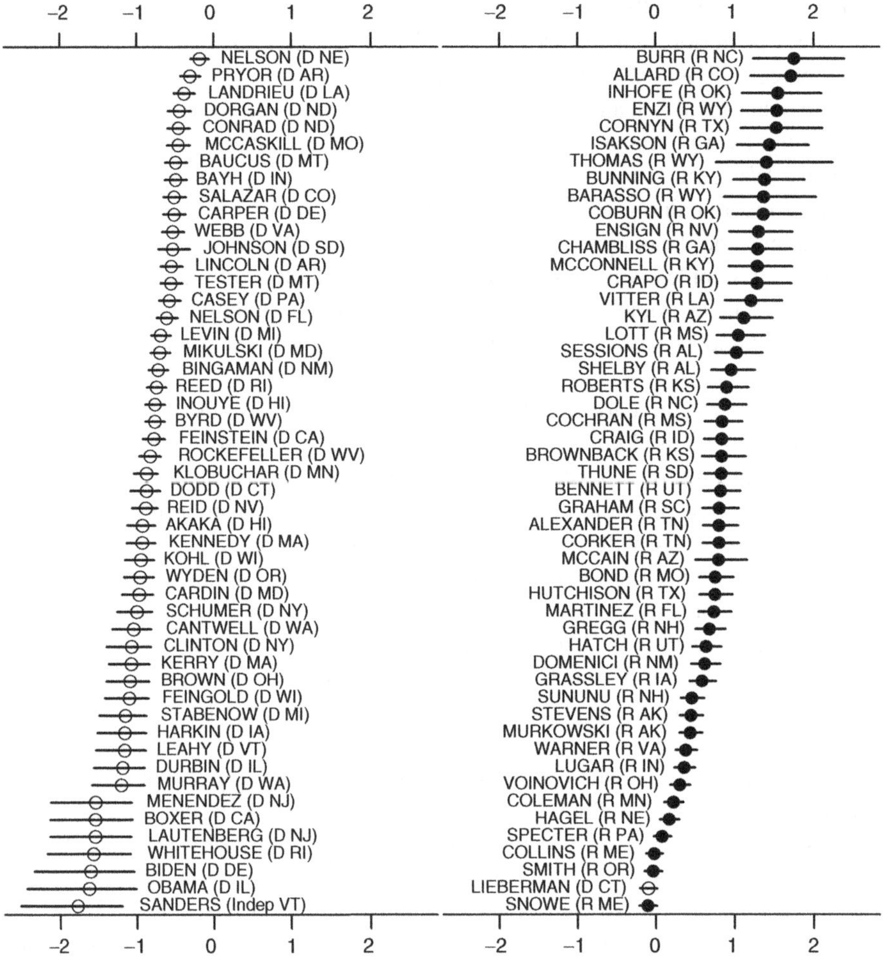

Figure 9.7 Posterior means and 95 % credible intervals for ideal points (ξ_i), using 99 *National Journal* key votes for 2007, Example 9.2. Solid plotting symbols indicate Republican senators.

Sensitivity to prior precision? To demonstrate the consequences of adjusting the weight given to the prior relative to the likelihood function, we analyze the set of 99 *National Journal* 'key' roll calls in 2007 using a one-dimensional model under four different priors: $\sigma_\beta^2 \in \{1, 5^2, 10^2, 25^2\}$. In each case we employ the mean zero, variance one, normalization for the ideal points ξ_i; with this normalization we set $\sigma_\xi^2 = 1$ such that the prior variance of the identified parameters $\tilde{\alpha}$ and $\tilde{\beta}$ is given by $\sigma_{\tilde{\beta}}^2 = \sigma_\xi^2 \sigma_\beta^2 = \sigma_\beta^2$. In this way, the prior is only a function of the σ_β^2 hyperparameter, which we adjust over the four different runs of the model.

For each set of priors we run the data-augmented Gibbs sampler for 1.1 million iterations, discarding the first 100 000 iterations, and saving every 100-th iteration of the remaining one million iterations. In each case the sampler is initialized with all

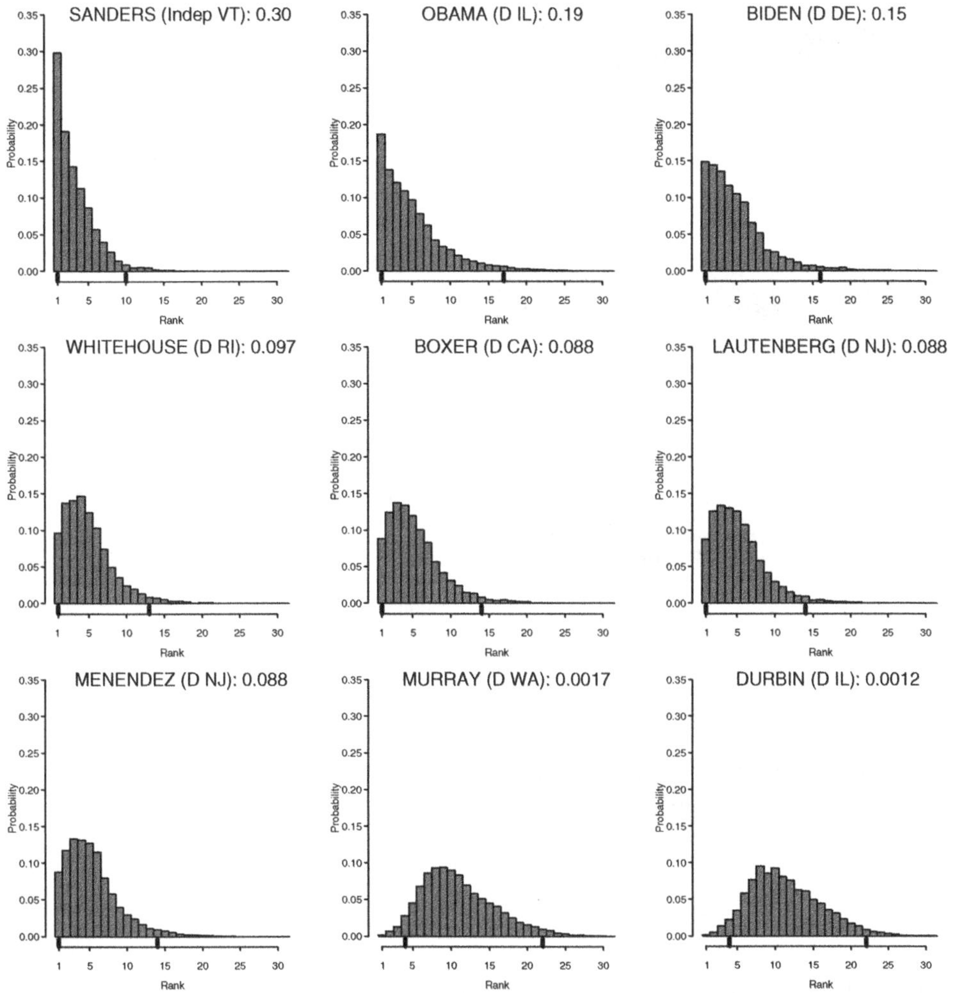

Figure 9.8 Posterior mass functions over order statistics of ideal points, nine 'most liberal' senators, using 99 *National Journal* key votes for 2007, Example 9.2. The number shown next to each name is the proportion of times we see the corresponding ideal point occupying rank one, over the 10 000 samples retained from the 1.1 million iteration run of the data-augmented Gibbs sampler.

Democratic senators at -1 and all Republicans at 1; start values for the bill parameters are generated by running a probit of each roll call on these initial values for the ideal points.

Figure 9.9 shows trace plots and autocorrelation functions for the output of the MCMC algorithm as it samples from marginal posterior density of Obama's ideal point, under the four different priors. As the prior information about $\tilde{\beta}$ and $\tilde{\alpha}$ becomes more vague (σ_β^2 gets larger), the performance of the MCMC algorithm degrades, with larger and more slowly decaying autocorrelations. This is because as the precision of the prior information about the bill parameters eases, the posterior densities for these parameters become more

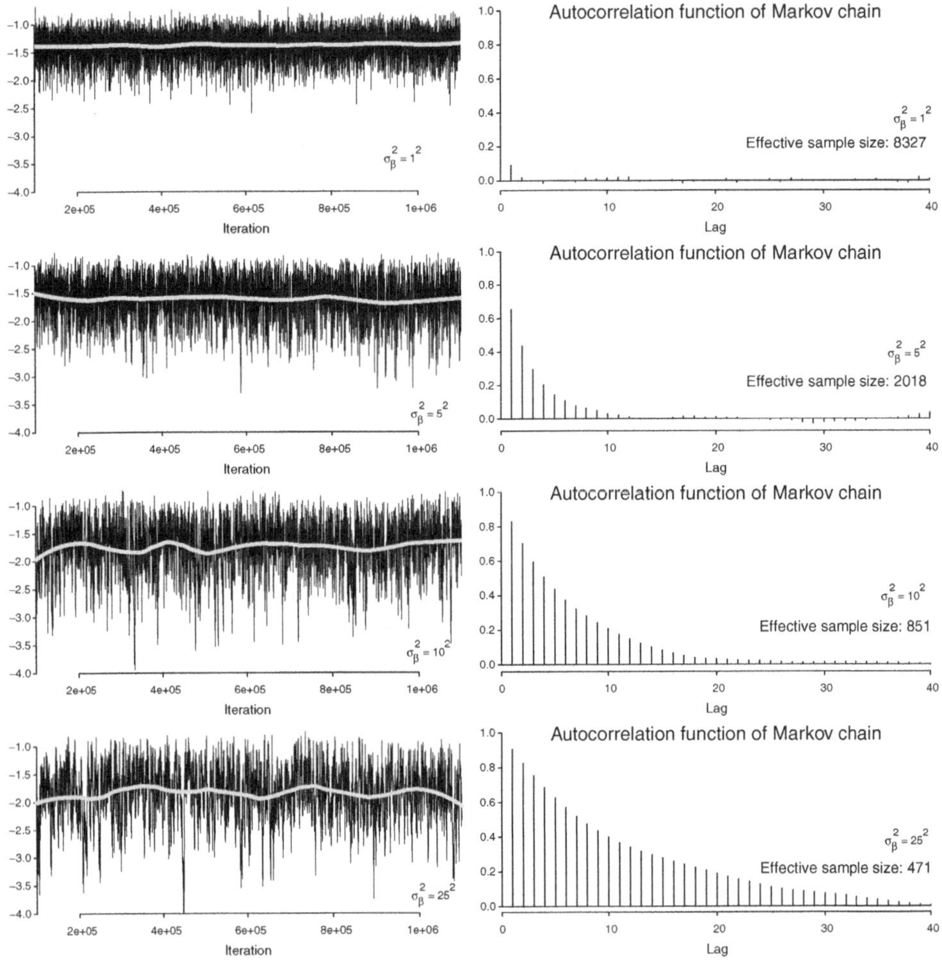

Figure 9.9 Trace plots and autocorrelation functions, output of Markov chain Monte Carlo algorithm for Obama's ideal point under different prior specifications, Example 9.2. In each instance, we impose the identifying restriction that the ideal points ξ_i have mean zero and variance one across senators, such that σ_β^2 is the only 'free' hyperparameter in the specification of the prior. For the trace plots, the gray line shows a moving average. The 'effective sample size' (Section 4.4.1) is a measure of the inefficiency of the MCMC algorithm.

diffuse; in turn, since the $\tilde{\alpha}$ and $\tilde{\beta}$ parameters contribute information about the ideal points, the posterior density for Obama's ideal point skews to the left, even with the identifying restriction that the ideal points have mean zero and variance one.

Figure 9.10 summarizes the marginal posterior density for Obama's ideal point in the panels on the left; the histograms summarize the 10 000 draws from the posterior density of Obama's ideal point that we retain from 1.1 million iteration run of the data-augmented Gibbs sampler. The phenomenon described above is clearly apparent, with less prior precision about the bill parameters working through the model and Obama's short and relatively-liberal voting history to skew his ideal point to the left. Over the four prior specifications for the bill parameters used here, the mean of the posterior density for Obama's ideal point moves about one half of a standard deviation of the ideal point distribution to the left, from -1.37 to -1.84; the increase in the left skew of the posterior density is also apparent, with the standard deviation of the posterior density for Obama's ideal point almost doubling as we increase the standard deviation of the prior density of the bill parameters by a factor of 25.

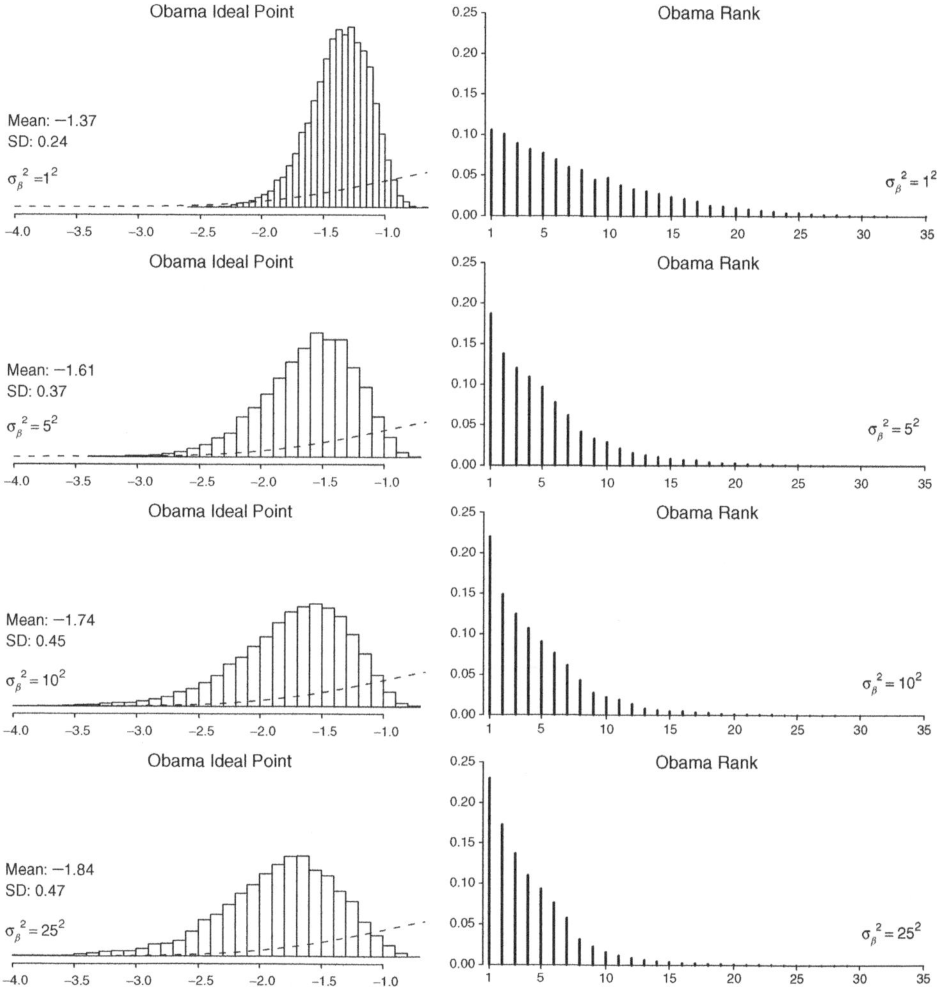

Figure 9.10 Marginal posterior densities for Obama's ideal point (left panels) and posterior mass functions for the rank of Obama's ideal point (right panels), under different prior specifications. In each instance, we impose the identifying restriction that the ideal points ξ_i have mean zero and variance one across senators, such that σ_β^2 is the only 'free' hyperparameter in the specification of the prior. The histograms on the right summarize 10,000 draws from the marginal posterior density for Obama's ideal point; the dotted line indicates the $N(0, 1)$ prior for the ideal point.

The impact of the change in the prior on the inferences we might draw about Obama's rank is summarized in the right column of panels of Figure 9.10. As we assume less about the bill parameters (increase their prior variance, σ_β^2), we become less certain about where to locate Obama's ideal point (left panels of Figure 9.10), but with the consequence that we become slightly more confident that Obama's ideal point occupies rank 1. The posterior mass function over the order statistic of Obama's ideal point is quite diffuse under the relatively informative prior given by setting $\sigma_\beta^2 = 1$ (top row of Figure 9.10); we assign just .11 probability to Obama being 'the most liberal senator' with this prior. But as we assume less about the bills, inducing the leftward drift and skew in the posterior density over Obama's ideal point, the probability that Obama occupies rank 1 increases to .19 ($\sigma_\beta^2 = 5^2$), .22 ($\sigma_\beta^2 = 10^2$) and .23 ($\sigma_\beta^2 = 25^2$).

Table 9.6 reports these posterior probabilities for the 8 senators with highest posterior probabilities of occupying rank 1. At no time is Obama the senator with the highest probability of being 'the most liberal'; Bernie Sanders (Indep VT) is always rated as 'more liberal' than any other Senator. Moreover, even with the least amount of prior information about the bills, no one Senator is unambiguously 'the most liberal'. For instance, Sanders' posterior probability of being 'the most liberal' never exceeds .30, with as many as 7 senators assigned a non-negligible posterior probability of occupying rank one.

Table 9.6 Posterior probabilities that a given Senator is the 'Most Liberal', under different prior variances for $\tilde{\beta}$ and $\tilde{\alpha}$. In each instance, we impose the identifying restriction that the ideal points ξ_i have mean zero and variance one across senators, such that σ_β^2 is the only 'free' hyperparameter in the specification of the prior. For each prior we list the senators with the 8 highest posterior probabilities of occupying rank 1 (having the lowest ideal point) in descending order.

$\sigma_\beta^2 = 1$		$\sigma_\beta^2 = 5^2$	
Sanders (Indep VT)	0.26	Sanders (Indep VT)	0.30
Whitehouse (D RI)	0.12	Obama (D IL)	0.19
Boxer (D CA)	0.12	Biden (D DE)	0.15
Lautenberg (D NJ)	0.12	Whitehouse (D RI)	0.10
Biden (D DE)	0.11	Boxer (D CA)	0.09
Menendez (D NJ)	0.11	Lautenberg (D NJ)	0.09
Obama (D IL)	0.11	Menendez (D NJ)	0.09
Leahy (D VT)	0.02	Murray (D WA)	0.00
$\sigma_\beta^2 = 10^2$		$\sigma_\beta^2 = 25^2$	
Sanders (Indep VT)	0.30	Sanders (Indep VT)	0.29
Obama (D IL)	0.22	Obama (D IL)	0.23
Biden (D DE)	0.15	Biden (D DE)	0.16
Whitehouse (D RI)	0.09	Whitehouse (D RI)	0.08
Boxer (D CA)	0.08	Lautenberg (D NJ)	0.08
Lautenberg (D NJ)	0.08	Boxer (D CA)	0.08
Menendez (D NJ)	0.07	Menendez (D NJ)	0.08
Murray (D WA)	0.00	Leahy (D VT)	0.00

It is hardly the case that the results 'fall apart' or are 'totally driven by the prior'. But it is apparent that as we assume less about the properties of the votes *a priori*, we wind up knowing relatively less about Obama's ideal point *a posteriori*. This is a subtle feature of the Bayesian IRT model deployed here. Nonetheless, we observe that for priors of the sort $\sigma_\beta^2 > 5^2$, the results are relatively robust; moving from the relatively 'tight' priors with $\sigma_\beta^2 = 1$ to $\sigma_\beta^2 = 5^2$ seems to produce the biggest change in the results, and even here these are not massive in any absolute or substantive sense.

In addition, only the estimated rank order of extremists exhibit any real sensitivity to changing the prior variance of the bill parameters. Recall that Obama not only missed a substantial number of votes, but that he also records a relatively extreme voting history. To illustrate the fact that the prior is influencing extremists more than those with a substantial number of missing votes, consider the case of McCain who, despite missing nearly 20 % more votes than Obama, is estimated to have nearly the same rank ordering regardless

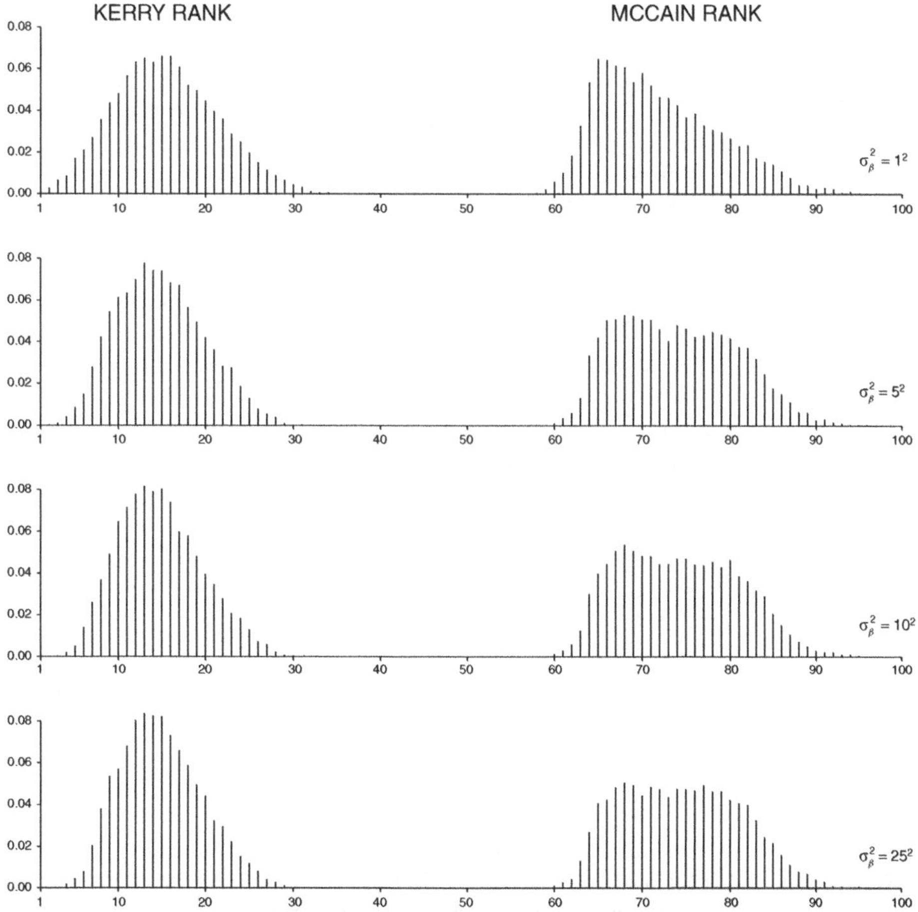

Figure 9.11 Posterior probability mass functions over ranks of Senators Kerry and McCain, Example 9.2.

of the prior variance of the bill parameters. Senator Kerry, supposedly the 'most liberal senator' according to the *National Journal* in 2003, misses only 3 % of the votes scored by the *National Journal* and the estimated rank order of his ideal point is insensitive to the choice of prior variance. Figure 9.11 presents the posterior mass functions for the order statistics of these senators' ideal points; there are some very modest changes in the posterior mass functions as we move from $\sigma_\beta^2 = 1$ to $\sigma_\beta^2 = 5^2$, but barely any discernible changes for successively larger values of σ_β^2.

In sum, to the extent that priors matter, they only matter for legislators with relatively one-sided voting histories. For almost all of the legislators the data dominates the prior. Given that most of the estimands of interest are relatively centrally located – e.g. medians of the floor, committees, and parties – so long as the roll calls have a reasonable number of cutpoints in this central region of the policy space, the choice of prior variance over the bill parameters is not substantively consequential. Moreover, it is trivial to replicate the analysis to assess the sensitivity of posterior inferences to the prior.

An informative prior via hierarchical modeling. When constructing their rankings in 2007, *National Journal* makes the extreme assumption that only votes cast in 2007 are relevant for estimating the ideology of senators in that year. We have also seen that these data are not particularly informative about the ideal points of some legislators. Recall that Table 9.5 suggests that the pattern of missing data in the *National Journal* key votes is not benign. Candidates for president missed a much larger percentage of votes than other senators. We now show how the Bayesian approach allows additional information can be used to supplement the analysis.

The most simple way to incorporate additional information is to pool roll calls across multiple time slices (in this case, years); we can then make the strong assumption that all ideal points are constant across time, effectively increasing the amount of information with which we make inferences about the ideal points. This is similar to the approach we adopted in our earlier work examining the 2003 *National Journal* ratings (Clinton, Jackman and Rivers 2004b).

Here we adopt a different approach, fitting a hierarchical model for the ideal points underlying the *National Journal* data. The model is very simple, exploiting two attributes of legislators: their party affiliations, and the political complexion of their respective state, as tapped by the share of the vote won by John Kerry in the 2004 presidential election. We let this information enter the model via the prior for the ideal points ξ_i with the following hierarchical model

$$\xi_i \sim N(\mu_\xi, \omega^2) \tag{9.19a}$$

$$\mu_\xi = \gamma_0 + R_i\gamma_1 + K_{s(i)}\gamma_2 \tag{9.19b}$$

$$\boldsymbol{\gamma} \sim N(\mathbf{0}, 10^2 \cdot \mathbf{I}_3) \tag{9.19c}$$

$$\omega \sim \text{Unif}(0, 1) \tag{9.19d}$$

where $\boldsymbol{\gamma} = (\gamma_0, \gamma_1, \gamma_2)'$ and ω^2 are hyperparameters, R_i is a binary indicator (1 if senator i is Republican, and 0 otherwise), and $K_{s(i)}$ is Kerry's share of the vote in state s. The last two lines of the hierarchical model specify vague priors over the hyperparameters.

This model is letting the observed characteristics R_i and $K_{s(i)}$ contribute information to the analysis, but with the vague priors on the hyperparameters reflecting uncertainty on our part as to how the information in R_i and $K_{s(i)}$ shapes inferences about ξ_i. That is, this model expresses an *a priori* belief that senators with identical values of R_i and $K_{s(i)}$ will have similar ideal points, but we are agnostic as to the precise mapping from the covariates R_i and $K_{s(i)}$ to ideal points (again, at least *a priori*).

As usual, we impose the identifying normalization that the ideal points have mean zero and variance one across legislators. This sets an upper bound on ω and results in transformations of the γ and ω parameters, above. If we transform from unidentified ideal points ξ_i to identified parameters $\tilde{\xi}_i = (\xi_i - c)/m$ then we have $\tilde{\gamma}_1 = \gamma_1/m$, and similarly for γ_2, $\tilde{\gamma}_0 = (\gamma_0 - c)/m$ and $\tilde{\omega} = \omega/m$. For the bill parameters we use the prior $(\beta_j, \alpha_j)' \sim N(\mathbf{0}, 5^2 \cdot \mathbf{I}_2), \forall j$. We implement this model in JAGS, which for this small data set can generate several thousand samples from the posterior density for this model quite quickly. The sampled values of the identified parameters reveal that the sampler in JAGS is generating a reasonably efficient exploration of the posterior density for this problem.

We compare the posterior densities over ideal points produced by this hierarchical model with the results of the non-hierarchical model fit with $\sigma_\beta^2 = 5^2$. Figure 9.12 presents the two sets of estimates, with lines connecting them so as to highlight the 'shrinkage' we usually see as a consequence of fitting a hierarchical model. The hierarchical model tends to pull together the ideal points of senators with similar covariate values, particularly when information in the roll call data is not particularly rich (e.g. extremists and/or senators with relatively short voting histories). The identifying restriction that the ideal points have mean zero and variance one means that we will not see big differences between the hierarchical and non-hierarchical models with respect to the Bayes estimates of the ideal points.

Figure 9.13 highlights the gains of utilizing the information in the covariates by comparing the standard deviations of the two sets of posterior densities (one obtained from the hierarchical model, vertical axis, the other from the non-hierarchical model, horizontal axis). The hierarchical model generally produces more precise estimates of the ideal points, especially for those ideal points estimated relatively imprecisely by

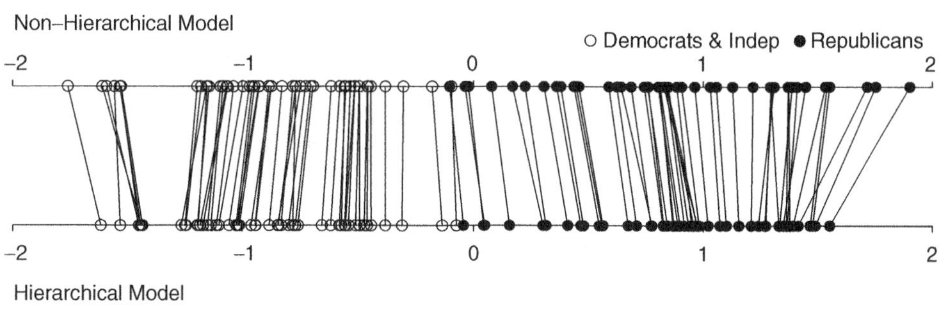

Figure 9.12 Comparison of estimated ideal points (means of the respective marginal posterior densities), hierarchical and non-hierarchical models fit to 99 *National Journal* key votes of 2007, Example 9.2.

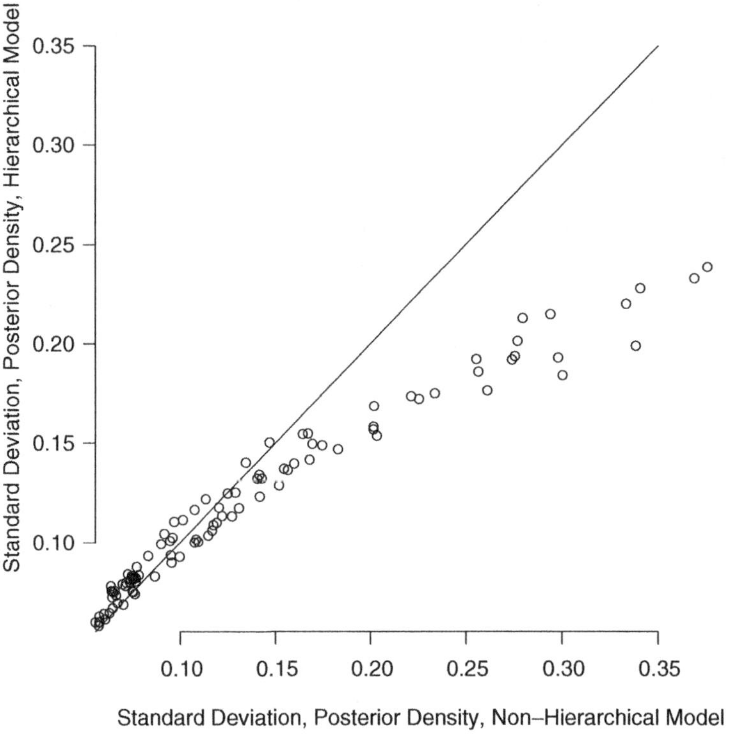

Figure 9.13 Comparison of standard deviations of posterior densities, hierarchical and non-hierarchical models fit to 99 *National Journal* key votes of 2007, Example 9.2. The solid line is at 45 degrees.

the non-hierarchical model. Note what we have accomplished here: politically relevant sources of information about the ideal points have entered the analysis, but not in a particularly heavy-handed way (we deployed vague priors for the hyperparameters), leading to a considerable boost in the precision of the inferences we draw about ideal points and quantities that are functions of the ideal points (rank orderings etc).

Also of interest are the $\tilde{\gamma}_1$ and $\tilde{\gamma}_2$ parameters, the hyperparameters attaching to the Republican indicator variables R_i and the Kerry vote shares $K_{s(i)}$; the posterior mean (and 95 % HPD interval) for $\tilde{\gamma}_1$ is 1.51 (1.35, 1.69) and −.032 (−.042, −.021) for $\tilde{\gamma}_2$. Consistent with the strong separation of ideal points by party, these parameter estimates suggest that on average, senators from the same state but from different parties will have ideal points that differ by 1.51 units, or 1.5 standard deviations of the ideal point distribution. Moreover, these two variables account for much of the variation in the ideal points: the posterior density for the 'residual standard deviation' $\tilde{\omega}$ has mean .35 (.30, .41).

The two-parameter IRT model has been extended in several directions in political science applications. Multidimensional models are routinely fit in the roll call setting (e.g. Jackman 2001; Londregan 2000; Poole and Rosenthal 1997). IRT models for ordinal responses are sometimes deployed in educational testing where they are known as models for 'graded responses'; these models are helpful when analyzing Likert-type ordinal survey responses or when experts assign ordinal ratings to countries, to party manifestos, to candidates, etc. Examples include: an analysis of the ordinal ratings produced by members of graduate admissions committees, so as to generate better measures of applicant quality (Jackman 2004); analysis of a mixture of ordinal and continuous indicators of country-level political and economic risk (Quinn 2004); and analysis of the Polity indicators of country-year levels of democracy (Treier and Jackman 2008). A non-parametric IRT model for ordinal data appears in van Schur (1992).

9.4 Dynamic measurement models

We now consider measurement models in a dynamic setting. There is a long history in the social sciences of analyzing panel studies of public opinion; i.e. surveys that generate repeated measures from survey respondents on various attitudinal items. The longitudinal dimension in these studies is typically short, and variants of the popular Wiley-Wiley (1970) model are often deployed to analyze data of this sort; a review of recent applications in political science appears in Jackman (2008a).

Here we consider a model well-suited to Bayesian analysis, for data spanning a longer longitudinal dimension. A typical application involves analyzing aggregated polling data, with inference for a dynamic, latent variable (or set of latent variables) being the primary goal of the analysis. Examples in political science include the work of Stimson and his co-workers on macro-partianship and dynamic representation (Erikson, MacKuen and Stimson 2002; Stimson, Mackuen and Erikson 1995) and, more generally, Stimson's (1991; 2004) work on the dynamics of public opinion. Contrast work by Donald Green and various co-workers making use of dynamic measurement models to examine the stability of partisanship in the American electorate: e.g. Green, Palmquist and Schickler (1998), Green, Gerber and Boef (1999), Green and Yoon (2002) and the recapitulation in Green, Palmquist and Schickler (2002, ch.4). Beck (1990) also used a dynamic measurement model to estimate the relationship between macro-economic indicators and presidential approval, using a survey measure of the latter. Baum and Kernell (2001) also use a similar model in investigating presidential approval for Franklin Roosevelt.

The model used in these applications is often called a *state-space* model, and has its home in electrical engineering and signal processing, with most social science applications in macroeconomics (e.g. Durbin and Koopman 2001). An observational or measurement equation relates the indicators $\mathbf{y}_t$ to the latent state $\boldsymbol{\xi}_t$:

$$\mathbf{y}_t = \mathbf{F}_t\boldsymbol{\xi}_t + \mathbf{W}_t\boldsymbol{\delta}_t + \boldsymbol{\epsilon}_t \tag{9.20}$$

where $\mathbf{F}_t$ and $\boldsymbol{\delta}_t$ are unknown parameters, $\mathbf{W}_t$ is a matrix of exogenous variables and $\boldsymbol{\epsilon}_t$ is a mean zero disturbance with variance-covariance matrix $\boldsymbol{\Sigma}_t$. The latent variable evolves via the transition equation for the model is

$$\boldsymbol{\xi}_t = \mathbf{G}_t\boldsymbol{\xi}_{t-1} + \mathbf{Z}_t\boldsymbol{\gamma}_t + \mathbf{u}_t, \tag{9.21}$$

$t = 2, \ldots, T$, where $\mathbf{G}_t$ and $\boldsymbol{\gamma}_t$ are unknown parameters, $\mathbf{Z}_t$ is a vector of exogenous variables, and $\mathbf{u}_t$ is a mean zero disturbance with variance-covariance matrix $\boldsymbol{\Omega}_t$.

The primary goal of analysis is inference for the latent state variable $\boldsymbol{\xi}_t$. As written, the model is quite general, allowing either the latent state vector $\boldsymbol{\xi}_t$ or the observational vector $\mathbf{y}_t$ to be a vector or a scalar. This model goes by different names in different parts of the statistics literature: Bayesians refer to this model as dynamic linear model (or DLM), for which West and Harrison (1997) is the standard reference, but see also Carter and Kohn (1994) and Frühwirth-Schnatter (2006); in classical or frequentist statistics, this model is called a Kalman filter (e.g., Harvey 1989). The model encompasses special cases that sometimes go by distinct names in the literature: e.g. the DYMIMIC (dynamic, multiple indicators, multiple cause) model of Watson and Engle (1980; 1983) is a special case. Independent normal densities are often assumed for the two sets of disturbances, which makes the model relatively easy to compute and plausible for many kinds of continuous $\mathbf{y}_t$. Two special cases are:

1. A *local level* or 'random walk' model, setting $\mathbf{G}_t = \mathbf{I}, \forall\, t$, such that we replace Equation 9.21 above with

$$\boldsymbol{\xi}_t = \boldsymbol{\xi}_{t-1} + \mathbf{u}_t. \tag{9.22}$$

2. A *local linear* model, given by setting $\boldsymbol{\xi}_t = (\boldsymbol{\mu}_t, \boldsymbol{\nu}_t)'$, where $\boldsymbol{\mu}_t$ is a latent state and $\boldsymbol{\nu}_t$ are trend coefficients: i.e.

$$\boldsymbol{\mu}_t = \boldsymbol{\mu}_{t-1} + \boldsymbol{\nu}_{t-1} + \boldsymbol{\eta}_t \tag{9.23}$$

$$\boldsymbol{\nu}_t = \boldsymbol{\nu}_{t-1} + \boldsymbol{\zeta}_t \tag{9.24}$$

Gaussian distributions for the level and trend innovations are convenient simplifying assumptions: i.e. $\boldsymbol{\eta}_t \sim N(\mathbf{0}, \boldsymbol{\Sigma}_{\eta_t})$ and $\boldsymbol{\zeta}_t \sim N(\mathbf{0}, \boldsymbol{\Sigma}_{\zeta_t})$. The *smooth local linear* model results by setting $\boldsymbol{\Sigma}_{\eta_t} = \mathbf{0}, \forall\, t$. With these definitions, the transition equation in 9.21 becomes

$$\boldsymbol{\xi}_t = \begin{bmatrix} \boldsymbol{\mu}_t \\ \boldsymbol{\nu}_t \end{bmatrix} = \begin{bmatrix} \mathbf{I} & \mathbf{I} \\ \mathbf{0} & \mathbf{I} \end{bmatrix} \begin{bmatrix} \boldsymbol{\mu}_{t-1} \\ \boldsymbol{\nu}_{t-1} \end{bmatrix} + \begin{bmatrix} \boldsymbol{\eta}_t \\ \boldsymbol{\zeta}_t \end{bmatrix}, \tag{9.25}$$

i.e.

$$\mathbf{G}_t = \begin{bmatrix} \mathbf{I} & \mathbf{I} \\ \mathbf{0} & \mathbf{I} \end{bmatrix}, \forall\, t.$$

Quite simple versions of the general model sketched in Equations 9.20 and 9.21 are typically deployed in political science applications. First, it is often the case that the latent variable is a scalar (e.g. when tracking support or levels of partisanship for one party, or presidential approval). There are often no exogenous variables $\mathbf{W}_t$ in the measurement equation, with the effect of covariates usually posited as working on the latent variable; contrast Jackman (2005a) and Beck, Rosenthal and Jackman (2006) where the measurement equation includes house effects (see Example 9.3, below). Simple dynamics are often posited for the latent variable, say a local-level/random-walk model, with $\mathbf{G}_t = 1 \forall\, t$. The effects of covariates $\mathbf{Z}_t$ are often made time-invariant, as is the innovation variance term (i.e. $\boldsymbol{\Omega}_t = \boldsymbol{\Omega}$).

9.4.1 State-space models for 'pooling the polls'

The special context of tracking public opinion with an ensemble of polls offers some further simplifications. For ease of exposition, consider the case where we are interested in a scalar latent state: e.g. presidential approval (the proportion of the population approving of the way the president is 'handling his job'), or the proportion of the electorate who intending to vote for a political party. The model we use here assumes that polls have two sources of error:

1. sampling error, $\boldsymbol{\epsilon}_t$, with covariance matrix $\boldsymbol{\Sigma}_t$. Given a poll with a large sample, the sampling variances of each y are well approximated by an observable function of the polls' sample size and its estimate; i.e. if poll i reports y_i as its estimate of a population proportion, with a sample size of n_i we set $V(y_i) = y_i(1 - y_i)/n_i$. Note also that sampling error can be considered independent across polls such that $\boldsymbol{\Sigma}_t$ is diagonal.
2. house effects, operationalized via the $\boldsymbol{\delta}_t$ parameters in equation 9.20; house effects are generally considered time-invariant biases specific to survey house $j = 1, \ldots, m$, and so we drop the t subscript and consider $\boldsymbol{\delta}$.

At time t, we have data from m_t polls, $0 < m_t \leq m$, and so the dimensions of $\mathbf{y}_t$, $\mathbf{F}_t$ and $\boldsymbol{\epsilon}_t$ and $\boldsymbol{\Sigma}_t$ vary accordingly. Let $\mathbf{W}_t$ be a set of binary indicators that vary over time, depending on which polls are generating data at time t, such that the appropriate house effects $\boldsymbol{\delta}$ enter the model for time t. Note also that given the assumptions above, $\mathbf{F}_t$ is a mere 'housekeeping' matrix of ones or zeros, with dimensions depending on the number of polls generating data at time t (m_t).

With these assumptions and simplifications, for the local level transition model, we have

$$\underset{(m_t \times 1)}{\mathbf{y}_t} = \underset{(m_t \times 1)}{\mathbf{F}_t} \underset{(1 \times 1)}{\xi_t} + \underset{(m_t \times m)}{\mathbf{W}_t} \underset{(m \times 1)}{\boldsymbol{\delta}} + \underset{(m_t \times 1)}{\boldsymbol{\epsilon}_t} \tag{9.26}$$

$$\xi_t = \xi_{t-1} + u_t \tag{9.27}$$

where $\mathbf{F}_t = \boldsymbol{\iota}_{m_t}$ is a m_t-by-1 unit vector and $\mathbf{W}_t = (W_{jkt})$ is a m_t-by-m matrix of binary indicators with $W_{jkt} = 1$ if polling house $k = 1, \ldots, m$ is the jth survey house in the field at time t, where $j = 1, \ldots, m_t$. Note also that in this case $\boldsymbol{\Omega}_t$ reduces to the scalar ω_t^2 which we usually assume is a time-invariant quantity ω^2. With the additional assumption of normality for both the sampling errors $\boldsymbol{\epsilon}_t$ and the innovations u_t we have

$$\mathbf{y}_t | \xi_t, \mathbf{W}_t, \boldsymbol{\delta} \sim N(\mathbf{F}_t \xi_t + \mathbf{W}_t \boldsymbol{\delta}, \boldsymbol{\Sigma}_t) \tag{9.28}$$

$$\xi_t | \xi_{t-1} \sim N(\xi_{t-1}, \omega^2). \tag{9.29}$$

For the case of the model with a local linear trend we have $\boldsymbol{\xi}_t = (\mu_t, \nu_t)'$ and

$$\mathbf{y}_t = \underset{(m_t \times 1)}{\mathbf{F}_t} \underset{(1 \times 2)}{[1 \; 0]} \underset{(2 \times 1)}{\begin{bmatrix} \mu_t \\ \nu_t \end{bmatrix}} + \mathbf{W}_t \, \boldsymbol{\delta} + \boldsymbol{\epsilon}_t \tag{9.30}$$

$$= \underset{(m_t \times 2)}{[\boldsymbol{\iota} \; \mathbf{0}]} \underset{(2 \times 1)}{\boldsymbol{\xi}_t} + \mathbf{W}_t \, \boldsymbol{\delta} + \boldsymbol{\epsilon}_t \tag{9.31}$$

$$= \mathbf{H}_t \boldsymbol{\xi}_t + \mathbf{W}_t \boldsymbol{\delta} + \boldsymbol{\epsilon}_t. \tag{9.32}$$

The transition equation for the local linear trend model is

$$\boldsymbol{\xi}_t = \underset{(2\times 2)}{\begin{bmatrix} 1 & 1 \\ 0 & 1 \end{bmatrix}} \boldsymbol{\xi}_{t-1} + \underset{(2\times 1)}{\mathbf{u}_t}, \qquad \text{or equivalently,} \tag{9.33}$$

$$\begin{bmatrix} \mu_t \\ \nu_t \end{bmatrix} = \begin{bmatrix} 1 & 1 \\ 0 & 1 \end{bmatrix} \begin{bmatrix} \mu_{t-1} \\ \nu_{t-1} \end{bmatrix} + \begin{bmatrix} \eta_t \\ \zeta_t \end{bmatrix}, \tag{9.34}$$

with $\boldsymbol{\Sigma}_\eta$ and $\boldsymbol{\Sigma}_\zeta$ reducing to the time-invariant scalars σ^2_η and σ^2_ζ, respectively, and $\mathbf{G}_t = \begin{bmatrix} 1 & 1 \\ 0 & 1 \end{bmatrix} \forall\, t$. One of the great strengths of the model is that it can easily handle the case where the indicators are missing (partially, or completely) for any given time point. In such a case inference for $\boldsymbol{\xi}_t$ can proceed with the transition model in Equation 9.21 serving as a (posterior) predictive model; in the case where the indicators $\mathbf{y}_t$ are available, both they and lagged and future values of $\boldsymbol{\xi}$ contribute to inferences for $\boldsymbol{\xi}_t$. This is especially useful in political science settings, where the indicators are typically survey aggregates of some kind, and it is quite typical for surveys to have spotty temporal coverage.

It should also be noted that the model parameters are not identified, since any additive combination of estimated levels of latent support $\boldsymbol{\xi}_t$ and house effects $\boldsymbol{\delta}$ can fit the observed data for a particular day, $\mathbf{y_t}$. In the special context of *ex post* analysis of election polling we exploit the fact that on day T ('Election Day'), ξ_T is observed as the election result. It is also helps if all survey houses were in the field close to Election Day and are reasonably prevalent elsewhere in the data, generating a rich 'overlapping generations' structure in the observed poll results. In addition, imposing an *a priori* upper bound on the innovation variance parameter ω^2, limits the extent to which differences between the final pre-election polls and the Election Day result can be accounted for by innovations in the latent state variable $\boldsymbol{\xi}_t$.

9.4.2 Bayesian inference

Bayesian inference for this problem requires computing the joint posterior density of the unknown parameters, given observables. Consider the relatively simple local level model in Equations 9.26 and 9.27. The observables are the survey estimates $\mathbf{Y} = \{\mathbf{y}_t\}$, the sampling variances $\boldsymbol{\Sigma} = \{\boldsymbol{\Sigma}_t\}$, and the indicators $\mathbf{W} = \{\mathbf{W}_t\}$. The unknowns are the latent states $\boldsymbol{\xi} = (\xi_1, \ldots, \xi_T)'$, the house effects $\boldsymbol{\delta}$ and the innovation variance parameter ω^2. By Bayes Rule,

$$p(\boldsymbol{\xi}, \boldsymbol{\delta}, \omega^2 | \mathbf{Y}, \boldsymbol{\Sigma}, \mathbf{W}) \propto p(\mathbf{Y}|\boldsymbol{\Sigma}, \mathbf{W}, \boldsymbol{\xi}, \boldsymbol{\delta}, \omega^2) p(\boldsymbol{\xi}, \boldsymbol{\delta}, \omega^2) \tag{9.35}$$

where $p(\mathbf{Y}|\boldsymbol{\Sigma}, \mathbf{W}, \boldsymbol{\xi}, \boldsymbol{\delta}, \omega^2)$ is the joint density of the observed polling data and $p(\boldsymbol{\xi}, \boldsymbol{\delta}, \omega^2)$ is the joint prior density over the unknowns.

The form of the joint density of the observed poll results (the likehood) is straightforward. Recall the assumptions of normality for the sampling errors $\boldsymbol{\epsilon}_t$ and conditional independence across polls given ξ_t. Let $j = 1, \ldots, m_t$ index the polls available for analysis at time t. Then

$$p(\mathbf{y}_t|\xi_t, \boldsymbol{\delta}, \mathbf{W}_t, \boldsymbol{\Sigma}_t) = \prod_{j=1}^{m_t} \phi\left(\frac{y_j - \xi_t - \mathbf{w}_j\boldsymbol{\delta}}{\sigma_{y_j}}\right) \tag{9.36}$$

where ϕ is the normal probability density function, $\mathbf{w}_j$ is the j-th row of $\mathbf{W}_t$ and $\sigma_{y_j}^2 = y_j(1 - y_j)/n_j$ is the j-th diagonal element of $\boldsymbol{\Sigma}_t$, where n_j is the sample size of poll j. Note also that conditional on the ξ_t, the polling data are independent of the innovation variance parameter ω^2. The joint density for all the polling data is easily obtained, again using the assumption of independence across polls:

$$p(\mathbf{Y}|\boldsymbol{\Sigma}, \mathbf{W}, \boldsymbol{\xi}, \boldsymbol{\delta}, \omega^2) = \prod_{t=1}^{T} p(\mathbf{y}_t|\xi_t, \boldsymbol{\delta}, \mathbf{W}_t, \boldsymbol{\Sigma}_t) \tag{9.37}$$

Priors

The transition model in Equation 9.27 specifies a prior for the ξ_t. In the application in Example 9.3, below, ξ_1 and ξ_T are observed election results and not unknown parameters. Priors on the house effects $\boldsymbol{\delta}$ and the innovation variance parameter complete the specification. Here I use independent normal priors for the $\boldsymbol{\delta}$:

$$p(\boldsymbol{\delta}) = \prod_{j=1}^{J} \phi(\delta_j/\sigma_\delta) \tag{9.38}$$

with $\sigma_\delta = .05$. That is, *a priori* I posit $\delta_j \sim N(0, .05^2)$. Plus or minus 2 prior standard deviations for each house effect is plus or minus 10 percentage points, which encompasses a wide range of possible house effects, at least in the context of election polling. For the innovation variance parameter, we employ a proper uniform prior with respect to the standard deviation of the innovations. In the context of election polling we might use a prior such as

$$\omega \sim \text{Unif}(0, .01) \tag{9.39}$$

meaning that we assign zero probability to the event that ω is greater than a percentage point. Since t indexes campaign days in most election polling applications, at the upper bound of this prior, $\sigma_u = .01$, implying that 95 % of the day-to-day movements in ξ_t will lie within $\pm$ 2 percentage points. This prior operationalizes a belief that these kind of large, day-to-day movements in voter sentiment larger than this are rare, although not unheard of. Less restrictive priors can be used if desired.

Bayesian computation via Markov chain Monte Carlo

The number of unknowns is large, and so we use a Gibbs sampler to explore the joint posterior density of the model parameters. In deriving the conditional distributions that drive the Gibbs sampler we will make repeated use of Proposition 5.2 and the results on (conditionally) conjugate Bayesian analysis of normal data (Sections 2.4 and 2.5).

1. $\delta_j, j = 1, \ldots, m$. *A priori*, $\delta_j \sim N(0, \sigma_\delta^2)$, see Equation 9.38. The children of δ_j are all the poll estimates y_i published by polling organization j. Let i index the complete set of poll results available for analysis, t_i be the time at which poll i was in the field. Letting k index polling organizations, define $\mathcal{P}_k = \{i : j_i = k\}$ as the set of polls published by polling organization k. Then δ_k has a conditional distribution that is normal, with mean

$$\left[\sum_{i \in \mathcal{P}_k} \frac{y_i - \xi_{t_i}}{\sigma_{y_i}^2}\right] \cdot \left[\sum_{i \in \mathcal{P}_k} \frac{1}{\sigma_{y_i}^2} + \frac{1}{\sigma_\delta^2}\right]^{-1}$$

and variance

$$\left[\sum_{i \in \mathcal{P}_k} \frac{1}{\sigma_{y_i}^2} + \frac{1}{\sigma_\delta^2}\right]^{-1}.$$

2. ω^2. For tractability I work with $\tau = g(\omega) = \omega^{-2}$; conversely, $\omega = g^{-1}(\tau) = \tau^{-1/2}$. The uniform prior over ω in Equation 9.39 implies the restriction that $\omega^2 < .0001$ and in turn, the restriction that $\tau > 10\,000$. Moreover, using standard results on transformations of random variables (see Proposition B.7), the uniform prior on ω implies the following prior for τ:

$$\begin{aligned} f(\tau) &= \left|\frac{\partial \omega}{\partial \tau}\right| f_\omega\left[g^{-1}(\tau)\right] \\ &= \frac{1}{2}\tau^{-3/2} p(\tau), \end{aligned}$$

where $p(\tau) = 100$ if $\tau > 10\,000$ and 0 otherwise, and so $f(\tau) \propto \tau^{-3/2}$, $\tau > 10\,000$. The children of τ are simply the children of ω^2, the ξ_t parameters, $t = 2, \ldots, T$; see Equation 9.27. Thus, using the general formulation in Proposition 5.2, the conditional distribution of τ given the rest of the model $\mathcal{G}_{-\tau}$ is

$$\begin{aligned} f(\tau|\mathcal{G}_{-\tau}) &\propto \tau^{-3/2} \prod_{t=2}^{T} f(\xi_t; \xi_{t-1}, \tau)\, \mathcal{I}(\tau > 10\,000) \\ &= \tau^{-3/2} \prod_{t=2}^{T} \frac{1}{\sqrt{2\pi\omega^2}} \exp\left[\frac{-(\xi_t - \xi_{t-1})^2}{2\omega^2}\right] \mathcal{I}(\tau > 10\,000) \\ &= \tau^{-3/2} \prod_{t=2}^{T} \frac{\tau^{1/2}}{\sqrt{2\pi}} \exp\left[-\tau\frac{1}{2}(\xi_t - \xi_{t-1})^2\right] \mathcal{I}(\tau > 10\,000) \\ &\propto \tau^{(T-4)/2} \exp\left[-\tau\frac{1}{2}\sum_{t=2}^{T}(\xi_t - \xi_{t-1})^2\right] \mathcal{I}(\tau > 10\,000) \end{aligned}$$

which is a Gamma density (Definition B.34) over τ with parameters $(T-4)/2$ and $\frac{1}{2}\sum_{t=2}^{T}(\xi_t - \xi_{t-1})^2$, and where $\mathcal{I}$ is an indicator function, evaluating to one if its argument is true, and zero otherwise (i.e. constraining the density to have support only where $\tau > 10\,000$). The sampled τ can be transformed back into ω (i.e. $\omega = g^{-1}(\tau) = \tau^{-1/2}$), since that is the scale on which we wish to perform inference (i.e. prior beliefs about the magnitudes of day-to-day movements in ξ_t were formulated on the scale of the standard deviation of those movements, ω, not the inverse of the variance of the shocks τ).

3. $\boldsymbol{\xi} = (\xi_1, \ldots, \xi_T)'$. It is well known that it is extremely inefficient to sample from the conditional distributions of each ξ_t, $t = 1, \ldots, T-1$, given its neighbors in the DAG representation of the model (e.g. ξ_{t-1} is a parent of ξ_t which is a parent of ξ_{t+1}, and so on). For this reason, this class of model is usually best handled *not* in general purpose computer programs for MCMC-based Bayesian analysis such as `WinBUGS` or `JAGS`. That is, these programs parse each ξ_t as a distinct node in the DAG representation of the state-space model. My experience is that tens to hundreds of millions of iterations – taking several days of CPU time on even fast desktop computers – are required so as to satisfactorily explore the posterior density for problems with large T using `WinBUGS` or `JAGS`. Because temporally proximate components of the state vector $\boldsymbol{\xi}$ are so highly correlated with another, it is far more effective to sample from their joint distribution *en bloc*. In the following discussion we momentarily set aside the fact that we sometimes have "end point" constraints on ξ_1 and for ξ_T.

   The standard method for generating a sample from the conditional distribution of the entire state vector is the *filter-forward, sample backwards* (FFSB) algorithm, independently due to Carter and Kohn (1994) and Frühwirth-Schnatter (1994), which we now describe. We wish to sample from

$$\begin{aligned} f(\boldsymbol{\xi}|\mathcal{G}_{-\boldsymbol{\xi}}) &\propto f(\boldsymbol{\xi}|\mathbf{Y}, \boldsymbol{\delta}, \mathbf{W}, \boldsymbol{\Sigma}, \omega^2) \\ &= f(\xi_T|\mathbf{Y}, \boldsymbol{\delta}, \mathbf{W}, \boldsymbol{\Sigma}, \omega^2) \prod_{t=1}^{T-1} f(\xi_t|\xi_{t+1}, \mathbf{Y}, \boldsymbol{\delta}, \mathbf{W}, \boldsymbol{\Sigma}, \omega^2). \end{aligned}$$

   This factorization of the conditional distribution of $\boldsymbol{\xi}$ is suggestive, implying that if we could sample ξ_T, we could then condition on it so as to sample from the conditional distribution for ξ_{T-1}, and so on, backwards-sampling through the series. Given ξ_{t+1}, ξ_t is independent of anything else in the future, so

$$f(\xi_t|\xi_{t+1}, \mathbf{Y}, \boldsymbol{\delta}, \mathbf{W}, \boldsymbol{\Sigma}, \omega^2) = f(\xi_t|\xi_{t+1}, \mathbf{Y}^t, \boldsymbol{\delta}, \mathbf{W}^t, \boldsymbol{\Sigma}_{\boldsymbol{\epsilon}}^t, \omega^2) \tag{9.40}$$

   where the superscript t indicates elements of the corresponding object up to time t. We also exploit the fact that by Bayes Rule

$$\begin{aligned} &f(\xi_t|\xi_{t+1}, \mathbf{Y}^t, \boldsymbol{\delta}, \mathbf{W}^t, \boldsymbol{\Sigma}_{\boldsymbol{\epsilon}}^t, \omega^2) \\ &\quad \propto p(\xi_{t+1}|\xi_t, \mathbf{Y}^t, \boldsymbol{\delta}, \mathbf{W}^t, \boldsymbol{\Sigma}_{\boldsymbol{\epsilon}}^t, \omega^2) p(\xi_t|\mathbf{Y}^t, \boldsymbol{\delta}, \mathbf{W}^t, \boldsymbol{\Sigma}_{\boldsymbol{\epsilon}}^t, \omega^2). \end{aligned} \tag{9.41}$$

   The first term on the right hand side simplifies considerably, since the transition model in Equation 9.27 implies that given ξ_t, beliefs about ξ_{t+1} do not depend on any quantities prior to time $t+1$; i.e. the only way that current or past polling data influence beliefs about future $\xi_{t+r}, r > 0$, is through ξ_t. With the additional assumption of normality we have

$$p(\xi_{t+1}|\xi_t, \mathbf{Y}^t, \boldsymbol{\delta}, \mathbf{W}^t, \boldsymbol{\Sigma}^t_{\boldsymbol{\epsilon}}, \omega^2) = p(\xi_{t+1}|\xi_t, \omega^2) \equiv N(\xi_t, \omega^2). \tag{9.42}$$

The second term on the right hand side of Equation 9.41 is itself a posterior density, the posterior density of ξ_t given observables up through time t. Under our maintained assumptions of normality, the mean and variance of this density are sufficient statistics for this density; i.e. we seek

$$m_t = E(\xi_t|\mathbf{Y}^t, \boldsymbol{\delta}, \mathbf{W}^t, \boldsymbol{\Sigma}^t_{\boldsymbol{\epsilon}}, \omega^2) \quad \text{and} \tag{9.43}$$

$$C_t = V(\xi_t|\mathbf{Y}^t, \boldsymbol{\delta}, \mathbf{W}^t, \boldsymbol{\Sigma}^t_{\boldsymbol{\epsilon}}, \omega^2) \tag{9.44}$$

The Kalman filter is a recursive procedure for obtaining these conditional moments. For notational clarity, denote the second term on the right hand side of Equation 9.41 as $p(\xi_t|\mathbf{Y}^t)$. Given a stream of data up through $t-1$, we have

$$\xi_{t-1}|\mathbf{Y}^{t-1} \sim N(m_{t-1}, C_{t-1}). \tag{9.45}$$

Recall that the model in Equations 9.26 and 9.27 implies

$$\xi_t|\xi_{t-1} \sim N(\xi_{t-1}, \omega^2) \tag{9.46}$$

$$\mathbf{y}_t|\xi_t \sim N(\xi_t + \mathbf{W}_t\boldsymbol{\delta}, \boldsymbol{\Sigma}_{\boldsymbol{\epsilon}_t}). \tag{9.47}$$

Combining Equations 9.45 and 9.42 we have the the step-ahead forecast for ξ_t (or, equivalently, a prior for ξ_t) given information observed up through $t-1$,

$$\xi_t|\mathbf{Y}^{t-1} \sim N(m_{t-1}, C_{t-1} + \omega^2). \tag{9.48}$$

Now, to obtain the second term on the right hand side of Equation 9.41, we combine the 'prior' in Equation 9.48 with the likelihood in Equation 9.47 to yield

$$\xi_t|\mathbf{Y}^t \sim N(m_t, C_t) \tag{9.49}$$

where

$$m_t = \left[\left(\sum_{j\in\mathcal{P}_t} \frac{y_{jt} - \delta_j}{\sigma^2_{y_j}}\right) + \frac{m_{t-1}}{C_{t-1} + \omega^2}\right] \cdot \left[\left(\sum_{j\in\mathcal{P}_t} \frac{1}{\sigma^2_{y_j}}\right) + \frac{1}{C_{t-1} + \omega^2}\right]^{-1} \tag{9.50}$$

is the precision-weighted average of the likelihood contributions and prior and

$$C_t = \left[\left(\sum_{j\in\mathcal{P}_t} \frac{1}{\sigma^2_{y_j}}\right) + \frac{1}{C_{t-1} + \omega^2}\right]^{-1}, \tag{9.51}$$

where $\mathcal{P}_t$ is the set of survey houses in the field at time t. If there is no survey data y_t available at time t, then we set $m_t = m_{t-1}$ and $C_t = C_{t-1} + \omega^2$.

With this forward-filtering procedure supplying m_t and C_t, $t = 1, \ldots, T$, we then backwards sample from the conditional distribution in Equation 9.41, $f(\xi_t|\xi_{t+1}, \mathbf{Y}^t)$, $t = T-1, T-2, \ldots, 1$. For clarity, let $\tilde{\xi}_t$ represent the sampled values from this backwards sampling procedure. Then via Bayes Rule and our assumptions of normality, $\tilde{\xi}_t$ is sampled from $f(\xi_t|\tilde{\xi}_{t+1}, \mathbf{Y}^t)$, a normal density with mean

$$E(\xi_t|\tilde{\xi}_{t+1}, \mathbf{Y}^t) = \left(\frac{\tilde{\xi}_{t+1}}{\omega^2} + \frac{m_t}{C_t}\right) \cdot \left(\frac{1}{\omega^2} + \frac{1}{C_t}\right)^{-1}$$

and variance

$$V(\xi_t|\tilde{\xi}_{t+1}, \mathbf{Y}^t) = \left(\frac{1}{\omega^2} + \frac{1}{C_t}\right)^{-1},$$

$t = T-1, T-2, \ldots, 1$. For $t = T$, we sample $\tilde{\xi}_T$ from a $N(m_T, C_T)$ density. The sampled values values $\tilde{\boldsymbol{\xi}} = (\tilde{\xi}_1, \ldots, \tilde{\xi}_T)'$ constitute a draw from the conditional distribution of $\boldsymbol{\xi}$.

We now examine the algorithm at work with an example.

■ Example 9.3

'Pooling the polls' over an election campaign; the 2007 Australian Federal election. In earlier work (Jackman 2005b) I considered a simple state space model for tracking public opinion over an election campaign, using the simple local level model with 'house effects' described above. Here we deploy the model and algorithm described above to examine Australian public opinion between the 2004 and 2007 elections.

Priors. We require priors for the initial state, ξ_1, the house effect parameters $\boldsymbol{\delta}$ and the innovation variance ω^2. For the house-effects parameters δ_j I use a vague normal prior centered on zero, $\delta_j \sim N(0, d^2)$ with d an arbitrary large constant, so that the data will dominate inferences for these parameters. Since the δ_j are house-specific shifts on the scale of the observed survey proportions, it is reasonable to posit that $\delta_j = .15$ would be a 'very large' house effect (i.e. survey organization j is biased by 15 percentage points, on average). Accordingly, a prior distribution that had a 95 % interval spanning $-.15$ to $.15$ would formalize the *a priori* belief that we are reasonably ignorant about the magnitudes of house effects, but that they are likely to be no larger than plus or minus 15 percentage points. In turn, this corresponds to $d^2 = (.15/2)^2 = .005625$. That is, given the scale of the data being modeled, a vague normal prior can be specified with any $d > .075$. For ω^2 we will use a uniform prior over the standard deviation of the innovations, ω; see the discussion accompanying equation 9.39.

Data. Here I use all 239 published polls provided by the 'Big Four' Australian pollsters (Newspoll, Nielsen, Morgan and Galaxy) between the October 9 2004 and November 24 2007 Australian Federal elections. Sample sizes for the polls range from roughly

500 to 2600, with a median of 1156 respondents; although most polls use some form of post-stratification weighting to adjust for biases that might result from non-response, we take the stated sample size of the poll as the effective sample size, n; i.e. the polls typically do not report sample sizes adjusted for design effects or weighting.

Polls are assigned to a specific day, equal to the median day of the poll's field period (rounded down for polls with an even number of field days). With Election Day 2004 constituting 'day one' (i.e. $t = 1$), there are $T = 1142$ days in the period under study. Morgan uses two modes of surveying – face-to-face and phone – which we treat as two separate 'houses' and we estimate a δ_j parameter for each, for a total of $J = 5$ survey houses in the analysis. Galaxy contributes 10 polls to the analysis in the second half of 2007; Morgan contributes 94 face-to-face polls and 15 phone surveys; there are 78 polls from Newspoll in the analysis, and Nielsen contributes 42 polls.

For each poll we note the proportion of respondents reporting an intention to vote for the Australian Labor Party (ALP) in a House of Representatives election. The ALP won the 2007 election, receiving 43.4 % of 'first preferences' (Australia has a single transferable voting system in its House of Representatives), up from 37.6 % in the 2004 election; these values supply known values for ξ_T and ξ_1, respectively. Alternatively, the constrained values of ξ_t can be treated as opinion polls with arbitrarily massive sample sizes and free of any house effect.

Bayesian computation. I implemented the Gibbs sampling algorithm described in Section 9.4.2 in R. The summaries reported here are based on a run of 250 000 iterations, thinned by a factor of 10, to leave 25 000 iterations for analysis. The Markov chain appears to converge on the posterior density almost instantly, and so we dispense with a burn-in period (as is often the case with MCMC algorithms, the issue here is not so much 'convergence' but the efficiency with which the algorithm explores the posterior density). The house effect parameters (δ_j) exhibit slow mixing, due to strong dependence in the model between these parameters and the latent states ξ_t; i.e. since we model $E(y_i) = \xi_{t(i)} + \delta_{j(i)}$, as the δ_j parameters drift up (or down) in the course of the MCMC iterations, the ξ_i parameters can drift in the opposite direction, subject to the identifying constraints from the fixed endpoints ξ_1 and ξ_T. The high *a posteriori* correlations among the δ_j are clearly apparent in Figure 9.14, which presents a series of pairwise summaries of the exploration of the posterior density of these parameters (bivariate trace plots shown in the panels above the diagonal, scatterplots in the panels below the diagonal). This suggests the MCMC sampler would benefit from an over-parameterization of the sort we have exploited numerous times throughout the book (e.g. Examples 6.9 and 7.7).

Results. Numerical summaries of the marginal posterior densities of the house effects and the innovation standard deviation ω are presented in Table 9.7. The largest house effect is for the face-to-face polls conducted by Morgan; the point estimate of the house effect is 2.7 percentage points, which is very large relative to the classical sampling error accompanying these polls. The 95 % HPD for this house effect, δ_2, is bounded well away from zero and we are extremely confident in the direction of the Morgan face-to-face house effect. Galaxy has the next largest house effect, in absolute terms, suggesting that its polls underestimate support for Labor by just over a percentage point. A similar result holds for Newspoll, except the house effect runs in the other direction, with a 1.2 percentage point overestimate of Labor support. Nielsen's house effect is closer to zero,

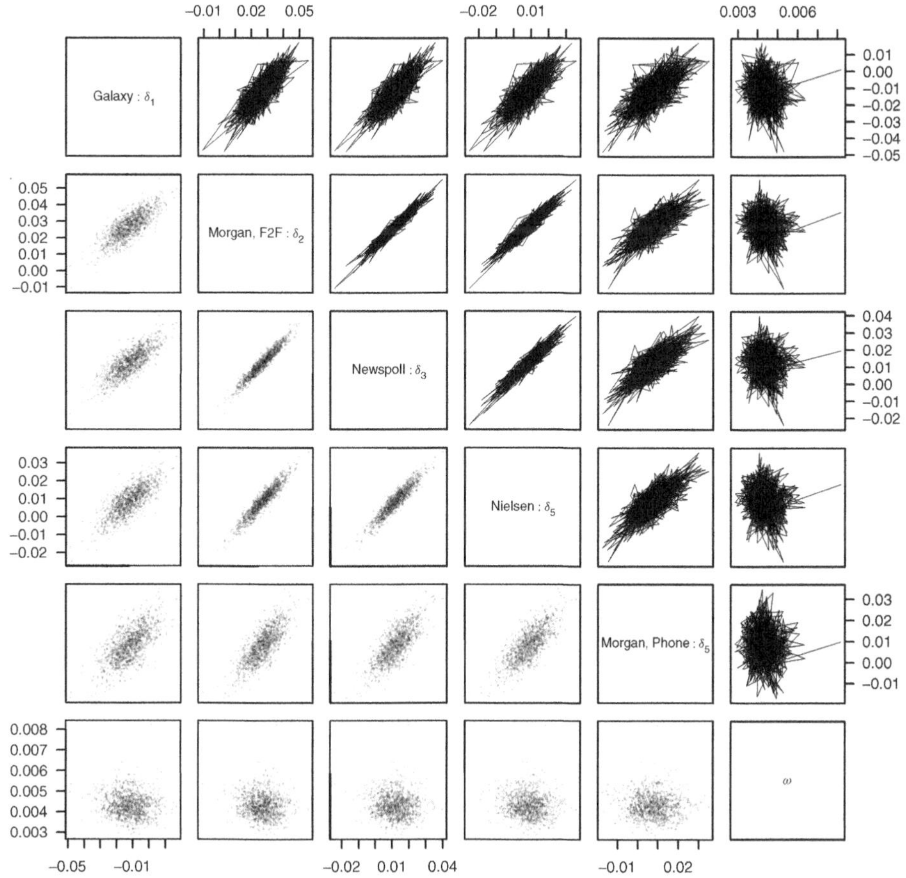

Figure 9.14 Pairwise traceplots (upper triangle of plots) and posterior densities (scatterplots, lower triangle of plots) for house-effect parameters (δ_j) and the standard deviation of the daily innovations ω, Example 9.3.

less than a percentage point in magnitude and the Morgan phone polls are smaller again. This disparity in the house effects for the two Morgan polling modes is quite striking.

The standard deviation of the day-to-day innovations, ω, is estimated to be less than half a percentage point. Since the model is fit on a daily time scale, this result implies a reasonably high level of volatility in Australian vote intentions. For instance, conditional on the Bayes estimate of ω and the normal model we have assumed here, 5 % of the day-to-day shifts in vote intentions have magnitudes greater than .86 of a percentage point. Marginalizing with respect to the posterior uncertainty in ω we can induce a posterior density over r-day-ahead forecasts; i.e. the local-level model holds that $E(\xi_{t+r}|\xi_t) = \xi_t \forall r > 0$, but with the conditional or forecast variance $V(\xi_{t+r}|\xi_t)$ an increasing function of r and the standard deviation of the (Gaussian) innovations ω. Over a $r = 7$ day time horizon, the 95 % HPD interval for the quantity $\nabla_r = \xi_{t+r} - \xi_t$ spans ± 2.3 percentage points, integrating over the posterior uncertainty in ω. This is extremely large, suggesting that in about 1 week out of 20, or about 2 or 3 weeks a year, we can expect to see swings in voter support greater than 2.3 percentage points in

Table 9.7 Summaries of marginal posterior densities, house effects (δ) and the innovation standard deviation (ω), Example 9.3. The summaries have been multiplied by 100 so as to be interpretable on the percentage point scale. Positive numbers indicate an over-estimate of Labor support. The posterior tail probabilities reported in the column on the right are computed by noting the proportion of times we observe $\delta_j > 0$ in the MCMC output.

	Mean	95 % HPD	$\Pr(\delta_j > 0\|\mathbf{y})$
δ_1, Galaxy	−1.2	[−3.1, 0.6]	.10
δ_2, Morgan (face to face)	2.7	[1.0, 4.3]	>.999
δ_3, Newspoll	1.2	[−0.5, 2.8]	.90
δ_4, Nielsen	0.9	[−0.8, 2.5]	.80
δ_5, Morgan (phone)	0.8	[−1.0, 2.3]	.80
ω	0.43	[0.33, 0.54]	

magnitude. Looking at the raw data (see Figure 9.15), we suspect that a lot of the apparent volatility here is being driven by some extremely abrupt changes in vote intentions, that are not well fit by the constant innovation variance model. A richer model would specify different volatility regimes for these data, a specification that is used frequently when fitting similar models to data on stock returns, exchange rates and the like (e.g. Chib, Nardari and Shephard 2002; Rachev *et al* 2008).

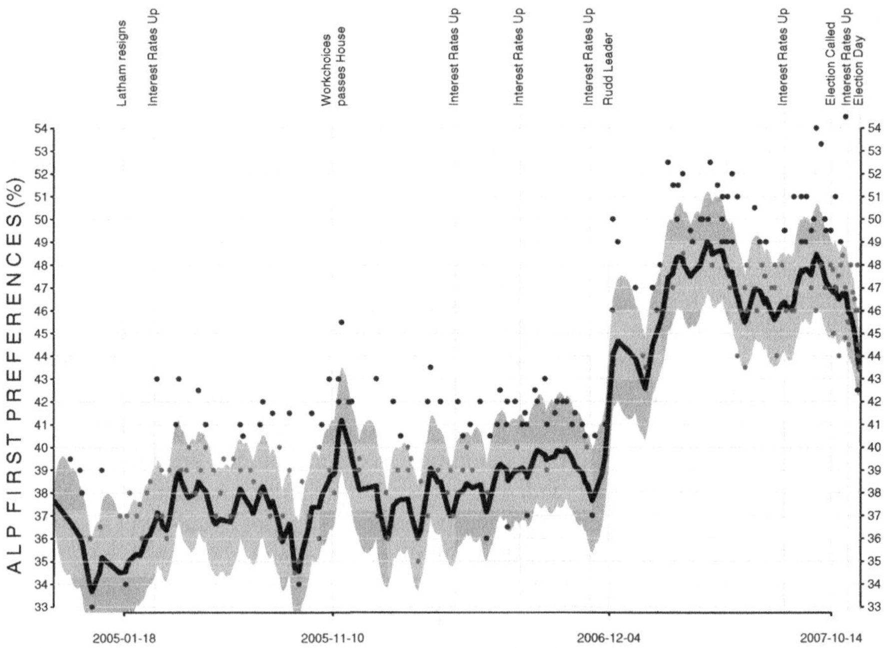

Figure 9.15 Time series, Bayes estimates of latent variable ξ_t, Example 9.3. The shaded region shows a pointwise 95 % credible interval. Solid points indicate poll results (y).

Figure 9.15 shows the Bayes estimates of the $T = 1142$ daily ξ_t, presented as a time series, with pointwise 95 % HPDs shown with the shaded area. The 239 individual polls are shown as small circles, and are considerably dispersed relative to one another and the ξ_t recovered by the model. Politically relevant events are indicated with the vertical lines and corresponding labels. The raw data and the recovered ξ_t show a jump in Labor support with the switch in leadership from Kim Beazley to Kevin Rudd in December of 2006. Labor's share of first preference vote intentions increased from a band between 35 % to 42 % to roughly 46 % to 52 % of first preference vote intentions. With the Greens drawing approximately anywhere from 3 % to 9 % of first preferences, and approximately 80 % of Green preferences flowing to Labor, published estimates of Labor's share of the two-party preferred vote were consistently above 53 % under Rudd's tenure as leader. There is some moderate evidence of a slow trend towards Labor prior to the leadership change, but this is dwarfed by the jump in Labor support associated with the change in leadership.

On the three year time scale displayed in Figure 9.15, the 2007 election is pushed up against the right hand edge of the graph. We focus on period between September 1 2007 and the November 24 election in Figure 9.16, but retain the vertical dimension and scale used in Figure 9.15. We also label the poll estimates so as to identify the corresponding survey house. The results indicate a trend back towards the Liberal/National Party government over the course of the campaign, with the election result, $\xi_T = .434$, some 3.6 percentage points below the 47 % level of ξ_t we estimate when the election was announced on October 14. The slow fall in Labor support appears to slightly accelerate

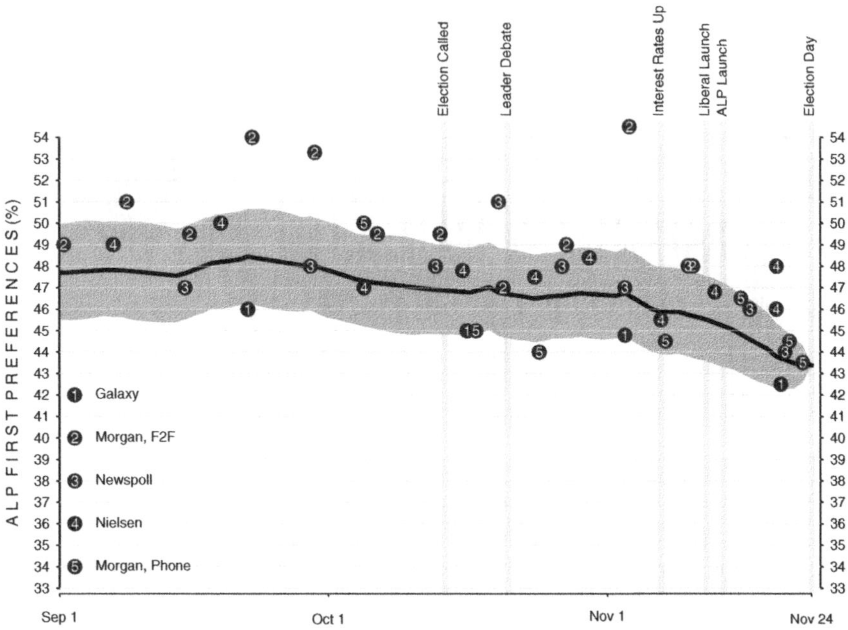

Figure 9.16 Time series, Bayes estimates of latent variable ξ_t, focusing on the 2007 election campaign, Example 9.3. The shaded region shows a pointwise 95 % credible interval. Solid points indicate poll results (y).

in the last two weeks of the campaign. The effect of the endpoint constraint is clearly visible in Figure 9.16, with the variances of the marginal posterior densities of the ξ_t tending towards zero as $t \to T$; this posterior variance is in fact zero for ξ_T.

The dispersion of the polls across survey houses is also apparent in Figure 9.15, with house-specific biases apparent in Figure 9.16. Some of the over-estimates produced by the Morgan face-to-face polls are quite striking, and we can also discern the tendency of the Galaxy polls to underestimate Labor support.

Finally, we can express the precision of the marginal posterior density of each ξ_t in terms of the sample size of a daily poll that would yield the equivalent amount of precision. That is, if $\xi_t^* \equiv E(\xi_t|\mathbf{y})$ is the Bayes estimate of ξ_t then we solve for n_t such that $v_t = V(\xi_t) = \xi_t^*(1 - \xi_t^*)/n_t$, where $V(\xi_t)$ is the variance of the marginal posterior density of ξ_t, which we recover from the Gibbs sampler. Note that the endpoint constraints on ξ_1 and ξ_T imply that $v_1 = v_T = 0$; in turn, for $t \to 1$ or $t \to T$, $v_t \to 0$ and so $n_t \to \infty$.

We plot the resulting series of implied sample sizes n_t in Figure 9.17, using a double-log scale for the n_t. The model produces day-by-day estimates of ξ_t with as much precision as if we had fielded a poll with 1500 to 2000 respondents every day. In no small part this is because of the somewhat strong assumption in the locally constant transition model; the assumption that $E(\xi_t) = \xi_{t-1}$ means that we assume/know a reasonable amount about ξ_t before we see any of the polls fielded at time t or thereafter.

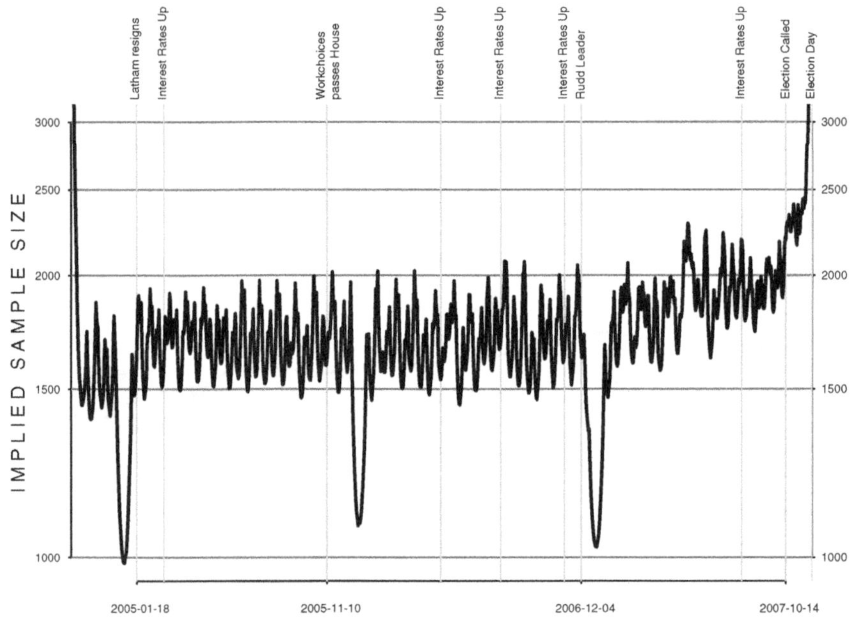

Figure 9.17 Time series, size of daily polls with precision equal to that of marginal posterior densities of each ξ_t, Example 9.3. For each t, we compute $n_t = \xi_t^*(1 - \xi_t^*)/v_t$, where ξ_t^* is the mean of the posterior density of ξ_i and v_t is the variance of posterior density. Note the asymptotes in the n_t series, since the constraints on ξ_1 and ξ_T imply that v_1 and v_T are zero and that n_1 and n_T are infinite. A double-log scale is used on the vertical axis.

Note that the precision of the recovered ξ_t increases as the tempo of polling increases during the 2007 election campaign, to generate the equivalent level of precision as if we had interviewed over 2000 respondents per day. There are obvious cycles in the n_t series, corresponding to the timing of the various polls, and hence the availability of more or less information on which to base inferences for the ξ_t. In particular, the three December-January periods spanned by these data see very little polling (the Christmas and New Year holidays coincide with summer in the southern hemisphere) and the precision of the ξ_t falls away in those periods.

Problems

9.1 Consider the five indicators of social attributes of the American states we examined in Example 9.1; the data are available in the `state.x77` data frame in `R`.

1. Estimate the factor loadings and uniquenesses for a one factor model using a maximum likelihood factor analysis program (e.g., the `factanal` program in `R`).

2. Note that almost all implementations of factor analysis are actually models for the covariance matrix of the indicators, $\boldsymbol{\Sigma}$, fitting the model $\boldsymbol{\Sigma} = \boldsymbol{\Lambda}\boldsymbol{\Phi}\boldsymbol{\Lambda}' + \boldsymbol{\Psi}$, where $\boldsymbol{\Lambda}$ is a p-by-k matrix of unknown factor loadings, $\boldsymbol{\Phi}$ is an identity matrix when fitting orthonormal factors and $\boldsymbol{\Psi}$ is typically a p-by-p diagonal matrix of measurement error variances. Critically, the n-by-k matrix of latent variables, $\boldsymbol{\xi}$, does not appear as a parameter in this implementation of the factor analysis model. Nonetheless, we can recover a predictive density for the latent variables $\boldsymbol{\xi}_i, i = 1, \ldots, n$, via the following argument. Suppose we have mean-deviated the indicators $\mathbf{y}_i$ and imposed the identifying restrictions that the latent variables $\boldsymbol{\xi}_i$ have mean zero and are orthonormal, i.e. $V(\boldsymbol{\xi}_i) = \boldsymbol{\Phi} = \mathbf{I}_k$; further, we assume that both $\mathbf{y}_i$ and $\boldsymbol{\xi}_i$ follow normal distributions. Then $\mathbf{y}_i \sim N(\boldsymbol{\Lambda}\boldsymbol{\xi}_i, \boldsymbol{\Psi})$ and $\boldsymbol{\xi}_i \sim N(\mathbf{0}, \mathbf{I}_k)$. Deduce the conditional density of $\boldsymbol{\xi}_i|\mathbf{y}_i, \boldsymbol{\Lambda}, \boldsymbol{\Psi}$.
   Hint: see Proposition B.2 in the discussion accompanying Definition B.31, or any serious treatment of the factor analytic model; e.g. Lawley and Maxwell (1971), Mardia, Kent and Bibby (1979 §9.7), Anderson (2003, §14.7), Wansbeek and Meijer (2000, 163–4) or Lee (2007, 83).

3. Using the result you derived in the previous question, compute $E(\xi_i|\mathbf{y}_i, \hat{\boldsymbol{\Lambda}}, \hat{\boldsymbol{\Psi}}), i = 1, \ldots, n$, for the one-factor model fit to the state-level social attributes data, where $\hat{\boldsymbol{\Lambda}}$ and $\hat{\boldsymbol{\Psi}}$ are the maximum likelihood estimates of $\boldsymbol{\Lambda}$ and $\boldsymbol{\Psi}$, respectively. Note that for one-factor model fit to the social attributes data, the matrix $\boldsymbol{\Lambda}$ is a vector, $\boldsymbol{\lambda}$; if you are using the `factanal` command in `R`, $\boldsymbol{\lambda}$ is stored in the `loadings` component of the object returned by `factanal`. Further, `R` users can form the diagonal matrix of error variances, $\boldsymbol{\Psi}$, by passing the `uniquenesses` component of the `factanal` object to the matrix constructor function `diag`.

4. Verify that (at least up to a scale factor) these estimates of ξ_i are the same as the 'factor scores' generated by 'regression scoring' from your factor analysis program. For the latter, `R` users should consult the documentation for the `factanal` function.

5. Replicate the Bayesian analysis presented in Example 9.1. Compare the means of the posterior densities of ξ_i from the Bayesian analysis with the factor scores you computed in the previous two questions. What differences do you see, if any?

6. In part 2, you derived the conditional density for ξ_i given the indicators $\mathbf{y}_i$ and maximum likelihood estimates of the factor analysis model. Use this result to compute the standard deviation of each ξ_i given $\mathbf{y}_i$ and the MLEs of $\boldsymbol{\lambda}$ and $\boldsymbol{\Psi}$.

7. Compare the standard deviations you computed in the previous question with the standard deviations of the marginal posterior densities of the ξ_i from the Bayesian analysis. How do these differ and why?

9.2 Again, consider the five indicators of social attributes of the American states we examined in Example 9.1 and in Problem 9.1.

1. Randomly select $J = 4$ observations. Letting j index these J observations, set j of the indicators to missing, $j = 1, \ldots, 4$ (i.e. one observation will have one indicator missing, another will have two, etc).

2. Re-run the Bayesian analysis on the data subject to the missingness-at-random generated in the previous question.

3. Compare the posterior densities of the latent traits ξ_i from this analysis with the posterior densities for ξ_i obtained with the complete data set. How do these posterior densities differ?

4. Compare the posterior densities of the γ parameters from the analysis with the MAR data with the results obtained with the complete data set.

9.3 In §9.2.5 we considered a hierarchical version of the factor analytic model, i.e.

$$\begin{aligned}
y_{ij} &\sim N(\mu_{ij}, \omega_j^2), i = 1, \ldots, n; j = 1, \ldots, J \\
\mu_{ij} &= \gamma_{j1} + \gamma_{j2}\xi_i \\
\xi_i &\sim N(\alpha_{S(i)}, 1) \\
\alpha_r &\sim N(0, 1), r \in \{0, 1\} \\
\boldsymbol{\gamma}_j &\sim N(\mathbf{g}_0, \mathbf{G}_0),
\end{aligned}$$

where $\mathbf{g}_0 = \mathbf{0}$ and $\mathbf{G}_0 = \kappa \cdot \mathbf{I}_2$, where κ is an arbitrarily large constant. We imposed the identifying restriction $\tilde{\xi}_i = (\xi_i - c)/m$, where $c = \bar{\xi}$ and $m = \text{sd}(\xi)$; i.e. we normalized the ξ_i to have mean zero and standard deviation one, denoting the identified version of the latent traits as $\tilde{\xi}_i$. Similarly, we transform α_r to the identified parameters $\tilde{\alpha}_r = (\alpha_r - c)/m$, $r \in \{0, 1\}$. We also make the transformations from γ to $\tilde{\gamma}$:

$$\begin{aligned}
\tilde{\gamma}_{j1} &= \gamma_{j1} + \gamma_{j2}\, c/m, \\
\tilde{\gamma}_{j2} &= \gamma_{j2}\, m.
\end{aligned}$$

But note that the priors used here are with respect to the unidentified parameters ξ_i, α_r and $\boldsymbol{\gamma}_j = (\gamma_{j1}, \gamma_{j2})'$. What are the priors these transformations imply for the identified parameters $\tilde{\xi}_i$, $\tilde{\alpha}_r$ and $\tilde{\boldsymbol{\gamma}}_j$?

9.4 The data set used in Example 9.2 is small (at least by the standards of roll call analysis and IRT modeling) and general purpose Bayesian analysis programs such as `WinBUGS` and `JAGS` can generate a reasonable exploration of the posterior density for the 'non-hierarchical' version of the model in a reasonable amount of time, without resorting to the data-augmented, Gibbs sampling approach described in Section 9.3. The necessary `WinBUGS/JAGS` program is rather simple:

`JAGS code`

```
model{
      for(i in 1:N){
            for(j in 1:M){
                  probit(pi[i,j]) <- beta[j,1]*xi[i] - beta[j,2]
                  y[i,j] ~ dbern(pi[i,j])
            }
      }

     ## non-exchangeable/non-hierarchical prior
     for(i in 1:N){
           xi[i] ~ dnorm(0,1)
     }

     ## bill parameters
     for(j in 1:M){
           beta[j,1:2] ~ dmnorm(b0[1:2],B0[1:2,1:2])
     }
}
```

Note that the constants `N` and `M` need to be passed to the program, along with the user-supplied hyper-parameters `b0` and `B0` (the prior mean vector and precision matrix of the bill parameters, respectively). The data are available as the object `nj06` in the `R` package `pscl`.

Fit the model. Impose local identification with the normalization $\tilde{\xi}_i^{(t)} = (\xi_i^{(t)} - c^{(t)})/m^{(t)}$ where t indexes MCMC iterations and $c^{(t)} = \bar{\xi}^{(t)}$ and $m^{(t)} = \text{sd}(\xi^{(t)})$; an offsetting set of normalizations are also required for the `betastar` parameters, see the discussion in Example 9.2. You can impose these identifying restrictions either *ex post* in `R` (an example of 'post-processing' the MCMC output) or have `WinBUGS/JAGS` make the identifying normalizations 'on-the-fly' by adding the necessary lines to the code given above. Initialize the sampler by setting $\xi_i^{(0)}$ for Democrats to be -1 and to +1 for Republicans.

As far as you can tell, what sampling methods are being used by `WinBUGS` and/or `JAGS` to sample from the conditional densities of the ξ_i parameters? Comment on the efficiency of the MCMC algorithm. How many MCMC iterations appear to be necessary so as to generate valid characterizations of the marginal posterior densities of the identified ideal points, $\tilde{\xi}_i$?

9.5 Try using a general purpose Bayesian analysis program such as `WinBUGS` or `JAGS` to fit the locally-constant DLM we examined in Example 9.3. The data are available as the data frame `AustralianElectionPolling` in the `R` package `pscl`. It is extremely simple to represent the locally-constant DLM in `WinBUGS/JAGS`: e.g.

JAGS code

```
model{
        ## observational model
        for(i in 1:NPOLLS){
             mu[i] <- xi[day[i]] + delta[org[i]]
                    y[i] ~ dnorm(mu[i],prec[i])
        }
        ## transition model
        for(t in 2:NPERIODS){
             xi[t] ~ dnorm(xi[t-1],tau)
       }

       ## priors for standard deviations
       omega ~ dunif(0,kappa)
       tau <- 1/pow(omega,2)

       ## priors for house effects
       delta[1:NHOUSES] ~ dmnorm(d0[1:NHOUSES],D0[1:NHOUSES,1:NHOUSES])
}
```

When working with this program we pass `xi` to the program as a vector of 'data', set to NA save for `xi[1]` and `xi[1142]` where we impose the endpoint constraints (see the discussion in Example 9.3). The constants `NPOLLS`, `NPERIODS` and `NHOUSES` are also passed to the program, along with the user-supplied hyperparameters `d0`, `D0` and `kappa`. Comment on the performance of the MCMC algorithm used by `WinBUGS` and/or `JAGS` when fitting this model. Which parameters 'mix' better than others? How many iterations will be needed to generate valid characterizations of the marginal posterior densities of the slowest mixing parameters?

9.6 Consider the local-constant DLM we fit in Example 9.3, generating a daily series of ALP voting intentions $\{\xi_t\}$, $\xi_t \in [0, 1]$ for the period between the 2004 and 2007 Australian Federal elections, with the endpoint constraints given by the respective election outcomes $\xi_1 = .376$ and $\xi_T = .434$. Fit the model without *any* data other than the endpoint constraints. What does the recovered series of Bayes estimates of ξ_t look like? What happens to the credible intervals for the ξ_t over the length of the series?

9.7 In Example 9.3 we exploited the fact that the election results in 2004 and 2007 supply 'endpoints' for the series $\{\xi_t\}$, allowing us to set ξ_1 and ξ_T to known constants. How could we fit the model without these constraints? What constraints might need to be applied so as to identify the model parameters?

9.8 Suppose we have a DLM with a locally-constant transition equation $\xi_t \sim N(\xi_{t-1}, \omega^2)$ with $\omega^2 \sim \text{inverse-Gamma}(30/2, 3/2)$. If ξ_i and ω are on a scale of percentage points, comment on what this density for ω^2 is saying about the volatility of the series $\{\xi_t\}$. To investigate this further, consider the following simulation exercise. Set $\xi_0 = 50$. Use Monte Carlo methods to characterize the predictive density of ξ_r, $r = 1, 2, \ldots, 7$. Graph the limits of the 95 % HDR for ξ_r as a function of r. Comment on what you see. Repeat the exercise but instead fix ω^2 at its mean value.

Part IV

Appendices

Appendix A

Working with vectors and matrices

Here I provide a practically oriented review of some definitions and properties of vectors and matrices that appear throughout the book. In statistical analysis at the level presented in this book, matrices and vectors are no more than useful mathematical representations of data and/or parameters (e.g. vectors as multiple observations or parameters; matrices as multiple observations on multiple variables). A mathematically rigorous treatment of the material below is beyond our scope here.

Definition A.1 (Matrix). *A two-dimensional array of numbers is known as a matrix, conventionally denoted with a bold, uppercase letter (e.g.* $\mathbf{M}$*). A* m*-by-*n *matrix has* m *rows and* n *columns. If* $i \in \{1, \ldots, m\}$ *and* $j \in \{1, \ldots n\}$ *then* M_{ij} *is the element of* $\mathbf{M}$ *in the* i*-th row and* j*-th column.*

Definition A.2 (Vector). *A vector is a special case of a matrix, conventionally denoted with a bold lowercase letter (e.g.* $\mathbf{v}$*) for which either the number of rows or the number of columns is one. A row vector of length* k *is a 1-by-*k *matrix, while a column vector of length* k *is a* k*-by-1 matrix.*

Remark. A unit vector is a vector whose elements are all 1, often written as $\boldsymbol{\iota}$. A zero (or null) vector/matrix is a vector/matrix whose elements are all zero, written as $\mathbf{0}$.

Definition A.3 (Matrix addition). *Matrices conform for the operation of addition provided they have the same number of rows and columns. For instance, if* $\mathbf{A}$ *and* $\mathbf{B}$ *are* m*-by-*n *matrices, then the matrix* $\mathbf{C} = \mathbf{A} + \mathbf{B}$ *is a* m*-by-*n *matrix with* $C_{ij} = A_{ij} + B_{ij}$.

Definition A.4 (Scalar multiplication). *A scalar* c *times a* m*-by-*n *matrix* $\mathbf{X}$ *yields a* m*-by-*n *matrix* $\mathbf{Y} = c\mathbf{X}$ *where* $Y_{ij} = cX_{ij}$.

Bayesian Analysis for the Social Sciences S. Jackman

Definition A.5 (Matrix multiplication). *Let* $\mathbf{A}$ *be a* n*-by-*m *matrix and let* $\mathbf{B}$ *be a* m*-by-*p *matrix. Then* $\mathbf{C} = \mathbf{AB}$ *is a* n*-by-*p *matrix with* $C_{ik} = \sum_{j=1}^{m} A_{ij} B_{jk}$.

Remark.

- Note that the matrix product $\mathbf{C} = \mathbf{AB}$ is only defined if $\mathbf{A}$ and $\mathbf{B}$ conform for the operation of multiplication, i.e. if $\mathbf{A}$ has as many columns as $\mathbf{B}$ has rows. Accordingly, in general, matrix multiplication is not commutative: i.e. in general, $\mathbf{AB}$ does not equal $\mathbf{BA}$ and indeed, one or both of these matrix products may not be defined.
- The conventional mathematical definition of a matrix is as a set of coefficients that define a linear transformation from one vector space to another. With matrix multiplication now defined, this understanding of a matrix can be elaborated. Suppose $\mathbf{y} \in \mathbb{R}^n$ and $\mathbf{x} \in \mathbb{R}^m$ and let $f : \mathbb{R}^m \mapsto \mathbb{R}^n$ be a linear transformation. Then for each unique f there is a unique n-by-m matrix $\mathbf{A}$ such that $\mathbf{y} = f(\mathbf{x}) = \mathbf{Ax}$.

Definition A.6 (Square matrix). *If* $\mathbf{A}$ *is a* n*-by-*n *matrix,* $\mathbf{A}$ *is a square matrix.*

Definition A.7 (Diagonal matrix). *A square matrix* $\mathbf{A}$ *is a diagonal matrix if* $A_{ij} = 0$, $\forall i \neq j$. *If* $\mathbf{x}$ *is a vector of length* n*, then* $\mathbf{A} = diag(\mathbf{x})$ *is a diagonal matrix with* $A_{ii} = x_i$, $i = 1, \ldots, n$.

Definition A.8 (Symmetric matrix). *A square matrix* $\mathbf{A}$ *is symmetric if* $A_{ij} = A_{ji}$, $\forall i \neq j$ *or equivalently, if* $\mathbf{A} = \mathbf{A}'$ *(see Definition A.11).*

Definition A.9 (Identity matrix). *A square,* n*-by-*n *diagonal matrix with ones on the diagonal is an identity matrix, denoted* $\mathbf{I}_n$*, so called since* $\mathbf{I}_n\mathbf{A} = \mathbf{A}$*, for any* n*-by-*p *matrix* $\mathbf{A}$.

Definition A.10 (Kronecker product). *If* $\mathbf{A}$ *is a* n*-by-*m *matrix and* $\mathbf{B}$ *is a* p*-by-*q *matrix then* $\mathbf{C} = \mathbf{A} \otimes \mathbf{B}$ *is the Kronecker product of* $\mathbf{A}$ *and* $\mathbf{B}$*, a* np*-by-*mq *matrix with entries formed by multiplying* $\mathbf{B}$ *by each scalar element of* $\mathbf{A}$*, i.e.*

$$\mathbf{C} = \mathbf{A} \otimes \mathbf{B} = \begin{bmatrix} A_{11}\mathbf{B} & A_{12}\mathbf{B} & \ldots & A_{1m}\mathbf{B} \\ A_{21}\mathbf{B} & A_{22}\mathbf{B} & \ldots & A_{2m}\mathbf{B} \\ \vdots & \vdots & \vdots & \vdots \\ A_{n1}\mathbf{B} & A_{n2}\mathbf{B} & \ldots & A_{nm}\mathbf{B} \end{bmatrix}.$$

Remark. If $\mathbf{A}$ is an identity matrix $\mathbf{I}_n$, then $\mathbf{C} = \mathbf{A} \otimes \mathbf{B}$ is a *block-diagonal* matrix formed by joining n copies of $\mathbf{B}$: i.e.

$$\mathbf{C} = \begin{bmatrix} \mathbf{B} & \mathbf{0} & \ldots & \mathbf{0} \\ \mathbf{0} & \mathbf{B} & \mathbf{0} & \mathbf{0} \\ \vdots & \vdots & \ddots & \vdots \\ \mathbf{0} & \mathbf{0} & \ldots & \mathbf{B} \end{bmatrix}.$$

Definition A.11 (Transpose of a matrix). *If* $\mathbf{A}$ *is a* n*-by-*m *matrix then the transpose of* $\mathbf{A}$ *is a* m*-by-*n *matrix, conventionally written* $\mathbf{A}'$*, where* $A'_{ij} = A_{ji}$, $\forall i \neq j$.

Remark. If $\mathbf{A}$ and $\mathbf{B}$ conform for multiplication, then $(\mathbf{AB})' = \mathbf{B}'\mathbf{A}'$.

Definition A.12 (Trace of a matrix). *The trace of a square n-by-n matrix $\mathbf{A}$ is the sum of its diagonal elements, conventionally denoted $tr(\mathbf{A})$; i.e. $tr(\mathbf{A}) = \sum_{i=1}^{n} A_{ii}$.*

Proposition A.1 (Properties of traces). *Let $\mathbf{A}$ and $\mathbf{B}$ be square matrices, and c be a scalar.*

1. $tr(\mathbf{A}) = tr(\mathbf{A}')$
2. $tr(\mathbf{A} + \mathbf{B}) = tr(\mathbf{A}) + tr(\mathbf{B})$.
3. $tr(c\mathbf{A}) = c \cdot tr(\mathbf{A})$.
4. $tr(\mathbf{AB}) = tr(\mathbf{BA})$, *provided* $\mathbf{AB}$ *is a square matrix.*

Proof. 1. Searle (1982, 27) 2. Searle (1982, 29) 3. Magnus and Neudecker (1999, 10). 4. Searle (1982, 45). $\triangleleft$

Definition A.13 (Linear independence). *The k vectors $\mathbf{v}_1, \ldots, \mathbf{v}_n$ are linearly independent if there does not exist a set of constants $c_1, \ldots, c_n$ such that*

$$c_1\mathbf{v}_1 + \ldots + c_n\mathbf{v}_n = \mathbf{0}.$$

Definition A.14 (Rank of a matrix). *The rank of a matrix is the number of linearly independent rows or columns of the matrix.*

Remark.

- The rank of a matrix is also equal to the dimension of the vector space spanned by the linear transformation defined by the matrix.
- The number of linearly independent rows in a matrix is also the number of linearly independent columns (Searle 1982, 169).
- A square matrix with rank less than its number of rows or columns has "less than full rank".

Definition A.15 (Determinant of a matrix). *The determinant of a square n-by-n matrix $\mathbf{A}$, conventionally designated $|\mathbf{A}|$ or $det(\mathbf{A})$, is*

$$|\mathbf{A}| = \sum_{p \in \mathcal{P}} s_p A_{1,p_1} A_{2,p_2} \ldots A_{n,p_n}$$

where

- *p indexes $\mathcal{P}$, the set of $n!$ permutations of the integers $1, \ldots, n$*
- *$s_p = (-1)^{v(p)}$, where $v(p)$ is the number of comparisons between elements i and j of permutation p where $i > j$ and $p_i < p_j$ (i.e. the number of times p differs from a strictly increasing sequence of the integers $1, \ldots, n$).*

Proposition A.2 (Properties of determinants).

1. *If* $\mathbf{A}$ *is a n-by-n matrix, and c is a scalar, then* $|c\mathbf{A}| = c^n|\mathbf{A}|$.
2. $|\mathbf{A}\,\mathbf{B}| = |\mathbf{A}| \cdot |\mathbf{B}|$.
3. $|\mathbf{A}'| = |\mathbf{A}|$

Proof. 1. Magnus and Neudecker (1999, 9). 2. Searle (1982, 96). 3. Searle (1982, 92). ◁

Definition A.16 (Singular matrix). *A square matrix* $\mathbf{A}$ *is singular if and only if its determinant is zero.*

Definition A.17 (Definiteness). *Consider a symmetric matrix* $\mathbf{A}$ *and the quadratic form* $\mathbf{x}'\mathbf{A}\mathbf{x}$.

- *If* $\mathbf{x}'\mathbf{A}\mathbf{x} > 0, \forall\ \mathbf{x} \neq \mathbf{0}$, *then* $\mathbf{A}$ *is positive definite.*
- *If* $\mathbf{x}'\mathbf{A}\mathbf{x} \geq 0, \forall\ \mathbf{x} \neq \mathbf{0}$ *with the equality holding for at least one* $\mathbf{x}$ *then* $\mathbf{A}$ *is positive semi-definite.*
- *If* $\mathbf{x}'\mathbf{A}\mathbf{x} < 0, \forall\ \mathbf{x} \neq \mathbf{0}$, *then* $\mathbf{A}$ *is negative definite.*
- *If* $\mathbf{x}'\mathbf{A}\mathbf{x} \leq 0, \forall\ \mathbf{x} \neq \mathbf{0}$ *with the equality holding for at least one* $\mathbf{x} \neq \mathbf{0}$ *then* $\mathbf{A}$ *is negative semi-definite.*

Remark.

- $\mathbf{A}$ is positive definite $\iff \exists\ \mathbf{M}$ such that $\mathbf{A} = \mathbf{M}\mathbf{M}'$, where $\mathbf{M}$ is non-singular.
- If $\mathbf{A}$ is positive definite and $\mathbf{B}$ is positive semi-definite, then $\mathbf{A} + \mathbf{B}$ is positive definite (provided $\mathbf{A}$ and $\mathbf{B}$ conform for addition).
- A necessary condition for a matrix $\mathbf{A}$ to be positive definite is that $\det(\mathbf{A}) > 0$.

Definition A.18 (Matrix inverse). *The inverse of a square matrix* $\mathbf{A}$ *is a matrix* $\mathbf{A}^{-1}$ *such that* $\mathbf{A}^{-1}\mathbf{A} = \mathbf{I}$.

Remark.

- The inverse of $\mathbf{A}$ exists if and only if $\det(\mathbf{A}) \neq 0$ (Searle 1982, 129).
- If $\mathbf{A}$ is not of full rank, then (by definition) there are linear dependencies among its rows/columns, in which case the determinant of $\mathbf{A}$ is zero and $\mathbf{A}^{-1}$ does not exist (i.e. $\mathbf{A}$ is not invertible). Hence,
  1. Singular matrices can not be inverted.
  2. Matrices of less than full rank can not be inverted.
  3. If $\mathbf{A}$ is positive definite then $\det(\mathbf{A}) > 0$ and hence $\mathbf{A}^{-1}$ exists.

- If two vector spaces $\mathcal{V}$ and $\mathcal{W}$ have the same dimension and $f : \mathcal{V} \mapsto \mathcal{W}$ is a linear map such that $f(\mathbf{x}) = \mathbf{A}\mathbf{x}$, then the inverse of $\mathbf{A}$, $\mathbf{A}^{-1}$, defines the linear map $f^{-1} : \mathcal{W} \mapsto \mathcal{V}$, if the inverse exists.

Proposition A.3 (Matrix inverses). *Assuming all inverses exist, then*

1. *(Inverse of a product).* $(\mathbf{AB})^{-1} = \mathbf{B}^{-1}\mathbf{A}^{-1}$.
2. *(Inverse of a sum). If* $\mathbf{A}$ *and* $\mathbf{B}$ *conform for addition, then*

$$(\mathbf{A} + \mathbf{B})^{-1} = \mathbf{A}^{-1}(\mathbf{A}^{-1} + \mathbf{B}^{-1})^{-1}\mathbf{B}^{-1}.$$

3. $(\mathbf{A} + \mathbf{BCD})^{-1} = \mathbf{A}^{-1} - \mathbf{A}^{-1}\mathbf{B}(\mathbf{C}^{-1} + \mathbf{D}\mathbf{A}^{-1}\mathbf{B})^{-1}\mathbf{D}\mathbf{A}^{-1}$, *where* $\mathbf{A}$ *is p-by-p,* $\mathbf{B}$ *is p-by-n,* $\mathbf{C}$ *is n-by-n and* $\mathbf{D}$ *is n-by-p.*
4. *(Determinant of an inverse).* $|\mathbf{A}^{-1}| = |\mathbf{A}|^{-1}$.

Proof. 1. Searle (1982, 130). 2 and 3. Horn and Johnson (1985, §0.7.4). 4. Searle (1982, 130). ◁

Proposition A.4 (Inverse of a partitioned matrix). *Consider the partitioned matrix*

$$\mathbf{M} = \begin{bmatrix} \mathbf{A} & \mathbf{B} \\ \mathbf{B}' & \mathbf{D} \end{bmatrix},$$

with $\mathbf{A}$ *and* $\mathbf{D}$ *square, non-singular matrices. Then*

$$\mathbf{M}^{-1} = \begin{bmatrix} \mathbf{A}^{-1} + \mathbf{F}\mathbf{E}^{-1}\mathbf{F}' & -\mathbf{F}\mathbf{E}^{-1} \\ -\mathbf{E}^{-1}\mathbf{F}' & \mathbf{E}^{-1} \end{bmatrix}$$

where $\mathbf{E} = \mathbf{D} - \mathbf{B}'\mathbf{A}^{-1}\mathbf{B}$ *and* $\mathbf{F} = \mathbf{A}^{-1}\mathbf{B}$.

Proof. Magnus and Neudecker (1999, 11); Horn and Johnson (1985, §0.7.3) ◁

Definition A.19 (Eigenvalues and eigenvectors). *A square n-by-n matrix* $\mathbf{A}$ *admits the relationship*

$$\mathbf{A}\mathbf{x} = \lambda\mathbf{x}$$

or equivalently, $(\mathbf{A} - \lambda)\mathbf{x} = \mathbf{0}$, *where* $\mathbf{x} \neq \mathbf{0}$ *is an eigenvector of* $\mathbf{A}$ *and* λ *is an eigenvalue of* $\mathbf{A}$. *Alternatively, the eigenvectors and eigenvalues of a square, n-by-n matrix* $\mathbf{A}$ *are the quantities* $\mathbf{X} = [\mathbf{x}_1, \ldots, \mathbf{x}_n]$ *and* $\boldsymbol{\lambda} = (\lambda_1, \ldots, \lambda_n)'$ *in the following equation:*

$$(\mathbf{A} - diag(\boldsymbol{\lambda}))\mathbf{X} = \mathbf{0}.$$

Remark.

- A n-by-n matrix $\mathbf{A}$ will have n eigenvalues, not necessarily unique, and not all of them may be real. For each eigenvalue, there is a corresponding eigenvector.

- The set of eigenvalues of $\mathbf{A}$, $\lambda(\mathbf{A})$, is sometimes referred to as the *spectrum* of $\mathbf{A}$; the *spectral radius* of $\mathbf{A}$ is the nonnegative real number $\rho(\mathbf{A}) = \max\{|\lambda| : \lambda \in \lambda(\mathbf{A})\}$.
- A matrix $\mathbf{A}$ is singular if and only if $0 \in \lambda(\mathbf{A})$ (e.g, Horn and Johnson 1985, 37).

Definition A.20 (Similar matrices). *A n-by-n matrix* $\mathbf{A}$ *is said to be similar to* $\mathbf{B}$ *if* $\mathbf{B} = \mathbf{S}^{-1}\mathbf{A}\mathbf{S}$.

Definition A.21 (Diagonalizable matrix). *If* $\mathbf{B}$ *is a diagonal matrix, similar to* $\mathbf{A}$*, then* $\mathbf{A}$ *is said to be diagonalizable.*

Proposition A.5 (Diagonalization and eigenstructure of a matrix). *A n-by-n matrix* $\mathbf{A}$ *is diagonalizable if and only if there exist n linearly independent vectors each of which is an eigenvector of* $\mathbf{A}$.

Proof. Horn and Johnson (1985, Theorem 1.3.7) ◁

Remark.

- The proposition actually provides a recipe for diagonalizing a matrix $\mathbf{A}$ via an eigen-decomposition of $\mathbf{A}$; i.e. using Definition A.21, we have $\mathbf{\Lambda} = \mathbf{X}\mathbf{A}\mathbf{X}^{-1}$, where $\mathbf{\Lambda}$ is a diagonal matrix containing the eigenvalues of $\mathbf{A}$ and $\mathbf{X}$ is the matrix formed by column-binding the eigenvectors of $\mathbf{A}$.
- If $\mathbf{A}$ is diagonalizable then the diagonal matrix to which it is similar must contain the eigenvalues of $\mathbf{A}$ (Horn and Johnson 1985, 47).

Proposition A.6 (Powers of a diagonalizable matrix). *Suppose* $\mathbf{A}$ *is diagonalizable, with* $\mathbf{\Lambda} = \mathbf{X}\mathbf{A}\mathbf{X}^{-1}$. *Then* $\mathbf{A}^t = \mathbf{X}\mathbf{\Lambda}^t\mathbf{X}^{-1}$.

Proof. $\mathbf{A} = \mathbf{X}\mathbf{\Lambda}\mathbf{X}^{-1}$. $\mathbf{A}^2 = \mathbf{X}\mathbf{\Lambda}\mathbf{X}^{-1}\mathbf{X}\mathbf{\Lambda}\mathbf{X}^{-1} = \mathbf{X}\mathbf{\Lambda}^2\mathbf{X}^{-1}$ and the result follows by induction. ◁

Appendix B

Probability review

B.1 Foundations of probability

Definition B.1 (σ-algebra). *Let Ω be a set of events ($\Omega \neq \emptyset$), and let $\mathcal{A}$ be a family of subsets of Ω. $\mathcal{A}$ is a σ-algebra over Ω, denoted $\sigma(\Omega)$, if and only if*

1. *$\emptyset \in \mathcal{A}$ (the null set is in $\mathcal{A}$)*
2. *for any subset $H \subseteq \Omega$, $H \in \mathcal{A} \Rightarrow H^C \in \mathcal{A}$, where $H^C = \Omega \backslash H$ is the complement of H.*
3. *if $H_1, H_2, \ldots$ is a collection of subsets of $\mathcal{A}$ then their countable union $\cup_{i=1}^{\infty} H_i \in \mathcal{A}$*

Remark. Properties 2 and 3 imply that if $H_1, H_2, \ldots$ is a collection of subsets of $\mathcal{A}$ then their countable intersection $\cap_{i=1}^{\infty} H_i \in \mathcal{A}$.

Definition B.2 (Borel σ-algebra). *The smallest σ-algebra over Ω is a Borel σ-algebra over Ω, written $\mathcal{B}(\Omega)$.*

Remark. The elements of a Borel σ-algebra are called Borel sets.

Definition B.3 (Measure). *If $\mathcal{A} = \sigma(\Omega)$, then a function $\mu : \mathcal{A} \mapsto [-\infty, \infty]$ is a measure if*

1. *$\mu(\emptyset) = 0$*
2. *$\mu(\cup_i H_i) = \sum_i \mu(H_i)$ for any sequence of pairwise disjoint sets, $H_1, H_2, \ldots$ in $\mathcal{A}$.*

Remark. A measure μ is strictly positive if $\mu(H) > 0\ \forall H \in \mathcal{A},\ H \neq \emptyset$.

Definition B.4 (Measureable function). *If $\mathcal{A} = \sigma(\Omega)$ and $\mathcal{V} = \sigma(\Psi)$ then a function $f : \Omega \mapsto \Psi$ is said to be measurable $\mathcal{A}/\mathcal{V}$.*

Remark. It is common to refer to a function that is measureable $\mathcal{A}/\mathcal{V}$ as being $\mathcal{A}$-measurable.

Definition B.5 (Measure space). *The tuple $(\Omega, \mathcal{A}, \mu)$ is a measure space.*

Definition B.6 (finite-measure, σ-finite measure).

1. *A measure μ on a σ-algebra $\mathcal{A}$ of Ω is finite if $\mu(\Omega) < \infty$.*
2. *A measure space $(\Omega, \mathcal{A}, \mu)$ is a finite measure space if μ is a finite measure.*
3. *A finite measure μ is σ-finite if Ω is the union of a countable collection of sets of finite measure, i.e. if $\Omega = \cup_{S \in \mathcal{S}}$, $\mu(S) < \infty\ \forall S \in \mathcal{S}$, $\mathcal{S} \subset \mathcal{A}$.*
4. *A measure space $(\Omega, \mathcal{A}, \mu)$ is a σ-finite measure space if μ is σ-finite.*

Definition B.7 (Dominating measure). *Let $\mathcal{M} = \{\mu_1, \mu_2, \ldots\}$ be a family of measures on $\mathcal{A} = \sigma(\Omega)$. The measure ν is a dominating measure with respect to $\mathcal{M}$ if $\nu(S) \geq \mu_j(S), \forall\ S \in \mathcal{A}, \mu_j \in \mathcal{M}$.*

Definition B.8 (Probability measure, probability space). *(1) A positive measure with the property $\mu(\Omega) = 1$ is a probability measure. (2) A measure space equipped with a probability measure is a probability space.*

Definition B.9 (Real-valued random variable). *Given a probability space $(\Omega, \mathcal{A}, \mu)$, a function $X : \Omega \mapsto \mathbb{R}$ such that*

$$\{\omega \in \Omega : X(\omega) \leq r\} \in \mathcal{A}, r \in \mathbb{R}$$

is a real-valued random variable.

Definition B.10 (Discrete random variable). *Given a probability space $(\Omega, \mathcal{A}, \mu)$, a function $X : \Omega \mapsto \mathcal{S}$ (where $\mathcal{S}$ is a countable subset of $\mathbb{R}$) such that*

$$\{\omega \in \Omega : X(\omega) = x\} \in \mathcal{A}, x \in \mathcal{S}$$

is a discrete (integer) valued random variable.

B.2 Probability densities and mass functions

Definition B.11 (Cumulative distribution function, real-valued random variable). *If X is a real-valued random variable then the cumulative distribution function (CDF) of X is defined as*

$$F(x) = P(X \leq x), \quad x \in \mathbb{R}$$

Remark. The CDF of a real-valued random variable has the following properties

1. $x_1 < x_2 \Rightarrow F(x_1) \leq F(x_2)$
2. $\lim_{x \to -\infty} F(x) = 0, \lim_{x \to \infty} F(x) = 1$
3. $F(x) = F(x^+)$, where $F(x^+) = \lim_{\substack{y \to x \\ y > x}} F(x)$ (i.e. continuous from the right).

Definition B.12 (Probability density function). *If x is a real-valued random variable then the probability density function of x is the function f such that*

$$F(x) = \int_{-\infty}^{x} f(t)dt, \forall x$$

where F is defined in Definition B.11.

Remark.

- When the derivative exists, $dF(x)/dx = f(x)$.
- In light of Definition B.11, we see that $f(x) \geq 0 \forall\ x$.

Definition B.13 (Improper density function). *Suppose $\mathbf{x} \in \mathcal{X} \subseteq \mathbb{R}^p$ has the density function $f(\mathbf{x})$. If $\int_{\mathcal{X}} f(\mathbf{x})d\mathbf{x} \neq 1$ then $f(\mathbf{x})$ is an improper density.*

Remark. A density function that is not improper is said to be proper. We try to refrain from using the term "probability density" in these cases.

Improper priors are often used to generate a posterior density that is proportional to the likelihood function: i.e. since by Bayes Rule $p(\boldsymbol{\theta}|\mathbf{y}) \propto p(\boldsymbol{\theta})p(\mathbf{y}|\boldsymbol{\theta})$, if $p(\boldsymbol{\theta}) \propto k$, where k is a constant, then the prior is absorbed into the constant of proportionality, and we obtain $p(\boldsymbol{\theta}|\mathbf{y}) \propto p(\mathbf{y}|\boldsymbol{\theta})$. For instance, see Proposition 2.9 and Example 2.14.

Definition B.14 (Probability mass function). *If X is a discrete random variable, taking values in a countable space $\mathcal{S} \subset \mathbb{R}$, then the probability mass function (pmf) $f : \mathcal{S} \mapsto (0, 1)$ assigns probability to the event $X = x, \forall x \in \mathcal{S}$ as follows:*

$$f(x) = \begin{cases} Pr(X = x) & x \in \mathcal{S}, \\ 0 & x \in \mathbb{R} \setminus \mathcal{S} \end{cases}$$

- For a discrete random variable X with a proper pmf $f(x)$, $\sum_{x \in \mathcal{S}} f(X = x) = 1$.
- Examples of pmfs include the Bernoulli, binomial (see Definition B.22), Poisson (Definition B.23), negative binomial (Definition B.24), geometric and hypergeometric probability mass functions.

Definition B.15 (Kernel of a probability density function or probability mass function). *Suppose $\mathbf{x} \in \mathcal{X} \subseteq \mathbb{R}^p$ has the density function $f(\mathbf{x})$, capable of being* *represented as a function of parameters* $\boldsymbol{\theta}$ *and a scaling constant,* k*; i.e.* $f(\mathbf{x}) = kg(\mathbf{x}; \boldsymbol{\theta})$. *Then*

$$f(\mathbf{x}) \propto g(\mathbf{x}; \boldsymbol{\theta})$$

where $g(\mathbf{x}; \boldsymbol{\theta})$ *is said to be the kernel of* $f(\mathbf{x})$.

- Note that the kernel of a pdf is not, in general, a proper density function. That is, since $f(\mathbf{x}) = kg(\mathbf{x}; \boldsymbol{\theta})$ and, by definition, $\int_{\mathcal{X}} f(\mathbf{x})d\mathbf{x} = 1$, then $\int_{\mathcal{X}} g(\mathbf{x})d\mathbf{x} = 1/k$. Thus, except for the special case of $k = 1$, the kernel $g(\mathbf{x})$ is not a proper pdf.
- If a positive function $g(\mathbf{x})$ is finitely integrable, then it can serve as the kernel of a proper pdf for $\mathbf{x}$. That is, if $\int_{\mathcal{X}} g(\mathbf{x})d\mathbf{x} = G, 0 < G < \infty$, then we can write $p(\mathbf{x}) = 1/Gg(\mathbf{x})$ where $p(\mathbf{x})$ is a proper pdf.

Definition B.16 (Likelihood function). *Consider data* $\mathbf{y} = (y_1, \ldots, y_n)'$ *with joint pdf (or pmf)* $p(\mathbf{y}|\boldsymbol{\theta})$. *When this density for the data is re-written as a function of* $\boldsymbol{\theta}$ *given* $\mathbf{y}$ *it is called the likelihood function and conventionally denoted* $\mathcal{L}(\boldsymbol{\theta}|\mathbf{y})$.

Definition B.17 (Identification; Rothenberg 1971). *Let* $\mathbf{y} \in \mathbb{R}^n$ *have density* p *indexed by a parameter vector* $\boldsymbol{\theta} \in \boldsymbol{\Theta} \subseteq \mathbb{R}^k$.

- *Two parameter values* $\boldsymbol{\theta}_1$ *and* $\boldsymbol{\theta}_2$ *are said to be observationally equivalent if* $p(\mathbf{y}|\boldsymbol{\theta} = \boldsymbol{\theta}_1) = p(\mathbf{y}|\boldsymbol{\theta} = \boldsymbol{\theta}_2)$.
- *A parameter value* $\boldsymbol{\theta}^0 \in \boldsymbol{\Theta}$ *is said to be (globally) identifiable if there is no other* $\boldsymbol{\theta} \in \boldsymbol{\Theta}$ *which is observationally equivalent.*
- *A parameter value* $\boldsymbol{\theta}^0$ *is said to be locally identifiable if there exists an open neighborhood of* $\boldsymbol{\theta}$ *containing no other* $\boldsymbol{\theta} \in \boldsymbol{\Theta}$ *which is observationally equivalent.*

Identification has an interesting status in the Bayesian framework, since there is nothing in Bayes Rule that requires that parameters be identified, in the sense given in Definition B.17 where identification is a property of the likelihood function. Alternative definitions and discussion from Bayesian perspectives appear in Kadane (1974) and Leamer (1978, §5.9).

Definition B.18 (Maximum likelihood estimate, MLE). *Let* $\hat{\boldsymbol{\theta}}(\mathbf{y})$ *be a parameter value at which a likelihood function* $\mathcal{L}(\boldsymbol{\theta}|\mathbf{y})$ *attains its maximum.* $\hat{\boldsymbol{\theta}}(\mathbf{y})$ *is said to be a maximum likelihood estimate (MLE) of* $\boldsymbol{\theta}$.

Remark. Nothing in this definition implies that a MLE will always exist or if it exists, that it is unique.

Proposition B.1 Invariance of MLEs. *Suppose* $\hat{\theta}$ *is the MLE of* θ. *Consider any function of* θ, $h(\theta)$. *The MLE of* $h(\theta)$ *is* $h(\hat{\theta})$.

Proof. Zehna (1966); Casella and Berger (2002, Theorem 7.2.10). ◁

Definition B.19 (Gamma function). *The Gamma function is defined as*

$$\Gamma(a) = \int_0^\infty t^{a-1} \exp(-t)dt, \quad \mathrm{Re}(a) > 0. \tag{B.1}$$

- $\Gamma(a+1) = a\Gamma(a)$.
- For $a = 0, 1, 2, \ldots, \Gamma(a+1) = a!$.
- The following result is helpful in deriving properties of the Gamma density (see Definition B.34, below):

$$\begin{aligned}\int_0^\infty t^{a-1} \exp(-bt)dt &= \int_0^\infty \frac{b^{a-1}}{b^{a-1}} t^{a-1} \exp(-z)\frac{dt}{dz}dz \\ &\quad \text{(i.e. setting } z = bt) \\ &= \frac{1}{b^{a-1}} \int_0^\infty z^{a-1} \exp(-z)dz \\ &= \frac{\Gamma(a)}{b^{a-1}}.\end{aligned}$$

Definition B.20 (Beta function). *The Beta function is*

$$\beta(a, b) = \frac{\Gamma(a)\Gamma(b)}{\Gamma(a+b)} = \int_0^1 t^{a-1}(1-t)^{b-1}dt. \tag{B.2}$$

A generalization of the Beta function is that for $\boldsymbol{\alpha} = (\alpha_1, \ldots, \alpha_k)$, then

$$\beta(\boldsymbol{\alpha}) = \frac{\prod_{i=1}^k \Gamma(\alpha_i)}{\Gamma\left(\sum_{i=1}^k \alpha_i\right)}, \quad \alpha_i > 0 \ \forall i = 1, \ldots, k. \tag{B.3}$$

This function appears as the normalizing constant in the Beta density (Definition B.28), the Dirichlet density (Definition B.29) and in Snedecor's F density (Definition B.39).

B.2.1 Probability mass functions for discrete random quantities

Definition B.21 (Bernoulli probability mass function). *If $x \in \{0, 1\}$ has a Bernoulli probability mass function, conventionally written $x \sim$ Bernoulli(θ), $\theta \in [0, 1]$, then*

$$p(x = 0) = 1 - \theta$$
$$p(x = 1) = \theta$$

- $E(x) = \theta$.
- $V(x) = \theta(1 - \theta)$.
- Binary data are often said to be the result of a "Bernoulli trial".

Definition B.22 (Binomial probability mass function). *If* $r \in \{0, 1, \ldots, n\}$ *has a binomial probability mass function, with success probability* $\theta \in [0, 1]$*, conventionally written* $r \sim$ *Binomial*(θ, n)*, then*

$$p(r) = \binom{n}{r} \theta^r (1 - \theta)^{n-r}.$$

- $\binom{n}{r}$ is the binomial coefficient, i.e.

  $$\binom{n}{r} = \frac{n!}{r!(n-r)!} = \frac{\Gamma(n+1)}{\Gamma(r+1)\Gamma(n-r+1)} = \frac{1}{(n+1)\beta(n-r+1, r+1)},$$

  using Definitions B.19 and B.20.
- The binomial pmf assigns probability over the outcome space $\mathcal{S} = \{0, 1, \ldots, n\}$, the sum of n independent Bernoulli trials, each with success probability θ.
- $E(r) = n\theta$.
- $V(r) = n\theta(1 - \theta)$.

Definition B.23 (Poisson probability mass function). *If* $x \in \{0, 1, 2, \ldots, \}$ *has a Poisson probability mass function with intensity parameter* $\lambda > 0$*, conventionally written* $x \sim$ *Poisson*(λ) *then*

$$p(x) = \frac{\lambda^x}{x!} \exp(-\lambda).$$

- $E(x) = V(x) = \lambda$.
- If $x \sim \text{Binomial}(\theta, n)$ for $x = 0, 1, \ldots, n$, then as $n \to \infty$ and $p \to 0$ such that $np = \lambda < \infty$, then the probability mass function for x converges to the Poisson pmf with intensity parameter λ (Poisson 1837).

Definition B.24 (Negative binomial probability mass function). *If* $x \in \{0, 1, 2, \ldots, \}$ *has a negative binomial probability mass function with success parameter* $\theta \in [0, 1]$ *and number of successes parameter* $y \in \{0, 1, 2, \ldots, \}$*, denoted* $x \sim NB(\theta, y)$ *then*

$$p(x) = \frac{\Gamma(x+y)}{\Gamma(y)x!} \theta^y (1 - \theta)^x$$

- A negative binomial pmf is the distribution over the number of failures in a series of Bernoulli trials with success probability θ until y successes are recorded. Note that the pmf is defined for non-integer values of y; when y is an integer, the negative binomial pmf is sometimes called the Pascal pmf.
- $E(x) = y(1 - \theta)/\theta$.

- When $y = 1$, x has a geometric pmf; in this case x the waiting time until the first failure in a series of independent Bernoulli trials each with probability θ.
- The negative binomial arises in congujate Bayesian analysis as the (prior/posterior) predictive pmf for counts y, where $y|\lambda \sim \text{Poisson}(\lambda)$, $\lambda \sim \text{Gamma}$. See Proposition 2.3.

Definition B.25 (Simplex). *The set*

$$\Delta^k = \left\{ (d_0, \ldots, d_k)' \in \mathbb{R}^{k+1} : \sum_{i=0}^{k} d_i = 1, d_i \geq 0 \,\forall\, i \right\}. \tag{B.4}$$

is known as a standard k-simplex.

For example, Δ^1 is the line connecting $(0, 1)$ and $(1, 0)$, points in $\mathbb{R}^2$; Δ^2 is enclosed by the equilateral triangle with vertices at $(0, 0, 1)$, $(0, 1, 0)$ and $(1, 0, 0)$, points in $\mathbb{R}^3$.

Definition B.26 (Multinomial probability mass function). *Consider the vector* $\mathbf{r} = (r_1, \ldots, r_J)'$ *where* $r_j \in \{0, 1, \ldots, n\}, \forall\, j$ *and* $\sum_{j=1}^{J} r_j = n$. *If* $\mathbf{r}$ *has a multinomial probability mass function, indexed by a parameter vector* $\boldsymbol{\lambda} \in \Delta^{J-1}$ *(the standard* $J - 1$ *simplex, see Definition B.25), then*

$$p(\mathbf{r}) = \frac{n!}{r_1! \ldots r_J!} \lambda_1^{r_1} \ldots \lambda_J^{r_J},$$

conventionally written as $\mathbf{r} \sim \textit{Multinomial}(\boldsymbol{\lambda}, n)$.

- $E(r_j) = n\lambda_j$.
- $V(r_j) = n\lambda_j(1 - \lambda_j)$.
- $\text{cov}(r_j, r_k) = -n\lambda_j\lambda_k$.
- If $n = 1$ then the multinomial pmf is sometimes called a "categorical" distribution, with one and only one element of $\mathbf{r}$ equal to one and the others all zero.

B.2.2 Probability density functions for continuous random quantities

Definition B.27 (Uniform density). *If* $x \in \mathbb{R}$ *has a uniform density on the interval* $[a, b] \subseteq \mathbb{R}$, *conventionally written* $x \sim \textit{Unif}(a, b)$, *then*

$$p(x) = \begin{cases} \dfrac{1}{b - a} & x \in [a, b] \\ 0 & x \notin [a, b] \end{cases}$$

- $E(x) = (a + b)/2$
- $V(x) = (b - a)^2/12$.

Definition B.28 (Beta density). *If $x \in [0, 1]$ has a Beta density with parameters $a > 0$ and $b > 0$, conventionally written as $x \sim Beta(a, b)$, then*

$$p(x) = \frac{1}{\beta(a, b)} x^{a-1}(1-x)^{b-1} = \frac{\Gamma(a+b)}{\Gamma(a)\Gamma(b)} x^{a-1}(1-x)^{b-1}.$$

- $E(x) = a/(a+b)$.
- $V(x) = ab/[(a+b)^2(a+b+1)]$.
- When $a, b > 1$, $p(x)$ has a unique mode at $(a-1)/(a+b-2)$.
- When $a = b = 1$, $x \sim \text{Uniform}(0, 1)$.

Definition B.29 (Dirichlet density). *Consider $\mathbf{x} \in \Delta^{k-1}$, the standard $k-1$ simplex (Definition B.25). If $\mathbf{x}$ has a Dirichlet density with parameter vector $\boldsymbol{\alpha} = (\alpha_1, \ldots, \alpha_k)'$, conventionally written as $\mathbf{x} \sim Dirichlet(\boldsymbol{\alpha})$, then*

$$p(\mathbf{x}) = \frac{1}{\beta(\boldsymbol{\alpha})} \prod_{i=1}^{k} x_i^{\alpha_i - 1}$$

where $\alpha_i > 0 \,\forall\, i = 1, \ldots, k$ and $\beta(\cdot)$ is the Beta function (Equation B.3 in Definition B.20).

- The Dirichlet is usefully considered a multivariate generalization of the Beta density. For instance, $x \sim \text{Unif}(0, 1) \Rightarrow x \sim \text{Beta}(1, 1)$ and $(x, 1-x) \sim \text{Dirichlet}(1, 1)$.
- If $\boldsymbol{\alpha} = (\alpha_1, \ldots, \alpha_k)' = (1, \ldots, 1)'$ then $p(\mathbf{x}) \equiv \text{Dirichlet}(\boldsymbol{\alpha})$ is a uniform density over Δ^{k-1}, with $p(\mathbf{x}) = 1/\beta(\boldsymbol{\alpha}) = \Gamma(k) = (k-1)!, \forall\ \mathbf{x} \in \Delta^{k-1}$. Suppose $k = 2$. Then $(x, 1-x) \sim \text{Dirichlet}(1, 1) \Rightarrow p(x, 1-x) = \Gamma(2) = 1! = 1$ and the density integrates to one over $\Delta^1 = [0, 1]$. Now suppose $k = 3$. Then $\mathbf{x} \sim \text{Dirichlet}(1, 1, 1) \Rightarrow p(\mathbf{x}) = \Gamma(3) = 2! = 2$; since Δ^2 is an equilateral triangle with area 1/2, we confirm that the uniform Dirichlet on Δ^2 integrates to 1.
- Since the support of the Dirichlet is a simplex there is a redundancy in the density, with $x_j = 1 - \sum_{i \neq j} x_i$.
- $E(x_i|\alpha_1, \ldots, \alpha_k) = \alpha_i/S$, where $S = \sum_{i=1}^{k} \alpha_i$.
- $V(x_i|\alpha_1, \ldots, \alpha_k) = \alpha_i(S - \alpha_i)/D$ and $\text{cov}(x_i, x_j|\alpha_1, \ldots, \alpha_k) = -\alpha_i\alpha_j/D$, where $D = S^2(S+1)$.
- If $\mathbf{x} \sim \text{Dirichlet}(\boldsymbol{\alpha})$, then the marginal density of any x_i is $\text{Beta}(\alpha_i, S - \alpha_i)$.

Definition B.30 (Normal density). *If x has a normal density with mean μ and σ^2, conventionally written as $x \sim N(\mu, \sigma^2)$, then*

$$p(x) = \frac{1}{\sqrt{2\pi\sigma^2}} \exp\left[\frac{-1}{2\sigma^2}(x-\mu)^2\right],$$

with $-\infty < x < \infty$, $-\infty < \mu < \infty$, $0 < \sigma^2 < \infty$. *The standard normal density results when* $\mu = 0$ *and* $\sigma^2 = 1$, *in which case* $p(x; \mu, \sigma^2) = \frac{1}{\sqrt{2\pi}} \exp\left[\frac{-x^2}{2}\right] \propto \exp\left[\frac{-x^2}{2}\right]$.

Definition B.31 (Multivariate normal density). *If* $\mathbf{x} \in \mathbb{R}^k$ *has a multivariate normal density, conventionally written* $\mathbf{x} \sim N(\boldsymbol{\mu}, \boldsymbol{\Sigma})$ *then*

$$p(\mathbf{x}) = (2\pi)^{-k/2} |\boldsymbol{\Sigma}|^{-1/2} \exp\left(-(\mathbf{x} - \boldsymbol{\mu})' \boldsymbol{\Sigma}^{-1} (\mathbf{x} - \boldsymbol{\mu})/2\right)$$

where $E(\mathbf{x}) = \boldsymbol{\mu}$ *and* $V(\mathbf{x}) = \boldsymbol{\Sigma}$ *provided* $\boldsymbol{\Sigma}$ *is positive definite.*

Proposition B.2 Marginal and conditional components of a multivariate normal density. *Suppose* $\mathbf{x} \in \mathbb{R}^k$ *can be partitioned as* $\mathbf{x} = (\mathbf{x}_1, \mathbf{x}_2)'$ *and that*

$$\mathbf{x} = \begin{bmatrix} \mathbf{x}_1 \\ \mathbf{x}_2 \end{bmatrix} \sim N\left(\begin{bmatrix} \boldsymbol{\mu}_1 \\ \boldsymbol{\mu}_2 \end{bmatrix}, \begin{bmatrix} \boldsymbol{\Sigma}_{11} & \boldsymbol{\Sigma}_{12} \\ \boldsymbol{\Sigma}_{21} & \boldsymbol{\Sigma}_{22} \end{bmatrix} \right).$$

where $\boldsymbol{\Sigma}_{22}$ *is a positive definite matrix. Then*

1. *the conditional density* $p(\mathbf{x}_1|\mathbf{x}_2)$ *is a normal density with mean*

   $$\boldsymbol{\mu}_{1|2} = \boldsymbol{\mu}_1 + \boldsymbol{\Sigma}_{12}\boldsymbol{\Sigma}_{22}^{-1}(\mathbf{x}_2 - \boldsymbol{\mu}_2)$$

   and variance-covariance matrix $\boldsymbol{\Sigma}_{11|2} = \boldsymbol{\Sigma}_{11} - \boldsymbol{\Sigma}_{12}\boldsymbol{\Sigma}_{22}^{-1}\boldsymbol{\Sigma}_{21}$.
2. *the marginal density* $p(\mathbf{x}_2) = \int p(\mathbf{x}_1, \mathbf{x}_2) d\mathbf{x}_1 \equiv N(\boldsymbol{\mu}_2, \boldsymbol{\Sigma}_{22})$.
3. $\mathbf{x}_1$ *and* $\mathbf{x}_2$ *are independent if and only if* $\boldsymbol{\Sigma}_{12} = \mathbf{0}$.

Proof. 1. Anderson (2003), Theorem 2.5.1. 2. Anderson (2003), Theorem 2.4.3. 3. Anderson (2003), Theorem 2.4.2. ◁

Remark. The matrix $\mathbf{B} = \boldsymbol{\Sigma}_{12}\boldsymbol{\Sigma}_{22}^{-1}$ is the matrix of regression coefficients resulting from the regression of $\mathbf{x}_1$ on $\mathbf{x}_2$.

Proposition B.3 Linear Transformation of Normal Random Variable. *If* $\mathbf{x} \sim N(\boldsymbol{\mu}, \boldsymbol{\Sigma})$ *then* $\mathbf{y} = \mathbf{Ax} + \mathbf{b}$ *is distributed normal with mean* $\mathbf{Ax} + \mathbf{b}$ *and variance-covariance matrix* $\mathbf{A\Sigma A}'$.

Proof. Anderson (2003), Theorems 2.4.4 and 2.4.5. ◁

Definition B.32 (Logistic density). *If* $x \in \mathbb{R}$ *follows a logistic density with mean parameter* $\mu \in \mathbb{R}$ *and scale parameter* $s > 0$, *then*

$$p(x) = \frac{\exp[-(x - \mu)/s]}{s\,(1 + \exp[-(x - \mu)/s])^2},$$

conventionally written as $x \sim Logistic(\mu, s)$.

- $E(x) = \mu$; the logistic density is also symmetric around μ and has a mode at μ (just like the normal density).
- $V(x) = \pi^2/3s^2$; i.e. $s = \sqrt{3/\pi} \Rightarrow V(x) = 1$.
- The logistic has heavier tails than the normal, with approximately the same kurtosis as a t density with 8 or 9 degrees of freedom.
- The CDF is $F(x) = (1 + \exp[-(x - \mu)/s])^{-1}$; the inverse CDF function is $F^{-1}(p) = \mu + s \ln\left(\frac{p}{1-p}\right)$, which is simply the log of the odds ratio for $p \in [0, 1]$ when $\mu = 0$ and $s = 1$.

Definition B.33 (Exponential density). *If $x > 0$ follows an exponential density with rate parameter $\alpha > 0$, then $p(x) = \alpha \exp(-\alpha x)$, conventionally written as $x \sim$ Exponential(α).*

- $E(x) = 1/\alpha$
- $V(x) = 1/\alpha^2$

Definition B.34 (Gamma density). *If $x > 0$ follows a Gamma density with shape parameter $a > 0$ and scale parameter $b > 0$, then*

$$p(x) = \frac{b^a}{\Gamma(a)} x^{a-1} \exp(-bx),$$

conventionally written as $x \sim$ Gamma(a, b).

- $E(x) = ab^{-1}$.
- $V(x) = ab^{-2}$.
- If $x \sim \text{Gamma}(\nu/2, 1/2)$ then $x \sim \chi^2_\nu$ (see Definition B.36).
- If $x \sim \text{Gamma}(1, b)$ then $x \sim \text{Exponential}(b)$.

Definition B.35 (inverse-Gamma density). *If $x > 0$ follows an inverse-Gamma density with shape parameter $a > 0$ and scale parameter $b > 0$, conventionally written as $x \sim$ inverse-Gamma(a, b), then*

$$p(x) = \frac{b^a}{\Gamma(a)} x^{-a-1} \exp\left(\frac{-b}{x}\right),$$

- $E(x) = \frac{b}{a-1}$ if $a > 1$.
- $V(x) = \frac{b^2}{(a-1)^2(a-2)}$ if $a > 2$.
- $p(x)$ has a mode at $b/(a+1)$.
- If $x \sim \text{inverse-Gamma}(a, b)$ then $1/x \sim \text{Gamma}(a, b)$.

- The typical use of the inverse-Gamma density in Bayesian analysis is as the conjugate prior for the variance parameter in a normal model, i.e. $\sigma^2 \sim \text{inverse-Gamma}(\nu_0/2, \nu_0\sigma_0^2/2)$; see Proposition 2.5 or Proposition 2.11 for examples. An improper prior $p(\sigma^2) \propto 1/\sigma^2$ results with $\nu_0 = 0$.

Definition B.36 (χ^2 density). *If $x > 0$ follows a χ^2 density with $\nu > 0$ degrees of freedom, conventionally written as $x \sim \chi^2_\nu$, then*

$$p(x) = \frac{(\frac{1}{2})^{\nu/2}}{\Gamma\left(\frac{\nu}{2}\right)} x^{\nu/2-1} e^{-x/2}$$

- $E(x) = \nu$
- $V(x) = 2\nu$
- If $\nu > 2$, then $p(x)$ has a mode at $\nu - 2$.
- If $x_j \stackrel{\text{iid}}{\sim} N(0, 1)$, $j = 1, \ldots, \nu$, then $y = \sum_{j=1}^{\nu} x_j^2 \sim \chi^2_\nu$.
- If $\mathbf{z} \sim N(\mathbf{a}, \mathbf{B})$, $\mathbf{z} \in \mathbb{R}^k$, then $(\mathbf{z} - \mathbf{a})'\mathbf{B}^{-1}(\mathbf{z} - \mathbf{a}) \sim \chi^2_k$.
- If $x \sim \chi^2_2$ then $x \sim \text{Exponential}(2)$.
- If $x \sim \text{inverse-Gamma}(\nu/2, 1/2)$ then $x \sim \text{inverse-}\chi^2_\nu$.
- If $\sigma^2 \sim \text{inverse-Gamma}(\nu/2, \nu s^2/2)$ then $\nu s^2 \sigma^{-2} \sim \chi^2_\nu$.

Definition B.37 (student-t density). *If x follows a (standardized) student-t density with $\nu > 0$ degrees of freedom, conventionally written $x \sim t_\nu$, then*

$$p(x) = \frac{\Gamma\left(\frac{\nu+1}{2}\right)}{\sqrt{\pi\nu}\,\Gamma\left(\frac{\nu}{2}\right)} \left(1 + \frac{x^2}{\nu}\right)^{-(\nu+1)/2}.$$

and has mean 0 and variance $\nu/(\nu - 2)$. In unstandardized form, the student-t density is

$$p(x) = \frac{\Gamma\left(\frac{\nu+1}{2}\right)}{\sqrt{\pi\nu}\;\sigma\;\Gamma\left(\frac{\nu}{2}\right)} \left(1 + \frac{1}{\nu}\left(\frac{x - \mu}{\sigma}\right)^2\right)^{-(\nu+1)/2},$$

and is conventionally written $x \sim t_\nu(\mu, \sigma^2)$, where μ is a location parameter, $\sigma > 0$ is a scale parameter, and $\nu > 0$ is a degrees of freedom parameter.

- The standardized version of the t-density is an unstandardized t-density with $\mu = 0$ and $\sigma = 1$.
- Provided $\nu > 1$, $E(x) = \mu$ and $V(x) = \frac{\nu}{\nu-2}\sigma^2$.
- As $\nu \to \infty$, $p(x)$ tends to the normal density.
- When $\nu = 1$, $p(x)$ is the Cauchy density.
- If $x \sim t_\nu$ and $y = x^2$, then $y \sim F_{1,\nu}$ (see Definition B.39).

Definition B.38 (Multivariate t density). *If $\mathbf{x} \in \mathbb{R}^k$ has a multivariate t density, conventionally written $\mathbf{x} \sim t_\nu(\boldsymbol{\mu}, \mathbf{V})$, then*

$$p(\mathbf{x}) = \frac{\Gamma((\nu+k)/2)}{\Gamma(\nu/2)\nu^{k/2}\pi^{k/2}} |\mathbf{V}|^{-1/2} \left(1 + \frac{1}{\nu}(\mathbf{x}-\boldsymbol{\mu})'\mathbf{V}^{-1}(\mathbf{x}-\boldsymbol{\mu})\right)^{-(\nu+k)/2}$$
$$\propto \left(1 + \frac{1}{\nu}(\mathbf{x}-\boldsymbol{\mu})'\mathbf{V}^{-1}(\mathbf{x}-\boldsymbol{\mu})\right)^{-(\nu+k)/2}$$

where $\boldsymbol{\mu} \in \mathbb{R}^k$ is a location parameter, $\mathbf{V}$ is a squared scale parameter (a k-by-k, symmetric, positive definite matrix) and $\nu > 0$ is a degrees of freedom parameter.

- Provided $\nu > 1$, $E(\mathbf{x}) = \boldsymbol{\mu}$.
- If $\nu > 2$, $\text{cov}(\mathbf{x}) = \frac{\nu}{\nu-2}\mathbf{V}$.

Proposition B.4 Marginal component of a multivariate t density. *Suppose $\mathbf{x} \sim t_\nu(\boldsymbol{\mu}, \mathbf{V})$, where $\mathbf{x} \in \mathbb{R}^k$ can be partitioned as $\mathbf{x} = (\mathbf{x}_1, \mathbf{x}_2)'$, where $\mathbf{x}_1$ is a m-by-1 vector and $\mathbf{x}_2$ is a n-by-1 vector (i.e. $k = m + n$), and $\boldsymbol{\mu}$ can be conformably partitioned as $\boldsymbol{\mu} = (\boldsymbol{\mu}_1, \boldsymbol{\mu}_2)'$ and*

$$\mathbf{V} = \begin{bmatrix} \mathbf{V}_{11} & \mathbf{V}_{12} \\ \mathbf{V}_{21} & \mathbf{V}_{22} \end{bmatrix},$$

where $\mathbf{V}_{11}$ is m-by-m and $\mathbf{V}_{22}$ is a n-by-n matrix, and $\mathbf{V}_{12}$ is a m-by-n matrix. Then the marginal density of $\mathbf{x}_1$ is a t density with location parameter $\boldsymbol{\mu}_1$, scale parameter $\mathbf{V}_{11}$ and degrees of freedom ν.

Proof. Tong (1990), Proposition 9.2.2. ◁

Definition B.39 (Snedecor F density). *If $x \sim \chi^2_p$ and $y \sim \chi^2_q$ are independent random variables then*

$$z = \frac{x/p}{y/q}$$

has a probability density function known as Snedecor's F density with p and q degrees of freedom, conventionally written $z \sim F_{p,q}$, and equal to

$$f(z) = \frac{1}{\beta(p/2, q/2)} \left(\frac{pz}{pz+q}\right)^{p/2} \left(1 - \frac{pz}{pz+q}\right)^{q/2} z^{-1}$$

for $z \geq 0$, $p, q > 0$ and where $\beta(\cdot, \cdot)$ is the Beta function given in Definition B.20.

Definition B.40 (Wishart density). *Let $\mathbf{W}$ be a p-by-p stochastic, symmetric, positive definite matrix. If the probability density function of $\mathbf{W}$ is*

$$p(\mathbf{W}) = \frac{|\mathbf{W}|^{\frac{n-p-1}{2}}}{2^{\frac{np}{2}} |\mathbf{V}| \Gamma_p\left(\frac{n}{2}\right)} \exp\left(-\frac{\text{tr}(\mathbf{V}^{-1}\mathbf{W})}{2}\right)$$

where Γ_p is the multivariate Gamma function

$$\Gamma_p\left(\frac{n}{2}\right) = \pi^{\frac{p(p-1)}{4}} \prod_{j=1}^{p} \Gamma\left(\frac{n+1-j}{2}\right),$$

then **W** *is said to have a Wishart density with $n > 0$ degrees of freedom and positive definite scale matrix* **V**, *conventionally written as* $\mathbf{W} \sim$ *Wishart*$(\mathbf{V}, n)$.

- $E(\mathbf{W}) = n\mathbf{V}$.
- Suppose $\mathbf{x}_i = (x_{i1}, \ldots, x_{ip})' \stackrel{\text{iid}}{\sim} N(\mathbf{0}, \mathbf{V})$, where $\mathbf{V}$ is a positive definite covariance matrix and $i = 1, \ldots, n$, with $\mathbf{X}$ the n-by-p matrix formed by stacking the n $\mathbf{x}_i$ vectors. Then $\mathbf{X}'\mathbf{X}$ is a p-by-p stochastic matrix with a Wishart$(\mathbf{V}, \mathbf{n})$ density.
- The Wishart density generalizes Gamma and χ^2 densities over precisions to the case of precision matrices, which arise when working with multivariate normal data.
- If $p = 1$ and $\mathbf{V} = 1$ then the Wishart reduces to a χ^2_n density, since in this case $\mathbf{W} = w = \sum_{i=1}^{n} x_i^2$, where $x_i \stackrel{\text{iid}}{\sim} N(0, 1)$; see Definition B.36.
- If $\mathbf{W} \sim$ Wishart$(\mathbf{V}, n)$ then the marginal distribution of any diagonal element of $\mathbf{W}$, $w_{jj}, j = 1, \ldots, p$ is a scaled χ^2 density; specifically, $w_{jj} \sim v_{jj}\chi^2_n$. See Anderson (2003, Theorem 7.3.4) or Mardia, Kent and Bibby (1979, Theorem 3.4.7).
- In Bayesian analysis, the Wishart density is the conjugate prior for a precision matrix $\mathbf{\Sigma}^{-1}$ given a multivariate normal likelihood for data $\mathbf{y}_i$, $i = 1, \ldots, n$. That is, if $\mathbf{y}_i \stackrel{\text{iid}}{\sim} N(\boldsymbol{\mu}, \mathbf{\Sigma})$, and $\mathbf{\Sigma}^{-1} \sim$ Wishart, then $p(\mathbf{\Sigma}^{-1}|\mathbf{Y})$ is also a Wishart density.
- Since $\Gamma((n + 1 - p)/2)$ is undefined for $n \leq p - 1$, the Wishart density is undefined for any integer $n < p$; but subtly, the Wishart distribution remains defined even with $n = p$.

Definition B.41 (inverse-Wishart density). *If a p-by-p matrix* $\mathbf{W} \sim$ *Wishart*$(\mathbf{V}, n)$ *then the density of a positive definite matrix* $\mathbf{S} = \mathbf{W}^{-1}$ *is*

$$p(\mathbf{S}) = \frac{|\mathbf{\Psi}|^{n/2}\,|\mathbf{S}|^{-(n+p+1)/2}}{2^{np/2}\,\Gamma_p\left(\frac{n}{2}\right)} \exp\left(-\frac{\text{tr}(\mathbf{\Psi}\mathbf{S}^{-1})}{2}\right),$$

an inverse-Wishart density with $n > 0$ degrees of freedom and positive-definite precision matrix $\mathbf{\Psi}$, *conventionally written as* $\mathbf{S} \sim$ *inverse-Wishart*$(\mathbf{\Psi}, n)$, *where Γ_p is the multivariate gamma function.*

- $E(\mathbf{S}) = (n - p - 1)^{-1}\mathbf{\Psi}$, $\mathbf{\Psi} = \mathbf{V}^{-1}$, $n > p + 1$.
- Some authors and computer programs parameterize the inverse-Wishart in terms of a scale matrix $\mathbf{V}$ instead of a precision $\mathbf{\Psi} = \mathbf{V}^{-1}$.

- In Bayesian analysis, the inverse-Wishart density is the conjugate prior for a covariance matrix $\boldsymbol{\Sigma}$ given a multivariate normal likelihood; see Proposition B.5.
- An improper, reference prior for a covariance matrix $\boldsymbol{\Sigma}$ can be considered a limiting version of the inverse-Wishart density: i.e. with $n = 0$ and $\boldsymbol{\Psi} = \mathbf{0}$, $p(\boldsymbol{\Sigma}) \propto |\boldsymbol{\Sigma}|^{-(p+1)/2}$.

Proposition B.5 Conjugate analysis of multivariate normal data. *Let* $\mathbf{y}_i \sim N(\boldsymbol{\mu}, \boldsymbol{\Sigma})$ *and suppose* $\boldsymbol{\theta} = (\boldsymbol{\mu}, \boldsymbol{\Sigma})$ *is assigned the prior density* $p(\boldsymbol{\theta}) = p(\boldsymbol{\mu}|\boldsymbol{\Sigma})p(\boldsymbol{\Sigma})$ *where* $p(\boldsymbol{\mu}|\boldsymbol{\Sigma}) \equiv N(\boldsymbol{\mu}_0, k^{-1}\boldsymbol{\Sigma})$ *and* $p(\boldsymbol{\Sigma}) \equiv$ *inverse-Wishart*$(\boldsymbol{\Psi}, n_0)$. *Then*

$$p(\boldsymbol{\mu}|\boldsymbol{\Sigma}, \mathbf{Y}) \sim N\left(\frac{n\bar{\mathbf{x}} + k\boldsymbol{\mu}_0}{n+k}, \frac{1}{n+k}\boldsymbol{\Sigma}\right)$$

$$p(\boldsymbol{\Sigma}|\mathbf{Y}) \sim \textit{inverse-Wishart}\left(\boldsymbol{\Psi} + n\mathbf{S} + \frac{nk}{n+k}(\bar{\mathbf{x}} - \boldsymbol{\mu}_0)(\bar{\mathbf{x}} - \boldsymbol{\mu}_0)', \; n + n_0\right)$$

where $\bar{\mathbf{x}} = n^{-1}\sum_{i=1}^{n} \mathbf{x}_i$ *and* $\mathbf{S} = n^{-1}\sum_{i=1}^{n}(\mathbf{x}_i - \bar{\mathbf{x}})(\mathbf{x}_i - \bar{\mathbf{x}})'$.

Proof. Anderson (2003, Theorem 7.7.3). ◁

Definition B.42 (Jacobian of a vector-valued function). *Let* $\mathbf{x} \in \mathcal{X} \subseteq \mathbb{R}^p$, $\mathbf{v} \in \mathcal{V} \subseteq \mathbb{R}^p$ *and* ϕ *be a map* $\phi : \mathcal{V} \mapsto \mathcal{X}$ *with continuous partial derivatives, i.e.* $\frac{\partial x_i}{\partial v_j}$ *exists and is continuous on* $\mathcal{V}$, $\forall\, i, j = 1, \ldots p$. *The quantity*

$$J\begin{pmatrix}\mathbf{x}\\ \mathbf{v}\end{pmatrix} = det\begin{bmatrix} \frac{\partial x_1}{\partial v_1} & \frac{\partial x_2}{\partial v_1} & \cdots & \frac{\partial x_p}{\partial v_1} \\ \frac{\partial x_1}{\partial v_2} & \frac{\partial x_2}{\partial v_2} & \cdots & \frac{\partial x_p}{\partial v_2} \\ \vdots & \vdots & \vdots & \vdots \\ \frac{\partial x_1}{\partial v_p} & \frac{\partial x_2}{\partial v_p} & \cdots & \frac{\partial x_p}{\partial v_p} \end{bmatrix}$$

is the Jacobian determinant of ϕ, *often simply referred to as the Jacobian of* ϕ.

Proposition B.6 Jacobian theorem. *Define* $\mathbf{x} \in \mathcal{X}$, $\mathbf{v} \in \mathcal{V}$, $\phi : \mathcal{V} \mapsto \mathcal{X}$, *and* J *as in Definition B.42. Then*

$$\int_{\mathcal{X}} g(\mathbf{x})d\mathbf{x} = \int_{\mathcal{V}} g\left(\phi(\mathbf{v})\right)\left|J\begin{pmatrix}\mathbf{x}\\ \mathbf{v}\end{pmatrix}\right| d\mathbf{v}$$

Proof. The proof is trivial for the special case of scalar quantities ($p = 1$). Then the propositon is $\int g(x)dx = \int g(\phi(v))dx/dvdv = \int g(x)dx$ via the chain rule of differential calculus and the facts that the determinant of a scalar is the scalar and $x = \phi(v)$. See Aliprantis and Burkinshaw (1981, Theorem 30.7) for a proof for the general case. ◁

Proposition B.7 Probability density functions under transformations. *Define* $\mathbf{x} \in \mathcal{X}$, $\mathbf{v} \in \mathcal{V}$, ϕ, *and* J *as in Definition B.42, and let* $f_{\mathbf{x}}$ *be the probability density function of* $\mathbf{x}$. *Then*

$$f_{\mathbf{v}} = f_{\mathbf{x}}\left(\phi(\mathbf{v})\right)\left|J\begin{pmatrix}\mathbf{x}\\ \mathbf{v}\end{pmatrix}\right|$$

is a probability density function for $\mathbf{v} \in \mathcal{V}$.

Proof. If $f_{\mathbf{v}}$ is a probability density function then $\Pr(\mathbf{v} \in \mathcal{A}) = \int_{\mathcal{A}} f_{\mathbf{v}} d\mathbf{v}$, $\mathcal{A} \subseteq \mathcal{V}$. In terms of $f_{\mathbf{x}}$, $\Pr(\mathbf{v} \in \mathcal{A}) = \int_{\phi(\mathcal{A})} f_{\mathbf{x}} d\mathbf{x}$. Note that $\phi^{-1}(\phi(\mathbf{v})) = \mathbf{v}$ (assuming the map ϕ is invertible). Then by Proposition B.6,

$$\Pr(\mathbf{v} \in \mathcal{A}) = \int_{\phi(\mathcal{A})} f_{\mathbf{x}} d\mathbf{x} = \int_{\mathcal{A}} f_{\mathbf{x}}(\phi(\mathbf{v})) \left| J\begin{pmatrix} \mathbf{x} \\ \mathbf{v} \end{pmatrix} \right| d\mathbf{v},$$

and the result follows. ◁

B.3 Convergence of sequences of random variables

Proposition B.8 Chebyshev inequality. *For any random variable Y with mean μ and variance $\sigma^2 < \infty$,*

$$Pr(|Y - \mu| \geq \lambda\sigma) \leq \frac{1}{\lambda^2}, \lambda > 0$$

Proof. Rao (1973, 95). ◁

Definition B.43 (Convergence of a sequence of random variables). *Let $Y_1, Y_2, \ldots$ be a sequence of random variables. If*

1. $\lim_{n\to\infty} Pr(|Y_n - Y| > \epsilon) = 0, \forall\, \epsilon > 0$ *then the sequence is said to converge to Y "in probability", conventionally abbreviated as $Y_n \xrightarrow{p} Y$ or $\text{plim}\, Y_n = Y$.*
2. $Pr\left(\lim_{n\to\infty} |Y_n - Y| = 0\right) = 1$ *then the sequence is said to converge to Y "almost surely", conventionally abbreviated as $Y_n \xrightarrow{a.s.} Y$.*
3. $\lim_{n\to\infty} E(Y_n - Y)^2 = 0$ *then the sequence is said to converege to Y "in quadratic mean", conventionally abbreviated as $Y_n \xrightarrow{q.m.} Y$.*

Proposition B.9 Relationships among modes of convergence. *Let $Y_1, Y_2, \ldots$ be a sequence of random variables.*

$Y_n \xrightarrow{q.m.} Y \Rightarrow Y_n \xrightarrow{p} Y.$
$Y_n \xrightarrow{a.s.} \Rightarrow Y_n \xrightarrow{p} Y.$

Proof. 1. The result follows from Chebyshev's inequality, Proposition B.8 (e.g. Rao 1973, 110). 2. Rao (1973, 110-111). ◁

Proposition B.10 Strong Law of Large Numbers (Kolmogorov). *Suppose that $Y_1, Y_2, \ldots$ is a sequence of iid random variables. Let $\bar{Y}_n = n^{-1} \sum_{i=1} Y_i$ and assume that $E(Y_i) = \mu$. Then $-\infty < E(Y_i) < \infty \iff \bar{Y}_n \xrightarrow{a.s.} \mu$.*

Proof. Rao (1973, 115). ◁

Appendix C

Proofs of selected propositions

C.1 Products of normal densities

The following propositions state results that are used frequently in Bayesian analysis, when using conjugate normal priors to analyze normal data. Recall from Definition B.30 that the kernel of a normal density for a random quantity θ involves a quadratic in θ, i.e. if $\theta \sim N(\theta_0, \sigma^2)$, then

$$p(\theta) \propto \exp\left[\frac{-(\theta - \theta_0)^2}{2\sigma^2}\right]. \tag{C.1}$$

When applying Bayes Rule with normal densities, we often have to multiply two expressions of the form in Equation C.1. Since $\exp(x) \times \exp(y) = \exp(x + y)$, applying Bayes Rule with normal densities involves summing two quadratic expressions of the sort appearing in the exponent of Equation C.1, to obtain an expression that is the kernel of another normal density for the random quantity θ (i.e. another quadratic in θ). The first proposition states the result for a scalar quantity x:

Proposition C.1 *The sum of two quadratics in x,*

$$f(x) = a(y - x)^2 + b(x - z)^2$$

yields the following sum of a quadratic in x and a term that does not involve x,

$$f(x) = (a + b)\left(x - \frac{ay + bz}{a + b}\right)^2 + \frac{ab}{a + b}(y - z)^2.$$

Bayesian Analysis for the Social Sciences S. Jackman

Proof. Via simple but tedious algebra

$$
\begin{aligned}
& a(y-x)^2 + b(x-z)^2 \\
&\quad = ay^2 + ax^2 - 2axy + bx^2 + bz^2 - 2bxz \\
&\quad = x^2(a+b) - 2x(ay+bz) + ay^2 + bz^2 \\
&\quad = (a+b)\left[x^2 + \frac{-2x(ay+bz) + ay^2 + bz^2}{a+b}\right] \\
&\quad = (a+b)\left[x^2 - \frac{2x(ay+bz)}{a+b} + \left(\frac{ay+bz}{a+b}\right)^2 - \left(\frac{ay+bz}{a+b}\right)^2 + \frac{ay^2+bz^2}{a+b}\right] \\
&\qquad \text{(i.e. "completing the square")} \\
&\quad = (a+b)\left[\left(x - \frac{ay+bz}{a+b}\right)^2 - \left(\frac{ay+bz}{a+b}\right)^2 + \frac{ay^2+bz^2}{a+b}\right] \\
&\quad = (a+b)\left(x - \frac{ay+bz}{a+b}\right)^2 - \frac{(ay+bz)^2}{a+b} + ay^2 + bz^2 \\
&\quad = (a+b)\left(x - \frac{ay+bz}{a+b}\right)^2 + \frac{-a^2y^2 - b^2z^2 - 2abyz + a^2y^2 + aby^2 + abz^2 + b^2z^2}{a+b} \\
&\quad = (a+b)\left(x - \frac{ay+bz}{a+b}\right)^2 + \frac{ab}{a+b}(y-z)^2.
\end{aligned}
$$

◁

Remark. A useful implication of Proposition C.1 is that if $f(x) = \exp[a(y-x)^2 + b(x-z)^2]$ then

$$f(x) \propto \exp\left[(a+b)\left(x - \frac{ay+bz}{a+b}\right)^2\right].$$

For quadratic forms in a vector $\mathbf{x}$, we have the following restatement of the proposition:

Proposition C.2 *Let* $\mathbf{x}, \mathbf{y}, \mathbf{z}$ *be vectors of length* k*, and* $\mathbf{A}$ *and* $\mathbf{B}$ *be positive definite, symmetric* k*-by-*k *matrices. Then the sum of two quadratics in* $\mathbf{x}$*,*

$$f(\mathbf{x}) = (\mathbf{y}-\mathbf{x})'\mathbf{A}(\mathbf{y}-\mathbf{x}) + (\mathbf{x}-\mathbf{z})'\mathbf{B}(\mathbf{x}-\mathbf{z})$$

yields the following sum of a quadratic in $\mathbf{x}$ *and a term that does not involve* $\mathbf{x}$*, i.e.*

$$f(\mathbf{x}) = (\mathbf{x}-\mathbf{c})'\mathbf{D}(\mathbf{x}-\mathbf{c}) + (\mathbf{y}-\mathbf{z})'\mathbf{E}(\mathbf{y}-\mathbf{z})$$

where $\mathbf{c} = (\mathbf{A}+\mathbf{B})^{-1}(\mathbf{A}\mathbf{y}+\mathbf{B}\mathbf{z})$, $\mathbf{D} = \mathbf{A}+\mathbf{B}$, *and* $\mathbf{E} = (\mathbf{A}^{-1}+\mathbf{B}^{-1})^{-1}$.

Proof. We make repeated use of the facts that if $\mathbf{P}$ and $\mathbf{Q}$ are symmetric k-by-k matrices, then $\mathbf{P} = \mathbf{P}'$ and $\mathbf{Q} = \mathbf{Q}'$ and so $(\mathbf{PQ})' = \mathbf{Q}'\mathbf{P}' = \mathbf{QP}$ which implies that $\mathbf{R} = \mathbf{PQ}$ is a square symmetric matrix, further implying that the order in which we multipy square symmetric matrices is immaterial (i.e. $\mathbf{PQ} = \mathbf{QP}$).

The proof is via the linear algebra analogue of the algebra used in the proof of Proposition C.1. Begin by expanding the quadratics:

$$\begin{aligned}
&(\mathbf{y}-\mathbf{x})'\mathbf{A}(\mathbf{y}-\mathbf{x}) + (\mathbf{x}-\mathbf{z})'\mathbf{B}(\mathbf{x}-\mathbf{z}) \\
&\quad = \mathbf{y}'\mathbf{A}\mathbf{y} - 2\mathbf{x}'\mathbf{A}\mathbf{y} + \mathbf{x}'\mathbf{A}\mathbf{x} + \mathbf{x}'\mathbf{B}\mathbf{x} - 2\mathbf{x}'\mathbf{B}\mathbf{z} + \mathbf{z}'\mathbf{B}\mathbf{z} \\
&\quad = \mathbf{x}'\mathbf{D}\mathbf{x} - 2\mathbf{x}'\mathbf{D}\mathbf{c} - \mathbf{y}'\mathbf{A}\mathbf{y} + \mathbf{z}'\mathbf{B}\mathbf{z}
\end{aligned}$$

using the definitions of $\mathbf{D}$ and $\mathbf{c}$ in the statement of the proposition, and noting that $\mathbf{Dc} = (\mathbf{A}+\mathbf{B})(\mathbf{A}+\mathbf{B})^{-1}(\mathbf{Ay}+\mathbf{Bz}) = (\mathbf{Ay}+\mathbf{Bz})$. Completing the square,

$$\begin{aligned}
&(\mathbf{y}-\mathbf{x})'\mathbf{A}(\mathbf{y}-\mathbf{x}) + (\mathbf{x}-\mathbf{z})'\mathbf{B}(\mathbf{x}-\mathbf{z}) \\
&\quad = \mathbf{x}'\mathbf{D}\mathbf{x} - 2\mathbf{x}\mathbf{D}\mathbf{c} + \mathbf{c}'\mathbf{D}\mathbf{c} - \mathbf{c}'\mathbf{D}\mathbf{c} + \mathbf{y}'\mathbf{A}\mathbf{y} + \mathbf{z}'\mathbf{B}\mathbf{z} \\
&\quad = (\mathbf{x}-\mathbf{c})'\mathbf{D}(\mathbf{x}-\mathbf{c}) - \mathbf{c}'\mathbf{D}\mathbf{c} + \mathbf{y}'\mathbf{A}\mathbf{y} + \mathbf{z}'\mathbf{B}\mathbf{z} \qquad \text{(C.2)}
\end{aligned}$$

Only the first term on the right-hand side of the last equality involves $\mathbf{x}$. The remaining terms can be manipulated by first exploiting a result stated in Definition A.18,

$$\mathbf{D}^{-1} = (\mathbf{A}+\mathbf{B})^{-1} = \mathbf{A}^{-1}(\mathbf{A}^{-1}+\mathbf{B}^{-1})^{-1}\mathbf{B}^{-1},$$

provided the matrix inverses $\mathbf{A}^{-1}$ and $\mathbf{B}^{-1}$ exist. Now let $\mathbf{E} = (\mathbf{A}^{-1}+\mathbf{B}^{-1})^{-1}$ such that $\mathbf{D}^{-1} = \mathbf{A}^{-1}\mathbf{E}\mathbf{B}^{-1}$, or alternatively $\mathbf{E} = \mathbf{A}\mathbf{D}^{-1}\mathbf{B} = \mathbf{A}\mathbf{B}\mathbf{D}^{-1} = \mathbf{A}\mathbf{B}(\mathbf{A}+\mathbf{B})^{-1}$. In addition,

$$\begin{aligned}
\mathbf{A}\mathbf{D}^{-1}\mathbf{A} &= \mathbf{A}\mathbf{A}^{-1}\mathbf{E}\mathbf{B}^{-1}\mathbf{A} = \mathbf{E}\mathbf{B}^{-1}\mathbf{A}, \\
\mathbf{B}\mathbf{D}^{-1}\mathbf{B} &= \mathbf{B}\mathbf{A}^{-1}\mathbf{E}\mathbf{B}^{-1}\mathbf{B} = \mathbf{B}\mathbf{A}^{-1}\mathbf{E}, \\
\mathbf{B}\mathbf{D}^{-1}\mathbf{A} &= \mathbf{B}\mathbf{A}^{-1}\mathbf{E}\mathbf{B}^{-1}\mathbf{A} = \mathbf{E}, \\
\mathbf{c}'\mathbf{D}\mathbf{c} &= (\mathbf{Ay}+\mathbf{Bz})'\mathbf{D}^{-1}\mathbf{D}\mathbf{D}^{-1}(\mathbf{Ay}+\mathbf{Bz}) \\
&= (\mathbf{y}'\mathbf{A}+\mathbf{z}'\mathbf{B})\mathbf{D}^{-1}(\mathbf{Ay}+\mathbf{Bz}) \\
&= \mathbf{y}'\mathbf{A}\mathbf{D}^{-1}\mathbf{A}\mathbf{y} + 2\mathbf{z}'\mathbf{B}\mathbf{D}^{-1}\mathbf{A}\mathbf{y} + \mathbf{z}'\mathbf{B}'\mathbf{D}^{-1}\mathbf{B}\mathbf{z} \\
&= \mathbf{y}'\mathbf{E}\mathbf{B}^{-1}\mathbf{A}\mathbf{y} + 2\mathbf{z}'\mathbf{E}\mathbf{y} + \mathbf{z}'\mathbf{B}\mathbf{A}^{-1}\mathbf{E}\mathbf{z}, \\
\mathbf{y}'\mathbf{A}\mathbf{y} &= \mathbf{y}'\mathbf{D}^{-1}\mathbf{D}\mathbf{A}\mathbf{y} = \mathbf{y}'\left[\mathbf{A}^{-1}\mathbf{E}\mathbf{B}^{-1}(\mathbf{A}+\mathbf{B})\right]\mathbf{A}\mathbf{y} \\
&= \mathbf{y}'\mathbf{E}\mathbf{B}^{-1}\mathbf{A}\mathbf{y} + \mathbf{y}'\mathbf{E}\mathbf{y} \text{ and} \\
\mathbf{z}'\mathbf{B}\mathbf{z} &= \mathbf{z}'\mathbf{D}^{-1}\mathbf{D}\mathbf{A}\mathbf{z} = \mathbf{z}'\left[\mathbf{A}^{-1}\mathbf{E}\mathbf{B}^{-1}(\mathbf{A}+\mathbf{B})\right]\mathbf{B}\mathbf{z} \\
&= \mathbf{z}'\mathbf{E}\mathbf{z} + \mathbf{z}'\mathbf{A}^{-1}\mathbf{E}\mathbf{B}\mathbf{z}.
\end{aligned}$$

With these results we find we can obtain a simple expression for the three quantities in Equation C.2 that are not functions of $\mathbf{x}$: i.e.

$$\mathbf{y}'\mathbf{A}\mathbf{y} + \mathbf{z}'\mathbf{B}\mathbf{z} - \mathbf{c}'\mathbf{D}\mathbf{c} = \mathbf{y}'\mathbf{E}\mathbf{y} + \mathbf{z}'\mathbf{E}\mathbf{z} - 2\mathbf{z}'\mathbf{E}\mathbf{y} = (\mathbf{y}-\mathbf{z})'\mathbf{E}(\mathbf{y}-\mathbf{z}).$$

◁

C.2 Conjugate analysis of normal data

Proposition C.3 (Conjugate prior for mean, normal data, variance known). *Let* $y_i \overset{iid}{\sim} N(\mu, \sigma^2)$, $i = 1, \ldots, n$, *with* σ^2 *known, and* $\mathbf{y} = (y_1, \ldots, y_n)'$. *If* $\mu \sim N(\mu_0, \sigma_0^2)$ *is the prior density for* μ, *then* μ *has posterior density*

$$\mu|\mathbf{y} \sim N\left(\frac{\mu_0\sigma_0^{-2} + \bar{y}\frac{n}{\sigma^2}}{\sigma_0^{-2} + \frac{n}{\sigma^2}}, \quad \left(\sigma_0^{-2} + \frac{n}{\sigma^2}\right)^{-1}\right).$$

Proof. The prior for μ is

$$p(\mu) = \frac{1}{\sqrt{2\pi\sigma_0^2}} \exp\left[\frac{-1}{2\sigma_0^2}(\mu - \mu_0)^2\right] \propto \exp\left[\frac{-1}{2\sigma_0^2}(\mu - \mu_0)^2\right] \qquad \text{(C.3)}$$

and the likelihood is

$$\mathcal{L}(\mu; \mathbf{y}, \sigma^2) = \prod_{i=1}^{n} \frac{1}{\sqrt{2\pi\sigma^2}} \exp\left[\frac{-1}{2\sigma^2}(y_i - \mu)^2\right]$$

$$\propto \exp\left[\frac{-1}{2\sigma^2}\sum_{i=1}^{n}(y_i - \mu)^2\right].$$

Via Bayes Rule, the posterior density is proportional to the prior density times the likelihood:

$$p(\mu|\mathbf{y}) \propto \exp\left[\frac{-1}{2}\left(\frac{1}{\sigma_0^2}(\mu - \mu_0)^2 + \frac{1}{\sigma^2}\sum_{i=1}^{n}(y_i - \mu)^2\right)\right].$$

Using the result in Proposition C.1,

$$\frac{1}{\sigma_0^2}(\mu - \mu_0)^2 + \frac{1}{\sigma^2}\sum_{i=1}^{n}(y_i - \mu)^2 = \left(\frac{1}{\sigma_0^2} + \frac{n}{\sigma^2}\right)\left(\mu - \frac{\frac{\mu_0}{\sigma_0^2} + \frac{n\bar{y}}{\sigma^2}}{\frac{1}{\sigma_0^2} + \frac{n}{\sigma^2}}\right)^2 + r,$$

where r is a set of terms that do not involve μ. Accordingly, the posterior density is

$$p(\mu|\mathbf{y}) \propto \exp\left[\frac{-1}{2}\left(\frac{1}{\sigma_0^2} + \frac{n}{\sigma^2}\right)\left(\mu - \frac{\frac{\mu_0}{\sigma_0^2} + \frac{n\bar{y}}{\sigma^2}}{\frac{1}{\sigma_0^2} + \frac{n}{\sigma^2}}\right)^2\right]$$

which is the kernel of a normal density for μ, with mean

$$\left(\frac{\mu_0}{\sigma_0^2} + \frac{n\bar{y}}{\sigma^2}\right)\left(\frac{1}{\sigma_0^2} + \frac{n}{\sigma^2}\right)^{-1}$$

and variance $\left(\frac{1}{\sigma_0^2} + \frac{n}{\sigma^2}\right)^{-1}$. ◁

Proposition C.4 (Conjugate priors for mean and variance, normal data). *Let* $y_i \stackrel{iid}{\sim} N(\mu, \sigma^2)$, $i = 1, \ldots, n$, *and let* $\mathbf{y} = (y_1, \ldots, y_n)'$. *If* $\boldsymbol{\theta} = (\mu, \sigma^2)'$ *has a normal/inverse-Gamma prior density with parameters* μ_0, n_0, ν_0 *and* σ_0^2, *then the posterior density of* $\boldsymbol{\theta}$ *is also a normal/inverse-Gamma density with parameters* μ_1, n_1, ν_1 *and* σ_1^2, *where*

$$\begin{aligned}
\mu_1 &= \frac{n_0\mu_0 + n\bar{y}}{n_0 + n} \\
n_1 &= n_0 + n \\
\nu_1 &= \nu_0 + n \\
\nu_1\sigma_1^2 &= \nu_0\,\sigma_0^2 + S + \frac{n_0\, n}{n_0 + n}(\mu_0 - \bar{y})^2
\end{aligned}$$

and where $S = \sum_{i=1}^n (y_i - \bar{y})^2$. *That is,*

$$\begin{aligned}
\mu|\,\sigma^2, \mathbf{y} &\sim N(\mu_1, \sigma^2/n_1) \\
\sigma^2|\,\mathbf{y} &\sim \textit{inverse-Gamma}\left(\frac{\nu_1}{2}, \;\frac{\nu_1\,\sigma_1^2}{2}\right)
\end{aligned}$$

Proof. The normal/inverse-Gamma prior density is $p(\mu, \sigma^2) = p(\mu|\sigma^2)p(\sigma^2)$ where

$$p(\mu|\sigma^2) = \frac{1}{\sqrt{2\pi\frac{\sigma^2}{n_0}}}\exp\left[\frac{-n_0}{2\sigma^2}(\mu - \mu_0)^2\right] \propto (\sigma^2)^{-1/2}\exp\left[\frac{-1}{2\sigma^2}n_0(\mu - \mu_0)^2\right], \tag{C.4}$$

and

$$p(\sigma^2) \propto (\sigma^2)^{-(\nu_0+2)/2}\exp\left(\frac{-\nu_0\,\sigma_0^2}{2\sigma^2}\right). \tag{C.5}$$

Multiplying Equations C.4 and C.5, the prior density is

$$\begin{aligned}
p(\mu, \sigma^2) &= p(\mu|\sigma^2)p(\sigma^2) \\
&\propto (\sigma^2)^{-(\nu_0+3)/2}\exp\left[\frac{-1}{2\sigma^2}\left(\nu_0\sigma_0^2 + n_0(\mu - \mu_0)^2\right)\right].
\end{aligned} \tag{C.6}$$

The likelihood is

$$\begin{aligned}
\mathcal{L}(\mu, \sigma^2; \mathbf{y}) &= p(\mathbf{y}|\mu, \sigma^2) = \prod_{i=1}^n p(y_i; \mu, \sigma^2) \\
&= \prod_{i=1}\frac{1}{\sqrt{2\pi\sigma^2}}\exp\left[\frac{-1}{2\sigma^2}(y_i - \mu)^2\right] \\
&\propto (\sigma^2)^{-n/2}\exp\left[\frac{-1}{2\sigma^2}\left(\sum_{i=1}^n (y_i - \mu)^2\right)\right] \\
&= (\sigma^2)^{-n/2}\exp\left[\frac{-1}{2\sigma^2}\left(\sum_{i=1}^n y_i^2 + n\mu^2 - 2\mu\sum_{i=1}^n y_i\right)\right].
\end{aligned} \tag{C.7}$$

Note that

$$
\begin{aligned}
\sum_{i=1}^{n} y_i^2 + n\mu^2 - 2\mu \sum_{i=1}^{n} y_i &= \sum_{i=1}^{n} y_i^2 + n\mu^2 - 2\mu n\bar{y} \\
&= \sum_{i=1}^{n} y_i^2 + n\mu^2 - 2\mu n\bar{y} + n\bar{y}^2 + n\bar{y}^2 - 2n\bar{y}^2 \\
&= \sum_{i=1}^{n} y_i^2 + \sum_{i=1}^{n} \bar{y}^2 - 2n\bar{y}^2 + n\mu^2 + n\bar{y}^2 - 2\mu n\bar{y} \\
&= \sum_{i=1}^{n} y_i^2 + \sum_{i=1}^{n} \bar{y}^2 - \sum_{i=1}^{n} 2\bar{y}y_i + n\mu^2 + n\bar{y}^2 - 2\mu n\bar{y} \\
&= \sum_{i=1}^{n} (y_i - \bar{y})^2 + n(\bar{y} - \mu)^2 = S + n(\bar{y} - \mu)^2
\end{aligned}
$$

where $S = \sum_{i=1}^{n} (y_i - \bar{y})^2$ is the sum of the squared mean deviations of y_i. Thus the likelihood in C.7 can be rewritten as

$$
\mathcal{L}(\mu, \sigma^2; \mathbf{y}) \propto (\sigma^2)^{-n/2} \exp\left[\frac{-1}{2\sigma^2}\left(S + n(\bar{y} - \mu)^2\right)\right]. \tag{C.8}
$$

By Bayes Rule, the posterior density for (μ, σ^2) is proportional to the prior density in Equation C.6 times the likelihood in Equation C.8:

$$
\begin{aligned}
p(\mu, \sigma^2|\mathbf{y}) &\propto p(\mu|\sigma^2)p(\sigma^2)p(\mathbf{y}|\mu, \sigma^2) \\
&\propto (\sigma^2)^{-(\nu_0+3)/2} \exp\left[\frac{-1}{2\sigma^2}\left(\nu_0\sigma_0^2 + n_0(\mu - \mu_0)^2\right)\right] \\
&\quad \times (\sigma^2)^{-n/2} \exp\left[\frac{-1}{2\sigma^2}\left(S + n(\bar{y} - \mu)^2\right)\right] \\
&= (\sigma^2)^{-(\nu_0+n+3)/2} \left[\frac{-1}{2\sigma^2}\left(\nu_0\sigma_0^2 + S + n_0(\mu - \mu_0)^2 + n(\bar{y} - \mu)^2\right)\right].
\end{aligned} \tag{C.9}
$$

By Proposition C.1,

$$
n_0(\mu - \mu_0)^2 + n(\bar{y} - \mu)^2 = (n_0 + n)\left(\mu - \frac{n_0\mu_0 + n\bar{y}}{n_0 + n}\right)^2 + \frac{n_0 n}{n_0 + n}(\mu_0 - \bar{y})^2.
$$

Substitute this result into the expression for the posterior density in Equation C.9, to yield

$$
\begin{aligned}
p(\mu, \sigma^2|\mathbf{y}) &\propto (\sigma^2)^{-(\nu_0+n+3)/2} \\
&\quad \times \exp\left[\frac{-1}{2\sigma^2}\left(\nu_0\sigma_0^2 + S + \frac{n_0 n}{n_0 + n}(\mu_0 - \bar{y})^2 + (n_0 + n)\left(\mu - \frac{n_0\mu_0 + n\bar{y}}{n_0 + n}\right)^2\right)\right]
\end{aligned}
$$

$$= (\sigma^2)^{-1/2} \exp\left[\frac{-(n_0+n)}{2\sigma^2}\left(\mu - \frac{n_0\mu_0 + n\bar{y}}{n_0+n}\right)^2\right]$$

$$\times \quad (\sigma^2)^{-\left(\frac{\nu_0}{2}+\frac{n}{2}+1\right)} \exp\left[\frac{-\nu_0\,\sigma_0^2 - S - \frac{n_0 n}{n_0+n}(\mu_0 - \bar{y})^2}{2\sigma^2}\right]. \qquad \text{(C.10)}$$

We now show that this joint posterior density can be factored as the product of a conditional posterior density for μ and a marginal posterior density for σ^2; i.e. $p(\mu, \sigma^2|\mathbf{y}) = p(\mu|\sigma^2, \mathbf{y})p(\sigma^2|\mathbf{y})$.

The first term in Equation C.10,

$$(\sigma^2)^{-1/2} \exp\left[\frac{-(n_0+n)}{2\sigma^2}\left(\mu - \frac{n_0\mu_0 + n\bar{y}}{n_0+n}\right)^2\right],$$

is the kernel of a normal density over μ:

$$\mu|\sigma^2, \mathbf{y} \sim N\left(\frac{n_0\mu_0 + n\bar{y}}{n_0+n}, \quad \frac{\sigma^2}{n_0+n}\right).$$

This means that the second term in Equation C.10 is (up to a factor of proportionality) $p(\sigma^2|\mathbf{y})$. In fact, the second term in Equation C.10 is the kernel of an inverse-Gamma density (see Definition B.35), such that the marginal posterior density of σ^2 is

$$p(\sigma^2|\mathbf{y}) \propto (\sigma^2)^{-\left(\frac{\nu_0}{2}+\frac{n}{2}+1\right)} \exp\left[\frac{-1}{2\sigma^2}\left(\nu_0\,\sigma_0^2 + S + \frac{n_0\,n}{n_0+n}(\mu_0 - \bar{y})^2\right)\right], \qquad \text{(C.11)}$$

or equivalently, $\sigma^2|\mathbf{y} \sim \text{inverse-Gamma}(\nu_1/2,\ \nu_1\,\sigma_1^2/2)$, where $\nu_1 = \nu_0 + n$ and $\nu_1\sigma_1^2 = \nu_0\sigma_0^2 + S + \frac{n_0 n}{n_0+n}(\mu_0 - \bar{y})^2$. ◁

Proposition C.5 (Conjugate prior for the variance parameter, normal data, mean known). *Let* $y_i \overset{iid}{\sim} N(\mu, \sigma^2), i = 1, \ldots, n$, *with* μ *known. If* σ^2 *has a inverse-Gamma prior density with parameters* ν_0 *and* σ_0^2 *then the posterior density of* σ^2 *is also a inverse-Gamma density with parameters* ν_1 *and* σ_1^2, *where*

$$\nu_1 = \nu_0 + n,$$
$$\nu_1\sigma_1^2 = \nu_0\sigma_0^2 + S \text{ and}$$
$$S = \sum_{i=1}^{n}(y_i - \mu)^2.$$

Proof. The inverse-Gamma prior for σ^2 is (see Definition B.35)

$$p(\sigma^2) \propto (\sigma^2)^{-(\nu_0+2)/2} \exp\left(\frac{-\nu_0\sigma_0^2}{2\sigma^2}\right),$$

and the likelihood for the data is

$$p(\mathbf{y}|\mu,\sigma^2) \propto (\sigma^2)^{-n/2} \exp\left[\frac{-S}{2\sigma^2}\right].$$

Via Bayes Rule,

$$\begin{aligned} p(\sigma^2|\mathbf{y}) &\propto p(\sigma^2)p(\mathbf{y}|\sigma^2) \\ &\propto (\sigma^2)^{-(\nu_0+2)/2} \exp\left(\frac{-\nu_0\sigma_0^2}{2\sigma^2}\right)(\sigma^2)^{-n/2} \exp\left[\frac{-\sum_{i=1}^n (y_i-\mu)^2}{2\sigma^2}\right] \\ &\propto (\sigma^2)^{-(\nu_0+n+2)/2} \exp\left(\frac{-(\nu_0\sigma_0^2+S)}{2\sigma^2}\right), \end{aligned}$$

where $S = \sum_{i=1}^n (y_i - \mu)^2$. We recognize the last expression as the kernel of an inverse-Gamma density (again, see Definition B.35) with parameters $\nu_1 = \nu_0 + n$ and $\nu_1\sigma_1^2 = \nu_0\sigma_0^2 + S$. ◁

Proposition C.6 (*t* density as scale mixture of normal densities). *If $\mu|\sigma^2 \sim N(m, \sigma^2/n)$, where $\sigma^2 \sim$ inverse-Gamma$(\nu/2, \nu s^2/2)$, then the marginal density of μ is a student-t density with location parameter m, scale parameter $\sqrt{s^2/n}$ and ν degrees of freedom.*

Proof. Begin by noting that the marginal density of μ is

$$p(\mu) = \int_0^\infty p(\mu|\sigma^2)p(\sigma^2)d\sigma^2.$$

Substituting the kernels of the normal density for $\mu|\sigma^2$ (see Definition B.30) and the inverse-Gamma density for σ^2 (see Definition B.35), then, up to a factor of proportionality,

$$\begin{aligned} p(\mu) &\propto \int_0^\infty \underbrace{(\sigma^2)^{-\frac{1}{2}} \exp\left(-\frac{n(\mu-m)^2}{2\sigma^2}\right)}_{\text{kernel of } p(\mu|\sigma^2)} \underbrace{(\sigma^2)^{\frac{-\nu-2}{2}} \exp\left(\frac{-1}{2\sigma^2}\nu s^2\right)}_{\text{kernel of } p(\sigma^2)} d\sigma^2 \\ &= \int_0^\infty (\sigma^2)^{\frac{-\nu-3}{2}} \exp\left[\frac{-1}{2\sigma^2}\left(\nu s^2 + n(\mu-m)^2\right)\right] d\sigma^2. \end{aligned}$$

Let $D = \nu s^2 + n(\mu - m)^2$ and re-write the integral via a change of variables to $z = D/(2\sigma^2)$, noting that $\sigma^2 = \frac{1}{2}Dz^{-1} \propto Dz^{-1}$ and $d\sigma^2/dz \propto Dz^{-2}$. That is,

$$\begin{aligned} p(\mu) &\propto \int_0^\infty (\sigma^2)^{\frac{-\nu-3}{2}} \exp(-z)\frac{d\sigma^2}{dz}dz \\ &\propto \int_0^\infty (Dz^{-1})^{\frac{-\nu-3}{2}} \exp(-z)Dz^{-2}dz \\ &= D^{-(\nu+1)/2} \int_0^\infty z^{\frac{\nu+1}{2}-1} \exp(-z)dz. \end{aligned}$$

The integral in the line above is a Gamma function (see Definition B.19) and evaluates to $\Gamma(\frac{\nu+1}{2})$, which is not a function of μ. Thus,

$$\begin{aligned} p(\mu) &\propto D^{-(\nu+1)/2} \\ &= \left[\nu s^2 + n(\mu - m)^2\right]^{-(\nu+1)/2} \\ &= \left[1 + \frac{1}{\nu}\frac{(\mu - m)^2}{s^2/n}\right]^{-(\nu+1)/2} \end{aligned}$$

which we recognize as the kernel of a (unstandardized) student-t density (see Definition B.37), with location parameter m, scale parameter $\sqrt{s^2/n}$ and ν degrees of freedom. $\triangleleft$

Proposition C.7 (Marginal Posterior Density of the Normal Mean, Conjugate Prior). *Under the conditions of Proposition C.4, the marginal posterior density of μ is a student-t density, with location parameter $\mu_1 = (n_0\mu_0 + n\bar{y})/n_1$, scale parameter $\sqrt{\sigma_1^2/n_1}$ and $\nu_1 = \nu_0 + n$ degrees of freedom, where $n_1 = n_0 + n$, $\sigma_1^2 = S_1/\nu_1$,*

$$S_1 = \nu_0\,\sigma_0^2 + S + \frac{n_0\,n}{n_1}(\bar{y} - \mu_0)^2,$$

$S = \sum_{i=1}^n (y_i - \bar{y})^2$ and $\bar{y} = n^{-1}\sum_{i=1}^n y_i$.

Proof. From Proposition C.4,

$$\begin{aligned} \mu|\sigma^2, \mathbf{y} &\sim N(\mu_1, \sigma^2/n_1), \\ \sigma^2|\mathbf{y} &\sim \text{inverse-Gamma}(\nu_1/2,\ \nu_1\sigma_1^2/2). \end{aligned}$$

The result follows from Proposition C.6. $\triangleleft$

Remark. As Lee (2004, 63–64) points out, it is more usual to express the result in Proposition C.7 in terms of a standardized student-t density. That is, let

$$t = \frac{\mu - \mu_1}{\sigma_1/\sqrt{n_1}}.$$

The change of variables from μ to t is simple. Following Proposition B.7, we seek

$$f_t = f_\mu(\phi(t))\left|J\begin{pmatrix}\mu \\ t\end{pmatrix}\right|$$

where $f_t = p(t|\mathbf{y})$, $f_\mu \propto p(\mu|\mathbf{y})$ is derived above, and $|J(\cdot)| = |d\mu/dt|$ is the Jacobian of the transformation. Note that $\phi(t) = t\sqrt{\sigma_1^2/n_1} + \mu_1$ and the Jacobian is a constant with respect to t. Thus

$$p(t|\mathbf{y}) \propto \left[1 + \frac{1}{\nu_1}\frac{(t\sqrt{\sigma_1^2/n_1} + \mu_1 - \mu_1)^2}{\sigma_1^2/n_1}\right]^{-(\nu_1+1)/2}$$

$$\propto \left[1 + \frac{t^2}{\nu_1}\right]^{-(\nu_1+1)/2}$$

which is the kernel of a standardized t density with ν_1 degrees of freedom. That is,

$$\frac{\mu - \mu_1}{\sqrt{\sigma_1^2/n_1}} \sim t_{\nu_1}.$$

Proposition C.8 (Posterior predictive density, normal data, conjugate priors, mean and variance unknown). *If* $y_i \stackrel{iid}{\sim} N(\mu, \sigma^2), i = 1, \ldots, n,$ *with prior beliefs over* $\boldsymbol{\theta} = (\mu, \sigma^2)' \in \boldsymbol{\Theta} \subseteq \mathbb{R} \times \mathbb{R}^+$ *represented with a normal/inverse-Gamma density with parameters* $\mu_0, n_0, \nu_0, \nu_0\sigma_0^2$, *then the posterior predictive density for a future observation* $\tilde{y}$, $p(\tilde{y}|\mathbf{y})$, *is a student-t density with location parameter* $\mu_1 = (n_0\mu_0 + n\bar{y})/n_1$, *scale parameter* $\sigma_1\sqrt{(n_1+1)/n_1}$ *and* $\nu_1 = n + \nu_0$ *degrees of freedom, where* $n_1 = n_0 + n$, $\sigma_1^2 = S_1/\nu_1$,

$$S_1 = \nu_0\,\sigma_0^2 + S + \frac{n_0\,n}{n_0 + n}(\mu_0 - \bar{y})^2,$$

$S = \sum_{i=1}^n (y_i - \bar{y})^2$ *and* $\bar{y} = n^{-1}\sum_{i=1}^n y_i$.

Proof.

$$\begin{aligned} p(\tilde{y}|\mathbf{y}) &= \int_{\boldsymbol{\Theta}} p(\tilde{y}|\boldsymbol{\theta}, \mathbf{y})p(\boldsymbol{\theta}|\mathbf{y})d\boldsymbol{\theta} \\ &= \int_0^\infty \int_{-\infty}^\infty p(\tilde{y}|\mu, \sigma^2, \mathbf{y})p(\mu|\sigma^2, \mathbf{y})p(\sigma^2|\mathbf{y})d\mu\, d\sigma^2 \\ &= \int_0^\infty \left[\int_{-\infty}^\infty p(\tilde{y}|\mu, \sigma^2)p(\mu|\sigma^2, \mathbf{y})d\mu\right] p(\sigma^2|\mathbf{y})d\sigma^2, \end{aligned} \qquad \text{(C.12)}$$

noting that if the y_i are iid, then $p(\tilde{y}|\mu, \sigma^2, \mathbf{y}) = p(\tilde{y}|\mu, \sigma^2)$, and specifically $\tilde{y}|\mu, \sigma^2 \sim N(\mu, \sigma^2)$. We also make use of the fact that $p(\mu, \sigma^2|\mathbf{y}) = p(\mu|\sigma^2, \mathbf{y})p(\sigma^2|\mathbf{y})$, and specifically, from Proposition C.4,

$$\begin{aligned} \mu|\sigma^2, \mathbf{y} &\sim N(\mu_1, \sigma^2/n_1) \\ \sigma^2|\mathbf{y} &\sim \text{inverse-Gamma}(\nu_1/2,\ \nu_1\sigma_1^2/2). \end{aligned}$$

Now consider the inner integral in Equation C.12:

$$\begin{aligned} &\int_{-\infty}^\infty p(\tilde{y}|\mu, \sigma^2)p(\mu|\sigma^2, \mathbf{y})d\mu = p(\tilde{y}|\sigma^2, \mathbf{y}) \\ &\quad \propto \int_{-\infty}^\infty \exp\left[\frac{-(\tilde{y} - \mu)^2}{2\sigma^2}\right] \exp\left[\frac{-(\mu - \mu_1)^2}{2\sigma^2/n_1}\right] d\mu \\ &\quad = \int_{-\infty}^\infty \exp\left[\frac{-1}{2\sigma^2}\left((\tilde{y} - \mu)^2 + n_1(\mu - \mu_1)^2\right)\right] d\mu \end{aligned}$$

and via Proposition C.1

$$= \exp\left[\frac{-1}{2\sigma^2}\frac{n_1}{n_1+1}(\tilde{y}-\mu_1)^2\right]\int_{-\infty}^{\infty}\exp\left[\frac{-1}{2\sigma^2}(n_1+1)\left(\mu-\frac{\tilde{y}+n_1\mu_1}{1+n_1}\right)\right]d\mu.$$

The second exponential term is the kernel of a normal density over μ and so integrates to a constant, and so

$$p(\tilde{y}|\sigma^2,\mathbf{y}) = \int_{-\infty}^{\infty} p(\tilde{y}|\mu,\sigma^2)p(\mu|\sigma^2,\mathbf{y})d\mu \propto \exp\left[\frac{-1}{2\sigma^2}\frac{n_1}{n_1+1}(\tilde{y}-\mu_1)^2\right],$$

or equivalently, $\tilde{y}|\sigma^2,\mathbf{y} \sim N(\mu_1, \sigma^2\frac{n_1+1}{n_1})$. Thus,

$$p(\tilde{y}|\mathbf{y}) \propto \int_0^{\infty}\exp\left[\frac{-1}{2\sigma^2}\frac{n_1}{n_1+1}(\tilde{y}-\mu_1)^2\right]p(\sigma^2|\mathbf{y})d\sigma^2$$

where $p(\sigma^2|\mathbf{y})$ is the inverse-Gamma density given in Equation C.11. This integral is of the same type as that considered in Proposition C.6. The result of that proposition means that

$$p(\tilde{y}|\mathbf{y}) \propto \left[1+\frac{1}{\nu_1}\frac{(\tilde{y}-\mu_1)^2}{\sigma_1^2}\frac{n_1}{n_1+1}\right]^{-(\nu_1+1)/2}$$

which is the kernel of a student-t density (see Definition B.37), with location parameter μ_1, scale parameter $\sigma_1\sqrt{\frac{n_1+1}{n_1}}$ and degrees of freedom parameter ν_1. ◁

Proposition C.9 (Posterior predictive density, normal data with mean and variance unknown, and an improper, reference prior). *Let $y_i \stackrel{iid}{\sim} N(\mu,\sigma^2)$, $i=1,\ldots,n$, $\mathbf{y}=(y_1,\ldots,y_n)'$ with prior beliefs over the unknown parameters $\boldsymbol{\theta}=(\mu,\sigma^2)'$ represented by the improper reference prior $p(\mu,\sigma^2)\propto 1/\sigma^2$. Then the posterior predictive density for a future observation $\tilde{y}$, $p(\tilde{y}|\mathbf{y})$ is a student-t density, with location parameter $\bar{y}$, scale parameter $s\sqrt{\frac{n+1}{n}}$ and degrees of freedom parameter $n-1$, where*

$$s^2 = (n-1)^{-1}\sum_{i=1}^{n}(y-\bar{y})^2 \qquad \textit{and}$$

$$\bar{y} = n^{-1}\sum_{i=1}^{n} y_i.$$

Proof. Following the proof of Proposition C.8,

$$\begin{aligned}p(\tilde{y}|\mathbf{y}) &= \int_0^{\infty}\int_{-\infty}^{\infty} p(\tilde{y}|\mu,\sigma^2)p(\mu|\sigma^2,\mathbf{y})p(\sigma^2|\mathbf{y})d\mu d\sigma^2\\ &= \int_0^{\infty}\left[\int_{-\infty}^{\infty} p(\tilde{y}|\mu,\sigma^2)p(\mu|\sigma^2,\mathbf{y})d\mu\right]p(\sigma^2|\mathbf{y})d\sigma^2.\end{aligned}$$

Note that $\tilde{y}|\mu, \sigma^2 \sim N(\mu, \sigma^2)$ and from Proposition 2.9, Equation 2.18, $\mu|\sigma^2 \sim N(\bar{y}, \sigma^2/n)$. Thus, the inner integral in the previous equation is

$$\int_{-\infty}^{\infty} p(\tilde{y}|\mu, \sigma^2) p(\mu|\sigma^2, \mathbf{y}) d\mu$$

$$\propto \int_{-\infty}^{\infty} \exp\left[\frac{-(\tilde{y}-\mu)^2}{2\sigma^2}\right] \exp\left[\frac{-(\mu-\bar{y})^2}{2\sigma^2/n}\right] d\mu$$

and via Proposition C.1

$$= \exp\left[\frac{-1}{2\sigma^2}\frac{n}{n+1}(\tilde{y}-\bar{y})^2\right] \int_{-\infty}^{\infty} \exp\left[\frac{-1}{2\sigma^2}(n+1)\left(\mu - \frac{\tilde{y}+n\bar{y}}{1+n}\right)\right] d\mu.$$

The second exponential term is the kernel of a normal density over μ and so integrates to a constant, and so

$$p(\tilde{y}|\sigma^2, \mathbf{y}) \propto \exp\left[\frac{-1}{2\sigma^2}\frac{n}{n+1}(\tilde{y}-\bar{y})^2\right],$$

or equivalently, $\tilde{y}|\sigma^2, \mathbf{y} \sim N(\bar{y}, \sigma^2 \frac{n+1}{n})$. From Proposition 2.9, Equation 2.19,

$$\sigma^2|\mathbf{y} \sim \text{inverse-Gamma}\left(\frac{n-1}{2}, \frac{s^2(n-1)}{2}\right),$$

and hence the result of Proposition C.6 can be applied to yield $\tilde{y}|\mathbf{y} \sim t_{n-1}(\bar{y}, s\sqrt{\frac{n+1}{n}})$. ◁

Proposition C.10 (Conditionally-Conjugate Prior for the Normal Mean and Variance). *Let* $y_i \overset{iid}{\sim} N(\mu, \sigma^2)$, $i = 1, \ldots, n$. *If* $p(\mu, \sigma^2) = p(\mu)p(\sigma^2)$, *where* $p(\mu) \equiv N(\mu_0, \omega_0^2)$ *and* $p(\sigma^2) \equiv$ *inverse-Gamma*$(\nu_0/2, \nu_0\sigma_0^2/2)$ *then*

$$\mu|(\sigma^2, \mathbf{y}) \sim N(\mu_1, \omega_1^2) \quad \textit{and} \quad \sigma^2|(\mu, \mathbf{y}) \sim \textit{inverse-Gamma}\left(\frac{\nu_1}{2}, \frac{\nu_1\sigma_1^2}{2}\right)$$

where

$$\mu_1 = \frac{\frac{n}{\sigma^2}\bar{y} + \frac{1}{\omega_0^2}\mu_0}{\frac{n}{\sigma^2} + \frac{1}{\omega_0^2}}, \qquad \omega_1^2 = \left(\frac{n}{\sigma^2} + \frac{1}{\omega_0^2}\right)^{-1},$$

$\bar{y} = n^{-1}\sum_{i=1}^n y_i$, $\nu_1 = \nu_0 + n$, $\nu_1\sigma_1^2 + S$ *and* $S = \sum_{i=1}^n (y_i - \mu)^2$.

Proof. $p(\mu|\mathbf{y}, \sigma^2) \propto p(\mu)p(\mathbf{y}|\mu, \sigma^2)$ and the result of Proposition C.3 applies. $p(\sigma^2|\mathbf{y}, \mu) \propto p(\sigma^2)p(\mathbf{y}|\mu, \sigma^2)$ and the result of Proposition C.5. applies. ◁

Proposition C.11 (t-prior for the normal mean). *Let* $y_i \overset{iid}{\sim} N(\mu, \sigma^2), i = 1, \ldots, n$ *and* $\mathbf{y} = (y_1, \ldots, y_n)'$, *with* μ *and* σ^2 *unknown. If* $p(\mu)$ *is a student t density and* $p(\sigma^2) \equiv$ *inverse-Gamma*$(\frac{a}{2}, \frac{a\sigma_0^2}{2})$, *then the marginal posterior density for* μ *is proportional to the product of two t densities.*

Proof. A more general result (for the case of t-priors over regression coefficients) was stated by Leamer (1978, Theorem 3.10). We adopt the same approach for the simpler case of a normal mean. The marginal posterior density for μ is

$$p(\mu|\mathbf{y}) \propto \int_0^\infty p(\mathbf{y}|\mu, \sigma^2)p(\sigma^2)p(\mu)d\sigma^2 = p(\mu)\int_0^\infty p(\mathbf{y}|\mu, \sigma^2)p(\sigma^2)d\sigma^2, \quad \text{(C.13)}$$

since the prior $p(\mu)$ is not a function of σ^2. Let $g(\mu) = \int_0^\infty p(\mathbf{y}|\mu, \sigma^2)p(\sigma^2)d\sigma^2$. Since $p(\mu)$ is a t density the proof requires demonstrating that $g(\mu)$ is a t density.

In Proposition C.4 we showed that the likelihood for this problem is

$$p(\mathbf{y}|\mu, \sigma^2) \propto (\sigma^2)^{-n/2} \exp\left[\frac{-1}{2\sigma^2}\left(S + n(\bar{y} - \mu)^2\right)\right]$$

where $S = \sum_{i=1}^n (y_i - \bar{y})^2$ and $\bar{y} = \sum_{i=1}^n y_i/n$. The inverse-Gamma prior over σ^2 is

$$p(\sigma^2) \propto (\sigma^2)^{-(a+2)/2} \exp\left(\frac{-a\sigma_0^2}{2\sigma^2}\right)$$

and so

$$g(\mu) = \int_0^\infty (\sigma^2)^{-(n+a+2)/2} \exp\left[\frac{-1}{2\sigma^2}\left(a\sigma_0^2 + S + n(\bar{y} - \mu)^2\right)\right] d\sigma^2.$$

As in Proposition C.6 we evaluate this integral by setting $D = a\sigma_0^2 + S + n(\bar{y} - \mu)^2$ and make the change of variables to $z = D/2\sigma^2$, with $\sigma^2 \propto Dz^{-1}$ and $d\sigma^2/dz \propto Dz^{-2}$. Recall also that we seek to express this integral as a function of μ, and so terms not involving μ can be absorbed into a constant of proportionality. Thus,

$$\begin{aligned}
g(\mu) &= \int_0^\infty (\sigma^2)^{-(n+a+2)/2} \exp(-z)\frac{d\sigma^2}{dz}dz \\
&\propto \int_0^\infty (Dz^{-1})^{-(n+a+2)/2} \exp(-z) Dz^{-2} dz \\
&= D^{-(n+a)/2} \int_0^\infty z^{(n+a+2)/2 - 2} \exp(-z) dz \\
&= D^{-(n+a)/2} \int_0^\infty z^{(n+a)/2 - 1} \exp(-z) dz \\
&\propto D^{-(n+a)/2},
\end{aligned}$$

since $\int_0^\infty z^{(n+a)/2-1}\exp(-z)dz = \Gamma[(n+a)/2]$ which is not a function of μ (see Definition B.19). Substituting for D,

$$\begin{aligned}
g(\mu) &\propto \left[1 + \frac{n(\bar{y} - \mu)^2}{S + a\sigma_0^2}\right]^{-(n+a)/2} \\
&= \left[1 + \frac{1}{\nu_1}\left(\frac{(\bar{y} - \mu)}{\tau}\right)^2\right]^{-(\nu_1+1)/2}
\end{aligned}$$

where $\nu_1 = (n + a - 1)$ and $\tau^2 = (S + a\sigma_0^2)/(n\nu_1)$. We now recognize $g(\mu)$ as the kernel of an unstandardized student t density (see Definition B.37) with location parameter $\bar{y}$, scale parameter $= \tau$ and degrees of freedom parameter ν_1. ◁

Proposition C.12 (Conjugate priors for regression parameters and variance, normal regression model). *Let* $y_i|\mathbf{x}_i \overset{iid}{\sim} N(\mathbf{x}_i\boldsymbol{\beta}, \sigma^2)$, $i = 1, \ldots, n$ *and* $\mathbf{y} = (y_1, \ldots, y_n)'$, *where* $\mathbf{x}_i$ *is a 1-by-k vector of predictors,* $\boldsymbol{\beta}$ *is a k-by-1 vector of unknown regression parameters and* σ^2 *is an unknown variance parameter. If* $\boldsymbol{\beta}|\sigma^2 \sim N(\mathbf{b}_0, \sigma^2\mathbf{B}_0)$ *is the conditional prior density for* $\boldsymbol{\beta}$ *given* σ^2, *and* $\sigma^2 \sim$ *inverse-Gamma*$(\nu_0/2, \nu_0\sigma_0^2/2)$ *is the prior density for* σ^2, *then*

$$\boldsymbol{\beta}|\sigma^2, \mathbf{y}, \mathbf{X} \sim N(\mathbf{b}_1, \sigma^2\mathbf{B}_1),$$
$$\sigma^2|\mathbf{y}, \mathbf{X} \sim \textit{inverse-Gamma}(\nu_1/2,\ \nu_1\sigma_1^2/2)$$

where

$$\mathbf{b}_1 = (\mathbf{B}_0^{-1} + \mathbf{X}'\mathbf{X})^{-1}(\mathbf{B}_0^{-1}\mathbf{b}_0 + \mathbf{X}'\mathbf{X}\hat{\boldsymbol{\beta}})$$
$$\mathbf{B}_1 = (\mathbf{B}_0^{-1} + \mathbf{X}'\mathbf{X})^{-1},$$
$$\nu_1 = \nu_0 + n \qquad \textit{and}$$
$$\nu_1\sigma_1^2 = \nu_0\sigma_0^2 + S + r,$$

and where

$$\hat{\boldsymbol{\beta}} = (\mathbf{X}'\mathbf{X})^{-1}\mathbf{X}'\mathbf{y},$$
$$S = (\mathbf{y} - \mathbf{X}\hat{\boldsymbol{\beta}})'(\mathbf{y} - \mathbf{X}\hat{\boldsymbol{\beta}}), \qquad \textit{and}$$
$$r = (\mathbf{b}_0 - \hat{\boldsymbol{\beta}})'(\mathbf{B}_0 + (\mathbf{X}'\mathbf{X})^{-1})^{-1}(\mathbf{b}_0 - \hat{\boldsymbol{\beta}}).$$

Proof. The prior density is $p(\boldsymbol{\beta}, \sigma^2) = p(\boldsymbol{\beta}|\sigma^2)p(\sigma^2)$ where (see Definition B.31)

$$p(\boldsymbol{\beta}|\sigma^2) = (2\pi)^{-k/2}|\sigma^2\mathbf{B}_0|^{-1/2}\exp\left(\frac{-1}{2\sigma^2}(\boldsymbol{\beta} - \mathbf{b}_0)'\mathbf{B}_0^{-1}(\boldsymbol{\beta} - \mathbf{b}_0)\right)$$
$$\propto (\sigma^2)^{-k/2}\exp\left(\frac{-1}{2\sigma^2}(\boldsymbol{\beta} - \mathbf{b}_0)'\mathbf{B}_0^{-1}(\boldsymbol{\beta} - \mathbf{b}_0)\right),$$

since if $\mathbf{B}_0$ is a k-by-k matrix, then $|\sigma^2\mathbf{B}_0| = (\sigma^2)^k|\mathbf{B}_0|$; see Proposition A.2. The prior for σ^2 is an inverse-Gamma density (see Definition B.35), i.e.

$$p(\sigma^2) \propto (\sigma^2)^{-(\nu_0+2)/2}\exp\left(\frac{-\nu_0\sigma_0^2}{2\sigma^2}\right).$$

The likelihood is

$$p(\mathbf{y}|\mathbf{X}, \boldsymbol{\beta}, \sigma^2) = \prod_{i=1}^{n} p(y_i|\boldsymbol{\beta}, \sigma^2) \propto (\sigma^2)^{-n/2}\exp\left[\frac{-(\mathbf{y} - \mathbf{X}\boldsymbol{\beta})'(\mathbf{y} - \mathbf{X}\boldsymbol{\beta})}{2\sigma^2}\right]. \qquad \text{(C.14)}$$

Note that

$$\begin{aligned}
(\mathbf{y}-\mathbf{X}\boldsymbol{\beta})'(\mathbf{y}-\mathbf{X}\boldsymbol{\beta}) &= (\mathbf{y}'-\boldsymbol{\beta}'\mathbf{X}')(\mathbf{y}-\mathbf{X}\boldsymbol{\beta}) \\
&= \mathbf{y}'\mathbf{y}-\mathbf{y}'\mathbf{X}\boldsymbol{\beta}-\boldsymbol{\beta}'\mathbf{X}'\mathbf{y}+\boldsymbol{\beta}'\mathbf{X}'\mathbf{X}\boldsymbol{\beta} \\
&= \boldsymbol{\beta}'\mathbf{X}'\mathbf{X}\boldsymbol{\beta}-2\boldsymbol{\beta}'\mathbf{X}'\mathbf{y}+\mathbf{y}'\mathbf{y}-S+S, \\
\text{where} \quad S &= (\mathbf{y}-\mathbf{X}\hat{\boldsymbol{\beta}})'(\mathbf{y}-\mathbf{X}\hat{\boldsymbol{\beta}}) \\
&= \mathbf{y}'\mathbf{y}-2\hat{\boldsymbol{\beta}}'\mathbf{X}'\mathbf{y}+\hat{\boldsymbol{\beta}}'\mathbf{X}'\mathbf{X}\hat{\boldsymbol{\beta}}, \quad \text{and so} \\
(\mathbf{y}-\mathbf{X}\boldsymbol{\beta})'(\mathbf{y}-\mathbf{X}\boldsymbol{\beta}) &= \boldsymbol{\beta}'\mathbf{X}'\mathbf{X}\boldsymbol{\beta}-2\boldsymbol{\beta}'\mathbf{X}'\mathbf{y}+2\hat{\boldsymbol{\beta}}'\mathbf{X}'\mathbf{y}-\hat{\boldsymbol{\beta}}'\mathbf{X}'\mathbf{X}\hat{\boldsymbol{\beta}}+S \\
&= \boldsymbol{\beta}'\mathbf{X}'\mathbf{X}\boldsymbol{\beta}+2(\hat{\boldsymbol{\beta}}-\boldsymbol{\beta})'\mathbf{X}'\mathbf{X}(\mathbf{X}'\mathbf{X})^{-1}\mathbf{X}'\mathbf{y}-\hat{\boldsymbol{\beta}}'\mathbf{X}'\mathbf{X}\hat{\boldsymbol{\beta}}+S \\
&= \boldsymbol{\beta}'\mathbf{X}'\mathbf{X}\boldsymbol{\beta}+2\hat{\boldsymbol{\beta}}'\mathbf{X}'\mathbf{X}\hat{\boldsymbol{\beta}}-2\boldsymbol{\beta}'\mathbf{X}'\mathbf{X}\hat{\boldsymbol{\beta}}-\hat{\boldsymbol{\beta}}'\mathbf{X}'\mathbf{X}\hat{\boldsymbol{\beta}}+S \\
&= \boldsymbol{\beta}'\mathbf{X}'\mathbf{X}\boldsymbol{\beta}+\hat{\boldsymbol{\beta}}'\mathbf{X}'\mathbf{X}\hat{\boldsymbol{\beta}}-2\boldsymbol{\beta}'\mathbf{X}'\mathbf{X}\hat{\boldsymbol{\beta}}+S \\
&= (\boldsymbol{\beta}-\hat{\boldsymbol{\beta}})'\mathbf{X}'\mathbf{X}(\boldsymbol{\beta}-\hat{\boldsymbol{\beta}})+S.
\end{aligned}$$

Substituting into the expression for the likelihood in Equation C.14 yields

$$p(\mathbf{y}|\boldsymbol{\beta},\sigma^2,\mathbf{X}) \propto (\sigma^2)^{-n/2}\exp\left[\frac{-1}{2\sigma^2}\left((\boldsymbol{\beta}-\hat{\boldsymbol{\beta}})'\mathbf{X}'\mathbf{X}(\boldsymbol{\beta}-\hat{\boldsymbol{\beta}})+S\right)\right], \tag{C.15}$$

and so via Bayes Rule the posterior density is

$$\begin{aligned}
p(\boldsymbol{\beta},\sigma^2|\mathbf{y},\mathbf{X}) &\propto \underbrace{p(\boldsymbol{\beta}|\sigma^2)p(\sigma^2)}_{\text{prior}}\ \underbrace{p(\mathbf{y}|\boldsymbol{\beta},\sigma^2)}_{\text{likelihood}} \\
&\propto (\sigma^2)^{-(\nu_0+n+2)/2}\exp\left[\frac{-1}{2\sigma^2}(\nu_0\sigma_0^2+S)\right] \qquad \text{(C.16)} \\
&\times\ (\sigma^2)^{-k/2}\exp\left[\frac{-1}{2\sigma^2}\left((\boldsymbol{\beta}-\mathbf{b}_0)'\mathbf{B}_0^{-1}(\boldsymbol{\beta}-\mathbf{b}_0)+(\boldsymbol{\beta}-\hat{\boldsymbol{\beta}})'\mathbf{X}'\mathbf{X}(\boldsymbol{\beta}-\hat{\boldsymbol{\beta}})\right)\right].
\end{aligned}$$

By Proposition C.2,

$$(\boldsymbol{\beta}-\mathbf{b}_0)'\mathbf{B}_0^{-1}(\boldsymbol{\beta}-\mathbf{b}_0)+(\boldsymbol{\beta}-\hat{\boldsymbol{\beta}})'\mathbf{X}'\mathbf{X}(\boldsymbol{\beta}-\hat{\boldsymbol{\beta}}) = (\boldsymbol{\beta}-\mathbf{b}_1)'\mathbf{B}_1^{-1}(\boldsymbol{\beta}-\mathbf{b}_1)+r$$

where

$$\begin{aligned}
\mathbf{b}_1 &= (\mathbf{B}_0^{-1}+\mathbf{X}'\mathbf{X})^{-1}(\mathbf{B}_0^{-1}\mathbf{b}_0+\mathbf{X}'\mathbf{X}\hat{\boldsymbol{\beta}}) \\
\mathbf{B}_1 &= (\mathbf{B}_0^{-1}+\mathbf{X}'\mathbf{X})^{-1} \qquad \text{and} \\
r &= (\mathbf{b}_0-\hat{\boldsymbol{\beta}})'(\mathbf{B}_0+(\mathbf{X}'\mathbf{X})^{-1})^{-1}(\mathbf{b}_0-\hat{\boldsymbol{\beta}}).
\end{aligned}$$

Substituting into the expression for the posterior density in Equation C.16 yields

$$\begin{aligned}
p(\boldsymbol{\beta},\sigma^2|\mathbf{y},\mathbf{X}) &\propto (\sigma^2)^{-(\nu_0+n+2)/2}\exp\left[\frac{-1}{2\sigma^2}(\nu_0\sigma_0^2+S+r)\right] \\
&\quad \times\ (\sigma^2)^{-k/2}\exp\left[\frac{-1}{2\sigma^2}(\boldsymbol{\beta}-\mathbf{b}_1)'\mathbf{B}_1^{-1}(\boldsymbol{\beta}-\mathbf{b}_1)\right].
\end{aligned}$$

The second term is the kernel of a multivariate normal density over $\boldsymbol{\beta}$, with mean vector $\mathbf{b}_1$ and variance-covariance matrix $\sigma^2\mathbf{B}_1$ (see Definition B.31), and hence proportional to $p(\boldsymbol{\beta}|\mathbf{y}, \sigma^2)$, the conditional posterior density for $\boldsymbol{\beta}$. The first term is the kernel of an inverse-Gamma density over σ^2, with shape parameter $(\nu_0 + n)/2$ and scale parameter $(\nu_0\sigma_0^2 + S + r)/2$ (see Definition B.35) and hence proportional to $p(\sigma^2|\mathbf{y})$, the marginal posterior density for σ^2. ◁

Proposition C.13 (Multivariate t density as a scale mixture of multivariate normal densities). *If $\boldsymbol{\mu}|\sigma^2 \sim N(\mathbf{m}, \sigma^2\mathbf{V})$, where $\boldsymbol{\mu}, \mathbf{m} \in \mathbb{R}^k$, $\mathbf{V}$ is a symmetric, positive-definite, k-by-k matrix, and $\sigma^2 \sim$ inverse-Gamma$(\nu/2, \nu s^2/2)$, then the marginal density of $\boldsymbol{\mu}$ is a multivariate t density with location parameter $\mathbf{m}$, squared scale parameter $s^2\mathbf{V}$, and degrees of freedom ν.*

Proof. As with Proposition C.6, begin by noting that marginal density of $\boldsymbol{\mu}$ is $p(\boldsymbol{\mu}) = \int_0^\infty p(\boldsymbol{\mu}|\sigma^2)p(\sigma^2)d\sigma^2$. Substituting the (conditional) multivariate normal density for $\boldsymbol{\mu}|\sigma^2$ (see Definition B.31) and the marginal inverse-Gamma density for σ^2 (see Definition B.35), then, up to a factor of proportionality

$$p(\boldsymbol{\mu}) \propto \int_0^\infty \underbrace{(\sigma^2)^{-k/2}|\mathbf{V}|^{-1/2}\exp\left[\frac{-1}{2\sigma^2}(\boldsymbol{\mu}-\mathbf{m})'\mathbf{V}^{-1}(\boldsymbol{\mu}-\mathbf{m})\right]}_{\text{from} p(\boldsymbol{\mu}|\sigma^2)}$$
$$\times \underbrace{(\sigma^2)^{-(\nu+2)/2}\exp\left[\frac{-1}{2\sigma^2}\nu s^2\right]}_{\text{from} p(\sigma^2)} d\sigma^2,$$

recalling that $|\sigma^2\mathbf{V}| = (\sigma^2)^k|\mathbf{V}|$ (see Definition A.15). After gathering terms,

$$p(\boldsymbol{\mu}) \propto \int_0^\infty (\sigma^2)^{-(\nu+k+2)/2}\exp\left[\frac{-1}{2\sigma^2}\left(\nu s^2 + (\boldsymbol{\mu}-\mathbf{m})'\mathbf{V}^{-1}(\boldsymbol{\mu}-\mathbf{m})\right)\right]d\sigma^2.$$

As in Proposition C.6, we set $D = \nu s^2 + (\boldsymbol{\mu}-\mathbf{m})'\mathbf{V}^{-1}(\boldsymbol{\mu}-\mathbf{m})$ and perform the intergration with respect to $z = D/2\sigma^2$, noting that $\sigma^2 = \frac{D}{2z}$ and $d\sigma^2 = \frac{-D}{2z^2}dz$. Then

$$p(\boldsymbol{\mu}) \propto \int \left(\frac{D}{2z}\right)^{-(\nu+k+2)/2}\frac{-D}{2z^2}\exp(-z)\,dz$$
$$\propto D^{-(\nu+k)/2}\int z^{(\nu+k)/2}\exp(-z)dz \quad \propto \quad D^{-(\nu+k)/2}$$

since $\int z^{(\nu+k)/2}\exp(-z)dz = \Gamma(\frac{\nu+k}{2})$, which is not a function of $\boldsymbol{\mu}$ (see Definition B.19). Thus,

$$p(\boldsymbol{\mu}) \propto \left(\nu s^2 + (\boldsymbol{\mu}-\mathbf{m})'\mathbf{V}^{-1}(\boldsymbol{\mu}-\mathbf{m})\right)^{-(\nu+k)/2}$$
$$= \left(1 + \frac{1}{\nu}(\boldsymbol{\mu}-\mathbf{m})'(s^2\mathbf{V})^{-1}(\boldsymbol{\mu}-\mathbf{m})\right)^{-(\nu+k)/2}$$

which is the kernel of a multivariate t density (see Definition B.38) with location parameter $\mathbf{m}$, squared scale parameter $s^2\mathbf{V}$ and ν degrees of freedom. ◁

Proposition C.14 (Marginal posterior density, conjugate priors, normal regression). *Assume the conditions of Proposition C.12. Then the marginal posterior density of* $\boldsymbol{\beta}$, $p(\boldsymbol{\beta}|\mathbf{y}, \mathbf{X})$, *is a multivariate t density with location parameter* $\mathbf{b}_1$, *squared scale parameter* $\sigma_1^2\mathbf{B}_1$ *and* ν_1 *degrees of freedom.*

Proof. From Proposition C.12,

$$\boldsymbol{\beta}|\sigma^2, \mathbf{y}, \mathbf{X} \sim N(\mathbf{b}_1, \sigma^2\mathbf{B}_1)$$

$$\sigma^2|\mathbf{y}, \mathbf{X} \sim \text{inverse-Gamma}(\nu_1/2,\ \nu_1\sigma_1^2/2)$$

Note that $p(\boldsymbol{\beta}|\mathbf{y}, \mathbf{X}) = \int_0^\infty p(\boldsymbol{\beta}|\sigma^2, \mathbf{y}, \mathbf{X})p(\sigma^2|\mathbf{y}, \mathbf{X})d\sigma^2$, where the latter two densities are given above. The result follows via Proposition C.13. ◁

Proposition C.15 (Posterior predictive density, normal regression, conjugate priors). *Assume the conditions of Proposition C.12. Then the posterior predictive density for q new observations* $\tilde{\mathbf{y}}$ *given predictors* $\tilde{\mathbf{X}}$ *(a q-by-k matrix) is a multivariate t density with location* $\tilde{\mathbf{X}}\mathbf{b}_1$, *squared scale parameter* $\sigma_1^2(\tilde{\mathbf{X}}\mathbf{B}_1\tilde{\mathbf{X}}' + \mathbf{I}_q)$ *and* ν_1 *degrees of freedom, where* $\mathbf{b}_1$ *and* $\mathbf{B}_1$ *are as defined in Proposition C.12.*

Proof. We proceed as in Proposition C.8, where we considered the predictive density for a new observation generated under a simple normal model $y_i \sim N(\mu, \sigma^2)$. In the regression context the conditioning on $\mathbf{X}$ and $\tilde{\mathbf{X}}$ introduces some minor complications, but otherwise the situation is analogous.

Letting $\boldsymbol{\theta} = (\boldsymbol{\beta}, \sigma^2)' \in \boldsymbol{\Theta} \subseteq \mathbb{R}^k \times \mathbb{R}^+$, the posterior predictive density for $\tilde{\mathbf{y}}$ is

$$p(\tilde{\mathbf{y}}|\tilde{\mathbf{X}}, \mathbf{y}, \mathbf{X}) = \int_{\boldsymbol{\Theta}} p(\tilde{\mathbf{y}}|\tilde{\mathbf{X}}, \mathbf{y}, \mathbf{X}, \boldsymbol{\theta})p(\boldsymbol{\theta}|\mathbf{y}, \mathbf{X}, \tilde{\mathbf{X}})d\boldsymbol{\theta}.$$

A maintained assumption is that conditional on $\boldsymbol{\theta}$ and $\tilde{\mathbf{X}}$, $\tilde{\mathbf{y}}$ is independent of the data $\mathbf{y}$ and $\mathbf{X}$, and thus $p(\tilde{\mathbf{y}}|\tilde{\mathbf{X}}, \mathbf{y}, \mathbf{X}, \boldsymbol{\theta}) = p(\tilde{\mathbf{y}}|\tilde{\mathbf{X}}, \boldsymbol{\theta})$. In addition, without $\tilde{\mathbf{y}}$, $\tilde{\mathbf{X}}$ contains no information about $\boldsymbol{\theta}$ beyond that in $\mathbf{y}$ and $\mathbf{X}$, and so $p(\boldsymbol{\theta}|\mathbf{y}, \mathbf{X}, \tilde{\mathbf{X}}) = p(\boldsymbol{\theta}|\mathbf{y}, \mathbf{X})$. Thus,

$$\begin{aligned}
p(\tilde{\mathbf{y}}|\tilde{\mathbf{X}}, \mathbf{y}, \mathbf{X}) &= \int_{\boldsymbol{\Theta}} p(\tilde{\mathbf{y}}|\tilde{\mathbf{X}}, \boldsymbol{\theta})p(\boldsymbol{\theta}|\mathbf{y}, \mathbf{X})d\boldsymbol{\theta} \\
&= \int_0^\infty \int_{\mathbb{R}^k} p(\tilde{\mathbf{y}}|\tilde{\mathbf{X}}, \boldsymbol{\beta}, \sigma^2)p(\boldsymbol{\beta}|\sigma^2, \mathbf{y}, \mathbf{X})p(\sigma^2|\mathbf{y}, \mathbf{X})\, d\boldsymbol{\beta}\, d\sigma^2 \\
&= \int_0^\infty \left[\int_{\mathbb{R}^k} p(\tilde{\mathbf{y}}|\tilde{\mathbf{X}}, \boldsymbol{\beta}, \sigma^2)p(\boldsymbol{\beta}|\sigma^2, \mathbf{y}, \mathbf{X})d\boldsymbol{\beta}\right] p(\sigma^2|\mathbf{y}, \mathbf{X})d\sigma^2. \quad \text{(C.17)}
\end{aligned}$$

Consider the expression being integrated inside the square brackets, $p(\tilde{\mathbf{y}}|\tilde{\mathbf{X}}, \boldsymbol{\beta}, \sigma^2)p(\boldsymbol{\beta}|\sigma^2, \mathbf{y}, \mathbf{X})$. The first term is a normal density, i.e. $\tilde{\mathbf{y}}|\tilde{\mathbf{X}}, \boldsymbol{\beta}, \sigma^2 \sim N(\tilde{\mathbf{X}}\boldsymbol{\beta}, \sigma^2)$ and the second density is simply the conditional posterior density for $\boldsymbol{\beta}$, i.e. $\boldsymbol{\beta}|\sigma^2, \mathbf{y}, \mathbf{X} \sim N(\mathbf{b}_1, \sigma^2\mathbf{B}_1)$ where $\mathbf{b}_1$ and $\mathbf{B}_1$ are as defined in Proposition C.12. Thus,

$$\begin{aligned} &p(\tilde{\mathbf{y}}|\tilde{\mathbf{X}}, \boldsymbol{\beta}, \sigma^2)p(\boldsymbol{\beta}|\sigma^2, \mathbf{y}, \mathbf{X}) \\ &\propto \exp\left[\frac{-1}{2\sigma^2}(\tilde{\mathbf{y}} - \tilde{\mathbf{X}}\boldsymbol{\beta})'(\tilde{\mathbf{y}} - \tilde{\mathbf{X}}\boldsymbol{\beta})\right] \exp\left[\frac{-1}{2\sigma^2}(\mathbf{b}_1 - \boldsymbol{\beta})'\mathbf{B}_1^{-1}(\mathbf{b}_1 - \boldsymbol{\beta})\right] \\ &= \exp\left[\frac{-1}{2\sigma^2}\left((\tilde{\mathbf{y}} - \tilde{\mathbf{X}}\boldsymbol{\beta})'(\tilde{\mathbf{y}} - \tilde{\mathbf{X}}\boldsymbol{\beta}) + (\mathbf{b}_1 - \boldsymbol{\beta})'\mathbf{B}_1^{-1}(\mathbf{b}_1 - \boldsymbol{\beta})\right)\right]. \end{aligned} \tag{C.18}$$

The term in the outer parentheses is the sum of two quadratic forms in $\boldsymbol{\beta}$. Completing the square in $\boldsymbol{\beta}$ (Proposition C.2), yields

$$\begin{aligned} &(\tilde{\mathbf{y}} - \tilde{\mathbf{X}}\boldsymbol{\beta})'(\tilde{\mathbf{y}} - \tilde{\mathbf{X}}\boldsymbol{\beta}) + (\mathbf{b}_1 - \boldsymbol{\beta})'\mathbf{B}_1^{-1}(\mathbf{b}_1 - \boldsymbol{\beta}) \\ &\quad = (\boldsymbol{\beta} - \mathbf{c})'\mathbf{D}(\boldsymbol{\beta} - \mathbf{c}) - \mathbf{c}'\mathbf{D}\mathbf{c} + \tilde{\mathbf{y}}'\tilde{\mathbf{y}} + \mathbf{b}_1'\mathbf{B}^{-1}\mathbf{b}_1 \end{aligned} \tag{C.19}$$

where $\mathbf{c} = \mathbf{D}^{-1}(\tilde{\mathbf{X}}'\tilde{\mathbf{y}} + \mathbf{B}_1^{-1}\mathbf{b}_1)$ and $\mathbf{D}^{-1} = (\tilde{\mathbf{X}}'\tilde{\mathbf{X}} + \mathbf{B}_1^{-1})^{-1}$. By Proposition A.3, part 3,

$$\mathbf{D}^{-1} = \mathbf{B}_1 - \mathbf{B}_1\tilde{\mathbf{X}}'(\tilde{\mathbf{X}}\mathbf{B}_1\tilde{\mathbf{X}}' + \mathbf{I}_q)^{-1}\tilde{\mathbf{X}}\mathbf{B}_1 = \mathbf{B}_1(\mathbf{I}_k - \tilde{\mathbf{X}}'\mathbf{P}^{-1}\tilde{\mathbf{X}}\mathbf{B}_1),$$

where

$$\mathbf{P}^{-1} = (\tilde{\mathbf{X}}\mathbf{B}_1\tilde{\mathbf{X}}' + \mathbf{I}_q)^{-1}. \tag{C.20}$$

Note that

$$\begin{aligned} \mathbf{B}_1^{-1}\mathbf{D}^{-1}\mathbf{B}_1^{-1} &= \mathbf{B}_1^{-1}\left[\mathbf{B}_1(\mathbf{I}_k - \tilde{\mathbf{X}}'\mathbf{P}^{-1}\tilde{\mathbf{X}}\mathbf{B}_1)\right]\mathbf{B}_1^{-1} \\ &= \mathbf{B}_1^{-1} - \tilde{\mathbf{X}}'\mathbf{P}^{-1}\tilde{\mathbf{X}} \end{aligned}$$

and $\mathbf{D}^{-1}\mathbf{B}_1^{-1} = \mathbf{B}_1(\mathbf{I}_k - \tilde{\mathbf{X}}'\mathbf{P}^{-1}\tilde{\mathbf{X}}\mathbf{B}_1)\mathbf{B}_1^{-1} = \mathbf{I}_k - \mathbf{B}_1\tilde{\mathbf{X}}'\mathbf{P}^{-1}\tilde{\mathbf{X}}$. Thus,

$$\begin{aligned} \mathbf{c}'\mathbf{D}\mathbf{c} &= (\tilde{\mathbf{X}}'\tilde{\mathbf{y}} + \mathbf{B}_1^{-1}\mathbf{b}_1)'\mathbf{D}^{-1}(\tilde{\mathbf{X}}'\tilde{\mathbf{y}} + \mathbf{B}_1^{-1}\mathbf{b}_1) \\ &= \tilde{\mathbf{y}}'\tilde{\mathbf{X}}\mathbf{D}^{-1}\tilde{\mathbf{X}}'\tilde{\mathbf{y}} + 2\tilde{\mathbf{y}}'\tilde{\mathbf{X}}\mathbf{D}^{-1}\mathbf{B}_1^{-1}\mathbf{b}_1 + \mathbf{b}_1'\mathbf{B}_1^{-1}\mathbf{D}^{-1}\mathbf{B}_1^{-1}\mathbf{b}_1, \end{aligned}$$

and so

$$\begin{aligned} &\tilde{\mathbf{y}}'\tilde{\mathbf{y}} + \mathbf{b}_1'\mathbf{B}_1^{-1}\mathbf{b}_1 - \mathbf{c}'\mathbf{D}\mathbf{c} \\ &= \tilde{\mathbf{y}}'\tilde{\mathbf{y}} + \mathbf{b}_1'\mathbf{B}_1^{-1}\mathbf{b}_1 - \tilde{\mathbf{y}}'\tilde{\mathbf{X}}\mathbf{D}^{-1}\tilde{\mathbf{X}}'\tilde{\mathbf{y}} - 2\tilde{\mathbf{y}}'\tilde{\mathbf{X}}\mathbf{D}^{-1}\mathbf{B}_1^{-1}\mathbf{b}_1 - \mathbf{b}_1'\mathbf{B}_1^{-1}\mathbf{D}^{-1}\mathbf{B}_1^{-1}\mathbf{b}_1 \\ &= \tilde{\mathbf{y}}'(\mathbf{I}_q - \tilde{\mathbf{X}}\mathbf{D}^{-1}\tilde{\mathbf{X}}')\tilde{\mathbf{y}} + \mathbf{b}_1'\tilde{\mathbf{X}}'\mathbf{P}^{-1}\tilde{\mathbf{X}}\mathbf{b}_1 - 2\tilde{\mathbf{y}}'\tilde{\mathbf{X}}(\mathbf{I}_k - \mathbf{B}_1\tilde{\mathbf{X}}'\mathbf{P}^{-1}\tilde{\mathbf{X}})\mathbf{b}_1 \end{aligned}$$

Since $(\mathbf{I}_q - \tilde{\mathbf{X}}\mathbf{D}^{-1}\tilde{\mathbf{X}}') = \mathbf{P}^{-1}$, we have $\tilde{\mathbf{y}}'(\mathbf{I}_q - \tilde{\mathbf{X}}\mathbf{D}^{-1}\tilde{\mathbf{X}}')\tilde{\mathbf{y}} = \tilde{\mathbf{y}}'\mathbf{P}^{-1}\tilde{\mathbf{y}}$, using the definition of $\mathbf{P}^{-1}$ in Equation C.20. Now complete the square to obtain

$$\begin{aligned} \tilde{\mathbf{y}}'\tilde{\mathbf{y}} + \mathbf{b}_1'\mathbf{B}_1^{-1}\mathbf{b}_1 - \mathbf{c}'\mathbf{D}\mathbf{c} &= \tilde{\mathbf{y}}'\mathbf{P}^{-1}\tilde{\mathbf{y}} + \mathbf{b}_1'\tilde{\mathbf{X}}'\mathbf{P}^{-1}\tilde{\mathbf{X}}\mathbf{b}_1 - 2\tilde{\mathbf{y}}'\mathbf{P}^{-1}\tilde{\mathbf{X}}\mathbf{b}_1 \\ &\quad + 2\tilde{\mathbf{y}}'\mathbf{P}^{-1}\tilde{\mathbf{X}}\mathbf{b}_1 - 2\tilde{\mathbf{y}}'\tilde{\mathbf{X}}'\mathbf{b}_1 + 2\tilde{\mathbf{y}}'\tilde{\mathbf{X}}\mathbf{B}_1\tilde{\mathbf{X}}'\mathbf{P}^{-1}\tilde{\mathbf{X}}\mathbf{b}_1. \end{aligned}$$

The terms in the second line on the right-hand side of this equality sum to a null matrix: i.e.

$$2\tilde{\mathbf{y}}'\mathbf{P}^{-1}\tilde{\mathbf{X}}\mathbf{b}_1 - 2\tilde{\mathbf{y}}'\tilde{\mathbf{X}}'\mathbf{b}_1 + 2\tilde{\mathbf{y}}'\tilde{\mathbf{X}}\mathbf{B}_1\tilde{\mathbf{X}}'\mathbf{P}^{-1}\tilde{\mathbf{X}}\mathbf{b}_1 = 2\tilde{\mathbf{y}}'(\mathbf{P}^{-1} - \mathbf{I}_q + \tilde{\mathbf{X}}\mathbf{B}_1\tilde{\mathbf{X}}'\mathbf{P}^{-1})\tilde{\mathbf{X}}\mathbf{b}_1$$

but $\tilde{\mathbf{X}}\mathbf{B}_1\tilde{\mathbf{X}}' = \mathbf{P} - \mathbf{I}_q$ (see the definition of $\mathbf{P}^{-1}$ in Equation C.20), which yields

$$\mathbf{P}^{-1} - \mathbf{I}_q + \tilde{\mathbf{X}}\mathbf{B}_1\tilde{\mathbf{X}}'\mathbf{P}^{-1} = \mathbf{P}^{-1} - \mathbf{I}_q + (\mathbf{P} - \mathbf{I}_q)\mathbf{P}^{-1} = \mathbf{0}.$$

Thus,

$$\tilde{\mathbf{y}}'\tilde{\mathbf{y}} + \mathbf{b}_1'\mathbf{B}_1^{-1}\mathbf{b}_1 - \mathbf{c}'\mathbf{D}\mathbf{c} = (\tilde{\mathbf{y}} - \tilde{\mathbf{X}}\mathbf{b}_1)'\mathbf{P}^{-1}(\tilde{\mathbf{y}} - \tilde{\mathbf{X}}\mathbf{b}_1). \tag{C.21}$$

Substituing this result into Equation C.19, we obtain

$$\begin{aligned} &(\tilde{\mathbf{y}} - \tilde{\mathbf{X}}\boldsymbol{\beta})'(\tilde{\mathbf{y}} - \tilde{\mathbf{X}}\boldsymbol{\beta}) + (\mathbf{b}_1 - \boldsymbol{\beta})'\mathbf{B}_1^{-1}(\mathbf{b}_1 - \boldsymbol{\beta}) \\ &= (\boldsymbol{\beta} - \mathbf{c})'\mathbf{D}(\boldsymbol{\beta} - \mathbf{c}) + (\tilde{\mathbf{y}} - \tilde{\mathbf{X}}\mathbf{b}_1)'\mathbf{P}^{-1}(\tilde{\mathbf{y}} - \tilde{\mathbf{X}}\mathbf{b}_1), \end{aligned}$$

and substituting into Equation C.18,

$$\begin{aligned} &p(\tilde{\mathbf{y}}|\tilde{\mathbf{X}}, \boldsymbol{\beta}, \sigma^2)p(\boldsymbol{\beta}|\sigma^2, \mathbf{y}, \mathbf{X}) \\ &\propto \exp\left(\frac{-1}{2\sigma^2}(\tilde{\mathbf{y}} - \tilde{\mathbf{X}}\mathbf{b}_1)'\mathbf{P}^{-1}(\tilde{\mathbf{y}} - \tilde{\mathbf{X}}\mathbf{b}_1)\right)\exp\left(\frac{-1}{2\sigma^2}(\boldsymbol{\beta} - \mathbf{c})'\mathbf{D}(\boldsymbol{\beta} - \mathbf{c}).\right) \end{aligned}$$

The second term on the right-hand side is the kernel of a multivariate normal density for $\boldsymbol{\beta}$, while the first term is the kernel of a multivariate normal density for $\tilde{\mathbf{y}}$, and is not a function of $\boldsymbol{\beta}$ (see Definition B.31). Thus, substituting in Equation C.17,

$$\begin{aligned} p(\tilde{\mathbf{y}}|\tilde{\mathbf{X}}, \mathbf{y}, \mathbf{X}) &= \int_0^\infty \left[\int_{\mathbb{R}^k} p(\tilde{\mathbf{y}}|\tilde{\mathbf{X}}, \boldsymbol{\beta}, \sigma^2)p(\boldsymbol{\beta}|\sigma^2, \mathbf{y}, \mathbf{X})d\boldsymbol{\beta}\right] p(\sigma^2|\mathbf{y}, \mathbf{X})d\sigma^2 \\ &\propto \int_0^\infty p(\tilde{\mathbf{y}}|\tilde{\mathbf{X}}, \mathbf{b}_1, \sigma^2)\left[\int_{\mathbb{R}^k} p(\boldsymbol{\beta}|\sigma^2, \mathbf{y}, \tilde{\mathbf{y}}, \mathbf{X}, \tilde{\mathbf{X}})d\boldsymbol{\beta}\right] p(\sigma^2|\mathbf{y}, \mathbf{X})d\sigma^2 \\ &\propto \int_0^\infty p(\tilde{\mathbf{y}}|\tilde{\mathbf{X}}, \mathbf{b}_1, \sigma^2)p(\sigma^2|\mathbf{y}, \mathbf{X})d\sigma^2, \end{aligned} \tag{C.22}$$

since the integration with respect $\boldsymbol{\beta}$ evaluates to a constant, which is absorbed into the proportionality constant, and where $p(\tilde{\mathbf{y}}|\tilde{\mathbf{X}}, \mathbf{b}_1, \sigma^2)$ is the multivariate normal density for $\tilde{\mathbf{y}}$ (the conditional posterior density for $\tilde{\mathbf{y}}$, given σ^2), i.e. $\tilde{\mathbf{y}}|\tilde{\mathbf{X}}, \sigma^2 \sim N(\tilde{\mathbf{X}}\mathbf{b}_1, \sigma^2\mathbf{P})$, and $p(\sigma^2|\mathbf{y}, \mathbf{X})$ is the marginal posterior density for σ^2. By conjugacy, $p(\sigma^2|\mathbf{y}, \mathbf{X})$ is an inverse-Gamma density (see Proposition C.12), i.e. $\sigma^2|\mathbf{y}, \mathbf{X} \sim \text{inverse-Gamma}(\nu_1/2, \nu_1\sigma_1^2/2)$.

Performing the integration with respect to σ^2 in Equation C.22 completes the proof. The result of Proposition C.13 applies, such that the resulting marginal posterior density for $\tilde{\mathbf{y}}$ is a multivariate t density, with location parameter $\tilde{\mathbf{X}}\mathbf{b}_1$, squared scale parameter $\sigma_1^2\mathbf{P}$, and ν_1 degrees of freedom. ◁

Proposition C.16 *Assume the conditions of Proposition C.12, assigning the parameters* $(\boldsymbol{\beta}, \sigma^2)'$ *the improper prior density* $p(\boldsymbol{\beta}, \sigma^2) \propto 1/\sigma^2$. *Further assume that* $\mathbf{X}'\mathbf{X}$ *is non-singular. Then*

$$\boldsymbol{\beta}|\sigma^2, \mathbf{y}, \mathbf{X} \sim N\left(\hat{\boldsymbol{\beta}}, \sigma^2(\mathbf{X}'\mathbf{X})^{-1}\right)$$

$$\sigma^2|\mathbf{y}, \mathbf{X} \sim \textit{inverse-Gamma}\left(\frac{n-k}{2}, \frac{S}{2}\right)$$

$$\boldsymbol{\beta}|\mathbf{y}, \mathbf{X} \sim t_{n-k}\left(\hat{\boldsymbol{\beta}}, S/(n-k)(\mathbf{X}'\mathbf{X})^{-1}\right)$$

where $\hat{\boldsymbol{\beta}} = (\mathbf{X}'\mathbf{X})^{-1}\mathbf{X}'\mathbf{y}$ *and* $S = (\mathbf{y} - \mathbf{X}\hat{\boldsymbol{\beta}})'(\mathbf{y} - \mathbf{X}\hat{\boldsymbol{\beta}})$.

Proof. Multiply the likelihood in Equation C.15 (given in Proposition C.12) by the prior density $p(\boldsymbol{\beta}, \sigma^2) \propto 1/\sigma^2$ to obtain the posterior density

$$p(\boldsymbol{\beta}, \sigma^2|\mathbf{y}, \mathbf{X}) \propto (\sigma^2)^{-(n+2)/2} \exp\left[\frac{-1}{2\sigma^2}\left((\boldsymbol{\beta} - \hat{\boldsymbol{\beta}})'\mathbf{X}'\mathbf{X}(\boldsymbol{\beta} - \hat{\boldsymbol{\beta}}) + S\right)\right].$$

The posterior density factors as the product of a conditional posterior density for $\boldsymbol{\beta}$ given σ^2 and a marginal posterior density for σ^2: i.e.

$$p(\boldsymbol{\beta}, \sigma^2|\mathbf{y}, \mathbf{X}) \propto \underbrace{(\sigma^2)^{-(n+2)/2} \exp\left(\frac{-S}{2\sigma^2}\right)}_{\text{kernel of } p(\sigma^2|\mathbf{y},\mathbf{X})} \underbrace{\exp\left(\frac{-1}{2\sigma^2}(\boldsymbol{\beta} - \hat{\boldsymbol{\beta}})'\mathbf{X}'\mathbf{X}(\boldsymbol{\beta} - \hat{\boldsymbol{\beta}})\right)}_{\text{kernel of } p(\boldsymbol{\beta}|\sigma^2,\mathbf{y},\mathbf{X})}. \tag{C.23}$$

The last term is the kernel of a multivariate normal density, i.e. via Definition B.31 we have

$$p(\boldsymbol{\beta}|\sigma^2, \mathbf{y}, \mathbf{X}) \propto \exp\left(\frac{-1}{2\sigma^2}(\boldsymbol{\beta} - \hat{\boldsymbol{\beta}})'\mathbf{X}'\mathbf{X}(\boldsymbol{\beta} - \hat{\boldsymbol{\beta}})\right)$$

or, equivalently, $\boldsymbol{\beta}|\sigma^2, \mathbf{y}, \mathbf{X} \sim N(\hat{\boldsymbol{\beta}}, \sigma^2(\mathbf{X}'\mathbf{X})^{-1})$. We obtain the marginal posterior density for σ^2 by integrating Equation C.23 with respect to $\boldsymbol{\beta} \in \mathbf{B} \subseteq \mathbb{R}^k$:

$$\begin{aligned} p(\sigma^2|\mathbf{y}, \mathbf{X}) &\propto \int_{\mathbf{B}} p(\boldsymbol{\beta}, \sigma^2|\mathbf{y}, \mathbf{X})d\boldsymbol{\beta} \\ &\propto (\sigma^2)^{-(n+2)/2} \exp\left(\frac{-S}{2\sigma^2}\right) \int_{\mathbf{B}} \exp\left(\frac{-1}{2\sigma^2}(\boldsymbol{\beta} - \hat{\boldsymbol{\beta}})'\mathbf{X}'\mathbf{X}(\boldsymbol{\beta} - \hat{\boldsymbol{\beta}})\right) d\boldsymbol{\beta}. \end{aligned}$$

Via Definition B.31,

$$\begin{aligned} \int_{\mathbf{B}} \exp\left(\frac{-1}{2\sigma^2}(\boldsymbol{\beta} - \hat{\boldsymbol{\beta}})'\mathbf{X}'\mathbf{X}(\boldsymbol{\beta} - \hat{\boldsymbol{\beta}})\right) d\boldsymbol{\beta} &= \left[(2\pi)^{-k/2} \left|\sigma^2(\mathbf{X}'\mathbf{X})^{-1}\right|^{-1/2}\right]^{-1} \\ &= (2\pi)^{k/2}(\sigma^2)^{k/2}|\mathbf{X}'\mathbf{X}|^{-1/2}, \end{aligned}$$

and thus

$$p(\sigma^2|\mathbf{y}, \mathbf{X}) \propto (\sigma^2)^{-(n-k+2)/2} \exp\left(\frac{-S}{2\sigma^2}\right)$$

or equivalently (via Definition B.35), $\sigma^2|\mathbf{y}, \mathbf{X} \sim$ inverse-Gamma $((n-k)/2, S/2)$. The claim that the marginal posterior density of $\boldsymbol{\beta}$ is a multivariate-t density follows via Proposition C.13. ◁

C.3 Asymptotic normality of the posterior density

Here I provide a informal re-statement of the fact that under a wide set of conditions, posterior densities tend to normal densities as the amount of data available for analysis becomes arbitrarily plentiful (i.e. as $n \to \infty$). This "heuristic proof" appears in numerous places in the literature: e.g. O'Hagan (2004, 73), Williams (2001, 204) and Gelman *et al.* (2004, 587); the following discussion is based on Bernardo and Smith (1994, 287).

To prove that $p(\boldsymbol{\theta}|\mathbf{y})$ tends to a normal distribution as $n \to \infty$, the general strategy is to first take a Taylor series expansion of the posterior distribution. Then, after ignoring higher order terms in the expansion that disappear asymptotically (subject to regularity conditions), we have something recognizable as a normal distribution. So, consider a parameter vector $\boldsymbol{\theta} \in \mathbb{R}^k$. Bayes Theorem tells us that a posterior distribution is proportional to a prior times a likelihood, or $p(\boldsymbol{\theta}|\mathbf{y}) \propto p(\boldsymbol{\theta})p(\mathbf{y}|\boldsymbol{\theta})$, which can be re-written as

$$p(\boldsymbol{\theta}|\mathbf{y}) \propto \exp\left(\log(p(\boldsymbol{\theta}) + \log p(\mathbf{y}|\boldsymbol{\theta})\right)$$

At the maximum of the log prior and the log likelihood, we have

$$\frac{\partial \log p(\boldsymbol{\theta})}{\partial \boldsymbol{\theta}} = \mathbf{0} \quad \text{and} \quad \frac{\partial \log p(\mathbf{y}|\boldsymbol{\theta})}{\partial \boldsymbol{\theta}} = \mathbf{0},$$

respectively. Let $\boldsymbol{\theta}_0$ denote the value of $\boldsymbol{\theta}$ that maximizes the prior and $\tilde{\boldsymbol{\theta}}_n$ denote the MLE of $\boldsymbol{\theta}$. Taylor series expansions around the respective log maxima yield

$$\log p(\boldsymbol{\theta}) = \log p(\boldsymbol{\theta}_0) - \frac{1}{2}(\boldsymbol{\theta} - \boldsymbol{\theta}_0)'\mathbf{Q}_0(\boldsymbol{\theta} - \boldsymbol{\theta}) + r_0$$

$$\log p(\mathbf{y}|\boldsymbol{\theta}) = \log p(\mathbf{y}|\hat{\boldsymbol{\theta}}_n) - \frac{1}{2}(\boldsymbol{\theta} - \hat{\boldsymbol{\theta}}_n)'\mathbf{Q}_n(\boldsymbol{\theta} - \hat{\boldsymbol{\theta}}_n) + r_n$$

where r_0 and r_n are higher-order terms and

$$\mathbf{Q}_0 = -\left.\frac{\partial^2 \log p(\boldsymbol{\theta}_0)}{\partial \boldsymbol{\theta} \partial \boldsymbol{\theta}'}\right|_{\boldsymbol{\theta}=\boldsymbol{\theta}_0} \quad \text{and} \quad \mathbf{Q}_n = -\left.\frac{\partial^2 \log p(\mathbf{y}|\boldsymbol{\theta})}{\partial \boldsymbol{\theta} \partial \boldsymbol{\theta}'}\right|_{\boldsymbol{\theta}=\hat{\boldsymbol{\theta}}_n}.$$

Note that the leading terms in the respective Taylor series expansions are not functions of $\boldsymbol{\theta}$. Thus, via the result in Proposition C.2,

$$p(\boldsymbol{\theta}|\mathbf{y}) \propto \exp\left(-\frac{1}{2}(\boldsymbol{\theta}-\boldsymbol{\theta}_p)'\mathbf{Q}_p(\boldsymbol{\theta}-\boldsymbol{\theta}_p)\right), \tag{C.24}$$

where $\mathbf{Q}_p = \mathbf{Q}_0 + \mathbf{Q}_n$ and $\boldsymbol{\theta}_p = \mathbf{Q}_p^{-1}(\mathbf{Q}_0\boldsymbol{\theta}_0 + \mathbf{Q}_n\hat{\boldsymbol{\theta}}_n)$. The right-hand side of Equation C.24 is recognizable as the kernel of a multivariate normal distribution with mean vector $\boldsymbol{\theta}_p$ and variance-covariance matrix $\mathbf{Q}_p^{-1}$; see Definition B.31.

References

Achen, Christopher. 1978. "Measuring representation." *American Journal of Political Science* 22:475–510.

Agresti, Alan. 2002. *Categorical Data Analysis*. Second ed. Hoboken, New Jersey: John Wiley & Sons, Inc.

Agresti, Alan and Barbara Finlay. 1997. *Statistical Methods for the Social Sciences*. Third ed. Upper Saddle River, New Jersey: Prentice Hall.

Aitchison, J. 1986. *The Statistical Analysis of Compositional Data*. London: Chapman & Hall.

Aitchison, J. and S. Silvey. 1957. "The generalization of probit analysis to the case of multiple responses." *Biometrika* 44:131–140.

Albert, James. 1992. "Bayesian estimation of normal ogive item response curves using Gibbs sampling." *Journal of Educational Statistics* 17:251–269.

Albert, James and Siddhartha Chib. 1993. "Bayesian analysis of binary and polychotomous response data." *Journal of the American Statistical Association* 88:669–79.

Aliprantis, Charalambos D. and Owen Burkinshaw. 1981. *Principles of Real Analysis*. New York: North Holland.

Altham, Patricia M. E. 1969. "Exact Bayesian analysis of a 2 $\times$ 2 contingency table, and Fisher's 'Exact' Significance Test." *Journal of the Royal Statistical Society, Series B* 31:261–269.

Alvarez, R. Michael and Jonathan Nagler. 1995. "Economics, issues and the Perot candidacy: Voter choice in the 1992 Presidential election." *American Journal of Political Science* 39:714–44.

Amemiya, Takeshi. 1985. *Advanced Econometrics*. Cambridge: Harvard University Press.

Anderson, T. W. 2003. *An Introduction to Multivariate Statistical Analysis*. Third ed. Hoboken, New Jersey: John Wiley & Sons, Inc.

Ardia, David, Lennart F. Hoogerheide and Herman K. van Dijk. 2008. *The AdMit Package: Adaptive Mixture of Student-t Distributions*. version 1-00.04. URL: http://cran.at.r-project.org/web/packages/AdMit/index.html

Arnold, B. C. and S. J. Press. 1989. "Compatible conditional distributions." *Journal of the American Statistical Association* 84:152–156.

Ashenfelter, Orley. 1994. "Report on expected absentee ballots." Typescript. Department of Economics, Princeton University.

Ashenfelter, Orley, Phillip Levine and David Zimmerman. 2003. *Statistics and Econometrics: Methods and Applications*. New York: John Wiley & Sons, Inc.

Baker, Frank B. and Seock-Ho Kim. 2004. *Item Response Theory: Parameter Estimation Techniques*. Second ed. New York: Dekker.

Barnett, Vic. 1999. *Comparative Statistical Inference*. Third ed. Chichester: John Wiley & Sons, Ltd.

Barone, Michael, Richard E. Cohen and Grant Ujifusa. 2002. *The Almanac of American Politics, 2002*. Washington, D.C.: National Journal Group.

Bartels, Larry M. 1991. "Constituency opinion and Congressional policy making: The Reagan defense buildup." *American Political Science Review* 85:457–474.

Bates, Douglas, Martin Maechler and Bin Dai. 2008. *lme4: Linear mixed-effects models using S4 classes*. R package version 0.999375-22. URL: http://lme4.r-forge.r-project.org/

Baum, Matthew A. and Samuel Kernell. 2001. "Economic class and popular support for Franklin Roosevelt in war and peace." *Public Opinion Quarterly* 65:198–229.

Bayes, Thomas. 1763. "An essay towards solving a problem in the doctrine of chances." *Philosophical Transactions of the Royal Society* 53:370–418.

Bayes, Thomas. 1958. "An essay towards solving a problem in the doctrine of chances." *Biometrika* 45:293–315.

Beck, Nathaniel. 1990. "Estimating dynamic models using Kalman filtering." In *Political Analysis*, ed. James A. Stimson. Vol. 1 Ann Arbor: University of Michigan Press pp. 121–156.

Beck, Nathaniel, Howard Rosenthal and Simon Jackman. 2006. "Presidential approval: the case of George W. Bush." Presented to the Annual Meeting of the Society for Political Methodology, University of California, Davis.

Bennett, James E., Amy Racine-Poon and Jon C. Wakefield. 1996. "MCMC for nonlinear hierarchical models." In *Markov Chain Monte Carlo in Practice*, ed. W. R. Gilks, S. Richardson and D. J. Spiegelhalter. London: Chapman & Hall.

Berger, James O. 1985. *Statistical Decision Theory and Bayesian Analysis*. Second ed. New York: Springer-Verlag.

Berger, James O. 2003. "Could Fisher, Jeffreys and Neyman have agreed on testing?" *Statistical Science* 18:1–12.

Berger, James O. and Thomas Sellke. 1987. "Testing a point null hypothesis: the irreconcilability of p values and evidence." *Journal of the American Statistical Association* 82:112–122.

Berk, Richard A., Bruce Western and Robert E. Weiss. 1995. "Statistical inference for apparent populations." *Sociological Methodology* 25:421–458.

Bernardo, José and Adrian F. M. Smith. 1994. *Bayesian Theory*. Chichester: John Wiley & Sons, Ltd.

Besag, J., P. J. Green, D. Higdon and K. Mengersen. 1995. "Bayesian computation and stochastic systems (with discussion)." *Statistical Science* 10:3–41.

Besag, Julian. 1974. "Spatial interaction and the statistical analysis of lattice systems (with discussion)." *Journal of the Royal Statistical Society, Series B* 41:143–168.

Besag, Julian and Peter J. Green. 1993. "Spatial statistics and Bayesian computation." *Journal of the Royal Statistical Society, Series B* 55:25–37.

Birkhoff, George D. 1931. "Proof of the Ergodic Theorem." *Proceedings of the National Academy of Sciences of the United States of America* 17:656–660.

Blackwell, D. and L. Dubins. 1962. "Merging of opinions with increasing information." *Annals of Mathematical Statistics* 33:882–886.

Bloch, Daniel A. and Geoffrey S. Watson. 1967. "A Bayesian study of the multinomial distribution." *Annals of Mathematical Statistics* 38:1423–1435.

Bock, R. Darrell and Murray Aitkin. 1981. "Marginal maximum likelihood estimation of item parameters: Application of an EM algorithm." *Psychometrika* 46:443–459.

Box, George E. P. and George C. Tiao. 1973. *Bayesian Inference in Statistical Analysis*. New York: John Wiley & Sons, Inc.

Bradley, Ian and Ronald L. Meek. 1986. *Matrices and Society*. Harmondsworth, Middlesex: Penguin.

Breiman, Leo. 1968. *Probability*. Reading, Massachusetts: Addison-Wesley.

Brémaud, Pierre. 1999. *Markov Chains: Gibbs Fields, Monte Carlo Simulation, and Queues*. New York: Springer.

Breusch, T. S., J. C. Robertson and A. H. Welsh. 1997. "The Emperor's New Clothes: A critique of the multivariate *t* regression model." *Statistica Neerlandica* 51:269–286.

Brooks, Stephen P. and Andrew Gelman. 1998. "General methods for monitoring convergence of iterative simulations." *Journal of Computational and Graphical Statistics* 7:434–455.

Brooks, Stephen P. and Gareth O. Roberts. 1998. "Assessing convergence of Markov chain Monte Carlo algorithms." *Statistics and Computing* 8:319–335.

Cameron, A. Colin and Pravin K. Trivedi. 1998. *Regression Analysis of Count Data*. Cambridge, United Kingdom: Cambridge University Press.

Carlin, Bradley P. and Thomas A. Louis. 2000. *Bayes and Empirical Bayes Methods for Data Analysis*. Second ed. London: CRC Press.

Carter, C. K. and R. Kohn. 1994. "On Gibbs sampling for state space models." *Biometrika* 81:541–553.

Casella, George. 1985. "An introduction to empirical Bayes data analysis." *The American Statistician* 39:83–87.

Casella, George and Edward I. George. 1992. "Explaining the Gibbs sampler." *The American Statistician* 46:167–74.

Casella, George and Roger L. Berger. 2002. *Statistical Inference*. Second ed. Pacific Grove, California: Duxbury.

Chacon, R. V. and Donald S. Ornstein. 1960. "A general ergodic theorem." *Illinois Journal of Mathematics* 4:153–160.

Chamberlain, Gary and Michael Rothschild. 1981. "A note on the probability of casting a decisive vote." *Journal of Econometric Theory* 25.

Chan, Kung Sik and Charles J. Geyer. 1994. 'Discussion of "Markov chains for exploring posterior distributions".' *The Annals of Statistics* 22:1747–1758.

Chib, Siddhartha and Edward Greenberg. 1995. "Understanding the Metropolis-Hastings algorithm." *The American Statistician* 49:327–335.

Chib, Siddhartha and Edward Greenberg. 1997. "Analysis of multivariate probit models." *Biometrika* 85:347–361.

Chib, Siddhartha, Federico Nardari and Neil Shephard. 2002. "Markov chain Monte Carlo methods for stochastic volatility models." *Journal of Econometrics* 108:281–316.

Clifford, P. 1993. "Discussion on the meeting on the Gibbs sampler and other Markov chain Monte Carlo methods." *Journal of the Royal Statistical Society, Series B* 55:53–102.

Clinton, Joshua D., Simon Jackman and Douglas Rivers. 2004a. "The statistical analysis of roll call data." *American Political Science Review* 98:355–370.

Clinton, Joshua D., Simon Jackman and Douglas Rivers. 2004b. "'The Most Liberal Senator'?: Analyzing and interpreting Congressional roll calls." *PS: Political Science and Politics* 37:805–811.

Cogburn, R. 1972. "The central limit theorem for Markov processes." In *Proceedings of the Sixth Berkeley Symposium on Mathematical Statistics and Probability, Volume 2: Probability Theory*, ed. Lucien M. Le Cam, Jerzy Neyman and Elizabeth L. Scott. University of California Press pp. 485–512.

Cole, Wade M. 2006. "When all else fails: International adjudication of human rights abuse claims, 1976-1999." *Social Forces* 84:1909–1935.

Congdon, Peter. 2003. *Applied Bayesian Modelling*. Chichester: John Wiley & Sons, Ltd.

Congdon, Peter. 2005. *Bayesian Models for Categorical Data*. Chichester: John Wiley & Sons, Ltd.

Congdon, Peter. 2007. *Bayesian Statistical Modelling*. Second ed. Chichester: John Wiley & Sons, Ltd.

Corley, Pamela C., Robert M. Howard and David C. Nixon. 2005. "The Supreme Court and opinion content: The use of the Federalist Papers." *Political Research Quarterly* 58:329–340.

Cowles, Mary K. 1996. "Accelerating Monte Carlo Markov chain convergence for cumulative-link generalized linear models." *Statistics and Computing* 6:101–111.

Cowles, Mary Kathryn and Bradley P. Carlin. 1996. "Markov Chain Monte Carlo convergence diagnostics: A comparative review." *Journal of the American Statistical Association* 91:883–904.

Damien, P., J. Wakefield and S. Walker. 1999. "Gibbs sampling for Bayesian non-conjugate and hierarchical models by using auxiliary variables." *Journal of the Royal Statistical Society, Series B* 61:331–344.

Daston, Lorraine. 1988. *Classical Probability in the Enlightenment*. Princeton: Princeton University Press.

Davis, Otto A., Melvin J. Hinich and Peter C. Ordeshook. 1970. "An expository development of a mathematical model of the electoral process." *American Political Science Review* 64:426–48.

de Finetti, B. 1931. "Funcione caratteristica di un fenomeno aleatorio." *Atti della Reale Accademia Nazionale dei Lincii* 4:251–299.

de Finetti, B. 1937. "La prévision: ses lois logiques, ses sources subjectives." *Annales de l'Institut Henri Poincaré* 7:1–68. Translated as "Foresight: Its logical laws, its subjective sources" in Kyburg, Henry E. and Howard E. Smokler (eds). (1980). *Studies in Subjective Probability*, Second ed. Krieger. Huntington, New York.

de Finetti, B. 1938. "Sur la condition de "équivalence partielle"." *Actualités Scientifiques et Industrielles* 739:5–18. Translated in R. Jeffrey (eds), *Studies in Inductive Logic and Probability*, Vol 2. Berkeley: University of California Press.

de Finetti, B. 1974, 1975. *Theory of Probability*. Chichester: John Wiley & Sons, Ltd. Volumes 1 and 2.

de Finetti, B. 1980a. "Foresight: Its logical laws, its subjective sources." In *Studies in Subjective Probability*, ed. Henry E. Kyburg and Howard E. Smokler. Second ed. Huntington, New York: Krieger. Reprinted from *Annales de l'Institut Henri Poincaré* (1937).

de Finetti, B. 1980b. "Probability: Beware of falsification." In *Studies in Subjective Probability*, ed. Henry E. Kyburg and Howard E. Smokler. Second ed. Huntington, New York: Krieger. Reprinted from *Scientia* (1977) 111:283–303.

DeGroot, M. H. and M. M. Rao. 1963. "Bayes estimation with convex loss." *Annals of Mathematical Statistics* 34:839–46.

Dellaportas, P. and Adrian F. M. Smith. 1993. "Bayesian inference for generalised linear and proportional hazards models via Gibbs sampling." *Applied Statistics* 42:443–460.

Delli Carpini, Michael X. and Scott Keeter. 1996. *What Americans Know about Politics and Why It Matters*. New Haven: Yale University Press.

Dempster, A. P., N. M. Laird and D. B. Rubin. 1977. "Maximum likelihood from incomplete data via the *EM* algorithm." *Journal of the Royal Statistical Society, Series B* 39:1–38.

Devroye, Luc. 1986. *Non-Uniform Random Variate Generation*. New York: Springer-Verlag.

Diaconis, Persi. 1977. "Finite forms of de Finetti's Theorem on exchangeability." *Synthese* 31:271–281.

Diaconis, Persi. 2005. "Exchangeability and de Finetti's Theorem." Lecture Notes. Stanford: Department of Statistics, Stanford University. http://www-stat.stanford.edu/cgates/PERSI/courses/stat_121/lectures/ex/.

Diaconis, Persi. 2009. "The Markov chain Monte Carlo revolution." *Bulletin of the American Mathematical Society* 46:179–205.

Diaconis, Persi and D. Ylvisaker. 1979. "Conjugate priors for exponential families." *Annals of Statistics* 7:269–281.

Diaconis, Persi and David Freedman. 1980a. "de Finetti's generalizations of exchangeability." In *Studies in Inductive Logic and Probability, Volume Two*, ed. Richard Jeffrey. Berkeley, California: University of California Press. Chapter 11.

Diaconis, Persi and David Freedman. 1980b. "Finite exchangeable sequences." *Annals of Probability* 8:745–764.

Diaconis, Persi and David Freedman. 1981. "Partial exchangeability and sufficiency." In *Proceedings of the Indian Statistical Institute Golden Jubilee International Conference on Statistics: Applications and New Directions*. Calcutta: Indian Statistical Institute pp. 205–236.

Diaconis, Persi and David Freedman. 1986a. "On the consistency of Bayes estimates (with discussion)." *Annals of Statistics* 14:1–67.

Diaconis, Persi and David Freedman. 1986b. "On inconsistent Bayes estimates of location." *Annals of Statistics* 14:68–87.

Diaconis, Persi, Kshitij Khare and Laurent Saloff-Coste. 2008. "Gibbs sampling, exponential families and orthogonal polynomials." *Statistical Science* 23:151–178.

Diaconis, Persi, M. L. Eaton and S. L. Lauritzen. 1992. "Finite de Finetti theorems in linear models and multivariate analysis." *Scandinavian Journal of Statistics* 19:289–316.

Dickey, James M. 1975. "Bayesian alternatives to the F-test and least squares estimates in the normal linear model." In *Studies in Bayesian Econometrics and Statistics*, ed. Stephen E. Fienberg and Arnold Zellner. Amsterdam: North-Holland pp. 515–554.

Dickey, James M. 1977. "Is the tail area useful as an approximate Bayes factor?" *Journal of the American Statistical Association* 72:138–142.

Durbin, James and Siem Jan Koopman. 2001. *Time Series Analysis by State Space Methods*. Oxford: Oxford University Press.

Eckhardt, Roger. 1987. "Stan Ulam, John von Neumann, and the Monte Carlo Method." *Los Alamos Science* Special Issue: 131–137.

Edwards, A. 2004. "Comment on Bellhouse, David R. 'The Reverand Thomas Bayes FRS: A biography to celebrate the tercentenary of his birth'." *Statistical Science* 19:34–37.

Edwards, R. G. and A. D. Sokal. 1988. "Generalization of the Fortuin-Kasteleyn-Swendsen-Wang representation and Monte Carlo algorithm." *Physical Review D* 38:2009–2012.

Efron, Bradley. 1986. "Why isn't everyone a Bayesian?" *The American Statistician* 40:1–11.

Efron, Bradley and Carl Morris. 1975. "Data analysis using Stein's estimator and its generalizations." *Journal of the American Statistical Association* 70:311–319.

Eicker, F. 1963. "Asymptotic normality and consistency of the least squares estimator of linear regressions." *Annals of Mathematical Statistics* 34:447–456.

Enelow, J. and Melvin J. Hinich. 1984. *The Spatial Theory of Voting: An Introduction*. New York: Cambridge University Press.

Engle, R. F., D. F. Hendry and J. F. Richard. 1983. "Exogeneity." *Econometrica* 51:277–304.

Engle, Robert and Mark Watson. 1980. "A time domain approach to dynamic factor and MIMIC models." *Les Cahiers du Seminaries d'Econometrie* 22:109–125.

Erikson, Robert S. 1990. "Roll calls, reputations, and representation in the U.S. Senate." *Legislative Studies Quarterly* 15:623–642.

Erikson, Robert S., Michael MacKuen and James A. Stimson. 2002. *The Macro Polity*. New York: Cambridge University Press.

Fair, R. 1990. "The effect of economic events on votes for the President: 1988 update." Cowles Foundation, Yale University.

Feller, W. 1968. *An Introduction to Probability Theory and its Applications*. Third ed. New York: John Wiley & Sons, Inc.

Fienberg, Stephen E. 2006. "When did Bayesian inference become 'Bayesian'?" *Bayesian Analysis* 1:1–40.

Fisher, R. A. 1922. "On the mathematical foundations of theoretical statistics." *Philosophical Transactions of the Royal Society, Series A* 222:309–368.

Fisher, R. A. 1935. "The logic of inductive inference (with discussion)." *Journal of the Royal Statistical Society* 98:39–82.

Fisher, R. A. 1950. *Contributions to Mathematical Statistics*. New York: John Wiley, & Sons, Inc.

Fisher, Ronald A. 1925. *Statistical Methods for Research Workers*. London: Oliver and Boyd.

Frühwirth-Schnatter, Sylvia. 1994. "Data augmentation and dynamic linear models." *Journal of Time Series Analysis* 15:183–202.

Frühwirth-Schnatter, Sylvia. 2006. *Finite mixture and Markov Switching Models*. New York: Springer.

Frühwirth-Schnatter, Sylvia and Rudolf Frühwirth. 2007. "Auxiliary mixture sampling with applications to logistic models." *Computational Statistics and Data Analysis* 51:3509–3528.

Galavotti, Maria Carla. 2005. *Philosophical Introduction to Probability*. Stanford, California: Center for the Study of Language and Information, Stanford University.

Gamerman, Dani and Hedibert F. Lopes. 2006. *Markov Chain Monte Carlo: Stochastic Simulation for Bayesian Inference*. Second ed. Chapman & Hall.

Geddes, Barbara. 1990. "How the cases you choose affect the answers you get: Selection bias in comparative politics." *Political Analysis* 2:131–150.

Gelfand, Alan E. and Adrian F. M. Smith. 1990. "Sampling based approaches to calculating marginal densities." *Journal of the American Statistical Association* 85:398–409.

Gelfand, Alan E., Susan E. Hills, Amy Racine-Poon and Adrian F. M. Smith. 1990. "Illustration of Bayesian inference in normal data models using Gibbs sampling." *Journal of the American Statistical Association* 85:972–985.

Gelman, Andrew. 2006. "Prior distributions for variance parameters in hierarchical models." *Bayesian Analysis* 1:514–534.

Gelman, Andrew, David A. van Dyk, Zaiying Huang and W. John Boscardin. 2007. "Using redundant parameters to fit hierarchical models." *Journal of Computational and Graphical Statistics* 17:95–122.

Gelman, Andrew and Donald B. Rubin. 1992. "Inference from iterative simulation using multiple sequences." *Statistical Sciences* 7:457–511.

Gelman, Andrew and Gary King. 1990. "Estimating the consequences of electoral redistricting." *Journal of the American Statistical Association* 85:274–82.

Gelman, Andrew, G. O. Roberts and W. R. Gilks. 1995. "Efficient Metropolis jumping rules." In *Bayesian Statistics*, ed. J. M. Bernardo, J. O. Berger, A. P. Dawid and A. F. M. Smith. Vol. 5 Oxford: Oxford University Press.

Gelman, Andrew and Jennifer Hill. 2007. *Data Analysis Using Regression and Multilevel/Hierarchical Models*. New York: Cambridge University Press.

Gelman, Andrew, John B. Carlin, Hal S. Stern and Donald B. Rubin. 2004. *Bayesian Data Analysis*. Second ed. Boca Raton, Florida: Chapman & Hall.

Geman, S. and D. Geman. 1984. "Stochastic relaxation, Gibbs distributions and the Bayesian restoration of images." *IEEE Transactions on Pattern Analysis and Machine Intelligence* 6:721–41.

Geweke, J. 1992. "Evaluating the accuracy of sampling-based approaches to the calculation of posterior moments (with discussion)." In *Bayesian Statistics 4*, ed. J. M. Bernardo, J. O. Berger, A. P. Dawid and A. F. M. Smith. Oxford: Oxford University Press pp. 169–193.

Geweke, John. 2005. *Contemporary Bayesian Econometrics and Statistics*. Hoboken, New Jersey: John Wiley & Sons, Inc.

Geweke, John, Michael Keane and David Runkle. 1994. "Alternative computational approaches to inference in the multinomial probit model." Staff Report, Research Department, Federal Reserve Bank of Minneapolis.

Geyer, Charles J. 2005. *mcmc: Markov Chain Monte Carlo*. R package version 0.5-1. URL: http://www.stat.umn.edu/geyer/mcmc/

Geyer, C. J. and E. A. Thompson. 1995. "Annealing Markov chain Monte Carlo with applications to ancestral inference." *Journal of the American Statistical Association* 90:909–920.

Gilks, W. R., D. G. Clayton, D. J. Spiegelhalter, N. G. Best, A. J. McNeil, L. D. Sharples and A. J. Kirby. 1993. "Modelling complexity: Applications of Gibbs sampling in medicine." *Journal of the Royal Statistical Society, Series B* 55:39–52.

Gilks, W. R. and P. Wild. 1992. "Adaptive rejection sampling for Gibbs sampling." *Applied Statistics* 41:337–348.

Gilks, W. R. 1992. "Derivative-free adaptive rejection sampling for Gibbs sampling." In *Bayesian Statistics 4*, ed. J. M. Bernardo, J. O. Berger, A. P. Dawid and A. F. M. Smith. Oxford: Clarendon pp. 641–649.

Gilks, W. R. 1996. "Full conditional distributions." In *Markov chain Monte Carlo in practice*, ed. W. R. Gilks, S. Richardson and D. J. Spiegelhalter. London: Chapman & Hall pp. 75–88.

Gilks, W. R., N. G. Best and K. K. C. Tan. 1995. "Adaptive rejection Metropolis sampling within Gibbs sampling." *Applied Statistics* 44:455–472.

Golombok, Susan and Fiona Tasker. 1996. "Do parents influence the sexual orientation of their children?: Findings From a Longitudinal Study of Lesbian Families." *Developmental Psychology* 32:3–11.

Good, I. J. 1988. "The interface between statistics and philosophy of science." *Statistical Science* 3:386–397.

Gowa, Joanne. 1998. "Politics at the water's edge: Parties, voters, and the use of force abroad." *International Organization* 52:307–324.

Graham, Alexander. 1987. *Nonnegative Matrices and Applicable Topics in Linear Algebra*. Chichester: Ellis Horwood.

Green, Donald P., Alan S. Gerber and Suzanna L. De Boef. 1999. "Tracking opinion over time: A method for reducing sampling error." *Public Opinion Quarterly* 63:178–192.

Green, Donald P., Bradley L. Palmquist and Eric Schickler. 1998. "Macropartisanship: A replication and critique." *American Political Science Review* 92:883–899.

Green, Donald P., Bradley L. Palmquist and Eric Schickler. 2002. *Partisan Hearts and Minds: Political Parties and the Social Identities of Voters*. New Haven, Conn.: Yale University Press.

Green, Donald P. and David H. Yoon. 2002. "Reconciling individual and aggregate evidence concerning partisan stability: Applying time series models to panel survey data." *Political Analysis* 10:1–24.

Green, Donald P. and Lynn Vavreck. 2008. "Analysis of cluster-randomized experiments: A comparison of alternative estimation approaches." *Political Analysis* 16:138–152.

Green, P. J. 1996. "MCMC in image analysis." In *Markov chain Monte Carlo in Practice*, ed. W. R. Gilks, S. Richardson and D. J. Spiegelhalter. London: Chapman & Hall pp. 381–400.

Guihenneuc-Jouyaux, C. and C. Robert. 1998. "Finite Markov chain convergence results and MCMC convergence assessment." *Journal of the American Statistical Association* 93:1055–1067.

Gurr, Ted Robert and Keith Jaggers. 1996. "Polity III: Regime change and political authority, 1800-1994." Computer file, Inter-university Consortium for Political and Social Research, Ann Arbor, MI.

Hacking, Ian. 1975. *The Emergence of Probability*. Cambridge: Cambridge University Press.

Hacking, Ian. 2001. *An Introduction to Probability and Inductive Logic*. Cambridge: Cambridge University Press.

Hall, A. 1873. "On an Experimental Determination of Pi." *The Messenger of Mathematics* 2.

Harvey, Andrew C. 1989. *Forecasting, Structural Time Series Models and the Kalman Filter*. New York: Cambridge University Press.

Hastings, W. K. 1970. "Monte Carlo sampling methods using Markov chains, and their applications." *Biometrika* 57:97–109.

Heath, David and William Sudderth. 1976. "De Finetti's Theorem on exchangeable variables." *The American Statistician* 30:188–189.

Heidelberger, P. and P. D. Welch. 1983. "Simulation run length control in the presence of an initial transient." *Operations Research* 31:1109–44.

Hewitt, Edwin and Leonard J. Savage. 1955. "Symmetric Measures on Cartesian products." *Transactions of the American Mathematical Society* 80:470–501.

Hibbs, Douglas. 1987. *The American Political Economy: Macroeconomics and Electoral Politics in the United States*. Cambridge: Harvard University Press.

Higdon, David M. 1998. "Auxiliary variable methods for Markov chain Monte Carlo with applications." *Journal of the American Statistical Association* 93:585–595.

Hill, Bruce M. 1965. "Inference about variance components in the one-way model." *Journal of the American Statistical Association* 60:806–825.

Hobert, James P. and George Casella. 1996. "The effect of improper priors on Gibbs sampling in hierarchical linear models." *Journal of the American Statistical Association* 91:1461–1473.

Hoff, Peter, Adrian E. Raftery and Mark S. Handcock. 2002. "Latent space approaches to social network analysis." *Journal of the American Statistical Association* 97:1090–1098.

Holmes, C. C. and L. Held. 2006. "Bayesian auxiliary variable models for binary and multinomial regression." *Bayesian Analysis* 1:145–168.

Hoogerheide, Lennart F., Johan F. Kaashoek and Herman K. van Dijk. 2007. "On the shape of posterior densities and credible sets in instrumental variable regression models with reduced rank: An application of flexible sampling methods using neural networks." *Journal of Econometrics* 139:154–180.

Horn, Roger A. and Charles R. Johnson. 1985. *Matrix Analysis*. Cambridge: Cambridge University Press.

Hosmer, David W. and Stanley Lemeshow. 2000. *Applied Logistic Regression*. Second ed. New York: John Wiley & Sons, Inc.

Howell, William G. and David E. Lewis. 2002. "Agencies by presidential design." *Journal of Politics* 64:1095–1114.

Howson, Colin and Peter Urbach. 1993. *Scientific Reasoning: the Bayesian Approach*. Second ed. Chicago: Open Court.

Huber, John and Ronald Inglehart. 1995. "Expert interpretations of party space and party locations in 42 societies." *Party Politics* 1:73–111.

Huber, Peter J. 1967. "The behavior of maximum likelihood estimation under nonstandard conditions." In *Proceedings of the Fifth Berkeley Symposium on Mathematical Statistics and Probability*, ed. L. M. Le Cam and J. Neyman. Vol. 1 Berkeley: University of California Press pp. 221–223.

Imai, Kosuke and David A. van Dyk. 2005a. "A Bayesian analysis of the multinomial probit model using marginal data augmentation." *Journal of Econometrics* 124:311–334.

Imai, Kosuke and David A. van Dyk. 2005b. "MNP: R package for fitting the multinomial probit model." *Journal of Statistical Software* 14:1–32.

Isaac, Larry and Lars Christiansen. 2002. "How the civil rights movement revitalized labor militancy." *American Sociological Review* 67:722–746.

Jackman, Simon. 1994. "Measuring electoral bias: Australia, 1949–1993." *British Journal of Political Science* 24:319–57.

Jackman, Simon. 2000. "Estimation and inference are missing data problems: Unifying social science statistics via Bayesian simulation." *Political Analysis* 8:307–332.

Jackman, Simon. 2001. "Multidimensional analysis of roll call data via Bayesian simulation: Identification, estimation, inference and model checking." *Political Analysis* 9:227–241.

Jackman, Simon. 2004. "What do we learn from graduate admissions committees?: A multiple-rater, latent variable model, with incomplete discrete and continuous indicators." *Political Analysis* 12:400–424.

Jackman, Simon. 2005a. "Incumbency advantage and candidate quality." In *Mortgage Nation: the 2004 Australian Election*, ed. Marian Simms and John Warhurst. Perth: API Network/Edith Cowan University Press.

Jackman, Simon. 2005b. "Pooling the polls over an election campaign." *Australian Journal of Political Science* 40:499–517.

Jackman, Simon. 2008a. "Measurement." In *The Oxford Handbook of Political Methodology*, ed. Janet Box-Steffensmeier, Henry Brady and David Collier. Oxford University Press.

Jackman, Simon. 2008b. *pscl: Classes and Methods for R Developed in the Political Science Computational Laboratory, Stanford University*. Stanford, California: Department of Political Science, Stanford University. R package version 1.00.

Jeffreys, H. 1961. *Theory of Probability*. Third ed. Oxford: Clarendon Press.

Johnson, Norman L., Samuel Kotz and N. Balakrishnan. 1994. *Continuous Univariate Distributions*. Vol. 1 second ed. New York: John Wiley & Sons, Inc.

Johnson, Norman L., Samuel Kotz and N. Balakrishnan. 1995. *Continuous Univariate Distributions*. Vol. 2 second ed. New York: John Wiley & Sons, Inc.

Johnson, Valen E. and James H. Albert. 1999. *Ordinal Data Modeling*. New York: Springer-Verlag.

Jones, Galin L. and James P. Hobert. 2001. "Honest exploration of intractable probability distributions via Markov chain Monte Carlo." *Statistical Science* 16:312–334.

Kadane, Joseph B. 1974. "The role of identification in Bayesian theory." In *Studies in Bayesian Econometrics and Statistics: In Honor of Leonard J. Savage*, ed. S. E. Fienberg and A. Zellner. Amsterdam: North-Holland pp. 175–191.

Karp, Jeffrey A. and Susan A. Banducci. 2001. "Absentee voting, mobilization, and participation." *American Politics Research* 29:183–195.

Katz, Jonathan N. and Gary King. 1999. "A statistical model for multiparty electoral data." *American Political Science Review* 93:15–32.

Keith, Bruce, Jenny Sundra Layne, Nicholas Babchuk and Kurt Johnson. 2002. "The context of scientific achievement: Sex status, organizational environments, and the timing of publication on scholarship outcomes." *Social Forces* 80:1253–1281.

Kendall, M. G. and A. Stuart. 1950. "The law of cubic proportion in election results." *The British Journal of Sociology* 1:183–196.

Kerman, Jouni. 2007. *Umacs: Universal Markov chain sampler*. R package version 0.924.

Kim, S., Neil Shephard and Siddhartha Chib. 1998. "Stochastic volatility: Likelihood inference and comparison with ARCH Models." *Review of Economic Studies* 65:361–393.

King, Gary and Andrew Gelman. 1991. "A unified model of evaluating electoral systems and redistricting plans." *American Journal of Political Science* 38:514–54.

King, Gary, Robert O. Keohane and Sidney Verba. 1994. *Designing Social Inquiry*. Princeton: Princeton University Press.

Kjellberg, Anders. 1983. *Facklig Organisering i Tolv Länder*. Lund: Arkiv.

Kleibergen, F. R. and E. Zivot. 2003. "Bayesian and classical approaches to instrumental variables regression." *Journal of Econometrics* 114:29–72.

Kleibergen, F. R. and Herman K. van Dijk. 1998. "Bayesian simultaneous equations analysis using reduced rank structures." *Econometric Theory* 14:701–743.

Knorr-Held, L. 1997. *Hierarchical Modelling of Discrete Longitudinal Data: Applications of Markov Chain Monte Carlo*. Munich: Herbert Utz Verlag.

Koch, Gary. 1983. "Intraclass correlation coefficient." In *Encyclopedia of Statistical Sciences*, ed. Samuel Kotz and Norman L. Johnson. New York: John Wiley & Sons, Inc.

Kolmogorov, A. N. 1933. *Grundbegriffe der Wahrscheinlichkeitsrechnung*. Berlin: Springer.

Kolmogorov, A. N. 1956. *Information Theory and Statistics*. New York: Chelsea.

Krauth, Werner. 2006. *Statistical Mechanics: Algorithms and Computations*. Oxford: Oxford University Press.

Krehbiel, Keith and Douglas Rivers. 1988. "The analysis of committee power: An application to senate voting on the minimum wage." *American Journal of Political Science* 32:1151–1174.

Kubrin, Charis E. and Ronald Weitzer. 2003. "Retaliatory homicide: Concentrated disadvantage and neighborhood culture." *Social Problems* 50:157–180.

Ladha, Krishna. 1991. "A spatial model of voting with perceptual error." *Public Choice* 78:43–64.

Lancaster, Tony. 2000. "The incidental parameter problem since 1948." *Journal of Econometrics* 95:391–413.

Lancaster, Tony. 2004. *An Introduction to Modern Bayesian Econometrics*. London: Blackwell.

Landau, David P. and Kurt Binder. 2000. *A Guide to Monte Carlo Simulations in Statistical Physics*. New York: Cambridge University Press.

Laplace, Pierre-Simon. 1774. "Mémoire sur la probabilité des causes par les évènemens." *Mémoires de mathématique et de physique presentés à l'Académie royale des sciences, par divers savs & lûs dans ses assemblées* 6:621–656. Reprinted in *Oeuvres complètes de Laplace*. Gauthier-Villars: Paris.

Laplace, Pierre-Simon. 1825. *Essai philosophique sur les probabilités*. Fifth ed. Paris: Bachelier. English translation by Andrew I. Dale, with notes. Springer-Verlag: New York. 1995.

Lawley, D. N. and A. E. Maxwell. 1971. *Factor Analysis as a Statistical Method*. Second ed. London: Butterworths.

Leamer, Edward. 1978. *Specification Searches: Ad Hoc Inference with Nonexperimental Data*. New York: John Wiley & Sons, Inc.

Lee, Peter M. 2004. *Bayesian Statistics: An Introduction*. Third ed. London: Hodder Arnold.

Lee, Sik-Yum. 2007. *Structural Equation Modeling: A Bayesian Approach*. Chichester, England: John Wiley & Sons, Ltd.

Leonard, Thomas and John S. J. Hsu. 1999. *Bayesian Methods*. Cambridge: Cambridge University Press.

Levendusky, Matthew S., Jeremy C. Pope and Simon Jackman. 2008. "Measuring district preferences with implications for the study of U.S. elections." *Journal of Politics* 70:736–753.

Levin, David A., Yuval Peres and Elizabeth L. Wilmer. 2008. *Markov Chains and Mixing Times*. Providence, Rhode Island: American Mathematical Society.

Lewis-Beck, Michael and Tom Rice. 1983. "Localism in presidential elections: The home-state advantage." *American Journal of Political Science* 27:548–556.

Lewis-Beck, Michael and Tom Rice. 1992. *Forecasting Elections*. Washington, D.C.: Congressional Quarterly Press.

Lindley, Dennis V. 1964. "The Bayesian analysis of contingency tables." *Annals of Mathematical Statistics* 35:1622–1643.

Lindley, Dennis V. 1965. *Introduction to Probability and Statistics from a Bayesian Viewpoint*. Cambridge: Cambridge University Press.

Lindley, Dennis V. 1985. *Making Decisions*. Second ed. London: John Wiley & Sons, Ltd.

Lindley, Dennis V. and L. D. Phillips. 1976. "Inference for a Bernoulli process (a Bayesian view)." *The American Statistician* 30:112–119.

Lindley, Dennis V. and Melvin R. Novick. 1981. "The role of exchangeability in inference." *Annals of Statistics* 9:45–58.

Little, Roderick J. A. and Donald B. Rubin. 2002. *Statistical Analysis with Missing Data*. Second ed. New York: John Wiley & Sons, Inc.

Liu, J. S. and Y. N. Wu. 1999. "Parameter expansion scheme for data augmentation." *Journal of the American Statistical Association* 94:1264–1274.

Liu, J. S. 1996. "Peskun's theorem and a modified discrete-state Gibbs sampler." *Biometrika* 83:681–682.

Liu, Jun S. 2001. *Monte Carlo Strategies in Scientific Computing*. New York: Springer.

Londregan, John. 2000. *Legislative Institutions and Ideology in Chile's Democratic Transition*. New York: Cambridge University Press.

Long, J. Scott. 1990. "The origins of sex differences in science." *Social Forces* 68:1297–1316.

Long, J. Scott. 1997. *Regression Models for Categorical and Limited Dependent Variables*. Thousand Oaks, California: Sage.

Lumley, Thomas. 2004. "Analysis of complex survey samples." *Journal of Statistical Software* 9:1–19.

Magnus, Jan R. and Heinz Neudecker. 1999. *Matrix Differential Calculus with Applications in Statistics and Econometrics*. Hoboken: John Wiley & Sons, Inc.

Magnusson, Arni and Ian Stewart. 2007. *scapeMCMC: MCMC diagnostic plots*. R package version 1.0-3. URL: http://students.washington.edu/arnima/s/

Mardia, Kanti V., John T. Kent and John M. Bibby. 1979. *Multivariate Analysis*. San Diego: Academic Press.

Marsaglia, G. 1977. "The squeeze method for generating gamma variables." *Computers and Mathematics with Applications* 3:321–325.

Martin, Andrew D. and Kevin M. Quinn. 2002. "Dynamic ideal point estimation via Markov chain Monte Carlo for the U.S. Supreme Court, 1953-1999." *Political Analysis* 10:134–153.

Martin, Andrew D., Kevin M. Quinn and Jong Hee Park. 2009. *MCMCpack: Markov Chain Monte Carlo (MCMC) Package*. R package version 0.9-6. URL: http://mcmcpack.wustl.edu

Matsumoto, Makoto and Takuji Nishimura. 1998. "Mersenne twister: a 623-dimensionally equidistributed uniform pseudo-random number generator." *ACM Trans. Model. Comput. Simul.* 8:3–30.

McCullagh, P. and J. A. Nelder. 1989. *Generalized Linear Models*. Second ed. London: Chapman & Hall.

McCulloch, Robert E., Nicholas G. Polson and Peter E. Rossi. 1998. "A Bayesian analysis of the multinomial probit model with fully identified parameters." Typescript. Graduate School of Business, University of Chicago.

McCulloch, Robert E. and Peter E. Rossi. 1994. "An exact likelihood analysis of the multinomial probit model." *Journal of Econometrics* 64:207–40.

McFadden, Daniel. 1974. "Conditional logit analysis of qualitative choice behavior." In *Frontiers in Econometrics*, ed. P. Zarembka. New York: Academic Press pp. 105–142.

McKelvey, Richard and William. Zavoina. 1975. "A statistical model for the analysis of ordinal level variables." *Journal of Mathematical Sociology* 4:103–120.

McLachlan, Geoffrey J. and David Peel. 2000. *Finite mixture models*. New York: John Wiley & Sons, Inc.

Meng, Xiao-Li and David A. van Dyk. 1997. "The *EM* algorithm – an old folk song sung to a fast new tune (with discussion)." *Journal of the Royal Statistical Society Series B* 59:511–567.

Meng, Xiao-Li and David A. van Dyk. 1999. "Seeking efficient data augmentation schemes via conditional and marginal augmentation." *Biometrika* 86:301–320.

Metropolis, N., A. W. Rosenbluth, M. N. Rosenbluth, A. H. Teller and E. Teller. 1953. "Equations of state calculations by fast computing machines." *Journal of Chemical Physics* 21:1087–91.

Metropolis, N. and S. Ulam. 1949. "The Monte Carlo method." *Journal of the American Statistical Association* 44:335–341.

Meyn, S. P. and R. L. Tweedie. 1993. *Markov Chains and Stochastic Stability*. London: Springer-Verlag.

Meyn, S. P. and R. L. Tweedie. 2009. *Markov Chains and Stochastic Stability*. Second ed. New York: Cambridge University Press.

Miller, Warren E. and Donald E. Stokes. 1963. "Constituency influence in Congress." *American Political Science Review* 57:45–56.

Miller, Warren E., Donald R. Kinder, Steven J. Rosenstone and the National Election Studies. 1999. *National Election Studies, 1992: Pre-/Post-Election Study*. Ann Arbor, Michigan: Center for Political Studies, University of Michigan.

Mira, Antonietta and Luke Tierney. 2002. "Efficiency and convergence properties of slice samplers." *Scandinavian Journal of Statistics* 29:1–12.

Morris, Martina. 1993. "Telling tails explain the discrepancy in sexual partner reports." *Nature* 365:437–440.

Murray, G. D. 1977. 'Comment on "Maximum Likelihood from Incomplete Data Via the EM Algorithm" by A. P. Dempster, N. M. Laird, and D. B. Rubin.' *Journal of the Royal Statistical Society, Series B* 39:27–28.

Nagler, Jonathan. 1991. "The effect of registration laws and education on U.S. voter turnout." *American Political Science Review* 85:1393–1405.

Natarajan, R. and R. E. Kass. 2000. "Reference Bayesian methods for generalized linear mixed models." *Journal of the American Statistical Association* 95:227–237.

Neal, Radford. 2003. "Slice sampling (with discussion)." *Annals of Statistics* 31:705–767.

Neal, Radford. 1997. Markov chain Monte Carlo methods based on 'slicing' the density function. Technical Report No. 9722. Department of Statistics, University of Toronto.

Neyman, J. and E. L. Scott. 1948. "Consistent estimates based on partially consistent observations." *Econometrica* 16:1–32.

Niemi, Richard G. and Patrick Fett. 1986. "The swing ratio: an explanation and an assessment." *Legislative Studies Quarterly* 11:75–90.

Niemi, Richard G. and Simon Jackman. 1991. "Bias and responsiveness in state legislative districting." *Legislative Studies Quarterly* 16:183–202.

Norris, J. R. 1997. *Markov Chains*. Cambridge: Cambridge University Press.

Nummelin, E. 1984. *General Irreducible Markov Chains and Non-Negative Operators*. Cambridge: Cambridge University Press.

Nurminen, Markku and Pertti Mutanen. 1987. "Exact Bayesian analysis of two proportions." *Scandinavian Journal of Statistics* 14:67–77.

O'Hagan, Anthony. 2004. *Bayesian Inference*. Vol. 2B of *Kendall's Advanced Theory of Statistics* Second ed. London: Arnold.

Oliver, J. Eric. 1996. "The effects of eligibility restrictions and party activity on absentee voting and overall turnout." *American Journal of Political Science* 40:498–513.

Patterson, H. D. and R. Thompson. 1971. "Recovery of inter-block information when block sizes are unequal." *Biometrika* 58:545–554.

Patterson, Samuel. C. and Gregory A. Caldeira. 1985. "Mailing in the vote: Correlates and consequences of absentee voting." *American Journal of Political Science* 29:766–787.

Patz, Richard J. and Brian W. Junker. 1999. "A straightforward approach to Markov chain Monte Carlo methods for item response models." *Journal of Educational and Behavioral Statistics* 24:146–178.

Petris, Giovanni and Luca Tardella. 2006. *HI: Simulation from distributions supported by nested hyperplanes*. R package version 0.3.

Pham-Gia, Thu and Noyan Turkkan. 1993. "Bayesian analysis of the difference of two proportions." *Communications in Statistics – Theory and Methods* 22:1755–1771.

Pham-Gia, Thu and Noyan Turkkan. 2002. "The product and quotient of general beta distributions." *Statistical Papers* 43:537–550.

Plummer, Martyn. 2009a. "JAGS Version 1.03 manual." Typescript.

Plummer, Martyn. 2009b. *rjags: Bayesian graphical models using MCMC*. R package version 1.0.3-8. URL: http://mcmc-jags.sourceforge.net

Plummer, Martyn, Nicky Best, Kate Cowles and Karen Vines. 2008. *Coda: Output Analysis and Diagnostics for MCMC*. R package version 0.13-2.

Poisson, S. D. 1837. *Recherches sur la probabilité des jugements en matière criminelle et en matière civile, précédées des regles générales du calcul des probabilitiés*. Paris: Bachelier, Imprimeur-Librarie pour les Mathematiques, la Physique, etc.

Poole, Keith, Jeffrey Lewis, James Lo and Royce Carroll. 2007. *wnominate: WNOMINATE Roll Call Analysis Software*. R package version 0.93.

Poole, Keith T. 2005. *Spatial Models of Parliamentary Voting*. New York: Cambridge University Press.

Poole, Keith T. and Howard Rosenthal. 1997. *Congress: A Political-Economic History of Roll Call Voting*. New York: Oxford University Press.

Press, S. James. 2003. *Subjective and Objective Bayesian Statistics*. Second ed. Hoboken, New Jersey: John Wiley & Sons, Inc.

Pryor, Frederic. 1973. *Property and Industrial Organization in Communist and Capitalist Countries*. Bloomington: Indiana University Press.

Przeworksi, Adam and Henry Tuene. 1970. *The Logic of Comparative Social Inquiry*. New York: Wiley.

Quinn, Kevin. 2004. "Bayesian factor analysis for mixed ordinal and continuous responses." *Political Analysis* 12:338–353.

R Development Core Team. 2009. *R: A Language and Environment for Statistical Computing*. Vienna, Austria: R Foundation for Statistical Computing. ISBN 3-900051-07-0. URL: http://www.R-project.org

Rachev, Svetlozar T., John S. Hsu, Biliana S. Bagasheva and Frank J. Fabozzi. 2008. *Bayesian Methods in Finance*. Hoboken, New Jersey: John Wiley & Sons, Inc.

Raftery, Adrian E. and S. M. Lewis. 1992a. "How many iterations of the Gibbs sampler?" In *Bayesian Statistics 4*, ed. J. M. Bernardo, J. Berger, A. P. Dawid and A. F. M. Smith. Oxford: Oxford University Press pp. 641–49.

Raftery, Adrian E. and S. M. Lewis. 1992b. "One long run with diagnostics: Implementation strategies for Markov chain Monte Carlo." *Statistical Science* 7:493–497.

Raftery, Adrian E. and Steven M. Lewis. 1996. "Implementing MCMC." In *Markov Chain Monte Carlo in Practice*, ed. W. R. Gilks, S. Richardson and D. J. Spiegelhalter. London: Chapman & Hall pp. 115–130.

Ramsey, F. P. 1931. "Truth and probability." In *The Foundations of Mathematics and other Logical Essays*, ed. R. B. Braithwaite. London: Humanities Press. Written in 1926.

Rao, C. Radhakrishna. 1973. *Linear Statistical Inference and Its Applications*. Second ed. New York: John Wiley & Sons, Inc.

Raudenbush, Stephen W. and Anthony S. Bryk. 2002. *Hierarchical Linear Models*. Second ed. Newbury Park, California: Sage.

Reckase, Mark D. 1997. "The past and future of multidimensional item response theory." *Applied Psychological Measurement* 21:25–36.

Rényi, A. 1970. *Probability Theory*. New York: American Elsevier.

Revuz, D. 1975. *Markov Chains*. Amsterdam: North Holland.

Ripley, Brian. 1987. *Stochastic Simulation*. New York: John Wiley & Sons, Inc.

Rivers, Douglas. 2003. "Identification of multidimensional item-response models." Typescript. Department of Political Science, Stanford University.

Robert, Christian P. 1994. 'Discussion of "Markov chains for exploring posterior distributions".' *The Annals of Statistics* 22:1742–1747.

Robert, Christian P. 1995. "Simulation of truncated normal variables." *Statistics and Computing* 5:121–125.

Robert, Christian P. 1996. "Mixtures of distributions: inference and estimation." In *Markov Chain Monte Carlo in Practice*, ed. W. R. Gilks, S. Richardson and D. J. Spiegelhalter. London: Chapman & Hall pp. 441–464.

Robert, Christian P. 2001. *The Bayesian Choice*. Second ed. New York: Springer.

Robert, Christian P. and George Casella. 2004. *Monte Carlo Statistical Methods*. Second ed. New York: Springer.

Roberts, Gareth O. 1996. "Markov chain concepts related to sampling algorithms." In *Markov Chain Monte Carlo in Practice*, ed. W. R. Gilks, S. Richardson and D. J. Spiegelhalter. London: Chapman & Hall pp. 45–57.

Roberts, Gareth O. and Adrian F. M. Smith. 1994. "Simple conditions for the convergence of the Gibbs sampler and Metropolis-Hastings algorithms." *Stochastic Processes and Their Applications* 49:207–216.

Roberts, Gareth O., Andrew Gelman and W. R. Gilks. 1997. "Weak convergence and optimal scaling of random walk Metropolis algorithms." *Annals of Applied Probability* 7:110–120.

Roberts, Gareth O. and Jeffrey S. Rosenthal. 1998. "Markov chain Monte Carlo: some practical implications of theoretical results (with discussion)." *Canadian Journal of Statistics* 26:5–32.

Roberts, Gareth O. and Jeffrey S. Rosenthal. 1999. "Convergence of slice sampler Markov chains." *Journal of the Royal Statistical Society, Series B* 61:643–60.

Roberts, Gareth O. and Jeffrey S. Rosenthal. 2002. "The polar slice sampler." *Stochastic Models* 18:257–280.

Rodriguez, Paulino Perez. 2007. *ars: Adaptive Rejection Sampling*. R package version 0.2; original C++ code from Arnost Komarek.

Rogosa, David R. and H. M. Saner. 1995. "Longitudinal data analysis examples with random coefficient models." *Journal of Educational and Behavioral Statistics* 20:149–170.

Rosenthal, Jeffrey S. 2006. *A First Look at Rigorous Probability Theory*. Second ed. Singapore: World Scientific.

Rossi, Peter and Rob McCulloch. 2008. *bayesm: Bayesian Inference for Marketing/Micro-econometrics*. R package version 2.2-2. URL: http://faculty.chicagogsb.edu/peter.rossi/research/bsm.htm

Rothenberg, Thomas J. 1971. "Identification in parametric models." *Econometrica* 39:577–591.

Rubin, Donald B. 1976. "Inference and missing data." *Biometrika* 63:581–592.

Rubin, Donald B. 1987. *Multiple Imputation for Nonresponse in Surveys*. New York: John Wiley & Sons, Inc.

Savage, L. J. 1961. "The subjective basis of statistical practice." Manuscript. University of Michigan.

Savage, L. J. 1962. "Bayesian statistics." In *Recent Developments in Information and Decision Processes*, ed. Robert E. Machol and Paul Gray. New York: Macmillan and Co.

Searle, Shayle R. 1982. *Matrix Algebra Useful for Statistics*. New York: John Wiley & Sons, Inc.

Segré, Emilio. 1980. *From X-Rays to Quarks: Modern Physicists and Their Discoveries*. San Francisco: W. H. Freeman and Company. Translation of *Personaggi e scoperte nella fisica contemporanea*.

Sekhon, Jasjeet S. 2005. "Making inference from 2 × 2 tables: The inadequacy of the Fisher exact test for observational data and a Bayesian alternative." Typescript. Survey Research Center, University of California, Berkeley.

Seneta, E. 1981. *Non-negative Matrices and Markov Chains*. New York: Springer-Verlag.

Sinclair, D. E. and R. Michael Alvarez. 2004. "Who overvotes, who undervotes, using punchcards?: Evidence from Los Angeles County." *Political Research Quarterly* 57:15–25.

Sing, Tobias, Oliver Sander, Niko Beerenwinkel and Thomas Lengauer. 2005. "ROCR: visualizing classifier performance in R." *Bioinformatics* 21:3940–3941.

Skocpol, Theda. 1979. *States and Social Revolutions: A Comparative Analysis of France, Russia, and China*. Cambridge: Cambridge University Press.

Smith, Adrian F. M. 1984. "Present positions and potential developments: Some personal views: Bayesian statistics." *Journal of the Royal Statistical Society, Series A* 147:245–259. Special issue commemorating the 150th Anniversary of the Royal Statistical Society.

Smith, Adrian F. M. and G. O. Roberts. 1993. "Bayesian computation via the Gibbs Sampler and related Markov chain Monte Carlo methods (with discussion)." *Journal of the Royal Statistical Society, Series B* 55:3–23.

Smith, Brian J. 2007. "boa: An R Package for MCMC output convergence assessment and posterior inference." *Journal of Statistical Software* 21:1–37.

Spiegelhalter, David J., Andrew Thomas, Nicky G. Best and W. R. Gilks. 1996. *BUGS: Bayesian Inference Using Gibbs Sampling, Version 0.5 (version ii)*. Cambridge, UK: MRC Biostatistics Unit.

Spiegelhalter, David J., Andrew Thomas, Nicky G. Best and Dave Lunn. 2003. *WinBUGS User Manual Version 1.4*. Cambridge, UK: MRC Biostatistics Unit.

Spiegelhalter, David J., Andrew Thomas and Nicky G. Best. 1996. "Computation on Bayesian graphical models." In *Bayesian Statistics 5*, ed. J. M. Bernardo, J. O. Berger, A. P. Dawid and Adrian F. M. Smith. Oxford University Press pp. 407–425.

Spiegelhalter, David J. and S. L. Lauritzen. 1990. "Sequential updating of conditional probabilities on directed graphical structures." *Networks* 20:579–605.

Stein, C. 1955. Inadmissibility of the usual estimator for the mean of a multivariate normal distribution. In *Proceedings of the Third Berkeley Symposium of Mathematical Statistics and Probability*. Vol. 1 Berkeley: University of California Press pp. 197–206.

Stephens, John D. 1979. *The Transition from Capitalism to Socialism*. Urbana: University of Illinois Press.

Stephens, John D. and Michael Wallerstein. 1991. "Industrial concentration, country size and trade union membership." *American Political Science Review* 85:941–953.

Stigler, Stephen M. 1982. "Thomas Bayes' Bayesian inference." *Journal of the Royal Statistical Society, Series A* 145:250–258.

Stigler, Stephen M. 1986a. *The History of Statistics: The Measurement of Uncertainty before 1900*. Cambridge, Massachusetts: Belknap Press of Harvard University Press.

Stigler, Stephen M. 1986b. "Laplace's 1774 memoir on inverse probability." *Statistical Science* 1:359–378.

Stigler, Stephen M. 1999. *Statistics on the Table: The History of Statistical Concepts and Methods*. Cambridge: Harvard University Press.

Stimson, James A. 1991. *Public Opinion in America: Moods, Cycles, and Swings*. Boulder: Westview.

Stimson, James A. 2004. *Tides of Consent: How Public Opinion Shapes American Politics*. New York: Cambridge University Press.

Stimson, James A., Michael B. Mackuen and Robert S. Erikson. 1995. "Dynamic representation." *American Political Science Review* 89:543–565.

Sturtz, Sibylle, Uwe Ligges and Andrew Gelman. 2005. "R2WinBUGS: A Package for Running WinBUGS from R." *Journal of Statistical Software* 12:1–16. URL: http://www.jstatsoft.org

Takane, Yoshio and Jan de Leeuw. 1987. "On the relationship between item response theory and factor analysis of discretized variables." *Psychometrika* 52:393–408.

Tanner, Martin A. 1996. *Tools for Statistical Inference: Methods for the Exploration of Postrior Distributions and Likelihood Functions*. Third ed. New York: Springer-Verlag.

Tanner, Martin A. and Wing Hung Wong. 1987. "The calculation of posterior distributions by data augmentation." *Journal of the American Statistical Association* 82:528–40.

Thomas, Andrew, David J. Spiegelhalter and W. R. Gilks. 1992. "BUGS: A program to perform Bayesian inference using Gibbs sampling." In *Bayesian Statistics 4*, ed. J. M. Bernardo, J. O. Berger, A. P. Dawid and Adrian F. M. Smith. Oxford: Clarendon Press pp. 837–842.

Thompson, W. A. 1962. "The problem of negative estimates of variance components." *Annals of Mathematical Statistics* 33:273–289.

Tierney, Luke. 1991. "Exploring posterior distributions using Markov chains." In *Computer Science and Statistics: Proceedings of the 23rd Symposium on the Interface*, ed. E. Keramidas. Fairfax Station: Interface Foundation pp. 563–570.

Tierney, Luke. 1994. "Markov chains for exploring posterior distributions (with discussion)." *Annals of Statistics* 22:1701–62.

Tierney, Luke. 1996. "Introduction to general state-space Markov chain theory." In *Markov Chain Monte Carlo in Practice*, ed. W. R. Gilks, S. Richardson and D. J. Spiegelhalter. London: Chapman & Hall pp. 59–74.

Tierney, Luke. 1997. "Markov chain Monte Carlo algorithms." In *Encyclopedia of the Statistical Sciences*, ed. Samuel Kotz, Campbell B. Read and David L. Banks. Vol. 1 (Update) New York: John Wiley & Sons, Inc. pp. 392–399.

Tierney, Luke and Antonietta Mira. 1999. "Some adaptive Monte Carlo methods for Bayesian inference." *Statistics in Medicine* 18:2507–15.

Titterington, D. M, Adrian F. M. Smith and U. E. Makov. 1985. *Statistical Analysis of Finite Mixture Distributions*. New York: John Wiley & Sons, Inc.

Todhunter, I. 1865. *A History of the Mathematical Theory of Probability*. London: Macmillan. Reprinted by Chelsea, New York, 1962.

Tomz, Michael, Joshua A. Tucker and Jason Wittenberg. 2002. "An easy and accurate regression model for multiparty electoral data." *Political Analysis* 10:66–83.

Tong, Y. L. 1990. *The Multivariate Normal Distribution*. New York: Springer-Verlag.

Treier, Shawn and Simon Jackman. 2008. "Democracy as a latent variable." *American Journal of Political Science* 52:201–217.

Tufte, Edward R. 1973. "The relationship between seats and votes in two-party systems." *American Political Science Review* 67:540–554.

van der Linden, W. J. and R. K. Hambleton, eds. 1997. *Handbook of Modern Item Response Theory*. New York: Springer-Verlag.

van der Vaart, A. W. 1998. *Asymptotic Statistics*. Cambridge, United Kingdom: Cambridge University Press.

van Dyk, David A. and Xiao-Li Meng. 2001. "The art of data augmentation (with discussion)." *Journal of Computational and Graphical Statistics* 10:1–111.

van Schur, Wijbrandt H. 1992. "Nonparametric unidimensional unfolding for multicategory data." In *Political Analysis*, ed. John R. Freeman. Vol. 4 Ann Arbor: University of Michigan Press pp. 41–74.

Venables, William N. and Brian D. Ripley. 2002. *Modern Applied Statistics with S-PLUS*. Fourth ed. New York: Springer-Verlag.

Venn, J. 1866. *The Logic of Chance*. London: Macmillan. Reprinted by Chelsea, New York, 1962.

von Mises, R. 1957. *Probability, Statistics, and Truth*. New York: Academic Press.

von Neumann, J. 1951. "Various techniques used in connection with random digits." *National Bureau of Standards Applied Mathematics Series* 12:36–38.

Wallerstein, Michael. 1989. "Union organization in advanced industrial democracies." *American Political Science Review* 83:481–501.

Wang, C. S., J. J. Rutledge and D. Gianola. 1993. "Marginal inferences about variance components in a mixed linear model using Gibbs sampling." *Genetique, Selection, Evolution* 25:41–62.

Wansbeek, Tom and Erik Meijer. 2000. *Measurement Error and Latent Variables in Econometrics*. Amsterdam: North-Holland.

Watson, Mark W. and Robert F. Engle. 1983. "Alternative algorithms for the estimation of dynamic factor, MIMIC, and time-varying coefficient models." *Journal of Econometrics* 15:385–400.

Weisberg, Herbert I. 1972. "Bayesian comparison of two ordered multinomial populations." *Biometrics* 28:859–867.

West, M. 1995. "Some statistical issues in paleoclimatology (with discussion)." In *Bayesian Statistics*, ed. J. M. Bernardo, J. O. Berger, A. P. Dawid and A. F. M. Smith. Oxford: Oxford University Press pp. 461–484.

West, Mike and Jeff Harrison. 1997. *Bayesian Forecasting and Dynamic Models*. New York: Springer-Verlag.

Western, Bruce. 1998. "Causal heterogeneity in comparative research: A Bayesian hierarchical modelling approach." *American Journal of Political Science* 42:1233–1259.

Western, Bruce and Simon Jackman. 1994. "Bayesian inference for comparative research." *American Political Science Review* 88:412–23.

White, H. 1980. "A heteroskedastic consistent covariance matrix estimator and a direct test of heteroskedasticity." *Econometrica* 48:817–838.

Wichura, Michael J. 1988. "Algorithm AS 241: The percentage points of the normal distribution." *Applied Statistics* 37:477–484.

Wild, P. and W. R. Gilks. 1993. "Algorithm AS 287: Adaptive rejection sampling from log-concave density functions." *Applied Statistics* 42:701–709.

Wilensky, Harold L. 1981. "Leftism, Catholicism, democratic corporatism: The role of political parties in recent welfare state development." In *The Development of Welfare States in Europe and America*, ed. Peter Flora and Arnold J. Heidenheimer. New Brunswick: Transaction Books.

Wiley, David E. and James A. Wiley. 1970. "The estimation of measurement error in panel data." *American Sociological Review* 35:112–117.

Williams, David. 2001. *Weighing the Odds: A Course in Probability and Statistics*. Cambridge, United Kingdom: Cambridge University Press.

Wood, G. R. 1992. "Binomial mixtures and finite exchangeability." *Annals of Probability* 20:1167–1173.

X-5 Monte Carlo Team. 2003. MNCP – A general Monte Carlo N Particle transport code, Version 5. User Manual. No. LA-UR-03-1987. Los Alamos National Laboratory. Volume 1: Overview and Theory. Revised 10/3/05.

Zabell, Sandy L. 1988. "Buffon, Price, and Laplace: Scientific attribution in the 18th Century." *Archive for History of Exact Sciences* 39:173–181.

Zabell, Sandy L. 1989a. "R. A. Fisher on the history of inverse probability." *Statistical Science* 4:247–263.

Zabell, Sandy L. 1989b. "The rule of succession." *Erkenntnis* 31:283–321.

Zehna, Peter W. 1966. "Invariance of maximum likelihood estimators." *Annals of Mathematical Statistics* 37:744.

Zellner, Arnold. 1962. "An efficient method of estimating seemingly unrelated regressions and tests for aggregation bias." *Journal of the American Statistical Association* 57:348–368.

Zellner, Arnold. 1971. *An Introduction to Bayesian Inference in Econometrics*. New York: Wiley.

Zellner, Arnold and Peter E. Rossi. 1984. "Bayesian analysis of dichotomous quantal response models." *Journal of Econometrics* 25:365–393.

Topic Index

Author Index

WILEY SERIES IN PROBABILITY AND STATISTICS

Established by WALTER A. SHEWHART AND SAMUEL S. WILKS

Editors: *David J. Balding, Noel A. C. Cressie, Garrett M. Fitzmaurice, Harvey Goldstein, Geert Molenberghs, David W. Scott, Adrian F.M. Smith, Ruey S. Tsay, Sanford Weisberg*

Editors Emeriti: *Vic Barnett, J. Stuart Hunter, David G. Kendall, Jozef L. Teugels*

The **Wiley Series in Probability and Statistics** is well established and authoritative. It covers many topics of current research interest in both pure and applied statistics and probability theory. Written by leading statisticians and institutions, the titles span both state-of-the-art developments in the field and classical methods.

Reflecting the wide range of current research in statistics, the series encompasses applied, methodological and theoretical statistics, ranging from applications and new techniques made possible by advances in computerized practice to rigorous treatment of theoretical approaches.

This series provides essential and invaluable reading for all statisticians, whether in academia, industry, government, or research.

ABRAHAM and LEDOLTER · Statistical Methods for Forecasting
AGRESTI · Analysis of Ordinal Categorical Data
AGRESTI · An Introduction to Categorical Data Analysis
AGRESTI · Categorical Data Analysis, Second Edition
ALTMAN, GILL and McDONALD · Numerical Issues in Statistical Computing for the Social Scientist
AMARATUNGA and CABRERA · Exploration and Analysis of DNA Microarray and Protein Array Data
ANDĚL · Mathematics of Chance
ANDERSON · An Introduction to Multivariate Statistical Analysis, Third Edition
* ANDERSON · The Statistical Analysis of Time Series
ANDERSON, AUQUIER, HAUCK, OAKES, VANDAELE and WEISBERG · Statistical Methods for Comparative Studies
ANDERSON and LOYNES · The Teaching of Practical Statistics
ARMITAGE and DAVID (editors) · Advances in Biometry
ARNOLD, BALAKRISHNAN and NAGARAJA · Records
* ARTHANARI and DODGE · Mathematical Programming in Statistics
* BAILEY · The Elements of Stochastic Processes with Applications to the Natural Sciences
BALAKRISHNAN and KOUTRAS · Runs and Scans with Applications
BALAKRISHNAN and NG · Precedence-Type Tests and Applications
BARNETT · Comparative Statistical Inference, Third Edition
BARNETT · Environmental Statistics: Methods & Applications
BARNETT and LEWIS · Outliers in Statistical Data, Third Edition
BARTOSZYNSKI and NIEWIADOMSKA-BUGAJ · Probability and Statistical Inference
BASILEVSKY · Statistical Factor Analysis and Related Methods: Theory and Applications
BASU and RIGDON · Statistical Methods for the Reliability of Repairable Systems
BATES and WATTS · Nonlinear Regression Analysis and Its Applications
BECHHOFER, SANTNER and GOLDSMAN · Design and Analysis of Experiments for Statistical Selection, Screening and Multiple Comparisons
BEIRLANT, GOEGEBEUR, SEGERS, TEUGELS and DE WAAL · Statistics of Extremes: Theory and Applications
BELSLEY · Conditioning Diagnostics: Collinearity and Weak Data in Regression
BELSLEY, KUH and WELSCH · Regression Diagnostics: Identifying Influential Data and Sources of Collinearity

*Now available in a lower priced paperback edition in the Wiley Classics Library.

BENDAT and PIERSOL · Random Data: Analysis and Measurement Procedures, Third Edition
BERNARDO and SMITH · Bayesian Theory
BERRY, CHALONER and GEWEKE · Bayesian Analysis in Statistics and Econometrics: Essays in Honor of Arnold Zellner
BHAT and MILLER · Elements of Applied Stochastic Processes, Third Edition
BHATTACHARYA and JOHNSON · Statistical Concepts and Methods
BHATTACHARYA and WAYMIRE · Stochastic Processes with Applications
BIEMER, GROVES, LYBERG, MATHIOWETZ and SUDMAN · Measurement Errors in Surveys
BILLINGSLEY · Convergence of Probability Measures, Second Edition
BILLINGSLEY · Probability and Measure, Third Edition
BIRKES and DODGE · Alternative Methods of Regression
BISWAS, DATTA, FINE and SEGAL · Statistical Advances in the Biomedical Sciences: Clinical Trials, Epidemiology, Survival Analysis, and Bioinformatics
BLISCHKE and MURTHY (editors) · Case Studies in Reliability and Maintenance
BLISCHKE and MURTHY · Reliability: Modeling, Prediction and Optimization
BLOOMFIELD · Fourier Analysis of Time Series: An Introduction, Second Edition
BOLLEN · Structural Equations with Latent Variables
BOLLEN and CURRAN · Latent Curve Models: A Structural Equation Perspective
BOROVKOV · Ergodicity and Stability of Stochastic Processes
BOSQ and BLANKE · Inference and Prediction in Large Dimensions
BOULEAU · Numerical Methods for Stochastic Processes
BOX · Bayesian Inference in Statistical Analysis
BOX · R. A. Fisher, the Life of a Scientist
BOX and DRAPER · Empirical Model-Building and Response Surfaces
* BOX and DRAPER · Evolutionary Operation: A Statistical Method for Process Improvement
BOX · Improving Almost Anything *Revised Edition*
BOX, HUNTER and HUNTER · Statistics for Experimenters: An Introduction to Design, Data Analysis and Model Building
BOX, HUNTER and HUNTER · Statistics for Experimenters: Design, Innovation and Discovery, Second Edition
BOX and LUCEÑO · Statistical Control by Monitoring and Feedback Adjustment
BRANDIMARTE · Numerical Methods in Finance: A MATLAB-Based Introduction
BROWN and HOLLANDER · Statistics: A Biomedical Introduction
BRUNNER, DOMHOF and LANGER · Nonparametric Analysis of Longitudinal Data in Factorial Experiments
BUCKLEW · Large Deviation Techniques in Decision, Simulation and Estimation
CAIROLI and DALANG · Sequential Stochastic Optimization
CASTILLO, HADI, BALAKRISHNAN and SARABIA · Extreme Value and Related Models with Applications in Engineering and Science
CHAN · Time Series: Applications to Finance
CHARALAMBIDES · Combinatorial Methods in Discrete Distributions
CHATTERJEE and HADI · Regression Analysis by Example, Fourth Edition
CHATTERJEE and HADI · Sensitivity Analysis in Linear Regression
CHERNICK · Bootstrap Methods: A Practitioner's Guide
CHERNICK and FRIIS · Introductory Biostatistics for the Health Sciences
CHILÉS and DELFINER · Geostatistics: Modeling Spatial Uncertainty
CHOW and LIU · Design and Analysis of Clinical Trials: Concepts and Methodologies, Second Edition
CLARKE · Linear Models: The Theory and Application of Analysis of Variance
CLARKE and DISNEY · Probability and Random Processes: A First Course with Applications, Second Edition
* COCHRAN and COX · Experimental Designs, Second Edition
CONGDON · Applied Bayesian Modelling

*Now available in a lower priced paperback edition in the Wiley Classics Library.

CONGDON · Bayesian Models for Categorical Data
CONGDON · Bayesian Statistical Modelling, Second Edition
CONOVER · Practical Nonparametric Statistics, Second Edition
COOK · Regression Graphics
COOK and WEISBERG · An Introduction to Regression Graphics
COOK and WEISBERG · Applied Regression Including Computing and Graphics
CORNELL · Experiments with Mixtures, Designs, Models and the Analysis of Mixture Data, Third Edition
COVER and THOMAS · Elements of Information Theory
COX · A Handbook of Introductory Statistical Methods
* COX · Planning of Experiments
CRESSIE · Statistics for Spatial Data, Revised Edition
CSÖRGÖ and HORVÁTH · Limit Theorems in Change Point Analysis
DANIEL · Applications of Statistics to Industrial Experimentation
DANIEL · Biostatistics: A Foundation for Analysis in the Health Sciences, Sixth Edition
* DANIEL · Fitting Equations to Data: Computer Analysis of Multifactor Data, Second Edition
DASU and JOHNSON · Exploratory Data Mining and Data Cleaning
DAVID and NAGARAJA · Order Statistics, Third Edition
* DEGROOT, FIENBERG and KADANE · Statistics and the Law
DEL CASTILLO · Statistical Process Adjustment for Quality Control
DEMARIS · Regression with Social Data: Modeling Continuous and Limited Response Variables
DEMIDENKO · Mixed Models: Theory and Applications
DENISON, HOLMES, MALLICK and SMITH · Bayesian Methods for Nonlinear Classification and Regression
DETTE and STUDDEN · The Theory of Canonical Moments with Applications in Statistics, Probability and Analysis
DEY and MUKERJEE · Fractional Factorial Plans
DILLON and GOLDSTEIN · Multivariate Analysis: Methods and Applications
DODGE · Alternative Methods of Regression
* DODGE and ROMIG · Sampling Inspection Tables, Second Edition
* DOOB · Stochastic Processes
DOWDY, WEARDEN and CHILKO · Statistics for Research, Third Edition
DRAPER and SMITH · Applied Regression Analysis, Third Edition
DRYDEN and MARDIA · Statistical Shape Analysis
DUDEWICZ and MISHRA · Modern Mathematical Statistics
DUNN and CLARK · Applied Statistics: Analysis of Variance and Regression, Second Edition
DUNN and CLARK · Basic Statistics: A Primer for the Biomedical Sciences, Third Edition
DUPUIS and ELLIS · A Weak Convergence Approach to the Theory of Large Deviations
EDLER and KITSOS (editors) · Recent Advances in Quantitative Methods in Cancer and Human Health Risk Assessment
* ELANDT-JOHNSON and JOHNSON · Survival Models and Data Analysis
ENDERS · Applied Econometric Time Series
ETHIER and KURTZ · Markov Processes: Characterization and Convergence
EVANS, HASTINGS and PEACOCK · Statistical Distribution, Third Edition
FELLER · An Introduction to Probability Theory and Its Applications, Volume I, Third Edition, Revised; Volume II, Second Edition
FISHER and VAN BELLE · Biostatistics: A Methodology for the Health Sciences
FITZMAURICE, LAIRD and WARE · Applied Longitudinal Analysis
* FLEISS · The Design and Analysis of Clinical Experiments
FLEISS · Statistical Methods for Rates and Proportions, Second Edition
FLEMING and HARRINGTON · Counting Processes and Survival Analysis
FULLER · Introduction to Statistical Time Series, Second Edition

*Now available in a lower priced paperback edition in the Wiley Classics Library.

FULLER · Measurement Error Models
GALLANT · Nonlinear Statistical Models.
GEISSER · Modes of Parametric Statistical Inference
GELMAN and MENG (editors) · Applied Bayesian Modeling and Casual Inference from Incomplete-data Perspectives
GEWEKE · Contemporary Bayesian Econometrics and Statistics
GHOSH, MUKHOPADHYAY and SEN · Sequential Estimation
GIESBRECHT and GUMPERTZ · Planning, Construction and Statistical Analysis of Comparative Experiments
GIFI · Nonlinear Multivariate Analysis
GIVENS and HOETING · Computational Statistics
GLASSERMAN and YAO · Monotone Structure in Discrete-Event Systems
GNANADESIKAN · Methods for Statistical Data Analysis of Multivariate Observations, Second Edition
GOLDSTEIN and WOLFF · Bayes Linear Statistics, Theory & Methods
GOLDSTEIN and LEWIS · Assessment: Problems, Development and Statistical Issues
GREENWOOD and NIKULIN · A Guide to Chi-Squared Testing
GROSS, SHORTLE, THOMPSON and HARRIS · Fundamentals of Queueing Theory, fourth Edition
GROSS, SHORTLE, THOMPSON and HARRIS · Solutions Manual to Accompany Fundamentals of Queueing Theory, fourth Edition
* HAHN and SHAPIRO · Statistical Models in Engineering
HAHN and MEEKER · Statistical Intervals: A Guide for Practitioners
HALD · A History of Probability and Statistics and their Applications Before 1750
HALD · A History of Mathematical Statistics from 1750 to 1930
HAMPEL · Robust Statistics: The Approach Based on Influence Functions
HANNAN and DEISTLER · The Statistical Theory of Linear Systems
HARTUNG, KNAPP and SINHA · Statistical Meta-Analysis with Applications
HEIBERGER · Computation for the Analysis of Designed Experiments
HEDAYAT and SINHA · Design and Inference in Finite Population Sampling
HEDEKER and GIBBONS · Longitudinal Data Analysis
HELLER · MACSYMA for Statisticians
HERITIER, CANTONI, COPT and VICTORIA-FESER · Robust Methods in Biostatistics
HINKELMANN and KEMPTHORNE · Design and Analysis of Experiments, Volume 1: Introduction to Experimental Design
HINKELMANN and KEMPTHORNE · Design and analysis of experiments, Volume 2: Advanced Experimental Design
HOAGLIN, MOSTELLER and TUKEY · Exploratory Approach to Analysis of Variance
HOAGLIN, MOSTELLER and TUKEY · Exploring Data Tables, Trends and Shapes
* HOAGLIN, MOSTELLER and TUKEY · Understanding Robust and Exploratory Data Analysis
HOCHBERG and TAMHANE · Multiple Comparison Procedures
HOCKING · Methods and Applications of Linear Models: Regression and the Analysis of Variance, Second Edition
HOEL · Introduction to Mathematical Statistics, Fifth Edition
HOGG and KLUGMAN · Loss Distributions
HOLLANDER and WOLFE · Nonparametric Statistical Methods, Second Edition
HOSMER and LEMESHOW · Applied Logistic Regression, Second Edition
HOSMER and LEMESHOW · Applied Survival Analysis: Regression Modeling of Time to Event Data
HUBER · Robust Statistics
HUBERTY · Applied Discriminant Analysis
HUBERY and OLEJNKI · Applied MANOVA and Discriminant Analysis, 2nd Edition
HUNT and KENNEDY · Financial Derivatives in Theory and Practice, Revised Edition
HURD and MIAMEE · Periodically Correlated Random Sequences: Spectral Theory and Practice

*Now available in a lower priced paperback edition in the Wiley Classics Library.

HUSKOVA, BERAN and DUPAC · Collected Works of Jaroslav Hajek – with Commentary
HUZURBAZAR · Flowgraph Models for Multistate Time-to-Event Data
IMAN and CONOVER · A Modern Approach to Statistics
JACKSON · A User's Guide to Principle Components
JOHN · Statistical Methods in Engineering and Quality Assurance
JOHNSON · Multivariate Statistical Simulation
JOHNSON and BALAKRISHNAN · Advances in the Theory and Practice of Statistics: A Volume in Honor of Samuel Kotz
JOHNSON and BHATTACHARYYA · Statistics: Principles and Methods, Fifth Edition
JOHNSON and KOTZ · Distributions in Statistics
JOHNSON and KOTZ (editors) · Leading Personalities in Statistical Sciences: From the Seventeenth Century to the Present
JOHNSON, KOTZ and BALAKRISHNAN · Continuous Univariate Distributions, Volume 1, Second Edition
JOHNSON, KOTZ and BALAKRISHNAN · Continuous Univariate Distributions, Volume 2, Second Edition
JOHNSON, KOTZ and BALAKRISHNAN · Discrete Multivariate Distributions
JOHNSON, KOTZ and KEMP · Univariate Discrete Distributions, Second Edition
JUDGE, GRIFFITHS, HILL, LU TKEPOHL and LEE · The Theory and Practice of Econometrics, Second Edition
JUREČKOVÁ and SEN · Robust Statistical Procedures: Asymptotics and Interrelations
JUREK and MASON · Operator-Limit Distributions in Probability Theory
KADANE · Bayesian Methods and Ethics in a Clinical Trial Design
KADANE and SCHUM · A Probabilistic Analysis of the Sacco and Vanzetti Evidence
KALBFLEISCH and PRENTICE · The Statistical Analysis of Failure Time Data, Second Edition
KARIYA and KURATA · Generalized Least Squares
KASS and VOS · Geometrical Foundations of Asymptotic Inference
KAUFMAN and ROUSSEEUW · Finding Groups in Data: An Introduction to Cluster Analysis
KEDEM and FOKIANOS · Regression Models for Time Series Analysis
KENDALL, BARDEN, CARNE and LE · Shape and Shape Theory
KHURI · Advanced Calculus with Applications in Statistics, Second Edition
KHURI, MATHEW and SINHA · Statistical Tests for Mixed Linear Models
* KISH · Statistical Design for Research
KLEIBER and KOTZ · Statistical Size Distributions in Economics and Actuarial Sciences
KLUGMAN, PANJER and WILLMOT · Loss Models: From Data to Decisions
KLUGMAN, PANJER and WILLMOT · Solutions Manual to Accompany Loss Models: From Data to Decisions
KOTZ, BALAKRISHNAN and JOHNSON · Continuous Multivariate Distributions, Volume 1, Second Edition
KOTZ and JOHNSON (editors) · Encyclopedia of Statistical Sciences: Volumes 1 to 9 with Index
KOTZ and JOHNSON (editors) · Encyclopedia of Statistical Sciences: Supplement Volume
KOTZ, READ and BANKS (editors) · Encyclopedia of Statistical Sciences: Update Volume 1
KOTZ, READ and BANKS (editors) · Encyclopedia of Statistical Sciences: Update Volume 2
KOVALENKO, KUZNETZOV and PEGG · Mathematical Theory of Reliability of Time-Dependent Systems with Practical Applications
KOWALSI and TU · Modern Applied U-Statistics
KROONENBERG · Applied Multiway Data Analysis
KULINSKAYA, MORGENTHALER and STAUDTE · Meta Analysis: A Guide to Calibrating and Combining Statistical Evidence
KUROWICKA and COOKE · Uncertainty Analysis with High Dimensional Dependence Modelling

*Now available in a lower priced paperback edition in the Wiley Classics Library.

KVAM and VIDAKOVIC · Nonparametric Statistics with Applications to Science and Engineering
LACHIN · Biostatistical Methods: The Assessment of Relative Risks
LAD · Operational Subjective Statistical Methods: A Mathematical, Philosophical and Historical Introduction
LAMPERTI · Probability: A Survey of the Mathematical Theory, Second Edition
LANGE, RYAN, BILLARD, BRILLINGER, CONQUEST and GREENHOUSE · Case Studies in Biometry
LARSON · Introduction to Probability Theory and Statistical Inference, Third Edition
LAWLESS · Statistical Models and Methods for Lifetime Data, Second Edition
LAWSON · Statistical Methods in Spatial Epidemiology, Second Edition
LE · Applied Categorical Data Analysis
LE · Applied Survival Analysis
LEE · Structural Equation Modelling: A Bayesian Approach
LEE and WANG · Statistical Methods for Survival Data Analysis, Third Edition
LEPAGE and BILLARD · Exploring the Limits of Bootstrap
LEYLAND and GOLDSTEIN (editors) · Multilevel Modelling of Health Statistics
LIAO · Statistical Group Comparison
LINDVALL · Lectures on the Coupling Method
LIN · Introductory Stochastic Analysis for Finance and Insurance
LINHART and ZUCCHINI · Model Selection
LITTLE and RUBIN · Statistical Analysis with Missing Data, Second Edition
LLOYD · The Statistical Analysis of Categorical Data
LOWEN and TEICH · Fractal-Based Point Processes
MAGNUS and NEUDECKER · Matrix Differential Calculus with Applications in Statistics and Econometrics, Revised Edition
MALLER and ZHOU · Survival Analysis with Long Term Survivors
MALLOWS · Design, Data and Analysis by Some Friends of Cuthbert Daniel
MANN, SCHAFER and SINGPURWALLA · Methods for Statistical Analysis of Reliability and Life Data
MANTON, WOODBURY and TOLLEY · Statistical Applications Using Fuzzy Sets
MARCHETTE · Random Graphs for Statistical Pattern Recognition
MARKOVICH · Nonparametric Analysis of Univariate Heavy-Tailed Data: Research and practice
MARDIA and JUPP · Directional Statistics
MARKOVICH · Nonparametric Analysis of Univariate Heavy-Tailed Data: Research and Practice
MARONNA, MARTIN and YOHAI · Robust Statistics: Theory and Methods
MASON, GUNST and HESS · Statistical Design and Analysis of Experiments with Applications to Engineering and Science, Second Edition
MCCULLOCH and SERLE · Generalized, Linear and Mixed Models
MCFADDEN · Management of Data in Clinical Trials
MCLACHLAN · Discriminant Analysis and Statistical Pattern Recognition
MCLACHLAN, DO and AMBROISE · Analyzing Microarray Gene Expression Data
MCLACHLAN and KRISHNAN · The EM Algorithm and Extensions
MCLACHLAN and PEEL · Finite Mixture Models
MCNEIL · Epidemiological Research Methods
MEEKER and ESCOBAR · Statistical Methods for Reliability Data
MEERSCHAERT and SCHEFFLER · Limit Distributions for Sums of Independent Random Vectors: Heavy Tails in Theory and Practice
MICKEY, DUNN and CLARK · Applied Statistics: Analysis of Variance and Regression, Third Edition
* MILLER · Survival Analysis, Second Edition
MONTGOMERY, JENNINGS and KULAHCI · Introduction to Time Series Analysis and Forecasting Solutions Set

*Now available in a lower priced paperback edition in the Wiley Classics Library.

MONTGOMERY, PECK and VINING · Introduction to Linear Regression Analysis, Fourth Edition
MORGENTHALER and TUKEY · Configural Polysampling: A Route to Practical Robustness
MUIRHEAD · Aspects of Multivariate Statistical Theory
MULLER and STEWART · Linear Model Theory: Univariate, Multivariate and Mixed Models
MURRAY · X-STAT 2.0 Statistical Experimentation, Design Data Analysis and Nonlinear Optimization
MURTHY, XIE and JIANG · Weibull Models
MYERS and MONTGOMERY · Response Surface Methodology: Process and Product Optimization Using Designed Experiments, Second Edition
MYERS, MONTGOMERY and VINING · Generalized Linear Models. With Applications in Engineering and the Sciences
† NELSON · Accelerated Testing, Statistical Models, Test Plans and Data Analysis
† NELSON · Applied Life Data Analysis
NEWMAN · Biostatistical Methods in Epidemiology
OCHI · Applied Probability and Stochastic Processes in Engineering and Physical Sciences
OKABE, BOOTS, SUGIHARA and CHIU · Spatial Tesselations: Concepts and Applications of Voronoi Diagrams, Second Edition
OLIVER and SMITH · Influence Diagrams, Belief Nets and Decision Analysis
PALTA · Quantitative Methods in Population Health: Extentions of Ordinary Regression
PANJER · Operational Risks: Modeling Analytics
PANKRATZ · Forecasting with Dynamic Regression Models
PANKRATZ · Forecasting with Univariate Box-Jenkins Models: Concepts and Cases
PARDOUX · Markov Processes and Applications: Algorithms, Networks, Genome and Finance
PARMIGIANI and INOUE · Decision Theory: Principles and Approaches
* PARZEN · Modern Probability Theory and Its Applications
PEÑA, TIAO and TSAY · A Course in Time Series Analysis
PIANTADOSI · Clinical Trials: A Methodologic Perspective
PORT · Theoretical Probability for Applications
POURAHMADI · Foundations of Time Series Analysis and Prediction Theory
POWELL · Approximate Dynamic Programming: Solving the Curses of Dimensionality
PRESS · Bayesian Statistics: Principles, Models and Applications
PRESS · Subjective and Objective Bayesian Statistics, Second Edition
PRESS and TANUR · The Subjectivity of Scientists and the Bayesian Approach
PUKELSHEIM · Optimal Experimental Design
PURI, VILAPLANA and WERTZ · New Perspectives in Theoretical and Applied Statistics
PUTERMAN · Markov Decision Processes: Discrete Stochastic Dynamic Programming
QIU · Image Processing and Jump Regression Analysis
RAO · Linear Statistical Inference and its Applications, Second Edition
RAUSAND and HØYLAND · System Reliability Theory: Models, Statistical Methods and Applications, Second Edition
RENCHER · Linear Models in Statistics
RENCHER · Methods of Multivariate Analysis, Second Edition
RENCHER · Multivariate Statistical Inference with Applications
RIPLEY · Spatial Statistics
RIPLEY · Stochastic Simulation
ROBINSON · Practical Strategies for Experimenting
ROHATGI and SALEH · An Introduction to Probability and Statistics, Second Edition
ROLSKI, SCHMIDLI, SCHMIDT and TEUGELS · Stochastic Processes for Insurance and Finance
ROSENBERGER and LACHIN · Randomization in Clinical Trials: Theory and Practice
ROSS · Introduction to Probability and Statistics for Engineers and Scientists

†Now available in a lower priced paperback edition in the Wiley-Interscience Paperback Series.
*Now available in a lower priced paperback edition in the Wiley Classics Library.

ROSSI, ALLENBY and MCCULLOCH · Bayesian Statistics and Marketing
ROUSSEEUW and LEROY · Robust Regression and Outline Detection
ROYSTON and SAUERBREI · Multivariable Model - Building: A Pragmatic Approach to Regression Anaylsis based on Fractional Polynomials for Modelling Continuous Variables
RUBIN · Multiple Imputation for Nonresponse in Surveys
RUBINSTEIN · Simulation and the Monte Carlo Method, Second Edition
RUBINSTEIN and MELAMED · Modern Simulation and Modeling
RYAN · Modern Engineering Statistics
RYAN · Modern Experimental Design
RYAN · Modern Regression Methods
RYAN · Statistical Methods for Quality Improvement, Second Edition
SALEH · Theory of Preliminary Test and Stein-Type Estimation with Applications
SALTELLI, CHAN and SCOTT (editors) · Sensitivity Analysis
* SCHEFFE · The Analysis of Variance
SCHIMEK · Smoothing and Regression: Approaches, Computation and Application
SCHOTT · Matrix Analysis for Statistics
SCHOUTENS · Levy Processes in Finance: Pricing Financial Derivatives
SCHUSS · Theory and Applications of Stochastic Differential Equations
SCOTT · Multivariate Density Estimation: Theory, Practice and Visualization
* SEARLE · Linear Models
SEARLE · Linear Models for Unbalanced Data
SEARLE · Matrix Algebra Useful for Statistics
SEARLE and WILLETT · Matrix Algebra for Applied Economics
SEBER · Multivariate Observations
SEBER and LEE · Linear Regression Analysis, Second Edition
SEBER and WILD · Nonlinear Regression
SENNOTT · Stochastic Dynamic Programming and the Control of Queueing Systems
* SERFLING · Approximation Theorems of Mathematical Statistics
SHAFER and VOVK · Probability and Finance: Its Only a Game!
SILVAPULLE and SEN · Constrained Statistical Inference: Inequality, Order and Shape Restrictions
SINGPURWALLA · Reliability and Risk: A Bayesian Perspective
SMALL and MCLEISH · Hilbert Space Methods in Probability and Statistical Inference
SRIVASTAVA · Methods of Multivariate Statistics
STAPLETON · Linear Statistical Models
STAUDTE and SHEATHER · Robust Estimation and Testing
STOYAN, KENDALL and MECKE · Stochastic Geometry and Its Applications, Second Edition
STOYAN and STOYAN · Fractals, Random and Point Fields: Methods of Geometrical Statistics
STREET and BURGESS · The Construction of Optimal Stated Choice Experiments: Theory and Methods
STYAN · The Collected Papers of T. W. Anderson: 1943–1985
SUTTON, ABRAMS, JONES, SHELDON and SONG · Methods for Meta-Analysis in Medical Research
TAKEZAWA · Introduction to Nonparametric Regression
TAMHANE · Statistical Analysis of Designed Experiments: Theory and Applications
TANAKA · Time Series Analysis: Nonstationary and Noninvertible Distribution Theory
THOMPSON · Empirical Model Building
THOMPSON · Sampling, Second Edition
THOMPSON · Simulation: A Modeler's Approach
THOMPSON and SEBER · Adaptive Sampling
THOMPSON, WILLIAMS and FINDLAY · Models for Investors in Real World Markets
TIAO, BISGAARD, HILL, PEÑA and STIGLER (editors) · Box on Quality and Discovery: with Design, Control and Robustness

*Now available in a lower priced paperback edition in the Wiley Classics Library.

TIERNEY · LISP-STAT: An Object-Oriented Environment for Statistical Computing and Dynamic Graphics
TSAY · Analysis of Financial Time Series
UPTON and FINGLETON · Spatial Data Analysis by Example, Volume II: Categorical and Directional Data
VAN BELLE · Statistical Rules of Thumb
VAN BELLE, FISHER, HEAGERTY and LUMLEY · Biostatistics: A Methodology for the Health Sciences, Second Edition
VESTRUP · The Theory of Measures and Integration
VIDAKOVIC · Statistical Modeling by Wavelets
VINOD and REAGLE · Preparing for the Worst: Incorporating Downside Risk in Stock Market Investments
WALLER and GOTWAY · Applied Spatial Statistics for Public Health Data
WEERAHANDI · Generalized Inference in Repeated Measures: Exact Methods in MANOVA and Mixed Models
WEISBERG · Applied Linear Regression, Second Edition
WELSH · Aspects of Statistical Inference
WESTFALL and YOUNG · Resampling-Based Multiple Testing: Examples and Methods for p-Value Adjustment
WHITTAKER · Graphical Models in Applied Multivariate Statistics
WINKER · Optimization Heuristics in Economics: Applications of Threshold Accepting
WONNACOTT and WONNACOTT · Econometrics, Second Edition
WOODING · Planning Pharmaceutical Clinical Trials: Basic Statistical Principles
WOODWORTH · Biostatistics: A Bayesian Introduction
WOOLSON and CLARKE · Statistical Methods for the Analysis of Biomedical Data, Second Edition
WU and HAMADA · Experiments: Planning, Analysis and Parameter Design Optimization
WU and ZHANG · Nonparametric Regression Methods for Longitudinal Data Analysis: Mixed-Effects Modeling Approaches
YANG · The Construction Theory of Denumerable Markov Processes
YOUNG, VALERO-MORA and FRIENDLY · Visual Statistics: Seeing Data with Dynamic Interactive Graphics
ZACKS · Stage-Wise Adaptive Designs
* ZELLNER · An Introduction to Bayesian Inference in Econometrics
ZELTERMAN · Discrete Distributions: Applications in the Health Sciences
ZHOU, OBUCHOWSKI and McCLISH · Statistical Methods in Diagnostic Medicine

*Now available in a lower priced paperback edition in the Wiley Classics Library.

Printed and bound by CPI Group (UK) Ltd, Croydon, CR0 4YY

12/01/2025

14624505-0001